W9-AVJ-691

Student Solutions Manual

to accompany

Technical Mathematics

and

Technical Mathematics
With Calculus

Second Edition

John C. Peterson
with
Alan Herweyer

Chattanooga State Technical College

Delmar Publishers

I(T)Pˉ An International Thomson Publishing Company

Albany • Bonn • Boston • Cincinnati • Detroit • London • Madrid • Melbourne
Mexico City • New York • Pacific Grove • Paris • San Francisco • Singapore • Tokyo
Toronto • Washington

NOTICE TO THE READER

Publisher does not warrant or guarantee any of the products described herein or perform any independent analysis in connection with any of the product information contained herein. Publisher does not assume, and expressly disclaims, any obligation to obtain and include information other than that provided to it by the manufacturer.

The reader is expressly warned to consider and adopt all safety precautions that might be indicated by the activities herein and to avoid all potential hazards. By following the instructions contained herein, the reader willingly assumes all risks in connection with such instructions.

The publisher makes no representation or warranties of any kind, including but not limited to, the warranties of fitness for particular purpose or merchantability, nor are any such representations implied with respect to the material set forth herein, and the publisher takes no responsibility with respect to such material. The publisher shall not be liable for any special, consequential, or exemplary damages resulting, in whole or part, from the readers' use of, or reliance upon, this material.

COPYRIGHT © 1997
By Delmar Publishers
a division of International Thomson Publishing Inc.

The ITP logo is a trademark under license.

Printed in the United States of America

For more information, contact:

Delmar Publishers
3 Columbia Circle, Box 15015
Albany, New York 12212-5015

International Thomson Publishing Europe
Berkshire House 168-173
High Holborn
London, WC1V 7AA
England

Thomas Nelson Australia
102 Dodds Street
South Melbourne, 3205
Victoria, Australia

Nelson Canada
1120 Birchmont Road
Scarborough, Ontario
Canada, M1K 5G4

International Thomson Editores
Campos Eliseos 385, Piso 7
Col Polanco
11560 Mexico D F Mexico

International Thomson Publishing GmbH
Konigswinterer Strasse 418
53227 Bonn
Germany

International Thomson Publishing Asia
221 Henderson Road
#05-10 Henderson Building
Singapore 0315

International Thomson Publishing Japan
Hirakawacho Kyowa Building, 3F
2-2-1 Hirakawacho
Chiyoda-ku, Tokyo 102
Japan

All rights reserved. Certain portions copyright 1994. No part of this work covered by the copyright hereon may be reproduced or used in any form or by any means—graphic, electronic, or mechanical, including photocopying, recording, taping, or information storage and retrieval systems—without the written permission of the publisher.

3 4 5 6 7 8 9 10 XXX 01 00

ISBN: 0-8273-7237-X

Contents

Online Services

Delmar Online
To access a wide variety of Delmar products and services on the World Wide Web,
point your browser to:
      http://www.delmar.com
      or email: info@delmar.com

thomson.com
To access International Thomson Publishing's
home site for information on more than 34 publishers
and 20,000 products, point your browser to:
      http://www.thomson.com
      or email: findit@kiosk.thomson.com

A service of I(T)P®

1

The Real Number System

☰ 1.1 SOME SETS OF NUMBERS

1. 15 is a natural number, whole number, integer, rational number, and a real number

3. $\frac{-\sqrt{7}}{8}$ is an irrational number and a real number

5. $|15| = 15$

7. $\left|\frac{-\sqrt{7}}{8}\right| = \frac{\sqrt{7}}{8}$

9. $|4 - 7| = |-3| = 3$

11.

$\frac{4}{7}$

13.

$\frac{-8}{3}$

15. $2 < 3; 2 - 3 = -1$ negative

17. $-4 < 7; -4 - 7 = -11$ negative

19. $-3 > -8; -3 - (-8) = 5$ positive

21. $-\frac{2}{3} < -\frac{1}{2}; -\frac{2}{3} - \left(-\frac{1}{2}\right) = -\frac{4}{6} - \left(-\frac{3}{6}\right) = -\frac{1}{6}$ negative

23. $0.7 > 0.5; 0.7 - 0.5 = 0.2$ positive

25. $|2.3| < |-4.1|$ because $|2.3| - |-4.1| = 2.3 - 4.1 = -1.8$ which is negative

27. The reciprocal of -5 is $\frac{1}{-5} = -\frac{1}{5}$

29. The reciprocal of $\frac{17}{3}$ is $\frac{1}{17/3} = 1 \times \frac{3}{17} = \frac{3}{17}$

31. $-5, -|4|, -\frac{2}{3}, \frac{-1}{3}, \frac{16}{3}, |-8|$

33. (a) $\frac{1}{32}$ in., (b) $\frac{1}{8}$ in., (c) $\frac{1}{16} = 0.0625$ in.

35. $|-19.4 - (-16.8)| = |-2.6| = 2.6$ so 2.6 V

37. $\frac{5}{7}$ (Sheila), $\frac{2}{3}$ (Hazel), $\frac{3}{5}$ (José), $\frac{9}{16}$ (Robert), and $\frac{4}{9}$ (Lamar)

39. Helen began at mile marker 157. If the mile markers were increasing, then after one hour she was at mile marker $157 + 64 = 221$. If the mile markers were decreasing, then after one hour she was at mile marker $157 - 64 = 93$.

☰ 1.2 BASIC LAWS OF REAL NUMBERS

1. Commutative law for addition (CA)

5. Associative law for addition (AA)

3. Commutative law for multiplication (CM)

7. Identity element for multiplication

9. Additive inverses

13. Additive inverse of 91 is -91

19. $\frac{2}{\sqrt{2}} = \frac{2}{\sqrt{2}} \frac{\sqrt{2}}{\sqrt{2}} = \frac{2\sqrt{2}}{2} = \sqrt{2}$

11. Associative law for multiplication

15. $-\sqrt{2}$

21. $16 - 8 \div 4 = 16 - 2 = 14$

17. Multiplicative inverse of $\frac{1}{2}$ is 2

23. $24 + 3 - 10 \div 5 \times 8 + 2 = 24 + 3 - 2 \times 8 + 2 = 24 + 3 - 16 + 2 = 27 - 16 + 2 = 11 + 2 = 13$

25. $(7 - 2) - (3 + 8 - 7) = 5 - (11 - 7) = 5 - 4 = 1$

27. $7 \times 3 + 5 \times 2 = 21 + 10 = 31$

29. $\{-[5 - (8 - 4) + (3 - 7)] - (4 - 2)\} = -[5 -$ $4 + (-4)] - 2 = -[1 + (-4)] - 2 = -[-3] - 2 = 3 - 2 = 1$

31. Press $\boxed{(-)}$ 39 $\boxed{-}$ 72. The result is -111.

33. Press 192 $\boxed{-}$ $\boxed{(-)}$ 78. The result is 270.

35. (a) Press 243 $\boxed{+}$ $\boxed{(}$ $\boxed{(-)}$ 691 $\boxed{+}$ 89 $\boxed{)}$. The result is -359.

 (b) Press $\boxed{(}$ 243 $\boxed{+}$ $\boxed{(-)}$ 691 $\boxed{)}$ $\boxed{+}$ 89. The result is -359.

37. (a) Press $\boxed{(}$ $\boxed{(-)}$ 342 $\boxed{+}$ $\boxed{(-)}$ 18 $\boxed{)}$ $\boxed{\times}$ 91. The result is -32760.

 (b) Press $\boxed{(-)}$ 342 $\boxed{\times}$ 91 $\boxed{+}$ $\boxed{(-)}$ 18 $\boxed{\times}$ 91. The result is -32760.

≡ 1.3 BASIC OPERATIONS WITH REAL NUMBERS

1. $27 + (+23)$. Same sign $27 + 23 = 50$; both positive so, $+50$

3. $27 + (-13)$. Different signs $|27| - |-13| = 27 - 13 = 14$; larger positive so, 14

5. $7 - 16 = 7 + (-16)$. Different signs $|-16| - |7| = 16 - 7 = 9$; larger negative so, -9

7. $-8 - 16 = -8 + (-16)$. Same sign $|-8| + |-16| = 8 + 16 = 24$; both negative so, -24

9. $-37 - (-49) = -37 + (+49)$. Different signs $|49| - |37| = 49 - 37 = 12$; larger positive so, $+12$

11. $(-3)(-5)$; $|-3| \cdot |-5| = 3 \cdot 5 = 15$. Same sign so $+15$

13. $-38 \div 4$; $|-38| \div |4| = 38 \div 4 = \frac{38}{4} = \frac{19}{2}$. Different signs so $-\frac{19}{2}$

15. $\frac{3}{4} + \frac{-5}{8}$; $\frac{3}{4} = \frac{6}{8}$; $\frac{6}{8} + \frac{-5}{8} = \frac{6 + (-5)}{8} = \frac{1}{8}$

17. $\frac{-9}{5} + \frac{7}{3}$ common denominator is 15; $\frac{-9}{5} + \frac{7}{3} = \frac{-27}{15} + \frac{35}{15} = \frac{-27 + 35}{15} = \frac{8}{15}$

19. $\frac{2}{5} + \frac{-1}{4} = \frac{8}{20} + \frac{-5}{20} = \frac{8 + (-5)}{20} = \frac{3}{20}$

21. $\frac{3}{8} - \frac{-1}{4} = \frac{3}{8} - \frac{-2}{8} = \frac{3 - (-2)}{8} = \frac{5}{8}$

23. $\frac{-9}{10} - \frac{2}{3} = \frac{-27}{30} - \frac{20}{30} = \frac{-27 - 20}{30} = \frac{-47}{30} = -1\frac{17}{30}$

25. $\frac{5}{32} - \frac{1}{8} = \frac{5}{32} - \frac{4}{32} = \frac{5 - 4}{32} = \frac{1}{32}$

27. $\frac{-2}{3} \times \frac{4}{5} = \frac{-2 \times 4}{3 \times 5} = \frac{-8}{15} = -\frac{8}{15}$

29. $\frac{-1}{8} \times \frac{-3}{4} = \frac{-1 \times (-3)}{8 \times 4} = \frac{3}{32}$

31. $-\frac{4}{3} \times \frac{5}{2} = \frac{-4 \times 5}{3 \times 2} = \frac{-20}{6} = -\frac{10}{3}$

 or $\frac{-2 \times 2 \times 5}{3 \times 2} = -3\frac{1}{3}$

33. $-\frac{3}{4} \div \frac{-5}{8} = -\frac{3}{4} \times \frac{8}{-5} = \frac{-3 \times 4 \times 2}{4 \times (-5)} = \frac{-6}{-5} = \frac{6}{5} = 1\frac{1}{5}$

35. $-\frac{3}{5} \div 4 = \frac{-3}{5} \times \frac{1}{4} = \frac{-3 \times 1}{5 \times 4} = \frac{-3}{20} = -\frac{3}{20}$

37. $-\frac{7}{5} \div \left(-\frac{5}{7}\right) = -\frac{7}{5} \times \frac{-7}{5} = \frac{-7 \times (-7)}{5 \times 5} = \frac{49}{25} = 1\frac{24}{25}$

39. $\left(-2 + \frac{2}{3}\right) \times \frac{-1}{2} - \left[\frac{3}{2} \div (-3)\right] + \frac{7}{3} = \frac{-4}{3} \times \frac{-1}{2} - \left(\frac{-1}{2}\right) + \frac{7}{3} = \frac{-4 \times (-1)}{3 \times 2} - \left(\frac{-1}{2}\right) + \frac{7}{3} = \frac{2}{3} - \left(\frac{-1}{2}\right) + \frac{7}{3} =$
$\frac{4}{6} - \frac{-3}{6} + \frac{7}{3} = \frac{7}{6} + \frac{7}{3} = \frac{7}{6} + \frac{14}{6} = \frac{21}{6} = \frac{7}{2} = 3\frac{1}{2}$

41. Add: $17\frac{1}{2} + 15\frac{7}{8} + 29\frac{3}{4} + 15\frac{3}{8} = 17\frac{4}{8} + 15\frac{7}{8} + 29\frac{6}{8} + 15\frac{3}{8} = 76\frac{20}{8} = 78\frac{4}{8} = 78\frac{1}{2} = 6'6\frac{1}{2}''$

43. We add the voltages of the batteries: $30 + 15 + (-12) + 24 = 57$. So, the total voltage is 57 V.

45. (a) $4\frac{5}{8} + 3\frac{1}{5} = \frac{37}{8} + \frac{16}{5} = \frac{185}{40} + \frac{128}{40} = \frac{313}{40} = 7\frac{33}{40}$ miles;

 (b) We need to add the distance between the second and third checkpoints to the answer for (a).
 $7\frac{33}{40} + 3\frac{3}{8} = \frac{313}{40} + \frac{27}{8} = \frac{313}{40} + \frac{135}{40} = \frac{448}{40} = 11\frac{8}{40} = 11\frac{1}{5}$ miles;

 (c) Using the answer from (b), we subtract that answer from the total length of the race.
 $14\frac{1}{2} - 11\frac{1}{5} = \frac{29}{2} - \frac{56}{5} = \frac{145}{10} - \frac{112}{10} = \frac{33}{10} = 3\frac{3}{10}$ miles

47. We want $37 \times 2'3\frac{1}{2}''$. There are $12''$ in 1 foot, so there are $2 \times 12'' = 24''$ in 2 feet. Thus, $2'3\frac{1}{2}'' = 24'' + 3\frac{1}{2}'' = 27\frac{1}{2}''$. Now, $37 \times 27\frac{1}{2} = \frac{37}{1} \times \frac{55}{2} = \frac{2035}{2} = 1017\frac{1}{2}'' = 84'9\frac{1}{2}''$

49. We subtract the two smaller lengths from the total length. $7\frac{5}{8} - 3\frac{3}{8} - 3\frac{3}{8} = \frac{61}{8} - \frac{27}{8} - \frac{27}{8} = \frac{34}{8} - \frac{27}{8} = \frac{7}{8}$
The missing length is $\frac{7}{8}''$.

51. **(a)** $-6.5 - 14.7 = -21.2$ V
(b) The negative sign indicates that the voltage dropped 21.2 V.

1.4 LAWS OF EXPONENTS

1. $5^3 = 5 \times 5 \times 5 = 125$

3. $\left(\frac{2}{3}\right)^{-1} = \frac{2^{-1}}{3^{-1}} = \frac{1/2}{1/3} = \frac{1}{2} \times \frac{3}{1} = \frac{3}{2}$

5. $(-4)^2 = (-4)(-4) = 16$

7. $\frac{7}{7^3} = \frac{7^1}{7^3} = 7^{1-3} = 7^{-2} = \frac{1}{7^2} = \frac{1}{49}$

9. $3^2 \cdot 3^4 = 3^{2+4} = 3^6$

11. $2^4 \cdot 2^3 \cdot 2^5 = 2^{4+3+5} = 2^{12}$

13. $\frac{2^5}{2^3} = 2^{5-3} = 2^2$

15. $(2^3)^2 = 2^{3 \times 2} = 2^6$

17. $(x^4)^5 = x^{4 \times 5} = x^{20}$

19. $(a^{-2}b)^{-3} = a^{-2 \times (-3)}b^{-3} = a^6 b^{-3} = \frac{a^6}{b^3}$

21. $\left(\frac{x}{4}\right)^3 = \frac{x^3}{4^3}$

23. $\left(\frac{a^2 b}{c^3}\right)^4 = \frac{a^{2 \times 4} b^4}{c^{3 \times 4}} = \frac{a^8 b^4}{c^{12}}$

25. $x^{-7} = \frac{1}{x^7}$

27. $\left(\frac{1}{5}\right)^3 = \frac{1}{5^3}$

29. $\frac{7^3}{7^8} = 7^{3-8} = 7^{-5} = \frac{1}{7^5}$

31. $\frac{a^2 y^3}{a^5 y^7} = a^{2-5} y^{3-7} = a^{-3} y^{-4} = \frac{1}{a^3} \frac{1}{y^4} = \frac{1}{a^3 y^4}$

33. $\frac{a^2 p^5 y^3}{a^6 p^5 y} = a^{2-6} p^{5-5} y^{3-1} = a^{-4} p^0 y^2 = \frac{y^2}{a^4}$

35. $(pr^2)^{-1} = \frac{1}{pr^2}$

37. $\left(\frac{4y^3}{5^2}\right)^{-1} = \frac{5^2}{4y^3}$

39. $\left(\frac{2b^2}{y^5}\right)^{-3} = \frac{2^{-3} b^{-6}}{y^{-15}} = \frac{y^{15}}{2^3 b^6}$ or $\left(\frac{2b^2}{y^5}\right)^{-3} = \left(\frac{y^5}{2b^2}\right)^3 = \frac{y^{15}}{8b^6}$

41. $(-b^4)^6 = (-b^4)(-b^4)(-b^4)(-b^4)(-b^4)(-b^4) = b^{24}$

43. First we simplify $\left((2\pi fC)^{-1}\right)^2$ as $(2\pi fC)^{-2}$. Next, we compute $(2\pi fC)^{-2}$. Substituting the given values of f and C and using the $\boxed{\pi}$ key on the calculator, we obtain $(2\pi fC)^{-2} = 0.030159289^{-2} = 1099.405205$. To this, add 40^2, with the result of 2699.405205. Then $\sqrt{2699.405205} = 51.95580049$ gives the answer of $51.96 \ \Omega$ when rounded off to 2 decimal places.

45. Substituting the given values, we obtain
$$R = 6.5 + \left(\frac{1}{5} + \frac{1}{6} + \frac{1}{7.5}\right)^{-1} = 6.5 + \left(\frac{6+5}{(5)(6)} + \frac{1}{7.5}\right)^{-1} = 6.5 + \left(\frac{11}{30} + \frac{1}{7.5}\right)^{-1} = 6.5 + \left(\frac{11}{30} + \frac{4}{30}\right)^{-1} =$$
$$6.5 + \left(\frac{11+4}{30}\right)^{-1} = 6.5 + \left(\frac{15}{30}\right)^{-1} = 6.5 + \left(\frac{1}{2}\right)^{-1} = 6.5 + 2 = 8.5$$
The total resistance is 8.5 Ω.
Note: On a calculator such as a TI-8x, you could have keyed this as

$6.5 \ \boxed{+} \ \boxed{(} \ 1 \ \boxed{\div} \ 5 \ \boxed{+} \ 1 \ \boxed{\div} \ 6 \ \boxed{+} \ 1 \ \boxed{\div} \ 7.5 \ \boxed{)} \ \boxed{x^{-1}} \ \boxed{\text{ENTER}}$

47. **(a)**   For -7^2, press $\boxed{(-)}$ 7 $\boxed{\wedge}$ 2 $\boxed{\text{ENTER}}$. The result should be -49. For $(-7)^2$, press $\boxed{(}$ $\boxed{(-)}$ 7 $\boxed{)}$ $\boxed{\wedge}$ 2 $\boxed{\text{ENTER}}$. The result should be 49. Note that you could have used the $\boxed{x^2}$ key rather than the $\boxed{\wedge}$ 2 keystroke combination.
(b)   Different
(c)   The notation -7^2 is a short way of writing $(-1)7^2 = (-1)49 = -49$
The notation $(-7)^2$ means $(-7)(-7) = 49$.

▤ 1.5 SIGNIFICANT DIGITS AND ROUNDING OFF

1. Exact

3. Approximate

5. Approximate

7. 3

9. 3

11. 1

13. 4

15. (a) 6.05
    (b) 6.05

17. (a) 5.01
    (b) 0.027

19. (a) 27,0̃00
    (b) 27,00̃0

21. (a) 86
    (b) 0.2

23. (a) 140.070
    (b) 140.070

25. (a) 10
    (b) 14
    (c) 14.4

27. (a) 7
    (b) 7.0
    (c) 7.04

29. (a) 400
    (b) 40̃0

31. (a) 300
    (b) 310
    (c) 305

33. (a) 10
    (b) 14
    (c) 14.4

35. (a) 7
    (b) 7.0
    (c) 7.04

37. (a) 400
    (b) 40̃0
    (c) 403

39. (a) 300

(c) 403

(b) 30̃0
(c) 305

41. (a) 90
    (b) 89.9
    (c) 89.899

43. (a) 240
    (b) 237.3
    (c) 237.302

45. (a) 440
    (b) 438.0
    (c) 437.998

47. (a) 80
    (b) 78.7
    (c) 78.671

49. Length: Absolute error is measured value minus true value, so $23.72 - 24 = -0.28$ mm. Relative error is absolute error divided by true value, so $\frac{-0.28}{24} = -0.0117$. Percent error is relative error $\times 100$, so $-0.0117 \times 100 = -1.17\%$
Width: Absolute error; $8.35 - 8 = 0.35$ mm. Relative error; $\frac{0.35}{8} = 0.0438$. Percent error; $0.0438 \times 100 = 4.38\%$
Thickness: Absolute error; $2.98 - 3 = -0.02$ mm. Relative error; $\frac{-0.02}{3} = -0.0067$. Percent error; $-0.0067 \times 100 = -0.67\%$

51. 99.37   **53.** 1618   **55.** 1020   **57.** 17

59. To find the total length, we need to multiply $37 \times 132 \times 0.072$. The product is 351.648, which is 351.65 when rounded to 5 significant digits.

61. The cutter removes 18×0.086 mm $= 1.548$ mm with each revolution. The total of 597.3 revolutions will remove $597.3 \times 1.548 = 924.6204$ mm. Rounded off to 4 significant digits, this is 924.6 mm.

▤ 1.6 SCIENTIFIC AND ENGINEERING NOTATION

1. 4.2×10^4

3. 3.8×10^{-4}

5. $9.807\,00 \times 10^9$

7. 9.70×10^{-5}

9. 4.3×10^0

11. 74×10^3

13. 470×10^{-6}

15. $9.807\,00 \times 10^9$

17. 53.10×10^{-6}

19. 5.6×10^0

21. 4500

23. 40500000

25. 0.000063

27. 72

29. 75 000

31. 47 500 000 000

33. 0.000 0392 0

35. 83.15

37. $(7.6 \times 10^5) \times (2.04 \times 10^{10}) = 15.504 \times 10^{15} = 1.5504 \times 10^{16}$

39. $(3.5 \times 10^{-5}) \times (7.6 \times 10^{-7}) = 26.6 \times 10^{-12} = 2.66 \times 10^{-11}$

41. $(8.4 \times 10^8) \times (3.5 \times 10^{-4}) = 29.4 \times 10^4 = 2.94 \times 10^5$

43. $(7.04 \times 10^4) \times (3.2 \times 10^{-6}) = 22.528 \times 10^{-2} = 2.2528 \times 10^{-1}$

45. $(2.88 \times 10^{10}) \div (2.4 \times 10^5) = 1.2 \times 10^5$

47. $(3.75 \times 10^5) \div (1.5 \times 10^8) = 2.5 \times 10^{-3}$

49. $(3.2 \times 10^{-3}) \div (1.6 \times 10^{-7}) = 2 \times 10^4$

51. $(3.6 \times 10^{-7}) \div (2 \times 10^{-4}) = 1.8 \times 10^{-3}$

53. $(8.76 \times 10^6)(2.46 \times 10^8)(6.4 \times 10^9) = 137.91744 \times 10^{6+8+9} = 137.91744 \times 10^{23} = 1.3791744 \times 10^{25} \approx 1.38 \times 10^{25}$

55. $(2.5 \times 10^5)(6.3 \times 10^8) \div (6.3 \times 10^{-9}) = 2.5 \times 10^{5+8-(-9)} = 2.5 \times 10^{22}$

57. $(5.2 \times 10^{-6})(4.8 \times 10^8) \div (6.4 \times 10^{-12}) = 3.9 \times 10^{-6+8-(-12)} = 3.9 \times 10^{14}$

59. 675 (answers may vary)

61. $(3 \times 10'')(3.7 \times 10^{-6}) = 11.1 \times 10^5 = 1.11 \times 10^6$ mm

63. $(1.6606 \times 10^{-27})(1.2 \times 10^1)(1.4 \times 10^7) = 2.789808 \times 10^{-27+1+7} = 2.789808 \times 10^{-19}$ kg

65. One neutron; approximately $(1.6750 \times 10^{-27}) \div (9.1095 \times 10^{-31}) = 1.84 \times 10^3$ times heavier

67. $X_L = 2\pi f L = 2\pi(18 \times 10^6)(2.5 \times 10^{-3}) \approx 282.74 \times 10^3$. The inductive reactance is about 282.74×10^3 Ω.

69. Easier to express very large and very small numbers. Historically, to use slide rules, numbers were first written in scientific notation.

≡ 1.7 ROOTS

1. 5

3. 12

5. 2

7. −3

9. 2

11. $\frac{2}{3}$

13. 0.2

15. −0.1

17. 3

19. 8.32

21. $\sqrt[3]{5}\sqrt[3]{25} = \sqrt[3]{5 \cdot 25} = \sqrt[3]{5^3} = 5$

23. $\sqrt[4]{8}\sqrt[4]{9} = \sqrt[4]{8 \cdot 9} = \sqrt[4]{72}$

25. $\frac{\sqrt{75}}{\sqrt{3}} = \sqrt{\frac{75}{3}} = \sqrt{25} = 5$

27. $\frac{\sqrt[3]{5}}{\sqrt[3]{40}} = \sqrt[3]{\frac{5}{40}} = \sqrt[3]{\frac{1}{8}} = \frac{1}{2}$

29. $\sqrt[3]{2^3} + \sqrt[4]{5^4} = 2 + 5 = 7$

31. $\sqrt[3]{\left(\frac{2}{3}\right)^3} - \sqrt[4]{\left(\frac{1}{3}\right)^4} = \frac{2}{3} - \frac{1}{3} = \frac{1}{3}$

33. $\sqrt{5^{2/3}} = \left(\sqrt{5^{1/3}}\right)^2 = 5^{1/3}$

35. $\sqrt[3]{16^{3/4}} = \left(\sqrt[3]{16^{1/4}}\right)^3 = 16^{1/4} = \sqrt[4]{16} = 2$

37. $16^{3/4} = \sqrt[4]{16^3} = 2^3 = 8$

39. $(-8)^{2/3} = (-2)^2 = 4$

41. $8^{-2/3} = \left(\sqrt[3]{8}\right)^{-2} = 2^{-2} = \frac{1}{2^2} = \frac{1}{4}$

43. $(-27)^{4/3} = (\sqrt[3]{-27})^4 = (-3)^4 = 81$

45. $\sqrt{\frac{81}{(8)(0.01)}} = \frac{\sqrt{81}}{\sqrt{8}\sqrt{0.01}} = \frac{9}{2\sqrt{2}(0.1)} = \frac{9}{0.2\sqrt{2}}$

47. $\sqrt[3]{\frac{(0.125)^3\sqrt{144}}{3/2}} = \frac{\sqrt[3]{(0.125)^3}\sqrt[3]{\sqrt{144}}}{\sqrt[3]{3/2}} = \frac{\sqrt[3]{(0.125)^3}\sqrt[3]{12}}{\frac{\sqrt[3]{3}}{\sqrt[3]{2}}} = 0.125\sqrt[3]{\frac{12 \cdot 2}{3}} = 0.125\sqrt[3]{8} = 0.125 \times 2 = 0.25$

49. $\lambda = \frac{3 \times 10^8}{1.435 \times 10^8} = 2.091$ m

51. $v_L = \sqrt{\frac{K}{\rho}} = \sqrt{\frac{2.1 \times 10^9}{1000}} = \sqrt{\frac{2.1 \times 10^9}{10^3}} = \sqrt{2.1 \times 10^{9-3}} = \sqrt{2.1 \times 10^6} \approx 1449.1377$ m/s

53. (a) $\sqrt{81} = 9$ and $-\sqrt{81} = -9$
    (b) 9

(c) $\sqrt[3]{-27} = -3$ and $-\sqrt{327} = 3$

(d) -3

55. (a) 4

(b) 4

(c) yes

(d) Both represent the cube root of 64.

CHAPTER 1 REVIEW

1. (a) Integers, rational numbers, real numbers
(b) Rational numbers, real numbers
(c) Irrational numbers, real numbers

2. (a) $\sqrt{42}$
(b) 16
(c) $\frac{5}{8}$

3. (a) $\frac{3}{2}$
(b) $\frac{-1}{8}$
(c) -5

4. (a) 5
(b) -17
(c) $-4\sqrt{2}$

5. (a) Commutative law for addition
(b) Distributive law
(c) Additive identity
(d) Multiplicative inverse

6. 64

7. -44

8. 53

9. $37 - (-61) = 37 + 61 = 98$

10. -32

11. $\frac{2}{3} + \frac{-5}{6} = \frac{4}{6} + \frac{-5}{6} = \frac{4-5}{6} = -\frac{1}{6}$

12. $\frac{4}{5} - \frac{5}{6} = \frac{24}{30} - \frac{25}{30} = -\frac{1}{30}$

13. -3

14. -3

15. $-\frac{2}{5}$

16. $3\frac{3}{4} \times -4\frac{1}{3} = \frac{15}{4} \times \frac{-13}{3} = \frac{5}{4} \times \frac{-13}{1} = -\frac{65}{4}$

17. $\frac{2}{3} \div \frac{1}{4} = \frac{2}{3} \times \frac{4}{1} = \frac{8}{3}$

18. $\frac{1}{5} \div \frac{-2}{15} = \frac{1}{5} \times \frac{15}{-2} = \frac{1}{1} \times \frac{3}{-2} = -\frac{3}{2}$

19. $2\frac{1}{2} \div 3\frac{1}{4} = \frac{5}{2} \div \frac{13}{4} = \frac{5}{2} \times \frac{4}{13} = \frac{5}{1} \times \frac{2}{13} = \frac{10}{13}$

20. $-4\frac{1}{3} \div -3\frac{1}{6} = -\frac{13}{3} \div -\frac{19}{6} = -\frac{13}{3} \times -\frac{6}{19} = \frac{26}{19}$

21. $2^5 = 2 \times 2 \times 2 \times 2 \times 2 = 32$

22. $(-3)^4 = (-3)(-3)(-3)(-3) = 81$

23. $(-4)^3 = (-4)(-4)(-4) = -64$

24. $8^{1/3} = \sqrt[3]{8} = 2$

25. $4^{1/2} = \sqrt{4} = 2$

26. $(-64)^{1/3} = \sqrt[3]{-64} = -4$

27. $2^5 \cdot 2^3 = 2^{5+3} = 2^8$

28. $3^5 \cdot 3^4 = 3^{5+4} = 3^9$

29. $2^{-3} \cdot 2^5 = 2^{-3+5} = 2^2$

30. $16^{-4} \cdot 16^{-3} = 16^{-7} = \frac{1}{16^7}$

31. $(4^3)^5 = 4^{3 \cdot 5} = 4^{15}$

32. $\left(2^{-3}\right)^4 = 2^{(-3)4} = 2^{-12} = \frac{1}{2^{12}}$

33. $\left(4^{1/3}\right)^3 = 4^1 = 4$

34. $\left(5^{1/4}\right)^{2/3} = 5^{1/4 \cdot 2/3} = 5^{1/6}$

35. $\frac{a^2 b^3}{ab^4} = \frac{a^{2-1}}{b^{4-3}} = \frac{a}{b}$

36. $\frac{x^2 y^3 z}{x^3 yz} = \frac{y^{3-1}}{x^{3-2}} = \frac{y^2}{x}$

37. $\left(ax^2\right)^{-2} = a^{-2}x^{-4} = \frac{1}{a^2 x^4}$ or $\left(ax^2\right)^{-2} = \frac{1}{\left(ax^2\right)^2} = \frac{1}{a^2 x^4}$

38. $\frac{\left(by^{-2}\right)^2}{\left(cy^{-4}\right)^{1/4}} = \frac{b^2 y^{-4}}{c^{1/4} y^{-1}} = \frac{b^2}{c^{1/4} y^3}$

39. (a) 2.37
(b) 2.37

40. (a) 2.02
(b) 0.002

41. (a) both
(b) 0.7

42. (a) both
(b) 0.0021

43. (a) 7.4

(b) 7.35
(c) 7.4
(d) 7.35

(b) 2.05
(c) 2.1
(d) 2.05

51. 29.08×10^{15}

52. 753×10^{21}

53. 75×10^{-24}

46. **(a)** 4.0
(b) 4.03
(c) 4.0
(d) 4.03

54. 193×10^{-18}

44. **(a)** 18
(b) 18.3
(c) 18.3
(d) 18.29

55. 12

56. -4

57. 5

47. 3.71×10^{11}

48. 2.54×10^{15}

58. $\frac{6}{11}$

45. **(a)** 2.1

49. 2.4×10^{-11}

50. 4.91×10^{-20}

59. **(a)** An order of 418 items is $\frac{418}{12} \approx 34.83$ dozen. The discount is given only for a full dozen, thus the manufacturer gets a discount on 34 dozen items. The total amount of the discount is $34 \times 12 \times 0.14 = 57.12$, or \$57.12.
(b) The total cost before the discount of the 418 items is $418 \times 13.79 = 5764.22$. Subtracting the discount of \$57.12, you get $5764.22 - 57.12 = 5707.10$. Thus, the total cost of the order is \$5,707.10.

60. **(a)** Substituting the given data in the above formula, we have
$$E = \frac{3}{2}kT$$
$$= \frac{3}{2}(8.617\ 33 \times 10^{-5})(295)$$
$$\approx 0.0381316853.$$

The average kinetic energy of particles at room temperature ($T = 295$ K) is about 0.038 eV.
(b) Substituting the given data in the above formula, we have
$$E = \frac{3}{2}kT$$
$$= \frac{3}{2}(8.617\ 33 \times 10^{-5})(5.714 \times 10^{3})$$
$$\approx 0.7385913543.$$
The average kinetic energy of particles at the surface of the sun ($T = 5.714 \times 10^3$ K) is about 0.739 eV.

CHAPTER 1 TEST

1. **(a)** $\left| \frac{4}{3} \right| = \frac{4}{3}$
(b) $\left| \frac{1}{2} - \frac{5}{8} \right| = \left| \frac{4}{8} - \frac{5}{8} \right| = \left| -\frac{1}{8} \right| = \frac{1}{8}$
(c) $|-6| = 6$

2. $-\frac{7}{3}$

3. $\frac{5}{8}$

4. 111

5. -112

6. -70

7. $\frac{5}{3} + 5\frac{2}{3} = \frac{5}{3} + \frac{17}{3} = \frac{22}{3}$ or $7\frac{1}{3}$

8. $\frac{7}{4} - \left(-\frac{3}{5} \right) = \frac{35}{20} + \frac{12}{20} = \frac{47}{20}$

9. $-4\frac{1}{2} \times 2\frac{1}{3} = -\frac{9}{2} \times \frac{7}{3} = -\frac{21}{2}$

10. $-\frac{5}{7} \div \frac{15}{28} = -\frac{5}{7} \times \frac{28}{15} = -\frac{4}{3}$

11. $-2\frac{1}{3} \div -4\frac{5}{6} = -\frac{7}{3} \div -\frac{29}{6} = -\frac{7}{3} \times -\frac{6}{29} = \frac{14}{29}$

12. $(-5)^3 = (-5)(-5)(-5) = -125$

13. $4^{3/2} = \left(\sqrt{4} \right)^3 = 2^3 = 8$

14. $2^6 \cdot 2^{-4} = 2^{6-4} = 2^2 = 4$

15. $\left(5^{1/4} \right)^8 = 5^{(1/4)8} = 5^2 = 25$

16. $3^{5/2} \div 3^{-3/2} = 3^{5/2 - (-3/2)} = 3^{5/2 + 3/2} = 3^4 = 81$

17. $\frac{a^3 b^5}{a^4 b^2} = \frac{b^{5-2}}{a^{4-3}} = \frac{b^3}{a}$

18. 4.516

19. 0.000 51

20. 4.75×10^{13}

21. $\sqrt{\frac{49}{25}} = \frac{\sqrt{49}}{\sqrt{25}} = \frac{7}{5}$

22. 92.5×10^{12}

CHAPTER
2
Algebraic Concepts and Operations

≡ 2.1 ADDITION AND SUBTRACTION

1. 4 is a constant; x and y are variables

3. 8 and π are constants; r is a variable

5. $3x^3$ and $4x$

7. $(2x^3)(5y)$, $\sqrt{3}ab$, and $-\frac{7a}{b}$

9. $-1, 5, x,$ and $\frac{1}{y}$

11. 47

13. $\frac{\pi a}{4}$, if a is a constant

15. $3x^2y$ and $17x^2y$

17. $(x+y)^2$ and $5(x+y)^2$

19. $a+b-c$ and $-2(a+b-c)$

21. $4x + 7x = (4+7)x = 11x$

23. $3z - z = 3z - 1z = (3-1)z = 2z$

25. $8x + 9x^2 - 2x = 8x - 2x + 9x^2 = (8-2)x + 9x^2 = 6x + 9x^2$ or $9x^2 + 6x$

27. $10w + w^2 - 8w^2 = 10w + (1-8)w^2 = 10w - 7w^2$ or $-7w^2 + 10w$

29. $ax^2 + a^2x + ax^2 = ax^2 + ax^2 + a^2x = 2ax^2 + a^2x$

31. $7xy^2 - 5x^2y + 4xy^2 = 7xy^2 + 4xy^2 - 5x^2y = 11xy^2 - 5x^2y$

33. $(a+6b) - (a-6b) = a + 6b - a + 6b = a - a + 6b + 6b = (1-1)a + (6+6)b = 0a + 12b = 12b$

35. $(2a^2 + 3b) + (2b + 4a) = 2a^2 + 3b + 2b + 4a = 2a^2 + (3+2)b + 4a = 2a^2 + 5b + 4a$

37. $(4x^2 + 3x) - (2x^2 - 3x) = 4x^2 + 3x - 2x^2 + 3x = (4-2)x^2 + (3+3)x = 2x^2 + 6x$

39. $2(6y^2 + 7x) = 2 \cdot 6y^2 + 2 \cdot 7x = 12y^2 + 14x$

41. $-3(4b - 2c) = -3 \cdot 4b + (-3)(-2c) = -12b + 6c$

43. $4(a+b) + 3(b+a) = 4a + 4b + 3b + 3a = 7a + 7b$

45. $3(x+y) - 2(x+y) = 3x + 3y - 2x - 2y = (3-2)x + (3-2)y = x + y$

47. $2(a+b+c) + 3(a+b-c) = 2a + 2b + 2c + 3a + 3b - 3c = 5a + 5b - c$

49. $3[2(x+y)] = 3[2x + 2y] = 6x + 6y$

51. $3(a+b) + 4(a+b) - 2(a+b) = 3a + 3b + 4a + 4b - 2a - 2b = 5a + 5b$ or $(3+4-2)(a+b) = 5(a+b) = 5a + 5b$

53. $2(a+b+c) + 3(a-b+c) + (a-b-c) = 2a + 2b + 2c + 3a - 3b + 3c + a - b - c = 6a - 2b + 4c$

55. $2(x+3y) - 3(x-2y) + 5(2x-y) = 2x + 6y - 3x + 6y + 10x - 5y = 9x + 7y$

57. $3(x+y-z) - 2(3x+2y-z) - 3(x-y+4z) = 3x + 3y - 3z - 6x - 4y + 2z - 3x + 3y - 12z = -6x + 2y - 13z$

59. $(x+y) - 3(x-z) + 4(y+4z) - 2(x+y-3z) = x + y - 3x + 3z + 4y + 16z - 2x - 2y + 6z = -4x + 3y + 25z$

61. $x + [3x + 2(x+y)] = x + [3x + 2x + 2y] = x + [5x + 2y] = x + 5x + 2y = 6x + 2y$

63. $y - [2z - 3(y+z) + y] = y - [2z - 3y - 3z + y] = y - [-z - 2y] = y + z + 2y = 3y + z$

65. $[2x + 3(x+y) - 2(x-y) + y] - 2x = [2x + 3x + 3y - 2x + 2y + y] - 2x = [3x + 6y] - 2x = x + 6y$

67. $-\{-[2a - (3b+a)]\} = -\{-[2a - 3b - a]\} = -\{-[a - 3b]\} = -\{-a + 3b\} = a - 3b$

69. $5a - 2\{4[a + 2(4a+b) - b] + a\} - a = 5a - 2\{4[a + 8a + 2b - b] + a\} - a = 5a - 2\{4[9a + b] + a\} - a = 5a - 2\{36a + 4b + a\} - a = 5a - 2\{37a + 4b\} - a = 5a - 74a - 8b - a = -70a - 8b$

71. $p + \frac{1}{2}p + \frac{2}{3}p = \frac{6}{6}p + \frac{3}{6}p + \frac{4}{6}p = \frac{6+3+4}{6}p = \frac{13}{6}p$

73. Follow the process in Example 2.7. The common denominator is C_1C_2. This produces the following result:
$$\frac{1}{C_T} = \frac{C_2}{C_1C_2} + \frac{C_1}{C_1C_2} = \frac{C_2+C_1}{C_1C_2}$$

≡ 2.2 MULTIPLICATION

1. $\left(a^2x\right)\left(ax^2\right) = a^{2+1} \cdot x^{1+2} = a^3x^3$

3. $(3ax)\left(2ax^2\right) = 3 \cdot 2 \cdot a^{1+1}x^{1+2} = 6a^2x^3$

5. $\left(2xw^2z\right)\left(-3x^2w\right) = 2(-3)x^{1+2}w^{2+1}z = -6x^3w^3z$

7. $(3x)(4ax)\left(-2x^2b\right) = 3 \cdot 4(-2)x^{1+1+2}ab = -24x^4ab$

9. $2(5y-6) = 2 \cdot 5y + 2(-6) = 10y - 12$

11. $-5(4w-7) = -5(4w) + (-5)(-7) = -20w + 35$ or $35 - 20w$

13. $3x(7y+4) = 3x(7y) + 3x(4) = 21xy + 12x$

15. $-5t(-3+t) = (-5t)(-3) - 5t(t) = 15t - 5t^2$

17. $\frac{1}{2}a(4a-2) = \frac{1}{2}a(4a) + \frac{1}{2}a(-2) = 2a^2 - a$

19. $2x(3x^2 - x + 4) = 2x(3x^2) + 2x(-x) + 2x(4) = 6x^3 - 2x^2 + 8x$

21. $4y^2(-5y^2 + 2y - 5 + 3y^{-1} - 6y^2) = 4y^2\left(-5y^{-2}\right) + 4y^2(2y) + 4y^2(-5) + 4y^2\left(3y^{-1}\right) - 4y^2\left(-6y^{-2}\right) - 20y^4 + 8y^3 - 20y^2 + 12y - 24$

23. $(a+b)(a+c) = (a+b)a + (a+b)c = a^2 + ab + ac + bc$ or by FOIL $a^2 + ac + ab + bc$

25. $(x+5)(x^2+6) = x^3 + 6x + 5x^2 + 30$ or $x^3 + 5x^2 + 6x + 30$

27. $(2x+y)(3x-y) = 6x^2 - 2xy + 3xy - y^2 = 6x^2 + xy - y^2$

29. $(2a-b)(3a-2b) = 6a^2 - 4ab - 3ab + 2b^2 = 6a^2 - 7ab + 2b^2$

31. $(b-1)(2b+5) = 2b^2 + 5b - 2b - 5 = 2b^2 + 3b - 5$

33. $\left(7a^2b + 3c\right)\left(8a^2b - 3c\right) = 56a^4b^2 - 21a^2bc + 24a^2bc - 9c^2 = 56a^4b^2 + 3a^2bc - 9c^2$

35. $(x+4)(x-4) = x^2 - 16$ (Difference of Squares)

37. $(p-6)(p+6) = p^2 - 36$

39. $(ax+2)(ax-2) = a^2x^2 - 4$

41. $\left(2r^2 + 3x\right)\left(2r^2 - 3x\right) = 4r^4 - 9x^2$

43. $\left(5a^2x^3 - 4d\right)\left(5a^2x^3 + 4d\right) = 25a^4x^6 - 16d^2$

45. $\left(\frac{2}{3}pa^2f + \frac{3}{4}tb^3\right)\left(-\frac{2}{3}pa^2f + \frac{3}{4}tb^3\right) = \left(\frac{3}{4}tb^3 + \frac{2}{3}pa^2f\right)\left(\frac{3}{4}tb^3 - \frac{2}{3}pa^2f\right) = \frac{9}{16}t^2b^6 - \frac{4}{9}p^2a^4f^2$

47. $(x+y)^2 = x^2 + 2xy + y^2$ (Square of a Binomial)

49. $(x-5)^2 = x^2 - 10x + 25$

51. $(a+3)^2 = a^2 + 6a + 9$

53. $(2a+b)^2 = 4a^2 + 4ab + b^2$

55. $(3x-2y)^2 = 9x^2 - 12xy + 4y^2$

57. $4x(x+4)(3x-2) = 4x\left(3x^2 - 2x + 12x - 8\right) = 4x\left(3x^2 + 10x - 8\right) = 12x^3 + 40x^2 - 32x$

59. $(x+y-z)(x-y+z) = (x+y-z)(x) + (x+y-z)(-y) + (x+y-z)(z) = x^2 + xy - xz - xy - y^2 + yz + xz + yz - z^2 = x^2 - y^2 - z^2 + 2yz$ or $[x+(y-z)][x-(y-z)] = x^2 - (y-z)^2 = x^2 - y^2 - z^2 + 2yz$

61. Think of $(a+2)^3$ as $(a+2)^2(a+2)$.

$(a+2)^3 = (a+2)^2(a+2) = \left(a^2 + 4a + 4\right)(a+2) = \left(a^2 + 4a + 4\right)(a) + \left(a^2 + 4a + 4\right)(2) = \left(a^3 + 4a^2 + 4a\right) + \left(2a^2 + 8a + 8\right) = a^3 + 6a^2 + 12a + 8$

63. $2[n(n+2)+n] = 2\left[n^2 + 2n + n\right] = 2\left[n^2 + 3n\right] = 2n^2 + 6n$

65. $\frac{1}{2}(y_2 - y_1)(y_2 + y_1) = \frac{1}{2}\left(y_2^2 - y_1^2\right) = \frac{1}{2}y_2^2 - \frac{1}{2}y_1^2$

67. $x = 3(t+4)(t+1) = 3\left(t^2 + 5t + 4\right) = 3t^2 + 15t + 12$

≡ 2.3 DIVISION

1. x^7 by $x^3 = \frac{x^7}{x^3} = x^{7-3} = x^4$

3. $2x^6$ by $x^4 = \frac{2x^6}{x^4} = 2x^2$

5. $12y^5$ by $4y^3 = \frac{12y^5}{4y^3} = \frac{12}{4} \cdot \frac{y^5}{y^3} = 3y^2$

7. $-45ab^2$ by $15ab = \frac{-45ab^2}{15ab} = -3b$

9. $33xy^2z$ by $3xyz = \frac{33xy^2z}{3xyz} = 11y$

11. $96a^2xy^3$ by $-16axy^2 = \frac{96a^2xy^3}{-16axy^2} = -6ay$

13. $144c^3d^2f$ by $8cf = \frac{144c^3d^2f}{8cf} = 18c^2d^2$

15. $9np^3$ by $-15n^3p^2 = \frac{9np^3}{-15n^3p^2} = \frac{3p}{-5n^2}$ or $\frac{-3p}{5n^2} = -\frac{3}{5}\frac{p}{n^2}$

17. $8abcdx^2y$ by $14adxy^2 = \frac{8abcdx^2y}{14adxy^2} = \frac{4bcx}{7y}$

19. $2a^3 + a^2$ by $a = \frac{2a^3+a^2}{a} = \frac{2a^3}{a} + \frac{a^2}{a} = 2a^2 + a$

21. $36b^4 - 18b^2$ by $9b = \frac{36b^4-18b^2}{9b} = \frac{36b^4}{9b} - \frac{18b^2}{9b} = 4b^3 - 2b$

23. $42x^2 + 28x$ by $7 = \frac{42x^2+28x}{7} = \frac{42x^2}{7} + \frac{28x}{7} = 6x^2 + 4x$

25. $34x^5 - 51x^2$ by $17x^2 = \frac{34x^5-51x^2}{17x^2} = \frac{34x^5}{17x^2} - \frac{51x^2}{17x^2} = 2x^3 - 3$

27. $24x^6 - 8x^4$ by $-4x^3 = \frac{24x^6-8x^4}{-4x^3} = \frac{24x^6}{-4x^3} - \frac{8x^4}{-4x^3} = -6x^3 + 2x$

29. $5x^2y + 5xy^2$ by $xy = \frac{5x^2y+5xy^2}{xy} = \frac{5x^2y}{xy} + \frac{5xy^2}{xy} = 5x + 5y$

31. $10x^2y + 15xy^2$ by $5xy = \frac{10x^2y+15xy^2}{5xy} = \frac{10x^2y}{5xy} + \frac{15xy^2}{5xy} = 2x + 3y$

33. $ap^2q - 2pq$ by $pq = \frac{ap^2q-2pq}{pq} = \frac{ap^2q}{pq} - \frac{2pq}{pq} = ap - 2$

35. $a^2bc + abc$ by $abc = \frac{a^2bc+abc}{abc} = \frac{a^2bc}{abc} + \frac{abc}{abc} = a + 1$

37. $9x^2y^2z - 3xyz^2$ by $-3xyz = \frac{9x^2y^2z}{-3xyz} - \frac{3xyz^2}{-3xyz} = -3xy + z$

39. $b^3x^2 + b^3$ by $-b = \frac{b^3x^2}{-b} + \frac{b^3}{-b} = -b^2x^2 - b^2$

41. $x^2y + xy - xy^2$ by $xy = \frac{x^2y}{xy} + \frac{xy}{xy} - \frac{xy^2}{xy} = x + 1 - y$

43. $18x^3y^2z - 24x^2y^3z$ by $-12x^2yz = \frac{18x^3y^2z}{-12x^2yz} - \frac{24x^2y^3z}{-12x^2yz} = -\frac{3}{2}xy + 2y^2$

45.
$$
\begin{array}{r}
x + 4 \\
x + 3 \,\overline{)\, x^2 + 7x + 12} \\
\underline{x^2 + 3x} \\
4x + 12 \\
\underline{4x + 12} \\
0
\end{array}
$$

47.
$$
\begin{array}{r}
x - 1 \\
x - 2 \,\overline{)\, x^2 - 3x + 2} \\
\underline{x^2 - 2x} \\
-x + 2 \\
\underline{-x + 2} \\
0
\end{array}
$$

49.
$$
\begin{array}{r}
x - 1 \\
x + 2 \,\overline{)\, x^2 + x - 2} \\
\underline{x^2 + 2x} \\
-x - 2 \\
\underline{-x - 2} \\
0
\end{array}
$$

51.
$$
\begin{array}{r}
2a + 1 \\
3a + 7 \,\overline{)\, 6a^2 + 17a + 7} \\
\underline{6a^2 + 14a} \\
3a + 7 \\
\underline{3a + 7} \\
0
\end{array}
$$

53.
$$
\begin{array}{r}
4y + 2 \\
2y - 3 \,\overline{)\, 8y^2 - 8y - 6} \\
\underline{8y^2 - 12y} \\
4y - 6 \\
\underline{4y - 6} \\
0
\end{array}
$$

55.
$$
\begin{array}{r}
x - 2 \\
x^2 + 2x - 1 \,\overline{)\, x^3 + 0x^2 - 5x + 2} \\
\underline{x^3 + 2x^2 - x} \\
-2x^2 - 4x + 2 \\
\underline{-2x^2 - 4x + 2} \\
0
\end{array}
$$

57.
$$
\begin{array}{r}
2a - 1 \\
3a^2 - 2a + 4 \,\overline{)\, 6a^3 - 7a^2 + 10a - 4} \\
\underline{6a^3 - 4a^2 + 8a} \\
-3a^2 + 2a - 4 \\
\underline{-3a^2 + 2a - 4} \\
0
\end{array}
$$

59.
$$
\begin{array}{r}
2x^2 + x - 1 \\
2x - 1 \,\overline{)\, 4x^3 + 0x^2 - 3x + 4} \\
\underline{4x^3 - 2x^2} \\
2x^2 - 3x + 4 \\
\underline{2x^2 - x} \\
-2x + 4 \\
\underline{-2x + 1} \\
3
\end{array}
$$

61.
$$
\begin{array}{r}
r^2 - 3r \\
r + 2 \,\overline{)\, r^3 - r^2 - 6r + 5} \\
\underline{r^3 + 2r^2} \\
-3r^2 - 6r + 5 \\
\underline{-3r^2 - 6r} \\
5
\end{array}
$$

63.
$$
\begin{array}{r}
x^3 + 3x^2 + 9x + 27 \\
x - 3 \,\overline{)\, x^4 \qquad\qquad\quad - 81} \\
\underline{x^4 - 3x^3} \\
3x^3 \qquad\qquad - 81 \\
\underline{3x^3 - 9x^2} \\
9x^2 \qquad - 81 \\
\underline{9x^2 - 27x} \\
27x - 81 \\
\underline{27x - 81} \\
0
\end{array}
$$

65.

$$
\begin{array}{r}
4x^3 \ -3x^2 \ -2x \ +6 \\
3x^2+x-2\,\overline{)12x^5-5x^4-14x^3+23x^2+8x-12} \\
\underline{12x^5+4x^4-\ 8x^3} \\
-9x^4-\ 6x^3+23x^2 \\
\underline{-9x^4-\ 3x^3+\ 6x^2} \\
-3x^3+17x^2+8x \\
\underline{-3x^3-\ \ x^2+2x} \\
18x^2+6x-12 \\
\underline{18x^2+6x-12} \\
0
\end{array}
$$

67.

$$
\begin{array}{r}
x+y \\
x-y\,\overline{)x^2\qquad -y^2} \\
\underline{x^2-xy} \\
xy-y^2 \\
\underline{} \\
0
\end{array}
$$

69.

$$
\begin{array}{r}
w^2+wz+z^2 \\
w-z\,\overline{)w^3\qquad\quad -z^3} \\
\underline{w^3-w^2z} \\
w^2z\qquad -z^3 \\
\underline{w^2z-wz^2} \\
wz^2-z^3 \\
\underline{wz^2-z^3} \\
0
\end{array}
$$

71.

$$
\begin{array}{r}
x^2+y^2 \\
x+y\,\overline{)x^3+x^2y+xy^2+y^3} \\
\underline{x^3+x^2y} \\
0+xy^2+y^3 \\
\underline{xy^2+y^3} \\
0
\end{array}
$$

73.

$$
\begin{array}{r}
c^2d^2+2cd+4 \\
cd-2\,\overline{)c^3d^3\qquad\qquad -8} \\
\underline{c^3d^3-2c^2d^2} \\
2c^2d^2\qquad -8 \\
\underline{2c^2d^2-4cd} \\
4cd-8 \\
\underline{4cd-8} \\
0
\end{array}
$$

75.

$$
\begin{array}{r}
x\ -y \\
x-y\,\overline{)x^2-2xy+y^2} \\
\underline{x^2-\ xy} \\
-xy+y^2 \\
\underline{-xy+y^2} \\
0
\end{array}
$$

77.

$$
\begin{array}{r}
p^2r\ -2p\ +3r^2 \\
5p-r\,\overline{)5p^3r-\ p^2r^2-10p^2+2pr+15pr^2-3r^3} \\
\underline{5p^3r-\ p^2r^2} \\
0-10p^2+2pr+15pr^2-3r^3 \\
\underline{-10p^2+2pr} \\
0+15pr^2-3r^3 \\
\underline{0+15pr^2-3r^3} \\
0
\end{array}
$$

79.

$$
\begin{array}{r}
a\ +d\ +4 \\
a-3d-1\,\overline{)a^2-2ad-3d^2+3a-13d-8} \\
\underline{a^2-3ad\qquad\ -a} \\
ad-3d^2+4a-13d-8 \\
\underline{ad-3d^2\qquad -d} \\
4a-12d-8 \\
\underline{4a-12d-4} \\
-4
\end{array}
$$

81.

$$
\begin{array}{r}
af \\
a-f\,\overline{)a^2f-af^2} \\
\underline{a^2f-af^2} \\
0
\end{array}
$$

83.

$$
\begin{array}{r}
a\ +\ b\ -c \\
a-b+c\,\overline{)a^2\qquad\quad -b^2+2bc-c^2} \\
\underline{a^2-ab+ac} \\
ab-ac-b^2+2bc-c^2 \\
\underline{ab\qquad -b^2+\ bc} \\
-ac\qquad +\ bc-c^2 \\
\underline{-ac\qquad +\ bc-c^2} \\
0
\end{array}
$$

85.

$$
\begin{array}{r}
a^2 - a + 1 \\
a^2 + a + 2 \overline{\smash{\big)}\ a^4 + 0a^3 + 2a^2 - a + 2} \\
\underline{a^4 + a^3 + 2a^2} \\
-a^3 + 0a^2 - a + 2 \\
\underline{-a^3 - a^2 - 2a} \\
a^2 + a + 2 \\
\underline{a^2 + a + 2} \\
0
\end{array}
$$

87. Reciprocal is $\frac{R_2R_3 + R_1R_3 + R_1R_2}{R_1R_2R_3} = \frac{R_2R_3}{R_1R_2R_3} +$

$\frac{R_1R_3}{R_1R_2R_3} + \frac{R_1R_2}{R_1R_2R_3} = \frac{1}{R_1} + \frac{1}{R_2} + \frac{1}{R_3}$

89. $V_2 = V_1\left(1 + \frac{T_2 - T_1}{T_1}\right) = V_1\left(1 + \frac{T_2}{T_1} - 1\right) =$

$V_1\left(\frac{T_2}{T_1}\right) = \frac{V_1T_2}{T_1}$

≡ 2.4 SOLVING EQUATIONS

1. $x - 7 = 32;\ (x - 7) + 7 = 32 + 7;\ x = 39$

3. $a + 13 = 25;\ (a + 13) - 13 = 25 - 13;\ a = 12$

5. $25 + c = 10;\ 25 + c - 25 = 10 - 25;\ c = -15$

7. $4.3 + w = 8.7;\ 4.3 + w - 4.3 = 8.7 - 4.3;\ w = 4.4$

9. $4x = 18;\ \frac{4x}{4};\ x = \frac{9}{2}$ or $4\frac{1}{2}$

11. $-3w = 24;\ \frac{-3w}{-3} = \frac{24}{-3};\ w = -8$

13. $12a = 18;\ \frac{12a}{12} = \frac{18}{12};\ a = \frac{3}{2}$

15. $21c = -14;\ \frac{21c}{21} = \frac{-14}{21};\ c = -\frac{2}{3}$

17. $\frac{p}{3} = 5;\ 3 \cdot \frac{p}{3} = 5 \cdot 3;\ p = 15$

19. $\frac{t}{4} = -6;\ 4 \cdot \frac{t}{4} = -6 \cdot 4;\ t = -24$

21. $4a + 3 = 11;\ 4a + 3 - 3 = 11 - 3;\ 4a = 8;\ \frac{4a}{4} = \frac{8}{4};\ a = 2$

23. $7 - 8d = 39;\ 7 - 8d - 7 = 39 - 7;\ -8d = 32;\ \frac{-8d}{-8} = \frac{32}{-8};\ d = -4$

25. $4x - 3 = -37;\ 4x - 3 + 3 = -37 + 3;\ 4x = -34;\ \frac{4x}{4} = \frac{-34}{4};\ x = -\frac{17}{2}$

27. $2.3w + 4.1 = 13.3;\ 2.3w + 4.1 - 4.1 = 13.3 - 4.1;\ 2.3w = 9.2;\ \frac{2.3w}{2.3} = \frac{9.2}{2.3};\ w = 4$

29. $2x + 5x = 28;\ 7x = 28;\ \frac{7x}{7} = \frac{28}{7};\ x = 4$

31. $3x = 7 - 10;\ 3x = -3;\ \frac{3x}{3} = \frac{-3}{3};\ x = -1$

33. $3a + 2(a + 5) = 45;\ 3a + 2a + 10 = 45;\ 5a + 10 = 45;\ 5a = 35;\ a = 7$

35. $4(6 + c) - 5 = 21;\ 24 + 4c - 5 = 21;\ 4c + 19 = 21;\ 4c = 2;\ c = \frac{1}{2}$

37. $2(p - 4) + 3p = 16;\ 2p - 8 + 3p = 16; 5p - 8 = 16;\ 5p = 24;\ p = \frac{24}{5}$

39. $3x = 2x + 5;\ 3x - 2x = 5;\ x = 5$

41. $4w = 6w + 12;\ 4w - 6w = 12;\ -2w = 12;\ w = -6$

43. $9a = 54 + 3a;\ 9a - 3a = 54;\ 6a = 54;\ a = 9$

45. $\frac{5x}{2} = \frac{4x}{3} - 7;\ 6\left(\frac{5x}{2}\right) = \left(\frac{4x}{3} - 7\right); 3 \cdot 5x = \frac{6 \cdot 4x}{3} - 6 \cdot 7; 15x = 8x - 42;\ 15x - 8x = -42; 7x = -42;\ x = -6$

47. $\frac{6p}{5} = \frac{3p}{2} + 4;\ 10\frac{6p}{5} = 10\left(\frac{3p}{2} + 4\right); 2 \cdot 6p = 5 \cdot 3p + 40;\ 12p = 15p + 40;\ -3p = 40;\ p = -\frac{40}{3}$

49. $8n - 4 = 5n + 14;\ 8n - 5n = 14 + 4;\ 3n = 18; n = 6$

51. $7r + 3 = 11r - 21;\ 7r - 11r = -21 - 3; -4r = -24;\ r = 6$

53. $9t + 6 = 3t - 5;\ 9t - 3t = -5 - 6;\ 6t = -11; t = -\frac{11}{6}$

55. $\frac{6x - 3z}{2} = \frac{7x + 2}{3};\ 6\left(\frac{6x - 3}{2}\right); 6\left(\frac{7x + 2}{3}\right); 3(6x - 3) = 2(7x + 2);\ 18x - 9 = 14x + 4; 18x - 14x = 4 + 9;\ 4x = 13;\ x = \frac{13}{4}$

57. $\frac{3t + 4}{4} = \frac{2t - 5}{2};\ 8\left(\frac{3t + 4}{4}\right) = 8\left(\frac{2t - 5}{2}\right); 2(3t + 4) = 4(2t - 5);\ 6t + 8 = 8t - 20; 6t - 8t = -20 - 8;\ -2t = -28;\ t = 14$

59. $3(x + 5) = 2x - 3;\ 3x + 15 = 2x - 3; 3x - 2x = -3 - 15;\ x = -18$

61. $5(w - 7) = 2w + 4;\ 5w - 35 = 2w + 4; 5w - 2w = 4 + 35;\ 3w = 39;\ w = 13$

63. $\frac{x}{2} + \frac{x}{3} - \frac{x}{4} = 2;\ 12\left(\frac{x}{2} + \frac{x}{3} - \frac{x}{4}\right) = 12 \cdot 2; 12\frac{x}{2} + \frac{12x}{3} - \frac{12x}{4} = 24;\ 6x + 4x - 3x = 24; 7x = 24;\ x = \frac{24}{7}$

65. $\frac{4(a-3)}{5} = \frac{3(a+2)}{4}$; $20\left(\frac{4(a-3)}{5}\right) = 20\left(\frac{3(a+2)}{4}\right)$; $4(4(a-3)) = 5(3(a+2))$; $16a - 48 = 15a + 30$; $16a - 15a = 30 + 48$; $a = 78$

67. Solve $ax + b = 3ax$ for x; $ax - 3ax = -b$; $x(a - 3a) = -b$; $x(-2a) = -b$; $x = \frac{-b}{-2a} = \frac{b}{2a}$

69. Solve $ax - 3a + x = 5a$ for a; $ax + x = 8a$; $ax - 8a = -x$; $a(x-8) = -x$; $a = \frac{-x}{x-8} = \frac{x}{x-8}$ or $\frac{x}{8-x}$

71. $\frac{3}{x} + \frac{4}{x} = 3$; $x\left(\frac{3}{x} + \frac{4}{x}\right) = x \cdot 3$; $3 + 4 = 3x$; $7 = 3x$; $3x = 7$; $x = \frac{7}{3}$

73. $\frac{3}{4p} + \frac{1}{p} = \frac{7}{4}$; $4p\left(\frac{3}{4p} + \frac{1}{p}\right) = 4p \cdot = \frac{7}{4}$; $3 + 4 = 7p$; $7 = 7p$; $p = 1$

75. $\frac{1}{x+1} - \frac{2}{x-1} = 0$; $(x+1)(x-1)\left[\frac{1}{x+1} - \frac{2}{x-1}\right] = 0(x+1)(x-1)$; $(x-1) - 2(x+1) = 0$; $x - 1 - 2x - 2 = 0$; $-x - 3 = 0$; $-x = 3$; $x = -3$

77. $\frac{3}{2x} = \frac{1}{x+5}$; $(x+5)2x \cdot \frac{3}{2x} = (x+5)2x\left(\frac{1}{x+5}\right)$; $3(x+5) = 2x$; $3x + 15 = 2x$; $3x - 2x = -15$; $x = -15$

79. $\frac{2x+1}{2x-1} = \frac{x-1}{x-3}$; $(x-3)(2x+1) = (x-1)(2x-1)$; $2x^2 - 5x - 3 = 2x^2 - 3x + 1$; $2x^2 - 2x^2 - 5x + 3x = 1 + 3$; $-2x = 4$; $x = -2$

81. A common denominator of these expressions is 12. Multiplying by 12 and simplifying produces the following:
$(12)\left(\frac{6b-5a}{3}\right) + (12)\left(\frac{5a+b}{2}\right) = (12)\left(\frac{5(a+2b)}{4}\right) + (12)5$; $4(6b - 5a) + 6(5a + b) = 15(a + 2b) + 60$; $24b - 20a + 30a + 6b = 15a + 30b + 60$; $30b + 10a = 15a + 30b + 60$; $-5a = 60$; $a = -12$

83. Multiplying by a common denominator of $6x$ produces the following:
$(6x)\left(\frac{2z+a}{3x}\right) - (6x)\left(\frac{9z+a}{6x}\right) = (6x)\left(\frac{z-a}{2x}\right) + (6x)\left(\frac{4a}{3x}\right)$; $2(2z+a) - (9z+a) = 3(z-a) + 2(4a)$; $4z + 2a - 9z - a = 3z - 3a + 8a$; $-5z + a = 3z + 5a$; $a - 5a = 3z + 5z$; $-4a = 8z$; $a = -2z$

85. $F - 32 = \frac{9}{5}C$; $\frac{9}{5}C = F - 32$; $\frac{5}{9} \cdot \frac{9}{5}C = \frac{5}{9}(F - 32)$; $C = \frac{5}{9}(F - 32)$

≡ 2.5 APPLICATIONS OF EQUATIONS

1. $\frac{79+85+74+x}{4} = 80$; $\frac{238+x}{4} = 80$; $238 + x = 320$; $x = 82$

3. $\frac{85+82+x}{3} = 75$; $167 + x = 225$; $x = 58$; minimum of 60

5. $.80w = 920$; $w = \frac{920}{.8} = \$1150$; $1150 - 920 = \$230$

7. $0.30c = 1839$; $c = \frac{1839}{0.30}$; $c = \$6130$

9. $a =$ amount at 7.5%; $(4500 - a) =$ amount at 6%; $0.075 \times a + .06(4500 - a) = 303$; $0.075a + 270 - .06a = 303$; $0.015a = 33$; $a = \frac{33}{.015} = \$2200$ at 7.5%; $4500 - 2200 = \$2300$ at 6%

11. $40 \times 8.50 + x(1.5)8.50 = 429.25$
$340 + 12.75x = 429.25$
$12.75x = 89.25$
$x = 80.25 \div 12.75 = 6.29$ h

13. $d = rt$; $d = 38$ mph $\times 7$ h $= 266$ mi

15.
$t =$ days of ships
$380 - 12 \times (t + 1) = 80t$
$380 - 12t - 12 = 80t$
$368 = 80t + 12t$
$368 = 92t \frac{368}{92} = t$
$t = 4$ days
$80 \times 4 = 320$ km

17. h = h together $\frac{1}{6} + \frac{1}{4} = \frac{1}{h}$; $12h = LCD 12h \cdot \frac{1}{6} + 12h \cdot \frac{1}{4} = 12h \cdot \frac{1}{h}2h + 3h = 125h = 12h = \frac{12}{5} = 2\frac{2}{5}$; 2 hours 24 minutes

19. $\frac{1}{4} + \frac{1}{2} = \frac{1}{h}$; $4h \cdot \frac{1}{4} + 4h \cdot \frac{1}{2} = 4h \cdot \frac{1}{h}$; $h + 2h = 4$; $3h = 4$; $h = \frac{4}{3} = 1$ h 20 min

21. $50 \times 0.86 = (50 + w) \times 0.40$; $43 = 20 + 0.40$; $23 = .4w$; $\frac{23}{.4} = w$; 57.5 mL $= w$

23. $x =$ amount of 35%; $750 - x =$ amount of 75%; $0.35x + 0.75(750 - x) = 0.60 \times 750$; $0.35x + 562.5 - 0.75x = 450$; $-0.4x = -112.5$; $x = 281.25$ of 35% copper; $750 - 281.25 = 468.75$ of 75% copper.

25. $850 - 500 = 350$ lb; $500x = 350(20 - x)$; $500x = 7000 - 350x$; $(500 + 350)x = 7000$; $850x = 7000$; $x = 8.24$ ft from 500 lb end

27. $x =$ distance from left
$25x = 15(12 - x)$
$25x = 180 - 15x$
$40x = 180$
$x = 180 \div 40$
$x = 4.5$ ft from the left end

29. Let d be the location of the center of gravity in cm from the right end. The torque to the left of the

center of gravity is $4 + 2(8 - d)$. To the right of the center of gravity, it is $2d$. Thus, $4 + 2(8 - d) = 2d$ or $20 = 4d$ and $d = 5$. The center of gravity is 5 in. from the right end.

31. Let n represent the number of ccs (cm^3) of pure alcohol that the nurse must add.
$$100\%(n) + 60\%(10) = 90\%(n + 10)$$
$$n + 6 = 0.9n + 9$$
$$0.1n = 3$$
$$n = 30$$

The nurse must add 30 cc of pure alcohol.

CHAPTER 2 REVIEW

1. $8y - 5y = (8 - 5)y = 3y$

2. $4z + 15z = (4 + 15)z = 19z$

3. $7x - 4x + 2x - 8 = (7 - 4 + 2)x - 8 = 5x - 8$

4. $-9a + 4a - 3a + 2 = (-9 + 4 - 3)a + 2 = -8a + 2$

5. $(2x^2 + 3x + 4) + (5x^2 - 3x + 7) = (2 + 5)x^2 + (3 - 3)x + (4 + 7) = 7x^2 + 11$

6. $(3y^2 - 4y - 3) + (5y - 3y^2 + 6) = (3 - 3)y^2 + (-4 + 5)y + (-3 + 6) = y + 3$

7. $2(8x + 4) = 2 \cdot 8x + 2 \cdot 4 = 16x + 8$

8. $-3(4a - 2) = -3 \cdot 4a - 3(-2) = -12a + 6$

9. $-3(x - 1) = -3x + 1$ or $1 - 3x$

10. $-(2z + 5) = -2z - 5$

11. $(4x^2 + 3x) - (2x - 5x^2 + 2) = 4x^2 + 3x - 2x + 5x^2 - 2 = 9x^2 + x - 2$

12. $(7y^2 + 6y) - (6y - 7y^2) + 2y - 5 = 7y^2 + 6y - 6y + 7y^2 + 2y - 5 = 14y^2 + 2y - 5$

13. $2(a + b) - 3(a - b) + 4(a + b) = 2a + 2b - 3a + 3b + 4a + 4b = 3a + 9b$

14. $6(c - d) - 4(d - c) + 2(c + d) = 6c - 6d - 4d + 4c + 2c + 2d = 12c - 8d$

15. $(ax^2)(a^2x) = (aa^2)(x^2x) = a^{1+2}x^{2+1} = a^3x^3$

16. $(cy^3)(dy) = cdy^{3+1} = cdy^4$

17. $(9ax^2)(3x) = (9 \cdot 3)(ax^{2+1}) = 27ax^3$

18. $(6cy^2z)(2cz^3) = 6 \cdot 2c^{1+1}y^2z^{1+3} = 12c^2y^2z^4$

19. $4(5x - 6) = 4 \cdot 5x - 4 \cdot 6 = 20x - 24$

20. $3(12y - 5) = 3 \cdot 12y - 3 \cdot 5 = 36y - 15$

21. $2x(4x - 5) = 2x \cdot 4x - 2x \cdot 5 = 8x^2 - 10x$

22. $3a(6a + a^2) = 3a \cdot 6a + 3a \cdot a^2 = 18a^2 + 3a^3$

23. $(a + 4)(a - 4) = a^2 - 16$ (Difference of Squares)

24. $(x - 9)(x + 9) = x^2 - 81$ (Difference of Squares)

25. $(2a - b)(3a - b) = 6a^2 - 2ab - 3ab + b^2 = 6a^2 - 5ab + b^2$ FOIL

26. $(4x + 1)(3x - 7) = 12x^2 - 28x + 3x - 7 = 12x^2 - 25x - 7$ FOIL

27. $(3x^2 + 2)(2x^2 - 3) = 6x^4 - 9x^2 + 4x^2 - 6 = 6x^4 - 5x^2 - 6$ FOIL

28. $(4a^3 + 2)(6a - 3) = 24a^4 - 12a^3 + 12a - 6$ FOIL

29. $(x + 2)^2 = x^2 + 2 \cdot x \cdot 2 + 2^2 = x^2 + 4x + 4$
Binomial Squared

30. $(3 - y)^2 = 3^2 - 2 \cdot 3 \cdot y + y^2 = 9 - 6y + y^2$
Binomial Squared

31. $5x(3x - 4)(2x + 1) = 5x[6x^2 + 3x - 8x - 4] = 5x[6x^2 - 5x - 4] = 30x^3 - 25x^2 - 20x$

32. $6a(4a + 3)(a - 2) = 6a[4a^2 - 8a + 3a - 6] = 6a[4a^2 - 5a - 6] = 24a^3 - 30a^2 - 36a$

33. $a^5 \div a^2 = \frac{a^5}{a^2} = a^{5-2} = a^3$

34. $x^7 \div x^3 = \frac{x^7}{x^3} = x^{7-3} = x^4$

35. $8a^2 \div 2a = \frac{8a^2}{2a} = \frac{8}{2} \cdot a^{2-1} = 4a$

36. $27b^3 \div 3b = \frac{27b^3}{3b} = \frac{27}{3} \cdot b^{3-1} = 9b^2$

37. $45a^2x^3 \div -5ax = \frac{45a^2x^3}{-5ax} = -9ax^2$

38. $52b^4c^2 \div -4b^2 = \frac{52b^4c^2}{-4b^2} = -13b^2c^2$

39. $(36x^2 - 16x) \div 2x = \frac{36x^2 - 16x}{2x} = \frac{36x^2}{2x} - \frac{16x}{2x} =$
$18x - 8$

40. $(39b^3 \dotplus 52b^5) \div 13b^2 = \frac{39b^3 + 52b^5}{13b^2} = \frac{39b^3}{13b^2} + \frac{52b^5}{13b^2} =$
$3b + 4b^3$

41.
$$\begin{array}{r} x - 4 \\ x+3 \overline{)\, x^2 - x - 12} \\ \underline{x^2 + 3x} \\ -4x - 12 \\ \underline{-4x - 12} \\ 0 \end{array}$$

42.
$$\begin{array}{r} x + 5 \\ x-6 \overline{)\, x^2 - x - 30} \\ \underline{x^2 - 6x} \\ 5x - 30 \\ \underline{5x - 30} \\ 0 \end{array}$$

43.
$$\begin{array}{r} x^2 + 3x + 9 \\ x-3 \overline{)\, x^3 \qquad\quad - 27} \\ \underline{x^3 - 3x^2} \\ 3x^2 \quad - 27 \\ \underline{3x^2 - 9x} \\ 9x - 27 \\ \underline{9x - 27} \\ 0 \end{array}$$

44.
$$\begin{array}{r} 4a^2 - 8a + 16 \\ 2a+4 \overline{)\, 8a^3 \qquad\qquad + 64} \\ \underline{8a^3 + 16a^2} \\ -16a^2 \quad + 64 \\ \underline{-16a^2 - 32a} \\ 32a + 64 \\ \underline{32a + 64} \\ 0 \end{array}$$

45.
$$\begin{array}{r} x - y \\ x+y \overline{)\, x^2 \quad - y^2} \\ \underline{x^2 + xy} \\ -xy - y^2 \\ \underline{-xy - y^2} \\ 0 \end{array}$$

46.
$$\begin{array}{r} y - a \\ y+a \overline{)\, y^2 \quad - a^2} \\ \underline{y^2 + ya} \\ -ya - a^2 \\ \underline{-ya - a^2} \\ 0 \end{array}$$

47.
$$\begin{array}{r} x - 2y \\ x^2-y^2 \overline{)\, x^3 - 2x^2y - xy^2 + 2y^3} \\ \underline{x^3 \qquad\quad - xy^2} \\ -2x^2y \quad + 2y^3 \\ \underline{-2x^2y \qquad + 2y^3} \\ 0 \end{array}$$

48.
$$\begin{array}{r} a^2 + 4ab + b^2 \\ a-b \overline{)\, a^3 + 3a^2b + 2ba - 3ab^2 - b^3} \\ \underline{a^3 - a^2b} \\ 4a^2b + 2ba - 3ab^2 - b^3 \\ \underline{4a^2b \qquad - 4ab^2} \\ 2ba + ab^2 - b^3 \\ \underline{ab^2 - b^3} \\ 2ba \end{array}$$

49. $x + 9 - 9 = 47 - 9;\ x = 38$

50. $y - 19 + 19 = -32 + 19;\ y = -13$

51. $\frac{2x}{2} = \frac{15}{2};\ x = \frac{15}{2}$

52. $\frac{-3y}{-3} = \frac{14}{-3};\ y = -\frac{14}{3}$

53. $4 \cdot \frac{x}{4} = 4 \cdot 9;\ x = 36$

54. $\frac{y}{3} = -7;\ 3\frac{y}{3} = 3(-7);\ y = -21$

55. $4x - 3 = 17;\ 4x - 3 + 3 = 17 + 3;\ 4x = 20;\ \frac{4x}{4} = \frac{20}{4};\ x = 5$

56. $7 + 8y = 23;\ 7 + 8y - 7 = 23 - 7;\ 8y = 16;\ \frac{8y}{8} = \frac{16}{8};\ y = 2$

57. $3.4a - 7.1 = 8.2;\ 3.4a - 7.1 + 7.1 = 8.2 + 7.1;\ 3.4a = 15.3;\ a = \frac{15.3}{3.4} = 4.5$

58. $6.2b + 19.1 = 59.4;\ 6.2b = 59.4 - 19.1;\ 6.2b = 40.3;\ b = \frac{40.3}{6.2} = 6.5$

59. $4x + 3 = 2x;\ 4x - 2x = -3;\ 2x = -3;\ x = -\frac{3}{2}$

60. $7a - 2 = 2a;\ 7a - 2a = 2;\ 5a = 2;\ a = \frac{2}{5}$

61. $4b + 2 = 3b - 5;\ 4b - 3b = -5 - 2;\ b = -7$

62. $7c + 9 = 12c - 4;\ 7c - 12c = -4 - 9;\ -5c = -13;\ c\frac{13}{5}$

63. $\frac{4(x-3)}{3} = \frac{5(x+4)}{2};\ 6\left(\frac{4(x-3)}{3}\right) = 6\left(\frac{5(x+4)}{2}\right);\ 2 \cdot 4(x-3) = 3 \cdot 5(x+4);\ 8x - 24 = 15x + 60;\ -7x = 84;\ x = -12$

64. $\frac{3(y-7)}{5} = \frac{5(y+4)}{2};\ 10\left(\frac{3(y-7)}{5}\right) = 10\left(\frac{5(y+4)}{2}\right);\ 2 \cdot 3(y-7) = 5 \cdot 5(y+4);\ 6y - 42 = 25y + 100;\ -19y = 142;\ y = -\frac{142}{19}$

65. $\frac{2}{a} - \frac{3}{a} = 5;\ a\left(\frac{2}{a} - \frac{3}{a}\right) = a \cdot 5;\ 2 - 3 = 5a;\ -1 = 5a;\ a = -\frac{1}{5}$

66. $\frac{4}{x} + \frac{5}{x} = \frac{1}{8}$; $8x\left(\frac{4}{x} + \frac{5}{x}\right) = 8x \cdot \frac{1}{8}$; $32 + 40 = x$; $72 = x$

67. $\frac{3}{2a} = \frac{3}{a+2}$; $2a(a+2)\frac{3}{2a} = 2a(a+2) \cdot \frac{3}{a+2}$; $(a+2)3 = 2a \cdot 3$; $3a + 6 = 6a$; $6 = 3a$; $a = 2$

68. $\frac{9}{4b} = \frac{12}{b+4}$; $(b+4)9 = 4b(12)$; $9b + 36 = 48b$; $36 = 39b$; $b = \frac{36}{39} = \frac{12}{13}$

69. $\frac{68+70+74+x}{4} = 72$; $\frac{212+x}{4} = 72$; $212 + x = 72 \times 4$; $212 + x = 288$; $x = 288 - 212 = 76$

70. $p + 0.065p = \$342.93$; $1.065p = 342.93$; $p = 342.93 \div 1.065 = \$322$

71. $d = rt$; $\frac{2718}{755} = \frac{755t}{755}$; $t = 3.6$ hours or 3 hrs 36 min

72. $t =$ satellite time; $t - 1\frac{3}{4} =$ shuttle time; $330t = 430(t - 1.75)$; $330t = 430t - 752.5$; $-100t = -752.5$; $t = 7.525$ hours or at $7:31\frac{1}{2}$ min $= 7:31:30$

73. $40\,\text{kg} \times 50\% + 15 = 20 + 15 = 35$ kg lead; $\frac{35}{40+15} = \frac{35}{55} = 63.64\%$ lead

74. $460 - 320 = 140\text{N}$; $320x = 140(8 - x)$; $320x = 1120 - 140x$; $(320 + 140)x = 1120$; $460x = 1120x = 2.43$ m from 320 force

75. Use the formula $F = ma$. $F_1 = 1538.6$ N; $F_2 = 1107.4$ N; $m_1 = \frac{F_1}{a} = \frac{1538.6}{9.8} = 157$ kg; $m_2 = \frac{F_2}{a} = \frac{1107.4}{9.8} = 113$ kg

76. Let n represent the number of grams of the 40% solution of the medicine that the pharmacist must use. Then $20 - n$ is the number of grams of the 70% solution that must be used.
$$40\%(n) + 70\%(20 - n) = 52\%(20)$$
$$0.4n + 14 - 0.7n = 10.4$$
$$-0.3n = -3.6$$
$$n = 12$$

The pharmacist should use 12 g of the 40% solution and $20 - 12 = 8$ g of the 70% solution.

CHAPTER 2 TEST

1. $5x^2 + (2 - 5)x = 5x^2 - 3x$

2. $(4a^3 - 2b) - (3b + a^3) = 4a^3 - 2b - 3b - a^3 = (4a^3 - a^3) - (2b + 3b) = 3a^3 - 5b$

3. $4x + [3(x + y - 2) - 5(x - y)] = 4x + [3x + 3y - 6 - 5x + 5y] = (4x + 3x - 5x) + (3y + 5y) - 6 = 2x + 8y - 6$

4. $\left(4xy^3z\right)\left(\frac{1}{2}xy^{-2}z^2\right) = 4\left(\frac{1}{2}\right)x^{1+1}y^{3-2}z^{1+2} = 2x^2yz^3$

5. $(2b - 3)(2b + 3) = (2b)^2 - 3^2 = 4b^2 - 9$

6. $(x^3 + 3x) \div x$; $\frac{x^3 + 3x}{x} = \frac{x^3}{x} + \frac{3x}{x} = x^{3-1} + 3x^{1-1} = x^2 + 3x^0 = x^2 + 3$

7. $\left(6x^5 + 4x^3 - 1\right) \div 2x^2$; $\frac{6x^5 + 4x^3 - 1}{2x^2} = \frac{6x^5}{2x^2} + ; \frac{4x^3}{2x^2} - \frac{1}{2x^2}; = 3x^3 + 2x - \frac{1}{2x^2}$

8.
$$
\begin{array}{r}
3x^2 - 8x + 17 \\
x + 2 \overline{)\,3x^3 - 2x^2 + x - 3} \\
\underline{3x^3 + 6x^2} \\
-8x^2 + x - 3 \\
\underline{-8x^2 - 16x} \\
17x - 3 \\
\underline{17x + 34} \\
-37
\end{array}
$$

9. $\frac{y}{3y+2} - \frac{4}{y-1} = \frac{y(y-1)}{(3y+2)(y-1)} - \frac{4(3y+2)}{(3y+2)(y-1)} = \frac{y(y-1)-4(3y+2)}{(3y+2)(y-1)} = \frac{y^2-y-12y-8}{(3y+2)(y-1)} = \frac{y^2-13y-8}{(3y+2)(y-1)}$

10. $5x - 8 = 3x$; $5x - 3x = 8$; $2x = 8$; $x = \frac{8}{2}$; $x = 4$

11. $\frac{7x+3}{2} - \frac{9x-12}{4} = 8$; $4\left(\frac{7x+3}{2}\right) - 4\left(\frac{9x-12}{4}\right) = 4 \cdot 8$; $14x + 6 - 9x + 12 = 32$; $5x + 18 = 32$; $5x = 14$; $x = \frac{14}{5}$

12. $p + 0.7p = 74.85$; $1.07p = 74.85$; $p = 74.85 \div 1.07$; $p = \$69.95$

13. Let n be the amount of original solution to be removed. Then $9 - n$ is the amount left. You want to end with 60% of 9 qt or 5.4 qt of antifreeze. The antifreeze left from the original solution is 50% of $9 - n$, so $50\%(9 - n) + n = 5.4$; $0.5(9 - n) + n = 5.4$; $4.5 - 0.5n + n = 5.4$; $0.5n = 0.9$; $n = 1.8$ qt

3

Geometry

≡ 3.1 LINES AND ANGLES

1. Using the proportion $\frac{D}{180°} = \frac{R}{\pi}$, we substitute $15°$ for D and solve for R. Thus, $\frac{15°}{180°} = \frac{R}{\pi}$ or $R = \frac{15°}{180°}\pi = \frac{1}{12}\pi = \frac{\pi}{12} \approx 0.2617994$.

3. Multiplying $210°$ by $\frac{\pi}{180°}$, we obtain $\frac{210°}{180°}\pi = \frac{7}{6}\pi = \frac{7\pi}{6} \approx 3.6651914$.

5. Multiplying $85.4°$ by $\frac{\pi}{180°}$ produces $\frac{85.4°}{180°}\pi = \frac{854}{1800}\pi = \frac{427}{900}\pi \approx 1.4905112$. Multiplying $85.4 \times 0.01745 = 1.49023$.

7. Multiplying $163.5°$ by $\frac{\pi}{180°}$ we obtain $\frac{163.5°}{180°}\pi = \frac{109}{120}\pi \approx 2.8536133$. Multiplying $163.5 \times 0.01745 = 2.853075$.

9. Multiplying by $\frac{180°}{\pi}$ produces $\frac{4\pi}{3} \cdot \frac{180°}{\pi} = 240°$.

11. Multiplying by $\frac{180°}{\pi}$ yields $(1.3\pi)\left(\frac{180°}{\pi}\right) = 1.3 \times 180° = 234°$.

13. Multiplying by $57.296°$ results in $14.324°$.

15. (a) To find the supplement of a $35°$ angle in degrees, subtract $35°$ from $180°$ with the result $180° - 35° = 145°$.
    (b) Converting $145°$ to radians, multiply by $\frac{\pi}{180°}$ and obtain $\frac{145}{180}\pi = \frac{29}{36}\pi \approx 2.5307274$.

17. Angle A and the given angle are supplementary, so $\angle A = 180° - 33° = 147°$. Angle B and the given angle are vertical angles and so they are congruent. Thus, $\angle B = 33°$.

19. Angle A and the given angle are alternate exterior angles and so congruent. Then, $\angle A = \frac{3\pi}{8}$. Angle B and $\angle A$ are supplementary, so $\angle B = \pi - \frac{3\pi}{8} = \frac{5\pi}{8}$.

21. Angle B and the given angle are opposite interior angles and so they are congruent. Thus, $\angle B = 70°30'$. Angle A and $\angle B$ are supplementary, so $\angle A = 180° - \angle B = 180° - 70°30' = 109°30'$.

23. Since corresponding segments of transversals that intersect parallel lines are proportional, we have $\frac{AB}{BC} = \frac{DE}{EF}$. Substituting the given information in this proportion we obtain $\frac{AB}{3} = \frac{8}{5}$. Solving this proportion we obtain $AB = \frac{3 \cdot 8}{5} = \frac{24}{5} = 4.8$.

25. The three angles form a straight angle. We have $\angle A + 85° + \angle B = 180°$ and so $\angle A + \angle B = 95°$. Since $\angle A \cong \angle B$, we have $\angle A = \frac{95°}{2} = 47.5°$.

27. Here $\omega = 2\pi f$ and, since $f = 60$, $\omega = 2\pi(60) = 120\pi$ rad/s.

29. Angle x is the supplement of $28°$. Hence, $x = 180° - 28° = 152°$.

≡ 3.2 TRIANGLES

1. Since $x + 35° + 75° = 180°$, we have $x = 180° - (35° + 75°) = 180° = 110° = 70°$.

3. This is an isosceles triangle. Base angles (the angles opposite the congruent sides) of an isosceles triangle are congruent. So, x and the $55°$ angle are congruent and $x = 55°$. To find y, we add to obtain $55° + y + x = 180°$. Solving for y produces $y = 180° - 2(55°) = 180° - 110° = 70°$.

5. This is an equilateral triangle, and so it is also an equiangular triangle and each angle is $\frac{180°}{3} = 60°$. Thus, $x = y = 60°$.

7. Labeling the triangle as shown, we see that $\angle ABC$ is a vertical angle of $53.13°$. Since $\triangle ABC$ is a right triangle, $x + 53.13° = 90°$ and we have $x = 90° - 53.13° = 36.87°$.

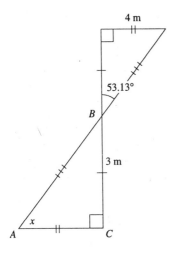

9. The perimeter $P = 6.5 + 7.2 + 9.7 = 23.4$ units. The area, $A = \frac{1}{2}bh$. Here $b = 7.2$ and $h = 6.5$, so $A = \frac{1}{2}bh = \frac{1}{2}(7.2)(6.5) = 23.4$ sq. units (or you may write this as 23.4 units^2).

11. $A = \frac{1}{2}bh = \frac{1}{2}(28)(9) = 126$ units^2. Notice that the dotted line $\overline{BD}$ is not part of the indicated triangle. To determine the perimeter, we need to find the length of $\overline{AC}$. We will do this using the Pythagorean Theorem and $\triangle ADC$. That means we need the length of $\overline{DC} = DB + 28$. Using the Pythagorean Theorem, we determine $BD = \sqrt{15^2 - 9^2} = \sqrt{144} = 12$, so $DC = 40$ and $AC = \sqrt{9^2 + 40^2} = \sqrt{1681} = 41$. Using this information, $P = AB + BC + CA = 15 + 28 + 41 = 84$ units.

13. **(a)** The amount of fencing is the same as the perimeter, or $23.2 + 47.6 + 62.5 = 133.3$ m.
(b) The area will provide the amount of sod that is needed. Using Hero's formula, we first determine the semiperimeter, $s = \frac{23.2 + 47.6 + 62.5}{2} = 66.65$. So, by Hero's formula, we find the area is $A =$

$\sqrt{66.65(66.65 - 23.2)(66.65 - 47.6)(66.65 - 62.5)}$
$= \sqrt{66.65(43.45)(19.05)(4.15)} = \sqrt{228,945.9742}$
≈ 478.5 m^2.

15. Assuming that the building is perpendicular to the ground, we have a right triangle with the ladder forming the hypotenuse. So, $5^2 + s^2 = 15^2$ or $s^2 = 15^2 - 5^2 = 200$ and $s = \sqrt{200}m = 10\sqrt{2}m \approx 14.121$ m.

17. The diagonal, d, is the hypotenuse of the right triangle formed by the two streets. So, $d = \sqrt{45^2 + 62^2} = \sqrt{5869} \approx 76.6$ ft.

19. **(a)** Let s represent the length of the top of the second beam. The figure below shows two similar triangles $\triangle ABC$ and $\triangle ADE$. Similar triangles have proportional sides and altitudes, so $\frac{s}{2400} = \frac{5700}{2600}$ and we obtain $s = 5261.54$ mm (5262 mm to 4 significant digits). Let t represent the length of the top of the third beam. Reasoning as above, we have $\frac{t}{2400} = \frac{8800}{2600}$ or $t = 8123$ mm.
(b) Reasoning as in (a), we have the proportion $\frac{b}{2400} = \frac{10\,000}{2600}$ or $b = 9231$ mm.

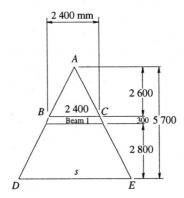

21. Since the total length of the conduit is 30 m, the length of the "bent" section is $30 - 8 - 10 = 12$ m. This bent section is the hypotenuse of a right triangle with one leg 5 m. Using the Pythagorean Theorem, we see that $x^2 + 5^2 = 12^2$ and so $x = \sqrt{119}\ m \approx 10.9$ m.

23. **(a)** The amount of pipe used is $97 + 62 + 53 = 212$ m.

(b) We need to find the length of the hypotenuse AB of a right triangle. One side of the triangle is $65 - 28 = 37$ m. The length of the other side is made up of three horizontal distances, two for the pipe sections on the sides of the ravine and the 62m section across the bottom. For the left-hand section, the horizontal distance is $\sqrt{97^2 - 65^2} = \sqrt{5184} = 72$ m. For the right-hand section, the horizontal distance is $\sqrt{53^2 - 28^2} = \sqrt{2025} = 45$ m. So, the length of the base of the right triangle is $72 + 62 + 45 = 179$ m and the length of the hypotenuse is $\sqrt{179^2 + 37^2} = \sqrt{33410} \approx 182.8$ m.

25. The triangle formed by the building and its shadow and the triangle formed by the meter stick and its shadow are similar. This leads to the proprotion $\frac{\text{height of building}}{\text{shadow of building}} = \frac{\text{height of meter stick}}{\text{shadow of meter stick}}$ or $\frac{\text{height of building}}{128.45 \text{ m}} = \frac{1 \text{ m}}{1.75 \text{ m}}$. Solving the proportion for the height of the building produces $\frac{128.45 \text{ m})(1 \text{ m})}{1.75 \text{ m}} = 73.4$ m.

3.3 OTHER POLYGONS

1. The figure is a square with $s = 15$ cm. $P = 4s = 4(15) = 60$ cm. $A = s^2 = 15^2 = 225$ cm^2.

3. This figure is a parallelogram with $a = 15\frac{1}{4}, b = 33$, and $h = 11\frac{1}{2}$, $P = 2(a + b) = 2(15\frac{1}{4} + 33) = 2(48\frac{1}{4}) = 96.5$ in. $A = bh = 33(11\frac{1}{2}) = 379\frac{1}{2} = 379.5$ in^2.

5. The figure for this exercise is a regular hexagon with $s = 11\frac{1}{2}$. Thus, $P = 6s = 6\left(11\frac{1}{2}\right) = 69$ in. Using the second formula for the area of regular hexagon that is given in Figure 3.27, we have $A = \frac{3\sqrt{3}}{2}s^2 = \frac{3\sqrt{3}}{2}\left(11\frac{1}{2}\right)^2 = \frac{3\sqrt{3}}{2}132.25 = 198.375\sqrt{3} \approx 343.6$ in^2.

7. Here we have a kite with $a = 10$ cm, $b = 12$ cm, $d_1 = 16$ cm, and $d_2 = 21$ cm. Thus, $P = 2(a + b) = 2(10 + 12) = 44$ cm. $A = \frac{1}{2}d_1d_2 = \frac{1}{2}(16)(21) = 168$ cm^2.

9. The figure is a parallelogram with $a = 15.7$ mm, $b = 21.2$ mm, and $h = 13.2$ mm. $P = 2(a + b) = 2(15.7 + 21.2) = 73.8$ mm. $A = bh = (21.2)(13.2) = 279.84$ mm^2.

11. Draw a line across one side of the "L" to make this into two rectangles. If the line is drawn as shown in the figure, one rectangle will be $26' \times 14'$ with an area of 364 ft^2 and the other rectangle will be $8' \times 10'$ with an area of 80 ft^2. Adding the two areas together, we obtain a total area of 444 ft^2.

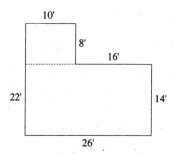

13. Following the hint, we think of this as a rectangle that measures 100 mm by 300 mm for an area of 30 000 mm^2. Each trapezoid has bases $b_1 = 240$ mm and $b_2 = 260$ mm with $h = 40$ mm. Thus, each trapezoid has an area of $A = \frac{1}{2}(b_1 + b_2)h = \frac{1}{2}(240 + 260)(40) = 10\,000$ mm^2. Thus, the area of the I-beam is $A = 30\,000 - 2(10\,000) = 30\,000 - 20\,000 = 10\,000$ mm^2.

15. Consulting Figure 3.27, we can see that the given dimension is $2h$ and so $h = \frac{7}{16}''$ and the area is $A = 2\sqrt{3}h^2 = 2\sqrt{3}\left(\frac{7}{16}\right)^2 = \frac{98}{256}\sqrt{3} \approx 0.663$ in^2. There are several ways we could determine s, but perhaps the easiest is to use the area formula $A = 3sh$. Since we know that $A = \frac{98}{256}\sqrt{3}$ and $h = \frac{7}{16}$, we find $s = A/3h = \left(\frac{98}{256}\sqrt{3}\right)/3\left(\frac{7}{16}\right) = \frac{7}{8\sqrt{3}} \approx 0.505$. Since the perimeter is $P = 6s = 6\frac{7}{8\sqrt{3}} = \frac{21}{4\sqrt{3}} \approx 3.031$ in.

17. **(a)** The cross-sectional area of this trapezoid is $A = \frac{1}{2}(85 + 40)12 = 750$ ft^2.

(b) For the number of linear feet of concrete, we need the length along the bottom and the two sides. One side is $13'$ and the bottom is $40'$. We need to determine the length of the right-hand side. That side is the hypotenuse of a right triangle with sides 12 and 40 and so is $\sqrt{1744} \approx 41.76$ ft. Thus, the linear feet is $13 + 40 + 41.76 = 94.76$ ft.

19. As in Exercise 11, we can draw a line to divide this into two rectangles. The area of one rectangle will be $5'' \times \frac{1''}{4} = \frac{5}{4}$ in^2 and the area of the other rectangle is $3\frac{3''}{4} \times \frac{1''}{4} = \frac{15}{16}$ in^2. Adding these two

areas together, we get the total area of $\frac{5}{4} + \frac{15}{16} = \frac{35}{16} = 2\frac{3}{16}$ in$^2 = 2.1875$ in^2 or 2.19 in^2.

21. Drawing a line across the top of the base of the "T" we divide the figure into two rectangles. The vertical stem of the "T" measures 18 mm by 391 mm and has an area of 7038 mm^2. The top of the "T" is a rectangle that measures 29 mm by 400 mm with an area of 11 600 mm^2. The total area of the "T" is $7038 + 11\,600 = 18638$ mm^2.

3.4 CIRCLES

1. $A = \pi r^2 = \pi(4)^2 = 16\pi$ cm^2; $C = 2\pi r = 2\pi(4) = 8\pi$ cm

3. $A = \pi r^2 = \pi(5)^2 = 25\pi$ in^2; $C = 2\pi r = 2\pi(5) = 10\pi$ in.

5. $A = \pi(14.2)^2 = 201.64\pi$ mm^2; $C = 2\pi(14.2) = 28.4\pi$ mm.

7. Here $d = 24.20$ mm, so $r = 12.10$ mm and $A = \pi(12.10)^2 = 146.41\pi$ mm^2; $C = 2\pi(12.10) = 24.20\pi$ mm

9. (a) This table top has a diameter, d, of 48 in and so, $r = 24$ in. The area, A, is $A = \pi(24)^2 = 576\pi$ in^2.
   (b) The amount of metal edging is the same as the circumference. $C = \pi d = \pi(48) = 48\pi$ in ≈ 151 in.

11. The diameter of this drum is 764 mm and so its circumference is $764\pi \approx 2400$ mm. If the rivets are 75 mm apart, then there are $2400 \div 75 = 32$ rivets.

13. Since $\theta = 110°$, we need to change it to radians. $\theta = 110° = \frac{11}{18}\pi$.
   (a) The arc length $s = r\theta = 85\left(\frac{11}{18}\pi\right) = 51\frac{17}{18}\pi \approx 163.2$ mm.
   (b) For the area, $A = \frac{1}{2}r^2\theta = \frac{1}{2}\left(85^2\right)\left(\frac{11}{18}\pi\right) \approx 6935.5$ mm^2.

15. (a) The amount of molding is the length of $\overset{\frown}{AB}$ + the length of $\overset{\frown}{BC}$ + the length of segment $\overline{AC}$.

Since $60° = \frac{\pi}{3}$, $m\overset{\frown}{AB} = 28''\left(\frac{\pi}{3}\right) = m\overset{\frown}{BC}$. So, the length of molding needed is $28''\left(\frac{\pi}{3}\right) + 28''\left(\frac{\pi}{3}\right) + 28'' \approx 86.643$ in.
(b) The area of this window is the area of the sector bounded by $\overset{\frown}{BC}$ and the sides $\overline{AB}$ and $\overline{AC}$ plus the area bounded by $\overline{AB}$ and $\overset{\frown}{AB}$. To find the area of this last figure, we will find the area of the sector bordered by $\overset{\frown}{AB}, \overline{BC}$ and $\overline{AC}$ and subtract the area of $\triangle ABC$. The area of each sector is $\frac{1}{2}(28'')^2\left(\frac{\pi}{3}\right) \approx 410.50$ in^2. Using Hero's formula, the area of $\triangle ABC$ is $\sqrt{42 \times 14^3} = 196\sqrt{3} \approx 339.48$. Subtracting this from the area of a sector produces $410.50 - 339.48 = 71.02$ and adding this to the area of a sector gives the desired answer, $410.50 + 71.02 = 481.52$ in^2.

17. (a) This is the area of a rectangle measuring $42'' \times 36''$ and two semicircles of radius $18''$. $A = 42 \times 36 + \pi(18)^2 = 1512 + 324\pi \approx 2529.876$ in^2.
   (b) The edging needed is $2(42'') + 2\pi(18'') = 84'' + 36\pi'' \approx 197.1$ in.

19. The area of a wire with a diameter of 8.42 mm is $4.21^2\pi = 17.7241\pi$. A wire with twice this area has an area of 35.4482π. If $\pi r^2 = 35.4482\pi$, then $r^2 = 35.4482$ and $r \approx 5.954$, so the diameter of the new wire is 11.91 mm.

21. (a) $3\frac{1}{7} - \pi \approx 0.00126$
   (b) $\frac{3\frac{1}{7} - \pi}{\pi}\% \approx 0.040\%$

3.5 GEOMETRIC SOLIDS

1. This is a cylinder with $r = 6''$ and $h = 18''$. $L = 2\pi rh = 2\pi(6'')(18'') = 216\pi$ in$^2 \approx 678.6$ in^2. $T = 2\pi r(r+h) = 2\pi(6'')(6''+18'') = 2\pi(6'')(24'')$

$= 288\pi$ in$^2 \approx 904.8$ in^2. $V = \pi r^2 h = \pi(6'')^2(18'') = 648\pi$ in$^3 \approx 2,035.8$ in^3.

3. This is a right circular cone with $r = 8$ mm, $h = 15$ mm, and $s = 17$ mm. $L = \pi r s = \pi(8 \text{ mm})(17 \text{ mm}) = 136\pi \text{ mm}^2 \approx 427.3 \text{ mm}^2$. $T = \pi r(r + s) = \pi(8 \text{ mm})(8 \text{ mm} + 17 \text{ mm}) = 200\pi \text{ mm}^2 \approx 628.3 \text{ mm}^2$. $V = \frac{1}{3}\pi r^2 h = \frac{1}{3}\pi (8 \text{ mm})^2(15 \text{ mm}) = 320\pi \text{ mm}^3 \approx 1005.3 \text{ mm}^3$.

5. This triangular prism has $h = 21$ in and $p = 10$ in $+9$ in $+ 16$ in $= 35$ in. Using Hero's formula, we see that $B \approx 40.91 \text{ in}^2$. $L = ph = (35 \text{ in})(21 \text{ in}) = 735 \text{ in}^2; T = L + 2B = 735 \text{ in}^2 + 2(40.91 \text{ in}^2) \approx 816.8 \text{ in}^2; V = Bh = (40.91 \text{ in}^2)(21 \text{ in}) \approx 859.1 \text{ in}^3$.

7. This is a sphere with radius 8 cm. There is no lateral surface area. $T = 4\pi r^2 = 4\pi(8 \text{ cm})^2 = 256\pi \text{ cm}^2 \approx 804.25 \text{ cm}^2$. $V = \frac{4}{3}\pi r^3 = \frac{4}{3}\pi(8 \text{ cm})^3 = \frac{2048}{3}\pi \text{ cm}^3 = 682.7\pi \text{ cm}^3 \approx 2144.7 \text{ cm}^3$.

9. The cross-section is a triangle on top of a rectangle. The area of the rectangle is 24 ft^2 (Note: $24' \times 12'' = 24' \times 1' = 24 \text{ ft}^2$) and the area of the triangle is $\frac{1}{2}(24')(6'') = \frac{1}{2}(24')(\frac{1}{2}'') = 6 \text{ ft}^2$. Thus, the total area of the cross-section is $24 + 6 = 30 \text{ ft}^2$. Multiplying this area by the length of the road gives a volume of $30 \times 5280 = 158,400 \text{ ft}^3$. Thus, there are $158,400 \div 27 \approx 5866.7 \text{ yd}^3$.

11. (a) The volume is $30(10)(12) = 3,600 \text{ ft}^3$.
(b) $T = 2[(30)(10) + (30)(12) + (10)(12)] = 2(300 + 360 + 120) = 1,560 \text{ ft}^2$

13. Here $r = 33$ mm.
(a) $V = \pi(33)^2(95) = 103,455\pi \approx 325,013.5 \text{ mm}^3$.
(b) Because the label overlaps 5 mm, we need to add 5 mm to the circumference in order to determine the area of the paper needed for the label. $L = (2\pi r + 5)h = [2\pi(33) + 5]95 = (66\pi + 5)95 = 6270\pi + 475 \approx 20\,172.79 \text{ mm}^2$.

15. $V = \frac{4}{3}\pi r^3 = \frac{4000}{3}\pi \text{ m}^3 \approx 1333.3\pi \text{ m}^3 \approx 4188.8 \text{ m}^3$.

17. In Exercise Set 3.3, Exercise #14, we determined that the cross-sectional area of this highway support is 1470 ft^2. Since it is $2'6'' = 2.5$ ft thick, the total volume of the support is $(1470 \text{ ft}^2)(2.5 \text{ ft}) = 3675 \text{ ft}^3 = \frac{3675}{27} \text{ yd}^3 \approx 136.1 \text{ yd}^3$.

19. When the piece of copper is joined along the straight edges, a cone is formed as shown in this figure below. To find the volume of this cone, we need the radius of the base and the height of the cone.

We can determine the radius of the base from the corcumference of the base. That circumference is the same as the arc length of the original sheet of copper. So, $C = 2\pi r = 85(110°)\left(\frac{\pi}{180°}\right)$ and $r = \frac{935}{36} \approx 25.97$. Using the Pythagorean Theorem, we get $h = \sqrt{85^2 - \left(\frac{935}{36}\right)^2} \approx 80.94$. So, $V = \frac{1}{3}\pi r^2 h \approx 18\,198.4\pi \approx 57\,172 \text{ mm}^3$.

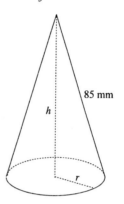

21. $L = \pi 12(1.5 + 4) = 66\pi \text{ in}^2 \approx 207.3 \text{ in}^2$.

23. Draw the square formed by the centers of the silos. Each side of the square is 6 m, so the square has an area of 36 m^2. What remains outside the square are 4 sectors, each of which is $\frac{3}{4}$ of the original circle (silo). The area of each sector is $\frac{1}{2}r^2\theta = \frac{1}{2}(3^2)\left(\frac{3\pi}{2}\right) = \frac{27\pi}{4}$. So, the shaded area in Figure 3.34 is $4\left(\frac{27\pi}{4}\right) + 36 = 27\pi + 36 \text{ m}^2$. The volume of grain is this area multiplied by the height of the silo, 10 m, or $270\pi + 360 \text{ m}^3 \approx 1208.23 \text{ m}^3$ of grain.

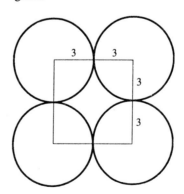

25. Here $h = 4$ ft and $r = 6$ ft, so $V = \frac{1}{3}\pi r^2 h = \frac{1}{3}(6^2)4 = 48\pi \approx 150.8 \text{ ft}^3$.

CHAPTER 3 REVIEW

1. Using the proportion $\frac{D}{180°} = \frac{R}{\pi}$, we substitute 27° for D and solve for R. Thus, $\frac{27°}{180°} = \frac{R}{\pi}$ or $R = \frac{27°}{180°}\pi = \frac{3}{20}\pi \approx 0.4712$.

2. Using the proportion $\frac{D}{180°} = \frac{R}{\pi}$, we substitute 212° for D and solve for R. Thus, $\frac{212°}{180°} = \frac{R}{\pi}$ or $R = \frac{212°}{180°}\pi = \frac{53}{45}\pi \approx 3.700$.

3. Multiplying by $\frac{180°}{\pi}$ produces 198°.

4. Multiplying by $\frac{180°}{\pi}$ produces $\frac{135°}{\pi} \approx 42.97°$.

5. The supplement of a 137° angle is $180° - 137° = 43°$.

6. The complement of a $\frac{\pi}{6}$ angle is $\frac{\pi}{2} - \frac{\pi}{6} = \frac{2\pi}{6} = \frac{\pi}{3}$.

7. Because $\angle A$ and the given angle are supplementary, $\angle A = 180° - 28° = 152°$. Angle A and $\angle B$ are corresponding angles formed by parallel lines cut (or intersected) by a transversal and so they are congruent. Thus, $\angle B = 152°$.

8. Angle B and the given angle are alternate exterior angles formed by parallel lines cut by a transversal and so they are congruent. Thus, $\angle B = \frac{7\pi}{16}$. Angle A and $\angle B$ are supplementary, so $\angle A = \pi - \angle B = \pi - \frac{7\pi}{16} = \frac{9\pi}{16}$.

9. Because corresponding segments are proportional, we have $\frac{2.7}{6.3} = \frac{AB}{7.7}$. Solving for AB produces $AB = \frac{(2.7)(7.7)}{6.3} = \frac{20.79}{6.3} = 3.3$.

10. Because the sum of the angles of a triangle is 180°, we have $x = 180° - (32° + 21°) = 180° - 53° = 127°$.

11. This is an isosceles triangle, so the base angles are congruent and $x = 180° - 2(75°) = 180° - 150° = 30°$.

12. Using the Pythagorean Theorem, the missing side is the hypotenuse, so its length is $\sqrt{120^2 + 209^2} = \sqrt{58081} = 241$.

13. In this right triangle, the hypotenuse is 37 so, using the Pythagorean Theorem, the length of the missing side is $\sqrt{37^2 - 16^2} = \sqrt{1113} \approx 33.36$.

14. The perimeter of this triangle is $P = 18 + 5 + 16 = 39$ units. Using Hero's formula, we determine that the semiperimeter is $s = \frac{P}{2} = \frac{39}{2} = 19.5$ units and so the area is
$$A = \sqrt{19.5(19.5) - 18)(19.5 - 5)(19.5 - 16)} = \sqrt{19.5(1.5)(14.5)(3.5)} = \sqrt{1484.4375} \approx 38.53 \text{ units}^2.$$

15. This rectangle has an area of $A = (24.5)(12) = 294$ units^2. The perimeter is $P = 2(l + w) = 2(24.5 + 12) = 2(36.5) = 73$ units.

16. This parallelogram has a perimeter of $P = 2(l + w) = 2(23 + 14) = 2(37) = 74$ units. The area is $A = bh = 23(10) = 230$ units^2.

17. The circumference of a circle is $C = 2\pi r = 2\pi(9) = 18\pi \approx 56.55$ units. The area is $A = \pi r^2 = \pi 9^2 = 81\pi \approx 254.5$ units^2.

18. For this circle, we are given the diameter of 21 units and so the radius is $r = 10.5$ units. $C = \pi d = 21\pi \approx 65.97$ units. The area is $A = \pi r^2 = \pi(10.5)^2 = 110.25\pi \approx 346.36$ units^2.

19. Using the Pythagorean Theorem, we find that the length of the hypotenuse is $\sqrt{56^2 + 33^2} = \sqrt{4225} = 65$. Thus, $P = 33 + 56 + 65 = 154$ units and, since this is a right triangle, $A = \frac{1}{2}(33)(56) = 924$ units^2.

20. Converting 110° to radians, we get $110° = \frac{11}{18}\pi$, so the arc length is $r\theta = 7\left(\frac{11}{18}\pi\right) = \frac{77}{18}\pi \approx 4.277\pi \approx 13.44$ units. Adding the lengths of the two straight sides (radii) we get a perimeter of $13.44 + 7 + 7 = 27.44$ units. The area is $A = \frac{1}{2}r^2\theta = \frac{1}{2}7^2\left(\frac{11}{18}\pi\right) = \frac{539}{36}\pi \approx 14.972\pi \approx 47.04$ units^2.

21. The perimeter of this trapezoid is $P = 15 + 12 + 10.6 + 29.5 = 67.1$ units. The area is $\frac{1}{2}(b_1 + b_2)h = \frac{1}{2}(12 + 29.5)(9) = \frac{1}{2}(41.5)(9) = 186.75$ units^2.

22. This is a rectangular prism. We will let $l = 15.5$, $w = 8.4$, and $h = 35$. $L = 2h(l + w) = 2(35)(15.5 + 8.4) = 2(35)(23.9) = 1673$ units^2. $T = 2(lw + lh + hw) = 2[(15.5)(8.4) + (15.5)(35) + (35)(8.4)] = 2[130.2 + 542.5 + 294] = 2(966.7) = 1933.4$ units^2. $V = lwh = (15.5)(8.4)(35) = 4557$ units^3.

23. This is a right circular cylinder with $r = 12.1$ and $h = 9.6$. $L = 2\pi rh = 2\pi(12.1)(9.6) = 232.32\pi \approx 729.85$ units^2. $T = 2\pi r(r + h) = 2\pi(12.1)(12.1 + 9.6) = 2\pi(12.1)(21.7) = 525.14\pi \approx 1649.78$

units^2. $V = \pi r^2 h = \pi(12.1)^2(9.6) = \pi(146.41)$ $(9.6) = 1405.54\pi \approx 4415.62$ units^3.

24. This figure is a right triangular prism with $p = 19 + 12 + 21 = 52$ units, $s = \frac{p}{2} = 26$ units, and $h = 17$ units. Using Hero's formula, we determine $B = \sqrt{26(26-21)(26-19)(26-12)} = \sqrt{26(5)(7)(14)} + \sqrt{12740} \approx 112.87$ units^2. $L = ph = (52)(17) = 884$ units^2. $T = ph + 2B = 884 + 2(112.87) = 1109.74$ units^2. $V = Bh \approx (112.87)(17) = 1918.79$ units^3.

25. This is a square pyramid with $p = 4(40) = 160$ units, $B = 40^2 = 1600$ units^2, $s = 29$ units, and $h = 21$ units. $L = \frac{1}{2}ps = \frac{1}{2}(160)(29) = 2320$ units^2. $T = \frac{1}{2}ps + B = L + B = 2320 + 1600$ units$^2 = 3920$ units^2. $V = \frac{1}{3}Bh = \frac{1}{3}(1600)(21) = 11,200$ units^3.

26. This figure is a right circular cone with $r = 8$ units and $h = 15$ units. Using the Pythagorean Theorem, we determine that $s = \sqrt{15^2 + 8^2} = \sqrt{225 + 64} = \sqrt{289} = 17$ units. $L = \pi r s = \pi(8)(17) = 136\pi \approx 427.26$ units^2. $T = \pi r(r + s) = \pi(8)(8 + 17) = 200\pi \approx 628.32$ units^2. $V = \frac{1}{3}\pi r^2 h = \frac{1}{3}\pi 8^2(15) = 320\pi \approx 1005.31$ units^3.

27. This is a frustum of a cone with $r_1 = 7$ units, $r_2 = 19$ units, $h = 16$ units, and $s = 20$ units. $L = \pi s(r_1 + r_2) = \pi(20)(7+19) = 520\pi \approx 1633.6$ units^2. $T = \pi[r_1(r_1 + s) + r_2(r_2 + s)] = \pi[(7)(7 + 20) + (19)(19 + 20)] = \pi(189 + 741) = 930\pi \approx 2921.68$ units^2. $V = \frac{1}{3}\pi h \left(r_1^2 + r_2^2 + r_1 r_2\right) = \frac{1}{3}\pi(16)\left(7^2 + 19^2 + (7)(19)\right) = \frac{1}{3}\pi(16)(49 + 361 + 133) = \frac{1}{3}\pi(16)(543) = 2,896\pi \approx 9098.05$ units^3.

28. This is a frustum of a square pyramid with $p_1 = 4(24) = 96$ units, $p_2 = 4(14) = 56$ units, $s = 13$ units, and $h = 12$ units. Since the bases are squares we can determine that $B_1 = 24^2 = 576$ units^2 and $B_2 = 14^2 = 196$ units^2. $L = \frac{1}{2}s(p_1 + p_2) = \frac{1}{2}(13)(96 + 56) = 988$ units^2. $T = \frac{1}{2}s(p_1 + p_2) + B_1 + B_2 = L + B_1 + B_2 = 988 + 576 + 196 = 1760$ units^2. $V = \frac{1}{3}h(B_1 + B_2 + \sqrt{B_1 B_2}) = \frac{1}{3}(12)\left(576 + 196 + \sqrt{(576)(196)}\right) = \frac{1}{3}(12)(576 + 196 + 336) = 4432$ units^3.

29. This sphere has a radius of 9 units. Because it is a sphere, it has no lateral area. $T = 4\pi r^2 = 4\pi 9^2 = 324\pi \approx 1017.88$ units^2. $V = \frac{4}{3}\pi r^3 = \frac{4}{3}\pi 9^3 = 972\pi \approx 3053.63$ units^3.

30. Using the Pythagorean Theorem, we can find the length of each of the four cables. The length of c_1, the cable that is attached 50 m above the ground, is $c_1 = \sqrt{75^2 + 50^2} \approx 90.14$ m. The length of c_2, the cable that is attached 100 m above the ground, is $c_2 = \sqrt{75^2 + 100^2} = 125.00$ m. c_3, the cable that is attached 150 m above the ground, has a length of $c_3 = \sqrt{75^2 + 150^2} \approx 167.71$ m. The length of c_4, the cable that is attached to the top of the antenna, is $c_4 = \sqrt{75^2 + 175^2} \approx 190.39$ m. One set of four cables has a total length of about $90.14 + 125.00 + 167.71 + 190.39 = 573.24$ m. The three sets have a total length of $3 \times 573.24 = 1,719.72$ m.

31. Assuming $\overline{AF}, \overline{BE}$, and $\overline{CD}$ are all parallel, we obtain the proportion $\frac{d}{100} = \frac{352.6}{82}$. Solving for d, we get $d = \frac{100 \times 352.6}{82} = 430$ ft.

32. (a) Only 95 cm of the strip are used to make washers. Each strip can make $\frac{95}{3.20} \approx 29.7$ washers. Do not round this up to 30 washers because the last 5 cm of each strip cannot be used. So, each strip will only produce 29 washers. Hence, to make 100 000 washers, a total of $\frac{100\,000}{29} \approx 3448.3$ strips are needed. Since 3448 strips will give a total of 99 992 washers, we must need 3449 strips.
(b) Each washer has a volume of $\pi(R^2 - r^2)h = \pi(1.6^2 - 0.525^2)(0.240) = 1.722378$ cm^3. The total metal in the 100 000 washers is 172 237.8 cm^3. (Actually, since the 3449 strips will produce 100 021 washers, a total of 172238 cm^3 will probably be used to produce washers.)
(c) Each strip contains $3.20 \times 100 \times 0.240 = 76.8$ cm^3 of metal. The total volume of metal in the 3,449 strips is $3,449 \times 76.8 = 264\,883.2$ cm^3. In (b), we found that 172 237.8 cm^3 of metal is actually used for the washers. Since 172 238.8 cm^3 is used to produce 100 000 washers, the rest must be scrap. Thus, there is a total of $264\,883.2 - 172\,237.8 = 92\,645.4$ cm^3 in scrap metal.

CHAPTER 3 TEST

1. $35\left(\frac{\pi}{180}\right) = \frac{35\pi}{180} = \frac{7\pi}{36} \approx 0.611$

2. $\left(\frac{7\pi}{15}\right)\left(\frac{180}{\pi}\right) = \frac{7\pi 180}{15\pi} = 84°$

3. $180° - 76° = 104°$

4. $133°$

5. $\frac{5}{7} = \frac{9}{\overline{AB}}$. Solving for $\overline{AB}$, we get $\overline{AB} = \frac{63}{5} = 12.6$ units.

6. Using the Pythagorean Theorem, with $c = 3.9$ and $b = 3.6$, we have $a^2 + 3.6^2 = 3.9^2$ or $a^2 = 3.9^2 = 3.6^2 = 15.21 - 12.96 = 2.25$. Since $\sqrt{2.25} = 1.5, a = 1.5$ cm.

7. This rectangle has $l = 36.4$ cm and $w = 14.3$ cm, so $P = 2(l + w) = 2(36.4 + 14.3) = 2(50.7)101.4$ cm.

8. This trapezoid has been divided into a rectangle and two triangles. The length of the bottom is $4 + 12 + 4 = 20$ in and so the perimeter is $P = 12 + 5 + 20 + 5 = 42$ in.

9. This is a right trapezoid, so $h = 10.4$ cm and the area is $A = \frac{1}{2}h(b_1 + b_2) = \frac{1}{2}(10.4)(21.6 + 32.4) = 280.8$ cm^2.

10. This circle has a radius of $r = 9$ in and so its area is $A = \pi r^2 = \pi 9^2 = 81\pi \approx 254.47$ in^2.

11. $\frac{14}{11.6} = \frac{x}{120}$. Solving for x produces $\frac{(14)(120)}{11.6} \approx 144.8$.

12. The circumference of the tire is $C = 2\pi r = \pi d = \pi(62)$ cm. To travel 25 m, the tire will have to make $\frac{25 \text{ m}}{\pi(62) \text{ cm}}$ revolutions. Converting 25 m to cm, we obtain 2500 cm. Substituting this in the previous statement, we get that the number of revolutions is $\frac{2500 \text{ cm}}{\pi(62) \text{ cm}} \approx 12.84$.

13. $V = lwh = (162)(52)(231) = 1\,945\,944$ mm$^2 = 1\,945\,944$ mm$^3 = 1945.944$ cm^3

14. $V = \frac{4}{3}\pi r^3$. We are given the diameter as $35'$, so $r = \frac{35}{2} = 17.5$ ft. Thus $V = \frac{4}{3}\pi(17.5)^3 \approx 7145.83333\pi \approx 22,449.29750$ feet.

15. The label covers the lateral surface area of the cylinder plus an overlap. The lateral surface area is $L = 2\pi rh = \pi dh$ and the amount needed for the overlap is $0.8h$. So, the area is $\pi(6.5)(9.5) + (0.8)(9.5) = 61.75\pi + 7.6 \approx 201.59$ cm^2.

CHAPTER
4
Functions and Graphs

≡ 4.1 FUNCTIONS

1. This is a function because each value of x produces just one value for y.

3. This is not a function of x because the same value of x can give two different values for y. For example, the ordered pairs (3, 2) and (3, −2) both satisfy the relation.

5. Notice that $\sqrt{x}$ is defined only when $x \geq 0$. Since $\sqrt{x}$ produces a unique number for each x, this is a function for $x \geq 0$.

7. Since the square of any real number is nonnegative, this equation will be defined only for those values of x where $2x - 7 \geq 0$ or when $x \geq \frac{7}{2}$. However, this is not a function since (4, 1) and (4, −1) both satisfy the equation.

9. Yes, because for each value of x there is just one value of y.

11. No, because when $x = 4$ there are two values of $y: -2$ and 2, or when $x = 1, y$ is 1 or −1.

x	-40	-30	-20	-10	0	10	20	30	40
y	-40	-22	-4	14	32	50	68	86	104

13. Domain: $\{-3, -2, -1, 0, 1, 2\}$;
Range: $\{-7, -5, -3, -1, 1, 3\}$

15. Domain: $\{-3, -2, -1, 0, 1, 2, 3\}$;
Range: $\{-25, -6, 1, 2, 3, 10, 25\}$

17. Domain and range are both all real numbers.

19. Domain is all real numbers except 5, that is the domain is $\{x : x \neq 5\}$. Rewrite $y = \frac{x}{x-5}$ as $y = 1 + \frac{5}{x-5}$. Since $\frac{x}{x-5}$ can never be 0, then $y = 1 + \frac{x}{x-5}$ can never be 1. Thus, the range is all real numbers except 1, that is the range is $\{y : y \neq 1\}$.

21. Domain is all nonnegative real numbers, $x \geq 0$. Range is all real numbers greater than or equal to -2, that is the range is $\{y : y \geq -2\}$.

23. Domain is all real numbers except 2 and -3, that is the domain is $\{x : x \neq 2, x \neq -3\}$.

25. Domain is all real numbers except -1 and 5, that is the domain is $\{x : x \neq -1, x \neq 5\}$.

27. Domain is all real numbers except -4 and 2, that is the domain is $\{x : x \neq -4, x \neq 2\}$.

29. Any function with a denominator of $x - 5$ will work as long as the numerator is defined for all real numbers. For example, $y = \frac{1}{x-5}$ is one answer. Other possible answers include $y = \frac{x}{x-5}$, $y = \frac{x-5}{x-5}$, and $y = -10(x-5)^2$.

31. Any function with a denominator of $(x+1)(x-2)$ will work as long as the numerator is defined for all real numbers. For example, $y = \frac{1}{(x+1)(x-2)}$ is one answer. Other possible answers include $y = \frac{x}{(x+1)(x-2)}, y = \frac{(x+1)(x-2)}{(x+1)(x-2)}, y = \frac{-10}{(x+1)(x-2)^2}$, and $y = \frac{3x(x+1)}{(x+1)^2(x-2)^5}$.

33. Any function with an even root will work if the quantity under the radical sign is nonnegative when $x \geq 5$. Since $x \geq 5$ is the same as $x - 5 \geq 0$, then $y = \sqrt{x-5}$ is one answer. Other possible answers include $y = \sqrt[4]{x-5}, y = -7\sqrt[4]{x-5}$, and $y = \sqrt[6]{x-5}$.

35. Any function with $x - 4$ in the denominator and with an even root where the quantity under the radical sign is nonnegative when $x \geq -1$. For example, $y = \frac{\sqrt{x+1}}{x-4}$ is one answer and $y = \frac{\sqrt{x+1}}{\sqrt[3]{x-4}}$ is another.

37. $f(x) = 3x - 2$, so $f(0) = 3(0) - 2 = 0 - 2 = -2$.

39. $f(-3) = 3(-3) - 2 = -9 - 2 = -11$.

41. Since $f(0) = -2$ and $f(-3) = -11$, then $f(0) + f(-3) = -2 + -11 = -13$.

43. $f(b+3) = 3(b+3) - 2 = 3b + 9 - 2 = 3b + 7.$

45. Since $g(x) = x^2 - 5x$, then $g(0) = 0^2 - 5(0) = 0 - 0 = 0.$

47. $g(2) = 2^2 - 5(2) = 4 - 10 = -6.$

49. $g\left(\frac{2}{5}\right) = \left(\frac{2}{5}\right)^2 - 5\left(\frac{2}{5}\right) = \left(\frac{4}{25}\right) - 2 = \frac{4}{25} - \frac{50}{25} - \frac{46}{25}$

51. $g(x-5) = (x-5)^2 - 5(x-5) = (x^2 - 10x + 25) - 5x + 25 = x^2 - 15x + 50.$

53. $g(4m^2) = (4m^2)^2 - 5(4m^2) = (4^2 m^4) - (20m^2) = 16m^4 - 20m^2.$

55. We have $F(x) = \frac{2-x}{x^2+2}$, and so $F(0) = \frac{2-0}{0^2+2} = \frac{2}{0+2} = \frac{2}{2} = 1.$

57. $F(1.5) = \frac{2-1.5}{(1.5)^2+2} = \frac{0.5}{2.25+2} = \frac{0.5}{4.25} = \frac{2}{17}.$

59. $F(2x) = \frac{2-2x}{(2x)^2+2} = \frac{2-2x}{4x^2+2} = \frac{2(1-x)}{2(2x^2+1)} = \frac{1-x}{2x^2+1}.$

61. $C(32) = \frac{5}{9}(32 - 32) = \frac{5}{9}(0) = 0.$

63. $C(-40) = \frac{5}{9}(-40 - 32) = \frac{5}{9}(-72) = -40.$

65. Explicit.

67. Implicit. Solving for y we get

$$y = x^2 y + x$$
$$y - x^2 y = x$$
$$y(1 - x^2) = x$$
$$y = \frac{x}{1-x^2}.$$

Thus, an explicit form is $y = \frac{x}{1-x^2}.$

69. Since $f(x,y) = 3x - 2y + 4$, then $f(1,0) = 3(1) - 2(0) + 4 = 3 - 0 + 4 = 7.$

71. $f(-1,2) = 3(-1) - 2(2) + 4 = -3 - 4 + 4 = -3.$

73. Here $g(x, y) = x^2 - y^2 + 2xy$, and so $g(-2,0) = (-2)^2 - 0^2 + 2(-2)(0) = 4 - 0 + 0 = 4.$

75. $g(5,4) = 5^2 - 4^2 + 2(5)(4) = 25 - 16 + 40 = 49.$

77. Since $h(x,y,z) = 3x - 4y - 2z + xyz$, then $h(-1, 2, 3) = 3(-1) - 4(2) - 2(3) + (-1)(2)(3) = -3 - 8 - 6 - 6 = -23.$

79. $h(3, 2, -1) = 3(3) - 4(2) - 2(-1) + (3)(2)(-1) = 9 - 8 + 2 - 6 = -3.$

81. If $r = 20, V = 16,000\pi$ ft^3, $S = 2400\pi$ ft^2; if $r = 30, V = 36,000\pi$ ft^3, $S = 4200\pi$ ft^2

83. Here $R(7) = 20 + 5(7 - 4) = 20 + 5(3) = 20 + 15 = 35$, so the cost of renting this chain saw is \$35.

85. Since, $P(d) = 62.4d$ and $d = 80$, then $P(80) = 62.4 \cdot 80 = 4,992$ lb/ ft^2.

4.2 OPERATIONS ON FUNCTIONS; COMPOSITE FUNCTIONS

1. Solving $y = x - 5$ for x, we obtain $x = y + 5$. Exchanging the x- and y-variables produces the inverse, $y = x + 5$.

3. Solving $y = 2x - 8$ for x, we obtain $x = \frac{y+8}{2}$. Exchanging the x- and y-variables produces the inverse, $y = \frac{x+8}{2}$. This can also be written as $y = \frac{1}{2}(x + 8)$ or $y = \frac{1}{2}x + 4$.

5. Solving $6y = 2x - 4$ for x, we obtain $2x = 6y + 4$, which simplifies to $x = 3y + 2$. Exchanging the x- and y-variables produces the inverse, $y = 3x + 2$.

7. Solving $y^2 = x - 5$ for x, we obtain $x = y^2 + 5$. Exchanging the x- and y-variables produces the inverse $y = x^2 + 5$.

9.

x	-3	-2	-1	0	1	2	3	4
$f(x)$	0	1	2	3	4	5	6	7
$g(x)$	-10	5	0	-1	5	1	-10	7
$(f+g)(x)$	-10	6	2	2	9	6	-4	14

11.

x	-3	-2	-1	0	1	2	3	4
$f(x)$	0	1	2	3	4	5	6	7
$g(x)$	-10	5	0	-1	5	1	-10	7
$(g-f)(x)$	-10	4	-2	-4	1	-4	-16	0

13.

x	-3	-2	-1	0	1	2	3	4
$f(x)$	0	1	2	3	4	5	6	7
$g(x)$	-10	5	0	-1	5	1	-10	7
$(f/g)(x)$	0	$\frac{1}{5}$	undefined	-3	$\frac{4}{5}$	5	$-\frac{3}{5}$	1

15.

x	-3	-2	-1	0	1	2	3	4
$g(x)$	-10	5	0	-1	5	1	-10	7
$(f \circ g)(x)$	-7	8	3	2	8	4	-7	10

17. The domain and range of f are all real numbers. The domain of g is $\{-3, -2, -1, 0, 1, 2, 3, 4\}$. The range of g is $\{-10, -1, 0, 1, 5, 7\}$.

19. **(a)** The domain of f/g is $\{-3, -2, 0, 1, 2, 3, 4\}$. Notice that -1 is not in the domain of f/g because the denominator of f/g is 0 when $x = -1$. By reading the bottom line of the table in the solution to Exercise 13, we can see that the range of f/g is $\{-3, -\frac{3}{5}, 0, \frac{1}{5}, 1, 5\}$.

(b) The domain of g/f is $\{-2, -1, 0, 1, 2, 3, 4\}$. Notice that -3 is not in the domain of g/f because the denominator g/f is 0 when $x = -3$. By reading the bottom line of the table in the solution to Exercise 14, we can see that the range of g/f is $\{-\frac{5}{3}, -\frac{1}{3}, 0, \frac{1}{5}, 1, \frac{5}{4}, 5\}$.

21. $f(+g)(x) = (x^2 - 1) + (3x + 5) = x^2 - 1 + 3x + 5 = x^2 + 3x + 4.$

23. $(f - g)(x) = (x^2 - 1) - (3x + 5) = x^2 - 1 - 3x - 5 = x^2 - 3x - 6.$

25. $(g - f)(x) = (3x + 5) - (x^2 - 1) = 3x + 5 - x^2 + 1 = 3x - x^2 + 6.$

27. $(f \cdot g)(x) = (x^2 = 1)(3x + 5) = 3x^3 + 5x^2 - 3x - 5.$

29. $(f/g)(x) = \left(\frac{f}{g}\right)(x) = \frac{f(x)}{g(x)} = \frac{x^2 - 1}{3x + 5}.$

31. $(g/f)(x) = \left(\frac{g}{f}\right)(x) = \frac{g(x)}{f(x)} = \frac{3x + 5}{x^2 - 1}.$

33. The domains of both f and g are all real numbers.

35. $(f \circ g)(x) = f(g(x)) = f(3x + 5) = (3x + 5)^2 - 1 = 9x^2 + 30x + 24.$

37. $(g \circ f)(x) = g(f(x)) = g(x^2 - 1) = 3(x^2 - 1) + 5 = 3x^2 - 3 + 5 = 3x^2 + 2.$

39. $(f + g)(x) = f(x) + g(x) = (3x - 1) + (3x^2 + x) = 3x^2 + 4x - 1.$

41. $f(-g)(x) = f(x) - g(x) = 3x - 1) - (3x^2 + x) = (3x - 1 - 3x^2 - x = -3x^2 + 2x - 1.$

43. $(g - f)(x) = g(x) - f(x) = (3x^2 + x) - (3x - 1) = 3x^2 + x - 3x_1 = 3x^2 - 2x + 1.$

45. $(f \cdot g)(x) = (3x - 1)(3x^2 + x) = 9x^3 + 3x^2 - x = 9x^3 - x.$

47. $(f/g)(x) = \frac{f(x)}{g(x)} = \frac{3x - 1}{3x^2 + x}.$

49. $(g/f)(x) = \frac{g(x)}{f(x)} = \frac{3x^2 + x}{3x - 1}.$

51. $(f \circ g)(x) = f(x(x)) = f(3x^2 + x) = 3(3x^2 + x) - 1 = 9x^2 + 3x - 1.$

53. $(g \circ g)(x) = g(f(x)) = g(3x - 1) = 3(3x - 1)^2 + (3x - 1) = 3(9x^2 - 6x + 1) + (3x - 1 = 27x^2 - 18x + 3 + 3x - 1 = 27x^2 - 15 + 2.$

55. The domains of both f and g are all real numbers.

57. **(a)** $C(n) = F(n) + V(n) = 7500 + 15n$ or $15n + 7500$;

(b) $C(100) = 15(100) + 7500 = 1500 + 7500 = \$9,000$;

(c) $C(1000) = 15(1000) + 7500 = 15000 + 7500 = \$22,500.$

59. **(a)** $P(n) = R(n) - C(n) = 90n - (30n + 275) = 60n - 275$;

(b) $P(50) = 60(50) - 275 = 3000 - 275 = \$2,725.$

61. $(P \circ n)(a) = P(7a + 4) = 300(7a + 4)^2 - 50(7a + 4) = 300(49a^2 + 56a + 16) - 50(7a + 4) = 14700a^2 + 16800a + 4800 - 350a - 200 = 14700a^2 + 16450a + 4600.$

4.3 RECTANGULAR COORDINATES

1.

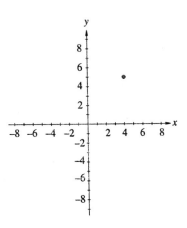

3.

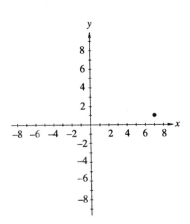

5.

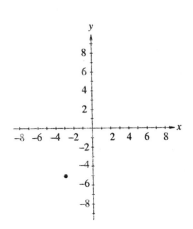

7.

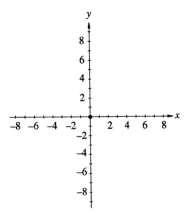

9.

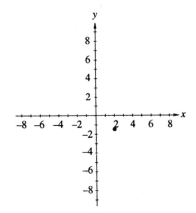

11. The fourth vertex will have the same x-coordinate as B, -1, and the same y-coordinate as C, -4. So its coordinates are $(-1, -4)$.

13. They are on a horizontal line and have the same y-coordinate.

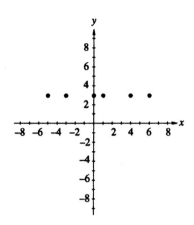

15. They all lie on the same straight line.

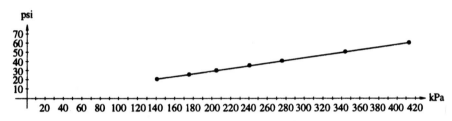

17. As in problem #13, these points all lie in a horizontal line through $y = -2$.

19. On the vertical line $x = -5$.

21. These points must all lie to the right of the vertical line $x = -3$.

23. To the right of the vertical line through the point $(1,0)$ and below the horizontal line through the point $(0, -2)$.

25. (a)

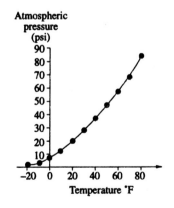

(b) 41.9 psi;
(c) 99.6 psi.

27. (a)

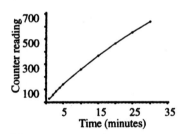

(b) around 317 min.;
(c) around 23.5 min.

≡ 4.4 GRAPHS

1.   (a)

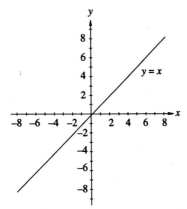

(b) x-intercept is 0; y-intercept is 0;
(c) Let $(x_1, y_1) = (0, 0)$ and $(x_2, y_2) = (2, 2)$; thus
$m = \frac{2-0}{2-0} = \frac{2}{2} = 1$.

3.   (a)

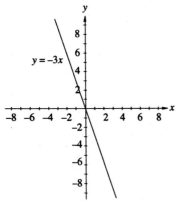

(b) x-intercept is 0; y-intercept is 0;
(c) Let $(x_1, y_1) = (0, 0)$ and $(x_2, y_2) = (1, -3)$;
then $m = \frac{-3-0}{1-0} = \frac{-3}{1} = -3$.

5.   (a)

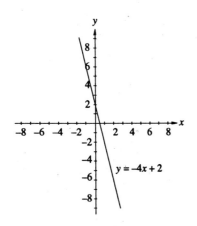

(b) x-intercept is $\frac{1}{2}$; y-intercept is 2;
(c) $m = \frac{0-2}{\frac{1}{2}-0} = \frac{-2}{\frac{1}{2}} = -4$.

7.   (a)

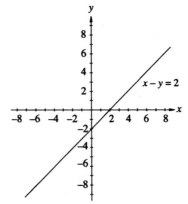

(b) x-intercept is 2; y-intercept is -2;
(c) $m = \frac{0-(-2)}{2-0} = \frac{2}{2} = 1$.

9.   (a)

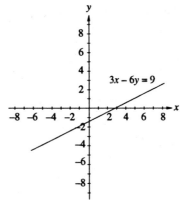

(b) x-intercept is 3; y-intercept is $-\frac{3}{2}$;

(c) $m = \dfrac{0 - \left(\frac{3}{2}\right)}{3 - 0} = \dfrac{\frac{3}{2}}{\frac{2}{3}} = \frac{1}{2}$.

11. **(a)**

x	-3	-2	-1	0	1	2	3
y	9	4	1	0	1	4	9

(b)

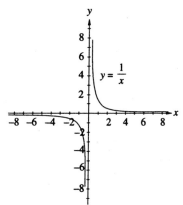

(c) Domain: all real numbers; range: $\{y{:}y \geq 0\}$.

13. **(a)**

x	-3	-2	-1	0	1	2	3
y	7	2	-1	-2	-1	2	7

(b)

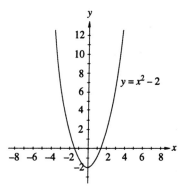

(c) Domain: all real numbers; range: $\{y{:}y \geq -2\}$.

15. **(a)**

x	-3	-2	-1	0	1	2	3	4	5	6
y	36	25	16	9	4	1	0	1	4	9

(b)

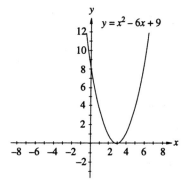

(c) Domain: all real numbers; range: $\{y{:}y \geq 0\}$.

17. **(a)**

x	-3	-2	-1	$-\frac{1}{2}$	$-\frac{1}{4}$	0	$\frac{1}{4}$	$\frac{1}{2}$	1	2	3
y	$-\frac{1}{3}$	$-\frac{1}{2}$	-1	-2	-4	undefined	4	2	1	$\frac{1}{2}$	$\frac{1}{3}$

(b)

(c) Domain: $\{x{:}x \neq 0\}$; range; $\{y{:}y \neq 0\}$.

19. (a)

x	-3	-2	-1	0	1	2	3
y	-27	-8	-1	0	1	8	27

(b)

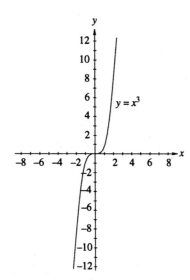

(c) Domain: all real numbers; range: all real numbers.

21.

x	0	0	5	-5	3	-3	3	-3	4	4	-4	-4
y	5	-5	0	0	4	4	-4	-4	3	-3	3	-3

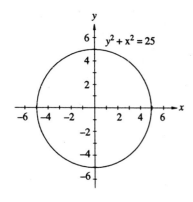

23.

x	-5	-4	-3	-2	-1	0	1	2
y	3	2	1	0	1	2	3	4

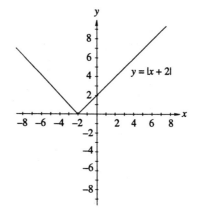

25.

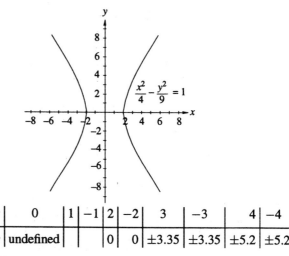

x	0	1	-1	2	-2	3	-3	4	-4
y	undefined			0	0	± 3.35	± 3.35	± 5.2	± 5.2

27. Function.

29. Function.

31. (a)

k	0	1	2	3	4	5	6	7	8	9
$p(k)$	25.0	8.33	5.00	3.57	2.78	2.27	1.92	1.67	1.47	1.32

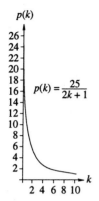

$$p(k) = \frac{25}{2k+1}$$

(b) Yes.

33. (a) $0 < p < \$4.65$;

(b)

p	0.25	0.50	0.75	1.00	1.25	1.50	1.75	2.00
$d(p)$	10.56	11	11.31	11.50	11.56	11.50	11.31	11

(c)

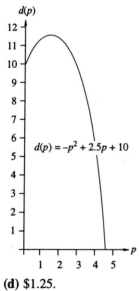

$$d(p) = -p^2 + 2.5p + 10$$

(d) $1.25.

35. No new solutions

☰ 4.5 GRAPHING CALCULATORS AND COMPUTER-AIDED GRAPHING (OPTIONAL)

Note: All graphs were generated on a TI-81 calculator. In each figure, the tick marks are 1 unit apart.

1. Graph $y = x$.

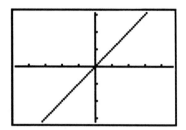

3. $y = -3x$.

5. $y = -4x + 2$.

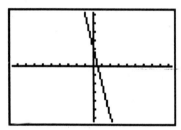

7. $x - y = 2$; graph as $y = x - 2$.

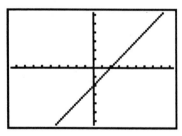

9. $3x - 6y = 9$; graph as $y = \frac{9-3x}{-6}$ or $y = \frac{1}{2}x - \frac{3}{2}$.

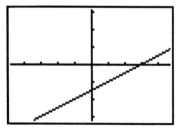

11. $y = x^2$.

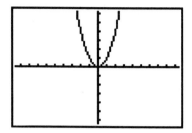

13. $y = x - 2$.

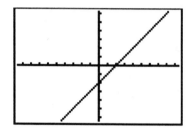

15. $y = x^2 - 6x + 9$.

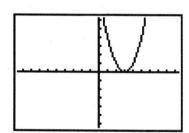

17. $y = \frac{1}{x}$.

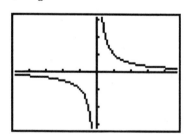

19. $y = x^3$.

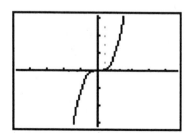

21. $y^2 + x^2 = 25$ or $y^2 = 25 - x^2$; graph both $y = \sqrt{25 - x^2}$ and $y = -\sqrt{25 - x^2}$.

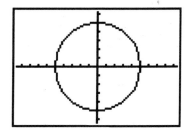

23. $y = |x + 2|$.

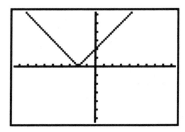

25. Here $\frac{x^2}{4} - \frac{y^2}{9} = 1$ can be written as $9x^2 - 4y^2 = 36$ or $-4y^2 = 36 - 9x^2$ or $4y^2 = 9x^2 - 36$ and so $2y = \pm\sqrt{9x^2 - 36}$; graph as $y = \frac{\pm\sqrt{9x^2 - 36}}{2}$; make sure you graph both.

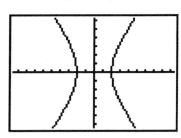

27.

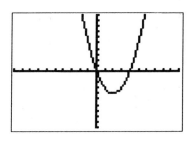

29.

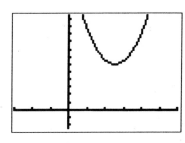

$[-6, 12, 2] \times [-5, 25, 2]$

31.

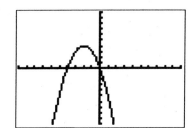

33.

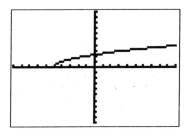

35.

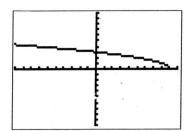

37.

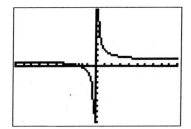

39.

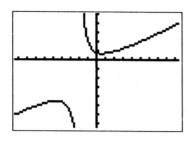

$[-9.4, 9.4, 1] \times [-10, 32, 2]$

41.

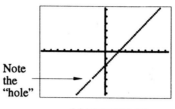

Note the "hole"

$[-9.4, 9.4] \times [-6.2, 6.2]$

4.6 USING GRAPHS TO SOLVE EQUATIONS

1. The root is $x = -\frac{5}{2}$.

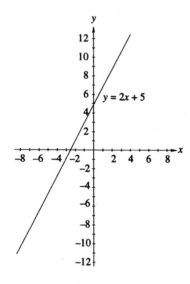

3. The root is $x = -3$ and $x = 3$.

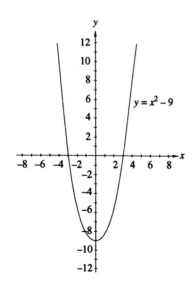

5. The roots are $x = 0$ and $x = 5$.

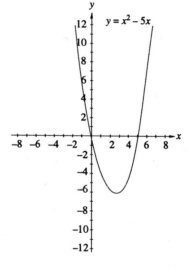

7. The roots are $x \approx -5.5414$ and $x \approx 0.5414$.

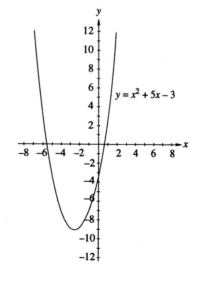

9. The roots are $x = -\frac{9}{4}$ and $x = \frac{6}{5}$.

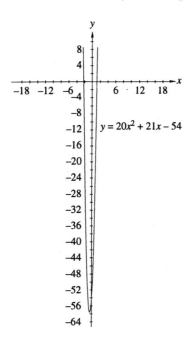

11. The root is $x = -1$.

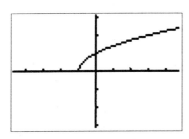

$$[-4.7, 4.7] \times [-3.1, 3.1]$$

13. Graph $y = \sqrt{x + 1} - 3$. The root is where the graph crosses the x-axis. The root is $x = 8$.

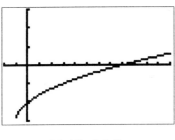

$$[-2, 12] \times [-3, 3]$$

15. Graph $y = \frac{x}{x+1} + 3$. The root is $x = -0.75$.

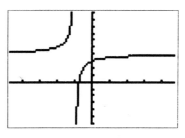

$$[-4.7, 4.7] \times [-6, 10]$$

17. Graph $y = \frac{x}{x-2} + x^2 - 2$. The root is $x \approx -1.2695$.

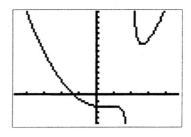

$$[-4.7, 4.7] \times [-1, 15]$$

19. (a)

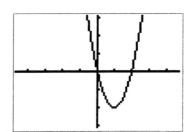

$$[-9.4, 9.2, 2] \times [-6.2, 6.2, 2]$$

(b) Domain: all real numbers; range $\{y : y \geq -4\}$.

21. (a)

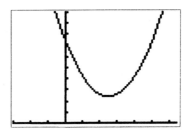

$$[-6, 12.8, 2] \times [0, 50, 5]$$

(b) Domain: all real numbers; range $\{y : y \geq 12\}$.

23. **(a)**

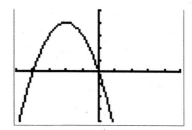

[–5, 5, 1] x [–4, 5, 1]

(b) Domain: all real numbers; range $\{y : y \leq 4\}$.

25. **(a)**

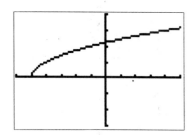

[–6, 5, 1] x [–3, 4, 1]

(b) Domain: $\{x : x \geq -5\}$; range $\{y : y \geq 0\}$.

27. **(a)**

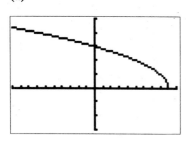

[–10, 10, 1] x [–3, 5, 1]

(b) Domain: $\{x : x \leq 9\}$; range $\{y : y \geq 0\}$.

29. **(a)**

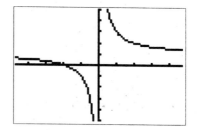

[–4.7, 4.7, 1] x [–5, 5, 1]

(b) Domain: $\{x : x \neq 0\}$; range $\{y : y \neq 1\}$

31. **(a)**

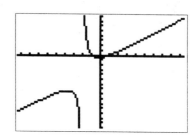

[–9.4, 9.4, 1] x [–15, 8, 1]

(b) Domain: $\{x : x \neq -2\}$; range $\{y : y < -7.46\}$ (decimal values are approximate).

33. **(a)**

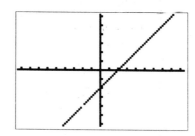

[–9.4, 9.4, 1] x [–6.2, 6.2, 1]

(b) Domain: $\{x : x \neq -2\}$; range $\{y : y \neq -4\}$.

35. Has an inverse function.

37. Does not have an inverse function.

39.

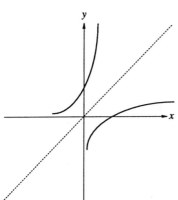

41.

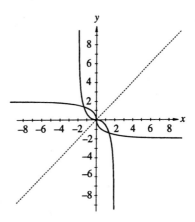

43. (a) 30%;
      (b) 60%;
      (c) 73%.

45. No.

 CHAPTER 4 REVIEW

1. (a)

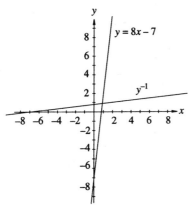

(b) Domain: all real numbers; Range: all real numbers; set $y = 0$ so $0 = 8x - 7$; $7 = 8x$; or $x = \frac{7}{8}$; x-intercept $\frac{7}{8}$; set $x = 0$; $y = 8(0) - 7 = -7$; y-intercept -7;
(c) Function;
(d) Has an inverse function.

2. (a)

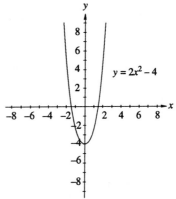

(b) Domain: all real numbers; Range: all real numbers greater than or equal to -4; set $y = 0$; $0 = 2x^2 - 4$; $2x^2 = 4$; $x = \pm\sqrt{2}$ x-intercept $\pm\sqrt{2}$; set $x = 0$; $y = 2(0)^2 - 4 = 0 - 4 = -4$; y-intercept -4;
(c) Function;
(d) Does not have an inverse function.

3.   **(a)**

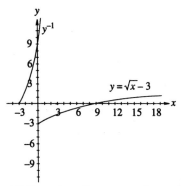

(b) Domain: all non-negative real numbers; $x \geq 0$; Range: all real numbers greater than or equal to -3. To find the x-intercept, set $y = 0$; Then, $0 = \sqrt{x} - 3$ or $\sqrt{x} = 3$ and so; $x = 3^2 = 9$ and the x-intercept $= 9$. To find the y-intercept, set $x = 0$. Then, $y - \sqrt{0} - 3 = -3$. and hence, the y-intercept $= -3$;
(c) Function;
(d) Has an inverse function.

4.   **(a)**

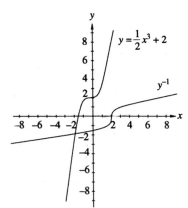

(b) Domain: all real numbers; Range: all real numbers; $0 = \frac{1}{2}x^3 + 2$; $\frac{1}{2}x^3 = -2$; $x^3 = -4$; $x = \sqrt[3]{-4}$; x-intercept $\sqrt[3]{-4}$; $y = \frac{1}{2}(0)^3 + 2 = 2$; y-intercept 2;
(c) Function;
(d) Has an inverse function.

5.   **(a)**

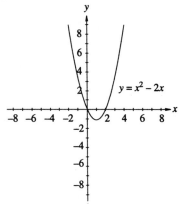

(b) Domain: all real numbers; Range: all real numbers greater than or equal to -1; $0 = x^2 - 2x$; $x(x - 2) = 0$, so $x = 0$ or $x = 2$; x-intercepts $0, 2$; $y = 0^2 - 2(0) = 0$; y-intercept 0;
(c) Function;
(d) Does not have an inverse function.

6.   **(a)**

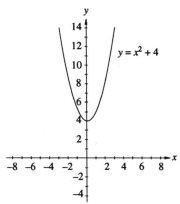

(b) Domain: all real numbers; Range: all real numbers 4 or larger; no x-intercept; y-intercept 4;
(c) Function;
(d) Does not have an inverse function.

7. $f(0) = 4(0) - 12 = 0 - 12 = -12$

8. $f(-2) = 4(-2) - 12 = -8 - 12 = -20$

9. $f(3) = 4(3) - 12 = 12 - 12 = 0$

10. $f(a) = 4a - 12$

11. $f(a - 2) = 4(a - 2) - 12 = 4a - 8 - 12 = 4a - 20$

12. $f(x + h) = 4(x + h) - 12 = 4x + 4h - 12$

13. $g(0) = \frac{0^2 - 9}{0^2 + 9} = \frac{-9}{9} = -1$

14. $f(3) = \frac{3^2 - 9}{3^2 + 9} = \frac{9 - 9}{9 + 9} = \frac{0}{18} = 0$

15. $g(-3) = \frac{(-3)^2 - 9}{(-3)^2 + 9} = \frac{9-9}{9+9} = \frac{0}{18} = 0$

16. $g(-2) = \frac{(-2)^2 - 9}{(-2)^2 + 9} = \frac{4-9}{4+9} = \frac{-5}{13} \approx -0.3846$

17. $g(4) = \frac{4^2 - 9}{4^2 + 9} = \frac{16-9}{16+6} = \frac{7}{25} = 0.28$

18. $g(-5) = \frac{(-5)^2 - 9}{(-5)^2 + 9} = \frac{25-9}{25+9} = \frac{16}{34} = \frac{8}{17} \approx 0.4706$

19.

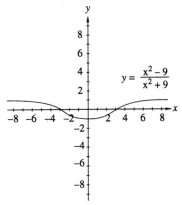

The zeros are -3 and 3.

20. $(f + g)(x) = f(x) + g(x) = 4x - 12 + \frac{x^2 - 9}{x^2 + 9}$

21. $(f + g)(3) = f(3) + g(3) = (4 \cdot 3 - 12) + \frac{3^2 - 9}{3^2 + 9} =$
$0 + 0 = 0$

22. $(f - g)(x) = f(x) - g(x) = 4x - 12 - \frac{x^2 - 9}{x^2 + 9}$

23. $(f - g)(-2) = f(-2) - g(-2) = 4(-2) - 12 - \frac{-5}{13} =$
$-20 + \frac{5}{13} = \frac{-260}{13} + \frac{5}{13} = \frac{-255}{13}$

24. $(f \cdot g)(x) = f(x) \cdot g(x) = (4x - 12)\left(\frac{x^2 - 9}{x^2 + 9}\right)$

25. $(f \cdot g)(0) = f(0) \cdot g(0) = (-12)(-1) = 12$

26. $(f/g)(x) = \frac{4x - 12}{\frac{x^2 - 9}{x^2 + 9}} = (4x - 12)\left(\frac{x^2 + 9}{x^2 - 9}\right) =$
$\frac{4(x-3)(x^2+9)}{(x-3)(x+3)} = \frac{4(x^2+9)}{x+3}$

27. $(f/g)(5) = \frac{4(5^2 + 9)}{5+3} = \frac{4(25+9)}{8} = \frac{1(34)}{2} = 17$

28. $(g/f)(x) = \frac{1}{(f/g)(x)} = \frac{x+3}{4(x^2+9)}$

29. $(g/f)(2) = \frac{2+3}{4(2^2+9)} = \frac{5}{4(4+9)} = \frac{5}{4(13)} = \frac{5}{52}$

30. $(f \circ g)(x) = f(g(x)) = f\left(\frac{x^2-9}{x^2+9}\right) = 4\left(\frac{x^2-9}{x^2+9}\right) - 12$

31. $(f \circ g)(4) = 4\left(\frac{4^2 - 9}{4^2 + 9}\right) - 12 = 4\left(\frac{16-9}{16+9}\right) - 12 =$
$4\left(\frac{7}{25}\right) - 12 = \frac{28}{25} - \frac{300}{25} = \frac{-272}{25}$

32. $(g \circ f)(x) = g(f(x)) = g(4x - 12) = \frac{(4x-12)^2 - 9}{(4x-12)^2 + 9} =$
$\frac{(16x^2 - 96x + 144) - 9}{(16x^2 - 96x + 144) + 9} = \frac{16x^2 - 96x + 135}{16x^2 - 96x + 153}$

33. $(g \circ f)(3) = \frac{16(3)^2 - 96(3) + 135}{16(3)^2 - 96(3) + 153} = \frac{144 - 288 + 135}{144 - 288 + 153} =$
$\frac{-9}{9} = -1$

34.

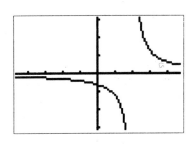

$[-9.4, 9.4] \times [-6.2, 6.2]$

35.

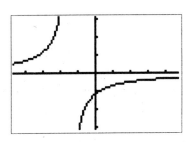

$[-9.4, 9.4] \times [-6.2, 6.2]$

36.

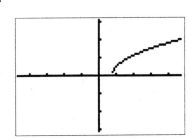

$[-9.4, 9.4] \times [-6.2, 6.2]$

37.

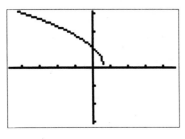

[–9.4, 9.4] x [–6.2, 6.2]

38.

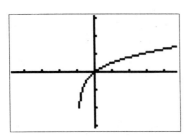

[–9.4, 9.4] x [–6.2, 6.2]

39. Solve for y to get two functions: $y = \sqrt{9 - x^2}$ and $y = -\sqrt{9 - x^2}$. When both of these are graphed on the same coordinate system you get the following figure.

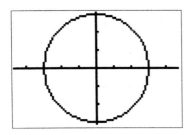

[–4.7, 4.7] x [–3.1, 3.1]

40. Solve for y to get two functions: $y = \sqrt{16 - 0.5x^2}$ and $y = -\sqrt{16 - 0.5x^2}$. When both of these are graphed on the same coordinate system, you get the following figure.

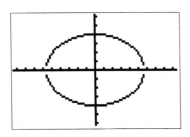

[–9.4, 9.4] x [–6.2, 6.2]

41. $4x + 7y = 0; 7y = -4x; y = \frac{-4}{7}x$; roots are $0 = -\frac{4}{7}x$ or $x = 0$.

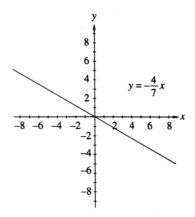

42. $x^2 - 20 = y$ or $y = x^2 - 20$; roots are $0 = x^2 - 20$; $x^2 = 20; x = \pm\sqrt{20} = \pm\sqrt{5} \approx \pm 4.4721$.

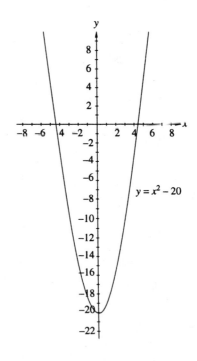

43. $2x^2 + 10x + 4 = 0$ or $2(x^2 + 5x + 2) = 0$ roots are about -0.44 and -4.50.

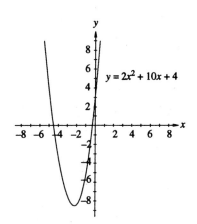

44. $y = 8x^3 - 20x^2 - 34x + 21$; roots: $x = -1.5, x = 0.5$, and $x = 3.5$.

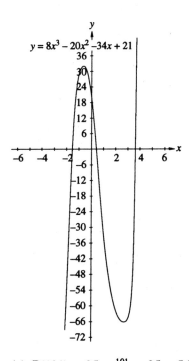

45. (a) $P(101) = 35 - \frac{101}{20} = 35 - 5.05 = \$29.95$;

$P(350) = 35 - \frac{350}{20} = 35 - 17.50 = \$17.50$;

$P(400) = 35 - \frac{400}{20} = 35 - 20 = \$15.00$;

(b) $R = D \times P = n\left(35 - \frac{n}{20}\right) = 35n - \frac{n^2}{20}$;

(c) $R = 101(29.95) = 3024.95$ or $R(n) = 35(101) - \frac{(101)^2}{20} = 3535 - 510.05 = 3024.95$;

$R(350) = 35(350) - \frac{350^2}{20} = 6125$;

$R(400) = 35(400) - \frac{400^2}{20} = 6000$

46. (a) $P(n) = \left(30n - \frac{n^2}{20}\right) - (550 + 10n) = 20n - \frac{n^2}{20} - 550$;

(b) $P(30) = 20(30) - \frac{30^2}{20} - 550 = \$5$;

$P(100) = 20(100) - \frac{100^2}{20} - 550 = \$950$; $P(150) = \$1325$; $P(300) = \$950$; $P(400) = -\$550$.

47.

t	0	1	2	3	4	5
$h(t)$	0	3.08	6.48	10.68	15.68	21.00

t	6	7	8	9	10
$h(t)$	25.68	28.28	26.88	19.08	2.00

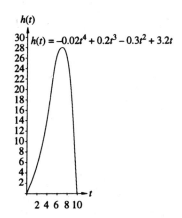

48. (a)

v	0	10	20	30	40	50	60	70
$s(v)$	0	14	36	66	104	150	204	266

(b) about 69.8 mph;

(c)

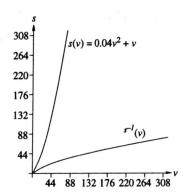

CHAPTER 4 TEST

1. $f(-2) = 7(-2) - 5 = -14 - 5 = -19$

2. $f(3 - a) = 7(3 - a) - 5 = 7 \cdot 3 - 7a - 5 = 21 - 7a - 5 = 16 - 7a$

3. $g(0) = \frac{0^2 - 2 \cdot 0 - 15}{0 + 3} = \frac{-15}{3} = -5$

4. $g(5) = \frac{5^2 - 2 \cdot 5 - 15}{5 + 3} = \frac{25 - 10 - 15}{8} = \frac{0}{8} = 0$

5.   **(a)**

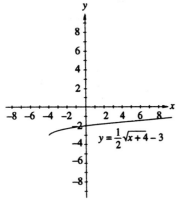

(b) This function is defined whenever $\sqrt{x + 4} \geq 0$ or when $x \geq -4$;
(c) The low point of the graph is when $x = -4$. Here the y-value is -3. So, the graph can assume all values of $y \geq -3$;
(d) The x-intercept is when $y = 0$. Solving $0 = \frac{1}{2}\sqrt{x + 4} - 3$, we get $3 = \frac{1}{2}\sqrt{x + 4}$ or $6 = \sqrt{x + 4}$. Squaring both sides we get $36 = x + 4$ or $x = 32$;
(e) The y-intercept is when $x = 0$. Solving $y = \frac{1}{2}\sqrt{0 + 4} - 3 = \frac{1}{2}\sqrt{4} - 3 = \frac{1}{2}2 - 3 = 1 - 3 = -2$.

6. $3x - 15 + \frac{x-5}{x+5} = \frac{(3x-15)(x+5)}{x+5} + \frac{x-5}{x+5} = \frac{3x^2 - 75}{x+5} + \frac{x-5}{x+5} = \frac{3x^2 + x - 80}{x+5}$

7. $3x - 15 - \frac{x-5}{x+5} = \frac{(3x-15)(x+5)}{x+5} - \frac{x-5}{x+5} = \frac{3x^2 - 75}{x+5} - \frac{x-5}{x+5} = \frac{3x^2 + x - 70}{x+5}$

8. $(3x - 15)\left(\frac{x-5}{x+5}\right) = \frac{(3x-15)(x-5)}{x+5} = \frac{3(x-5)(x-5)}{x+5} = \frac{3(x-5)^2}{x+5}$

9. $(3x - 15) \div \frac{x-5}{x+5} = (3x - 15)\left(\frac{x+5}{x-5}\right) = \frac{3(x-5)(x+5)}{x-5} = 3(x + 5) = 3x + 15$

10. $3\left(\frac{x-5}{x+5}\right) - 15 = \frac{3x-15}{x+5} - \frac{15x+75}{x+5} = \frac{3x-15-15x-75}{x+5} = \frac{-12x-90}{x+5} = -6\left(\frac{2x+15}{x+5}\right)$

11. $\frac{(3x-15)-5}{(3x-15)+5} = \frac{3x-20}{3x-10}$

12.

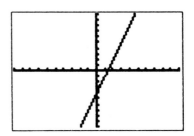

13.

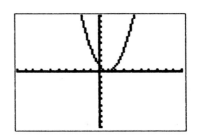

14.

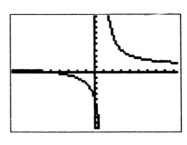

[–9.4, 9.4] x [–10, 10]

15.   Graph $y = \frac{x^2}{x+1} - 2x + 1$.

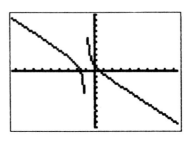

[–9.4, 9.4] x [–10, 10]

16.   Graph $y = \sqrt{x^2 - 1} - 3 + x$.

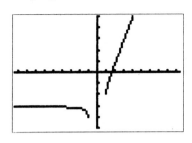

[–9.4, 9.4] x [–5, 5]

17.   **(a)**

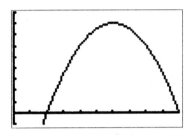

[0, 10, 1] x [–2, 18, 2]

(b) $r(3) = -3^2 + 12(3) - 20 = -9 + 36 - 20 = 7$.
So, 7 responses can be expected after 3 ms.
(c) Here $-s^2 + 12s - 20 = 16$ or $s^2 - 12s + 36 = 0$.
This factors as $(s - 6)^2 = 0$, so $s = 6$. It has been
6 ms since the nerve was stimulated.
(d) We need to solve $-s^2 + 12s - 20 = 12$ or
$s^2 - 12s + 32 = 0$. This factors as $(s-4)(s-8) = 0$,
so $s = 4$ or $s = 8$. It has been either 4 ms or 8
ms since the nerve was stimulated.

18.

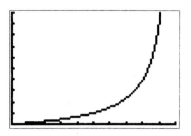

[0, 100, 10] x [0, 50, 5]

(b) $5,000,
(c) $45,000,
(d) $y = \frac{5x}{100-x}$. Solving for x, we have $y(100 - x) = 5x$ or $100y - xy = 5x$. Grouping x's
we obtain $5x + xy = 100y$, and factoring yields
$x(y + 5) = 100y$ and so, $x = \frac{100y}{y+5}$. Interchanging
x and y, and using function notation, we obtain
$C^{-1}(x) = \frac{100x}{x+5}$,
(e) 75%.

5

Systems of Linear Equations and Determinants

≡ 5.1 LINEAR EQUATIONS

1. $(2,5)$, $(3,8)$, $m = \frac{y_2 - y_1}{x_2 - x_1} = \frac{8-5}{3-2} = \frac{3}{1} = 3$

3. $(1,8)$, $(5,3)$, $m = \frac{3-8}{5-1} = \frac{-5}{4} = -\frac{5}{4}$

5. $(9,3)$, $(2,-7)$, $m = \frac{-7-3}{2-9} = \frac{-10}{-7} = \frac{10}{7}$

7. $\frac{5-5}{4-0} = \frac{0}{4} = 0$, so the slope is 0, or $m = 0$

9. $\dfrac{\frac{3}{4} - 17}{2 - 2} = \dfrac{\frac{-65}{4}}{0}$. You cannot divide by 0, and so the slope is undefined.

11. $m = 4$, point: $(5,3)$. Using $y - y_1 = m(x - x_1)$ we get $y - 3 = 4(x - 5)$

13. $m = \frac{2}{3}$, point: $(1,-5)$, $y - (-5) = \frac{2}{3}(x - 1)$ or $y + 5 = \frac{2}{3}(x - 1)$

15. Since $m = 0$, this is a horizontal line. All horizontal lines have the same y-coordinate, so the equation of this line is $y = -5$.

17. $m = -\frac{5}{3}$, point: $(2,0)$, $y - 0 = -\frac{5}{3}(x - 2)$ or $y = -\frac{5}{3}(x - 2)$

19. Since the slope m is undefined, this is a vertical line. All vertical lines have the same x-coordinate, so the equation of this line is $x = 7$.

21. points: $(1,5)$ and $(-3,2)$. First $m = \frac{2-5}{-3-1} = \frac{-3}{-4} = \frac{3}{4}$; then using $(-3,2)$ we get $y - 2 = \frac{3}{4}(x + 3)$; using $(1,5)$ we get $y - 5 = \frac{3}{4}(x - 1)$

23. $m = 0$, so this is a horizontal line. All horizontal lines have the same y-coordinate, so the equation of this line is $y = 3$.

25. $m = 2, b = 4$. Using $y = mx + b$ we get $y = 2x + 4$

27. $m = 5, b = -3, y = 5x - 3$

29. $y - 3x = 6$; adding $3x$ to both sides we get $y = 3x + 6$ so $m = 3, b = 6, x$-intercept $= -2$

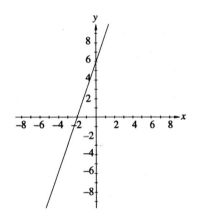

31. $2y - 5x = 8 \Rightarrow 2y = 5x + 8 \Rightarrow y = \frac{5}{2}x + 4; m = \frac{5}{2}; b = 4; x$-intercept $= -\frac{8}{5}$

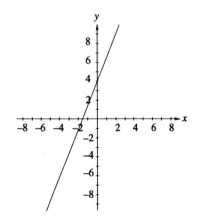

33. $5x - 2y - 10 = 0 \Rightarrow -2y = -5x + 10 \Rightarrow y = \frac{5}{2}x - 5$; $m = \frac{5}{2}$; $b = -5$; x-intercept $= 2$

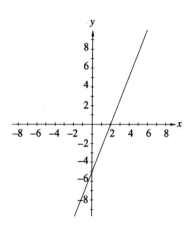

35. $x = -3y + 7$; $3y = -x + 7$; $y = -\frac{1}{3}x + \frac{7}{3}$; $m = -\frac{1}{3}$; $b = \frac{7}{3}$; x-intercept $= 7$

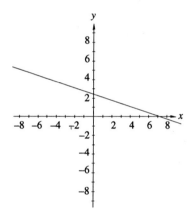

37. **(a)** This line is horizontal, so $m = 0$.
(b)

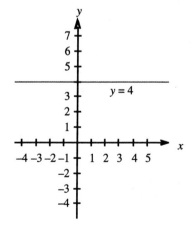

39. **(a)** This is a vertical line, so the slope is undefined.
(b)

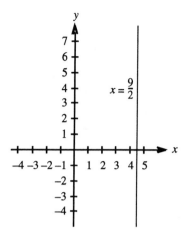

41. $s = md = \left(\frac{1780\pi}{60}\right)d$ so $m = \frac{1780\pi}{60} = \frac{178}{6}\pi = \frac{89\pi}{3} \approx 93.2$

43. **(a)** $F = kx = k(L - L_0)$ or $F = kL - kL_0$;
(b) $F = 4.5(L - 6)$ or $F = 4.5L - 27$
(c)

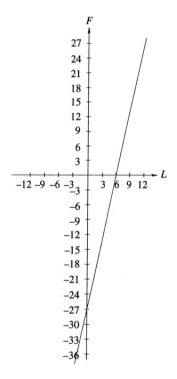

≡ 5.2 GRAPHICAL AND ALGEBRAIC METHODS FOR SOLVING TWO LINEAR EQUATIONS IN TWO VARIABLES

1.

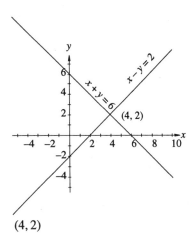

$(4, 2)$

3.

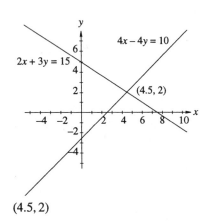

$(4.5, 2)$

5.

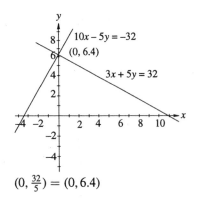

$(0, \frac{32}{5}) = (0, 6.4)$

7.

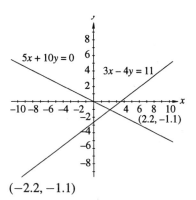

$(-2.2, -1.1)$

9. $\begin{cases} y = 3x - 4 & (1) \\ x + y = 8 & (2) \end{cases}$
Substituting $(3x - 4)$ for y in equation (2) yields $x + (3x - 4) = 8$. Solving $4x - 4 = 8; 4x = 12$, $x = 3$. Substituting $x = 3$ in (1) $\Rightarrow y = 3(3) - 4 = 9 - 4 = 5; (3, 5)$

11. $\begin{cases} y = -2x - 2 & (1) \\ 3x + 2y = 0 & (2) \end{cases}$
Substituting $-2x - 2$ for y in (2) $\Rightarrow 3x + 2(-2x - 2) = 0; 3x - 4x - 4 = 0; -x = 4$ or $x = -4$. Back substituting in (1) $y = -2(-4) - 2 = +8 - 2 = 6; (-4, 6)$

13. $\begin{cases} 2x + 5y = 6 & (1) \\ x - y = 10 & (2) \end{cases}$
Solving (2) for $x \Rightarrow x = 10 + y$. Substituting in (1) $\Rightarrow 2(10 + y) + 5y = 6; 20 + 2y + 5y = 6; 7y = -14; y = -2. x = 10 + (-2) = 8$. Solution $(8, -2)$

15. $\begin{cases} 2x + 3y = 3 & (1) \\ 6x + 4y = 15 & (2) \end{cases}$
Solving for x in (1) $\Rightarrow 2x = 3 - 3y$ or $x = \frac{3-3y}{2}$ (3). Substituting this in (2) yields $6(\frac{3-3y}{2}) + 4y = 15; 3(3 - 3y) + 4y = 15; 9 - 9y + 4y = 15; -5y = 6$ or $y = -\frac{6}{5}$ or -1.2. Substituting this for y in (3) yields $x = \frac{3-3(-1.2)}{2} = \frac{3+3.6}{2} = \frac{6.6}{2} = 3.3$. Solution $(3.3, -1.2)$

17. $\begin{cases} 4.8x - 1.3y = 16.9 \quad (1) \\ -7.2x - 2.8y = -9.2 \quad (2) \end{cases}$

Solving equation (1) for x, we obtain
$4.8x = 1.3y + 16.9$ or $x = 0.2708\bar{3}y$
$+ 3.5208\bar{3}$ (3)

Now, substituting for x in equation (2), we have

$$-7.2(0.2708\bar{3}y + 3.5208\bar{3}) - 2.8y = -9.2$$
$$-1.95y - 25.35 - 2.8y = -9.2$$
$$-4.75y = 16.15$$
$$y = -3.4$$

We still need to find the x-coordinate of the solution. Back-substituting into equation (3) yields $x = 0.2708\bar{3}(-3.4) + 3.5208\bar{3} = 2.6$. The solution is $(2.6, -3.4)$.

19. $\begin{cases} 4.2x + 3.7y = 10.79 \quad (1) \\ 6.5x - 0.3y = -15.24 \quad (2) \end{cases}$

Solving equation (1) for x, we obtain
$4.2x = -3.7y + 10.79$ or
$x = -0.88095231y + 2.569047619$ (3)

Substituting this result for x in equation (2) gives

$$6.5(-0.88095231y + 2.569047619) - 0.3y = -15.24$$
$$-5.726190476y + 16.69880952 - 0.3y = -15.24$$
$$-6.026190476y = -31.93880952$$
$$y = 5.3$$

Back-substituting into equation (3) yields $x = -0.880952381(5.3) + 2.569047619$ or $x = -2.1$. The solution is $(-2.1, 5.3)$.

21. $\begin{cases} x + y = 9 \quad (1) \\ x - y = 5 \quad (2) \end{cases}$

Adding $(1) + (2) \Rightarrow 2x = 14; x = 7$; Substituting in (1) yields $7 + y = 9; y = 2; (7, 2)$

23. $\begin{cases} -x + 3y = 5 \quad (1) \\ 2x + 7y = 3 \quad (2) \end{cases}$

Multiply (1) by $2 \Rightarrow -2x + 6y = 10$ (3). Adding $(3) + (2)$ yields $13y = 13$ or $y = 1$. Back substituting in (1) yields $-x + 3 = 5$ or $-x = 2$ and so we see that $x = -2$. The solution is $(-2, 1)$.

25. $\begin{cases} 3x - 2y = -15 \quad (1) \\ 5x + 6y = 3 (2) \end{cases}$

Multiply (1) by $3 \Rightarrow 9x - 6y = -45$ (3). Add $(3) + (2): 14x = -42 \Rightarrow x = -3$. Substituting $x = -3$ in (1) we get $3(-3) - 2y = -15 \Rightarrow -9 - 2y = -15; -2y = -6$, so $y = 3$ and the solution is $(-3, 3)$.

27. $\begin{cases} 3x - 5y = 37 \quad (1) \\ 5x - 3y = 27 \quad (2) \end{cases}$

Multiply (1) by $5 \Rightarrow 15x - 25y = 185$ (3). Multiply (2) by $3 \Rightarrow 15x - 9y = 81$ (4). Subtract $(3) - (4) \Rightarrow -16y = 104$ and so $y = -6.5$. Substituting in (1) $\Rightarrow 3x - 5(-6.5) = 37$ or $3x + 32.5 = 37 \Rightarrow x = 1.5$. The solution is $(1.5, -6.5)$.

29. $\begin{cases} 6.4x - 1.7y = 66.7 \quad (1) \\ -4.2x + 5.1y = -62.1 \quad (2) \end{cases}$

Multiply equation (1) by 3 to obtain $19.2x - 5.1y = 200.1$ (3). Adding equations (2) and (3), we get $15x = 138$ or $x = 9.2$. Now, substitute this value for x in equation (1) to get $6.4(9.2) - 1.7y = 66.7$ or $-1.7y = 7.82$, which means that $y = -4.6$. The solution is $(9.2, -4.6)$.

31. $\begin{cases} 2x + 3y = 5 \quad (1) \\ x - 2y = 6 (2) \end{cases}$

Solving (2) for x, we get $x = 2y + 6$. Substituting in (1) produces $2(2y + 6) + 3y = 5$ or $4y + 12 + 3y = 5; 7y = -7; y = -1; x = 2(-1) + 6 = -2 + 6 = 4; (4, -1)$

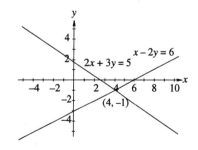

33. $\begin{cases} 8x + 3y = 13 & (1) \\ 3x + 2y = 11 & (2) \end{cases}$

$(1) \times 2 \Rightarrow 16x + 6y = 26\,(3).\,(2) \times 3 \Rightarrow 9x + 6y = 33\,(4).\,(3) - (4) \Rightarrow 7x = -7$ or $x = -1$. Substituting in (1) we get $8(-1) + 3y = 13 \Rightarrow -8 + 3y = 13$ and so $3y = 21$ or $y = 7$. The solution is $(-1, 7)$.

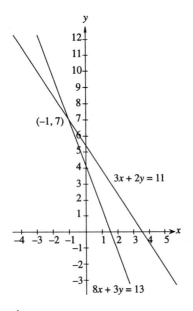

35. $\begin{cases} 10x - 9y = 18 & (1) \\ 6x + 2y = 1 & (2) \end{cases}$

$(1) \times 2 \Rightarrow 20x - 18y = 36\,(3); (2) \times 9 \Rightarrow 54x + 18y = 9\,(4)$. Adding $(3) + (4) \Rightarrow 74x = 45$ and so $x = \frac{45}{74} \approx 0.608$. Substituting in (1), we get $10\left(\frac{45}{74}\right) - 9y = 18$ or $\frac{450}{74} - 9(y = 18 \Rightarrow -9y = 18 - \frac{450}{74} = \frac{441}{37}$, and so $y = -\frac{1}{9}\left(\frac{441}{37}\right) = -\frac{49}{37} \approx -1.324$. The solution is $\left(\frac{45}{74}, -\frac{49}{37}\right) \approx (0.608, -1.324)$.

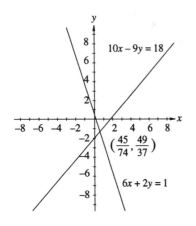

37. $\begin{cases} x - 9y = 0 & (1) \\ \frac{x}{3} = 2y + \frac{1}{3} & (2) \end{cases}$

$(2) \times (3) \Rightarrow x = 6y + 1$ or $x - 6y = 1\,(3)$. Subtracting: $(1) - (3) \Rightarrow -3y = -1$ or $y = \frac{1}{3}$. Substituting in (1), we obtain $x - 9\left(\frac{1}{3}\right) = 0$ or $x - 3 = 0$ and so $x = 3$. The solution is $\left(3, \frac{1}{3}\right)$.

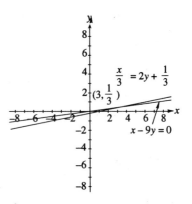

39. $\begin{cases} 4.9x + 1.7y = 10.6 & (1) \\ 3.6x - 1.2y = 14.4 & (2) \end{cases}$

Multiply equation (1) by 1.2 to obtain $5.88x + 2.04y = 12.72\,(3)$. Multiply equation (2) by 1.7 with the result $6.12x - 2.04y = 24.48\,(4)$. Now, add equations (3) and (4) to get $12x = 37.2$ or $x = 3.1$. Back substitution into equation (1) yields $4.9(3.1) + 1.7y = 10.6$ or $1.7y = -4.59$. Hence, $y = -2.7$ and the solution is $(3.1, -2.7)$.

41. Substituting $w + 8$ for L in the first equation yields $2(w + 8) + 2w = 36; 2w + 16 + 2w = 36 \Rightarrow 4w + 16 = 36$ or $4w = 20$ and so $w = 5$. $L = w + 8$ or $5 + 8 = 13$. Length is 13 km and the width is 5 km.

43. $2L + 2w = 45$ and $L = 3w$ are two equations that fit this situation. By substitution $2(3w) + 2w = 45 \Rightarrow 6w + 2w = 45$ or $8w = 45$ and so $w = 5.625$. Substituting, we get $L = 3(5.625) = 16.875$. Thus, the length is 16.875 km and the width is 5.625 km.

45. Multiplying (2) by $100 \Rightarrow 5x + 13y = 80,000$ and multiplying (1) by $5 \Rightarrow 5x + 5y = 50,000$. Subtracting yields $8y = 30,000$ and so $y = 3750$. Substituting in (1), we see that $x + 3750 = 10,000$ and so $x = 6,250$. As a result of these calculations, we have determined that we need 6,250 L of 5% gasohol and 3,750 L of 13% gasohol.

47. Equations that fit this situation are $8x = 5y$ (1) and $12x = 5(y + 3.2)$ (2). Solving (1) for x produces $x = \frac{5}{8}y$. Substituting into (2) yields $12\left(\frac{5}{8}y\right) = 5(y + 3.2)$; $\frac{15}{2}y = 5y + 16$; $2.5y = 16$ or $y = 6.4$. $x = \frac{5}{8} \times 6.4$ or $x = 4$. Thus, the unknown mass is 6.4 kg and the 8-kg force is at a distance of 4 m.

49. $8 = I_1(3) + I_2(5)$; $5 = 5(I_2) - 6(I_1 - I_2)$. The second equation equation simplifies to $5 = 5I_2 - 6I_1 + 6I_2$ or $5 = 11I_2 - 6I_1$. Taking the first equation times 2 yields $16 = 6I_1 + 10I_2$. Adding these two gives $21 = 21I_2$ or $I_2 = 1$. Replacing I_2 with 1 in the first equation you get $8 = 3I_1 + 5$ or $3 = 3I_1$ or $I_1 = 1$. Since $I_1 = I_2 + I_3$; $I_3 = I_1 - I_2$ or $I_3 = 1 - 1 = 0$.

51. Using $I_2 = I_1 + I_3$ is equivalent to $I_2 - I_1 = I_3$. Hence $E_2 = R_3 I_3 + R_2 I_2$ is equivalent to $E_2 = R_3(I_2 - I_1)$. Substituting the given values yields $6 = 8I_1 + 4I_2$ and $10 = 7(I_2 - I_1) + 4I_2$ or $10 = 11I_2 - 7I_1$. Taking the first equation times $11 \Rightarrow 66 = 88I_1 + 44I_2$. The second times 4 gives $40 = 44I_2 - 28I_1$. Subtracting gives $26 = 116I_1$ or $I_1 = \frac{26}{116} = 0.224$ A. Thus, $6 = 8(0.224) + 4I_2 \Rightarrow 6 = 1.792 + 4I_2 \Rightarrow 4.208 = 4I_2$. Thus, $I_2 = 1.052$ A; $I_3 = 1.052 - 0.224 = .828$ A.

5.3 ALGEBRAIC METHODS FOR SOLVING THREE LINEAR EQUATIONS IN THREE VARIABLES

1.
$$\begin{cases} 2x + y + z = 7 & (1) \\ x - y + 2z = 11 & (2) \\ 5x + y - 2z = 1 & (3) \end{cases}$$
Solving (2) for x yields $x = 11 + y - 2z$ (4). Substituting in (3) yields $5(11 + y - 2z) + y - 2z = 1$; $55 + 5y - 10z + y - 2z = 1$; or $6y - 12z = -54$ (5). Substituting (4) into (1) we get $2(11 + y - 2z) + z = 7$ or $22 + 2y - 4z + y + z = 7$ or $3y - 3z = -15$ (7). Substituting using (6) into (7) we get $3(2z - 9) - 3z = -15$; $6z - 27 - 3z = -15$; $3z = 12$; $z = 4$. Using (6) we get $y = 2 \cdot 4 - 9 = 8 - 9 = -1$. Using (4) we get $x = 11 + (-1) - 2(4) = 11 - 1 - 8 = 11 - 9 = 2$. The solution is $x = 2$; $y = -1$; $z = 4$.

3.
$$\begin{cases} 2x - y - z = -8 & (1) \\ x + y - z = -9 & (2) \\ x - y + 2z = 7 & (3) \end{cases}$$
Solving equation (1) for y we obtain $y = 8 + 2x - z$ (4). Substituting this into (2) yields $x + (8 + 2x - z) - z = -9$ which simplifies to $3x - 2z = -17$ (5). Again, using equation (4), we substitute into equation (3), which yields $x - (8 + 2x - z) + 2z = 7$. This simplifies to $-x + 3z = 15$ (6). Solving equation (6) for x, we have $x = 3z - 15$ (7). Using equation (7) and substituting into equation (5) yields $3(3z - 15) - 2z = -17$ or $9z - 45 - 2z = -17$ or $7z = 28$. Hence, $z = 4$. Back substituting into equation (7) we get $x = 3(4) - 15 = 12 - 15 = -3$. Now, back substitution into equation (4) yields $y = 8 + 2(-3) - 4 = 8 - 6 - 4 = -2$. The solution is $(-3, -2, 4)$.

5.
$$\begin{cases} 2x + y + z = 7 & (1) \\ x - y + 2z = 11 & (2) \\ 5x + y - 2z = 1 & (3) \end{cases}$$
Adding (1) + (2) yields $3x + 3z = 18$ (4). Adding (2) + (3) yields $6x = 12$ or $x = 2$ (5). Substituting 2 for x in (4) yields $3 \cdot 2 + 3z = 18$; $6 + 3z = 18$; $3z = 12$, $z = 4$. Substituting $z = 4$, $x = 2$ in (1) yields $2(2) + y + 4 = 7$ or $4 + y + 4 = 7$; $y = -1$. The solution is $x = 2$, $y = -1$, $z = 4$.

7.
$$\begin{cases} 2x - y - z = -8 & (1) \\ x + y - z = -9 & (2) \\ x - y + 2z = 7 & (3) \end{cases}$$
$(1) + (2) \Rightarrow 3x - 2z = -17$ (4). $(2) + (3) \Rightarrow 2x + z = -2$ (5). $2 \times (5) \Rightarrow 4x + 2z = -4$ (6). $(4) + (6) \Rightarrow 7x = -21$ or $x = -3$. Substituting into (5) yields $2(-3) + z = -2$ or $-6 + z = -2$ or $z = 4$. Substituting into (2) yields $(-3) + y - 4 = -9$ or $-7 + y = -9$ or $y = -2$. The solution is $x = -3$, $y = -2$, $z = 4$.

9.
$$\begin{cases} x + y + z = 2 & (1) \\ 8x - 2y + 4z = -3 & (2) \\ 6x - 4y - 3z = 3 & (3) \end{cases}$$
$2 \times (1) \Rightarrow 2x + 2y + 2z = 4$ (4). $(4) + (2) \Rightarrow 10x + 6z = 1$ (5). $4 \times (1) \Rightarrow 4x + 4y + 4z = 8$ (6). $(6) + (3) \Rightarrow 10x + z = 11$ (7). $(5) - (7) \Rightarrow 5z = -10$ or $z = -2$. Substituting in (7) yields $10x - 2 = 11$ or $10x = 13$ or $x = 1.3$. Substituting in (1) yields $1.3 + y - 2 = 2$; $y - 0.7 = 2$ so $y = 2.7$. The solution is $x = 1.3$, $y = 2.7$, $z = -2$.

11. $\begin{cases} 3x - y - 2z = 11 & (1) \\ -x + 3y + 2z = -1 & (2) \\ 2x - 2y - 4z = 17 & (3) \end{cases}$

$2 \times (2) \Rightarrow -2x + 6y + 4z = -2\,(4)$. Adding: $(4) + (3) \Rightarrow 4y = 15$ or $y = 3.75\,(5)$, and adding $(1)+(2) \Rightarrow 2x+2y = 10$ or $x+y = 5\,(6)$. Substituting 3.75 for y into (6) yields $x+3.75 = 5$; $x = 1.25$. Substituting into (1) yields $3(1.25)-(3.75)- 2z = 11$; $3.75 - 3.75 - 2z = 11$ or $-2z = 11$ and so $z = -5.5$. The solution is $x = 1.25, y = 3.75, z = -5.5$.

13. $\begin{cases} 2x + 3y + 3z = 9 & (1) \\ 5x - 2y + 8z = 6 & (2) \\ 4x - y + 5z = -1 & (3) \end{cases}$

To eliminate ys: $(3) \times 2 \Rightarrow 8x - 2y + 10z = -2\,(4)$. $(4) - (2) \Rightarrow 3x + 2z = -8\,(5)$. $(3) \times 3 \Rightarrow 12x - 3y + 15z = -3\,(6)$. $(1)+(6) \Rightarrow 14x+18z = 6\,(7)$. $9 \times (5) \Rightarrow 27x + 18z = -72\,(8)$. $(8) - (7) \Rightarrow 13x = -78$ or $x = -6$. Substituting in (5) yields $3(-6)+2z = -8$; $-18+2z = -8$; $2z = 10$; $z = 5$. Substituting in (1) yields $2(-6) + 3y + 3(5) = 9$; $-12 + 3y + 15 = 9$; $3y + 3 = 9$; $3y = 6$; $y = 2$. The solution is $x = -6, y = 2, z = 5$.

15. $\begin{cases} I_1 - I_2 + I_3 = 0 & (1) \\ 6I_1 + 6I_2 = 18 \Rightarrow I_1 + I_2 = 3 & (2) \\ 6I_2 + I_3 = 14 & (3) \end{cases}$

$(3)-(1) \Rightarrow -I_1+7I_2 = 14\,(4)$. Adding $(2)+(4) \Rightarrow 8I_2 = 17$; $I_2 = \frac{17}{8} = 2.125$. Substituting in (2) yields $I_1 + 2.125 = 3$ or $I_1 = 0.875$. Substituting in (3) yields $6(2.125) + I_3 = 14$; $12.75 + I_3 = 14$ or $I_3 = 1.25$. The values of the currents are $I_1 = 0.875$ A, $I_2 = 2.125$ A, $I_3 = 1.25$ A.

17. Substituting the coordinates of $P(5, 1)$ in the standard equation for a circle, we get $5^2 + 1^2 + 5a + b + c = 0 \Rightarrow 5a + b + c = -26\,(1)$. Next we substitute $x = -2$ and $y = -6$, the coordinates of Q, in the standard equation of a circle, with the result: $(-2)^2 + (-6)^2 - 2a - 6b + c = 0 \Rightarrow -2a - 6b + c = -40$ or $2a + 6b - c = 40\,(2)$. Finally, substituting the coordinates of $R(-1, -7)$ in the standard equation of a circle yields $(-1)^2 + (-7)^2 - a - 7b + c = 0$ or $-a - 7b + c = -50$ or $a + 7b - c = 50\,(3)$. We now proceed to solve this

system of three equations. We begin by adding $(1) + (2) \Rightarrow 7a + 7b = 14$ or $a + b = 2\,(4)$ and adding $(1) + (3) \Rightarrow 6a + 8b = 24$ or $3a + 4b = 12\,(5)$. Then, $3 \times (4) \Rightarrow 4a + 4b = 8\,(6)$ and subtracting $(6) - (5)$, we find that $a = -4$. Substituting -4 for a in $(4) \Rightarrow -4 + b = 2$; $b = 6$. Substituting into (1) yields $5(-4) + 6 + c = -26$ or $-20 + 6 + c = -26$; $c - 14 = -26$ and so $c = -12$ and so the equation of the circle through points P, Q, and R is $x^2 + y^2 - 4x + 6y - 12 = 0$.

19. Organizing the information as in Example 5.18, we get $\begin{cases} 7L + 6M + 8S = 64 & (1) \\ 6L + 3M + 1S = 33 & (2) \\ 4L + 2M + 2S = 26 & (3) \end{cases}$

$(2) \times 3 \Rightarrow 12L + 6M + 2S = 66\,(4)$. Subtracting $(4) - (3) \Rightarrow 8L + 4M = 40\,(5)$. Multiplying $4 \times (3) \Rightarrow 16L + 8M + 8S = 104\,(6)$ and then subtracting $(6) - (1) \Rightarrow 9L + 2M = 40\,(7)$. $2 \times (7) \Rightarrow 18L + 4M = 80\,(8)$. $(8) - (5) \Rightarrow 10L = 40$ or $L = 4$. Substituting into (5) yields $8(4) + 4M = 40$; $32 + 4M = 40$; $4M = 8$ and $M = 2$. Substituting into (2) yields $6(4) + 3(2) + S = 33$ and $S = 3$. So, the company needs 4 large, 2 medium and 3 small trucks.

21. The final 400 lb of fertilizer contains $0.16 \times 400 = 64$ lb of nitrogen, $0.3 \times 400 = 12$ lb of phosphorus, and $0.9 \times 400 = 36$ lb of potash. Let x be the amount, in pounds, of 10-12-15 fertilizer, y the amount of 10-0-6 and z the amount of the 30-6-15 fertilizer. This leads to the following system of equations.

$\begin{cases} 0.10x + 0.10y + 0.30z = & 64 & (1) \\ 0.12x + 0y + 0.06z = & 12 & (2) \\ 0.15x + 0.06y + 0.15z = & 36 & (3) \end{cases}$

$\begin{cases} (1) + -5(2) \Rightarrow -0.5x + 0.10y = & 4 & (4) \\ (1) + -2(3) \Rightarrow -0.2x = & -8 & (5) \end{cases}$

Equation (5) yields $x = \frac{-8}{-0.2} = 40$. Back-substitution into equation (4) produces $y = 240$. Finally, back-substitution into equation (1) gives $0.30z = 64 - 4 - 24 = 36$ or $z = \frac{36}{0.30} = 120$. The solution is 40 lb of the 10-12-15 fertilizer; 240 lb of the 10-0-5 fertilizer, and 120 lb of the 30-6-15 fertilizer.

≡ 5.4 DETERMINANTS

1. $\begin{vmatrix} 2 & 3 \\ 4 & -1 \end{vmatrix} = 2(-1) - 4(3) = -2 - 12 = -14$

3. $\begin{vmatrix} 5 & -1 \\ 8 & 1 \end{vmatrix} = 5(1) - 8(-1) = 5 + 8 = 13$

5. $\begin{vmatrix} 4 & 7 \\ -3 & 1 \end{vmatrix} = 4(1) - (-3)7 = 4 + 21 = 25$

7. $\begin{vmatrix} -9 & \frac{1}{2} \\ 2 & 1 \end{vmatrix} = (-9)(1) - (2)\frac{1}{2} = -9 - 1 = -10$

9. (a) 2,

(b) $\begin{vmatrix} 3 & -4 \\ -2 & 1 \end{vmatrix} = -5$,

(c) Since $1 + 2 = 3$ is odd, the cofactor is the negative of the minor or $-(-5) = 5$.

11. (a) 1,

(b) minor $= \begin{vmatrix} 4 & -1 \\ 3 & -4 \end{vmatrix} = -16 + 3 = -13$,

(c) $3 + 2$ is odd so the cofactor is $-(-13) = 13$.

13. (a) -2,

(b) minor $= \begin{vmatrix} 2 & -1 \\ 7 & -4 \end{vmatrix} = -8 + 7 = -1$,

(c) $3 + 1$ is even so cofactor = minor = -1.

15. $-3\begin{vmatrix} 2 & -1 \\ 1 & 1 \end{vmatrix} + 7\begin{vmatrix} 4 & -1 \\ -2 & 1 \end{vmatrix} - (-4)\begin{vmatrix} 4 & 2 \\ -2 & 1 \end{vmatrix} =$

$-3(3) + 7(2) + 4(8) = -9 + 14 + 32 = 37$

17. 0, rows 1 and 3 are identical.

19. 0, because row 3 is a multiple of row 1 (or because column 2 is a multiple of column 1).

21. Using the first row to evaluate, we obtain

$$\begin{vmatrix} 5 & 2 & 3 \\ 4 & -5 & -6 \\ -2 & 5 & -9 \end{vmatrix} = 5\begin{vmatrix} -5 & -6 \\ 5 & -9 \end{vmatrix} - 2\begin{vmatrix} 4 & -6 \\ -2 & -9 \end{vmatrix} + 3\begin{vmatrix} 4 & -5 \\ -2 & 5 \end{vmatrix}$$

$$= 5[9 - 5)(-9) - 5(-6)] - 2[4(-9) - (-2)(-6)]$$
$$+ 3[4(5) - (-2)(-5)]$$
$$= 5[45 + 30] - 2[-36 - 12] + 3[20 - 10]$$
$$= 5(75) - 2(-48) + 3(10)$$
$$= 375 + 96 + 30 = 501$$

23. $\begin{vmatrix} 4 & 3 & 9 \\ -4 & -6 & 16 \\ 2 & 3 & 2 \end{vmatrix} \overset{R1+R2}{=} \begin{vmatrix} 4 & 3 & 9 \\ 0 & -3 & 25 \\ 2 & 3 & 2 \end{vmatrix} \overset{-2R3+R1}{=}$

$\begin{vmatrix} 0 & -3 & 5 \\ 0 & -3 & 25 \\ 2 & 3 & 2 \end{vmatrix} \overset{-R2+R1}{=} \begin{vmatrix} 0 & 0 & -20 \\ 0 & -3 & 25 \\ 2 & 3 & 2 \end{vmatrix} =$

$-\begin{vmatrix} 2 & 3 & 2 \\ 0 & -3 & 25 \\ 0 & 0 & -20 \end{vmatrix} = -2(-3)(-20) = -120$

25. 252.86583

27. -0.057525

29. $\begin{vmatrix} r & s & t \\ a & b & c \\ x & y & z \end{vmatrix} = -(-5) = 5$

31. $\begin{vmatrix} a & b & c \\ 3r & 3s & 3t \\ -2x & -2y & -2z \end{vmatrix} = 3(-2)(-5) = 30$

33. $\begin{vmatrix} 1 & 2 & 3 & 4 \\ 5 & 0 & 7 & 0 \\ 9 & 10 & 11 & 12 \\ 13 & 14 & 15 & 16 \end{vmatrix} \overset{-3(R1)+R3}{\underset{-4(R1)+R4}{=}}$

$\begin{vmatrix} 1 & 2 & 3 & 4 \\ 5 & 0 & 7 & 0 \\ 6 & 4 & 2 & 0 \\ 9 & 6 & 3 & 0 \end{vmatrix} =$

0, since $R4 = 1.5 \times R3$.

35. Using a calculator, we obtain approximately -321.545

37. Using a calculator, we obtain 0.

39. See *Computer Programs* in main text.

41. See *Computer Programs* in main text.

≡ 5.5 USING CRAMER'S RULE TO SOLVE SYSTEMS OF LINEAR EQUATIONS

1. $D = \begin{vmatrix} 2 & 1 \\ 3 & -2 \end{vmatrix} = -4 - 3 = -7, D_x = \begin{vmatrix} 7 & 1 \\ -7 & -2 \end{vmatrix} =$
$-14 + 7 = -7, D_y = \begin{vmatrix} 2 & 7 \\ 3 & -7 \end{vmatrix} = -14 - 21 =$
$-35, x = \frac{-7}{-7} = 1, y = \frac{-35}{-7} = 5$

3. $D = \begin{vmatrix} 4 & 3 \\ 3 & -2 \end{vmatrix} = -8 - 9 = -17, D_x =$
$\begin{vmatrix} 4 & 3 \\ -14 & -2 \end{vmatrix} = -8 + 42 = 34, D_y = \begin{vmatrix} 4 & 4 \\ 3 & -14 \end{vmatrix} =$
$-56 - 12 = -68, x = \frac{34}{-17} = -2, y = \frac{-68}{-17} = 4$

5. $D = \begin{vmatrix} 1.2 & 3.7 \\ 4.3 & -5.2 \end{vmatrix} = -22.15, D_x = \begin{vmatrix} 9.1 & 3.7 \\ 8.3 & -5.2 \end{vmatrix} =$
$-78.03, D_y = \begin{vmatrix} 1.2 & 9.1 \\ 4.3 & 8.3 \end{vmatrix} = -29.17, x = \frac{-78.03}{-22.15} \approx$
$3.5228, y = \frac{-29.17}{-22.15} = 1.3169$

7. $D = \begin{vmatrix} 2.5 & 3.8 \\ .5 & .76 \end{vmatrix} = 0$, inconsistent system, no solution.

9. $D = \begin{vmatrix} -5.3 & 2.1 \\ 6.2 & -3.1 \end{vmatrix} = 3.41, D_x = \begin{vmatrix} 4.6 & 2.1 \\ 6 & -3.1 \end{vmatrix} =$
$-26.86, D_y = \begin{vmatrix} -5.3 & 4.6 \\ 6.2 & 6 \end{vmatrix} = -60.32, x =$
$\frac{-26.86}{3.41} \approx -7.8768, y = \frac{-60.32}{3.41} \approx -17.689$

11. $D = \begin{vmatrix} 6 & 2.5 \\ 13.8 & 5.75 \end{vmatrix} = 0$, inconsistent system, no solution.

13. $D = \begin{vmatrix} 4 & 3 \\ 3 & -2 \end{vmatrix} = -8 - 9 = -17, D_x =$
$\begin{vmatrix} 2 & 3 \\ -24 & -2 \end{vmatrix} = -4 + 72 = 68, D_y = \begin{vmatrix} 4 & 2 \\ 3 & -24 \end{vmatrix} =$
$-96 - 6 = -102, x = \frac{68}{-17} = -4, y = \frac{-102}{-17} = 6$

15. $D = \begin{vmatrix} 1 & -.5 & 1.5 \\ 1 & .5 & -3.75 \\ -3 & -2.5 & 4 \end{vmatrix} = -12.5,$

$D_x = \begin{vmatrix} 2 & -.5 & 1.5 \\ .25 & .5 & -3.75 \\ -.75 & -2.5 & 4 \end{vmatrix} = -16.03125,$

$D_y = \begin{vmatrix} 1 & 2 & 1.5 \\ 1 & .25 & -3.75 \\ -3 & -.75 & 4 \end{vmatrix} = 12.6875,$

$D_z = \begin{vmatrix} 1 & -.5 & 2 \\ 1 & .5 & .25 \\ -3 & -2.5 & -.75 \end{vmatrix} = -1.75,$

$x = \frac{-16.03125}{-12.5} = 1.2825, y = \frac{12.6875}{-12.5} = -1.015, z = \frac{-1.75}{-12.5} = 0.14$

17. First, we evaluate $D = -150, D_x = 315, D_y = -25.5, D_z = 652.5$, and $D_w = -1437$. Then, by Cramer's rule, we have $x = -2.1, y = 0.17, z = -4.35$, and $w = 9.58$.

19. See *Computer Programs* in main text.

21. First, we make the substitutions for t and s to obtain the following system of three equations.
$$\begin{cases} s_0 + 2v_0 + 2a = 212 \\ crs_0 + 5v_0 + \frac{25}{2}a = 156.5 \\ crs_0 + 7v_0 + \frac{49}{2}a = 70.5 \end{cases}$$

$D = \begin{vmatrix} 1 & 2 & 2 \\ 1 & 5 & 12.5 \\ 1 & 7 & 24.5 \end{vmatrix} = 15,$

$D_{s_0} = \begin{vmatrix} 212 & 2 & 2 \\ 156.5 & 5 & 12.5 \\ 70.5 & 7 & 24.5 \end{vmatrix} = 3000,$

$D_{v_0} = \begin{vmatrix} 1 & 212 & 2 \\ 1 & 156.5 & 12.5 \\ 1 & 70.5 & 24.5 \end{vmatrix} = 237,$

$D_a = \begin{vmatrix} 1 & 2 & 212 \\ 1 & 5 & 156.5 \\ 1 & 7 & 70.5 \end{vmatrix} = -147,$

$s_0 = \frac{3000}{15} = 200,$

$v_0 = \frac{237}{15} = 15.8, a = \frac{147}{15} = -9.8$. Thus, $s = 200 + 15.8t - 4.9t^2$.

23. Let A, B, and C represent the three distances. Then we get the equations $B = 90 + A$ and $C = 130 + B$, which leads to the system of three equations:

$$\begin{cases} -A + B = 90 \\ -B + C = 130 \\ A + B + C = 1900 \end{cases}$$

$$D = \begin{vmatrix} -1 & 1 & 0 \\ 0 & -1 & 1 \\ 1 & 1 & 1 \end{vmatrix} = 3,$$

$$D_A = \begin{vmatrix} 90 & 1 & 0 \\ 130 & -1 & 1 \\ 1900 & 1 & 1 \end{vmatrix} = 1590,$$

$$D_B = \begin{vmatrix} -1 & 90 & 0 \\ 0 & 130 & 1 \\ 1 & 1900 & 1 \end{vmatrix} = 1860,$$

$$D_C = \begin{vmatrix} -1 & 1 & 90 \\ 0 & -1 & 130 \\ 1 & 1 & 1900 \end{vmatrix} = 2250.$$

$A = \frac{1590}{3} = 530,$

$B = \frac{1860}{3} = 620, C = \frac{2250}{3} = 750.$ Hence, the car traveled 530 km on fuel A, 620 km on fuel B, and 750 km on fuel C.

25. **(a)** Let $A, B,$ and C represent the three companies. The 5 units of salt will satisfy $2A + 2B + C$. Similarly, for the sand we get the equation $2A + 2B + 5C = 6$ and for the inhibiting agent the equation $A + 2B + C = 4$. Thus, we have the system

$$\begin{cases} 2A + 2B + C = 5 \\ A + 2B + 5C = 6 \\ A + 2B + C = 4 \end{cases}$$

From this sytem, we get the determinants

$$D = \begin{vmatrix} 2 & 2 & 1 \\ 1 & 2 & 5 \\ 1 & 2 & 1 \end{vmatrix} = -8,$$

$$D_A = \begin{vmatrix} 5 & 2 & 1 \\ 6 & 2 & 5 \\ 4 & 2 & 1 \end{vmatrix} = -8,$$

$$D_B = \begin{vmatrix} 2 & 5 & 1 \\ 1 & 6 & 5 \\ 1 & 4 & 1 \end{vmatrix} = -10, \text{ and}$$

$$D_C = \begin{vmatrix} 2 & 2 & 5 \\ 1 & 2 & 6 \\ 1 & 2 & 4 \end{vmatrix} = -4. \text{ Thus, } A = \frac{-8}{-8} = 1, B = \frac{-10}{-8} = \frac{5}{4}, \text{ and } C = \frac{-4}{-8} = \frac{1}{2}. \text{ Thus, we have the proportion}$$

$1 : \frac{5}{4} : \frac{1}{2} = 4 : 5 : 2.$

(b) If we let x represent the basic number of units that must be purchased, then the ratio of $4 : 5 : 2$ can be thought of as $4x + 5x + 2x = 3,297,283,$ or $11x = 3,297,283.$ Thus, $x = \frac{3,297,283}{11} = 299,753.$ Thus, the city must purchase $4 \times 299,753 = 1,199,095$ units from Company A, it must purchase $5 \times 299,753 = 1,498,765$ units from Company B, and $2 \times 299,753 = 599,506$ units from Company C.

CHAPTER 5 REVIEW

1. $\begin{vmatrix} 2 & 5 \\ -3 & 6 \end{vmatrix} = 12 + 15 = 27$

2. $\begin{vmatrix} 4 & -7 \\ 3 & -6 \end{vmatrix} = -24 + 21 = -3$

3. $\begin{vmatrix} 5 & 9 \\ 8 & 1 \end{vmatrix} = 5 - 72 = -67$

4. $\begin{vmatrix} 3 & -12 \\ 0 & 5 \end{vmatrix} = 15 - 0 = 15$

5. $\begin{vmatrix} 6 & -2 \\ -4 & 3 \end{vmatrix} = 6 \times 3 - (-4)(-2) = 18 - 8 = 10$

6. $\begin{vmatrix} 8 & 9 \\ -1 & -2 \end{vmatrix} = -16 + 9 = -7$

7. We first get this determinant in triangular form.

$\begin{vmatrix} 9 & 2 & 1 \\ 3 & -4 & 6 \\ 7 & 2 & 1 \end{vmatrix} \xrightarrow{-R3+R1} \begin{vmatrix} 2 & 0 & 0 \\ 3 & -4 & 6 \\ 7 & 2 & 1 \end{vmatrix}$

$\xrightarrow{-6R3+R2} \begin{vmatrix} 2 & 0 & 0 \\ -39 & -16 & 0 \\ 7 & 2 & 1 \end{vmatrix}$

The determinant is now triangular, so $\begin{vmatrix} 9 & 2 & 1 \\ 3 & -4 & 6 \\ 7 & 2 & 1 \end{vmatrix} =$

$\begin{vmatrix} 2 & 0 & 0 \\ -39 & -16 & 0 \\ 7 & 2 & 1 \end{vmatrix} = 2(-16)(1) = -32.$

8. $\begin{vmatrix} 9 & -2 & 3 \\ 4 & -5 & 2 \\ 3 & 2 & 1 \end{vmatrix} = 9(-5)1 + (-2)2(3) + 3(4)2-$

$3(-5)3 - 2(2)(9) - 1(4)(-2) = -45 - 12 + 12 + 24 + 45 - 36 + 8 = -16$

9. (a) $m = \frac{1-3}{5-(-2)} = \frac{-2}{7}$,

   (b) $y - 1 = \frac{-2}{7}(x - 5)$ or $y - 3 = \frac{-2}{7}(x + 2)$

10. (a) $m = \frac{-2-(-4)}{3-8} = \frac{2}{-5} = \frac{2}{-5}$.

    (b) $y + 2 = \frac{-2}{5}(x - 3)$ or $y + 4 = \frac{-2}{5}(x - 8)$

11. (a) $m = \frac{4-(-1)}{-2-9} = \frac{5}{-11} = \frac{-5}{11}$,

    (b) $y - 4 = \frac{5}{11}(x + 2)$ or $y + 1 = \frac{-5}{11}(x - 9)$

12. (a) $m = \frac{-1-(-3)}{2-(-4)} = \frac{2}{6} = \frac{1}{3}$,

    (b) $y + 1 = \frac{1}{3}(x - 2)$ or $y + 3 = \frac{1}{3}(x + 4)$

13. $4x - 2y = 6, -2y = -4x + 6, y = 2x - 3$

14. $9x + 3 = 5, 3y = -9x, y = -3x + \frac{5}{3}$

15. $4x = 5y - 8, 5y = 4x + 8, y = \frac{4}{5}x + \frac{8}{5}$

16. $7x + 3y - 8 = 0, 3y = -7x + 8, y = \frac{-7}{3}x + \frac{8}{3}$

17.

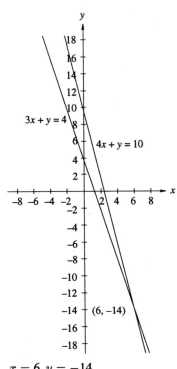

$3x + y = 4$

$4x + y = 10$

$(6, -14)$

$x = 6, y = -14$

18.

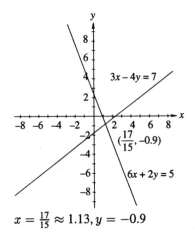

$3x - 4y = 7$

$\left(\frac{17}{15}, -0.9\right)$

$6x + 2y = 5$

$x = \frac{17}{15} \approx 1.13, y = -0.9$

19. $\begin{cases} 3x + y = 5 & (1) \\ 4x - y = 16 & (2) \end{cases}$

$y = 5 - 3x; 4x - (5 - 3x) = 16;$
$7x - 5 = 16; 7x = 21; x = 3, y = 5 - 3(3) =$
$5 - 9 = -4; (3, -4)$

20. $\begin{cases} 3x - 2y = 4 & (1) \\ x + 8y = 3 & (2) \end{cases}$

Solving (2) for x, we get $x = 3 - 8y$ (3). Substituting in (1) yields $3(3 - 8y) - 2y = 4$ or $9 - 24y - 2y = 4 \Rightarrow -26y = -5$ and so $y = \frac{5}{26} = 0.1923$. Substituting this value of y in (3) produces $x = 3 - 8(0.1923) = 1.4615$. The solution is $(1.4615, 0.1923)$.

21. $\begin{cases} 4x - y = 7 & (1) \\ 3x + 2y = 5 & (2) \end{cases}$

Solving (1) for y, we get $y = 4x - 7$ (3). Substituting this in (2) yields $3x + 2(4x - 7) = 5 \Rightarrow 3x + 8x - 14 = 5$ or $11x = 19$ and so, $x = \frac{19}{11} = 1.7273$. Substituting this value of x in (3), we see that $y = 4\left(\frac{19}{11}\right) - 7 = \frac{76}{11} - \frac{77}{11} = \frac{-1}{11} = -0.0909$. Thus, the solution is $\left(\frac{19}{11}, \frac{-1}{11}\right) = (1.7273, -0.0909)$.

22. $\begin{cases} 6x + y - 5 = 0 & (1) \\ 4y - 3x = -7 & (2) \end{cases}$

$y = 5 - 6x; 4(5 - 6x) - 3x = -7; 20 - 24x - 3x = -7; -27x = -27, x = 1; y = 5 - 6(1) = 5 - 6 = -1$

23. $\begin{cases} x - y = -2 & (1) \\ x + y = 8 & (2) \end{cases}$

Adding gives $2x = 6; x = 3; 3 - y = -2; -y = -5; y = 5$

24. $\begin{cases} 6x + 5y = 7 & (1) \\ 3x - 7y = 13 & (2) \end{cases}$

$(2) \times 2 \Rightarrow 6x - 14y = 26\,(3); (1) - (3) \Rightarrow 19y = -19; y = -1; 6x + 5(-1) = 7; 6x = 12; x = 2$

25. $\begin{cases} x + \frac{1}{2}y = 2 & (1) \\ 3x - y = 1 & (2) \end{cases}$

$(1) \times 2 \Rightarrow 2x + y = 4\,(3); (3) + (2) \Rightarrow 5x = 5; x = 1\,(1) - y = 1; 3 - y = 1; -y = -2; y = 2$

26. $\begin{cases} \frac{1}{6}x + \frac{1}{4}y = \frac{1}{3} & (1) \\ \frac{1}{4}x - \frac{1}{2}y = 1 & (2) \end{cases}$

$(1) \times 12 \Rightarrow 2x + 3y = 4\,(3); (2) \times 8 \Rightarrow 2x - 4y = 8\,(4); (3) - (4) \Rightarrow 7y = -4; y = -\frac{4}{7}; 2x + 3(-\frac{4}{7}) = 4; 2x - \frac{12}{7} = 4; 2x = \frac{40}{7}; x = \frac{20}{7}$

27. $\begin{cases} 3x + 2y + 3z = -7 & (1) \\ 5x - 3y + 2z = -4 & (2) \\ 7x + 4y + 5z = 2 & (3) \end{cases}$

Begin by multiplying (1) $\times 3 \Rightarrow 9x + 6y + 9z = -21$ (4). Next, multiply (2) $\times 2 \Rightarrow 10x - 6y + 4z = -8$ (5). Add: (4) + (5) $\Rightarrow 19x + 13z = -29$ (6). Now, multiply (1) $\times 2 \Rightarrow 6x + 4y + 6z = -14$ (7), and subtract (7) $-$ (3) $\Rightarrow -x + z = -16$ (8). If you now multiply $19 \times$ (8) $\Rightarrow -19x + 19z = -304$ (9), and add (9) + (6) $\Rightarrow 32z = -333$. As a result, we obtain $z = -\frac{333}{32} = -10.40625$. From (8), we obtain $x = z + 16$, and substituting for z yields $x = 16 - 10.40625 = 5.59375$. Substituting these values for x and z in (1) produces $3(5.59375) + 2y + 3(-10.40625) = -7 \Rightarrow 2y = -7 - 3(5.59375) + 3(10.40625)$ and so $2y = 7.4375$ or $y = 3.71875$. Thus, the solution is $x = 5.59375, y = 3.71875, z = -10.40625$.

28. $\begin{cases} x + y + z = 2.7 & (1) \\ 6x + 7y - 5z = -8.8 & (2) \\ 10x + 16y - 3z = -6.4 & (3) \end{cases}$

First, multiply (1) $\times 6 \Rightarrow 6x + 6y + 6z = 16.2$ (4), and subtract: (4) $-$ (2) $\Rightarrow -y + 11z = 25$ (5). Next, multiply (1) $\times 10 \Rightarrow 10x + 10y + 10z = 27$ (6), and subtract: (3) $-$ (6) $\Rightarrow 6y - 13z = -33.4$ (7). Finally, multiply (5) $\times 6 \Rightarrow -6y + 66z = 150$ (8), and add (7) + (8) $\Rightarrow 53z = 116.6$, and so $z = 2.2$. From (5), $y = 11z - 25 = 11(2.2) - 25 = -0.8$. From (1), $x = 2.7 - y - z = 2.7 - (-.8) - 2.2 = 1.3$, and the solution is $(1.3, -0.8, 2.2)$.

29. $D = \begin{vmatrix} 3 & -2 \\ 5 & 2 \end{vmatrix} = 6 + 10 = 16;$

$D_x = \begin{vmatrix} 4 & -2 \\ 12 & 2 \end{vmatrix} = 8 + 24 = 32;$

$D_y = \begin{vmatrix} 3 & 4 \\ 5 & 12 \end{vmatrix} = 36 - 20 = 16; x = \frac{32}{16} = 2;$

$y = \frac{16}{16} = 1$

30. $D = \begin{vmatrix} 4 & 3 \\ 2 & -5 \end{vmatrix} = -20 - 6 = -26;$

$D_x = \begin{vmatrix} 27 & 3 \\ -19 & -5 \end{vmatrix} = -78;$

$D_y = \begin{vmatrix} 4 & 27 \\ 2 & -19 \end{vmatrix} = -130;$

$x = \frac{-78}{-26} = 3; y = \frac{-130}{-26} = 5$

31. $D = \begin{vmatrix} 4 & -3 \\ 11 & 4 \end{vmatrix} = 16 + 33 = 49;$

$D_x = \begin{vmatrix} 9 & -3 \\ 7 & 4 \end{vmatrix} = 57; x = \frac{57}{49};$

$D_y = \begin{vmatrix} 4 & 9 \\ 11 & 7 \end{vmatrix} = 28 - 99 = -71; y = \frac{-71}{49}$

32. $D = \begin{vmatrix} 2 & -5 \\ 4 & 6 \end{vmatrix} = 12 + 20 = 32;$

$D_x = \begin{vmatrix} -8 & -5 \\ 17 & 6 \end{vmatrix} = 37; x = \frac{37}{32};$

$D_y = \begin{vmatrix} 2 & -8 \\ 4 & 17 \end{vmatrix} = 34 + 32 = 66; y = \frac{66}{32} = \frac{33}{16}$

33. $D = \begin{vmatrix} 5 & 3 & -2 \\ 3 & -4 & 3 \\ 1 & 6 & -4 \end{vmatrix} - 9;$

$D_x = \begin{vmatrix} 5 & 3 & -2 \\ 13 & -4 & 3 \\ -8 & 6 & -4 \end{vmatrix} = -18;$

$D_y = \begin{vmatrix} 5 & 5 & -2 \\ 3 & 13 & 3 \\ 1 & -8 & -4 \end{vmatrix} = 9;$

$D_z = \begin{vmatrix} 5 & 3 & 5 \\ 3 & -4 & 13 \\ 1 & 6 & -8 \end{vmatrix} = -9;$

$x = \frac{-18}{-9} = 2;$

$y = \frac{9}{-9} = -1; z = \frac{-9}{-9} = 1. (2, -1, 1)$

34. $D = \begin{vmatrix} 5 & -2 & 3 \\ 6 & -3 & 4 \\ -4 & 4 & -9 \end{vmatrix} = 15;$

$D_x = \begin{vmatrix} 6 & -2 & 3 \\ 10 & -3 & 4 \\ 4 & 4 & -9 \end{vmatrix} = 10;$

$D_y = \begin{vmatrix} 5 & 6 & 3 \\ 6 & 10 & 4 \\ -4 & 4 & -9 \end{vmatrix} = -110;$

$D_z = \begin{vmatrix} 5 & -2 & 6 \\ 6 & -3 & 10 \\ -4 & 4 & 4 \end{vmatrix} = -60;$

$x = \frac{10}{15} = \frac{2}{3};$

$y = \frac{-110}{15} = -7.3; z = \frac{-60}{15} = -4. (\frac{2}{3}, -7\frac{1}{3}, -4)$

35. Let x be the amount of Colombian Supreme and y the amount Mocha Java coffee. Then we obtain two equations (1) $x + y = 50$ and (2) $4.99x + 5.99y = 50 \times 5.39$ or $499x + 599y = 26950$. Multiplying (1) by 499 gives $499x + 499y = 24950$. Subtracting yields $100y = 2000$ or $y = 20$. This leaves 30 for x. Solution: 30 pounds of Colombian Supreme and 20 pounds of Mocha Java coffee.

36. Organizing the information in chart form we have

computer type	Chip A	Chip A
PCx	4	11
BCy	9	6

$4x + 9y = 670; 11x + 6y = 1055$. Using Cramer's Rule $D = \begin{vmatrix} 4 & 9 \\ 11 & 6 \end{vmatrix} = 24 - 99 = -75,$

$D_x = \begin{vmatrix} 670 & 9 \\ 1055 & 6 \end{vmatrix} = -5475,$

$D_y = \begin{vmatrix} 4 & 670 \\ 11 & 1055 \end{vmatrix} = -3150,$

$x = \frac{-5475}{-75} = 73, y = \frac{-3150}{-75} = 42.$ They should produce 73 PC and 42 BC computers.

37. Let x be the number of 1 room offices and y be the number of 2 room offices and z be the number of 3 room offices; then
$\begin{cases} x + y + z = 66 \\ x + 2y + 3z = 146 \\ 300x + 520y + 730z = 37,160 \end{cases}$
Using Cramer's Rule $D = \begin{vmatrix} 1 & 1 & 1 \\ 1 & 2 & 3 \\ 300 & 520 & 730 \end{vmatrix} = -10,$

$D_x = \begin{vmatrix} 66 & 1 & 1 \\ 146 & 2 & 3 \\ 37,160 & 520 & 730 \end{vmatrix} = -100,$

$D_y = \begin{vmatrix} 1 & 66 & 1 \\ 1 & 146 & 3 \\ 300 & 37,160 & 730 \end{vmatrix} = -320,$

$D_z = \begin{vmatrix} 1 & 1 & 66 \\ 1 & 2 & 146 \\ 300 & 520 & 37,160 \end{vmatrix} = -240, x = \frac{-100}{-10} = 10,$

$y = \frac{-320}{-10} = 32, z = \frac{-240}{-10} = 24.$ There are 10 small, 32 medium, and 24 large offices.

CHAPTER 5 TEST

1. (a) $m = \frac{6-2}{5-(-4)} = \frac{4}{9}$.

(b) $y - 2 = \frac{4}{9}(x+4)$ or $y - 6 = \frac{4}{9}(x-5)$; $y - 2 = \frac{4}{9}x + \frac{16}{9}$ and so $y = \frac{4}{9}x + \frac{34}{9}$

2. (a) $-5y = -6x + 12$; $y = \frac{6}{5}x - \frac{12}{5}$,

(b) slope: $m = \frac{6}{5}$,

(c) y-intercept is $-\frac{12}{5}$,

(d) $6x - 5y = 12$

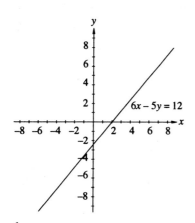

3. $\begin{cases} 2x + 3y = 1 & (1) \\ -3x + 6y = 16 & (2) \end{cases}$
$(1) \times 2 \Rightarrow 4x + 6y = 2$ (3), $(3) - (2) \Rightarrow 7x = -14$, and we see that $x = -2$. Substitution in (1) yields $2(-2) + 3y = 1$ or $-4 + 3y = 1 \Rightarrow 3y = 5$ and so, $y = \frac{5}{3}$. Thus, the solution is $\left(-2, \frac{5}{3}\right)$.

4. $\begin{vmatrix} 4 & -2 \\ 5 & 6 \end{vmatrix} = 4(6) - 5(-2) = 24 + 10 = 34$

5. $\begin{vmatrix} 4 & 2 & -1 \\ 3 & 0 & 4 \\ -3 & 1 & 1 \end{vmatrix}$. Using the 2nd column and minors,

$-2 \begin{vmatrix} 3 & 4 \\ -3 & 1 \end{vmatrix} - 1 \begin{vmatrix} 4 & -1 \\ 3 & 4 \end{vmatrix} = 2(15) - 1(19) = -30 - 19 = -49$

6. $\begin{cases} 2x + 3y = 5 \\ x - 4y = -14 \end{cases}$

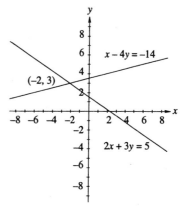

7. $\begin{cases} 4x + 3y = 9 & (1) \\ 2x + y = 2 & (2) \end{cases}$
From (2), we get $y = 2 - 2x$ (3). Substituting in (1) yields $4x + 3(2 - 2x) = 9 \Rightarrow 4x + 6 - 6x = 9$, or $-2x + 6 = 9$ and so, $x = -\frac{3}{2}$. Substituting in (3) produces $y = 2 - 2(-\frac{3}{2}) = 2 + 3 = 5$. Hence, the solution is $(-\frac{3}{2}, 5)$.

8. We have $D = \begin{vmatrix} 2 & -4 \\ -6 & 8 \end{vmatrix} = 16 - 24 = -8$,

$D_x = \begin{vmatrix} 2 & -4 \\ -9 & 8 \end{vmatrix} = 16 - 36 = -20$, and

$D_y = \begin{vmatrix} 2 & 2 \\ -6 & -9 \end{vmatrix} = -18 + 12 = -6$.

Using Cramer's rule, we see that $x = \frac{-20}{-8} = \frac{5}{2}$ and $y = \frac{-6}{-8} = \frac{3}{4}$. The solution is $\left(\frac{5}{2}, \frac{3}{4}\right)$.

9. Using the addition method on the system
$\begin{cases} -x + 3y - 2z = 7 & (1) \\ 3x + 4y - 7z = -8 & (2) \\ x + 2y + z = 2 & (3) \end{cases}$
we obtain
$(1) + (2) \Rightarrow 5y - z = 9$ \hfill (4)
$3(1) + (2) \Rightarrow 13y - 13z = 13$ or $y - z = 1$ (5)
$(4) - (5) \Rightarrow 4y = 8$ or $y = 2$
Back-substitution into equation (5) yields $z = 1$ and finally $x = -3$. So, the solution is $(-3, 2, 1)$.

10. Using Cramer's rule, $D = 2, D_{x_1} = 220, D_{x_2} = -246, D_{x_3} = -130$, and $D_{x_4} = -102$. This produces the solution $x_1 = 110, x_2 = -123, x_3 = -65$, and $x_4 = -51$.

11. $x =$ cost of 1st, $y =$ cost of 2nd, $z =$ cost of 3rd

$$\begin{cases} x + y + z = 60 & (1) \\ x = y + z & (2) \\ y = 2z + 3 & (3) \end{cases}$$

or (2) can be written as $x - y - z = 0\,(4)$. Adding $(1) + (4)$ we get $2x = 60$ or $x = 30$. Substituting this value for x in (2), we get $y + z = 30\,(5)$. Substituting $2z + 3$ for y in (5) yields $2z + 3 + z = 30$ or $3z = 27$, and so $z = 9$. Finally, substituting for z in (3), we see that $y = 2\,(9) + 3 = 18 + 3 = 21$. So, the cost of the first part is \$30, the cost of the second part \$21, and the cost of the third part is \$9.

12. By Kirchhoff's laws, we have the following equations

$$\begin{aligned} I_1 - I_2 - I_3 &= 0 & (1) \\ 6I_1 + 4I_2 &= 20 & (2) \\ 4I_2 - 2I_3 &= 10 & (3) \end{aligned}$$

Now, $6 \times (1) - (2) \Rightarrow -10I_2 - 6I_3 = -20$ or $5I_2 + 3I_3 = 10\,(4)$. Then, $2 \times (4) + 3 \times (3) \Rightarrow 22I_2 = 50$, and so $I_2 = \frac{25}{11}$. Back-substituting into equation (4) produces $I_3 = -\frac{5}{11}$ and, from equation (1), we find that $I_1 = I_2 + I_3 = \frac{20}{11}$. The solutions are $I_1 = \frac{20}{11}$ A, $I_2 = \frac{25}{11}$ A, and $I_3 = -\frac{5}{11}$ A.

CHAPTER
6
Ratio, Proportion, and Variation

≡ 6.1 RATIO AND PROPORTION

1. $1.38 for 16 bolts is 1.38 : 16

3. 86 L per km is 86 : 1

5. $\frac{9}{2} = \frac{4.5}{1}$ or 4.5 : 1

7. $\frac{23}{7} \approx \frac{3.2857}{1}$ or 3.2857 : 1

9. $2 \times 15 = 30; 3 \times 10 = 30$ so equal

11. $145 \times 5 = 725; 25 \times 29 = 725$ so equal

13. $8c = 7 \times 32; c = \frac{7 \times 32}{8}; c = 7 \times 4 = 28$

15. $124d = 62 \times 158; d = \frac{62 \times 158}{124} = 79$

17. $0.15a = 4 \times 0.16; a = \frac{4 \times 0.16}{0.15} \approx 4.267$

19. $8b = 20 \times 5.6; b = \frac{20 \times 5.6}{8} = 14$

21. $9x = 4 \times 4; x = \frac{4 \times 4}{9} \approx 1.78$

23. $\frac{2}{x} = \frac{5}{14} = \frac{9}{z}; 5x = 2 \times 14; x = 5.6; 5z = 9 \times 14; z = \frac{9 \times 14}{5} = 25.2$

25. 54 : 13 or $\approx$ 4.1538 : 1

27. A steering wheel that makes $4\frac{2}{3}$ complete turns turns a total of $4\frac{2}{3} \times 360° = \frac{14}{3} \times 360° = 1680°$. Thus, the steering ratio is $\frac{1680}{60} = \frac{28}{1}$ or 28 : 1

29. $\frac{5 \times 10^{-4}}{300} \approx 1.67 \times 10^{-6}$ F $= 1.67\mu$F

31. $\frac{8.7}{1} = \frac{96}{x}; x = \frac{96}{8.7} \approx 11.03$ or 11.03 cm^3

33. $\frac{80 \text{ cal}}{1 \text{ g}} = \frac{x \text{ cal}}{785 \text{ kg}} = \frac{x \text{ cal}}{785,000 \text{ g}}; x = 80 \times 785000; x = 62{,}800{,}000$ calories

35. $\frac{25.4 \text{ mm}}{1 \text{ in.}} = \frac{88.9 \text{ mm}}{x \text{ in.}}; x = \frac{88.9}{25.4} = 3.5; 3.5$ in.

37. We have the proportion $\frac{x \text{ L}}{14.2 \text{ gal}} = \frac{1 \text{L}}{0.2642 \text{ gal}}$. Multiplying by 14.2 gal produces 53.7472 L.

≡ 6.2 SIMILAR GEOMETRIC SHAPES

1. $\frac{a}{2} = \frac{3}{4} = \frac{b}{5}; 4a = 3 \times 2; a = 6 \div 4 = 1.5; 4b = 3 \times 5; b = 15 \div 4 = 3.75$

3. $\frac{a}{33} = \frac{24.3}{c} = \frac{22}{b}; 66a = 33 \times 34; a = 17; 34b = 22 \times 66; b \approx 42.71; 34c = 24.3 \times 66; c \approx 47.17$

5. $\frac{a}{3} = \frac{b}{4} = \frac{c}{5} = \frac{d}{5} = \frac{1.7}{2}; 2a = 3 \times 1.7; a = 2.55; 2b = 4 \times 1.7; b = 3.4; 2c = 5 \times 1.7; c = 4.25; d = c = 4.25$

7. $\frac{A}{20} = \frac{1^2}{(\frac{1}{8})^2}, \frac{1}{64}A = 20, A = 1280$ ft^2; 1280 ft$^2 = 144 \times 1280$ in.$^2 = 184{,}320$ in.^{2}

9. For a fixed length, the mass is proportional to the cross sectional area; $\frac{19^2}{38^2} = \frac{x}{22}, x = \frac{19^2 \times 22}{38^2} = 5.5$ kg

11. Amount of paint is proportional to area; $\frac{700}{x} = \frac{20^2}{35^2}; x = \frac{35^2 \times 700}{20^2} = 2\,143.75$ L

13. $\frac{3930}{x} = \frac{15^3}{5^3} = \frac{3375}{125}, x = \frac{125 \times 3930}{3375} \approx 145.56$ mm^3

15. $\frac{10^2}{2^2} = \frac{1256.64}{A}; A = \frac{4 \times 1256.64}{100} \approx 50.27$ cm^2, $\frac{10^3}{2^3} = \frac{4188.79}{V}; V = \frac{8 \times 4188.79}{1000} \approx 33.51$ cm^3

☰ 6.3 DIRECT AND INVERSE VARIATION

1. $R = kl$ or $\frac{R}{l} = k$

3. $A = kd^2$ or $\frac{A}{d^2} = k$

5. $IN = k$ or $I = \frac{k}{N}$

7. $k = \frac{r}{t} = \frac{4}{6} = \frac{2}{3}$

9. $d = kr^2; k = \frac{d}{r^2} = \frac{8}{4^2} = \frac{8}{16} = \frac{1}{2}$

11. $\frac{a_1}{d_1} = \frac{a_2}{d_2} \Rightarrow \frac{4}{8} = \frac{a}{18}; a = 4 \times 18 \div 8 = 9$

13. Note 1 min equals 60 s; $\frac{v_1}{t_1} = \frac{v_2}{t_2} \Rightarrow \frac{v}{60} = \frac{60}{20} \Rightarrow$ $v = \frac{60 \times 60}{20} = 180$ m/s

15. $\frac{n_1}{d_1^2} = \frac{n_2}{d_2^2} \Rightarrow \frac{n}{2^2} = \frac{10890}{10^2} \Rightarrow n = \frac{4 \times 10890}{100} = 435.6$
neutrons

17. (a) $V = kT \Rightarrow = V/T = \frac{25}{293.18} \approx 0.08527 =$ constant;
(b) $V = 0.08527 \times (85+273.18) \approx 30.542$ m^3; ΔV $= 30.542 - 25 = 5.542$ m$^3 =$ change in volume

19. $R = k\ell$ or $\frac{R_1}{l_1} = \frac{R_2}{l_2}; \frac{32}{1500} = \frac{R}{5000}$, $1500R = 32 \times$ $5000; R\frac{32 \times 5000}{1500} = 106.67 \ \Omega$

21. $P_1 V_1 = P_2 V_2; 40 \times 200 = P \times 50 \Rightarrow P = \frac{40 \times 200}{50} =$ $40 \times 4 = 160$ psi

23. $In = k$, $I = $ current, $n = $ number of turns or $I_1 n_1 = I_2 n_2; I \times 200 = 0.3 \times 50; I = \frac{0.3 \times 50}{200} -$ 0.075 A

25. $wr^2 = k$ or $w_1 r_1^2 = w_2 r_2^2; 180 \times 4000^2 = w \times$ $(4000 + 500)^2; w = \frac{180 \times 4000^2}{4500^2} \approx 142.22$ lb

☰ 6.4 JOINT AND COMBINED VARIATION

1. $K = kmv^2$

3. $f = \frac{kv}{l}$

5. $E = \frac{klI}{d^2}$

7. $a = kbc \Rightarrow 20 = k4(2); k = \frac{20}{4 \cdot 2} = \frac{20}{8} = 2.5$

9. $u = \frac{kv}{w}; 20 = \frac{k \cdot 5}{2}; k = \frac{20 \times 2}{5} = 8$

11. $r = kst; 10 = k \cdot 6 \cdot 5; k = \frac{10}{30} = \frac{1}{3}; r = \frac{1}{3} \cdot 3 \cdot 4 = 4$

13. $a = kxy^2; 9 = k \cdot 3 \cdot 5^2; k = \frac{9}{3 \cdot 5^2} = \frac{3}{25} = 0.12;$ $a = 0.12 \cdot 6 \times 15^2 = 162$

15. $p = \frac{kr\sqrt{t}}{w}; 9 = \frac{k18\sqrt{9}}{6}; k = \frac{9 \times 6}{18 \times 3} = 1; p = \frac{36\sqrt{4}}{2} =$ 36

17. (a) $A = kbh; 12 = k \cdot 3 \cdot 8; k = \frac{12}{24} = \frac{1}{2};$
(b) $A = \frac{1}{2} \cdot 10 \cdot 5 = 25$

19. $w = \frac{kV^2}{R}; 2 = \frac{k6^2}{18}; k = \frac{2 \times 18}{36} = 1; V^2 = wR =$ $50 \times 10 = 500; V = \sqrt{500} \approx 22.36$ V

21. $R = \frac{k\ell}{A}; 0.56 = \frac{k80}{2.5}; k = \frac{0.56 \times 2.5}{80} = 0.0175; 3 =$ $\frac{0.0175\ell}{0.1}; \ell = \frac{3 \times 0.1}{0.0175} \approx 17.14$ m

23. $H = \frac{kA(\Delta T)}{d}; 50 = \frac{k12(72-32)}{3}; k = 0.3125; H =$ $\frac{0.3125(72-10) \times 12}{3} = 77.5$ Btu/hr

25. $P = \frac{kT}{V}; 15 = \frac{k(70+460)}{5}; k = \frac{15 \times 5}{530} \approx 0.1415; 75 =$ $\frac{0.1415(200+460)}{V}; V = \frac{0.1415 \times 660}{75} = 1.2452$ or 1.25 ft^3

☰ CHAPTER 6 REVIEW

1. $3 : x = 4 : 6; 4x = 3 \times 6; 4x = 18; x = \frac{18}{4} =$ $\frac{9}{2} = 4.5$

2. $x : 5 = 3 : 15; 15x = 5 \times 3; x = 1$

3. $\frac{7}{9} = \frac{21}{d}; 7d = 21 \cdot 9; d = \frac{21 \cdot 9}{7} = 27$

4. $\frac{14}{6} = \frac{c}{27}; 6c = 14 \times 27; c = \frac{14 \times 27}{6} = 63$

5. $4 : 8 = 19 : x; 4x = 8 \times 19; x = \frac{8 \times 19}{4} = 38$

6. $x : 12 = 15 : 32; 32x = 12 \times 15; x = \frac{12 \times 15}{32} =$ 5.625

7. $7 : 24.5 : x = 8 : y : 42; \frac{7}{8} = \frac{24.5}{y} = \frac{x}{42};$ $8x = 7 \times 42; x = 36.75; 7y = 8 \times 24.5; y = 28$

8. $\frac{12.5}{x} = \frac{y}{47} = \frac{8}{5}; 8x = 5 \times 12.5; x = 7.8125; 5y = 8 \times 47; y = 75.2$

9. $TR = \frac{N_p}{N_s}; 25 = \frac{4000}{N_s}; N_s = \frac{4000}{25} = 160$

10. $\Delta S = kI; 12.50 = k6.5; k \approx 1.923; \Delta S = 1.923 \times 9.6 \approx \$18.46$

11. $\frac{s_1}{4} = \frac{s_2}{5} = \frac{180}{8}; S_1 = \frac{180 \times 4}{8} = 90$ mm; $S_2 = \frac{180 \times 5}{8} = 112.5$ mm

12. TMA$= \frac{90}{5} = 18$

13. $E = \frac{AMA}{TMA} = \frac{16}{18} = \frac{8}{9}$ or ≈ 0.889 or 88.9%

14. If rate of flow is constant, then $V_1 A_1 = V_2 A_2; A \propto d^2 \Rightarrow V_1 d_1^2 = V_2 d_2^2$;

(a) $3 \cdot 1^2 = (.5)^2 \cdot V_2; V_2 = \frac{3}{.25} = 12$ m/s;

(b) $30 d^2 = 3; d^2 = \frac{1}{10}; d = \sqrt{.1} \approx .316$ cm $= 3.16$ mm diameter

15. $5 \times .5^2 = V \times 2.5^2; V = \frac{5 \times 0.25}{6.25} = 0.2$ m/s

16. $W = kT^4; k = \frac{W}{T^4} = \frac{6.5 \times 10^7}{(5800)^4} = 5.7438 \times 10^{-8}$

17. $F = \frac{kC_1 C_2}{d^2}; k = \frac{Fd^2}{C_1 C_2} = \frac{7.192 \times 10^{-4}(.05)^2}{4 \times 10^{-9} \times 5 \times 10^{-8}} = 8.99 \times 10^9$

18. $M = \frac{kI_1}{d}; 4 \times 10^{-6} = \frac{k100}{5}; k = \frac{4 \times 10^{-6} \times 5}{100}; k = 2 \times 10^{-7}; M = \frac{2 \times 10^{-7} \times 150}{15} = 2 \times 10^{-6}T$

19. $2.42 = \frac{3 \times 10^8}{s}; s = \frac{3 \times 10^8}{2.42} \approx 1.24 \times 10^8$ m/s

▤ CHAPTER 6 TEST

1. $4 : 9 = x : 51; 9x = 4 \times 51; x = \frac{4 \times 51}{9} = \frac{68}{3} = 22\frac{2}{3} \approx 22.67$

2. $8 : x = 68 : 93.5; 68x = 8 \times 93.5; x = 11$

3. $\frac{a}{28} = \frac{12}{20}; 20a = 12 \times 28; a = 16.8$

4. $\frac{24}{42} = \frac{78}{d}; 24d = 78 \times 42; d = 136.5$

5. $6 : 8 : x = y : 14 : 24.5; \frac{6}{y} = \frac{8}{14} = \frac{x}{24.5}; 8y = 6 \times 14; y = 10.5; 14x = 8 \times 24.5; x = 14$

6. $\frac{a}{9} = \frac{b}{30} = \frac{75}{45}; 45a = 75 \times 9; a = 15; 45b = 75 \times 30; b = 50$

7. $E = kd^2$

8. $h = \frac{kr}{t^3}$

9. $\frac{a}{5} = \frac{b}{12} = \frac{71.5}{13}; 13a = 5 \times 71.5; a = 27.5; 13b = 12 \times 71.5; b = 66$

10. $3x + 2x = \frac{540}{2}; 5x = 270; x = 54; \ell = 3 \cdot 54 = 162$ cm; $w = 2 \cdot 54 = 108$ cm

11. $A = ks^2; 72 = k(12)^2$ and so $k = 0.5$. Thus, $A = (0.5)(18^2) = 162$ m^2 or $A_1 S_2^2 = A_2 S_1^2; A \cdot 12^2 = 72 \cdot 18^2; A = 162$ m^2

12. $d = \frac{k}{ph}$ and so $45 = \frac{k}{(80)(8)}$. Thus, $k = 28,800$ and $d = \frac{28,800}{(60)(10)} = 48$

CHAPTER

7

Factoring and Algebraic Fractions

≣ 7.1 SPECIAL PRODUCTS

1. $3(p + q) = 3p + 3q$

3. $3x(5 - y) = 15x - 3xy$

5. $(p + q)(p - q) = p^2 - q^2$

7. $(2x - 6p)(2x + 6p) = (2x)^2 - (6p)^2 = 4x^2 - 36p^2$

9. $(r + w)^2 = r^2 + 2rw + w^2$

11. $(2x + y)^2 = (2x)^2 + 2 \cdot 2x \cdot y + y^2 = 4x^2 + 4xy + y^2$

13. $\left(\frac{2}{3}x + 4bx\right)^2 = \left(\frac{2}{3}x\right)^2 + 2 \cdot \frac{2}{3}x(4b) + (4b)^2 = \frac{4}{9}x^2 = \frac{16}{3}xb + 16b^2$

15. $\left(2p - \frac{3}{4}r\right)^2 = (2p)^2 - 2 \cdot 2p \cdot \frac{3}{4}r + \left(\frac{3}{4}r\right)^2 = 4p^2 - 3pr + \frac{9}{16}r^2$

17. $(a + 2)(a + 3) = a^2 + (2 + 3)a + 2 \cdot 3 = a^2 + 5a + 6$

19. $(x - 5)(x + 2) = x^2 + (-5 + 2)x + (-5)(2) = x^2 - 3x - 10$

21. $(2a + b)(3a + b) = 2 \cdot 3a^2 + (2 + 3)ab + b^2 = 6a^2 + 5ab + b^2$

23. $(3x + 4)(2x - 5) = 3 \cdot 2x^2 + [3(-5) + 4 \cdot 2]x + 4(-5) = 6x^2 + [-15 + 8]x - 20 = 6x^2 - 7x - 20$

25. $(a + b)^3 = a^3 + 3a^2b + 3ab^2 + b^3$

27. $(x + 4)^3 = x^3 + 3 \cdot x^2 \cdot 4 + 3x \cdot 4^2 + 4^3 = x^3 + 12x^2 + 48x + 64$

29. $(2a - b)^3 = (2a)^3 - 3(2a)^2b + 3(2a)b^2 - b^3 = 8a^3 - 12a^2b + 6ab^2 - b^3$

31. $(3x - 2y)^3 = (3x)^3 - 3(3x)^2(2y) + 3(3x)(2y)^2 - (2y)^3 = 27x^3 - 54x^2y + 36xy^2 - 8y^3$

33. $(m + n)(m^2 - mn + n^2) = m^3 + n^3$

35. $(r - t)\left(r^2 + rt + t^2\right) = r^3 - t^3$

37. $(2x + b)\left(4x^2 - 2xb + b^2\right) = (2x + b)\left[(2x)^2 - 2x \cdot b + b^2\right] = (2x)^3 + b^3 = 8x^3 + b^3$

39. $(3a - d)(9a^2 + 3da + d^2) = (3a - d)\left[(3a)^2 + 3ad + d^2\right]; = (3a)^3 - d^3 = 27a^3 - d^3$

41. $3(a + 2)^2 = 3(a^2 + 4a + 4) = 3a^2 + 12a + 12$

43. $\frac{5r}{t}\left(t + \frac{r}{5}\right)^2 = \frac{5r}{t}\left(t^2 + \frac{2rt}{5} + \frac{r^2}{25}\right) = \frac{5r}{t} \cdot t^2 + \frac{5r}{t} \cdot \frac{2rt}{5} + \frac{5r}{t} \cdot \frac{r^2}{25} = 5rt + 2r^2 + \frac{r^3}{5t}$

45. $(x^2 - 6)(x^2 + 6) = x^4 - 6^2 = x^4 - 36$

47. $\left(\frac{3x}{y} - \frac{y}{x}\right)\left(\frac{3x}{y} + \frac{y}{x}\right)x^2y^2 = \left(\left(\frac{3x}{y}\right)^2 - \left(\frac{y}{x}\right)^2\right)x^2y^2 = \left(\frac{9x^2}{y^2} - \frac{y^2}{x^2}\right)x^2y^2 = 9x^4 - y^4$

49. $(x - y)^2 - (y - x)^2 = (x^2 - 2xy + y^2) - (y^2 - 2yx + x^2) = x^2 - 2xy + y^2 - y^2 + 2xy - x^2 = 0$

51. $(x + 3)(x - 3)^2 = (x + 3)(x^2 - 6x + 9) = x^3 - 6x^2 + 9x + 3x^2 - 18x + 27 = x^3 - 3x^2 - 9x + 27$

53. $r(r - t)^2 - t(t - r)^2 = r\left(r^2 - 2rt + t^2\right) - t\left(t^2 - 2rt + r^2\right) = r^3 - 2r^2 + rt^2 - t^3 + 2rt^2 - r^2t = r^3 - 3r^2t + 3rt^2 - t^3$

55. $(5 + 3x)(25 - 15x + 9x^2) = (5 + 3x)\left[5^2 - 5 \cdot 3x + (3x)^2\right] = 5^3 + (3x)^3 = 125 + 27x^3$

57. $[(x + y) - (w + z)]^2 = (x + y)^2 - 2(x + y)(w + z) + (x + z)^2$

59. $(x + y - z)(x + y + z) = [(x + y) - z][(x + y) + z] = (x + y)^2 - z^2 = x^2 + 2xy + y^2 - z^2$

61. $z^2 = R^2 + \left[x_L^2 - 2x_Lx_C + x_C^2\right] = R^2 + x_L^2 - 2x_Lx_C + x_C^2$

63. $a_c = \frac{(2r^2 - t)^2}{r} = \frac{4r^4 - 4t^3 + r^2}{r}$

65. $d = \left(\frac{P}{3EI}\right)\left(l_1^3 - l_2^3\right)$

≡ 7.2 FACTORING

1. $6x + 6 = 6(x + 1)$

3. $12a - 6 = 6(2a - 1)$

5. $4x - 2y + 8 = 2(2x - y + 4)$

7. $5x^2 + 10x + 15 = 5\left(x^2 + 2x + 3\right)$

9. $10x^2 - 15 = 5\left(2x^2 - 3\right)$

11. $4x^2 + 6x = 2x(2x + 3)$

13. $7b^2y + 28b = 7b(by + 4)$

15. $3ax + 6ax^2 - 2ax = ax(3 + 6x - 2) = ax(1 + 6x)$

17. $4ap^2 + 6a^2pq + 8apq^2 = 2ap(2p + 3aq + 4q^2)$

19. $a^2 - b^2 = (a + b)(a - b)$

21. $x^2 - 4 = x^2 - 2^2 = (x + 2)(x - 2)$

23. $y^2 - 81 = y^2 - 9^2 = (y + 9)(y - 9)$

25. $4x^2 - 9 = (2x)^2 - 3^2 = (2x + 3)(2x - 3)$

27. $9a^4 - b^2 = (3a^2)^2 - b^2 = (3a^2 + b)(3a^2 - b)$

29. $25a^2 - 49b^2 = (5a)^2 - (7b)^2 = (5a + 7b)(5a - 7b)$

31. $144 - 25b^4 = (12)^2 - \left(5b^2\right)^2$
$= \left(12 + 5b^2\right)\left(12 - 5b^2\right)$

33. $5a^2 - 125 = 5\left(a^2 - 25\right) = 5(a + 5)(a - 5)$

35. $28a^2 - 63b^4 = 7\left(4a^2 - 9b^4\right) = 7\left[(2a)^2 - \left(3b^2\right)^2\right]$
$= 7\left(2a + 3b^2\right)\left(2a - 3b^2\right)$

37. $a^4 - 81 = \left(a^2 + 9\right)\left(a^2 - 9\right) = \left(a^2 + 9\right)(a + 3)(a - 3)$

39. $16x^4 - 256y^4 = 16\left(x^4 - 16y^4\right) = 16\left(x^2 + 4y^2\right)\left(x^2 - 4y^2\right) = 16\left(x^2 + 4y^2\right)(x + 2y)(x - 2y)$

41. $\pi r \left(r + \sqrt{h^2 + r^2}\right)$

43. $\frac{1}{2}d\left(v_2^2 - v_1^2\right) = \frac{1}{2}d\left(v_2 + v_1\right)\left(v_2 - v_1\right)$

45. $\frac{(\omega_f - \omega_0)(\omega_f + \omega_0)}{2\theta}$

47. $W = \frac{1}{2}I\omega_2^2 - \frac{1}{2}I\omega_1^2$
$= \frac{1}{2}I\left(\omega_2^2 - \omega_1^2\right)$
$= \frac{1}{2}I\left(\omega_2 - \omega_1\right)\left(\omega_2 + \omega_1\right)$

≡ 7.3 FACTORING TRINOMIALS

1. $x^2 + 9x - 8$; $b^2 - 4ac = 113$; $\sqrt{113} \approx 10.6$; factors, but does not factor using rational numbers.

3. $3x^2 - 10x - 8$; $b^2 - 4ac = 196$; $\sqrt{196} = 14$; factors.

5. $5x^2 + 23x + 18$; $b^2 - 4ac = 169$; $\sqrt{169} = 13$; factors.

7. $x^2 + 7x + 10 = (x + 2)(x + 5)$

9. $x^2 - 12x + 27 = (x - 3)(x - 9)$

11. $x^2 - 27x + 50 - (x - 2)(x - 25)$

13. $x^2 - x - 2 = (x - 2)(x + 1)$

15. $x^2 - 3x - 10 = (x - 5)(x + 2)$

17. $r^2 + 10r + 25 = r^2 + 2 \cdot 5r + 5^2 = (r + 5)^2$

19. $a^2 + 22a + 121 = a^2 + 1 \cdot 11a + 11^2 = (a + 11)^2$

21. $f^2 = 30f + 225 = f^2 - 2 \cdot 15f + 15^2 = (f - 15)^2$

23. $6y^2 - 7y + 1 = (6y - 1)(y - 1)$

25. $7 \times 2 = 14$; $7 + 2 = 9$; $7t^2 + 9t + 2 = 7t^2 + 7t + 2t + 2$; $= 7t(t + 1) + 2(t + 1) = (7t + 2)(t + 1)$

27. $7(-5) = -35$; $-35 \times 1 = -35$; $-35 + 1 = -34$; $7b^2 - 34b - 5 = 7b^2 - 35b + b - 5 = 7b(b - 5) + 1(b - 5)(7b + 1)(b - 5)$

29. $4(-5) = -20$; $20 \times (-1) = -20$; $20 + (-1) = 19$; $4e^2 + 19e - 5 = 4e^2 + 20e - 3 - 5 = 4e(e + 5) - 1(e + 5) = (4e - 1)(e + 5)$

31. $3 \times 8 = 24$; $4 \times 6 = 24$; $4 + 6 = 10$; $3u^2 + 10u + 8 = 3u^2 + 6u + 4u + 8 = 3u(u + 2) + 4(u + 2) = (3u + 4)(u + 2)$

33. $9(-6) = -54$; $(-27)(2) = -54$; $(-27) + 2 = -25$; $9t^2 - 25t - 6 = 9t^2 - 27t + 2t - 6 = 9t(t - 3) + 2(t - 3) = (9t + 2)(t - 3)$

35. $6(-5) = -30$; $15(-2) = -30$; $15 + (-2) = 13$; $6x^2 + 13x - 5 = 6x^2 + 15x - 2x - 5 = 3x(2x + 5) - 1(2x + 5) = (3x - 1)(2x + 5)$

37. $15(-15) = -225$; $-25 \times 9 = -225$; $-25 + 9 = -16$; $15a^2 - 16a - 15 = 15a^2 - 25a + 9a - 15 = 5a(3a - 5) + 3(3a - 5) = (5a + 3)(3a - 5)$

39. $15 \times 15 = 225$; $25 \times 9 = 225$; $25 + 9 = 34$; $15e^2 + 34e + 15 = 15e^2 + 25e + 9e + 15 = 5e(3e + 5) + 3(3e + 5) = (5e + 3)(3e + 5)$

41. $10 \times 6 = 60$; $(-15) \times (-4) = 60$; $-15 + (-4) = -19$; $10x^2 - 19x + 6 = 10x^2 - 15x - 4x + 6 = 5x(2x - 3) - 2(2x - 3) = (5x - 2)(2x - 3)$

43. $3r^2 - 18r - 21 = 3(r^2 - 6r - 7) = 3(4 - 7)(r + 1)$

45. $49t^4 - 105t^3 + 14t^2 = 7t^2(7t^2 - 15t + 2) = 7t^2(7t - 1)(t - 2)$

47. $6(-10) = -60$; $(-15)4 = -60$; $-15 + 4 = -11$; $6x^2 - 11xy - 10y^2 = 6x^2 - 15xy + 4xy - 10y^2 = 3x(2x - 5y) + 2y(2x - 5y) = (3x + 2y)(2x - 5y)$

49. $8 \times (-9) = -72$; $-18(4) = -72$; $-18 + 4 = -14$; $8a^2 - 14ab - 9b^2 = 8a^2 - 18ab + 4ab - 9b^2 = 2a(4a - 9b) + b(4a - 9b) = (2a + b)(4a - 9b)$

51. $a^3 - b^3 = (a - b)\left(a^2 + ab + b^2\right)$

53. $8x^3 - 27 = (2x)^3 - 3^3 = (2x - 3) \times \left[(2x)^2 + 2x \cdot 3 + 3^2\right] = (2x - 3)\left(4x^2 + 6x + 9\right)$

55. $i = 0.7\left(t^2 - 3t - 4\right) = 0.7(t - 4)(t + 1)$

57. $0.0001n^2(n - 2000) - 3(n - 2000) = \left(0.0001n^2 - 3\right)(n - 2000) = 0.0001\left(n^2 - 30,000\right)(n - 2000)$

59. **(a)** $(6 + 2x)(10 + 2x) - 60 = 36$; $60 + 32x + 4x^2 - 60 - 36 = 0$; $4x^2 + 32x - 36 = 0$.
(b) $4\left(x^2 + 8x - 9\right) = 0$; $4(x + 9)(x - 2) = 0$

≡ 7.4 FRACTIONS

1. $\frac{7}{8}$ (by 5); $\frac{7}{8} \cdot \frac{5}{5} - \left(\frac{3}{5}\right)$ 40; $7 \times 40 = 280$; $8 \times 35 = 280$

3. $\frac{x}{y}$ (by a); $\frac{x}{y} \cdot \frac{a}{a} = \frac{ax}{ay}$. The cross product is $axy = axy$.

5. $\frac{x^2 y}{a}$ (by $3ax$); $\frac{3ax^3 y}{3a^2 x}$. The cross-product is $3a^2 x^3 y = 3a^2 x^3 y$.

7. $\frac{4}{x - y}$ (by $x + y$); $\frac{4}{(x - y)} \cdot \frac{(x + y)}{(x + y)} = \frac{4(x + y)}{x^2 - y^2} = \frac{4x + 4y}{x^2 - y^2}$. The cross-product is $4\left(x^2 - y^2\right) = (4x + 4y)(x - y) = 4x^2 - 4y^2$.

9. $\frac{a + b}{a - b}$ (by $a + b$); $\frac{(a + b)}{(a - b)} \cdot \frac{(a + b)}{(a + b)} = \frac{a^2 + 2ab + b^2}{a^2 - b^2}$. The cross-product is $(a + b)\left(a^2 - b^2\right) = a^3 + a^2 b - ab^2 - b^3 = (a - b)\left(a^2 + 2ab + b^2\right)$.

11. $\frac{38}{24}$ (by 2) $= \frac{19}{12}$; $12 \times 38 = 456$; $24 \times 19 = 456$

13. $\frac{3x^2}{12x}$ (by $3x$) $= \frac{x}{4}$ $4 \cdot 3x^2 = 12x^2$ $12x \cdot x = 12x^2$

15. $\frac{4(x + 2)}{(x + 2)(x - 3)}$ (by $x + 2$) $= \frac{4}{x - 3}$; $4(x + 2)(x - 3) = 4(x + 2)(x - 3)$.

17. $\frac{x^2 - 16}{x^2 + 8x + 16}$ (by $x + 4$) $= \frac{x^2 - 16}{x^2 + 8x + 10} = \frac{(x + 4)(x - 4)}{(x + 4)^2} = \frac{x - 4}{x + 4}$; $(x + 4)\left(x^2 - 16\right) = (x + 4)^2(x - 4)$

19. $\frac{4x^2}{12x} = \frac{4x \cdot x}{4x \cdot 3} = \frac{x}{3}$

21. $\frac{x^2 + 3x}{x^3 + 5x} = \frac{x(x + 3)}{x(x^2 + 5)} = \frac{x + 3}{x^2 + 5}$

23. $\frac{6m^2 - 3m^3}{9m + 18m^3} = \frac{3m^3(2 - m)}{9m(1 + 2m^2)} = \frac{m(2 - m)}{3(1 + 2m^2)}$ or $\frac{2m - m^2}{3 + 6m^2}$

25. $\frac{x^2 + 3x}{x^2 - 9} = \frac{x(x + 3)}{(x + 3)(x - 3)} = \frac{x}{x - 3}$

27. $\frac{2b^2 - 10b}{3b^2 - 75} = \frac{2b(b - 5)}{3(b + 5)(b - 5)} = \frac{2b}{3(b + 5)}$ or $\frac{2b}{3b + 15}$

29. $\frac{z^2 - 9}{z^2 - 6z + 9} = \frac{(z + 3)(z - 3)}{(z - 3)^2} = \frac{z + 3}{z - 3}$

31. $\frac{x^2 + 4x + 3}{x^2 + 7x + 12} = \frac{(x + 1)(x + 3)}{(x + 4)(x + 3)} = \frac{x + 1}{x + 4}$

33. $\frac{2x^2 + 9x + 4}{x^2 + 9x + 20} = \frac{(2x + 1)(x + 4)}{(x + 5)(x + 4)} = \frac{2x + 1}{x + 5}$

35. $\frac{12y^3 + 12y^2 + 3y}{6y^2 - 3y - 3} = \frac{3y(4y^2 + 4y + 1)}{3(2y^2 - y - 1)} = \frac{y(2y + 1)^2}{(2y + 1)(y - 1)} = \frac{y(2y + 1)}{y - 1}$

37. $\frac{x^3 y^6 - y^3 x^6}{2x^3 y^4 - 2x^4 y^3} = \frac{x^3 y^3\left(y^3 - x^3\right)}{2x^3 y^3(y - x)} = \frac{(y - x)\left(y^2 + yx + x^2\right)}{2(y - x)} = \frac{y^2 + xy + x^2}{2}$

39. $\frac{x^2 - y^2}{x + y} = \frac{(x + y)(x - y)}{(x + y)} = \frac{x - y}{1} = x - y$

41. $p = \frac{6whs + 6wh}{3s^2 + 6s + 3} = \frac{6wh(s + 1)}{3(s + 1)^2}$
$= \frac{6wh}{3(s + 1)}$
$= \frac{2wh}{s + 1}$

☰ 7.5 MULTIPLICATION AND DIVISION OF FRACTIONS

1. $\frac{2}{x} \cdot \frac{5}{6} = \frac{2 \times 5}{x \times y} = \frac{10}{xy}$

3. $\frac{4x^2}{5} \cdot \frac{3}{y^3} = \frac{4x^2 \cdot 3}{5y^3} = \frac{12x^2}{5y^3}$

5. $\frac{3}{x} \div \frac{7}{y} = \frac{3}{x} \times \frac{y}{7} = \frac{3y}{7x}$

7. $\frac{2x^2}{3} \div \frac{7y}{4x} = \frac{2x^2}{3} \times \frac{4x}{7y} = \frac{8x^3}{21y}$

9. $\frac{2x}{3y} \cdot \frac{5}{4x^2} = \frac{2 \cdot x \cdot 5}{3 \cdot y \cdot 2 \cdot 2 \cdot x \cdot x} = \frac{5}{3y2x} = \frac{5}{6xy}$

11. $\frac{3a^2b}{5d} \cdot \frac{25ad^2}{6b^2} = \frac{3a^2b \cdot 5 \cdot 5 \cdot add}{5d \cdot 3 \cdot 2 \cdot bb} = \frac{5a^3d}{2b}$

13. $\frac{3y}{5x} \div \frac{15x^2}{8xy} = \frac{3y}{5x} \times \frac{8xy}{15x^2} = \frac{3y8xy}{5x \cdot 3 \cdot 5xx} = \frac{8y^2}{25x^2}$

15. $\frac{3x^2y}{7p} \div \frac{15x^2p}{7y^2} = \frac{3x^2y}{7p} \times \frac{7y^2}{15x^2p} = \frac{3x^2y \cdot 7y^2}{7p \cdot 3 \cdot 5x^2p} = \frac{y^3}{5p^2}$

17. $\frac{4y+16}{5} \cdot \frac{15y}{3y+12} = \frac{4(y+4) \cdot 3 \cdot 5y}{5 \cdot 3(y+4)} = \frac{4y}{1} = 4y$

19. $\frac{a^2-b^2}{a+3b} \cdot \frac{5a+15b}{a+b} = \frac{(a+b)(a-b)5(a+3b)}{(a+3b)(a+b)} = 5(a-b)$
or $5a - 5b$

21. $\frac{x^2-100}{10} \div \frac{2x+10}{15} = \frac{x^2-100}{10} \times \frac{15}{2x+10} =$
$\frac{(x+10)(x-10) \cdot 3 \cdot 5}{2 \cdot 5 \cdot 2(x+5)} = \frac{3(x^2-100)}{4(x+5)}$

23. $\frac{4x^2-1}{9x-3x^2} \div \frac{2x+1}{x^2-9} = \frac{4x^2-1}{9x-3x^2} \times \frac{x^2-9}{2x+1} =$
$\frac{(2x+1)(2x-1)(x+3)(x-3)}{3x(3-x)(2x+1)} = -\frac{(2x-1)(x+3)}{3x}$

25. $\frac{a^2-8a}{a-8} \cdot \frac{a+2}{a} = \frac{a(a-8)(a+2)}{(a-8)a} = a+2$

27. $\frac{2a-b}{4a} \cdot \frac{2a-b}{4a^2-4ab+b^2} = \frac{(2a-b)^2}{4a(2a-b)^2} = \frac{1}{4a}$

29. $\frac{y^2}{x^2-1} \div \frac{y^2}{x-1} = \frac{y^2}{(x+1)(x-1)} \times \frac{x-1}{y^2} = \frac{1}{x+1}$

31. $\frac{2y^2-y}{4y^2-4y+1} \div \frac{y^2}{8y-4} = \frac{y(2y-1)}{(2y-1)^2} \cdot \frac{(4(2y-1))}{y^2} = \frac{4}{y}$

33. $\frac{x^2-3x+2}{x^2+5x+6} \cdot \frac{x+3}{3x-6} = \frac{(x-2)(x-1)}{(x+2)(x+3)} \cdot \frac{x+3}{3(x-2)} = \frac{x-1}{3(x+2)}$

35. $\frac{x^2+xy-6y^2}{x^2+6xy+8y^2} \cdot \frac{x^2-9xy+20y^2}{x^2-4xy-21y^2} = \frac{(x+3y)(x-2y)}{(x+4y)(x+2y)} \cdot$
$\frac{(x-4y)(x-5y)}{(x-7y)(x+3y)} = \frac{(x-2y)(x-4y)(x-5y)}{(x+4y)(x+2y)(x-7y)}$

37. $\frac{9x^2-25}{x^2+6x+9} \div \frac{3x+5}{x+3} = \frac{(3x+5)(3x-5)}{(x+3)^2} \times \frac{x+3}{3x+5} = \frac{3x-5}{x+3}$

39. $\frac{x^2+4xy+4y^2}{x^2-4y^2} \div \frac{x^2+xy-2y^2}{x^2-xy-2y^2} = \frac{(x+2y)^2}{(x+2y)(x-2y)} \times$
$\frac{(x-2y)(x+y)}{(x+2y)(x-y)} = \frac{x+y}{x-y}$

41. $\frac{x+y}{4x-4y} \div \left[\frac{(x+y)^2}{x^2-y^2} \cdot \frac{x^3-y^3}{x^3+y^3} \right] = \frac{x+y}{4(x-y)} \cdot \left[\frac{(x^2-y^2)(x^3+y^3)}{(x+y)^2(x^3-y^3)} \right]$

$= \frac{x+y}{4(x-y)} \cdot \left[\frac{(x-y)(x+y)(x+y)(x^2-xy+y^2)}{(x+y)^2(x-y)(x^2+xy+y^2)} \right]$

$= \frac{x+y}{4(x-y)} \cdot \frac{(x^2-xy+y^2)}{(x^2+xy+y^2)}$

$= \frac{x^3+y^3}{4(x^3-y^3)}$

43. $\frac{x^2-25}{5x^2-24x-5} \cdot \frac{2x^2+12x+2}{6x^2-12x} \div \frac{x^2+6x+1}{5x^2-9x-2}$

$= \frac{x^2-25}{5x^2-24x-5} \cdot \frac{2x^2+12x+2}{6x^2-12x} \cdot \frac{5x^2-9x-2}{x^2+6x+1}$

$= \frac{(x+5)(x-5)}{(5x+1)(x-5)} \cdot \frac{2(x^2+6x+1)}{6x(x-2)} \cdot \frac{(5x+1)(x-2)}{x^2+6x+1}$

$= \frac{x+5}{3x}$

45. $\left(\frac{2x^2-5x-3}{x^2-x-12} \div \frac{2x^2+5x+2}{3x+9} \right) \div \frac{x^2-9}{x^2-2x-8}$

$= \frac{(2x+1)(x-3)}{(x-4)(x+3)} \cdot \frac{3(x+3)}{(2x+1)(x+2)} \cdot \frac{(x-4)(x+2)}{(x-3)(x+3)}$

$= \frac{3}{x+3}$

47. $\dfrac{(a^3 - a'^3)/a^3}{(a^2 - a'a)/a^2} = \dfrac{a^3 - a'^3}{a^3} \cdot \dfrac{a^2}{a^2 - a'a}$

$= \dfrac{(a - a')(a^2 + aa' - a'^2)}{a(a - a')(a + a')}$

$= \dfrac{a^2 + aa' - a'^2}{a(a + a')}$

≡ 7.6 ADDITION AND SUBTRACTION OF FRACTIONS

1. $\dfrac{2}{7} + \dfrac{5}{7} = \dfrac{2+5}{7} = \dfrac{7}{7} = 1$

3. $\dfrac{7}{3} - \dfrac{5}{3} = \dfrac{7-5}{3} = \dfrac{2}{3}$

5. $\dfrac{1}{2} + \dfrac{1}{3} = \dfrac{3}{6} + \dfrac{2}{6} = \dfrac{3+2}{6} = \dfrac{5}{6}$

7. $\dfrac{4}{5} - \dfrac{2}{3} = \dfrac{12}{15} - \dfrac{10}{15} = \dfrac{12-10}{15} = \dfrac{2}{15}$

9. $\dfrac{1}{x} + \dfrac{5}{x} = \dfrac{1+5}{x} = \dfrac{6}{x}$

11. $\dfrac{4}{a} - \dfrac{3}{a} = \dfrac{4-3}{a} = \dfrac{1}{a}$

13. $\dfrac{2x}{y} + \dfrac{3x}{y} = \dfrac{2x+3x}{y} = \dfrac{5x}{y}$

15. $\dfrac{3r}{2t} + \dfrac{-r}{2t} - \dfrac{5r}{2t} = \dfrac{3r-r-5r}{2t} = \dfrac{-3r}{2t}$

17. $\dfrac{3}{x+2} + \dfrac{x}{x+2} = \dfrac{3+x}{x+2}$

19. $\dfrac{t}{t+1} - \dfrac{2}{t+1} = \dfrac{t-2}{t+1}$

21. $\dfrac{y-3}{x+2} + \dfrac{3+y}{x+2} = \dfrac{(y-3)+(3+y)}{x+2} = \dfrac{2y}{x+2}$

23. $\dfrac{x+2}{a+b} - \dfrac{x-5}{a+b} = \dfrac{(x+2)-(x-5)}{a+b} = \dfrac{7}{a+b}$

25. $\dfrac{2}{x} + \dfrac{3}{y} = \dfrac{2}{x} \cdot \dfrac{y}{y} + \dfrac{x}{x} \cdot \dfrac{3}{y} = \dfrac{2y}{xy} + \dfrac{3x}{xy} = \dfrac{2y+3x}{xy}$

27. $\dfrac{a}{b} - \dfrac{4}{d} = \dfrac{a}{b} \cdot \dfrac{d}{d} - \dfrac{4}{d} \cdot \dfrac{b}{b} = \dfrac{ad}{bd} - \dfrac{4b}{bd} = \dfrac{ad-4b}{bd}$

29. $\dfrac{3}{x(x+1)} + \dfrac{4}{x^2-1} = \dfrac{3}{x(x+1)} + \dfrac{4}{(x+1)(x-1)} = \dfrac{3(x-1)}{x(x+1)(x-1)} + \dfrac{4x}{x(x+1)(x-1)} = \dfrac{3x-3+4x}{x(x+1)(x-1)} = \dfrac{7x-3}{x(x+1)(x-1)}$

31. $\dfrac{2}{x^2-1} - \dfrac{4}{(x+1)^2} = \dfrac{2}{(x+1)(x-1)} - \dfrac{4}{(x+1)^2} = \dfrac{2(x+1)}{(x+1)^2(x-1)} - \dfrac{4(x-1)}{(x+1)^2(x-1)} = \dfrac{(2x+2)-(4x-4)}{(x+1)^2(x-1)} = \dfrac{-2x+6}{(x+1)^2(x-1)}$

33. $\dfrac{x}{x^2-11x+30} + \dfrac{2}{x^2-36} = \dfrac{x}{(x-6)(x-5)} + \dfrac{2}{(x-6)(x+6)} = \dfrac{x(x+6)+2(x-5)}{(x-6)(x+6)(x-5)} = \dfrac{x^2+6x+2x-10}{(x-6)(x+6)(x-5)} = \dfrac{x^2+8x-10}{(x-6)(x+6)(x-5)}$

35. $\dfrac{2}{x^2-x-6} - \dfrac{5}{x^2-4} = \dfrac{2}{(x-3)(x+2)} - \dfrac{5}{(x+2)(x-2)} = \dfrac{2(x-2)}{(x-3)(x+2)(x-2)} - \dfrac{5(x-3)}{(x-3)(x-2)(x+2)} = \dfrac{(2x-4)-(5x-15)}{(x-3)(x+2)(x-2)} = \dfrac{3x+11}{(x-3)(x+2)(x-2)}$

37. $\dfrac{x-1}{3x^2-13x+4} + \dfrac{3x+1}{4x-x^2} = \dfrac{x-1}{(3x-1)(x-4)} + \dfrac{3x+1}{x(4-x)} + \dfrac{x(x-1)}{x(3x-1)(x-4)} + \dfrac{(-1)(3x+1)(3x-1)}{x(x-4)(3x-1)} = \dfrac{(x^2-x)-(9x^2-1)}{x(x-4)(3x-1)} = \dfrac{-8x^2-x+1}{x(x-4)(3x-1)}$

39. $\dfrac{x-3}{x^2-1} + \dfrac{2x-7}{x^2+5x+4} = \dfrac{x-3}{(x+1)(x-1)} + \dfrac{2x-7}{(x+1)(x+4)} = \dfrac{(x-3)(x+4)}{(x+1)(x-1)(x+4)} + \dfrac{(2x-7)(x-1)}{(x+1)(x-1)(x+4)} = \dfrac{(x^2+x-12)+(2x^2-9x+7)}{(x+1)(x-1)(x+4)} = \dfrac{3x^2-8x-5}{(x+1)(x-1)(x+4)}$

41. $\dfrac{y+3}{y^2-y-2} - \dfrac{2y-1}{y^2+2y-8} = \dfrac{y+3}{(y-2)(y+1)} - \dfrac{2y-1}{(y+4)(y-2)} = \dfrac{(y+3)(y+4)}{(y-2)(y+1)(y+4)} - \dfrac{(2y-1)(y+1)}{(y-2)(y+1)(y+4)} = \dfrac{(y^2+7y+12)-(2y^2+y-1)}{(y-2)(y+1)(y+4)} = \dfrac{-y^2+6y+13}{(y-2)(y+1)(y+4)}$

43. $\dfrac{x}{(x^2+3)(x-1)} + \dfrac{3x^2}{(x-1)^2(x+2)} - \dfrac{x+2}{x^2+3} = $

$\dfrac{x(x-1)(x+2)+3x^2(x^2+3)-(x+2)(x+2)(x-1)^2}{(x^2+3)(x-1)^2(x+2)} = $

$\dfrac{(x^3+x^2-2x)+(3x^4+9x^2)-(x^4+2x^3-3x^2-4x+4)}{(x^2+3)(x-1)^2(x+2)} = $

$\dfrac{2x^4-x^3+13x^2+2x-4}{(x^2+3)(x-1)^2(x+2)}$

45. $\dfrac{1+\frac{2}{x}}{1-\frac{3}{x}} = \dfrac{\left(1+\frac{2}{x}\right)x}{\left(1-\frac{3}{x}\right)x} = \dfrac{x+2}{x-3}$

47. $\dfrac{x-1}{1+\frac{1}{x}} = \dfrac{(x-1)x}{\left(1+\frac{1}{x}\right)x} = \dfrac{x^2-x}{x+1} = \dfrac{x(x-1)}{x+1}$

49. $\dfrac{\frac{x}{x+y} - \frac{y}{x-y}}{\frac{x}{x+y} + \frac{y}{x-y}} = \dfrac{\left(\frac{x}{x+y} - \frac{y}{x-y}\right)}{\frac{x}{x+y} + \frac{y}{x-y}} \cdot$

$\dfrac{(x+y)(x-y)}{(x+y)(x-y)} = \dfrac{x(x-y)-y(x+y)}{x(x-y)+y(x+y)} = $

$\dfrac{x^2-xy-xy-y^2}{x^2-xy+xy+y^2} = \dfrac{x^2-2xy-y^2}{x^2+y^2}$

51. $\dfrac{1 + \frac{3}{x}}{1 + \frac{2}{x}} = \dfrac{\frac{x+3}{x}}{\frac{x+2}{x}} = \dfrac{x+3}{x} \times \dfrac{x}{x+2} = \dfrac{x+3}{x+2}$

53. $\dfrac{t-1}{t+\frac{1}{t}} = \dfrac{t-1}{\frac{t^2+1}{t}} = \dfrac{t-1}{1} \times \dfrac{t}{t^2+1} = \dfrac{t^2-t}{t^2+1}$

55. $\dfrac{\frac{x}{x-y} - \frac{y}{x+y}}{\frac{1}{x-y} + \frac{1}{x+y}} = \dfrac{\frac{x^2+xy}{x^2-y^2} - \frac{xy-y^2}{x^2-y^2}}{\frac{x+y+x-y}{x^2-y^2}} = \dfrac{\frac{x^2+y^2}{x^2-y^2}}{\frac{2x}{x^2-y^2}} =$

$\dfrac{x^2+y^2}{x^2-y^2} \times \dfrac{x^2-y^2}{2x} = \dfrac{x^2+y^2}{2x}$

57. $\dfrac{1}{R_1} + \dfrac{1}{R_2} = \dfrac{R_2}{R_1 R_2} + \dfrac{R_1}{R_1 R_2} = \dfrac{R_1 + R_2}{R_1 R_2}$

59. $\dfrac{1}{C_1} + \dfrac{1}{C_2} + \dfrac{1}{C_3} = \dfrac{C_2 C_3}{C_1 C_2 C_3} + \dfrac{C_1 C_3}{C_1 C_2 C_3} + \dfrac{C_1 C_2}{C_1 C_2 C_3} =$
$\dfrac{C_1 C_2 + C_1 C_3 + C_2 C_3}{C_1 C_2 C_3}$

61. $\left(\dfrac{m_1}{m_1 + m_2} - \dfrac{m_2}{m_1 + m_2} \right) v + 2 \left(\dfrac{m_2}{m_1 + m_2} \right) v$

$\qquad = \dfrac{vm_1}{m_1 + m_2} - \dfrac{vm_2}{m_1 + m_2} + \dfrac{2vm_2}{m_1 + m_2}$

$\qquad = \dfrac{vm_1 + vm_2}{m_1 + m_2}$

$\qquad = \dfrac{v(m_1 + m_2)}{m_1 + m_2} = v$

63. $\dfrac{\frac{V_1}{R_1} + \frac{V_2}{R_2} + \frac{V_3}{R_3}}{\frac{1}{R_1} + \frac{1}{R_2} + \frac{1}{R_3}}$. Multiply by $\dfrac{R_1 R_2 R_3}{R_1 R_2 R_3}$ to obtain

$\dfrac{V_1 R_2 R_3 + V_2 R_1 R_3 + V_3 R_1 R_2}{R_2 R_3 + R_1 R_3 + R_1 R_2}$.

▤ CHAPTER 7 REVIEW

1. $5x(x-y) = 5x^2 - 5xy$

2. $(3+x)^2 = 9 + 6x + x^2$

3. $(x-2y)^3 = x^3 - 3x^2(2y) + 3x(2y)^2 - (2y)^3 = x^3 - 6x^2 y + 12xy^2 - 8y^3$

4. $(x+y)(x-6) = x^2 - 6x + xy - 6y$ using FOIL

5. $(2x+3)(x-6) = 2x^2 - 12x + 3x - 18 = 2x^2 - 9x - 18$

6. $(x+7)(x-7) = x^2 - 7^2 = x^2 - 49$

7. $(x^2 - 5)(x^2 + 5) = (x^2)^2 - 5^2 = x^4 - 25$

8. $(7x-1)(x+5) = 7x^2 + 35x - x - 5 = 7x^2 + 34x - 5$

9. $(2+x)^3 = 2^3 + 3 \cdot 2^2 x + 3 \cdot 2x^2 + x^3 = 8 + 12x + 6x^2 + x^3$

10. $(x-7)^2 = x^2 - 2 \cdot x \cdot 7 + 7^2 = x^2 - 14x + 49$

11. $9 + 9y = 9(1+y)$

12. $x^2 - 4 = (x+2)(x-2)$

13. $7x^2 - 63 = 7(x^2 - 9) = 7(x+3)(x-3)$

14. $x^2 - 12x + 36 = x^2 - 2 \cdot x \cdot 6 + 6^2 = (x-6)^2$

15. $x^2 - 11x + 30 = (x-5)(x-6)$ or $(x-6)(x-5)$

16. $x^2 + 15x + 36 = (x+12)(x+3)$

17. $x^2 + 6x - 16 = (x+8)(x-2)$

18. $x^2 - 4x - 45 = (x+5)(x-9)$

19. $2(-9) = -18; \; -6 \cdot 3 = -18; \; -6 + 3 = -3;$ $2x^2 - 3x - 9 = 2x^2 - 6x + 3x - 9 = 2x(x-3) + 3(x-3) = (2x+3)(x-3)$

20. Begin by factoring out $2x$, getting $8x^3 + 6x^2 - 20x = 2x(4x^2 + 3x - 10)$. Then, for the $4x^2 + 3x - 10$ expression, we have $4(-10) = -40$; $8(-5) = -40; \; 8 + (-5) = 3; \; 2x(4x^2 + 3x - 10) = 2x[4x^2 + 8x - 5x - 10] = 2x[4x(x+2) - 5(x+2)] = 2x(4x-5)(x+2)$

21. $\dfrac{2x}{6y} = \dfrac{x}{3y}$

22. $\dfrac{7x^2 y}{9xy^2} = \dfrac{7x}{9y}$

23. $\dfrac{x^2 - 9}{(x+3)^2} = \dfrac{(x+3)(x-3)}{(x+3)^2} = \dfrac{x-3}{x+3}$

24. $\dfrac{x^2 - 4x - 45}{x^2 - 81} = \dfrac{(x-9)(x+5)}{(x-9)(x+9)} = \dfrac{x+5}{x+9}$

25. $\dfrac{x^3+y^3}{x^2+2xy+y^2} = \dfrac{(x+y)\left(x^2-xy+y^2\right)}{(x+y)^2} = \dfrac{x^2-xy+y^2}{x+y}$

26. $\dfrac{x^3-16x}{x^2+2x-8} = \dfrac{x(x+4)(x-4)}{(x+4)(x-2)} = \dfrac{x(x-4)}{x-2}$

27. $\dfrac{x^2}{y} \cdot \dfrac{3y^2}{7x} = \dfrac{3x^2y^2}{7xy} = \dfrac{3xy}{7}$

28. $\dfrac{x^2-9}{x+4} \cdot \dfrac{x^3-16x}{x-3} = \dfrac{(x+3)(x-3)(x)(x+4)(x-4)}{(x+4)(x-3)} = x(x+3)(x-4)$

29. $\dfrac{4x}{3y} \div \dfrac{2x^2}{6y} = \dfrac{4x}{3y} \cdot \dfrac{6y}{2x^2}\, \dfrac{4x\cdot3\cdot2y}{3y\cdot2x^2} = \dfrac{4}{x}$

30. $\dfrac{x^2-25}{x^2-4x} \div \dfrac{2x^2+2x-40}{x^3-x} = \dfrac{x^2-25}{x^2-4x} \cdot \dfrac{x^3-x}{2x^2+2x-40} =$
$\dfrac{(x+5)(x-5)x(x+1)(x-1)}{x(x-4)2(x+5)(x-4)} = \dfrac{(x-5)(x+1)(x-1)}{2(x-4)^2}$

31. $\dfrac{4x}{y} + \dfrac{3x}{y} = \dfrac{4x+3x}{y} = \dfrac{7x}{y}$

32. $\dfrac{4}{x-y} + \dfrac{6}{x+y} = \dfrac{4(x+y)}{x^2-y^2} + \dfrac{6(x-y)}{x^2-y^2} = \dfrac{4x+4y+6x-6y}{x^2-y^2} = \dfrac{10x-2y}{x^2-y^2}$

33. $\dfrac{3(x-3)}{(x+2)(x-5)^2} + \dfrac{4(x-1)}{(x+2)^2(x-5)} = \dfrac{3(x-3)(x+2)+4(x-1)(x-5)}{(x+2)^2(x-5)^2}$
$= \dfrac{3x^2-3x-18+4x^2-24x+20}{(x+2)^2(x-5)^2} = \dfrac{7x^2-27x+2}{(x+2)^2(x-5)^2}$

34. $\dfrac{8a}{b} - \dfrac{3}{b} = \dfrac{8a-3}{b}$

35. $\dfrac{x}{y+x} - \dfrac{x}{y-x} = \dfrac{x(y-x)-x(y+x)}{(y+x)(y-x)} = \dfrac{xy-x^2-xy-x^2}{y^2-x^2} =$
$\dfrac{-2x^2}{y^2-x^2}$ or $\dfrac{2x^2}{x^2-y^2}$ or $\dfrac{2x^2}{(x+y)(x-y)}$

36. $\dfrac{2(x+3)}{(x+1)^2(x+2)} - \dfrac{3(x-1)}{(x+1)(x+2)^2} = \dfrac{2(x+3)(x+2)-3(x-1)(x+1)}{(x+1)^2(x+2)^2}$
$= \dfrac{2x^2+10x+12-3x^2+3}{(x+1)^2(x+2)^2} = \dfrac{-x^2+10x+15}{(x+1)^2(x+2)^2}$

37. $\dfrac{x^2-5x-6}{(x+6)(x+2)} + \dfrac{x^2+7x+6}{(x+2)(x-6)} = \dfrac{\left(x^2-5x-6\right)(x-6)}{(x+6)(x+2)(x-6)} +$
$\dfrac{\left(x^2+7x+6\right)(x+6)}{(x+2)(x-6)(x+6)} = \dfrac{x^3-11x^2+24x+36+x^3+13x^2+48x+36}{(x+2)(x-6)(x+6)}$
$= \dfrac{2x^3+2x^2+72x+72}{(x+2)(x-6)(x+6)} = \dfrac{2\left(x^3+x^2+36x+36\right)}{(x+2)(x-6)(x+6)} =$
$\dfrac{2\left[x^2(x+1)+36(x+1)\right]}{(x+2)(x-6)(x+6)} = \dfrac{2\left(x^2+36\right)(x+1)}{(x+2)(x-6)(x+6)}$

38. $\dfrac{2x-1}{4x^2-12x+5} - \dfrac{x+1}{4x^2-4x-15} = \dfrac{2x-1}{(2x-5)(2x-1)} -$
$\dfrac{x+1}{(2x-5)(2x+3)} = \dfrac{(2x-1)(2x+3)}{(2x-5)(2x-1)(2x+3)} -$

$\dfrac{(x+1)(2x-1)}{(2x-5)(2x+3)(2x-1)} = \dfrac{\left(4x^2+4x-3\right)-\left(2x^2+x-1\right)}{(2x-5)(2x+3)(2x-1)}$
$= \dfrac{2x^2+3x-2}{(2x-5)(2x+3)(2x-1)} = \dfrac{(2x-1)(x+2)}{(2x-5)(2x+3)(2x-1)}$
$= \dfrac{x+2}{(2x-5)(2x+3)}$

39. $\dfrac{x^2-5x-6}{x^2+8x+12} \div \dfrac{x^2+7x+6}{x^2-4x-12} = \dfrac{(x+1)(x-6)}{(x+6)(x+2)} \cdot \dfrac{(x-6)(x+1)}{(x+6)(x+1)} =$
$\dfrac{(x-6)^2}{(x+6)^2}$

40. $\dfrac{x^2+x-2}{7a^2x^2-14a^2x+7a^2} \cdot \dfrac{14ax-28a}{1-2x+x^2} = \dfrac{(x+2)(x-1)}{7a^2(x-1)^2} \cdot \dfrac{14a(x-2)}{(x-1)^2} =$
$\dfrac{2(x+2)(x-2)}{a(x-1)^3} = \dfrac{2\left(x^2-4\right)}{a(x-1)^3}$

41. $\dfrac{\frac{1}{x} - \frac{1}{y}}{\frac{1}{x} + \frac{1}{y}} = \dfrac{\frac{y-x}{xy}}{\frac{y+x}{xy}} \cdot \dfrac{xy}{xy} = \dfrac{y-x}{y+x}$

42. $\dfrac{\frac{1}{x} + \frac{1}{y}}{x+y} = \dfrac{\frac{y+x}{xy}}{x+y} \cdot \dfrac{xy}{xy} = \dfrac{y+x}{(x+y)(xy)} = \dfrac{1}{xy}$

43. $\dfrac{\frac{1}{x} - \frac{1}{y}}{\frac{x-y}{xy}} = \dfrac{\frac{y-x}{xy}}{\frac{x-y}{xy}} \cdot \dfrac{xy}{xy} = \dfrac{y-x}{x-y} = -1$

44. $\dfrac{1-\frac{1}{x}}{x-2+\frac{1}{x}} = \dfrac{\frac{x-1}{x}}{\frac{x^2-2x+1}{x}} \cdot \dfrac{x}{x} = \dfrac{x-1}{x^2-2x+1}$
$= \dfrac{x-1}{(x-1)^2} = \dfrac{1}{x-1}$

45. $\dfrac{\frac{x}{1+x} - \frac{1-x}{x}}{\frac{x}{1+x} + \frac{1-x}{x}} = \dfrac{\frac{x^2-(1-x)(1+x)}{x(1+x)}}{\frac{x^2+(1+x)(1-x)}{x(1+x)}} \cdot \dfrac{x(1+x)}{x(1+x)} =$
$\dfrac{x^2-\left(1-x^2\right)}{x^2+1-x^2} = \dfrac{2x^2-1}{1} = 2x^2-1$

46. $\dfrac{x - \frac{xy}{x-y}}{\frac{x^2}{x^2-y^2}-1} = \dfrac{\frac{x(x-y)-xy}{x-y}}{\frac{x^2-(x^2-y^2)}{x^2-y^2}} \cdot \dfrac{(x-y)(x+y)}{(x-y)(x+y)} =$
$\dfrac{x(x+y)(x-y)-xy(x+y)}{x^2-\left(x^2-y^2\right)} =$
$\dfrac{x^3-xy^2-x^2y-xy^2}{y^2} = \dfrac{x^3-x^2y-2xy^2}{y^2} =$
$\dfrac{x\left(x^2-xy-2y^2\right)}{y^2} = \dfrac{x(x+y)(x-2y)}{y^2}$

CHAPTER 7 TEST

1. $(x+5)(x-3) = x^2 - 3x + 5x - 15 = x^2 + 2x - 15$

2. $(2x-3)(2x+3) = (2x)^2 - 3^2 = 4x^2 - 9$

3. $(3x^2 - 4)(2 - 5x) = 6x^2 - 15x^3 - 8 + 20x = -15x^3 + 6x^2 + 20x - 8$

4. $(x-4)^3 = (x-4)(x-4)^2 = (x-4)(x^2 - 8x + 16)$
$= x^3 - 8x^2 + 16x - 4x^2 + 32x - 64$
$= x^3 - 12x^2 + 48x - 64$

5. $2x^2 - 128 = 2\left(x^2 - 64\right) = 2(x+8)(x-8)$

6. $x^2 - 12x + 32 = (x-8)(x-4)$

7. $10 \times (-21) = -210$ which has factors 15 and -14. Hence, $10x^2 + x - 21 = 10x^2 + 15x - 14x - 21$
$= 5x(2x+3) - 7(x+3)$
$= (5x-7)(2x+3)$

8. $x^3 - 125 = x^3 - 5^3$. This is a difference of cubes and it factors as $(x-5)(x^2 + 5x + 25)$.

9. $\frac{x^2 - 25}{x^2 + 6x + 5} = \frac{(x+5)(x-5)}{(x+5)(x+1)} = \frac{x-5}{x+1}$

10. $\frac{3(a+b)^3 - x(a+b)}{a^2 - b^2} = \frac{(a+b)[3(a+b)^2 - x]}{(a+b)(a-b)} = \frac{3(a+b)^2 - x}{a-b}$

11. $\frac{3x}{x+2} \cdot \frac{x-1}{x+2} = \frac{3x(x-1)}{(x+2)(x+2)} = \frac{3x(x-1)}{(x+2)^2} = \frac{3x^2 - 3x}{x^2 + 4x + 4}$

12. $\frac{2x+6}{x-2} \div \frac{3x+9}{x^2-4} = \frac{2x+6}{x-2} \cdot \frac{x^2-4}{3x+9} = \frac{2(x+3)(x+2)(x-2)}{(x-2)3(x+3)} = \frac{2(x+2)}{3}$

13. $\frac{6}{x-5} + \frac{x^2 - 2x}{x-5} = \frac{x^2 - 2x + 6}{x-5}$

14. $\frac{2x}{x+3} - \frac{x+4}{x-2} = \frac{2x(x-2)}{(x+3)(x-2)} - \frac{(x+4)(x+3)}{(x+3)(x-2)} = \frac{(2x^2 - 4x) - (x^2 + 7x + 12)}{(x+3)(x-2)} = \frac{x^2 - 11x - 12}{(x+3)(x-2)}$

15. $\dfrac{x - \frac{1}{x}}{x - \frac{2}{x+1}} = \dfrac{\frac{x^2-1}{x}}{\frac{x^2+x-2}{x+1}} = \dfrac{(x^2-1)}{x} \cdot \dfrac{(x+1)}{x^2+x-2} = \dfrac{(x+1)(x-1)(x+1)}{x(x+2)(x-1)} = \dfrac{(x+1)^2}{x(x+2)}$

16. The LCD is $(2x+1)^2$. Hence, $\frac{1}{2x+1} - \frac{2}{4x^2 + 4x + 1} = \frac{2x+1-2}{(2x+1)^2} = \frac{2x-1}{(2x+1)^2}$.

17. $\dfrac{2d}{\frac{d}{r_1} + \frac{d}{r_2}} = \dfrac{2d}{\frac{d(r_1+r_2)}{r_1 r_2}} = \dfrac{2d}{1} \cdot \dfrac{r_1 r_2}{d(r_1+r_2)} = \dfrac{2r_1 r_2}{r_1 + r_2}$

18. $\frac{1}{R_1} + \frac{1}{R_2} + \frac{1}{R_3} = \frac{R_2 R_3}{R_1 R_2 R_3} + \frac{R_1 R_3}{R-1 R-2 R_3} + \frac{R_1 R_2}{R_1 R_2 R_3} = \frac{R_1 R_2 + R_1 R_3 + R_2 R_3}{R_1 R_2 R_3}$

CHAPTER

8

Fractional and Quadratic Equations

≡ 8.1 FRACTIONAL EQUATIONS

1. $\frac{x}{2} + \frac{x}{3} = \frac{1}{4}$; LCD = 12; $12\left(\frac{x}{2} + \frac{x}{3}\right) = 12 \cdot \frac{1}{4}$; $12 \cdot \frac{x}{2} + 12 \cdot \frac{x}{3} = 12 \cdot \frac{1}{4}$; $6x + 4x = 3$; $10x = 3$; $x = \frac{3}{10}$ or 0.3

3. $\frac{y}{2} + 3 = \frac{4y}{5}$; LCD = 10; $10\left(\frac{y}{2} + 3\right) = 10\frac{4y}{5}$; $10 \cdot \frac{y}{2} + 10 \cdot 3 = 10 \cdot \frac{4y}{5}$; $5y + 30 = 2 \cdot 4y$; $5y + 30 = 8y$; $30 = 3y$; $10 = y$

5. $\frac{x-1}{2} + \frac{x+1}{3} = \frac{x-1}{4}$; LCD= 12; $12 \cdot \frac{x-1}{2} + 12 \cdot \frac{x+1}{3} = 12 \cdot \frac{x-1}{4}$; $6(x-1) + 4(x+1) = 3(x-1)$; $6x - 6 + 4x + 4 = 3x - 3$; $10x - 2 = 3x - 3$; $7x = -1$; $x = -\frac{1}{7}$

7. $\frac{1}{x} + \frac{2}{x} = \frac{1}{3}$; LCD = 3x; $x \neq 0$; $3x \cdot \frac{1}{x} + 3x \cdot \frac{2}{x} = 3x \cdot \frac{1}{3}$; $3 + 6 = x$; $9 = x$

9. $\frac{7}{w-4} = \frac{1}{2w+5}$; LCD = $(-4)(2w+5)$; $\neq 4$; $w \neq -\frac{5}{2}$; $(w-4)(2w+5) \cdot \frac{7}{w-4} = (w-4)(2w+5) \cdot \frac{1}{2w+5}$; $(2w+5)7 = w-4$; $14w + 35 = w - 4$; $13w = -39$; $w = -3$

11. $\frac{2}{2x-1} = \frac{5}{x+5}$; LCD = $(2x-1)(x+5)$; $x \neq \frac{1}{2}$; $x \neq -5$; $(2x-1)(x+5) \cdot \frac{2}{2x-1} = (2x-1)(x+5) \cdot \frac{5}{x+5}$; $(x+5)2 = (2x-1)5$; $2x + 10 = 10x - 5$; $15 = 8x$; $x = \frac{15}{8}$

13. $\frac{4x}{x-3} - 1 = \frac{3x}{x+2}$; LCD = $(x-3)(x+2)$; $x \neq 3$; $x \neq -2$; $4x(x+2) - (x-3)(x+2) = 3x(x-3)$; $(4x^2 + 8x) - (x^2 - x - 6) = 3x^2 - 9x$; $3x^2 + 9x + 6 = 3x^2 - 9x$; $18x = -6$; $x = -\frac{1}{3}$

15. $\frac{4}{x+2} - \frac{3}{x-1} = \frac{5}{(x-1)(x+2)}$; LCD = $(x+2)(x-1)$; $x \neq -2$; $x \neq 1$; $(x+2)(x-1) \cdot \frac{4}{x+2} - (x+2)(x-1)$

$\cdot \frac{3}{x-1} = (x+2)(x-1) \cdot \frac{5}{(x+2)(x-1)}$; $4(x-1) - 3(x+2) = 5$; $4x - 4 - 3x - 6 = 5$; $x - 10 = 5$; $x = 15$

17. $\frac{x+1}{x+2} + \frac{x+3}{x-2} = \frac{2x^2+3x-5}{x^2-4}$; LCD = $(x+2)(x-2)$ or $x^2 - 4$; $x \neq 2$; $x \neq -2$; $(x+2)(x-2) \cdot \frac{x+1}{x+2} + (x+2)(x-2) \cdot \frac{x+3}{x-2} = (x^2-4) \cdot \frac{2x^2+3x-5}{x^2-4}$; $(x-2)(x+1) + (x+2)(x+3) = 2x^2 + 3x - 5$; $x^2 - x - 2 + x^2 + 5x + 6 = 2x^2 + 3x - 5$; $2x^2 + 4x + 4 = 2x^2 + 3x - 5$; $x = -9$

19. $\frac{3}{a+1} + \frac{a+1}{a-1} = \frac{a^2}{a^2-1}$; LCD = $(a+1)(a-1)$ or $a^2 - 1$; $a \neq 1$; $a \neq -1$; $3(a-1) + (a+1)(a+1) = a^2$; $3a - 3 + a^2 + 2a + 1 = a^2$; $5a - 2 = 0$; $a = \frac{2}{5}$

21. $\frac{2}{x-1} + \frac{5}{x+1} = \frac{4}{x^2-1}$; LCD = $(x-1)(x+1)$ or $x^2 - 1$; $x \neq 1$; $x \neq -1$; $2(x+1) + 5(x-1) = 4$; $2x + 2 + 5x - 5 = 4$; $7x - 3 = 4$; $x = 1$; no solution since x cannot equal 1.

23. $\frac{5x-2}{x-3} + \frac{4-5x}{x+4} = \frac{10}{x^2+x-12}$; LCD = $(x-3)(x+4)$ or $x^2 + x - 12$; $x \neq 3$; $x \neq -4$; $(5x-2)(x+4) + (4-5x)(x-3) = 10$; $(5x^2 + 18x - 8) + (-5x^2 + 19x - 12) = 10$; $37x - 20 = 10$; $37x = 30$; $x = \frac{30}{37}$

25. $\frac{5}{x} + \frac{3}{x+1} = \frac{x}{x+1} - \frac{x+1}{x}$; LCD = $x(x+1)$; $x \neq 0$; $x \neq -1$; $5(x+1) + 3x = x^2 - (x+1)(x+1)$; $5x + 5 + 3x = x^2 - (x^2 + 2x + 1)$; $8x + 5 = -2x - 1$; $10x - 6$; $x = -\frac{6}{10} = -\frac{3}{5}$

27. $\frac{2t-4}{2t+4} = \frac{t+2}{t+4}$; LCD = $(2t+4)(t+4)$; $t \neq -2$; $t \neq -4$; $(2t-4)(t+4) = (t+2)(2t+4)$; $2t^2 + 4t - 16 = 2t^2 + 8t + 8$; $4t - 16 = 8t + 8$; $-4t = 24$; $t = -6$

29. $\frac{3x+1}{x-1} - \frac{x-2}{x+3} = \frac{2x-3}{x+3} + \frac{4}{x-1}$; LCD = $(x-1)(x+3)$; $x \neq 1$; $x \neq -3$; $(3x+1)(x+3) - (x-2)(x-1) = (2x-3)(x-1) + 4(x+3)$; $3x^2 + 10x + 3 - (x^2 - $

$3x+2) = 2x^2 - 5x + 3 + 4x + 12; 2x^2 + 13x + 1 = 2x^2 - x + 15; 14x = 14; x = 1;$ no solution

31. $\frac{1}{r} + \frac{1}{s} = \frac{1}{t};$ LCD $= rst; rst \cdot \frac{1}{r} + rst \cdot \frac{1}{s} = rst \cdot \frac{1}{t};$ $st + rt = rs; rs - st = rt; s(r - t) = rt; s = \frac{rt}{r-t}$

33. $\frac{1}{R} = \frac{1}{R_1} + \frac{1}{R_2} + \frac{1}{R_3};$ LCD $= RR_1R_2R_3; R_1R_2R_3 = RR_2R_3 + RR_1R_3 + RR_1R_2; R(R_2R_3 + R_1R_3 + R_1R_2) = R_1R_2R_3; R = \frac{R_1R_2R_3}{R_2R_3 + R_1R_3 + R_1R_2}$

35. $V = 2\pi rh + 2\pi r^2; 2\pi rh = V - 2\pi r^2; h = \frac{V - 2\pi r^2}{2\pi r}$

37. $\frac{1}{f} = (n-1)\left(\frac{1}{R_1} + \frac{1}{R_2}\right);$ LCD $= fR_1R_2; R_1R_2 = (n-1)(fR_2 + fR_1); R_1R_2 - (n-1)(fR_2) = (n-1)fR_1; R_2(R_1 - fn + f) = (n-1)fR_1; R_2 = \frac{(n-1)fR_1}{R_1 - nf + f}$

39. First simplify $\frac{1}{R_2} - \frac{1}{R_1} = \frac{R_1 - R_2}{R_1R_2}.$ Now, solving for C, we get $C\left(\frac{R_1 - R_2}{R_1R_2}\right) = \frac{d}{9 \times 10^9}$ and so, $C = \frac{d}{9 \times 10^9} \cdot \frac{R_1R_2}{R_1 - R_2} = \frac{dR_1R_2}{(9 \times 10^9)(R_1 - R_2)}.$

41. $(n-1)\left(\frac{1}{r_1} - \frac{1}{r_2}\right) = \frac{(n-1)(r_1 - r_1)}{r_1r_2}.$ Hence, $f = \frac{r_1r_2}{(n-1)(r_2 - r_1)}.$

43. $\frac{h}{D-d} = 2$ is equivalent to $h = 2(D-d) = 2D - 2d.$ So, $2D = 2d + h$ or $D = \frac{2d+h}{2} = d + \frac{h}{2}.$

45. $\frac{1}{x} = \frac{1}{10} + \frac{1}{6} = \frac{6}{60} + \frac{10}{60} = \frac{16}{60}.$ Thus, $x = \frac{60}{16} = 3\frac{3}{4}.$ It will take $3\frac{3}{4}$h or 3 h 45 min to complete the payroll.

47. h = hours B is open; $\frac{h+1}{6} + \frac{h}{4} = 1;$ LCD = 12; $12 \cdot \frac{h+1}{6} + 12 \cdot \frac{h}{4} = 12; 2h + 2 + 3h = 12; 5h = 10; h = 2; h + 1 = 3.$ It takes a total of 3 hours.

49. $\frac{h+4}{18} + \frac{h}{12} = 1;$ LCD = 36; $2(h+4) + 3h = 36; 5h + 8 = 36; 5h = 28; h = \frac{28}{5} = 5.6$ hr together; Total time is $5.6 + 4 = 9.6$ hr

51. $\frac{1}{t} = \frac{1}{3} - \frac{1}{4} = \frac{4-3}{12} = \frac{1}{12}.$ Hence, $t = 12,$ so it will take 12 h to fill the tank.

≡ 8.2 QUADRATIC EQUATIONS AND FACTORING

1. Factoring $x^2 - 9 = 0$ we get $(x+3)(x-3) = 0.$ By the zero product rule we have $x + 3 = 0$ and $x = -3$ or $x - 3 = 0$ and so $x = 3.$

3. $x^2 + x - 6 = 0; (x+3)(x-2) = 0; x + 3 = 0 \Rightarrow x = -3$ or $x - 2 = 0 \Rightarrow x = 2$

5. $x^2 - 11x - 12 = 0; (x-12)(x+1) = 0; x - 12 = 0 \Rightarrow x = 12$ or $x + 1 = 0 \Rightarrow x = -1$

7. $x^2 + 2x - 8 = 0; (x+4)(x-2) = 0; x + 4 = 0 \Rightarrow x = -4$ or $x - 2 = 0 \Rightarrow x = 2$

9. $x^2 - 5x = 0; x(x-5) = 0; x = 0$ or $x - 5 = 0 \Rightarrow x = 5$

11. $x^2 + 12 = 7x; x^2 - 7x + 12 = 0; (x-3)(x-4) = 0; x - 3 = 0 \Rightarrow x = 3; x - 4 = 0 \Rightarrow x = 4$

13. $2x^2 - 3x - 14 = 0; (2x-7)(x+2) = 0; 2x - 7 = 0 \Rightarrow x = \frac{7}{2}; x + 2 = 0 \Rightarrow x = -2$

15. $2x^2 + 12 = 11x; 2x^2 - 11x + 12 = 0; (2x-3)(x-4) = 0; 2x - 3 = 0 \Rightarrow x = \frac{3}{2}; x - 4 = 0 \Rightarrow x = 4$

17. $3x^2 - 8x - 3 = 0; (3x+1)(x-3) = 0; 3x + 1 = 0 \Rightarrow x = -\frac{1}{3}; x - 3 = 0 \Rightarrow x = 3$

19. $4x^2 - 24x + 35 = 0; (2x-5)(2x-7) = 0; 2x - 5 = 0 \Rightarrow x = \frac{5}{2}; 2x - 7 = 0 \Rightarrow x = \frac{7}{2}$

21. $6x^2 + 11x - 35 = 0; (3x-5)(2x+7) = 0; 3x - 5 = 0 \Rightarrow x = \frac{5}{3}; 2x + 7 = 0 \Rightarrow x = -\frac{7}{2}$

23. $10x^2 - 17x + 3 = 0; (5x-1)(2x-3) = 0; 5x - 1 = 0 \Rightarrow x = \frac{1}{5}; 2x - 3 = 0 \Rightarrow x = \frac{3}{2}$

25. $6x^2 = 31x + 60; 6x^2 - 31x - 60 = 0; (3x-20)(2x+3) = 0; 3x - 20 = 0 \Rightarrow x = \frac{20}{3}; 2x + 3 = 0 \Rightarrow x = -\frac{3}{2}$

27. $(x-1)^2 = 4; x - 1 = \pm\sqrt{4}; x - 1 = \pm 2; x - 1 = 2 \Rightarrow x = 3; x - 1 = -2 \Rightarrow x = -1$

29. $(5x-2)^2 = 16; 5x - 2 = \pm\sqrt{16}; 5x - 2 = \pm 4; 5x - 2 = 4 \Rightarrow x = \frac{6}{5}; 5x - 2 = -4 \Rightarrow x = -\frac{2}{5}$

31. $\frac{x}{x+1} = \frac{x+2}{3x}; 3x \cdot x = (x+2)(x+1); 3x^2 = x^2 + 3x + 2; 2x^2 - 3x - 2 = 0; (2x+1)(x-2) = 0; x = -\frac{1}{2}$ or 2

33. $(x+5)^3 = x^3 + 1385.$ Expanding the left-hand side we can rewrite the equation as $x^3 + 15x^2 + 75x + 125 = x^3 + 1385.$ Combining like terms this becomes $15x^2+75x-1260 = 0$ or $x^2+5x-84 = 0$. This factors as $(x + 12)(x - 7) = 0$ and so $x = -12$ or $x = 7$.

35. $(x-4)^3-x^3 = -316$. Expanding the left-hand side we can rewrite the equation as $x^3 - 12x^2 + 48x - 64 - x^3 = -316$. Combining like terms we get $-12x^2+48x+252 = 0$ or $12x^2-48x-252 = 0$. Dividing by 12, this becomes $x^2-4x-21 = 0$ or $(x - 7)(x + 3) = 0$. Thus, $x = 7$ or $x = -3$.

37. $\frac{1}{x-3} + \frac{1}{x+4} = \frac{1}{12}$; $12(x + 4) + 12(x - 3) = (x - 3)(x + 4)$; $12x + 48 + 12x - 36 = x^2 + x - 12$; $24x + 12 = x^2 + x - 12$; $0 = x^2 - 23x - 24$; $(x - 24)(x + 1) = 0$; $x = 24$ or $x = -1$.

39. Multiplying $\frac{1}{x-1} + \frac{1}{x-2} = \frac{7}{12}$ by the LCD, $12(x - 1)(x - 2)$, we obtain $12(x - 2) + 12(x - 1) = 7(x-1)(x-2)$. Distributing produces $12x-24 + 12x - 12 = 7\left(x^2 - 3x + 2\right)$ which simplifies to:

$$24x - 36 = 7x^2 - 21x + 14$$
$$0 = 7x^2 - 45x + 50$$
$$0 = (7x - 10)(x - 5)$$

Thus, $x = \frac{10}{7}$, 5.

41. Multiplying $\frac{2}{x-24} + \frac{1}{x-9} = \frac{1}{6}$ by the LCD of $6(x - 4)(x - 9)$, we obtain $12(x - 9) + 6(x - 4) = (x - 4)(x - 9)$ or $18x - 132 = x^2 - 13x + 36$. This simplifies to $x^2 - 31x + 168 = 0$ and factors as $(x - 24)(x - 7) = 0$. So, $x = 24, 7$.

43. Multiplying $\frac{x}{x+1} - \frac{2x}{x+3} = -\frac{1}{15}$ by the LCD of $15(x + 1)(x + 3)$, we obtain $15x(x + 3) - 30x(x + 1) = -(x+1)(x+3)$ or $15x^2+45x-30x^2-30x = -x^2-4x-3$. This simplifies to $14x^2-19x-3 = 0$ and factors as $(7x + 1)(2x - 3) = 0$. Thus, $x = -\frac{1}{7}, \frac{3}{2}$.

45. $-16t^2 + 64t + 192 = 192$; $-16t^2 + 64t = 0$; $-16t(t - 4) = 0$; $t = 0$ or $t = 4$; we want $t = 4$ s

47. If we let ℓ represent the length in cm and w its width in cm, then we are given $\ell = w + 5$. The area of a rectangle is $\ell w = (w + 5)w = 104$. Multiplying, we get $w^2 + 5w - 104 = 0$, which factors as $(w + 13)(w - 8) = 0$. Thus, by the zero product rule, we have $w = -13$ or $w = 8$. Choose $w = 8$ cm and then $\ell = w + 5 = 8 + 5 = 13$ cm.

49. Pythagorean Theorem yields $(x - 7)^2 + x^2 = 13^2$; $x^2 - 14x + 49 + x^2 = 169$; $2x^2 - 14x - 120 = 0$; $x^2 - 7x - 60 = 0$; $(x - 12)(x + 5) = 0$; $x = 12$ or $x = -5$; cannot be negative so $x = 12$; $x - 7 = 5$. The rafters are 5 m and 12 m.

51. Since total area − inside area = area of concrete, we have the equation $(2x + 16)(2x + 10) - 160 = 192$ or $4x^2 + 52x + 160 - 160 = 192$. Collecting like terms, this becomes $4x^2 + 52x - 192 = 4\left(x^2 + 13x - 48\right) = 0$. This factors as $4(x + 16)(x - 3)$ and so, $x = -16$ or $x = 3$. The only answer that fits the situation is $x = 3.00$ in.

53. The radius of the pool is 12 m. Its area is 144π m^2. The area of the deck plus the pool is $(12 + w)^2\pi$. This leads to the equation $(12 + w)^2\pi - 144\pi = 432\pi$ or $(12 + w)^2 - 144 = 432$. Expanding, we get $144+24w+w^2-144 = 432$. Collecting terms, we can rewrite this as $w^2 + 24w - 432 = 0$. This factors as $(x + 36)(x - 12) = 0$. Thus, $x = -36$ or $x = 12$. The only answer that makes physical sense is $x = 12$. The width of the deck should be 12.0 m.

55. $\frac{1}{B} + \frac{1}{B+6} = \frac{1}{4}$ or $4(B + 6) + 4B = B(B + 6)$ or $8B + 24 = B^2 + 6B$. Collecting like terms, this simplifies to $B^2-2B-24 = 0$ or $(B-6)(B+4) = 0$. Thus, $B = 6$ or $B = -4$. Since $B > 0$, $B = 6$ h and $A = 12$ h.

57. Substituting the given values, we obtain $47.75 = 0.002837D^2(75) - 0.127248$. This simplifies to $0.212775D^2 = 47.877248$ or $D^2 = 225.01$. Thus, $D = \sqrt{225.01} \approx 15.00$ in.

≡ 8.3 COMPLETING THE SQUARE

1. $x^2+6x+8 = 0$; $x^2+6x+9 = -8+9$; $(x+3)^2 = 1$; $x + 3 = \pm 1$; $x = -3 + 1 = -2$ or $x = -3 - 1 = -4$

3. $x^2-10x = 11$; $x^2-10x+25 = 11+25$; $(x-5)^2 = 36$, $x - 5 = \pm 6$ and so, $x = 5 + 6 = 11$ or $x = 5 - 6 = -1$

5. $x^2+6x+3 = 0$; $x^2+6x+9 = -3+9$; $(x+3)^2 = 6$; $x + 3 = \pm\sqrt{6}$ and so $x = -3 \pm \sqrt{6}$

7. $x^2 - 5x + 5 = 0$; $x^2 - 5x + \frac{25}{4} = -5 + \frac{25}{4}$; $\left(x - \frac{5}{2}\right)^2 = \frac{5}{4}$; $x - \frac{5}{2} = \pm\frac{\sqrt{5}}{2}$; $x = \frac{5}{2} \pm \frac{\sqrt{5}}{2}$ or $x = \frac{5\pm\sqrt{5}}{2}$

9. $2x^2 - 6x - 10 = 0$; $x^2 - 3x - 5 = 0$; $x^2 - 3x + \frac{9}{4} = 5 + \frac{9}{4}$; $\left(x - \frac{3}{2}\right)^2 = \frac{29}{4}$; $x - \frac{3}{2} = \pm\frac{\sqrt{29}}{2}$; $x = \frac{3 \pm \sqrt{29}}{2}$

11. $4x^2 - 12x - 18 = 0$; $x^2 - 3x - \frac{9}{2} = 0$; $x^2 - 3x + \frac{9}{4} = \frac{9}{2} + \frac{9}{4}$; $\left(x - \frac{3}{2}\right)^2 = \frac{27}{4}$; $x - \frac{3}{2} = \frac{\pm\sqrt{27}}{2}$; $x = \frac{3 \pm \sqrt{27}}{2} = \frac{3 \pm 3\sqrt{3}}{2}$

13. $x^2 + 2kx + c = 0$; $x^2 + 2kx = -c$; $x^2 + 2kx + k^2 = -c + k^2$; $(x + k)^2 = -c + k^2$; $x + k = \pm\sqrt{k^2 - c}$; $x = -k \pm \sqrt{k^2 - c}$

15. $196 - 16t^2 = 0$; $16t^2 = 196$; $t^2 = 12.25$; $t = \pm\sqrt{12.25}$ cannot be negative so $\sqrt{12.25} = 3.5$ s

17. $q(1560 - 4q) = 29{,}600$; $-4q^2 + 1560q = 29{,}600$; $q^2 - 390q = -7400$; $q^2 - 390q + (195)^2 = -7400 +$

195^2; $(q - 195)^2 = 30625$; $q - 195 = \pm175$; $q = 195 + 175 = 370$ objects; or $q = 195 - 175 = 20$ objects

19. $P = 4w + 3\ell = 2700$; $3\ell = 2700 - 4w$; $\ell = 900 - \frac{4}{3}w$; $A = 2\ell w = 270{,}000$; $\ell w = 135{,}000$; $\left(900 - \frac{4}{3}w\right)w = 135{,}000$; $900w - \frac{4}{3}w^2 = 135{,}000$; $-\frac{4}{3}w^2 + 900w = 135{,}000$; $w^2 - 675w = -101{,}250$; $w^2 - 675w + \left(\frac{675}{2}\right)^2 = -101{,}250 + \left(\frac{675}{2}\right)^2$; $\left(w - \frac{675}{2}\right)^2 = 12656.25$; $w - 337.5 = \pm\sqrt{12656.25}$; $w - 337.5 = \pm112.5$; $w = 337.5 + 112.5 = 450 \Rightarrow \ell = 300$; or $w = 337.5 - 112.5 = 225 \Rightarrow \ell = 600$; 450 ft by 300 ft or 225 ft by 600 ft

≡ 8.4 THE QUADRATIC FORMULA

1. $x^2 + 3x - 4 = 0$; $x = \frac{-3 \pm \sqrt{(3)^2 - 4(1)(-4)}}{2 \cdot 1} = \frac{-3 \pm \sqrt{9 + 16}}{2} = \frac{-3 \pm \sqrt{25}}{2} = \frac{-3 \pm 5}{2}$; $\frac{-3 + 5}{2} = 1$; $\frac{-3 - 5}{2} = -4$; $\{1, -4\}$

3. $3x^2 - 5x - 2 = 0$; $x = \frac{-(-5) \pm \sqrt{(-5)^2 - 4(3)(-2)}}{2 \cdot 3} = \frac{5 \pm \sqrt{25 + 24}}{6} = \frac{5 \pm \sqrt{49}}{6} = \frac{5 \pm 7}{6}$; $\frac{5 + 7}{6} = \frac{12}{6} = 2$; $\frac{5 - 7}{6} = \frac{-2}{6} = -\frac{1}{3}$; $\left\{-\frac{1}{3}, 2\right\}$

5. $7x^2 + 6x - 1 = 0$; $x = \frac{-6 \pm \sqrt{6^2 - 4 \cdot 7(-1)}}{14} = \frac{-6 \pm \sqrt{36 + 28}}{14} = \frac{-6 \pm \sqrt{64}}{14} = \frac{-6 \pm 8}{14}$; $\frac{-6 + 8}{14} = \frac{2}{14} = \frac{1}{7}$; $\frac{-6 - 8}{14} = \frac{-14}{14} = -1$; $\left\{-1, \frac{1}{7}\right\}$

7. $2x^2 - 5x - 7 = 0$; $x = \frac{-(-5) \pm \sqrt{(-5)^2 - 4 \cdot 2(-7)}}{2 \cdot 2} = \frac{5 \pm \sqrt{25 + 56}}{4} = \frac{5 \pm \sqrt{81}}{4} = \frac{5 \pm 9}{4}$; $\frac{5 + 9}{4} = \frac{14}{4} = \frac{7}{2}$; $\frac{5 - 9}{4} = \frac{-4}{4} = -1$; $\left\{-1, \frac{7}{2}\right\}$

9. $3x^2 + 2x - 8 = 0$; $x = \frac{-2 \pm \sqrt{2^2 - 4(3)(-8)}}{2 \cdot 3} = \frac{-2 \pm \sqrt{4 + 96}}{6} = \frac{-2 \pm \sqrt{100}}{6} = \frac{-2 \pm 10}{6}$; $\frac{-2 + 10}{6} = \frac{8}{6} = \frac{4}{3}$; $\frac{-2 - 10}{6} = \frac{-12}{6} = -2$; $\left\{-2, \frac{4}{3}\right\}$

11. $9x^2 + 12x + 4 = 0$; $x = \frac{-12 \pm \sqrt{12^2 - 4 \cdot 9 \cdot 4}}{2 \cdot 9} = \frac{-12 \pm \sqrt{144 - 144}}{18} = \frac{-12}{18} = \frac{-2}{3}$ double root

13. $2x^2 - 3x - 1 = 0$; $x = \frac{-(-3) \pm \sqrt{(-3)^2 - 4 \cdot 2 \cdot (-1)}}{2 \cdot 2} = \frac{3 \pm \sqrt{9 + 8}}{4} = \frac{3 \pm \sqrt{17}}{4}$

15. $x^2 + 5x + 2 = 0$; $x = \frac{-5 \pm \sqrt{5^2 - 4 \cdot 2 \cdot 1}}{2 \cdot 1} = \frac{-5 \pm \sqrt{25 - 8}}{2} = \frac{-5 \pm \sqrt{17}}{2}$

17. $2x^2 + 6x - 3 = 0$; $x = \frac{-6 \pm \sqrt{6^2 - 4(2)(-3)}}{2 \cdot 2} = \frac{-6 \pm \sqrt{36 + 24}}{4} = \frac{-6 \pm \sqrt{60}}{4} = \frac{-6 \pm 2\sqrt{15}}{4} = \frac{2\left(-3 \pm \sqrt{15}\right)}{2 \cdot 2} = \frac{-3 \pm \sqrt{15}}{2}$

19. $x^2 - 2x - 7 = 0$; $x = \frac{2 \pm \sqrt{(-2)^2 - 4 \cdot 1(-7)}}{2} = \frac{2 \pm \sqrt{4 + 28}}{2} = \frac{2 \pm \sqrt{32}}{2} = \frac{2 \pm 4\sqrt{2}}{2} = 1 \pm 2\sqrt{2}$

21. $2x^2 - 3 = 0$; $x = \frac{0 \pm \sqrt{0 - 4(2)(-3)}}{2 \cdot 2} = \frac{\pm\sqrt{24}}{4} = \frac{\pm2\sqrt{6}}{4} = \frac{\pm\sqrt{6}}{2}$

23. $3x^2 + 4 = 0$; $x = \frac{0 \pm \sqrt{0^2 - 4(3)(4)}}{2 \cdot 3} = \frac{\pm\sqrt{-48}}{6}$ no real roots because the discriminant is -48

25. $\frac{2}{3}x^2 - \frac{1}{9}x + 3 = 0$; $x = \frac{\frac{1}{9} \pm \sqrt{\left(-\frac{1}{9}\right)^2 - 4 \cdot \frac{2}{3} \cdot 3}}{2 \cdot \frac{2}{3}} = \frac{\frac{1}{9} \pm \sqrt{\frac{1}{81} - 8}}{\frac{4}{3}} = \frac{\frac{1}{9} \pm \sqrt{\frac{-647}{81}}}{\frac{4}{3}}$ no real roots because the discriminant is $\frac{-647}{81}$

27. $0.01x^2 + 0.2x = 0.6$; $0.01x^2 + 0.2x - 0.6 = 0$; $x = \frac{-0.2 \pm \sqrt{.2^2 - 4 \cdot (0.01)(-0.6)}}{2 \cdot (0.01)} = \frac{-0.2 \pm \sqrt{0.064}}{0.02}$ or see the alternate solution.

Alternate Solution:

$0.01x^2 + 0.2x - 0.6 = 0$; $x^2 + 20x - 60 = 0$; $x = \frac{-20 \pm \sqrt{20^2 - 4(-60)}}{2} = \frac{-20 \pm \sqrt{400 + 240}}{2} = \frac{-20 \pm \sqrt{640}}{2} = \frac{-20 \pm 2\sqrt{160}}{2} = -10 \pm \sqrt{160} = -10 \pm 4\sqrt{10}$

29. $\frac{1}{4}x^2 + 3 = \frac{5}{2}x$; $\frac{1}{4}x^2 - \frac{5}{2}x + 3 = 0$ or $x^2 - 10x + 12 = 0$; $x = \frac{10 \pm \sqrt{(-10)^2 - 4 \cdot 12}}{2} = \frac{10 \pm \sqrt{52}}{2} = \frac{10 \pm 2\sqrt{13}}{2} = 5 \pm \sqrt{13}$

31. $1.2x^2 = 2x - 0.5 = 1.2x^2 - 2x + 0.5 = 0$; $12x^2 - 20x + 5 = 0$; $x = \frac{20 \pm \sqrt{(-20)^2 - 4(12)5}}{2 \cdot 12} = \frac{20 \pm \sqrt{400 - 240}}{24}$ $= \frac{20 \pm \sqrt{160}}{24} = \frac{20 \pm 4\sqrt{10}}{24} = \frac{5 \pm \sqrt{10}}{6}$ or $x = \frac{2 \pm \sqrt{(-2)^2 - 4(1.2)(0.5)}}{2 \cdot 1.2} = \frac{2 \pm \sqrt{4 - 2.4}}{2.4} = \frac{2 \pm \sqrt{1.6}}{2.4}$

33. $3x^2 + \sqrt{3}x - 7 = 0$; $x = \frac{-\sqrt{3} \pm \sqrt{\sqrt{3}^2 - 4(3)(-7)}}{2 \cdot 3} = \frac{-\sqrt{3} \pm \sqrt{3 + 84}}{6} = \frac{-\sqrt{3} \pm \sqrt{87}}{6}$

35. $\frac{x-3}{7} = 2x^2$; $x - 3 = 14x^2$; $14x^2 - x + 3 = 0$; $x = \frac{1 \pm \sqrt{(-1)^2 - 4 \cdot 14 \cdot 3}}{2} = \frac{1 \pm \sqrt{-167}}{2}$. Since the discriminant is negative, there are no real roots.

37. $\frac{2}{x-1} + 3 = \frac{-2}{x+1}$; LCD $= (x-1)(x+1)$. Multiplying by the LCD produces $2(x + 1) + 3(x - 1)(x + 1) = -2(x - 1)$; $2x + 2 + 3x^2 - 3 = -2x + 2$; $3x^2 + 4x - 3 = 0$; $\frac{-4 \pm \sqrt{4^2 - 4(3)(-3)}}{2(3)}$; $\frac{-4 \pm \sqrt{16 + 36}}{6}$, $\frac{-4 \pm \sqrt{52}}{6} = \frac{-4 \pm 2\sqrt{13}}{6} = \frac{-2 \pm \sqrt{13}}{3}$

39. $-4.9t^2 + 411 = 0$; $\frac{0 \pm \sqrt{0 - 4(-4.9)(411)}}{2(-4.9)} = \frac{\pm \sqrt{8055.6}}{-9.8} \approx 9.16$ s; only positive answer works

41. $\frac{-20 \pm \sqrt{20^2 - 4(-4.9)(411)}}{2(-4.9)} = \frac{-20 \pm 91.95}{-9.8} = \frac{-20 - 91.95}{-9.8} \approx +11.42$; $\frac{-20 + 91.95}{-9.8} \approx -7.34$; only positive answer fits the problem situation 11.42 s

43. $V = \ell \cdot w \cdot h = (1.5w - 6)(w - 6) \cdot 3 = 578$; $(1.5w^2 - 15w + 36)3 = 578$; $4.5w^2 - 45w + 108 = 578$; $4.5w^2 - 45w - 470 = 0$; $x = \frac{45 \pm \sqrt{(-45)^2 - 4(4.5)(-470)}}{2 \cdot 4.5}$

$= \frac{45 \pm \sqrt{10,485}}{9} \approx \frac{45 + 102.4}{9} = \frac{147.4}{9} \approx 16.38$; $w = 16.38$ cm; $\ell = 1.5(16.38) = 24.57$ cm.

45. $x(12 - 2x) = 16.875$; $12x - 2x^2 = 16.875$; $-2x^2 + 12x - 16.875 = 0$; $2x^2 - 12x + 16.875 = 0$; $x = \frac{12 \pm \sqrt{(-12)^2 - 4 \cdot (2)(16.875)}}{2 \cdot 2} = \frac{12 \pm 3}{4}$; $\frac{12 + 3}{4} = \frac{15}{4} = 3\frac{3}{4}$; $\frac{12 - 3}{4} = \frac{9}{4} = 2\frac{1}{4}$; $12 - 2 \cdot 3\frac{3}{4} = 4.5$; $12 - 2 \cdot 2\frac{1}{4} = 7.5$; 2.25 in. by 7.5 in.; or 3.75 in. by 4.5 in.

47. $x =$ amount of increase; new price is $30 + x$; number of customers is $1000 - 5x$; revenue $= (1000 - 5x)(30 + x)$; $(1000 - 5x)(30 + x) = 45,000$; $30000 + 850x - 5x^2 = 45,000$; $-5x^2 + 850x - 15,000 = 0$; $x^2 - 170x + 3,000 = 0$; $x = \frac{170 \pm \sqrt{(-170)^2 - 4(3,000)}}{2} = \frac{170 \pm 130}{2}$. Thus, one value produces $x = \frac{170 - 130}{2} = \frac{40}{2} = 20$. The other value yields $x = \frac{170 + 130}{2} = 150$; a \$20 increase will lose the fewest customers.

49. $610^2 = 300^2 + (X_L - 531)^2$; $372100 = 90000 + (X_L - 531)^2$; $282100 = (X_L - 531)^2$; $x_L - 531 = \pm\sqrt{282100} \approx \pm 531.1$; $x_L \approx 531 \pm 531.1$. Since x_L cannot be negative, we have $x_L = 1062.13 \, \Omega$

51. Since the total area − inside area = area of concrete, we have the equation $(2x + 16)(2x + 10) - 160 = 245$ which simplifies to $4x^2 + 52x - 245 = 0$. Using the quadratic formula we have $x = \frac{-52 \pm \sqrt{52^2 - 4(4)(-245)}}{2(4)} = \frac{-52 \pm \sqrt{6624}}{8} = \frac{-52 \pm 12\sqrt{46}}{8} = \frac{-13 \pm 3\sqrt{46}}{2}$. Since $x > 0$, we get $\frac{-13 + 3\sqrt{46}}{2} \approx 3.67$ in.

53. Substituting, we get $926 = 0.79D^2 - 2D - 4$. Collecting like terms produces $0.79D^2 - 2D - 930 = 0$. By the quadratic formula, $D = \frac{2 \pm \sqrt{(-2)^2 + 4(0.79)(930)}}{2(0.79)} \approx 35.6$ in. or -33.1 in. Since, $D > 0$, we ignore the negative answer.

55. See *Computer Programs* in main text.

57. (a) $3x^2 + 5x - 2 = (3x - 1)(x + 2)$ and so, $x = -2$ or $x = \frac{1}{3}$.

(b) $\frac{-5 \pm \sqrt{25 + 24}}{6} = \frac{-5 \pm 7}{6} \cdot \frac{-5 + 7}{6} = \frac{2}{6} = \frac{1}{3}$ and $\frac{-5 - 7}{6} = \frac{-12}{6} = -2$

CHAPTER 8 REVIEW

1. $\frac{x}{3} + \frac{x}{2} = 5$; LCD $= 6$; $6 \cdot \frac{x}{3} + 6 = 6 \cdot \frac{x}{2} = 6 \cdot 5 \Rightarrow 2x + 3x = 30 \Rightarrow 5x = 30$; $x = 6$

2. $\frac{2}{x} - \frac{3}{x} = \frac{1}{5}$; LCD $= 5x$, $x \neq 0$; $5 \cdot 2 - 5 \cdot 3 = x$; $10 - 15 = x$; $-5 = x$

3. $\frac{x-2}{4} - \frac{x+2}{5} = \frac{x}{2}$; LCD 20; $20\left(\frac{x-2}{4}\right) - 20\left(\frac{x+2}{5}\right) =$
$20 \cdot \frac{x}{2}$; $(5x-10)-(4x+8) = 10x$; $5x-10-4x-8 =$
$10x$; $x - 18 = 10x$; $-18 = 9x$; $x = -2$

4. $\frac{2}{x-1} + \frac{3}{x-2} = \frac{4}{x^2-3x+2}$; LCD $(x-1)(x-2) =$
$x^2 - 3x + 2$; $x \neq 1$; $x \neq 2$; $2(x-2)+3(x-1) = 4$;
$2x - 4 + 3x - 3 = 4$; $5x = 11$; $x = \frac{11}{5}$

5. $\frac{3x}{x-1} - \frac{3x+2}{x} = 3$; LCD $(x-1)(x) = x^2-x$; $3x(x) -$
$(3x+2)(x-1) = 3\left(x^2-x\right)$; $3x^2 - 3x^2 + x + 2 =$
$3x^2 - 3x$; $0 = 3x^2 - 4x - 2$; $\frac{4 \pm \sqrt{(-4)^2-4\cdot(3)(-2)}}{2\cdot 3} =$
$\frac{4\pm\sqrt{40}}{6} = \frac{4\pm 2\sqrt{10}}{2\cdot 3} = \frac{2\pm\sqrt{10}}{3}$

6. $\frac{4x}{2x-3} + \frac{1}{x-1} = \frac{2x+1}{x-1}$; We see that the LCD $=$
$(2x-3)(x-1)$ and that $x \neq \frac{3}{2}$ and $x \neq 1$. Multiplying by the LCD, we get $4x(x-1) + (2x-3) =$
$(2x+1)(2x-3)$ or $4x^2 - 4x + 2x - 3 = 4x^2 - 4x - 3$.
Collecting terms, we get $2x = 0$ and so, $x = 0$.

7. $A = 2lw + 2(l+w)h$; $Aa - 2lw = 2(l+w)h$;
$\frac{A-2lw}{2(l+w)} = h$

8. $A = 2lw + 2(l+w)h$; $A = 2lw + 2lh + 2wh$;
$2lw + 2lh = A - 2wh$; $l(2w+2h) = A - 2wh$;
$l = \frac{A-2wh}{2w+2h}$

9. $\frac{1}{f} = \frac{1}{p} + \frac{1}{q}$; LCD $= fpq$; $fpq \cdot \frac{1}{f} = fpq \cdot \frac{1}{p} + fpq \cdot \frac{1}{q}$;
$pq = fq + fp$; $pq = f(q+p)$; $f = \frac{pq}{p+q}$

10. $X = wL - \frac{1}{wc}$; $Xwc = wLwC - 1$; $Xwc =$
$w^2Lc - 1$; $w^2Lc - Xwc = 1$; $c\left(w^2L - Xw\right) = 1$;
$c = \frac{1}{w^2L-Xw}$ or $c = \frac{1}{w(wL-X)}$

11. $x^2 - 8x + 7 = 0$; $(x-7)(x-1) = 0$; $x - 7 = 0 \Rightarrow$
$x = 7$; $x - 1 = 0 \Rightarrow x = 1$

12. $x^2 + 4x + 3 = 0$; $(x+3)(x+1) = 0$; $x + 3 = 0 \Rightarrow$
$x = -3$; $x + 1 = 0 \Rightarrow x = -1$

13. $x^2 - 11x + 10 = 0$; $(x-10)(x-1) = 0$; $x - 10 =$
$0 \Rightarrow x = 10$; $x - 1 = 0 \Rightarrow x = 1$

14. $x^2 + 15x + 14 = 0$; $(x+14)(x+1) = 0$; $x + 14 =$
$0 \Rightarrow x = -14$; $x + 1 = 0 \Rightarrow x = -1$

15. $x^2 + 15x + 56 = 0$; $(x+8)(x+7) = 0$; $x + 7 =$
$0 \Rightarrow x = -7$; $x + 8 = 0 \Rightarrow x = -8$

16. $x^2 - 8x + 12 = 0$; $(x-6)(x-2) = 0$; $x - 6 =$
$0 \Rightarrow x = 6$; $x - 2 = 0 \Rightarrow x = 2$

17. $2x^2 - 5x + 3 = 0$; $(2x-3)(x-1) = 0$; $2x - 3 =$
$0 \Rightarrow x = \frac{3}{2}$; $x - 1 = 0 \Rightarrow x = 1$

18. $3x^2 + 10x + 7 = 0$; $(3x+7)(x+1) = 0$; $3x + 7 =$
$0 \Rightarrow x = -\frac{7}{3}$; $x + 1 = 0 \Rightarrow x = -1$

19. $6x^2 + 7x - 10 = 0$; $(6x-5)(x+2) = 0$; $6x - 5 =$
$0 \Rightarrow x = \frac{5}{6}$; $x + 2 = 0 \Rightarrow x = -2$

20. $8x^2 + 22x + 9 = 0$; $8 \times 9 = 72$; $18 \times 4 = 72$;
$18 + 4 = 22$; $8x^2 + 18x + 4x + 9 = 0$; $2x(4x +$
$9) + 1(4x + 9)$; $(2x+1)(4x+9) = 0$; $2x + 1 =$
$0 \Rightarrow x = -\frac{1}{2}$; $4x + 9 = 0 \Rightarrow x = -\frac{9}{4}$

21. $4x^2 - 9 = 0$; $(2x+3)(2x-3) = 0$; $2x + 3 = 0 \Rightarrow$
$x = -\frac{3}{2}$; $2x - 3 = 0 \Rightarrow x = \frac{3}{2}$

22. $9x^2 - 16 = 0$; $(3x+4)(3x-4) = 0$; $3x + 4 =$
$0 \Rightarrow x = -\frac{4}{3}$; $3x - 4 = 0 \Rightarrow x = \frac{4}{3}$

23. $x^2 + 4x - 5 = 0$; $x^2 + 4x + 4 = 5 + 4$; $(x+2)^2 = 9$;
$x + 2 = \pm 3$; $x = -2 + 3 = 1$; $x = -2 - 3 = -5$

24. $x^2 - 7x + 6 = 0$; $x^2 - 7x + \frac{49}{4} = -6 + \frac{49}{4}$;
$\left(x - \frac{7}{2}\right)^2 = \frac{25}{4}$; $x - \frac{7}{2} = \pm\frac{5}{2}$; $x = \frac{7}{2} + \frac{5}{2} = \frac{12}{2} = 6$;
$x = \frac{7}{2} - \frac{5}{2} = \frac{2}{2} = 1$

25. $x^2 + 19x = 11$; $x^2 + 19x + \left(\frac{19}{2}\right)^2 = 11 + \left(\frac{19}{2}\right)^2$;
$\left(x = \frac{19}{2}\right)^2 + \frac{44}{4} + \frac{361}{4} = \frac{405}{4}$; $x + \frac{19}{2} = \frac{\pm\sqrt{405}}{2} =$
$\frac{\pm 9\sqrt{5}}{2}$; $x = \frac{-19\pm 9\sqrt{5}}{2}$

26. $2x^2 + 7x = 15$, upon dividing by 2 becomes
$x^2 + \frac{7}{2}x = \frac{15}{2}$ and after completing the square
yields $x^2 + \frac{7}{2}x + \left(\frac{7}{4}\right)^2 = \frac{15}{2} + \left(\frac{7}{4}\right)^2 = \frac{120}{16} + \frac{49}{16}$
or $\left(x + \frac{7}{4}\right)^2 = \frac{169}{16}$, and after taking the square
roots of both sides produces $x + \frac{7}{4} = \pm\frac{13}{4}$ or $x =$
$-\frac{7}{4} + \frac{13}{4} = \frac{6}{4} = \frac{3}{2}$ and $x = \frac{-7}{4} - \frac{13}{4} = \frac{-20}{4} = -5$.

27. $3x^2 + 5x - 14 = 0$ or $3x^2 + 5x = 14$, and, after
dividing by 3 yields $x^2 + \frac{5}{3}x = \frac{14}{3}$. Completing
the square produces $x^2 + \frac{5}{3}x + \left(\frac{5}{6}\right)^2 = \frac{14}{3} + \left(\frac{5}{6}\right)^2$
or $\left(x + \frac{5}{6}\right)^2 = \frac{168}{36} + \frac{25}{36} = \frac{193}{36}$. Taking the square
roots of both sides produces $x + \frac{5}{6} = \pm\frac{\sqrt{193}}{6}$ and
so $x = \frac{-5\pm\sqrt{193}}{6}$.

28. $4x^2 + 2x - 5 = 0$ or $4x^2 + 2x = 5$, and, after dividing by 4 yields $x^2 + \frac{1}{2}x = \frac{5}{4}$. Completing the square produces $x^2 + \frac{1}{2}x + \frac{1}{16} = \frac{5}{4} + \frac{1}{16}$ or $\left(x + \frac{1}{4}\right)^2 = \frac{21}{16}$. Taking the square roots of both sides produces $x = -\frac{1}{4} \pm \frac{\sqrt{21}}{4}$ and so $x = \frac{-1 \pm \sqrt{21}}{4}$.

29. $x^2 - 8x + 5 = 0$; $x = \frac{8 \pm \sqrt{(-8)^2 - 4(5)}}{2} = \frac{8 \pm \sqrt{64 - 20}}{2} = \frac{8 \pm \sqrt{44}}{2} = \frac{8 \pm 2\sqrt{11}}{2} = 4 \pm \sqrt{11}$

30. $x^2 + 7x - 6 = 0$; $x = \frac{-7 \pm \sqrt{7^2 - 4(-6)}}{2} = \frac{-7 \pm \sqrt{49 + 24}}{2} = \frac{-7 \pm \sqrt{73}}{2}$

31. $2x^3 + 3x - 5 = 0$; $x = \frac{-3 \pm \sqrt{3^2 - 4(2)(-5)}}{2 \cdot 2} = \frac{-3 \pm \sqrt{9 + 40}}{4} = \frac{-3 \pm 7}{4}$; $\frac{-3 + 7}{4} = 1$; $\frac{-3 - 7}{4} = \frac{-5}{2}$

32. $2x^2 - 7x + 4 = 0$; $x = \frac{7 \pm \sqrt{(-7)^2 - 4(2)(4)}}{2 \cdot 2} = \frac{7 \pm \sqrt{49 - 32}}{4} = \frac{7 \pm \sqrt{17}}{4}$

33. $3x^2 + 2x - 4 = 0$; $x = \frac{-2 \pm \sqrt{2^2 - 4(3)(-4)}}{2 \cdot 3} = \frac{-2 \pm \sqrt{4 + 48}}{6} = \frac{-2 \pm \sqrt{52}}{6} = \frac{-2 \pm 2\sqrt{13}}{2 \cdot 3} = \frac{-1 \pm \sqrt{13}}{3}$

34. $4x^2 - 3x = 1$; $4x^2 - 3x - 1 = 0$; $x = \frac{3 \pm \sqrt{(-3)^2 - 4(4)(-1)}}{2 \cdot 4} = \frac{3 \pm \sqrt{9 + 16}}{8} = \frac{3 \pm 5}{8}$; so $x = \frac{3 + 5}{8} = 1$ or, $x = \frac{3 - 5}{8} = -\frac{1}{4}$

35. $5x^2 + 2 = 8x$; $5x^2 - 8x + 2 = 0$; $x = \frac{8 \pm \sqrt{(-8)^2 - 4(5)(2)}}{2 \cdot 5} = \frac{8 \pm \sqrt{64 - 40}}{2 \cdot 5} = \frac{8 \pm \sqrt{24}}{2 \cdot 5} = \frac{8 \pm 2\sqrt{6}}{2 \cdot 5} = \frac{2(4 \pm \sqrt{6})}{2 \cdot 5} = \frac{4 \pm 6}{5}$

36. $6x^2 + 2x = 3$; $6x^2 + 2x - 3 = 0$; $x = \frac{-2 \pm \sqrt{2^2 - 4(6)(-3)}}{2 \cdot 6} = \frac{-2 \pm \sqrt{4 + 72}}{12} = \frac{-2 \pm \sqrt{76}}{12} = \frac{-2 \pm 2\sqrt{19}}{2 \cdot 6} = \frac{-1 \pm \sqrt{19}}{6}$

37. $3x^2 - 8x + 10 = 0$; $x = \frac{8 \pm \sqrt{(-8)^2 - 4(3)(10)}}{2 \cdot 3} = \frac{8 \pm \sqrt{64 - 120}}{6} = \frac{8 \pm \sqrt{-56}}{6}$; no answer because the discriminant is negative

38. $8x^2 = 4x + 3$; $8x^2 - 4x - 3 = 0$; $x = \frac{4 \pm \sqrt{(-4)^2 - 4(8)(-3)}}{2 \cdot 8} = \frac{4 \pm \sqrt{16 + 96}}{16} = \frac{4 \pm \sqrt{112}}{16} = \frac{4 \pm 4\sqrt{7}}{16} = \frac{1 \pm \sqrt{7}}{4}$

39. $\frac{x}{x-1} + \frac{2}{x+1} = 3$; $(x+1)x + 2(x-1) = 3(x+1)(x-1)$; $x^2 + x + 2x - 2 = 3x^2 - 3$; $0 = 2x^2 - 3x - 1$; $x = \frac{3 \pm \sqrt{(-3)^2 - 4(2)(-1)}}{2 \cdot 2} = \frac{3 \pm \sqrt{9 + 8}}{4} = \frac{3 \pm \sqrt{17}}{4}$

40. $\frac{2}{x} - \frac{3}{x+2} = 4$; $2(x+2) - 3(x) = 4(x)(x+2)$; $2x + 4 - 3x = 4x^2 + 8x$; $0 = 4x^2 + 9x - 4$; $x = \frac{-9 \pm \sqrt{9^2 - 4(4)(-4)}}{2 \cdot 4} = \frac{-9 \pm \sqrt{81 + 64}}{8} = \frac{-9 \pm \sqrt{145}}{8}$

41. $12 = (i_1 + 0.4)^2 \cdot 50$; $(i_1 + 0.4)^2 = \frac{12}{50} = .24$; $(i_1 + 0.4) = \pm\sqrt{.24} \approx \pm.4899$; $i_1 \approx -0.4 + .4899$; $i_1 \approx 0.0899$ A; Only positive answers fit the problem

42. $\frac{h}{7} + \frac{h}{8} = 1$; $8h + 7h = 56$; $15h = 56$; $h = \frac{56}{15} \approx 3.73$ hr

43. $144 - 16t^2 = 0$; $9 - t^2 = 0$; $t^2 = 9$; $t = \pm\sqrt{9} = \pm 3$; only 3 fits the problem, so it takes 3 s for the stone to reach the bottom

44. $a^2 + (a - 6)^2 = 27^2$; $a^2 + a^2 - 12a + 36 = 729$; $2a^2 - 12a - 693 = 0$; $a = \frac{12 \pm \sqrt{(-12)^2 - 4(2)(-693)}}{2 \cdot 2} = \frac{12 \pm \sqrt{5688}}{4} \approx \frac{12 + 75.42}{4} \approx 21.86$. Then $a - 6 = 21.86 - 6 = 15.86$ and the lengths are 15.86 mm and 21.86 mm

CHAPTER 8 TEST

1. $\frac{x}{8} + \frac{3x}{4} = \frac{7}{2}$; LCD = 8; $8 \cdot \frac{x}{8} + 8 \cdot \frac{3x}{4} = 8 \cdot \frac{7}{2}$; $x + 6x = 28$; $7x = 28$; $x = 4$

2. $\frac{12}{x^2 - 9} = \frac{x}{x-3} - \frac{x}{x+3}$; LCD = $(x-3)(x+3) = x^2 - 9$; $x \neq 3$; $x \neq -3$; $12 = x(x+3) - x(x-3)$; $12 = x^2 + 3x - x^2 + 3x$; $12 = 6x$; $x = 2$

3. $x^2 - 4x - 5 = 0 \Rightarrow (x-5)(x+1) = 0$; $x - 5 = 0 \Rightarrow x = 5$; $x + 1 = 0 \Rightarrow x = -1$

4. $3x^2 + x - 2 = 0$; Using the quadratic formula produces $x = \frac{-1 \pm \sqrt{1^2 - 4(3)(-2)}}{2 \cdot 3} = \frac{-1 \pm \sqrt{1 + 24}}{6} = \frac{-1 \pm 5}{6}$. Thus, we see that $x = \frac{-1 - 5}{6} = -1$ or $x = \frac{-1 + 5}{6} = \frac{2}{3}$. However, the given equation factors, and so this could have been solved by factoring where $(x + 1)(3x - 2) = 0$. By the zero product rule, $x + 1 = 0 \Rightarrow x = -1$ or $3x - 2 = 0 \Rightarrow x = \frac{2}{3}$.

5. $16x^2 + 9 = 24x$; $16x^2 - 24x + 9 = 0$; $(4x - 3)^2 = 0$; $4x - 3 = 0 \Rightarrow x = \frac{3}{4}$ double root

6. $2x^2 + 3x + 3$; $x = \frac{-3 \pm \sqrt{3^2 - 4(2)(3)}}{2 \cdot 2} = \frac{-3 \pm \sqrt{-15}}{4}$; Discriminant is -15 so no solution

7. $x^2 + 3x + 1 = 0$; $\frac{-3 \pm \sqrt{3^2 - 4}}{2} = \frac{-3 \pm \sqrt{5}}{2}$

8. $x^2 - 8x + 16 = -5 + 16$; $(x-4)^2 = 11$; $x - 4 = \pm\sqrt{11}$; $x = 4 \pm \sqrt{11}$

9. Since $x^2 + 7x - 30 = (x + 10)(x - 3)$, the solutions are $x = -10$ and $x = 3$.

10. Since $3x^2 + 5x - 28 = (x + 4)(3x - 7)$, the solutions are $x = -4$ and $x = \frac{7}{3}$.

11. $V = 3D^2H - 5$ so, $V + 5 = 3D^2H$ and $H = \frac{V+5}{3D^2}$.

12. $V = 3D^2H - 5$ so $3D^2H = V + 5$ and $D^2 = \frac{V+5}{3H}$. Thus, $D = -\sqrt{\frac{V+5}{3H}}$ and $D = \sqrt{\frac{V+5}{3H}}$.

13. $l = w + 3$; $(w+3)w = 46.75$; $w^2 + 3w - 46.75 = 0$; $w = \frac{-3 \pm \sqrt{9 + 4(46.75)}}{2} = \frac{-3 \pm \sqrt{196}}{2}$. Since a negative value does not make sense in this situation, we have $w = \frac{-3 + 14}{2} = \frac{11}{2} = 5.5$. Thus, $w = 5.5$ m and $l = 8.5$ m.

14. $s = vt + \frac{at^2}{2}$; $s - \frac{at^2}{2} = vt$ or $vt = \frac{2s - at^2}{2}$ and so $v = \frac{2s - at^2}{2t}$

Trigonometric Functions

≡ 9.1 ANGLES, ANGLE MEASURE, AND TRIGONOMETRIC FUNCTIONS

1. $90° = 90° \times \frac{\pi}{180°} = \frac{\pi}{2}$ or 1.5708 or $90° \times 0.01745 = 1.5705$

3. $80° = 80° \times \frac{\pi}{180°} = \frac{4\pi}{9}$ or 1.3963 or $80° \times 0.01745 = 1.396$

5. $155° = 155° \times \frac{\pi}{180°} = \frac{31\pi}{36} = 2.7053$ or $155° \times 0.01745 = 2.70475$

7. $215° = -215° \times \frac{\pi}{180°} = 3.7525$ or $215° \times 0.01745 = 3.75175$

9. $2 \text{ rad} = 2 \cdot \frac{180°}{\pi} = \frac{360°}{\pi} \approx 114.592°$ or $2 \times 57.296° = 114.592°$

11. $1.5 = 1.5 \times \frac{180°}{\pi} \approx 85.944°$ or $1.5 \times 57.296 = 85.944°$

13. $\frac{\pi}{3} = \frac{\pi}{3} \cdot \frac{180°}{\pi} = 60°$

15. $-\frac{\pi}{4} = -\frac{\pi}{4} \cdot \frac{180°}{\pi} = -45°$

17. (c) $150° = 150° + 360° = 510°; 150° - 360° = -210°$

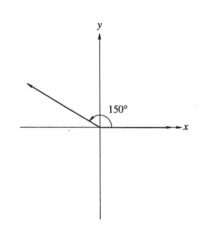

19. (c) $-135° = -135° + 360° = 225°; -135° - 360° = -495°$

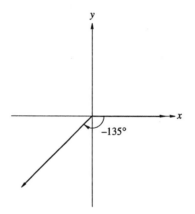

21. $r = \sqrt{4^2 + 3^2} = \sqrt{25} = 5; \sin\theta = \frac{3}{5}; \cos\theta = \frac{4}{5}; \tan\theta = \frac{3}{4}; \csc\theta = \frac{5}{3}; \sec\theta = \frac{5}{4}; \cot\theta = \frac{4}{3}$

23. $r = \sqrt{8^2 + (-15)^2} = 17; \sin\theta = \frac{-15}{17}; \cos\theta = \frac{8}{17}; \tan\theta = \frac{-15}{8}; \csc\theta = \frac{-17}{15}; \sec\theta = \frac{17}{8}; \cot\theta = \frac{-8}{15}$

25. $r = \sqrt{1^2 + 2^2} = \sqrt{5}; \sin\theta = \frac{2}{\sqrt{5}}; \cos\theta = \frac{1}{\sqrt{5}}; \tan\theta = 2; \csc\theta = \frac{\sqrt{5}}{2}; \sec\theta = \sqrt{5}; \cot\theta = \frac{1}{2}$

27. $r = \sqrt{10^2 + (-8)^2} = \sqrt{164} = 2\sqrt{41}; \sin\theta = -\frac{4}{\sqrt{41}}; \cos\theta = \frac{5}{\sqrt{41}}; \tan\theta = \frac{-4}{5}; \csc\theta = -\frac{\sqrt{41}}{4}; \sec\theta = \frac{\sqrt{41}}{5}; \cot\theta = \frac{-5}{4}$

29. $r = \sqrt{\left(\sqrt{11}\right)^2 + 5^2} = \sqrt{11 + 25} = \sqrt{36} = 6, \sin\theta = \frac{5}{6}; \cos\theta = \frac{\sqrt{11}}{6}; \tan\theta = \frac{5}{\sqrt{11}}; \csc\theta = \frac{6}{5}; \sec\theta = \frac{6}{\sqrt{11}}; \cot\theta = \frac{\sqrt{11}}{5}$

31. $r = 3$, $\sin\theta = 0$; $\cos\theta = 1$; $\tan\theta = 0$; $\csc\theta =$ Does not exist; $\sec\theta = 1$; $\cot\theta =$ Does not exist.

33. $r = 5$; $\sin\theta = 0$; $\cos\theta = -1$; $\tan\theta = 0$; $\csc\theta =$ Does not exist; $\sec\theta = -1$; $\cot\theta =$ Does not exist.

35. $x = 6$, $r = 10$, $y > 0$, $y^2 = r^2 - x^2 = 10^2 - 6^2 = 100 - 36 = 64$, $y = \pm\sqrt{64}$ so $y = 8$; $\sin\theta = \frac{8}{10} = \frac{4}{5}$; $\cos\theta = \frac{x}{r} = \frac{6}{10} = \frac{3}{5}$; $\tan\theta = \frac{4}{3}$; $\csc\theta = \frac{5}{4}$; $\sec\theta = \frac{5}{3}$; $\cot\theta = \frac{3}{4}$

37. $x = -20$, $r = 29$, $y > 0$, $y^2 = r^2 - x^2 = 29^2 - (-20)^2 = 841 - 400 = 441$, $y = \pm\sqrt{441} = \pm21$ so $y = 21$, $\sin\theta = \frac{21}{29}$; $\cos\theta = \frac{-20}{29}$; $\tan\theta = \frac{-21}{20}$; $\csc\theta = \frac{29}{21}$; $\sec\theta = \frac{-29}{20}$; $\cot\theta = \frac{-20}{21}$

39. $x = -7$, $r = 8$, $y < 0$, $y^2 = 8^2 - (-7)^2 = 64 - 49 = 15$, $y = \pm\sqrt{15}$ so $y = -\sqrt{15}$; $\sin\theta = -\frac{\sqrt{15}}{8}$; $\cos\theta = \frac{-7}{8}$; $\tan\theta = \frac{\sqrt{15}}{7}$; $\csc\theta = -\frac{8}{\sqrt{15}}$; $\sec\theta = \frac{-8}{7}$; $\cot\theta = \frac{7}{\sqrt{15}}$

9.2 VALUES OF THE TRIGONOMETRIC FUNCTIONS

1. $a = 5$, $c = 13$, $b = \sqrt{c^2 - a^2} = \sqrt{13^2 - 5^2} = \sqrt{169 - 25} = \sqrt{144} = 12$; $\sin\theta = \frac{5}{13}$; $\cos\theta = \frac{12}{13}$; $\tan\theta = \frac{5}{12}$; $\csc\theta = \frac{13}{5}$; $\sec\theta = \frac{13}{12}$; $\cot\theta = \frac{12}{5}$

3. $a = 1.2$, $c = 2$, $b = \sqrt{2^2 - 1.2^2} = \sqrt{4 - 1.44} = \sqrt{2.56} = 1.6$, $\sin\theta = \frac{1.2}{2} = \frac{12}{20} = \frac{3}{5}$, $\cos\theta = \frac{1.6}{2} = \frac{16}{20} = \frac{4}{5}$, $\tan\theta = \frac{3}{4}$; $\csc\theta = \frac{5}{3}$; $\sec\theta = \frac{5}{4}$; $\cot\theta = \frac{4}{3}$

5. $a = 1.4$, $b = 2.3$, $c = \sqrt{1.4^2 + 2.3^2} = \sqrt{1.96 + 5.29} = \sqrt{7.25}$; $\sin\theta = \frac{1.4}{\sqrt{7.25}}$; $\cos\theta = \frac{2.3}{\sqrt{7.25}}$; $\tan\theta = \frac{14}{23}$; $\csc\theta = \frac{\sqrt{7.25}}{1.4}$; $\sec\theta = \frac{\sqrt{7.25}}{2.3}$; $\cot\theta = \frac{23}{14}$

7. Since $\cos\theta = \frac{8}{17}$, the side adjacent is 8 and the hypotenuse is 17, and so the side opposite is 15. $\sin\theta = \frac{\text{opp}}{\text{hyp}} = \frac{15}{17}$; $\tan\theta = \frac{\text{opp}}{\text{adj}} = \frac{15}{8}$, $\csc\theta = \frac{\text{hyp}}{\text{opp}} = \frac{17}{15}$, $\sec\theta = \frac{\text{hyp}}{\text{adj}} = \frac{17}{8}$, $\cot\theta = \frac{\text{adj}}{\text{opp}} = \frac{8}{15}$

9. Since $\sin\theta = \frac{3}{5}$, the side opposite is 3 and the hypotenuse is 5, so side adjacent is 4. $\sin\theta = \frac{3}{5}$; $\cos\theta = \frac{4}{5}$; $\tan\theta = \frac{3}{4}$; $\csc\theta = \frac{5}{3}$; $\sec\theta = \frac{5}{4}$; $\cot\theta = \frac{4}{3}$

11. Here $\csc\theta = \frac{29}{20}$, so the hypotenuse is 29 and the side opposite is 20. Thus, the side adjacent is 21.

$\sin\theta = \frac{20}{29}$; $\cos\theta = \frac{21}{29}$; $\tan\theta = \frac{20}{21}$; $\csc\theta = \frac{29}{20}$; $\sec\theta = \frac{29}{21}$; $\cot\theta = \frac{21}{20}$

13. $\sin\theta = 0.866$, $\cos\theta = 0.5$, $\tan\theta = \frac{0.866}{0.5} = 1.732$, $\csc\theta = \frac{1}{0.866} \approx 1.155$, $\sec\theta = 2$, $\cot\theta = \frac{0.5}{0.866} \approx 0.577$

15. $\sin\theta = 0.085$, $\sec\theta = 1.004$, $\cos\theta = \frac{1}{1.004} \approx 0.996$, $\tan\theta \approx \frac{0.085}{0.996} \approx 0.085$, $\csc\theta = \frac{1}{0.085} \approx 11.765$, $\cot \approx \frac{0.996}{0.085} \approx 11.718$

17. $\sin 18.6° = 0.3189593$

19. $\tan 18.3° = 0.3307184$

21. $\tan 76°32' = \tan 76\frac{32}{60}° = 4.1760011$

23. $\cot 82.6° = \frac{1}{\tan 82.6°} = 0.1298773$

25. $\sin 0.25$ rad $= 0.247404$

27. $\tan 0.63$ rad $= 0.7291147$

29. $\cot 1.43$ rad $= \frac{1}{\tan 1.43} = 0.1417341$

31. $\csc 0.21$ rad $= \frac{1}{\sin 0.21} = 4.7970857$

33. $R = \frac{15^2}{9.8}\sin(2 \cdot 40) \approx 22.61$ m

35. $V_R = 5.8\cos 32° \approx 4.92$ V

37. See *Computer Programs* in main text.

9.3 THE RIGHT TRIANGLE

1. $\sin A = 0.732$, $A = 47.05°$

3. $\tan C = 4.671$, $C = 77.92°$

5. $\cos E = 0.839$, $E = 32.97°$

7. $\sec G = 3.421$, $G = 73.00°$

9. $\cot I = 0.539$, $I = 61.68°$

11. $\csc K = 4.761$, $K = 12.12°$

13. $B = 90 - 16.5 = 73.5$; $\sin A = \frac{a}{c}$, $c = \frac{a}{\sin A} = \frac{7.3}{\sin 16.5} \approx 25.7$, $\tan A = \frac{a}{b}$, $b = \frac{a}{\tan A} = \frac{7.3}{\tan 16.5} \approx$ 24.6. In summary, we have

$$a = \ 7.3 \quad A = 16.5°$$
$$b \approx 24.6 \quad B = 73.5°$$
$$c \approx 25.7 \quad C = 90°.$$

15. $B = 90 - 72.6 = 17.4$, $\sin A = \frac{a}{c}$, $a = c \sin A = 20 \sin 72.6 \approx 19.1$, $\cos A = \frac{b}{c}$, $b = 20 \cos 72.6 \approx 6.0$. In summary, we have

$$a \approx 19.1 \quad A = 72.6°$$
$$b \approx \ 6.0 \quad B = 17.4°$$
$$c = 20.0 \quad C = 90°.$$

17. $B = 90° - 43° = 47°$ and $\cos A = \frac{b}{c}$ which means that $c = \frac{34.6}{\cos 43} \approx 47.3$. Finally, $\tan A = \frac{a}{b}$, and so $a = b \tan A = 34.6 \tan 43 = 32.3$. Summarizing, we have

$$a \approx 32.3 \quad A = 43°$$
$$b = 34.6 \quad B = 47°$$
$$c \approx 47.3 \quad C = 90°.$$

19. We begin by determining that $B = \frac{\pi}{2} - 0.92 \approx 0.65$ rad. Next, we determine that $\sin A = \frac{a}{c}$ and so $c = \frac{a}{\sin A} = \frac{6.5}{\sin 0.92} \approx 8.2$. Finally, we find that $\tan A = \frac{a}{b}$ and so $b = \frac{6.5}{\tan 0.92} \approx 4.9$. In summary, we have the following results:

$$a = 6.5 \quad A = 0.92 \text{ rad}$$
$$b \approx 4.9 \quad B \approx 0.65 \text{ rad}$$
$$c \approx 8.2 \quad C = \frac{\pi}{2} \text{ rad}$$

21. We first determine that $B = \frac{\pi}{2} - 0.15 \approx 1.42$ rad. Next, we find that $\sin A = \frac{a}{c}$ and so $a = c \sin A = 18 \sin 0.15 \approx 2.7$. Finally, we have $\cos A = \frac{b}{c}$ which produces $b = 18 \cos 0.15 \approx 17.8$. Summarizing, we have:

$$a \approx \ 2.7 \quad A = 0.15 \text{ rad}$$
$$b \approx 17.8 \quad B \approx 1.42 \text{ rad}$$
$$c = 18.0 \quad C = \frac{\pi}{2} \text{ rad}$$

23. Using the given information, we find that $B \approx 1.57 - 1.41 \approx 0.16$ rad. Next, we see that $\tan A = \frac{a}{b}$ produces $a = 40 \tan 1.41 \approx 246.6$. Finally, we have $\cos A = \frac{b}{c}$, and so $c = \frac{40}{\cos 1.41} \approx 249.8$. In summary, we have the following results:

$$a \approx 246.6 \quad A = 1.41 \text{ rad}$$
$$b = \ 40.0 \quad B \approx 0.16 \text{ rad}$$
$$c \approx 249.8 \quad C = \frac{\pi}{2} \text{ rad}$$

25. $a = 9$, $b = 15$, $c = \sqrt{9^2 + 15^2} \approx 17.5$, $\tan A = \frac{a}{b} = \frac{9}{15}$, $A = \tan^{-1} \frac{9}{15} = 31.0°$ or 0.54 rad, $B \approx 90° - 31° = 59°$ or $B \approx 1.57 - 0.54 = 1.03$ rad

27. $b = 9.3$, $c = 18$, $a = \sqrt{18^2 - 9.3^2} \approx 15.4$, $\cos A = \frac{b}{c} = \frac{9.3}{18}$, $A \approx 58.9°$ or 1.03 rad, $B \approx 90° - 58.9° = 31.1°$ or $1.57 - 1.03 = 0.54$ rad

29. $a = 20$, $c = 30$, $b = \sqrt{30^2 - 20^2} \approx 22.4$, $\sin A = \frac{a}{c} = \frac{20}{30}$, $A \approx 41.8°$ or 0.73 rad, $B \approx 90° - 41.8° = 48.2°$ or $B \approx 1.57 - 0.73 = 0.84$ rad

31. $\tan \phi = \frac{12.3}{19.7}$, $\phi = \tan^{-1} \frac{12.3}{19.7} \approx 31.9792$ or about $32°$

33. $\sin 37.6° = \frac{700}{d}$, $d = \frac{700}{\sin 37.6°} \approx 1147$ m

35. $F_x = 10 \cos 30° \approx 8.7$ N, $F_y = 10 \sin 30° = 5$ N

37. $\tan \phi = \frac{(80.67)^2}{32 \cdot 1200} \approx 0.1695$, $\phi = \tan^{-1}(0.1695) \approx 9.62°$

39. $\cos 67° = \frac{x}{300}$, $x = 3.00 \cos 67° \approx 117.2$ m, $\sin 67° = \frac{d}{300}$ and so $d = 300 \sin 67° \approx 276.2$ m. So 117.2 m from intersection, 276.2 m from service station.

41. $\tan 67.3° = \frac{h}{25.75}$, so $h = 25.75 \tan 67.3° \approx 61.56$ ft

☰ 9.4 TRIGONOMETRIC FUNCTIONS OF ANY ANGLE

1. $87°$ is in Quadrant I, so $\theta_{ref} = 87°$

3. $137°$ is in Quadrant II, so $\theta_{ref} = 180° - 137° = 43°$

5. $\frac{9\pi}{8}$ rad is in Quadrant III, so $\theta_{ref} = \frac{9\pi}{8} - \pi = \frac{9\pi}{8} - \frac{8\pi}{8} = \frac{\pi}{8}$

7. 4.5 rad is in Quadrant III, so $\theta_{ref} = 4.5 - \pi \approx 4.5 - 3.14 = 1.36$

9. $518°$; $518° - 360° = 158°$

11. $-871°$; $3 \times 360° - 871° = 1080° - 871° = 209°$

13. 7.3 rad; $7.3 - 2\pi \approx 7.3 - 6.28 = 1.02$

15. -2.17 rad; $-2.17 + 2\pi \approx -2.17 + 6.28 = 4.11$

17. $165°$; Quadrant II

19. $-47°$; Quadrant IV

21. $250°$; Quadrant III

23. $98°$; Quadrant II

25. $\tan\theta$ is negative in Quadrants II and IV.

27. $\sec\theta$ is negative in Quadrants II and III.

29. $\csc\theta$ is negative and $\cos\theta$ is positive in Quadrant IV.

31. $\sin 105°$ is in Quadrant II, so the expression is positive.

33. $\tan 372°$ is in Quadrant I, so the expression is positive.

35. $\cos 1.93$ rad is in Quadrant II, so the expression is negative.

37. $\sin 215°$ is in Quadrant III, so the expression is negative.

39. $\sin 137° = 0.6820$

41. $\tan 293° = -2.3559$

43. $\tan 164.2° = -0.2830$

45. $\sin 2.4$ rad $= 0.6755$

47. $\tan 6.1$ rad $= -0.1853$

49. $\tan 1.37$ rad $= 4.9131$

51. $\sin 415.5° = 0.8241$

53. $\cot -87.4° = -0.0454$

55. $\csc 432.4° = -1.0491$

57. $\sin 6.5$ rad $= 0.2151$

59. $\tan 8.35$ rad $= -1.8479$

61. $\cos -9.43$ rad $= 0.9090$

63. $I = 32.65 \sin 132° \approx 24.26$ mA

65. The unlabeled angle at B is $180° - 123° - 20° = 37°$. Thus, $AB = \frac{96}{\sin 37°} \approx 159.52$; $BC = \frac{264}{\cos 20°} \approx 280.94$. The total length is $159.52 + 280.94 = 440.46$ mm.

67. Note that the drawing is not to scale. The ground that needs to be sodded is the shape of a trapezoid.

To find h, the height of the trapezoid, we divide the trapezoid into a rectangle and a right triangle. The leg of the right triangle that joins the base of the trapezoid is h. The other leg is $223 - 156 = 67$ ft. The angle opposite h is $19.7°$, so $h = 67 \tan 19.7° \approx 23.99$ ft. Thus, the area of the trapezoid is $A = \frac{1}{2}h(b_1 + b_2) = \frac{1}{2}(23.99)(223 + 156) = 4{,}546$ ft^2.

☰ 9.5 INVERSE TRIGNOMETRIC FUNCTIONS

1. $\sin\theta = \frac{1}{2}$ in Quadrants I and II.

3. $\cot\theta = -2$ in Quadrants II and IV.

5. $\sec\theta = 4.3$ in Quadrants I and IV.

7. $\csc\theta = -6.1$ in Quadrants III and IV.

9. $\sin\theta = \frac{1}{2}$; $\theta_r = \sin^{-1} 0.5 = 30.0°$; $\theta_I = 30.0°$, $\theta_{II} = 180° - 30° = 150°$

11. $\cot\theta = -2$; $\theta_r = \cot^{-1} 2 = \tan^{-1}\frac{1}{2} = 26.6°$; $\theta_{II} = 180° - \theta_r = 153.4°$; $\theta_{IV} = 360° - \theta_r = 333.4°$

13. $\sec\theta = 4.3$; $\theta_r = \sec^{-1} 4.3 = \cos^{-1}\frac{1}{4.3} = 76.6°$; $\theta_I = \theta_r = 76.6°$; $\theta_{IV} = 360 - \theta_r = 283.4°$

15. $\csc\theta = -6.1$; $\theta_r = \csc^{-1} 6.1 = \sin^{-1}\frac{1}{6.1} = 9.4°$; $\theta_{III} = 180° + \theta_r = 189.4°$; $\theta_{IV} = 360° - \theta_r = 350.6°$

17. $\sin\theta = 0.75$; $\theta_r = \sin^{-1} 0.75 = 0.85$; $\theta_I = \theta_r = 0.85$; $\theta_{II} = \pi - \theta_r = 2.29$

19. $\cot\theta = -0.4$; $\theta_r = \cot^{-1} 0.4 = \tan^{-1}\frac{1}{0.4} = 1.19$; $\theta_{II} = \pi - \theta_r = 1.95$; $\theta_{IV} = 2\pi - \theta_r = 5.09$

21. $\csc\theta = 4.3$; $\theta_r = \csc^{-1} 4.3 = \sin^{-1}\frac{1}{4.3} = 0.23$; $\theta_I = \theta_r = 0.23$; $\theta_{II} = \pi - \theta_r = 2.91$

23. $\sec\theta = 2.7$; $\theta_r = \sec^{-1} 2.7 = \cos^{-1}\frac{1}{2.7} = 1.19$; $\theta_I = \theta_r = 1.19$; $\theta_{IV} = 2\pi - \theta_r = 5.09$

25. $\arcsin 0.84 = 57.1°$

31. $\tan^{-1}(-0.64) = -32.6°$

37. $\sin^{-1} 0.95 = 1.25$

27. $\arctan 4.21 = 76.6°$

33. $\arccos(-0.33) = 1.91$

39. $\tan^{-1} 0.25 = 0.24$

29. $\sin^{-1} 0.32 = 18.7°$

35. $\arccos 0.29 = 1.28$

41. $\theta = \tan^{-1}\left(-\frac{33}{40}\right) \approx -0.6898 \text{ rad} \approx -39.52°$

43. $\theta = \sin^{-1}\left(\frac{4.635}{6.725}\right) \approx 43.57°$

45. See *Computer Programs* in main text.

47. See *Computer Programs* in main text.

≡ 9.6 APPLICATIONS OF TRIGONOMETRY

1. $47° = 47 \times \frac{\pi}{180} \text{ rad} \approx 0.8203 \text{ rad}$; $s = r\theta = 1050.250 \times 0.8203 \approx 861.520 \text{ m}$

3. $30 \text{ rpm} = 30 \cdot 2\pi \text{ rad/min} \approx 188.5 \text{ rad/min}$. $v = r\omega = 188.5 \times 12 = 2262 \text{ in./min} = 2262 \text{ in/min}\left(\frac{1 \text{ min}}{60 \text{ s}}\right) = 37.7 \text{ in/s}$

5. $s = r\theta = 15 \times \frac{5\pi}{8} \approx 29.45 \text{ cm}$

7. $A = \frac{1}{2}r^2\theta = \frac{1}{2}(15)^2\left(\frac{5\pi}{8}\right) \approx 220.89 \text{ cm}^2$

9. $\theta = 2 \times 0.07 = 0.14$; $s = r\theta = 1.2 \times 0.14 = 0.168 \text{ m}$

11. **(a)** $\omega = 45\frac{\text{rev}}{\text{min}} \times \frac{2\pi}{\text{rev}} \approx 282.74\frac{\text{rad}}{\text{min}}$ or $282.74 \frac{\text{rad}}{\text{min}}\frac{1 \text{ min}}{60 \text{ s}} \approx 4.712 \text{ rad/s}$; $r = 175.26/2 = 87.63 \text{ mm}$; $v = r\omega = 87.63 \times 282.74 \approx 24776.5 \text{ mm/min}$ or $v = 87.63 \times 4.712 \approx 412.9 \text{ mm/s}$; **(b)** 24 776.5 mm

13. **(a)** We use the given information to determine that $\cos D = (\sin 35°30')(\sin 40°40') + (\cos 35°30')(\cos 40°40')(\cos[-10°10'])$ and so $\cos D \approx 0.98624$ and so the angular distance between the cities is $D \approx 9.515\ 478°$. **(b)** To find the linear distance s between the cities, we use $s = r\theta$ with $\theta = 9.515\ 478°\left(\frac{\pi}{180°}\right) \approx$

0.1660764 rad. Using $r = 3960$ mi, we get $s = 3960(0.1660764) \approx 657.66$ mi.

15. $\frac{8}{\frac{1}{4}} \times \frac{\pi}{3} \times 2.5 = 83.77$. Since you cannot have a fraction of a byte, the answer is 83.

17. $850\frac{\text{rev}}{\text{min}} \cdot \frac{360°}{1 \text{ rev}} \cdot \frac{1 \text{ min}}{60 \text{ s}} = 5100°/\text{s}$

19. $v = r\omega = 4.2 \text{ m} \cdot \frac{20 \text{ rev}}{\text{min}} \cdot \frac{2\pi}{\text{rev}} \cdot \frac{1 \text{ min}}{60 \text{ s}} \approx 8.80 \text{ m/s}$

21. $\theta = 66.5° \times \frac{\pi}{180°} = 1.160\ 644 \text{ rad}$, $s = r\theta = 6370 \times 1.1606 = 7393.3 \text{ km}$

23. $\cos^2\theta = \frac{I}{I_m} = \frac{3}{4}$, so $\cos\theta = \sqrt{\frac{3}{4}}$, and so $\theta = \cos^{-1}\sqrt{\frac{3}{4}} = 30°$.

25. **(a)** $12 \text{ hr} = 12 \cdot 60 \text{ min} = 12 \cdot 60 \cdot 60 \text{ s} = 43200$ s. In 12 hr, a satellite completes 1 revolution, amounting to 2π radians. Its angular speed is thus $\omega = \frac{\theta}{t} = \frac{2\pi}{43200}$, or $\frac{\pi}{21600} \text{ rad/sec} \approx 0.000\ 145\ 444$ rad/s. **(b)** The radius of the orbit of a satellite is $3960 + 11,000 = 14,960$ mi. Thus, the linear speed of a satellite is given by $v = r\omega = (14,960)\left(\frac{\pi}{21600}\right)$, or about 2.175843801 mi/s.

≡ CHAPTER 9 REVIEW

1. $60° \cdot \frac{\pi}{180} = \frac{\pi}{3}$ or $60° \times 0.01745 = 1.047$

2. $198° \times 0.01745 = 3.4551$ or $198 \times \frac{\pi}{180} \approx 3.456$

3. $325 \times 0.01745 = 5.67125$ or $325 \times \frac{\pi}{180} \approx 5.672$

4. $180° \times \frac{\pi}{180} = \pi \approx 3.1416$

5. $-115° \times 0.01745 = -2.00675$; $-115° \times \frac{\pi}{180} \approx -2.007$

6. $435° \times 0.01745 = 7.59075$; $435 \cdot \frac{\pi}{180} \approx 7.592$

7. $\frac{3\pi}{4}\text{rad} = \frac{3\pi}{4} \cdot \frac{180°}{\pi} = 135°$

8. $1.10 \text{ rad} = 1.10 \times \frac{180°}{\pi} \approx 63.025$ or $1.10 \times 57.296 \approx 63.026$

9. $2.15 \text{ rad} = 2.15 \times 57.296 \approx 123.19°$ or $2.15 \times \frac{180°}{\pi} \approx 123.19°$

10. $\frac{7\pi}{3} \text{ rad} = \frac{7\pi}{3} \cdot \frac{180°}{\pi} = 420°$

11. $-4.31 \text{ rad} = -4.31 \times 57.296 \approx -246.95°$ or $-4.31 \times \frac{180°}{\pi} \approx -246.94°$

12. $5.92 \text{ rad} = 5.92 \times 57.296 \approx 339.19°$ or $5.92 \times \frac{180°}{\pi} \approx 339.19°$

13. (a) I,
(b) $60°$,
(c) $60° + 360 = 420°$; $60 - 360 = -300°$

14. (a) III,
(b) $\theta_r = 198 - 180 = 18°$,
(c) $198 + 360 = 558°$; $198 - 360 = -162°$

15. (a) IV,
(b) $\theta_r = 360 - 325 = 35°$,
(c) $325 + 360 = 685°$, $325 - 360 = 35°$

16. (a) Quadrantal angle between II and III,
(b) $0°$,
(c) $180 + 360 = 540°$; $180 - 360° = -180°$

17. (a) III,
(b) $\theta_r = 245 - 180 = 65°$,
(c) $-115 + 360 = 245°$; $-115 - 360° = -475°$

18. (a) I,
(b) $75°$,
(c) $435 + 360 = 795°$; $435 - 360 = 75°$; $75° - 360° = -285°$

19. (a) II,
(b) $\pi - \frac{3\pi}{4} = \frac{\pi}{4}$,
(c) $\frac{3\pi}{4} + 2\pi = \frac{11\pi}{4}$, $\frac{3\pi}{4} - 2\pi = \frac{-5\pi}{4}$

20. (a) I,
(b) 1.10,
(c) $1.10 + 2\pi = 1.10 + 6.28 = 7.38$; $1.10 - 6.28 = -5.18$

21. (a) II,
(b) $\pi - 2.15 = 3.14 - 2.15 = 0.99$,
(c) $2.15 + 2\pi = 2.15 + 6.28 = 8.43$; $2.15 - 2\pi = 2.15 - 6.28 = -4.13$

22. (a) I,
(b) $\frac{7\pi}{3} - 2\pi = \frac{\pi}{3}$,
(c) $\frac{7\pi}{3} + 2\pi = \frac{13\pi}{3}$, $\frac{7\pi}{3} - 2\pi = \frac{\pi}{3}$, or $\frac{7\pi}{3} - 4\pi = -\frac{5\pi}{3}$

23. (a) II,
(b) $4.31 - \pi = 1.17$,
(c) $-4.31 + 6.28 = 1.97$; $-4.31 - 6.28 = -10.59$

24. (a) IV,
(b) $2\pi - 5.92 = 6.28 - 5.92 = 0.36$,
(c) $5.92 - 2\pi = -0.36$; $5.92 + 6.28 = 12.20$

25. $r = \sqrt{3^2 + (-4)^2} = 5$; $\sin\theta = \frac{y}{r} = \frac{-4}{5}$; $\cos\theta = \frac{x}{r} = \frac{3}{5}$; $\tan\theta = \frac{y}{x} = \frac{-4}{3}$; $\csc\theta = \frac{r}{y} = \frac{-5}{4}$; $\sec\theta = \frac{r}{x} = \frac{5}{3}$; $\cot\theta = \frac{x}{y} = \frac{-3}{4}$

26. $r = \sqrt{5^2 + 12^2} = 13$; $\sin\theta = \frac{y}{r} = \frac{12}{13}$; $\cos\theta = \frac{x}{r} = \frac{5}{13}$; $\tan\theta = \frac{y}{x} = \frac{12}{5}$; $\csc\theta = \frac{r}{y} = \frac{13}{12}$; $\sec\theta = \frac{r}{x} = \frac{13}{5}$; $\cot\theta = \frac{x}{y} = \frac{5}{12}$

27. $r = \sqrt{(-20)^2 + 21^2} = \sqrt{841} = 29$; $\sin\theta = 21/29$, $\cos\theta = -20/29$, $\tan\theta = -21/20$, $\csc\theta = 29/21$, $\sec\theta = -29/20$, $\cot\theta = -20/21$

28. $r = \sqrt{(-4)^2 + (-7)^2} = \sqrt{16 + 49} = \sqrt{65}$; $\sin\theta = -7/\sqrt{65}$, $\cos\theta = -4/\sqrt{65}$, $\tan\theta = 7/4$, $\csc\theta = -\sqrt{65}/7$, $\sec\theta = -\sqrt{65}/4$, $\cot\theta = 4/7$

29. $r = \sqrt{7^2 + 1^2} = \sqrt{50} = 5\sqrt{2}$; $\sin\theta = \frac{1}{5\sqrt{2}}$; $\cos\theta = \frac{7}{5\sqrt{2}}$; $\tan\theta = \frac{1}{7}$; $\csc\theta = 5\sqrt{2}$; $\sec\theta = \frac{5\sqrt{2}}{7}$; $\cot\theta = 7$

30. $r = \sqrt{144 + 64} = \sqrt{208} = 4\sqrt{13}$; $\sin\theta = \frac{8}{4\sqrt{13}} = \frac{2}{\sqrt{13}}$; $\cos\theta = \frac{-12}{4\sqrt{13}} = \frac{-3}{\sqrt{13}}$; $\tan\theta = \frac{-8}{12} = \frac{-2}{3}$; $\csc\theta = \frac{\sqrt{13}}{2}$; $\sec\theta = -\frac{\sqrt{13}}{3}$; $\cot\theta = -\frac{3}{2}$

31. $b = \sqrt{17^2 - 8^2} = 15$; $\sin\theta = \frac{a}{c} = \frac{8}{17}$; $\cos\theta = \frac{b}{c} = \frac{15}{17}$; $\tan\theta = \frac{a}{b} = \frac{8}{15}$; $\csc\theta = \frac{c}{a} = \frac{17}{8}$; $\sec\theta = \frac{c}{b} = \frac{17}{15}$; $\cot\theta = \frac{b}{a} = \frac{15}{8}$

32. $c = \sqrt{5^2 + 12^2} = 13$; $\sin\theta = \frac{a}{c} = \frac{5}{13}$; $\cos\theta = \frac{b}{c} = \frac{12}{13}$; $\tan\theta = \frac{a}{b} = \frac{5}{12}$; $\csc\theta = \frac{c}{a} = \frac{13}{5}$; $\sec = \frac{c}{b} = \frac{13}{12}$; $\cot\theta = \frac{b}{a} = \frac{12}{5}$

33. $a = \sqrt{\left(\frac{40}{3}\right)^2 - 8^2} = \sqrt{\frac{1600}{9} - \frac{576}{9}} = \sqrt{\frac{1024}{9}} = \frac{32}{3}$; $\sin\theta = \frac{32/3}{40} = \frac{4}{5}$; $\cos\theta = \frac{8}{\frac{40}{3}} = \frac{24}{40} = \frac{3}{5}$; $\tan\theta = \frac{32/3}{8} = \frac{4}{3}$; $\csc\theta = \frac{5}{4}$; $\sec\theta = \frac{5}{3}$; $\cot = \frac{3}{4}$

34. $b = \sqrt{7.5^2 - 6^2} = 4.5$; $\sin\theta = \frac{6}{7.5} = .8$; $\cos\theta = \frac{4.5}{7.5} = 0.6$; $\tan\theta = \frac{6}{4.5} = 1.33$; $\csc\theta = \frac{7.5}{6} = 1.25$; $\sec\theta = \frac{7.5}{4.5} = 1.67$; $\cot\theta = \frac{4.5}{6} = 0.75$

35. $a = \sqrt{18.2^2 - 7^2} = 16.8$; $\sin\theta = \frac{16.8}{18.2} = \frac{168}{182} = \frac{84}{91}$; $\cos\theta = \frac{7}{18.2} = \frac{70}{182} = \frac{35}{91}$; $\tan\theta = \frac{16.8}{7} = \frac{168}{70} = \frac{84}{35}$; $\csc\theta = \frac{91}{84}$; $\sec\theta = \frac{91}{35}$; $\cot\theta = \frac{35}{84}$

36. $c = \sqrt{42^2 + 44.1^2} = 60.9$; $\sin\theta = \frac{42}{60.9} = \frac{420}{609}$; $\cos\theta = \frac{44.1}{60.9} = \frac{441}{609}$; $\tan\theta = \frac{42}{44.1} = \frac{420}{441} = \frac{20}{21}$; $\csc\theta = \frac{609}{420}$; $\sec\theta = \frac{609}{441}$; $\cot\theta = \frac{21}{20}$

37. $a = 12.8$; $c = 27.2$; $b = \sqrt{27.2^2 - 12.8^2} = 24$; $\cos\theta = \frac{24}{27.2}$; $\tan\theta = \frac{12.8}{24}$; $\csc\theta = \frac{27.2}{12.8}$; $\sec\theta = \frac{27.2}{24}$; $\cot\theta = \frac{24}{12.8}$

38. $a = 16$; $b = 16.8$; $c = \sqrt{16^2 + 16.8^2} = 23.2$; $\sin\theta = \frac{16}{23.2}$; $\cos\theta = \frac{16.8}{23.2}$; $\tan\theta = \frac{16}{16.8}$; $\csc\theta = \frac{23.2}{16}$; $\sec\theta = \frac{23.2}{16.8}$; $\cot\theta = \frac{16.8}{16}$

39. $c = 4$; $b = 2.5$; $a = \sqrt{4^2 - 2.5^2} \approx 3.12$; $\sin\theta = \frac{3.12}{4}$; $\cos\theta = \frac{2.5}{4}$; $\tan\theta = \frac{3.12}{2.5}$; $\csc\theta = \frac{4}{3.12}$; $\sec\theta = \frac{4}{2.5}$; $\cot\theta = \frac{2.5}{3.12}$

40. $\cos\theta = \frac{\sin\theta}{\tan\theta} = \frac{0.532}{0.628} \approx 0.847$; $\csc\theta = \frac{1}{0.532} \approx 1.880$; $\sec\theta = \frac{1}{0.847} \approx 1.181$; $\cot\theta = \frac{1}{0.628} \approx 1.592$

41. $\tan\theta = \frac{0.5}{0.866} \approx 0.577$; $\csc\theta = \frac{1}{0.5} = 2$; $\sec\theta = \frac{1}{0.866} \approx 1.155$; $\cot\theta = \frac{0.866}{0.5} = 1.732$

42. $\sin\theta = \frac{1}{1.364} \approx 0.733$; $\tan\theta \approx \frac{0.733}{0.680} \approx 1.078$; $\sec\theta = \frac{1}{0.680} \approx 1.471$; $\cot\theta \approx \frac{0.680}{0.733} \approx 0.928$

43. $\sin 45° = 0.7071$

44. $\cos 82.5° = 0.1305$

45. $\tan 213.5° = 0.6619$

46. $\sec(-81°) = 6.3925$

47. $\cos 2.3 \text{ rad} = -0.6663$

48. $\sin 4.75 \text{ rad} = -0.9993$

49. $\tan(-3.2) \text{ rad} = -0.0585$

50. $\csc 0.21 \text{ rad} = 4.7971$

51. $\sin\theta = 0.5$; $\theta_r = \sin^{-1} 0.5 = 30°$; $\theta_I = \theta_r = 30°$; $\theta_{II} = 180 - 30° = 150°$

52. $\tan\theta = 2.5$; $\theta_r = \tan^{-1} 2.5 = 68.2$; $\theta_I = 68.2$; $\theta_{III} = 180 + 68.2 = 248.2$

53. $\cos\theta = -0.75$; $\theta_r = \cos^{-1} 0.75 = 41.4$; $\theta_{II} = 180 - 41.4 = 138.6$; $\theta_{III} = 180 + 41.4 = 221.4$

54. $\csc\theta = 3.0$; $\theta_r = \sin^{-1}\frac{1}{3} = 19.5$; $\theta_I = 19.5$; $\theta_{II} = 180 - 19.5 = 160.5$

55. $\cos\theta = -0.5$; $\theta_r = \cos^{-1} 0.5 = 1.047 \approx 1.05$; $\theta_{II} = \pi - 1.05 = 3.14 - 1.05 = 2.09$; $\theta_{III} = \pi + 1.05 = 3.14 + 1.05 = 4.19$

56. $\sin\theta = 0.717$; $\theta_r = \sin^{-1} 0.717 = 0.80$; $\theta_I = 0.80$; $\theta_{II} = \pi - 0.80 = 3.14 - 0.80 = 2.34$

57. $\theta_r = \tan^{-1} 0.95 = 0.76$; $\theta_{II} = \pi - 0.76 = 3.14 - 0.76 = 2.38$; $\theta_{IV} = 2\pi - 0.76 = 6.28 - 0.76 = 5.52$

58. $\theta_r = \cos^{-1}\frac{1}{2.25} = 1.11$; $\theta_I = 1.11$; $\theta_{IV} = 2\pi - 1.11 = 6.28 - 1.11 = 5.17$

59. $\arcsin 0.866 = 60.0°$

60. $\arccos 0.5 = 60.0°$

61. $\arctan(-1) = -45.0°$

62. $\cos^{-1}(-0.707) = 135.0°$

63. $\sin^{-1} 0.385 = 22.6°$

64. $\cot^{-1}(3.5) = \tan^{-1}\frac{1}{3.5} = 15.9°$

65. $1.33 = \frac{\sin 30°}{\sin r}$, $\sin r = \frac{\sin 30°}{1.33} \approx 0.3759$; $r \approx \sin^{-1} 0.3759 \approx 22.1°$

66. $n = \frac{\sin 0.90}{\sin 0.55} \approx 1.50$

67. $I = 12.6\sin\frac{\pi}{5} \approx 7.41$ A

68. $\sin\theta = \frac{4.732}{7.3} \approx 0.6482$; $\theta \approx \sin^{-1} 0.6482 \approx 40.4°$

69. $\tan\theta = \frac{44}{50}$; $\theta = \tan^{-1}\frac{44}{50} \approx 41.3°$; $V = \sqrt{V_L^2 + V_R^2} = \sqrt{44^2 + 50^2} \approx 66.6$ V

70. $\frac{1}{2}(57 - 46) = 5.5$; $\tan 3° = \frac{5.5}{h}$; $h = \frac{5.5}{\tan 3°} \approx 104.9$ mm

71. We use the equation $s = r\theta$ with $\theta = 37° \times \frac{\pi}{180} \approx 0.6458$. Thus, we obtain $s \approx 900 \times 0.6458 \approx 581.2$ ft.

72. $300\frac{\text{rev}}{\text{min}} \cdot \frac{2\pi \text{ rad}}{\text{rev}} \cdot \frac{1 \text{ min}}{60 \text{ sec}} = 10\pi$ rad/s ≈ 31.416 rad/s; $v = r\omega = \frac{5.25 \text{ in.}}{2} \times 31.416 = 82.467$ in./s

73. $\omega = \frac{1 \text{ rev}}{365 \text{ days}} \cdot \frac{2\pi}{1 \text{ rev}} \cdot \frac{1 \text{ day}}{24 \text{ hr}} \approx 7.17 \times 10^{-4}\frac{\text{rad}}{\text{hr}}$; $v = r\omega = 9.3 \times 10^7 \times 7.17 \times 10^{-4} \approx 66{,}700$ mph

74. $\tan 68° = \frac{h}{15}$; $h = 15 \tan 68° \approx 37.1$ ft

75. Let $y = $ length of common side, then $\cos 36° = \frac{y}{110.5}$; $y = 110.5 \cos 36° = 89.4$; $\cos x = \frac{82.3}{89.4}$; $x = \cos^{-1} \frac{82.3}{89.4}$; $x = 23.0°$

76. $\tan \theta = \frac{56}{43}$; $\theta = \tan^{-1} \frac{56}{43} \approx 52.5°$

77. $\tan 53.7° = \frac{h}{150}$; $h = 150 \tan 53.7° \approx 204.2$ ft

78. Draw a segment parallel to the axis from the top of the smaller circle. This makes a right triangle with legs of 128.3 m and $(50.2 - 26.1) = 24.1$ m. Then $\tan \phi = \frac{24.1}{128.3}$; $\phi \approx 10.64$; $\theta = 2\phi \approx 21.3°$

79. $F_x = 3500 \sin 34.3° \approx 1972.3$ lb; $F_y = 3500 \cos 34.3$ 2891.3 lb

80. Let h be height of the tower and x be the distance from the base of the tower to the closer anchor. Then $\tan 23.7° = \frac{h}{x+35}$ and $\tan 36.4° = \frac{h}{x}$; $\tan 23.7° \approx 0.439$; $\tan 36.4 \approx 0.737$, so $0.439(x+35) = h$ and $0.737x = h$; by substitution $0.439(x+35) = 0.737x$; $0.439x + 15.365 = 0.737x$; $15.365 = 0.298x$; $x = \frac{15.365}{0.298} \approx 51.56$; $h = x \tan 36.4$; $h = 51.56 \tan 36.4 \approx 38.0$ m

CHAPTER 9 TEST

1. $50° \times \frac{\pi}{180°} = \frac{5\pi}{18}$ or $50 \times 0.01745 = 0.8725$

2. $\frac{4\pi}{3} = \frac{4\pi}{3} \cdot \frac{180°}{\pi} = \frac{4}{3} \cdot 180° = 4(60°) = 240°$

3. (a) III;
(b) $237° - 180° = 57°$

4. $r = \sqrt{(-5)^2 + 12^2} = 13$; $\sin \theta = \frac{12}{13}$; $\cos \theta = \frac{-5}{13}$; $\tan \theta = \frac{-12}{5}$; $\csc \theta = \frac{13}{12}$; $\sec \theta = \frac{-13}{5}$; $\cot \theta = \frac{-5}{12}$

5. $b = \sqrt{c^2 - a^2} = \sqrt{23.4^2 - 9^2} = \sqrt{466.56} = 21.6$;
(a) $\sin \theta = \frac{9}{23.4} \approx 0.38462$;
(b) $\tan \theta = \frac{9}{21.6} \approx 0.41667$

6. Here the side opposite $\theta = 30$ and the side adjacent to θ is 6.75. So, the hypotenuse $r = \sqrt{30^2 + 6.75^2} = 30.75$;
(a) $\tan \theta = \frac{1}{\cot \theta} = \frac{30}{6.75} \approx 4.44444$;
(b) $\sin \theta = \frac{30}{30.75} \approx 0.97561$

7. Using a calculator, we get
(a) 0.79864;
(b) 2.47509;
(c) -1.66164; (d) -0.49026 [In (d) make sure that your calculator is in radian mode.]

8. (a) $\theta_r = \tan^{-1} 1.2 \approx 50.2°$; $\theta_{III} = 180° + \theta_r = 230.2°$;
(b) $\theta_r = \sin^{-1} 0.72 = 46.1$; $\theta_{III} = 180° + \theta_r = 226.1°$; $\theta_{IV} = 360° - \theta_r = 313.9°$

9. $\tan 53° = \frac{h}{135}$; $h = 135 \tan 53° \approx 179.15$ ft

10. One revolution is π radians, so it rotates $3600 \times 2\pi = 7200\pi$ radians per minute. There are 60 seconds in one minute, so it rotates $7200\pi \div 60 = 120\pi$ radians per second.

CHAPTER

10

Vectors and Trigonometric Functions

≡ 10.1 INTRODUCTION TO VECTORS

1.

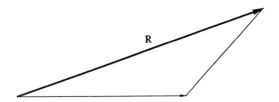

3.

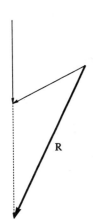

5.

7.

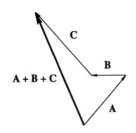

9.

11.

13.

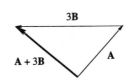

15.

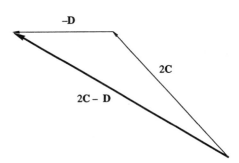

17.

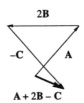

19.

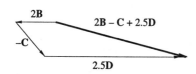

21. $P_x = 20 \cos 75° = 5.176$; $P_y = 20 \sin 75° = 19.319$

23. $P_x = 18.4 \cos 4.97 \text{ rad} = 4.688$; $P_y = 18.4 \sin 4.97 = -17.793$

25. $V_x = 9.75 \cos 16° = 9.372$; $V_y = 9.75 \sin 16° = 2.687$

27. $A = \sqrt{(-9)^2 + 12^2} = 15$; θ is in Quadrant II; $\theta_{Ref} = \tan^{-1} \frac{12}{9} = 52.13$; $\theta = 180° - 53.13° = 126.87°$

29. $C = \sqrt{8^2 + 15^2} = 17$; θ is in Quadrant I; $\theta_{Ref} = \tan^{-1} \frac{15}{8} = 61.93°$; $\theta = 61.93°$

31. $v = \sqrt{5^2 + 12^2} = 13 \text{ km/hr}$

33. $V_x = 976 \cos 72.4° = 295 \text{ N}$; $V_y = 976 \sin 72.4° = 930 \text{ N}$

35. Parallel to ramp $153 \times \sin 12° = 31.8$ lb. Perpendicular to ramp $153 \cos 12° = 149.7$ lb.

37. $V = \sqrt{15^2 + (8 - 17)^2} = 17.5 \text{ V}$; $\phi = \tan^{-1} \frac{-9}{5} \approx -30.96°$ or $30.96°$ from V_R towards $(V_L - V_C)$.

≣ 10.2 ADDING AND SUBTRACTING VECTORS

1. $R = |20.0 - 32.5| = 12.5$; $\theta_R =$ angle of larger so $180°$

3. $R = |121.7 - 86.9| = 34.8$; $\theta_R = 270°$ (angle of the larger magnitude)

5. $R = \sqrt{55^2 + 48^2} = 73.$; $\theta = \tan^{-1} \frac{55}{48} = 48.9$; since it's in Quadrant II $\theta_R = 180° - 48.9° = 131.1°$. (Note **C** is a vertical vector, **D** is a horizontal vector. $\theta_R = \tan^{-1} \frac{\text{vertical}}{\text{horizontal}}$)

7. $R = \sqrt{81.4^2 + 37.6^2} = 89.7$; $\theta_R = \tan^{-1} \frac{37.6}{81.4} = 24.79°$

9. $A = \sqrt{33^2 + 56^2} = 65$; $\theta_A = \tan^{-1} \frac{33}{56} = 59.49°$

11. $C = \sqrt{11.7^2 + 4.4^2} = 12.5$; $\theta_C = \tan^{-1} \frac{11.7}{4.4} = 20.61°$

13. $E = \sqrt{6.3^2 + 1.6^2} = 6.5$; $\theta_E = \tan^{-1} \frac{6.3}{1.6} = 14.25°$

15. $G = \sqrt{8.4^2 + 12.6^2} = 15.1$; $\theta_G = \tan^{-1} \frac{8.4}{12.6} = 56.31°$

17. From these values for R_x and R_y we can determine $R = \sqrt{R_x^2 + R_y^2} = 12.33$ and $\theta_R = \tan^{-1} \frac{R_y}{R_x} = 32.03°$

vector	horizontal component	vertical component
A	$A_x = 4 \cos 60° = 2.000$	$A_y = 4 \sin 60° = 3.464$
B	$B_x = 9 \cos 20° = 8.457$	$B_y = 9 \sin 20° = 3.078$
R	$R_x \qquad\qquad 10.457$	$R_y \qquad\qquad 6.542$

19. $R = \sqrt{R_x^2 + R_y^2} = 53.00$; $\theta = \tan^{-1} \frac{R_y}{R_x} = 178.11°$

vector	horizontal component	vertical component
C	$C_x = 28 \cos 120° = -14.000$	$C_y = 28 \sin 120° = 24.249$
D	$D_x = 45 \cos 210° = -38.971$	$D_y = 45 \sin 210° = -22.500$
R	$R_x \qquad\qquad -52.971$	$R_y \qquad\qquad 1.749$

21. $R = \sqrt{R_x^2 + R_y^2} = 88.50$; $\theta = \tan^{-1} \frac{R_y}{R_x} = 223.83°$

vector	horizontal component	vertical component
A	$A_x = 31.2 \cos 197.5° = -29.756$	$A_y = 31.2 \sin 197.5° = -9.382$
B	$B_x = 62.1 \cos 236.7° = -34.094$	$B_y = 62.1 \sin 236.7° = -51.904$
R	$R_x \qquad\qquad -63.850$	$R_y \qquad\qquad -61.286$

23. $R = \sqrt{R_x^2 + R_y^2} = 22.44$; $\theta = \tan^{-1} \frac{R_y}{R_x} = 348.90°$

vector	horizontal component	vertical component
E	$E_x = 12.52 \cos 46.4° = 8.634$	$E_y = 12.52 \sin 46.4° = 9.067$
F	$F_x = 18.93 \cos 315° = 13.386$	$F_y = 18.93 \sin 315° = -13.386$
R	$R_x \qquad\qquad 22.020$	$R_y \qquad\qquad -4.319$

25. $R = \sqrt{R_x^2 + R_y^2} = 13.04$; $\theta = \tan^{-1} \frac{R_y}{R_x} = 150.30°$

vector	horizontal component	vertical component
A	$A_x = 9.84 \cos 215.5° = -8.011$	$A_y = 9.84 \sin 215.5° = -5.714$
B	$B_x = 12.62 \cos 105.25° = -3.319$	$B_y = 12.62 \sin 105.25° = 12.176$
R	$R_x \qquad\qquad -11.330$	$R_y \qquad\qquad 6.462$

27. $R = \sqrt{R_x^2 + R_y^2} = 63.58;$

$\theta = \tan^{-1} \frac{R_y}{R_x} = 4.62°$

vector	horizontal component	vertical component
E	$\mathbf{E_x} = 42.0 \cos 3.4 = -40.606$	$\mathbf{E_y} = 42.0 \sin 3.4 = -10.733$
F	$\mathbf{F_x} = 63.2 \cos 5.3 = 35.036$	$\mathbf{F_y} = 63.2 \sin 5.3 = -52.599$
R	$\mathbf{R_x} -5.570$	$\mathbf{R_y} -63.332$

29. $R = \sqrt{R_x^2 + R_y^2} = 13.13;$

$\theta = \tan^{-1} \frac{R_y}{R_x} = 64.41°$

vector	horizontal component	vertical component
A	$\mathbf{A_x} = 15 \cos 25° = 13.59$	$\mathbf{A_y} = 15 \sin 25° = 6.34$
B	$\mathbf{B_x} = 20 \cos 120° = -10.00$	$\mathbf{B_y} = 20 \sin 120° = 17.32$
C	$\mathbf{C_x} = 12 \cos 280° = 2.08$	$\mathbf{C_y} = 12 \sin 280° = -11.82$
R	$\mathbf{R_x} 5.67$	$\mathbf{R_y} 11.84$

31. Draw the coordinate axes and label the angles as shown in the figure below. The origin is placed at the ring where the three cables meet, because the tension on all three cables acts on this ring.

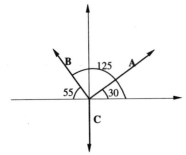

We use the following table listing the horizontal and vertical components of each vector.

vector	horizontal component	vertical component
A	$\mathbf{A_x} = A \cos 30°$	$\mathbf{A_y} = A \sin 30°$
B	$\mathbf{B_x} = B \cos 125°$	$\mathbf{B_y} = B \sin 125°$
C	$\mathbf{C_x} = C \cos 270° = 0$	$\mathbf{C_y} = C \sin 270° = -C = -215$
R	$\mathbf{R_x}$	$\mathbf{R_y}$

Since, $R_x = A \cos 30° + B \cos 125° = 0$ and $R_y = A \sin 30° + B \sin 125° - 215 = 0$, we are led to the system of two equations in two variables.

$$\begin{cases} 0.86603A - 0.57358B = 0 \\ 0.5A + 0.81915B = 215 \end{cases}$$

Using Cramer's rule, we determine that $A \approx 123.8$ N and $B = 186.9$ N. The tension in C is 215 kg $\times 9.8$ m/s$^2 \approx$ 2107 N.

33. Use a diagram similar to the one used to solve the previous exercise. This should lead to the following table listing the horizontal and vertical components of each vector.

vector	horizontal component	vertical component
A	$\mathbf{A_x} = A \cos 0° = A$	$\mathbf{A_y} = A \sin 0° = 0$
B	$\mathbf{B_x} = B \cos 143°$	$\mathbf{B_y} = B \sin 143°$
S	$\mathbf{S_x} = 235 \cos 270° = 0$	$\mathbf{S_y} = 235 \sin 270° = -235$
R	$\mathbf{R_x}$	$\mathbf{R_y}$

Here $R_x = A + B \cos 143° + 0 = 0$ and $R_y = 0 + B \sin 143° - 235 = 0$. From the second equation we obtain $0.6018B - 235 = 0$ and so, $B = \frac{235}{0.6018} \approx 390.5$. Substituting this value in the first equation produces $A + (390.5) \cos 143° = 0$ and this leads to $A \approx 311.9$. Thus, the brace exerts a force of 311.9 lb and the cable a force of 390.5 lb.

35. The components for the airplane are $A_x = 480 \cos 27° = 427.68$, $A_y = 480 \sin 27° = 217.92$; while the components for the wind are $W_x = 45 \cos -55° = 25.81$, $W_y = 45 \sin -55° = -36.86$. Using these, we can determine that the components for the resultant are $R_x = 427.68 + 25.81 = 453.49$, $R_y = 217.92 - 36.86 = 181.06$ and the length of the component produces a ground speed of $R = \sqrt{R_x^2 + R_y^2} = 488.3$ mph. To determine the course and drift angle, we first find that $\theta = \tan^{-1}\frac{R_y}{R_x} = 21.8°$. Hence the course is $90 - 21.8 = 68.2°$ and the drift angle is $68.2 - 63 = 5.2°$.

37. We will assume that one voltage, vector **A**, is horizontal and that the other voltage, **B**, makes an angle of 68° with the horizontal vector. We can form the following component table.

vector	horizontal component		vertical component	
A	$A_x = 196 \cos 0° =$	196.0	$A_y = 196 \sin 0° =$	0
B	$B_x = 196 \cos 68° \approx$	73.4	$B_y = 196 \sin 68° \approx$	181.7
R	R_x	269.4	R_y	181.7

Using the value of R_x and R_y, we see that

$$R = \sqrt{R_x^2 + R_y^2}$$
$$= \sqrt{269.4^2 + 181.7^2} \approx 324.9$$

The total voltage is about 324.9 V.

39. See *Computer Programs* in main text.

10.3 APPLICATIONS OF VECTORS

1. Let the 70 lb force be at 0° and the 50 lb force at 35°. Then
$R = \sqrt{R_x^2 + R_y^2} = 114.60$ and
$\theta_R = \tan^{-1}\frac{R_y}{R_x} = 14.49°$ from the 70 lb force.

vector	horizontal component		vertical component	
A	$A_x = 70 \cos 0° =$	70.000	$A_y = 70 \sin 0° =$	0.000
B	$B_x = 50 \cos 35° =$	40.958	$B_y = 50 \sin 35° =$	28.679
R	R_x	110.958	R_y	28.679

3. (a) Downward force is $25 \sin 55° = 20.48$ lb.
(b) Horizontal force is $25 \cos 55° = 14.34$ lb.
(c) for a 40° angle, Downward force $= 25 \sin 40° = 16.07$ lb, Horizontal force is $25 \cos 40° = 19.15$ lb.

5. The forward force is $F_f = 1700 \cos 18° = 1616.80$ N; and the sideward force is $F_s = 1700 \sin 18° = 525.33$ N.

7. Since the heading and wind are perpendicular, the ground speed is $\sqrt{370^2 + 40^2} = 372.16$ mi/h. The angle east of south is $\theta = \tan^{-1}\frac{40}{370} = 6.17°$. This is a compass direction of $180° - 6.17° = 173.83°$.

9. Parallel component is $1200 \sin 22° = 449.53$ kg. Perpendicular component is $1200 \cos 22° = 1112.62$ kg.

11. The angle the guy wire makes with the ground is $\theta = \tan^{-1}\frac{35}{27} = 52.352°$. Horizontal tension $T_x = 195 \cos 52.352° = 119.11$ lb. Vertical-downward $T_y = 195 \sin 52.352° = 154.40$ lb.

13. $I_C + I_R = I$; $I_C = I - I_R$; $I_C^2 = 9.6^2 - 7.5^2$, $I_C = 5.99$ or 6 A; $\phi = \tan^{-1}\frac{I_C}{I_R} = 38.66°$.

15. $I_L = \sqrt{I^2 - I_R^2} = \sqrt{12.4^2 - 6.3^2} = 10.68$;
$\theta = -\tan^{-1}\frac{I_L}{I_R} = -\tan^{-1}\frac{10.68}{6.3} = -59.46°$

17.

vector	horizontal component	vertical component
X_C	$X_{Cx} = 0$	$X_{Cy} = -60$
X_L	$X_{Lx} = 0$	$X_{Ly} = 40$
R	$R_x = 24$	$R_y = 0$
Z	$Z_x = 24$	$Z_y = -20$

$Z = \sqrt{Z_x^2 + Z_y^2} = \sqrt{24^2 + -20^2} = 31.24$;
$\theta = \tan^{-1}\frac{-20}{24} = -39.81°$

19. $I = \sqrt{I_x^2 + I_y^2} = 55.84$;
$\theta = \tan^{-1}\frac{I_y}{I_x} = \tan^{-1}\frac{6.342}{55.478} = 6.52°$

vector	horizontal component		vertical component	
I_A	$I_{Ax} = 20 \cos 35° =$	16.383	$I_{Ay} = 20 \sin 35° =$	11.472
I_B	$I_{Bx} = 15 \cos -20° =$	14.095	$I_{By} = 15 \sin -20° =$	-5.130
I_C	I_{Cx}	$= 25.000$	I_{Cy}	$= 0.000$
I	I_x	$= 55.478$	I_y	$= 6.342$

21. $V_x = 120$ m/s; $V_y = -9.8 \cdot 4$ m/s $= -39.2$ m/s;

$V = \sqrt{V_x^2 + V_y^2} = 126.24$ m/s;

$\theta = \tan^{-1} \frac{V_y}{V_x} = \tan^{-1} \frac{-39.2}{120} = -18.09°$

23. We will let the longer side be vector **A** and assume that it is horizontal. The other (shorter) side will be **B** and it makes an angle of 40° with the horizontal vector. Before we make the component table, we will convert $1'9''$ to $12 + 9 = 21''$ and $2'3''$ to $2 \cdot 12 + 3 = 24 + 3 = 27''$. We now form the following component table.

vector	horizontal component		vertical component	
A	$A_x = 27 \cos 0° =$	27.0	$A_y = 27 \sin 0° =$	0
B	$B_x = 21 \cos 40° \approx$	16.1	$B_y = 21 \sin 40° \approx$	13.5
R	R_x	43.1	R_y	13.5

Thus, the smallest rectangle that can be used is one that measures 43.1 in. $\times$ 13.5 in. or $3'7\frac{1}{10}'' \times 1'1\frac{1}{2}''$.

≡ 10.4 OBLIQUE TRIANGLES: LAW OF SINES

1. $A = 19.4°$; $B = 85.3°$; $c = 22.1$; $C = 180° - A - B = 180° - 19.4° - 85.3° = 75.3°$; $\frac{a}{\sin A} = \frac{c}{\sin C}$; $\frac{a}{\sin 19.4°} = \frac{(22.1)}{\sin 75.3°}$; $a = \frac{(22.1)\sin 19.4°}{\sin 75.3°} = 7.59$; $\frac{b}{\sin B} = \frac{c}{\sin C}$; $\frac{b}{\sin 85.3°} = \frac{22.1}{\sin 75.3°}$; $b = \frac{(22.1)\sin 85.3°}{\sin 75.3°} = 22.77$. Summary: $C = 75.3°$, $a = 7.59$, $b = 22.77$

3. $a = 14.2$; $b = 15.3$; $B = 97°$; $\frac{a}{\sin A} = \frac{b}{\sin B}$; $\sin A = \frac{a \sin B}{b} = \frac{14.2 \sin 97°}{15.3}$; $\sin A = .9211866$; $A = 67.1°$; Since B is obtuse, A must be acute. $C = 180° - A - B = 180° - 67.1° - 97° = 15.9°$; $\frac{c}{\sin C} = \frac{b}{\sin B}$; $c = \frac{15.3 \sin 15.9°}{\sin 97°} = 4.22$; $A = 67.1°$; $C = 15.9°$; $c = 4.22$

5. $A = 86.32°$, $a = 19.19$, $c = 18.42$; $\frac{c}{\sin C} = \frac{a}{\sin A}$; $\sin C = \frac{18.42 \sin 86.32}{19.19}$; $\sin C = 0.9578958$, $C_1 = 73.31°$; $C_2 = 180° - C_1 = 106.69°$; but $C_2 + A > 180°$ is not possible. $B = 180° - A - C = 180° - 86.32° - 73.31° = 20.37°$; $\frac{b}{\sin B} = \frac{a}{\sin A}$; $b = \frac{19.19 \sin 20.37}{\sin 86.32} = 6.69$; $C = 73.31°$; $B = 20.37°$; $b = 6.69$

7. $B = 39.4°$, $b = 19.4$, $c = 35.2$; $\frac{c}{\sin C} = \frac{b}{\sin B}$; $\sin C = \frac{c \sin B}{b} = \frac{35.2 \sin 39.4}{19.4}$; $\sin C = 1.15$. Since the range of the sine function is $[-1, 1]$, there is no solution, and so this is not a triangle.

9. $A = 45°$, $a = 16.3$, $b = 19.4$; $\frac{b}{\sin B} = \frac{a}{\sin A}$; $\sin B = \frac{b \sin A}{a} = \frac{19.4 \sin 45°}{16.3}$; $\sin B = 0.841587$; $B_1 = 57.31°$; $B_2 = 180° - B_1 = 122.69°$. $C_1 = 180° - A - B_1 = 180° - 45° - 57.31° = 77.69°$. $C_2 = 180° - A - B_2 = 180° - 45° - 122.69° = 12.31°$. $\frac{c_1}{\sin C_1} = \frac{a}{\sin A}$; $c_1 = \frac{16.3 \sin 77.69°}{\sin 45°} = 22.52$. $\frac{c_2}{\sin C_2} = \frac{a}{\sin A}$; $c_2 = \frac{16.3 \sin 12.31}{\sin 45} = 4.91$; $B_1 = 57.31°$; $C_1 = 77.69°$; $c_1 = 22.52$; $B_2 = 122.69°$; $C_2 = 12.31°$; $c_2 = 4.91$.

11. $a = 42.3$, $B = 14.3°$, $C = 16.9°$; $A = 180° - B - C = 180° - 14.3° - 16.9° = 148.8°$; $\frac{b}{\sin B} = \frac{a}{\sin A}$; $b = \frac{42.3 \sin 14.3}{\sin 148.8} = 20.17$; $\frac{c}{\sin C} = \frac{a}{\sin A}$; $c = \frac{42.3 \sin 16.9}{\sin 148.8} = 23.74$. Summary: $A = 148.8°$, $b = 20.17$, $c = 23.74$

13. $b = 19.4$, $c = 12.5$, $C = 35.6°$; $\frac{b}{\sin B} = \frac{c}{\sin C}$; $\sin B = \frac{b \sin C}{c} = \frac{19.4 \sin 35.6}{12.5}$; $\sin B = 0.9034549$; $B_1 = 64.62°$; $B_2 = 180° - B_1 = 115.38°$; $A_1 = 180° - B_1 - C = 180° - 64.62° - 35.6° = 79.78°$; $A_2 = 180° - B_2 - C = 180° - 115.38° - 35.6° = 29.02°$; $a_1 = \frac{c \sin A_1}{\sin C} = \frac{12.5 \sin 79.78}{\sin 35.6°} = 21.13$; $a_2 = \frac{c \sin A_2}{\sin C} = \frac{12.5 \sin 29.02}{\sin 35.6°} = 10.42$. Summary: $B_1 = 64.62$, $A_1 = 79.78$, $a_1 = 21.13$, or $B_2 = 115.38$, $A_2 = 29.02$, $a_2 = 10.42$.

15. $a = 19.7$, $b = 8.5$, $B = 78.4°$; $\sin A = \frac{a \sin B}{b} = \frac{19.7 \sin 78.4}{8.5} = 2.27$. Since 2.27 is not in the range of the sine function, there is no solution.

17. $b = 8.5$, $c = 19.7$, $C = 1.37$ rad; $\sin B = \frac{b \sin C}{c} = \frac{8.5 \sin 1.37}{19.7} = .4228$; $B_1 = 0.44$ rad; $B_2 = 3.14 - 0.44 = 2.7$; $A_1 = \pi - B_1 - C = 3.14 - .44 - 1.37 = 1.33$; $A_2 = \pi - B_2 - C = 3.14 - 2.7 - 1.37 = -0.93$ cannot be so only $B_1 + A_1$ are answers. $a = \frac{c \sin A}{\sin C} = \frac{19.7 \sin 1.33}{\sin 1.37} = 19.52$: $B = 0.44$, $A = 1.33$, $a = 19.52$.

19. $A = 0.47$ rad, $b = 195.4$, $C = 1.32$ rad; $B = \pi - A - C = 3.14 - 0.47 - 1.32 = 1.35$. $a = \frac{b \sin A}{\sin B} = \frac{195.4 \sin 0.47}{\sin 1.35} = 90.7$; $c = \frac{b \sin C}{\sin B} = \frac{195.4 \sin 1.32}{\sin 1.35} = 194.0$: $B = 1.35$, $a = 90.7$, $c = 194.0$.

21. The length between towers A and B is $AB = c = 360$ m. The angle at tower $B = 67.4°$ and at tower $A = 49.3°$. Hence, the angle at tower $C = 180° - A - B = 63.3°$. The distance between towers A and C is $AC = b = \frac{c \sin B}{\sin C} = \frac{360 \sin 67.4}{\sin 63.3} = 372.0$ m and the distance between towers B and C is $BC = a = \frac{c \sin A}{\sin C} + \frac{360 \sin 49.3}{\sin 63.3} = 305.5$ m.

23. Let C designate the present location of the airplane. Since the plane has a heading of $313°$, then $\angle A$ of $\triangle ABC$ is $43°$. We are given that the heading from B is $27°$ and so $B = 90° - 27° = 63°$. Using these two angles, we find that $C = 180° - 43° - 63° = 74°$. Now, using the Sine law, if the distance the plane has flown is b, then we have $\frac{b}{\sin B} = \frac{c}{\sin C}$. We were given that $c = 37$ mi, and so we have $b = \frac{c \sin B}{\sin C} = \frac{37 \sin 63°}{\sin 74°} \approx 34.3$ mi.

25. $c = 180° - 47° - 67° = 66°; CB = a = \frac{400 \sin 47°}{\sin 66°} = 320.22; \sin 67° = \frac{h}{a}; h = 320.22 \sin 67° = 294.77$ m.

27. Let h be the height of the building above the eye of the observer and d the distance of the observer from the building. Then, $\tan 57.8° = \frac{h}{d}$

and $\tan 67.4° = \frac{h+225}{d}$. Hence $d = \frac{h}{\tan 57.8°} = \frac{h+225}{\tan 67.4°}$. Solving $\frac{h}{\tan 57.8°} = \frac{h}{\tan 67.4°} + \frac{225}{\tan 67.4°}$ or $h\left(\frac{1}{\tan 57.8°} - \frac{1}{\tan 67.4°}\right) = \frac{225}{\tan 67.4°}$ or $h = 438.74$ ft or $438'9''$. Adding the height of the observer's eyes, we get $438'9'' + 5'7'' = 444'4''$.

29. Call the center of the circle C and the two adjacent holes A and B. Since the holes are equally spaced, $\angle ACB = \frac{360°}{12} = 30°$. A and B are on a circle centered at C, so $\overline{CA} = \overline{CB} = 16.40$ cm and $\triangle ABC$ is an isosceles triangle. Thus, $\angle CAB = \angle CBA = \frac{180° - 30°}{2} = 75°$. We can use the Sine Law.

$$\frac{AB}{\sin C} = \frac{AC}{\sin B}$$
$$AB = \frac{AC \sin C}{\sin B}$$
$$= \frac{16.40 \sin 30°}{\sin 75°} \approx 6.42$$

The holes are about 8.49 cm apart.

10.5 OBLIQUE TRIANGLES: LAW OF COSINES

1. $a = 9.3, b = 16.3, C = 42.3°$ SAS so Cosine Law. First we find c by using the cosine law: $c^2 = a^2 + b^2 - 2ab \cos C = 9.3^2 + 16.3^2 - (9.3)(16.3) \cos 42.3°$, $c^2 = 127.9386, c = 11.3$. Next find the angle opposite the short side, namely A using Law of Sines. $\frac{a}{\sin A} = \frac{c}{\sin C}; \sin A = \frac{a \sin C}{c} = \frac{9.3 \sin 42.3}{11.3}; \sin A = 0.553895, A = 33.6°, B = 180° - A - C = 180° - 42.3° - 33.6° = 104.1°$. (Note: If you use the Law of Sines to find B, you may not recognize that it is obtuse. That's why you should find the angle opposite the shorter side first since it must be acute.) $c = 11.31, A = 33.6°, B = 104.1°$.

3. $a = 47.85, B = 113.7°, c = 32.79; b^2 = a^2 + c^2 - 2ac \cos B = 47.85^2 + 32.79^2 - 2(47.85)(32.79) \cos 113.7° = 4626.1199$; hence, $b = 68.02$ and so, $\sin A = \frac{a \sin B}{b} = \frac{47.85 \sin 113.7°}{68.02} = 0.64414$. Thus, $A = 40.1°. C = 180° - A - B = 180° - 40.1° - 113.7° = 26.2°. b = 68.02, A = 40.1°, C = 26.2°$: (Note: Since B is obtuse, A and C must be acute).

5. $a = 29.43, b = 16.37, c = 38.62$: Use the alternate form of Law of Cosines to find one of the angles.

Choose C since it's the largest and if it's obtuse, the Law of Cosines will yield this answer directly. $\cos C = \frac{a^2 + b^2 - c^2}{2ab} = \frac{29.43^2 + 16.37^2 - 38.62^2}{2(29.43)(16.37)}; \cos C \approx -0.37093; C \approx 111.8°; \sin A = \frac{a \sin C}{c} = \frac{29.43 \sin 111.8°}{38.62} \approx 0.7075; A = 45.0°; B = 180° - A - C = 180° - 45.0° - 111.8° = 23.2°: A = 45.0°, B = 23.2°, C = 111.8°$.

7. $a = 63.92, B = 92.44°, c = 78.41; b^2 = a^2 + c^2 - 2ac \cos B = 63.92^2 + 78.41^2 - 2(63.92)(78.41) \cos 92.44° = 10660.645; b = 103.25. \sin A = \frac{a \sin B}{b} = \frac{63.92 \sin 92.44}{103.25} = .6185186; A = 38.21. C = 180° - A - B = 180° - 38.21° - 92.44° = 49.35°: b = 103.25, A = 38.21°, C = 49.35°$

9. $a = 4.527, b = 6.239, c = 8.635$: Find C first, $\cos C = \frac{a^2 + b^2 - c^2}{2ab} = \frac{4.527^2 + 6.239^2 - 8.635^2}{2(4.527)(6.239)}; \cos C = -0.268099; C = 105.55° \sin A = \frac{a \sin C}{c} = \frac{4.527 \sin 105.55°}{8.635} = 0.505072; A = 30.34. B = 180° - A - C = 180° - 30.34° - 105.55° = 44.11°: A = 30.34°, B = 44.11°, C = 105.55°$.

11. $a = 8.5, b = 15.8, C = 0.82$ rad; SAS, $c^2 = a^2 + b^2 - 2ab\cos C = 138.6454$; $c = 11.77$; $\sin A = \frac{a\sin C}{c} = .5280$; $A = 0.56$ rad. $B = \pi - A - C = 3.14 - 0.82 - 0.56 = 1.76$ rad: $c = 11.77, A = 0.56$ rad, $B = 1.76$ rad.

13. $a = 52.65, B = 1.98, c = 35.8$; $b^2 = a^2 + c^2 - 2ac\cos B = 5553.5624$; $b = 74.52$. $\sin C = \frac{c\sin B}{b} = .440744, C = 0.46$; $A = \pi - B - C = 3.14 - 1.98 - 0.46 = 0.70$: $b = 74.52, A = 0.70, C = 0.46$.

15. $a = 36.27, b = 24.55, c = 44.26$; $\cos C = \frac{a^2+b^2-c^2}{2ab} = -0.02287$; $C = 1.59$ rad. $\sin A = \frac{a\sin C}{c} = 0.8193$; $A = 0.96$. $B = \pi - A - C = 0.59$: $A = 0.96, B = 0.59, C = 1.59$

17. $a = 54.8, B = 1.625, c = 38.33$; SAS; $b^2 = a^2 + c^2 - 2ac\cos B = 4699.8253$; $b = 68.56$. $\sin A = \frac{a\sin B}{b} = 0.7981$; $A = 0.924$. $C = \pi - A - B = 3.142 - 0.924 - 1.625 = 0.593$: $b = 68.56, A = 0.924, C = 0.593$

19. $a = 2.317, b = 1.713, c = 1.525$; SSS, $\cos A = (b^2+c^2-a^2)/2bc = -0.020766, A = 1.59$; $\sin B =$

$\frac{b\sin A}{a} = .7392, B = 0.83$; $C = \pi - A - B = 3.14 - 1.59 - 0.83 = 0.72$: $A = 1.59, B = 0.83, C = 0.72$

21. From the diagram we know that $a = 90 + 34 = 124.7°$, $\overline{AC} = b = 6335, \overline{AB} = c = 12325, \overline{BC} = a$; $a^2 = b^2 + c^2 - 2bc\cos A = 6335^2 + 12525^2 - 2(6335)(12325)\cos 124.7° = 280\,935\,259.5$ and so $a = 16\,761$. Distance above the Earth's surface is $16\,761 - 6\,335 = 10\,426$ km.

23. The angle between the antenna and the hill is $90 + 12.37° = 102.37°$ If the length of the wire is ℓ then $\ell^2 = 40^2 + 75^2 - 2 \cdot 40 \cdot 75\cos 102.37°$; $\ell^2 = 8510.34$; $\ell = 92.25$ ft.

25. $R^2 = 12^2 + 23^2 - 2 \cdot 12 \cdot 23\cos 121.27° = 959.53$; $R = \sqrt{959.53} = 30.98$ or 31 N. $\theta = \sin^{-1}\left(\frac{23\sin 121.27}{31}\right) = 39.36°$

27. The other angle of the parallelogram is $180° - 74° = 106°$. By the Law of Cosines, $Z^2 = 56^2 + 38^2 - 2 \cdot 56 \cdot 38\cos 106° = 5753.11$. Taking the square root, we find that $Z = \sqrt{5753.11} \approx 75.8$ or about $76\,k\Omega$.

CHAPTER 10 REVIEW

1.

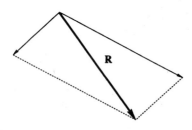

3.

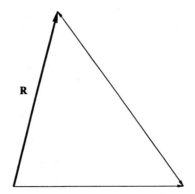

2.

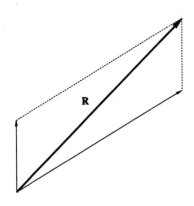

4.

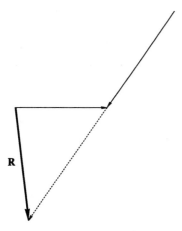

5. $P_x = 35\cos 67° = 13.68; P_y = 35\sin 67° = 32.22$

6. $P_x = 19.7\cos 237° = -10.73; P_y = 19.7\sin 237° = -16.52$

7. $P_x = 23.4\cos 172.4° = -23.19; P_y = 23.4\sin 172.4° = 3.09$

8. $P_x = 14.5\cos 338° = 13.44; P_y = 14.5\sin 338° = -5.43$

9. $A = \sqrt{A_x^2 + A_y^2} = \sqrt{16^2 + (-8)^2} = 17.89. \theta = \tan^{-1}\left(\frac{A_y}{A_x}\right) = \tan^{-1}\frac{-8}{16} = \tan^{-1}-\frac{1}{2} = -26.57°$ or $333.43°$

10. $B = \sqrt{B_x^2 + B_y^2} = \sqrt{(-27)^2 + 32^2} = 41.87;$ $\theta_R = \tan^{-1}\frac{32}{27} = 49.84$ Second Quadrant so $180° - 49.84° = 130.16°$

11. $A_x = 38\cos 15° = 36.71; A_y = 38\sin 15° = 9.84$

12. $B_x = 43.5\cos 127° = -26.18; A_y = 43.5\sin 127° = 34.74$

13. $C_x = 19.4\cos 1.25 = 6.12;$ $C_y = 19.4\sin 1.25 = 18.41$

14. $D_x = 62.7\cos 5.37 = 38.32; D_y = 62.7\sin 5.37 = -49.62$

15. $R = \sqrt{R_x^2 + R_y^2} = 50.41;$ $\theta = \tan^{-1}\frac{R_y}{R_x} = 20.69°$

vector	horizontal component	vertical component
A	$A_x = 19\cos 32° = 16.11$	$A_y = 19\sin 32° = 10.07$
B	$B_x = 32\cos 14° = 31.05$	$B_y = 32\sin 14° = 7.74$
R	R_x \qquad\qquad 47.16	R_y \qquad\qquad 17.81

16. $R = \sqrt{R_x^2 + R_y^2} = 36.96;$ $\theta_R = \tan^{-1}\frac{5.88}{36.49} = 9.15°$. Since θ is in 4th Quadrant $\theta = 360 - 9.15 = 350.85°$

vector	horizontal component	vertical component
C	$C_x = 24\cos 57° = 13.07$	$C_y = 24\sin 57° = 20.13$
D	$D_x = 35\cos 312° = 23.42$	$D_y = 35\sin 312° = -26.01$
R	R_x \qquad\qquad 36.49	R_y \qquad\qquad -5.88

17. $R = \sqrt{R_x^2 + R_y^2} = 72.77;$ $\theta_R = \tan^{-1}\frac{92}{72.76} = .01264. \theta$ in 2nd Quadrant so $\theta = \pi - \theta_R = 3.13$ rad

vector	horizontal component	vertical component
E	$E_x = 52.6\cos 2.53 = -43.07$	$E_y = 52.6\sin 2.53 = 30.20$
F	$F_x = 41.7\cos 3.92 = -29.69$	$F_y = 41.7\sin 3.92 = -29.28$
R	R_x \qquad\qquad -72.76	R_y \qquad\qquad 0.92

18. $R = \sqrt{R_x^2 + R_y^2} = 57.67;$ $\theta_R = \tan^{-1}\frac{57.57}{3.41} = 1.51. \theta$ is in 3rd Quadrant so $\theta = \pi + 1.51 = 4.65$

vector	horizontal component	vertical component
G	$G_x = 43.7\cos 4.73 = 0.77$	$G_y = 43.7\sin 4.73 = -43.69$
H	$H_x = 14.5\cos 4.42 = -4.18$	$H_y = 14.5\sin 4.42 = -13.88$
R	R_x \qquad\qquad -3.41	R_y \qquad\qquad -57.57

19. $a = 14, b = 32, c = 27$: SSS so use Law of Cosines to find the largest angle, namely $B \cos B = \frac{a^2 + c^2 - b^2}{2ac} = -0.1310; B = 97.5. \sin A = \frac{a\sin B}{b} = .4338; A = 25.7. C = 180° - A - B = 56.8°: A = 25.7°, B = 97.5°, C = 56.8°$

20. $a = 43, b = 52, B = 86.4°$: SSA so Law of Sines. $\sin A = \frac{a\sin B}{b} = .8253; A = 55.6. C = 180° - A - B = 38.0°. c = \frac{b\sin C}{\sin B} = 32.08.$

21. $b = 87.4, B = 19.57°, c = 65.3$: SSA, Sine Law. $\sin C = \frac{c\sin B}{b} = 0.25026; C = 14.49°. A = 180° - B - C = 145.94°. a = \frac{b\sin A}{\sin B} = 146.14. A = 145.94°, a = 146.14, C = 14.49°$

22. $A = 121.3°, b = 42.5, c = 63.7$: SAS, Cosine Law. $a^2 = b^2 + c^2 - 2bc\cos A = 8676.8762; a = 93.15. \sin B = \frac{b\sin A}{a} = 0.3898497; B = 22.9°; C = 180° - A - B = 35.8°: a = 93.15, B = 22.9°, C = 35.8°$

23. $a = 127.35, A = 0.12, b = 132.6$: SSA, Sine Law. $\sin B = \frac{b \sin A}{a} = 0.1246; B = 0.12, C = \pi - A - B = 3.14 - .12 - .12 = 2.90. c = \frac{a \sin C}{\sin A} = 254.51. B = 0.12, C = 2.90, c = 254.51$

24. $b = 84.3, c = 95.4, C = 0.85$: SSA so Sine Law. $\sin B = \frac{b \sin C}{c} = 0.6639; B = 0.73. A = \pi - B - C = 1.56. a = \frac{c \sin A}{\sin C} = 127.0: B = 0.73, A = 1.56, a = 127.0.$

25. $a = 67.9, b = 54.2, C = 2.21$: SAS so Cosine Law. $c^2 = a^2 + b^2 - 2ab \cos C = 11938.92; c = 109.3. \sin B = \frac{b \sin C}{c} = 0.39798; B = 0.41: A = \pi - B - C = 0.52: c = 109.3, A = 0.52, B = 0.41.$

26. $a = 53.1, b = 63.2, c = 74.3$: SSS so use the Cosine Law. $\cos C = (a^2 + b^2 - c^2)/2ab = 0.1927; C = 1.38. \sin A = \frac{a \sin C}{c} = 0.7017; A = 0.78. B = \pi - A - C = 0.98: A = 0.78, B = 0.98, C = 1.38$

27. $R = \sqrt{R_x^2 + R_y^2} = 2831.17$;
$\theta = \tan^{-1} \frac{R_y}{R_x} = 83.97°$.
Resultant force is 2831 kg directed at an angle of 83.97° to the axis of the vehicle.

vector	horizontal component	vertical component
A	$A_x = 1650 \cos 68° = 618.10$	$A_y = 1650 \sin 68° = 1529.85$
B	$B_x = 1325 \cos 76° = -320.55$	$B_y = 1325 \sin 76° = 1285.64$
R	R_x 297.55	R_y 2815.49

28. Call the alternate route a. Then by the Cosine Law we see that $a^2 = 3621^2 + 2342^2 - 2(3621)(2342) \cos 72.4°; a^2 = 13468181; a = 3669.9$ ft. Over the hill $3621 + 2342 = 5963; 2.3 \times 3669.9 = 8440.77$. Over the hill is less expensive.

29. Parallel component is $126.5 \sin 31.7° = 66.47$ lb. Perpendicular component $126.5 \cos 31.7° = 107.63$ lb.

30. $X_C + X_L = -72 + 52 = -20\Omega$ $\theta = \tan^{-1}\left(\frac{-20}{35}\right) = -29.74°$

31. $AB = c, AC = b = 73, A = 123.4, C = 42.1. B = 180° - A - C = 14.5. c = \frac{b \sin C}{\sin B} = 195.47$ m. The distance is 195.47 m.

32. By Cosine Law $d^2 = 1235^2 + 962^2 - 2(1235)(962) \cos 52.57°; d^2 = 1006471; d = 1003.23$ m

▣ CHAPTER 10 TEST

1. $V_x = 47 \cos 117° = -21.34$;
$V_y = 47 \sin 117° = 41.88$

2. $A = \sqrt{A_x^2 + A_y} = \sqrt{12.91^2 + (-14.36)^2} = 19.31$;
$\theta = \tan^{-1} \frac{-14.36}{12.91} = -48.04°$ or $311.96°$

3.

vector	horizontal component	vertical component
A	$A_x = 25 \cos 64° = 10.96$	$A_y = 25 \sin 64° = 22.47$
B	$B_x = 40 \cos 112° = -14.98$	$B_y = 40 \sin 112° = 37.09$
R	R_x -4.02	R_y 59.56

$R = \sqrt{R_x^2 + R_y^2} = 59.70; \theta = \tan^{-1} \frac{59.56}{-4.02} = 93.86°$

4. $b = \frac{a \sin B}{\sin A} = \frac{9.42 \sin 67.5}{\sin 35.6} = 14.95.$

5. $a^2 = b^2 + c^2 - 2bc \cos A = 4.95^2 + 6.24^2 - 2(4.95)(6.24) \cos 13.4; a^2 = 87.9743; a = 9.38.$

6. $A = 24°, b = 36.5$, and $C = 97°: B = 180° - A - C = 59°. a = \frac{b \sin A}{\sin B} = \frac{36.5 \sin 24°}{\sin 59°} = 17.32. c = $

$\frac{b \sin C}{\sin B} = \frac{36.5 \sin 97°}{\sin 59°} = 42.26. B = 59°, a = 17.32, c = 42.26.$

7. $c^2 = 30^2 + 36^2 - 2 \cdot 30 \cdot 36 \cos 97° = 2459.237782; c = 49.59$

CHAPTER

11

Graphs of Trigonometric Functions

≡ 11.1 SINE AND COSINE CURVES: AMPLITUDE AND PERIOD

1. Period $= \frac{2\pi}{2} = \pi$,
Amplitude $= |3| = 3$,
Frequency $= \frac{1}{\pi}$

3. Period $= 2\pi$,
Amplitude $= |2| = 2$,
Frequency $\frac{1}{2\pi}$

5. Period $= \frac{2\pi}{2\pi} = 1$,
Amplitude $= 8$,
Frequency 1

7. Period $= \frac{2\pi}{4} = \frac{\pi}{2}$,
Amplitude $= \frac{1}{2}$,
Frequency $\frac{2}{\pi}$

9. Period $= \frac{2\pi}{\frac{1}{2}} = 4\pi$,
Amplitude $= \frac{1}{3}$,
Frequency $\frac{1}{4\pi}$

11. Period $= \frac{2\pi}{\frac{1}{4}} = 8\pi$,
Amplitude $= |-3| = 3$,
Frequency $\frac{1}{8\pi}$

13. Period $= 2\pi$,
Amplitude $= |-\frac{1}{2}| = \frac{1}{2}$,
Frequency $\frac{1}{2\pi}$

15.

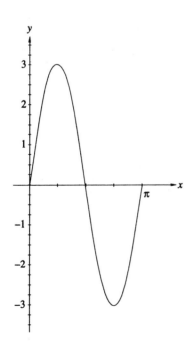

17.

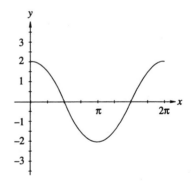

19.

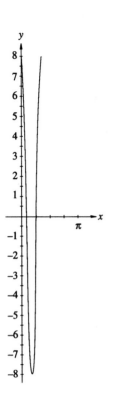

21.

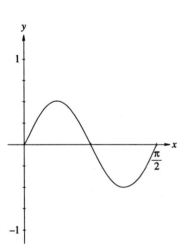

23.

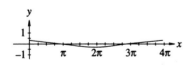

25.

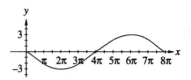

27.

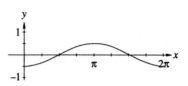

29. Period $= \frac{2\pi}{120\pi} = \frac{1}{60}s$, Amplitude $= 6.5$ A, Frequency 60 Hz

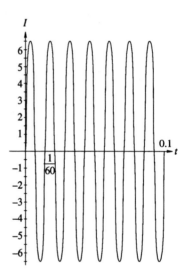

31. The given equation, $G_i = 1094\cos i$, is of the form $E = A\cos Bi$, where $A = 1094$ and $B = 1$.
(a) Amplitude: $|A| = |1094| = 1094$, period: $\frac{2\pi}{|B|} = \frac{2\pi}{|1|} = 2\pi$, frequency: $\frac{1}{\text{period}} = \frac{1}{2\pi}$
(b)

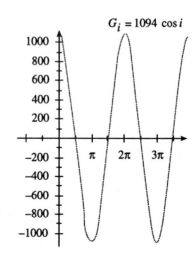

$G_i = 1094\cos i$

☰ 11.2 SINE AND COSINE CURVES: DISPLACEMENT OR PHASE SHIFT

1.   Amplitude $= 2$,
Period $= 2\pi$,
Phase Shift $= -\frac{\pi}{4}$ or $\frac{\pi}{4}$ left,
Vertical displacement $= 0$

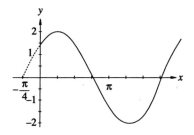

3.   Amplitude $= 2.5$,
Period $= \frac{2\pi}{3}$,
Phase Shift $= \frac{\pi}{3}$ or $\frac{\pi}{3}$ to the right,
Vertical displacement $= 0$

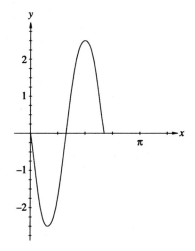

5.   Amplitude $= 6$,
Period $= \frac{360°}{1.5} = 240° = \frac{4\pi}{3}$,
Phase Shift $= -\frac{180°}{1.5} = -120° = -\frac{2\pi}{3}$ to the left,
Vertical displacement $= 0$

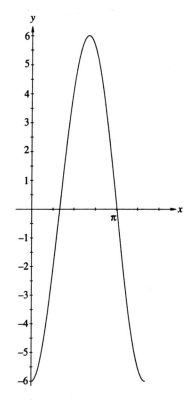

7.   Amplitude $= 2$,
Period $= \frac{360°}{3} = 120° = \frac{2\pi}{3}$,
Phase Shift $= \frac{90°}{3} = 30° = \frac{\pi}{6}$,
Vertical displacement $= 0$

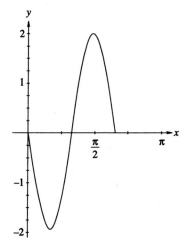

9.   Amplitude $= 4.5$

Period $= \frac{2\pi}{6} = \frac{\pi}{3}$,

Phase Shift $= \frac{4}{3}$,

Vertical displacement $= 0$

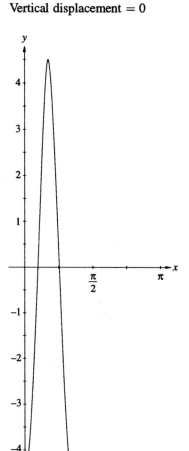

11.   Amplitude $= 0.5$,

Period $= \frac{2\pi}{\pi} = 2$,

Phase Shift $= -\frac{\pi}{8} \div \pi = -\frac{\pi}{8} \cdot \frac{1}{\pi} = -\frac{1}{8}$,

Vertical displacement $= 0$

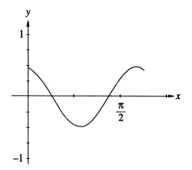

13.   Amplitude $= 0.75$,

Period $= \frac{2\pi}{\pi} = 2$,

Phase Shift $= \frac{\pi^2}{3} \div \pi = \frac{\pi^2}{3} \cdot \frac{1}{\pi} = -\frac{\pi}{3}$,

Vertical displacement $= 0$

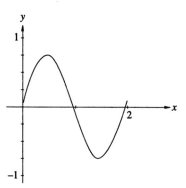

15.   Amplitude $= 2$,

Period $= 2$,

Phase Shift $= -\frac{1}{3\pi}$,

Vertical displacement $= 4$

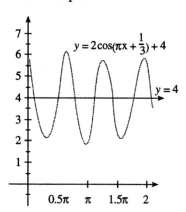

$$y = 2\cos\left(\pi x + \frac{1}{3}\right) + 4$$

$$y = 4$$

17.   Amplitude $= \frac{1}{2}$,

Period $= 2$,

Phase Shift $= -\frac{1}{\pi}$,

Vertical displacement $= \frac{3}{2}$

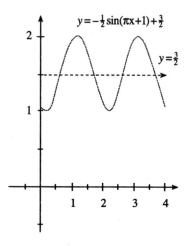

$$y = -\frac{1}{2}\sin(\pi x + 1) + \frac{3}{2}$$

$$y = \frac{3}{2}$$

19. Amplitude = 2,
Period = $180° = \pi$,
Phase Shift = $22.5°$,
Vertical displacement = 3.5

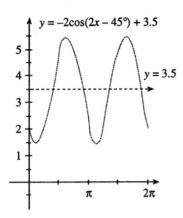

$y = -2\cos(2x - 45°) + 3.5$

$y = 3.5$

21. Amplitude = 65 A,
Period = $\frac{2\pi}{120\pi} = \frac{1}{60}$ s,
Phase Shift = $\dfrac{-\frac{\pi}{2}}{120\pi} = -\frac{1}{240}$ s

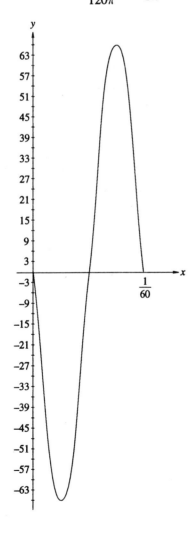

23. **(a)** The amplitude $A = \frac{1}{2}(11.8 - 6.2) = 2.8$. Since the period is 12.4 h, we have $12.4 = \frac{2\pi}{B}$ yields $B = \frac{\pi}{6.2} = \frac{5\pi}{31}$. Since high tide is at 4:00 A.M., no tide is $\frac{1}{4}$ period or $\frac{12.4}{4} = 3.1$ hr before or at $4 - 3.1 = 0.9$ hr. $\frac{C}{B} = 0.9$ and so, $C = 0.9B = \frac{0.9\pi}{6.2} = \frac{9\pi}{62}$. The average water height is $\frac{1}{2}(6.2+11.8) = 9$ m, so $D = 9$. Putting these all together, we get $d = 2.8\sin\left(\frac{\pi}{6.2}t - \frac{0.9\pi}{6.2}\right) + 9 = 2.8\sin\left(\frac{5\pi}{31}t - \frac{9\pi}{62}\right) + 9$.

(b)

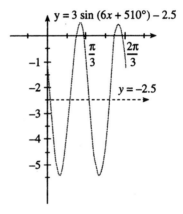

$y = 3\sin(6x + 510°) - 2.5$

$y = -2.5$

25. $A = \frac{1}{2}(72 - 11) = 30.5$ $D = \frac{1}{2}(72 + 11) = 41.5$. The period is 52 weeks so $52 = \frac{2\pi}{B}$ which leads to $B = \frac{2\pi}{52} = \frac{\pi}{26}$. The phase shift is 12 weeks, so $\frac{C}{B} = 12$ and so $C = 12\left(\frac{\pi}{26}\right) = \frac{6\pi}{13}$. Putting this all together, we get $T = 30.5\sin\left(\frac{\pi}{26}t - \frac{6\pi}{13}\right) + 41.5$.

(b)

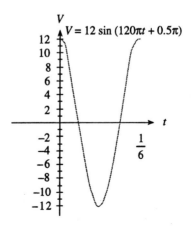

$V = 12\sin(120\pi t + 0.5\pi)$

11.3 COMPOSITE SINE AND COSINE CURVES

1.

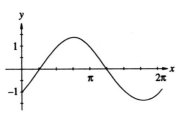

3.

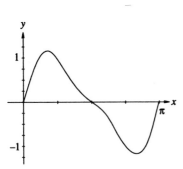

5.

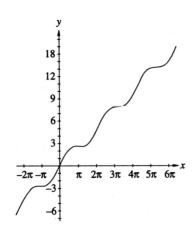

7.

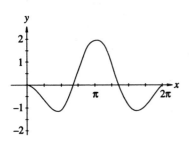

9.

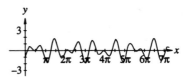

11.

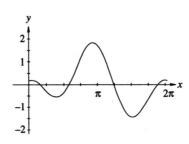

13.

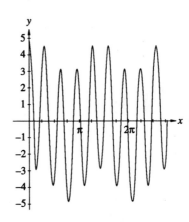

15.

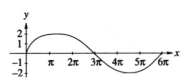

17.

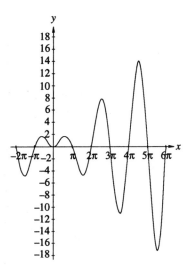

19. **(a)** Amplitude (maximum): 5, Period: π,

(b) Viewing window: one cycle: $\left[0, \pi, \frac{\pi}{4}\right] \times \left[-3, 7, 1\right]$, two cycles: $\left[0, 2\pi, \frac{\pi}{4}\right] \times \left[-3, 7, 1\right]$,

(c)

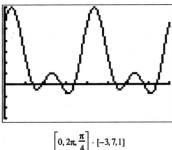

$\left[0, 2\pi, \frac{\pi}{4}\right] \cdot [-3, 7, 1]$

21. **(a)** Amplitude (maximum): 6, Period: The least common multiple of $\frac{2\pi}{3}$ and $\frac{2\pi}{5}$ is 2π,

(b) Viewing window: one cycle: $\left[0, 2\pi, \frac{\pi}{4}\right] \times \left[-9, 3, 1\right]$, two cycles $\left[0, 4\pi, \frac{\pi}{4}\right] \times [-9, 3, 1]$, and

(c)

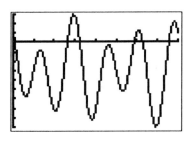

$\left[0, 2\pi, \frac{\pi}{4}\right] \cdot [-9, 3, 1]$

23. **(a)** Amplitude (maximum): 5.5, Period: Since the individual periods are 1 and 2π, and one is rational and the other irrational, they have no least common multiple. Hence, this function is not periodic.

(b) Viewing window: $\left[0, 4\pi, \frac{\pi}{2}\right] \times [-9, 2, 1]$,

(c)

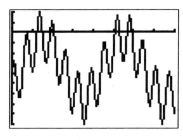

$\left[0, 4\pi, \frac{\pi}{2}\right] \cdot [-9, 2, 1]$

25.

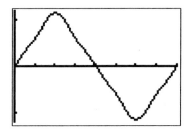

$\left[0, 2\pi, \frac{\pi}{4}\right] \cdot [0, 3, 1]$

≡ 11.4 GRAPHS OF THE OTHER TRIGONOMETRIC FUNCTIONS

1.

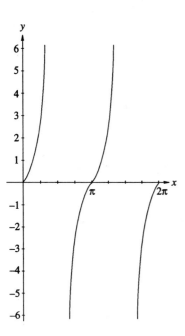

3.

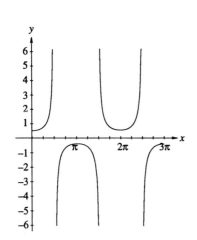

5.

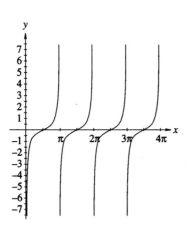

7.

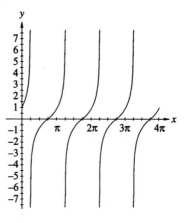

9.

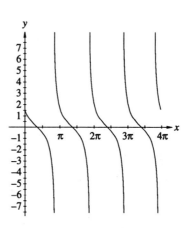

11.

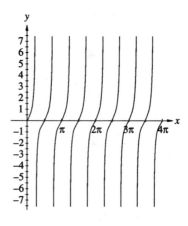

13.

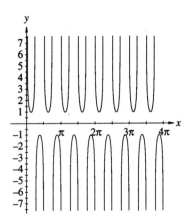

15.

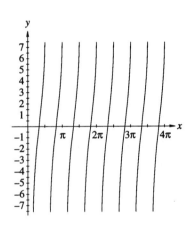

17.

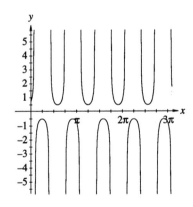

19. **(a)** $\left[0, 4\pi, \frac{\pi}{2}\right] \times [-8, 6, 2]$,

(b)

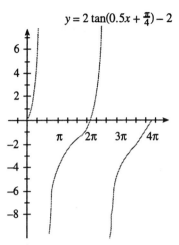

$$y = 2\tan(0.5x + \tfrac{\pi}{4}) - 2$$

21. For the given tapered hole,
$$d = -2h\tan\theta + D$$
$$= -2(8)\tan\theta + 4.5$$
$$= -16\tan\theta + 4.5$$

(a) If $\theta = 10°$, then $d = -16\tan 10° + 4.5 \approx$ 1.6788. Thus, the small-end diameter is about 1.68 cm.

(b) Here, we let $d = 0$ and solve the equation for θ.

$$0 = -16\tan\theta + 4.5$$
$$16\tan\theta = 4.5$$
$$\tan\theta = \frac{4.5}{16}$$
$$\theta = \tan^{-1}\left(\frac{4.5}{16}\right) \approx 15.71°$$

Thus, an angle of about 15.71° will produce a small-end diameter of 0 cm.

(c) The figure shows the graph of $d = -16\tan\theta + 4.5$ for $0° \leq \theta < 16°$, as drawn by a TI-82. Note that you need to put your calculator in degree mode to obtain this graph.

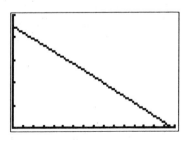

$[0, 16] \times [0, 5]$

☰ 11.5 APPLICATIONS OF TRIGONOMETRIC GRAPHS

1. The amplitude is 10 so $A = 10$. Since $f = 4, \omega = 2\pi f = 8\pi$. $y = A \sin \omega t = 10 \sin 8\pi t$.

3. Amplitude $A = 0.8$ m, $\omega = \frac{\pi}{3}$ rad/s. Since $y = 0$ when $t = 0$ there is no phase shift. Substituting into the form $y = A \sin \omega t$ we get $y = 0.8 \sin \frac{\pi}{3} t$.

5. $a = -g \sin \theta = -32 \sin 5° = -2.79$ ft/s^2.

7. $y = 8.5 \cos 2.8t$
(a) Amplitude is 8.5 cm,
(b) Period $= \frac{2\pi}{2.8} = \frac{\pi}{1.4} \approx 2.24$ s,
(c) $f = \frac{2.8}{2\pi} = \frac{1.4}{\pi} \approx 0.45$ Hz.

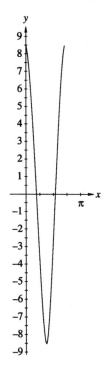

9. $I = 10 \sin 120\pi t$: Amplitude 10 A, Period $= \frac{2\pi}{120\pi} = \frac{1}{60}$ s; $f = \frac{120\pi}{2\pi} = 60$ Hz, $\omega = 2\pi f = 120\pi$ rad/s

11. If $\phi = \frac{\pi}{3}$ then $I = 6.8 \sin \left(160\pi t + \frac{\pi}{3} \right)$

13. $V = 220 \sin \left(80\pi t - \frac{\pi}{3} \right)$

15. Since cosine is the same as sine except for a phase shift of $\frac{\pi}{2}$ we have $V = 220 \cos \left(80\pi t - \frac{\pi}{3} + \frac{\pi}{2} \right) = 220 \cos \left(80\pi t + \frac{\pi}{6} \right)$

17. $y = 10^{-12} \sin 2\pi 10^{23} t$, frequency is 10^{23} Hz, amplitude is 10^{-12}

19. $I = I_{\max} \sin(2\pi f t + \phi) = 1.2 \sin(800\pi t + 37°)$ or $I = 1.2 \sin(800\pi t + 0.6458)$

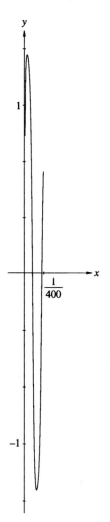

21. **(a)** Period: $\frac{2\pi}{2\pi \times 10^{18}} = 10^{-18}$ Hz, Amplitude: 10^{-10} m,

(b)

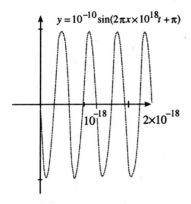

$$y = 10^{-10} \sin(2\pi x \times 10^{18} t + \pi)$$

☰ 11.6 PARAMETRIC EQUATIONS

1.

t	-3	-2	-1	0	1	2	3
x	-3	-2	-1	0	1	2	3
y	-9	-6	-3	0	3	6	9

Given $x = t$ and $y = 3t$, substituting x for t in the second equation yields $y = 3x$.

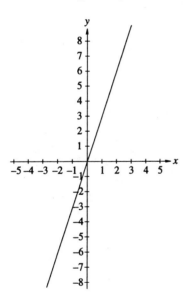

3.

t	-3	-2	-1	0	1	2	3
x	-3	-2	-1	0	1	2	3
y	$-\frac{1}{3}$	$-\frac{1}{2}$	-1	$*$	1	$\frac{1}{2}$	$\frac{1}{3}$

$*$ Not defined

Here $x = t$ and $y = \frac{1}{t}$. Direct substitution yields $y = \frac{1}{x}$.

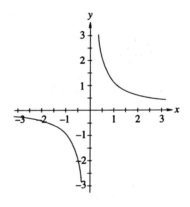

5.

t	-5	-4	-3	-2	-1	0	1	2	3	4	5
x	8	7	6	5	4	3	2	1	0	-1	-2
y	16	7	0	-5	-8	-9	-8	-5	0	7	16

Here $x = 3 - t$ and $y = t^2 - 9$. The first equation gives $t = 3 - x$. Substituting into the second equation yields $y = (3-x)^2 - 9 = 9 - 6x + x^2 - 9 = x^2 - 6x$ or $y = x^2 - 6x$.

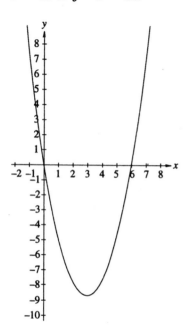

7.

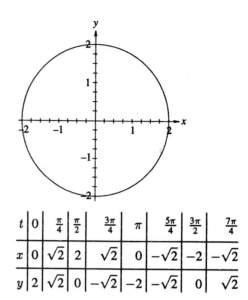

t	0	$\frac{\pi}{4}$	$\frac{\pi}{2}$	$\frac{3\pi}{4}$	π	$\frac{5\pi}{4}$	$\frac{3\pi}{2}$	$\frac{7\pi}{4}$
x	0	$\sqrt{2}$	2	$\sqrt{2}$	0	$-\sqrt{2}$	-2	$-\sqrt{2}$
y	2	$\sqrt{2}$	0	$-\sqrt{2}$	-2	$-\sqrt{2}$	0	$\sqrt{2}$

9.

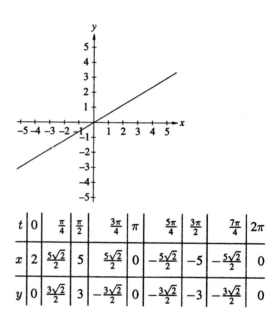

t	0	$\frac{\pi}{4}$	$\frac{\pi}{2}$	$\frac{3\pi}{4}$	π	$\frac{5\pi}{4}$	$\frac{3\pi}{2}$	$\frac{7\pi}{4}$	2π
x	2	$\frac{5\sqrt{2}}{2}$	5	$\frac{5\sqrt{2}}{2}$	0	$-\frac{5\sqrt{2}}{2}$	-5	$-\frac{5\sqrt{2}}{2}$	0
y	0	$\frac{3\sqrt{2}}{2}$	3	$-\frac{3\sqrt{2}}{2}$	0	$-\frac{3\sqrt{2}}{2}$	-3	$-\frac{3\sqrt{2}}{2}$	0

11.

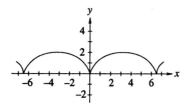

t	−5	−4	−3	−2	−1	0	1	2	3	4	5
x	−5.96	−4.76	−2.86	−1.09	−0.16	0	0.16	1.09	2.86	4.76	5.96
y	0.72	1.65	1.99	1.41	0.46	0	0.46	1.41	1.99	1.65	0.72

13.

t	0	$\frac{\pi}{6}$	$\frac{\pi}{4}$	$\frac{\pi}{3}$	$\frac{\pi}{2}$	$\frac{2\pi}{3}$	$\frac{3\pi}{4}$	$\frac{5\pi}{6}$	π	$\frac{7\pi}{6}$	$\frac{5\pi}{4}$	$\frac{4\pi}{3}$	$\frac{3\pi}{2}$	$\frac{5\pi}{3}$	$\frac{7\pi}{4}$	$\frac{11\pi}{6}$	2π
x	1	$\frac{2}{\sqrt{3}}$	$\sqrt{2}$	2	*	−2	$-\sqrt{2}$	$-\frac{2}{\sqrt{3}}$	−1	$-\frac{2}{\sqrt{3}}$	$-\sqrt{2}$	−2	*	2	$\sqrt{2}$	$\frac{2}{\sqrt{3}}$	1
y	*	4	$2\sqrt{2}$	$\frac{4}{\sqrt{3}}$	2	$\frac{4}{\sqrt{3}}$	$2\sqrt{2}$	4	*	−4	$-2\sqrt{2}$	$-\frac{4}{\sqrt{3}}$	−2	$-\frac{4}{\sqrt{3}}$	$-2\sqrt{2}$	−4	*

* Not defined

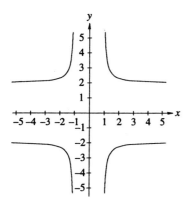

15. (a) 1:1;

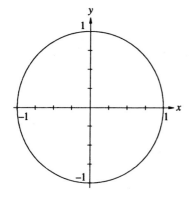

17. (a) 2:1;

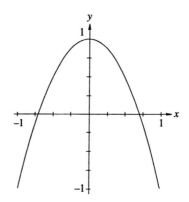

19.   **(a)** 1:4;

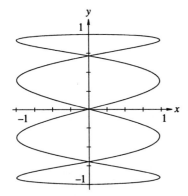

21.   **(a)** 2:3 or 1:1.5;

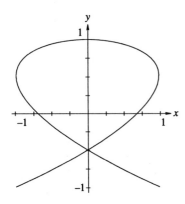

23.   **(a)** 2:5 or 1:2.5;

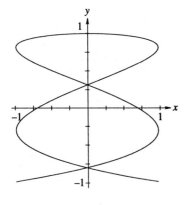

25.

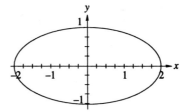

27.

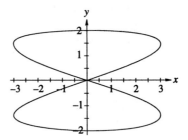

29.

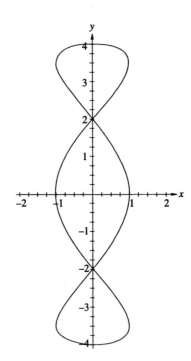

31. **(a)**

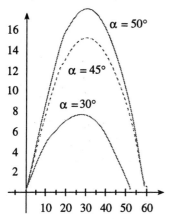

α = 50°
α = 45°
α = 30°

(b) Solving $x = (V \cos \alpha)t$ for t, we get $t = \frac{x}{V \cos \alpha}$. Substituting into the second equation, we get

$$y = (V \sin \alpha)\left(\frac{x}{V \cos \alpha}\right) - \frac{1}{2}g\left(\frac{x}{V \cos \alpha}\right)^2$$

$$= x \tan \alpha - \frac{gx^2}{2V^2} \sec^2 \alpha$$

(c) When the gravel hits the road $y = 0$ which means that $x \tan \alpha = \frac{gx^2}{2V^2} \sec^2 \alpha$. Solving for x, we get

$$x = \frac{2V^2 \tan \alpha}{g \sec^2 \alpha}$$

$$= \frac{2V^2 \left(\frac{\sin \alpha}{\cos \alpha}\right)}{g \left(\frac{1}{\cos^2 \alpha}\right)}$$

$$= \frac{2V^2 \sin \alpha \cos \alpha}{g}$$

$$= \frac{V^2 \sin 2\alpha}{g}$$

Note that the last step used the identity $2 \sin \cos \alpha = \sin 2\alpha$. $x = \frac{V^2 \sin 2\alpha}{g}$

≡ 11.7 POLAR COORDINATES

1.

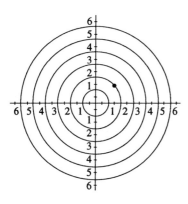

5.

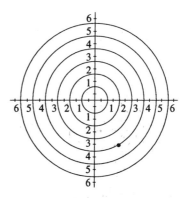

3.

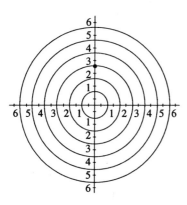

7.

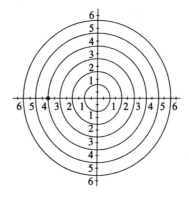

9.

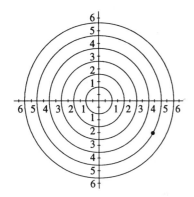

11.

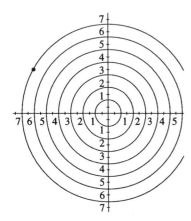

13. Using $x = r\cos\theta$, we find $x = 4\cos\frac{\pi}{3} = 4 \cdot \frac{1}{2} = 2$. Using $y = r\sin\theta$ yields $y = 4\sin\frac{\pi}{3} = 4 \cdot \frac{\sqrt{3}}{2} = 2\sqrt{3}$. Thus, the polar coordinates $\left(4, \frac{\pi}{3}\right)$ have the rectangular coordinates $\left(2, 2\sqrt{3}\right)$ or $(2, 3.46)$.

15. Here $x = r\cos\theta$ yields $x = 2\cos 135° = 2 \cdot \frac{-\sqrt{2}}{2} = -\sqrt{2}$, and $y = r\sin\theta$ produces $y = 2\sin 135° = 2 \cdot \frac{\sqrt{2}}{2} = \sqrt{2}$. Thus, the polar coordinates $(2, 135°)$ have the rectangular coordinates $(-\sqrt{2}, \sqrt{2})$ or $(-1.414, 1.414)$.

17. Here $x = -6\cos 20° = -5.64$ and $y = -6\sin 20° = -2.05$ with the result $(-5.64, -2.05)$.

19. Here $x = -2\cos 4.3 = 0.80$ and $y = -2\sin 4.3 = 1.83$ with the result $(0.80, 1.83)$.

21. $x = 3\cos -170° = -2.95$, $y = 3\sin -170° = -0.52$; $(-2.95, -0.52)$

23. $x = -6\cos(-2.5) = 4.81$, $y = -6\sin(-2.5) = 3.59$; $(4.81, 3.59)$

25. Here $r = \sqrt{x^2 + y^2} = \sqrt{4^2 + 4^2} = 4\sqrt{2} \approx 5.66$ and $\theta = \tan^{-1}\frac{4}{4} = \tan^{-1} 1 = 45°$ or $\frac{\pi}{4}$, so the rectangular coordinates $(4,4)$ have a polar coordinate $(4\sqrt{2}, 45°)$ or $(4\sqrt{2}, \frac{\pi}{4})$ or $(5.66, 45°)$.

27. Here $r = \sqrt{4^2 + 3^2} = 5$ and $\theta = \tan^{-1}\frac{3}{4} = 36.87° = 0.64$ rad, so the rectangular coordinates $(4, 3)$ have a polar coordinate $(5, 36.87°)$ or $(5, 0.64°)$.

29. Here $r = \sqrt{(-20)^2 + 21^2} = 29$, and $\theta_R = \tan^{-1}\frac{21}{20} = 46.40°$ or 0.81 rad. Since θ is in Quadrant II, $\theta = 180 - \theta_r = 133.60°$ or, $\theta = \pi - 0.81 = 2.33$, so the rectangular coordinates $(-20, 21)$ have a polar coordinate $(29, 133.60°)$ or $(29, 2.33)$.

31. $r = \sqrt{(-3)^2 + 4^2} = 5, \theta_R = \tan^{-1}\frac{4}{3} = 53.13° = 0.73$ rad, θ in Quadrant II so $180 - 53.13 = 126.87°$ or, $\theta = \pi - 0.93 = 2.21$; $(5, 126.87°)$ or $(5, 2.21)$

33. $r = \sqrt{(-7)^2 + (-10)^2} = 12.21, \theta_R = \tan^{-1}\frac{10}{7} = 55.01°$ or 0.96 rad. θ in Quadrant III so $180 + 55.01 = 235.01°$ or, $\theta = \pi + .96 = 4.102$; $(12.21, 235.01°)$ or $(12.21, 4.102)$

35. $r = \sqrt{2^2 + 9^2} = 9.22, \theta_R = \tan^{-1}\left(\frac{9}{2}\right) = 77.47°$ or 1.352; $(9.22, 77.47°)$ or $(9.22, 1.352)$

37.

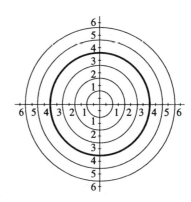

39.

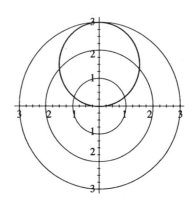

41.

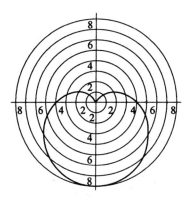

43.

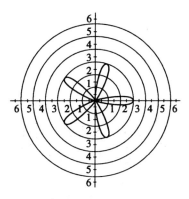

45.

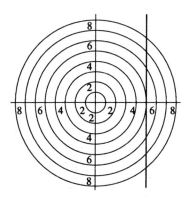

47.

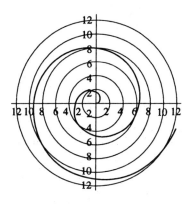

49.

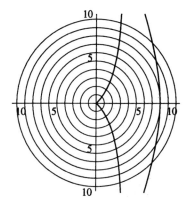

51.

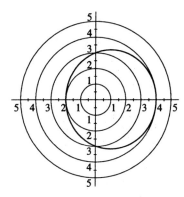

53.

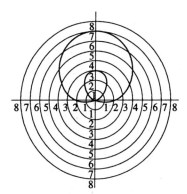

55.

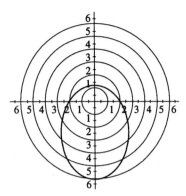

57.

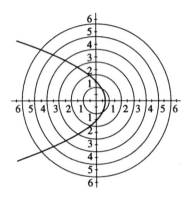

59.

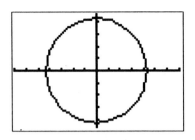

$[-7050, 7050, 1000] \times [-4650, 4650, 1000]$

61.

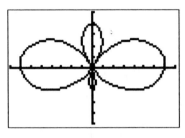

$[-8, 8, 1] \times [-5, 5, 1]$

CHAPTER 11 REVIEW

1. Period $= \frac{2\pi}{4} = \frac{\pi}{2}$,
Amplitude $= |8| = 8$,
Frequency $= \frac{2}{\pi}$, displacement or
Phase Shift $= 0$

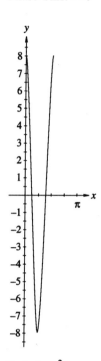

3. Period $= \frac{\pi}{3}$,
Amplitude $= \infty$,
Frequency $= \frac{3}{\pi}$, displacement or
Phase Shift $= 0$

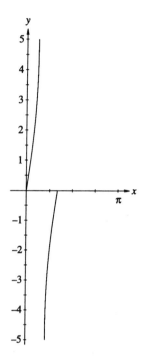

2. Period $= \frac{2\pi}{2} = \pi$,
Amplitude $= 3$,
Frequency $= \frac{1}{\pi}$, displacement or
Phase Shift $= 0$

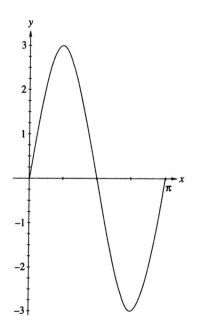

4. Period $= \frac{2\pi}{2} = \pi$,
Amplitude $= 3$,
Frequency $= \pi$, displacement $= -\frac{\frac{\pi}{2}}{2} = -\frac{\pi}{4}$

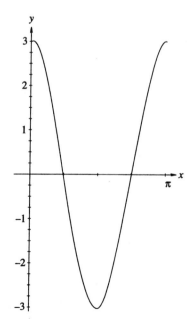

5.   Period $= \frac{2\pi}{3}$,

Amplitude $= \frac{1}{2}$,

Frequency $= \frac{3}{2\pi}$, displacement $= \frac{\frac{\pi}{3}}{3} = \frac{\pi}{9}$

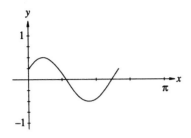

6.   Period $= \frac{2\pi}{2} = \pi$,

Amplitude $= \infty$,

Frequency $= \frac{1}{\pi}$, displacement $= \frac{\frac{\pi}{6}}{2} = \frac{\pi}{12}$

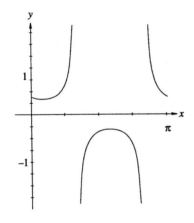

7.   Period $= \pi$,

Amplitude $= \infty$,

Frequency $= \frac{1}{\pi}$, displacement $= -\frac{\pi}{4}$

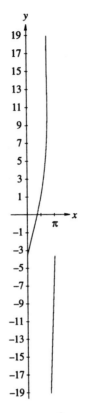

8.   Period $= \frac{2\pi}{\frac{1}{3}} = 6\pi$,

Amplitude $= \infty$,

Frequency $= \frac{1}{6\pi}$, displacement $= \frac{\frac{\pi}{5}}{\frac{1}{3}} = \frac{3\pi}{5}$

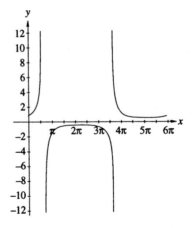

9. Period $= \dfrac{2\pi}{\frac{2}{3}} = 3\pi,$

Amplitude $= 2,$

Frequency $= \frac{1}{3\pi}$, displacement $= -\dfrac{\frac{\pi}{6}}{\frac{2}{3}} = -\frac{\pi}{4}$

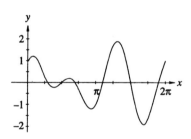

10. Period $= \dfrac{2\pi}{\frac{1}{2}} = 4\pi,$

Amplitude $= \frac{1}{4},$

Frequency $= \frac{1}{4\pi}$, displacement $= \dfrac{\frac{2\pi}{3}}{\frac{1}{2}} = \frac{4\pi}{3}$

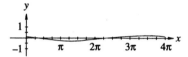

11.

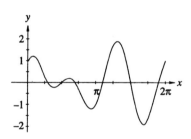

12.

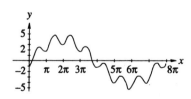

13.

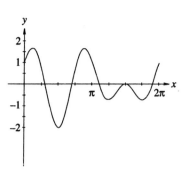

14.

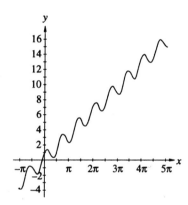

15.

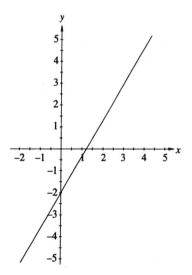

16.

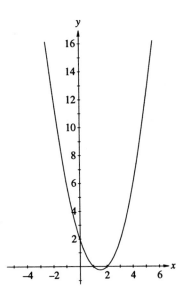

17.

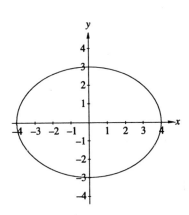

18.

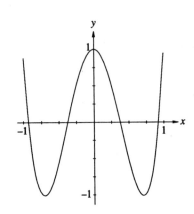

19.

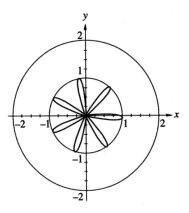

20.

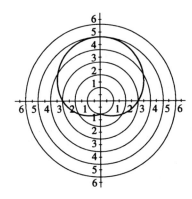

21. $y = A \sin 2\pi f t = 85 \sin 2\pi 5 t, y = 85 \sin 10\pi t$

22.   (a) $A = 1.7$ m,
   (b) Period $= \frac{2\pi}{3.4} = 1.85$ s,
   (c) $f = \frac{3.4}{2\pi} = 0.54$ Hz

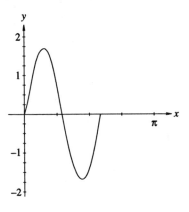

23. $V = V_{\max} \sin 2\pi ft$, $I = I_{\max} \sin(2\pi ft + \phi)$ **24.**
$= 4.8 \sin(2\pi 60t + \frac{\pi}{2})$, $I = 4.8 \sin\left(120\pi t + \frac{\pi}{2}\right)$

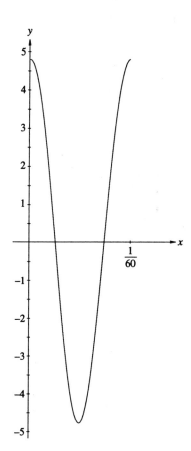

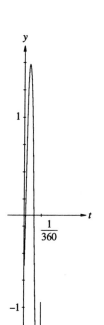

CHAPTER 11 TEST

1.   **(a)** Period $= \frac{2\pi}{5}$, amplitude $= |-3| = 3$, frequency, $= \frac{5}{2\pi}$, displacement $= 0$

(b) Period $= \frac{2\pi}{3}$, Amplitude $= 2.4$, frequency $= \frac{3}{2\pi}$, displacement $= \frac{\pi}{12}$

(c) Period $= \frac{\pi}{2}$, Amplitude $= \infty$, frequency $= \frac{2}{\pi}$, displacement $= -\frac{\pi}{10}$

2.

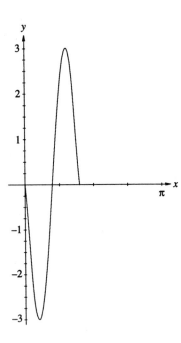

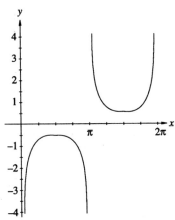

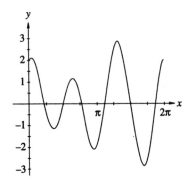

3.

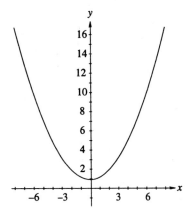

4.

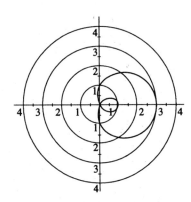

5. Using $x = r\cos\theta$, we find $x = 2\cos 55° = 1.147$, and using $y = r\sin\theta$ yields $y = 2\sin 55° = 1.638$. Thus, the polar coordinates $(2, 55°)$ have the rectangular coordinates $(1.147, 1.638)$.

6. Here $r = \sqrt{5^2 + (-12)^2} = 13$ and $\theta_R = \tan^{-1}\frac{\pm 12}{5}$ $= 67.38°$ or 1.176 rad. Since θ is Quadrant IV, we see that $\theta = 360 - 67.38 = 292.52°$, or $\theta = 2\pi - 1.175 = 5.107$. Thus, the rectangular coordinates $(5, -12)$ have polar coordinates $(13, 292.62°)$ or $(13, 5.107)$.

7. $y = I_{max}\sin(2\pi f t) = 5.7\sin(2\pi \cdot 60 \cdot t), y = 5.7\sin 120\pi t$

12

Exponents and Radicals

☰ 12.1 FRACTIONAL EXPONENTS

1. $25^{1/2} = \sqrt{25} = 5$

3. $64^{1/3} = \sqrt[3]{64} = 4$

5. $25^{-1/2} = \frac{1}{25^{1/2}} = \frac{1}{\sqrt{25}} = \frac{1}{5}$

7. $32^{-1/5} = \frac{1}{\sqrt[5]{32}} = \frac{1}{2}$

9. $27^{2/3} = \sqrt[3]{27}^2 = 3^2 = 9$

11. $125^{2/3} = \sqrt[3]{125}^2 = 5^2 = 25$

13. $16^{-3/4} = \frac{1}{\sqrt[4]{16}^3} = \frac{1}{2^3} = \frac{1}{8}$

15. $(-8)^{-1/3} = \frac{1}{\sqrt[3]{-8}} = -\frac{1}{2}$

17. $\left(\frac{1}{8}\right)^{1/3} = \frac{1}{\sqrt[3]{8}} = \frac{1}{2}$

19. $\left(\frac{1}{16}\right)^{-5/4} = \sqrt[4]{16}^5 = 2^5 = 32$

21. $3^2 \cdot 3^5 = 3^{2+5} = 3^7$

23. $7^6 7^{-2} = 7^{6-2} = 7^4$

25. $x^4 x^6 = x^{4+6} = x^{10}$

27. $y^6 y^{-4} = y^{6-4} = y^2$

29. $\left(9^5\right)^2 = 9^{5 \times 2} = 9^{10}$

31. $\left(x^7\right)^3 = x^{7 \times 3} = x^{21}$

33. $(xy)^5 = x^5 y^5$

35. $(ab)^{-5} = \frac{1}{(ab)^5} = \frac{1}{a^5 b^5}$

37. $\frac{x^{10}}{x^2} = x^{10-2} = x^8$

39. $\frac{x^2}{x^8} = \frac{1}{x^{8-2}} = \frac{1}{x^6}$

41. $x^{1/2} x^{3/2} = x^{1/2 + 3/2} = x^2$

43. $r^{3/4} r = r^{3/4+1} = r^{7/4}$

45. $a^{1/2} a^{1/3} = a^{1/2+1/3} = a^{3/6+2/6} = a^{5/6}$

47. $d^{2/3} d^{-1/4} = d^{2/3-1/4} = d^{8/12-3/12} = d^{5/12}$

49. $\frac{a^2 b^5}{a^5 b^2} = \frac{b^{5-2}}{a^{5-2}} = \frac{b^3}{a^3}$

51. $\frac{r^5 s^2 t}{t r^3 s^5} = r^{5-3} s^{2-5} t^{1-1} = r^2 s^{-3} t^0 = \frac{r^2}{s^3}$

53. $\frac{(xy^2 z)^4}{x^4 (yz^2)^2} = \frac{x^4 y^8 z^4}{x^4 y^2 z^4} = y^6$

55. $\left(\frac{a}{b^2}\right)^3 \left(\frac{a}{b^3}\right)^2 = \frac{a^3}{b^6} \cdot \frac{a^2}{b^6} = \frac{a^5}{b^{12}}$

57. $\frac{(xy^2 b^3)^{1/2}}{(x^{1/4} b^4 y)^2} = \frac{x^{1/2} y b^{3/2}}{x^{1/2} b^8 y^2} = \frac{1}{b^{8-3/2} y} = \frac{1}{b^{13/2} y}$

59. $\left(\frac{2x}{p^2}\right)^{-2} \left(\frac{p}{4}\right)^{-1} = \frac{p^4}{4x^2} \cdot \frac{4}{p} = \frac{p^3}{x^2}$

61. $\frac{(x^2 y^{-1} z)^{-2}}{(xy)^{-4}} = \frac{(xy)^4}{(x^2 y^{-1} z)^2} = \frac{x^4 y^4}{x^4 y^{-2} z^2} = \frac{y^6}{z^2}$

63. $\frac{(27x^6 y^3)^{-1/3}}{(16x^4 y^{12})^{-1/4}} = \frac{(16x^4 y^{12})^{1/4}}{(27x^6 y^3)^{1/3}} = \frac{2xy^3}{3x^2 y} = \frac{2y^2}{3x}$

65. $\left(\frac{x^2}{y^3}\right)^{-1/5} \left(\frac{y^2}{x^5}\right)^{1/3} = \left(\frac{y^3}{x^2}\right)^{1/5} \left(\frac{y^2}{x^5}\right)^{1/3} = \left(\frac{y^3}{x^2}\right)^{3/15}$
$\left(\frac{y^2}{x^5}\right)^{5/15} = \left(\frac{y^9}{x^6}\right)^{1/15} \left(\frac{y^{10}}{x^{25}}\right)^{1/15} = \left(\frac{y^{19}}{x^{31}}\right)^{1/15} = \frac{y^{19/15}}{x^{31/15}}$

67. $\left(\frac{9x^3}{t^5}\right)^{-1/2} \left(\frac{8t^2}{x^5}\right)^{-1/3} = \left(\frac{t^5}{9x^3}\right)^{1/2} \left(\frac{x^5}{8t^2}\right)^{1/3} = \frac{t^{5/2}}{3x^{3/2}} \cdot \frac{x^{5/3}}{2t^{2/3}} = \frac{t^{15/6} x^{10/6}}{6x^{9/6} t^{4/6}} = \frac{t^{11/6} x^{1/6}}{6}$

69. $\frac{(9x^4 y^{-6})^{-3/2}}{(8x^{-6} y^3)^{-2/3}} = \frac{(8x^{-6} y^3)^{2/3}}{(9x^4 y^{-6})^{3/2}} = \frac{2^2 x^{-4} y^2}{3^3 x^6 y^{-9}} = \frac{4y^{11}}{27x^{10}}$

71. $8.3^{2/3} \approx 4.0993852$

73. $92.47^{5/7} \approx 25.368006$

75. $(-81.52)^{2/7} = \left((-81.42)^2\right)^{1/7} \approx 3.5162154$

77. $(432.61)^{1/4} \approx 4.5606226$

79. Using a calculator: $\left((-32.35)\wedge 7 \boxed{x^{-1}}\right)\wedge \boxed{(-)}$
$3 = -0.2253777406$

81. $9.8(4.75)^2 \approx 221.1125$ or 221 m

83. $P = 5^5 \left(\frac{15}{0.5}\right)^{5/3} = 905\ 146.3 \text{ N/m}^2$

$T = 315 \left(\frac{15}{0.5}\right)^{5/3} = 91\ 238.747 \text{ K}$

85. Substituting for α, we see that the exponent becomes $\frac{\alpha}{1-\alpha} = \frac{1.4}{1-1.4} = \frac{1.4}{-0.4} = -\frac{7}{2}$. Hence, we get

$E = 1 - \left(\frac{p_1}{p_2}\right)^{-7/2} = 1 - \left(\frac{p_2}{p_1}\right)^{7/2}$

☰ 12.2 LAWS OF RADICALS

1. $\sqrt[3]{16} = \sqrt[3]{2^4} = \sqrt[3]{2^3}\sqrt[3]{2^1} = 2\sqrt[3]{2}$

3. $\sqrt{45} = \sqrt{9}\sqrt{5} = 3\sqrt{5}$

5. $\sqrt[3]{y^{12}} = y^{12/3} = y^4$

7. $\sqrt[5]{a^7} = \sqrt[5]{a^5 a^2} = a\sqrt[5]{a^2}$

9. $\sqrt{x^2 y^7} = \sqrt{x^2}\sqrt{y^6}\sqrt{y} = xy^3\sqrt{y}$

11. $\sqrt[4]{a^5 b^3} = \sqrt[4]{a^4}\sqrt[4]{ab^3} = a\sqrt[4]{ab^3}$

13. $\sqrt[3]{8x^4} = \sqrt[3]{2^3 x^3 x} = 2x\sqrt[3]{x}$

15. $\sqrt{27x^3 y} = \sqrt{3^2 x^2 3xy} = 3x\sqrt{3xy}$

17. $\sqrt[3]{-8} = \sqrt[3]{(-2)^3} = -2$

19. $\sqrt[3]{a^2 b^4}\sqrt[3]{ab^5} = \sqrt[3]{a^3 b^9} = ab^3$

21. $\sqrt[4]{p^3 q^2 r^6}\sqrt[4]{pq^6 r} = \sqrt[4]{p^4 q^8 r^7} = pq^2 r\sqrt[4]{r^3}$

23. $\sqrt[3]{\frac{8x^3}{27}} = \frac{\sqrt[3]{2^3 x^3}}{\sqrt[3]{3^3}} = \frac{2x}{3}$

25. $\sqrt[5]{\frac{x^5 y^{10}}{z^5}} = \frac{\sqrt[5]{x^5 y^{10}}}{\sqrt[5]{z^5}} = \frac{xy^2}{z}$

27. $\sqrt[3]{\frac{16x^3 y^2}{z^6}} = \frac{\sqrt[3]{2^3 \cdot 2x^3 y^2}}{\sqrt[3]{z^6}} = \frac{2x\sqrt[3]{2y^2}}{z^2}$

29. $\sqrt{\frac{64x^3 y^4}{9z^4 p^2}} = \frac{\sqrt{8^2 x^2 xy^4}}{\sqrt{3^2 z^4 p^2}} = \frac{8xy^2\sqrt{x}}{3z^2 p}$

31. $\sqrt{\frac{16}{3}} = \frac{\sqrt{16}}{\sqrt{3}} \cdot \frac{\sqrt{3}}{\sqrt{3}} = \frac{4\sqrt{3}}{3}$

33. $\sqrt[3]{\frac{27}{4}} = \frac{\sqrt[3]{3^3}}{\sqrt[3]{2^2}} \cdot \frac{\sqrt[3]{2}}{\sqrt[3]{2}} = \frac{3\sqrt[3]{2}}{2}$

35. $\sqrt{\frac{25}{2x}} = \frac{\sqrt{25}}{\sqrt{2x}} \cdot \frac{\sqrt{2x}}{\sqrt{2x}} = \frac{5\sqrt{2x}}{2x}$

37. $\sqrt[4]{\frac{81}{32z^2}} = \frac{\sqrt[4]{3^4}}{\sqrt[4]{2^5 z^2}} \cdot \frac{\sqrt[4]{2^3 z^2}}{\sqrt[4]{2^3 z^2}} = \frac{3\sqrt[4]{8z^2}}{4z}$

39. $\sqrt[3]{\frac{16x^2 y}{x^5}} = \sqrt[3]{\frac{16y}{x^3}} = \frac{\sqrt[3]{(2^3)2y}}{\sqrt[3]{x^3}} = \frac{2\sqrt[3]{2y}}{x}$

41. $\sqrt[3]{\frac{-8x^3 yz}{27b^2 z^4}} = \sqrt[3]{\frac{-8x^3 y}{27b^2 z^3}} = \frac{\sqrt[3]{(-2)^3 x^3 y}}{\sqrt[3]{3^3 b^2 z^3}} \cdot \frac{\sqrt[3]{b}}{\sqrt[3]{b}} = \frac{-2x\sqrt[3]{by}}{3bz}$

43. $\sqrt{4 \times 10^4} = \sqrt{4} \times \sqrt{10^4} = 2 \times 10^2 = 200$

45. $\sqrt{25 \times 10^3} = \sqrt{25}\sqrt{10^1 10^2} = 5\sqrt{10} \times 10 = 50\sqrt{10}$

47. $\sqrt{4 \times 10^7} = \sqrt{4}\sqrt{10(10^6)} = 2\sqrt{10} \times 10^3 = 2000\sqrt{10}$

49. $\sqrt[3]{1.25 \times 10^{10}} = \sqrt[3]{125 \times 10^8} = \sqrt[3]{125}\sqrt[3]{10^2}\sqrt[3]{10^6}$
$= 5\sqrt[3]{100} \times 10^2 = 500\sqrt[3]{100}$ or $\sqrt[3]{1.25 \times 10^{10}} =$
$\sqrt[3]{1.25 \times 10 \times 10^9} = \sqrt[3]{12.5} \times 10^3$

51. $\sqrt{\frac{x}{y} + \frac{y}{x}} = \sqrt{\frac{x^2}{xy} + \frac{y^2}{xy}} = \sqrt{\frac{x^2 + y^2}{xy}} = \frac{\sqrt{x^2 + y^2}}{\sqrt{xy}} \cdot$
$\frac{\sqrt{xy}}{\sqrt{xy}} = \frac{\sqrt{xy(x^2 + y^2)}}{xy} = \frac{\sqrt{x^3 y + xy^3}}{xy}$

53. $\sqrt{a^2 + 2ab + b^2} = \sqrt{(a+b)^2} = |a + b|$

55. $\sqrt{\frac{1}{a^2} + \frac{1}{b}} = \sqrt{\frac{b}{a^2 b} + \frac{a^2}{a^2 b}} = \sqrt{\frac{b+a^2}{a^2 b}} =$
$\frac{\sqrt{b+a^2}}{\sqrt{a^2 b}} \cdot \frac{\sqrt{b}}{\sqrt{b}} = \frac{\sqrt{b^2 + a^2 b}}{ab}$

57. $\sqrt[6]{\sqrt[3]{27x^2}}$. Multiply the indexes to get $\sqrt[18]{27x^2}$.

59. Multiply the indexes to get $\sqrt[21]{-624.2x^{15} y^{10} z}$.

61. $f = \frac{1}{2L}\sqrt{\frac{T}{\mu}} = \frac{1}{2L} \cdot \frac{\sqrt{T}}{\sqrt{\mu}} \cdot \frac{\sqrt{\mu}}{\sqrt{\mu}} = \frac{1}{2L} \cdot \frac{\sqrt{T\mu}}{\mu} = \frac{1}{2L\mu} \cdot \sqrt{T\mu}$

63. $z = \frac{1}{\sqrt{\frac{1}{x^2} + \frac{1}{R^2}}} = \frac{1}{\sqrt{\frac{R^2 + x^2}{x^2 R^2}}} = \sqrt{\frac{x^2 R^2}{R^2 + x^2}} =$
$\frac{xR\sqrt{R^2 + x^2}}{R^2 + x^2}$

65. First, we simplify the expression under the radical to obtain $1 + \sqrt{\frac{m+2h}{m}}$ which equals $1 + \frac{m^2 + 2mh}{m^2}$
or $1 + \frac{\sqrt{m^2 + 2mh}}{m} = \frac{m + \sqrt{m^2 + 2h}}{m}$

≡ 12.3 BASIC OPERATIONS WITH RADICALS

1. $2\sqrt{3} + 5\sqrt{3} = 7\sqrt{3}$

3. $\sqrt[3]{9} + 4\sqrt[3]{9} = 5\sqrt[3]{9}$

5. $2\sqrt{3} + 4\sqrt{2} + 6\sqrt{3} = 8\sqrt{3} + 4\sqrt{2}$

7. $\sqrt{5} + \sqrt{20} = \sqrt{5} + 2\sqrt{5} = 3\sqrt{5}$

9. $\sqrt{7} - \sqrt{28} = \sqrt{7} - 2\sqrt{7} = -\sqrt{7}$

11. $\sqrt{60} - \sqrt{\frac{5}{3}} = 2\sqrt{15} - \sqrt{\frac{5}{3}}\sqrt{\frac{3}{3}} = 2\sqrt{15} - \frac{\sqrt{15}}{3} =$
$\frac{6\sqrt{15}}{3} - \frac{\sqrt{15}}{3} = \frac{5\sqrt{15}}{3}$

13. $\sqrt{\frac{1}{2}} - \sqrt{\frac{9}{2}} = \frac{\sqrt{2}}{2} - \frac{3\sqrt{2}}{2} = \frac{-2\sqrt{2}}{2} = -\sqrt{2}$

15. $\sqrt{x^3 y} + 2x\sqrt{xy} = x\sqrt{xy} + 2x\sqrt{xy} = 3x\sqrt{xy}$

17. $\sqrt[3]{24p^2 q^4} + \sqrt[3]{3p^8 q} = \sqrt[3]{8 \cdot 3p^2 q^3 q} + \sqrt[3]{3p^6 p^2 q} =$
$2q\sqrt[3]{3p^2 q} + p^2\sqrt[3]{3p^2 q} = (2q + p^2)\sqrt[3]{3p^2 q}$

19. $\sqrt{\frac{x}{y^3}} - \sqrt{\frac{y}{x^3}} = \sqrt{\frac{xy}{y^4}} - \sqrt{\frac{xy}{x^4}} = \left(\frac{1}{y^2} - \frac{1}{x^2}\right)\sqrt{xy}$
or $\frac{x^2 - y^2}{x^2 y^2}\sqrt{xy}$

21. $a\sqrt{\frac{x}{3a}} + b\sqrt{\frac{a}{3b}} = \frac{a\sqrt{3ab3a}}{3a} + \frac{b\sqrt{3ab}}{3b} = \frac{\sqrt{3ab}}{3} +$
$\frac{\sqrt{3ab}}{3} = \frac{2\sqrt{3ab}}{3}$

23. $\sqrt{5}\sqrt{8} = \sqrt{40} = 2\sqrt{10}$

25. $\sqrt{3x}\sqrt{5x} = \sqrt{15x^2} = x\sqrt{15}$

27. $\left(\sqrt{4x}\right)^3 = \left(2\sqrt{x}\right)^3 = 2^3\sqrt{x^3} = 8\sqrt{x^3} = 8x\sqrt{x}$

29. $\sqrt{\frac{5}{8}} \cdot \sqrt{\frac{9}{10}} = \sqrt{\frac{5 \cdot 9}{8 \cdot 10}} = \sqrt{\frac{9}{16}} = \frac{3}{4}$

31. $\sqrt{2}(\sqrt{x} + \sqrt{2}) = \sqrt{2x} + 2$

33. $(\sqrt{x} + \sqrt{y})^2 = \sqrt{x}^2 + 2\sqrt{xy} + \sqrt{y}^2 = x + 2\sqrt{xy} + y$

35. $\sqrt[3]{\frac{5}{2}} \cdot \sqrt[3]{\frac{2}{7}} = \sqrt[3]{\frac{5 \cdot 2}{2 \cdot 7}} = \sqrt[3]{\frac{5}{7}} \cdot \sqrt[3]{\frac{7^2}{7^2}} = \frac{\sqrt[3]{245}}{7}$

37. $\left(\sqrt{a} + \sqrt{b}\right)\left(\sqrt{a} - \sqrt{b}\right) = \sqrt{a}^2 - \sqrt{b}^2 = |a| - |b|$

39. $\sqrt[3]{x}\sqrt{x} = x^{1/3} x^{1/2} = x^{2/6} x^{3/6} = x^{5/6} = \sqrt[6]{x^5}$

41. To obtain a common index, rewrite $\sqrt[3]{49x^2}\sqrt[4]{3x}$
as $\left(49x^2\right)^{1/3} (3x)^{1/4} = \left(49x^2\right)^{4/12}(3x)^{3/12} =$
$\left(49^4 x^8\right)^{1/12} \left(3^3 x^3\right)^{1/12} = \left(49^4 3^3 x^{11}\right)^{1/12} =$
$\sqrt[12]{49^4 3^3 x^{11}} = \sqrt[12]{3^3 49^4 x^{11}} = \sqrt[12]{155,649,627 x^{11}}$

43. $\frac{\sqrt{32}}{\sqrt{2}} = \sqrt{\frac{32}{2}} = \sqrt{16} = 4$

45. $\frac{\sqrt[3]{4b^2}}{\sqrt[3]{16b}} = \sqrt[3]{\frac{4b^2}{16b}} = \sqrt[3]{\frac{b}{4}}\sqrt[3]{\frac{2}{2}} = \frac{\sqrt[3]{2b}}{2}$

47. $\frac{\sqrt{3x^2}}{\sqrt[3]{9x}} = \frac{\sqrt[6]{(3x^2)^3}}{\sqrt[6]{(9x)^2}} = \frac{\sqrt[6]{27x^6}}{\sqrt[6]{81x^2}} = \sqrt[6]{\frac{27x^6}{81x^2}} = \sqrt[6]{\frac{x^4}{3}} \cdot \frac{\sqrt[6]{3^5}}{\sqrt[6]{3^5}} =$
$\frac{\sqrt[6]{243x^4}}{3}$

49. $\frac{\sqrt[3]{16a^2 b}}{\sqrt[4]{2ab^3}} = \frac{\sqrt[12]{(16a^2 b)^4}}{\sqrt[12]{(2ab^3)^3}} = \sqrt[12]{\frac{(16a^2 b)^4}{(2ab^3)^3}} = \sqrt[12]{\frac{(2^4 a^2 b)^4}{(2ab^3)^3}} =$
$\sqrt[12]{\frac{2^{16} a^8 b^4}{2^3 a^3 b^9}} = \sqrt[12]{\frac{2^{16} a^8 b^4}{2^3 a^3 b^9}} = \sqrt[12]{\frac{2^{13} a^5}{b^5}} = 2\sqrt[12]{\frac{2a^5}{b^5}} =$
$2\sqrt[12]{\frac{2a^5}{b^5}} \cdot \frac{\sqrt[12]{b^7}}{\sqrt[12]{b^7}} = 2\frac{\sqrt[12]{2a^5 b^7}}{b}$

51. $\frac{\sqrt[3]{4}}{\sqrt{2}} = \frac{\sqrt[3]{4}}{\sqrt{2}} \cdot \frac{\sqrt{2}}{\sqrt{2}} = \frac{\sqrt[3]{2}^2 \sqrt{2}}{2} = \frac{\sqrt[6]{2^4 \cdot 2^3}}{2} = \frac{\sqrt[6]{2^7}}{2} = \sqrt[6]{2}$

53. $\frac{1}{x + \sqrt{5}} \cdot \frac{x - \sqrt{5}}{x - \sqrt{5}} = \frac{x - \sqrt{5}}{x^2 - 5}$

55. $\frac{\sqrt{5} - \sqrt{3}}{\sqrt{5} + \sqrt{3}} \cdot \frac{\sqrt{5} - \sqrt{3}}{\sqrt{5} - \sqrt{3}} = \frac{5 - 2\sqrt{15} + 3}{5 - 3} = \frac{8 - 2\sqrt{15}}{2} = \frac{2(4 - \sqrt{15})}{2}$
$= 4 - \sqrt{15}$

57. $\frac{\sqrt{x+1}}{\sqrt{x-1}} + \frac{\sqrt{x-1}}{\sqrt{x+1}} = \frac{\sqrt{x+1}\sqrt{x-1}}{\sqrt{x-1}\sqrt{x-1}} + \frac{\sqrt{x-1}\sqrt{x+1}}{\sqrt{x+1}\sqrt{x+1}} = \frac{\sqrt{x^2-1}}{x-1} +$
$\frac{\sqrt{x^2-1}}{x+1} = \frac{(x+1)\sqrt{x^2-1}}{x^2-1} + \frac{(x-1)\sqrt{x^2-1}}{x^2-1} = \frac{2x\sqrt{x^2-1}}{x^2-1}, x \neq$
$1, -1$

59. $\frac{\sqrt{x+y}}{\sqrt{x-y} - \sqrt{x}} \cdot \frac{\sqrt{x-y} + \sqrt{x}}{\sqrt{x-y} + \sqrt{x}} = \frac{\sqrt{x^2-y^2} + \sqrt{x^2+xy}}{x-y-x} =$
$-\frac{\sqrt{x^2-y^2} + \sqrt{x^2+xy}}{y}, y \neq 0$

61. $\frac{-b + \sqrt{b^2 - 4ac}}{2a} + \frac{-b - \sqrt{b^2 - 4ac}}{2a} = \frac{-2b}{2a} = \frac{-b}{a}$

63. **(a)** $R = \frac{R_1 R_2}{R_1 + R_2} = \frac{x^{3/2}\sqrt{x}}{x^{3/2} + \sqrt{x}} = \frac{x^{3/2} x^{1/2}}{x^{3/2} + x^{1/2}} = \frac{x^2}{\sqrt{x^3} + \sqrt{x}} \cdot$
$\frac{\sqrt{x^3} - \sqrt{x}}{\sqrt{x^3} - \sqrt{x}} = \frac{x^2 x^{3/2} - x^2 x^{1/2}}{x^3 - x} = \frac{x^{7/2} - x^{5/2}}{x^3 - x} = \frac{x^{5/2}(x-1)}{x(x^2-1)} =$
$\frac{x \cdot x^{3/2}(x-1)}{x(x+1)(x-1)} = \frac{x^{3/2}}{x+1}$

(b) if $x = 20$; $\frac{20^{3/2}}{20+1} = \frac{89.44279191}{21} = 4.2591771 \approx$
$4.259 \, \Omega$

65. $\left(\frac{-b + \sqrt{b^2 - 4ac}}{2a}\right)\left(\frac{-b - \sqrt{b^2 - 4ac}}{2a}\right) = \frac{b^2 - \left(\sqrt{b^2 - 4ac}\right)^2}{4a^2}$
$= \frac{b^2 - (b^2 - 4ac)}{4a^2} = \frac{4ac}{4a^2} = \frac{c}{a}$

12.4 EQUATIONS WITH RADICALS

1. $\sqrt{x+3}=5$; $\left(\sqrt{x+3}\right)^2=5^2$; $x+3=25$; $x=22$. Check: $\sqrt{22+3}=\sqrt{25}=5$. Solution is 22.

3. $\sqrt{2x+4}-7=0$; $\sqrt{2x+4}=7$; $\left(\sqrt{2x+4}\right)^2=7^2$; $2x+4=49$; $2x=45$; $x=\frac{45}{2}$. Check:
$$\sqrt{2\cdot\frac{45}{2}+4}-7=\sqrt{45+4}-7=\sqrt{49}-7=0.$$
Solution is $\frac{45}{2}$ or 22.5.

5. $\sqrt{x^2+24}=x-4$; $\left(\sqrt{x^2+24}\right)^2=(x-4)^2$; $x^2+24=x^2-8x+16$; $24=-8x+16$; $8=-8x$; $x=-1$. Check: $\sqrt{(-1)^2+24}=\sqrt{1+24}=\sqrt{25}=5$; $-1-4=-5$ but $5\neq-5$ so no solution.

7. $(y+12)^{1/3}=3$; $\left((y+12)^{1/3}\right)^3=3^3$; $y+12=27$; $y=15$. Check: $(15+12)^{1/3}=27^{1/3}=3$. Solution is 15.

9. $\left(\frac{x}{2}+1\right)^{1/2}=3$; $\left[\left(\frac{x}{2}+1\right)^{1/2}\right]^2=3^2$; $\frac{x}{2}+1=9$; $\frac{x}{2}=8$; $x=16$. Check: $\left(\frac{16}{2}+1\right)^{1/2}=(8+1)^{1/2}=9^{1/2}=3$. Solution is 16.

11. $(x-1)^{3/2}=27$; $\left((x-1)^{3/2}\right)^2=27^2$; $(x-1)^3=729$; $(x-1)=\sqrt[3]{729}=9$; $x=10$. Check: $(10-1)^{3/2}=9^{3/2}=3^3=27$. Solution is 10.

13. $\sqrt{x+1}=\sqrt{2x-1}$; $\sqrt{x+1}^2=\sqrt{2x-1}^2$; $x+1=2x-1$; $-x=-2$; $x=2$. Check: $\sqrt{2+1}=\sqrt{3}$; $\sqrt{2\cdot2-1}=\sqrt{4-1}=\sqrt{3}$. Solution is 2.

15. $\sqrt{x+1}+\sqrt{x-5}=7$; $\sqrt{x+1}=7-\sqrt{x-5}$; $\sqrt{x+1}^2=(7-\sqrt{x-5})^2$; $x+1=49-14\sqrt{x-5}+(x-5)$; $x+1=x-14\sqrt{x-5}+44$; $-43=-14\sqrt{x-5}$; $\frac{43}{14}=\sqrt{x-5}$; $\left(\frac{43}{14}\right)^2=\sqrt{x-5}^2$; $\frac{1849}{196}=$

$x-5$; $x=\frac{2829}{196}$. Check: $\sqrt{\frac{2829}{196}+1}+\sqrt{\frac{2829}{196}-5}=$
$\sqrt{\frac{3025}{196}}+\sqrt{\frac{1849}{196}}=\frac{55}{14}+\frac{43}{14}=\frac{98}{14}=7.$
Solution is $\frac{2829}{196}\approx14.43$.

17. $\sqrt{x-1}+\sqrt{x+5}=2$; $\sqrt{x-1}=2-\sqrt{x+5}$; $\sqrt{x-1}^2=(2-\sqrt{x+5})^2$; $x-1=4-4\sqrt{x+5}+(x+5)$; $-10=-4\sqrt{x+5}$; $\frac{5}{2}=\sqrt{x+5}$; $\left(\frac{5}{2}\right)^2=\sqrt{x+5}^2$; $\frac{25}{4}=x+5$; $x=\frac{25}{4}-\frac{20}{4}=\frac{5}{4}$. Check: $\sqrt{\frac{5}{4}-1}+\sqrt{\frac{5}{4}+5}=\sqrt{\frac{1}{4}}+\sqrt{\frac{25}{4}}=\frac{1}{2}+\frac{5}{2}=\frac{6}{2}=3$. Does not check so there is no solution.

19. $\sqrt{\frac{1}{x}}=\sqrt{\frac{4}{3x-1}}$; square both sides to get $\frac{1}{x}=\frac{4}{3x-1}$; $3x-1=4x$; $-1=x$. This does not check as it gives negative radicands so no solution.

21. Here $v=\sqrt{\mu g_s R}$. Squaring both sides give $v^2=\mu_s g R$ and so $R=\frac{v^2}{\mu_s g}$.

23. We are given $\frac{d_A}{d_B}-\frac{\sqrt{I_A}}{\sqrt{I_B}}$. Substituting the given data produces $\frac{11}{20}=\frac{3.5I_B-15}{\sqrt{I_B}}$. Now, square both sides to get $\frac{121}{400}=\frac{3.5I_B-15}{I_B}$ and cross-multiply, with the result $121I_B=1400I_B-6000$ or $-1279I_B=-6000$ and so, $I_B=\frac{6000}{1279}\approx4.691$. Since $I_A=3.5I_B-15$, we have $I_A=3.5\times4.691-15=1.419$. The final result is $I_A=1.419, I_B=4.691$.

25. Substituting for t, we have $8.6=\frac{1}{4}\sqrt{s}$. Multiply by 4 to get $\sqrt{s}=34.4$. Squaring both sides, we get $s=34.4^2=1,183.36$ ft.

CHAPTER 12 REVIEW

1. $49^{1/2}=\sqrt{49}=7$

2. $81^{1/4}=\sqrt[4]{81}=3$

3. $25^{3/2}=\sqrt{25}^3=5^3=125$

4. $64^{2/3}=\sqrt[3]{64}^2=4^2=16$

5. $9^{-3/2}=\frac{1}{9^{3/2}}=\frac{1}{\sqrt{9}^3}=\frac{1}{3^3}=\frac{1}{27}$

6. $125^{-4/3}=\frac{1}{125^{4/3}}=\frac{1}{\sqrt[3]{125}^4}=\frac{1}{5^4}=\frac{1}{625}$

7. $5^4 5^9=5^{13}$

8. $13^6 13^{-9}=13^{6-9}=13^{-3}=\frac{1}{13^3}$

9. $(2^4)^8=2^{32}$

10. $(5^3)^6=5^{18}$

11. $(x^2y^3)^4=x^8y^{12}$

12. $(y^6x)^4=y^{24}x^4$

13. $\left(\frac{a^3}{ya^4}\right)^{-5}=\left(\frac{1}{ya}\right)^{-5}=(ya)^5=y^5a^5$

14. $\left(\frac{a^5}{ab^4}\right)^3=\left(\frac{a^4}{b^4}\right)^3=\frac{a^{12}}{b^{12}}$ or $\left(\frac{a}{b}\right)^{12}$

15. $x^{1/2}x^{1/3} = x^{3/6}x^{2/6} = x^{5/6}$

16. $y^{2/3}y^{3/4} = y^{8/12}y^{9/12} = y^{17/12}$

17. $\dfrac{(xy^2a)^5}{x^4(ya^3)^2} = \dfrac{x^5y^{10}a^5}{x^4y^2a^6} = \dfrac{xy^8}{a}$

18. $\dfrac{(ab^3c^2)^4}{(a^2bc)^3} = \dfrac{a^4b^{12}c^8}{a^6b^3c^3} = \dfrac{b^9c^5}{a^2}$

19. $\sqrt[3]{-27} = -3$

20. $\sqrt[4]{81^3} = \sqrt[4]{81}^3 = 3^3 = 27$

21. $\sqrt{a^2b^5} = \sqrt{a^2}\sqrt{b^4}\sqrt{b} = ab^2\sqrt{b}$

22. $\sqrt[4]{a^3b^7} = \sqrt[4]{b^4}\sqrt[4]{a^3b^3} = b\sqrt[4]{a^3b^3}$

23. $\sqrt[3]{\dfrac{-8x^3y^6}{z}} = \dfrac{-2xy^2}{\sqrt[3]{z}} = \dfrac{-2xy^2\sqrt[3]{z^2}}{\sqrt[3]{z}\sqrt[3]{z^2}} = \dfrac{-2xy^2\sqrt[3]{z^2}}{z}$ or $\dfrac{-2xy^2}{z}\sqrt[3]{z^2}$

24. $\sqrt[4]{\dfrac{32a^6b^8}{cd^2}} = \sqrt[4]{\dfrac{2^5a^6b^8}{cd^2}} \cdot \sqrt[4]{\dfrac{c^3d^2}{c^3d^2}} = \dfrac{2ab^2\sqrt[4]{2a^2c^3d^2}}{cd} = \dfrac{2ab^2}{cd}\sqrt[4]{2a^2c^3d^2}$

25. $\sqrt{\dfrac{a}{b^2} - \dfrac{b}{a^2}} = \sqrt{\dfrac{a^3}{a^2b^2} - \dfrac{b^3}{a^2b^2}} = \dfrac{\sqrt{a^3-b^3}}{\sqrt{a^2b^2}} = \sqrt{\dfrac{a^3-b^3}{ab}}$

26. $\sqrt{a^2+8a+16} = \sqrt{(a+4)^2} = a+4$

27. $2\sqrt{5}+6\sqrt{5} = 8\sqrt{5}$

28. $3\sqrt{7}-6\sqrt{7} = -3\sqrt{7}$

29. $3\sqrt{6}+5\sqrt{6}-2\sqrt{3} = 8\sqrt{6}-2\sqrt{3}$

30. $\sqrt{6}+\sqrt{24} = \sqrt{6}+2\sqrt{6} = 3\sqrt{6}$

31. $\sqrt[4]{16a^3b}+\sqrt[4]{81a^7b} = 2\sqrt[4]{a^3b}+3a\sqrt[4]{a^3b}; (2+3a)\sqrt[4]{a^3b}$

32. $\sqrt[4]{32x^5} - x^2\sqrt[4]{2x} = 2x\sqrt[4]{2x} - x^2\sqrt[4]{2x}; = (2x - x^2)\sqrt[4]{2x}$

33. $a\sqrt{\dfrac{a}{7b}} - b\sqrt{\dfrac{b}{7a}} = \dfrac{a}{7b}\sqrt{7ab} - \dfrac{b}{7a}\sqrt{7ab} = \dfrac{a^2}{7ab}\sqrt{7ab} - \dfrac{b^2}{7ab}\sqrt{7ab} = \dfrac{(a^2-b^2)\sqrt{7ab}}{7ab}$

34. $\sqrt{\dfrac{x-2}{x+2}} + \sqrt{\dfrac{x+2}{x-2}} = \dfrac{\sqrt{x-2}\sqrt{x+2}}{x+2} + \dfrac{\sqrt{x+2}\sqrt{x-2}}{x-2} = \dfrac{\sqrt{x^2-4}}{x+2} + \dfrac{\sqrt{x^2-4}}{x-2} = \dfrac{(x-2)\sqrt{x^2-4}}{(x-2)(x+2)} + \dfrac{(x+2)\sqrt{x^2-4}}{(x+2)(x-2)} = \dfrac{x\sqrt{x^2-4}-2\sqrt{x^2-4}}{x^2-4} + \dfrac{x\sqrt{x^2-4}+2\sqrt{x^2-4}}{x^2-4} = \dfrac{2x\sqrt{x^2-4}}{x^2-4}$

35. $\sqrt{3}\sqrt{5} = \sqrt{15}$

36. $\sqrt{7}\sqrt{6} = \sqrt{42}$

37. $\sqrt{2x}\sqrt{7x} = \sqrt{14x^2} = x\sqrt{14}$

38. $\sqrt[3]{5y}\sqrt[3]{25y} = \sqrt[3]{125y^2} = 5\sqrt[3]{y^2}$

39. $\left(\sqrt{a} + \sqrt{2b}\right)\left(\sqrt{a} - \sqrt{2b}\right) = \sqrt{a}^2 - \sqrt{2b}^2 = a - 2b$

40. $\left(\sqrt{x} - 3\sqrt{y}\right)\left(\sqrt{x} + 3\sqrt{y}\right) = \sqrt{x}^2 - \left(3\sqrt{y}\right)^2 = x - 9y$

41. $\dfrac{\sqrt[3]{2a^2}}{\sqrt[3]{16a}} = \sqrt[3]{\dfrac{2a^2}{16a}} = \sqrt[3]{\dfrac{a}{8}} = \dfrac{\sqrt[3]{a}}{2}$

42. $\dfrac{\sqrt[3]{5a^2b}}{\sqrt[3]{25ab^2}} = \sqrt[3]{\dfrac{5a^2b}{25ab^2}} = \sqrt[3]{\dfrac{a}{5b}} \cdot \sqrt[3]{\dfrac{5^2b^2}{5^2b^2}} = \dfrac{\sqrt[3]{25ab^2}}{5b}$

43. $\sqrt{x-1} = 9$ so $x - 1 = 81$ and $x = 82$. Check: $\sqrt{82-1} = \sqrt{81} = 9$. Solution is 82.

44. $\sqrt{2x-7} - 5 = 0$ or $\sqrt{2x-7} = 5$. Squaring both sides gives $2x - 7 = 25$ or $2x = 32$, and so $x = 16$. Check: $\sqrt{2 \cdot 16 - 7} - 5; = \sqrt{32 - 7} - 5 = \sqrt{25} - 5 = 5 - 5 = 0$. Solution is 16.

45. Here $\sqrt{x^2-7} = 2x-4$. Squaring yields $\sqrt{x^2-7}^2 = (2x-4)^2$ or $x^2 - 7 = 4x^2 - 16x + 16$ and so $0 = 3x^2 - 16x + 25$. The discriminant, $b^2 - 4ac = -44$, is negative, so there is no solution.

46. Multiplying $\dfrac{3x}{4} - 1 = 25x^2 - 30x + 9$ by 4 gives $3x - 4 = 100x^2 - 120x + 36$, or $100x^2 - 123x + 40 = 0$. The discriminant, $b^2 - 4ac = -871$, so there is no solution.

47. Here $\sqrt{x+1} + \sqrt{x-5} = 8$ can be written as $\sqrt{x+1} = 8 - \sqrt{x-5}$. Squaring both sides produces $\sqrt{x+1}^2 = (8 - \sqrt{x-5})^2$ or $x + 1 = 64 - 16\sqrt{x-5} + (x-5)$ which simplifies to $-58 = -16\sqrt{x-5}$ or $\dfrac{58}{16} = \sqrt{x-5}$ or $3.625 = \sqrt{x-5}$. Squaring again yields $3.625^2 = x - 5$ or $13.140625 = x - 5$ and so $x = 18.140625$. This answer does check so $x \approx 18.14$.

48. Here $\sqrt{x+3} + \sqrt{x-2} = 20$ can be written as $\sqrt{x+3} = 20 - \sqrt{x-2}$. Squaring both sides gives $\sqrt{x+3}^2 = (20 - \sqrt{x-2})^2$ or $x + 3 = 400 - 40\sqrt{x-2} + (x-2)$ and so $-395 = -40\sqrt{x-2}$ or $\sqrt{x-2} = 9.875$. Squaring again gives $x - 2 = 9.875^2$ or $x - 2 = 97.51562$ and so, $x = 99.51562$. This answer does check so the solution is about 99.52.

49. Square both sides to get $\dfrac{1}{x} = \dfrac{3x}{4} - 1 = \dfrac{3x-4}{4}$. Multiplying by $4x$, we get $4 = 3x^2 - 4x$ or $3x^2 - 4x - 4 = 0$. This factors as $(3x + 2)(x - 2) = 0$ and so $x = -\dfrac{2}{3}$ or $x = 2$. We see that $-\dfrac{2}{3}$ yields a negative radicand so does not check. Thus, the only solution is 2.

50. Here $\sqrt{\frac{3}{x-1}} = \sqrt{\frac{4}{x+1}}$. Square both sides to get $\frac{3}{x-1} = \frac{4}{x+1}$ and so $3(x+1) = 4(x-1)$ or $3x+3 = 4x-4$ which simplifies to $x = 7$. Check: $\sqrt{\frac{3}{7-1}} = \sqrt{\frac{3}{6}} = \sqrt{\frac{1}{2}}; \sqrt{\frac{4}{7+1}} = \sqrt{\frac{4}{8}} = \sqrt{\frac{1}{2}}$. Solution is 7.

51. $q(t) = q_0 2^{-t/12.5}$
   (a) $q(10) = 50 \cdot 2^{-10/12.5} = 28.71745887$; or about 28.72 mg;
   (b) $q(20) = 50 \cdot 2^{-20/12.5} = 16.49384888$ or 16.49 mg.

52. **(a)** $v = \sqrt{\frac{2KE}{m}} = \frac{\sqrt{2KEm}}{m}$;
   (b) $v = \sqrt{\frac{2KE}{m}}; v^2 = \frac{2KE}{m}; m = \frac{2KE}{v^2}$

53. By the Pythagorean theorem $v^2 = v_x^2 + v_y^2$ or $v^2 = (3t+2)^2 + (4t-1)^2 = 9t^2 + 12t + 4 + 16t^2 - 8t + 1; v^2 = 25t^2 + 4t + 5; v = \sqrt{25t^2 + 4t + 5}$

54. From the diagram and the given information we have that $AB = 12, AC = x$ and let $BC = y$. Then $y - x = 5$ and $x^2 + 12^2 = y^2; y = 5 + x$ by substitution $x^2 + 144 = (5+x)^2; x^2 + 144 = 25 + 10x + x^2; 10x = 119; x = 11.9$ m; $y = 5 + 11.16 = 16.9$ m. $AC = 11.9$ m, $BC = 16.9$ m

---

CHAPTER 12 TEST

1. $8^5 8^{-7} = 8^{-2} = \frac{1}{8^2}$

2. $\left(3^5\right)^4 = 3^{20}$

3. $\left(\frac{x^2}{xy^3}\right)^5 = \left(\frac{x}{y^3}\right)^5 = \frac{x^5}{y^{15}}$

4. $x^{1/4} x^{4/3} = x^{3/12} x^{16/12} = x^{19/12}$

5. $\sqrt[3]{-64} = \sqrt[3]{(-4)^3} = -4$

6. $\sqrt[4]{81} = \sqrt[4]{3^4} = 3$

7. $\sqrt{x^4 y^3} = \sqrt{x^4}\sqrt{y^2}\sqrt{y} = x^2 y \sqrt{y}$

8. $\sqrt[3]{\frac{-27x^6 y}{z^2}} = \frac{\sqrt[3]{-27x^6 yz}}{\sqrt[3]{z^3}}$
   $= \frac{-3x^2}{z}\sqrt[3]{yz}$

9. $\sqrt{x^2 + 6x + 9} = \sqrt{(x+3)^2}$
   $= x + 3$

10. $3\sqrt{6} + 5\sqrt{24} = 3\sqrt{6} + 10\sqrt{6}$
   $= 13\sqrt{6}$

11. $\sqrt{2x}\sqrt{6x} = \sqrt{12x^2} = 2\sqrt{3}x$

12. $(3\sqrt{x} + \sqrt{5y})(3\sqrt{x} - \sqrt{5y}) = (3\sqrt{x})^2 - \sqrt{5y}^2 = 9x - 5y$

13. $\frac{\sqrt[3]{4x^2 y}}{\sqrt[3]{2x^2 y^2}} = \sqrt[3]{\frac{4x^2 y}{2x^2 y^2}} = \sqrt[3]{\frac{2}{y}}\sqrt[3]{\frac{y^2}{y^2}} = \frac{\sqrt[3]{2y^2}}{y}$

14. $\sqrt{x+3} = 6$. Square both sides to get the result $x + 3 = 36$ or $x = 33$. Check: $\sqrt{33+3} = \sqrt{36} = 6$. Solution is 33.

15. $\sqrt{2x+1} = 5$. Squaring both sides produces the result $2x + 1 = 25$ or $2x = 24$ and so $x = 12$. Check: $\sqrt{2 \cdot 12 + 1} = \sqrt{24 + 1} = \sqrt{25} = 5$. Solution is 12.

16. $\sqrt{2x^2 - 7} = x + 3$. Squaring both sides gives the result $\sqrt{2x^2 - 7}^2 = (x+3)^2$ or $2x^2 - 7 = x^2 + 6x + 9$ and so $x^2 - 6x - 16 = 0$. This factors as $(x-8)(x+2) = 0$, so $x = 8$ or $x = -2$. Check: $\sqrt{2 \cdot 8^2 - 7} = \sqrt{2 \cdot 64 - 7} = \sqrt{128 - 7} = \sqrt{121} = 11; 8 + 3 = 11; 8$ works;

$\sqrt{2(-2)^2 - 7} = \sqrt{2(4) - 7} = \sqrt{8 - 7} = 1; -2 + 3 = 1$ also, so -2 works. Solutions are 8 and -2.

17. $\sqrt{\frac{5}{x+3}} = \sqrt{\frac{4}{x-3}}$. Square both sides to get $\frac{5}{x+3} = \frac{4}{x-3}$ and so $5(x-3) = 4(x+3)$ or $5x - 15 = 4x + 12$ which simplifies to $x = 27$. Check: $\sqrt{\frac{5}{27+3}} = \sqrt{\frac{5}{30}} = \sqrt{\frac{1}{6}}; \sqrt{\frac{4}{27-3}} = \sqrt{\frac{4}{24}} = \sqrt{\frac{1}{6}}$. Solution is 27.

18. **(a)** $d = (1.12 \times 10^{-2})\sqrt[3]{F\ell}$ or $\sqrt[3]{F\ell} = \frac{d}{1.12 \times 10^{-2}} = 8.92857 \times 10^1 d$. Cubing both sides produces $F\ell = (8.92857 \times 10^1 d)^3 = 7.12 \times 10^5 d^3$ and so $F = \frac{7.12 \times 10^5 d^3}{\ell}$,
   (b) $F = \frac{7.12 \times 10^5 (1.2)^3}{5} = 246,067.2$ kg

CHAPTER
13
Exponential and Logarithmic Functions

≡ 13.1 EXPONENTIAL FUNCTIONS

1. $3^{\sqrt{2}} = 4.7288$

3. $\pi^3 = 31.0063$

5. $\left(\sqrt{3}\right)^{\sqrt{5}} = 3.4154$

7.

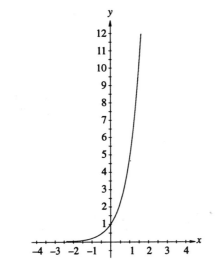

x	-3	-2	-1	0	1	2	3	4	5
$f(x)$	$\frac{1}{64}$	$\frac{1}{16}$	$\frac{1}{4}$	1	4	16	64	256	1024

9.

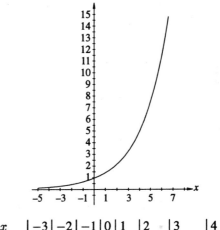

x	-3	-2	-1	0	1	2	3	4	5
$h(x)$	$\frac{8}{27}$	$\frac{4}{9}$	$\frac{2}{3}$	1	1.5	2.25	3.375	5.0625	7.59375

11.

x	-3	-2	-1	0	1	2	3	4	5
$f(x)$	27	9	3	1	$\dfrac{1}{3}$	$\dfrac{1}{9}$	$\dfrac{1}{27}$	$\dfrac{1}{81}$	$\dfrac{1}{243}$

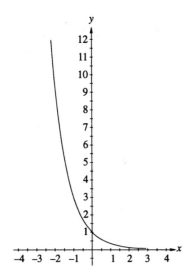

13.

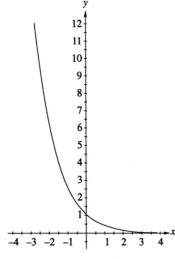

x	-3	-2	-1	0	1	2	3	4	5
$h(x)$	13.824	5.760	2.4	1	0.417	0.174	0.072	0.030	0.0126

15.

x	-3	-2	-1	0	1	2	3	4
$f(x)$	0.064	0.192	0.577	1.732	5.196	15.588	46.765	140.296

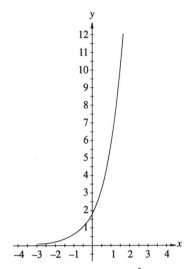

17. (a) $S = 1000(1 + 0.06)^5 = \$1338.23$,

(b) $S = 1000\left(1 + \frac{0.06}{2}\right)^{2 \times 5} = 1000(1.03)^{10} = \$1343.92$,

(c) $S = 1000\left(1 + \frac{0.06}{4}\right)^{4 \times 5} = 1000(1.015)^{20} = \$1346.86$,

(d) $S = 1000\left(1 + \frac{0.06}{12}\right)^{12 \times 5} = 1000(1.005)^{60} = \$1348.85$.

19. First Bank: $S = 1000\left(1 + \frac{0.06}{2}\right)^{2 \times 5} = 1343.92$.

Second Bank: $S = 1000\left(1 + \frac{0.06}{12}\right)^{12 \times 5} = 1348.85$. Since $1348.85 - 1343.92 = \$4.93$, you will earn \$4.93 more at the second bank.

21. (a) $200,000 = Q_0 2^2$ or $200,000 = 4Q_0$ and so $Q_0 = \frac{200,000}{4} = 50,000$.

(b) $Q = Q_0 2^4 = 50,000 \cdot 2^4 = 50,000 \cdot 16 = 800,000$

(c) We want $100,000 = 50,000 \cdot 2^t$ or $2 = 2^t$. Thus, $t = 1$, and the doubling time is 1 hr.

(d) We want $8 \cdot 50,000 = 50,000 \cdot 2^t$ or $8 = 2^t$. Since $8 = 2^3 = 2^t$, we see that $t = 3$ and it takes 3 hr for Q to become eight times as large as Q_0.

23. (a) When $t = 0$, you get $400 = Q_0 3^0$ and so, $Q_0 = 400$.

(b) Here, $t = 2.5$ gives $1,200 = 400 \cdot 3^{2.5k}$ or $3 = 3^{2.5k}$. Hence, $2.5k = 1$ and $k = \frac{1}{2.5} = \frac{2}{5}$.

(c) $Q = 400 \cdot 3^{(2/5)5} = 400 \cdot 3^2 = 400 \cdot 9 = 3,600$.

25.

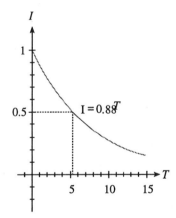

13.2 THE EXPONENTIAL FUNCTION e^x

1. $e^3 = 20.085\,537$

3. $e^{4.65} = 104.584\,986$

5. $e^{-4} = 0.018\,316$

7. $e^{-2.75} = 0.063\,928$

9.

x	-2	-1	0	1	2	3	4
$f(x)$	0.5413	1.4715	4	10.8731	29.5562	80.3421	218.3926

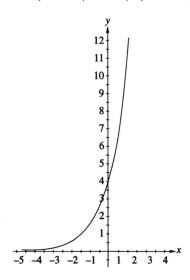

11.

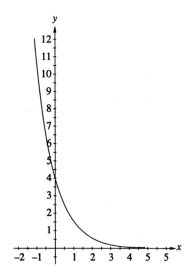

x	-4	-3	-2	-1	0	1	2
$h(x)$	218.3926	80.3421	29.5562	10.8731	4	1.4715	0.5413

13. $Q = 125e^{-0.375t}$
(a) when $t = 0$, we get $e^0 = 1$ so Q is 125 mg
(b) when $t = 1$, we have $Q = 125e^{-0.375} = 85.91$ mg
(c) when $t = 16$, we obtain $Q = 125e^{-0.375 \times 16} = 0.3098$ mg or 0.31 mg

15. Initial amount is 5000. $y = ce^{kt}$. $15{,}000 = 5000e^{20kt}; 3 = e^{20kt}$.
(a) when $t = 10$, $y = 5000e^{k10} = 5000 \left(e^{20k}\right)^{1/2} = 5000 \cdot 3^{1/2} = 8{,}660$.
(b) when $t = 30$, $y = 5000e^{30k} = 5000 \left(e^{20k}\right)^{1.5} = 5000 \times 3^{1.5} = 25{,}981$.

(c) when $t = 3$ days $= 3 \times 24$ hr $= 72$ hr. $72 \div 20 = 3.6$, $y = 5000e^{72k} = 5000 \left(e^{20k}\right)^{3.6} = 5000 \cdot 3^{3.6} = 260{,}980$.

17. Method 1. Use the compound interest formula $S = P(1 + r)^t$
(a) $S = 200{,}000(1.07^5 = 280{,}510$.
(b) $S = 2000{,}000(1.07)^{10} = 393{,}430$.

Method 2. Use the growth formula $y = ce^{kt}$ with $c = 200{,}000$ and $e^k = 1.07$.
(a) $y = 200{,}000e^{5k} = 200{,}000(e^k)^5 = 200{,}000(1.07)^5 = 280{,}510$.

(b) $y = 200{,}000e^{10k} = 200{,}000(e^k)^{10} = 200{,}000(1.07)^{10} = 393{,}430.$

19. Use $T = T_m + (T_0 - T_m)e^{-kt}$, with $T = 150°$, $T_0 = 180°$, $T_m = 60°$, and $t = 20$ min. This gives $150 = 60 + (180 - 60)e^{-k \cdot 20}$, or $90 = (120)e^{-k20}$, and so $e^{-k20} = 0.75$. When $t = 1$ hr $= 60$ min, we have $60 = 20 \times 3$. $T = 60 + (180 - 60)e^{-k \cdot 60} = 60 + 120\left(e^{-k \cdot 20}\right)^3$ or $T = 60 + 120(0.75)^3 = 60 + 50.625 \approx 110.6°$F.

21. Using the formula for Newton's law of cooling $T = T_m + (T_0 - T_m)e^{-kt}$, with $T = 70$ when $t = 25$, $T_0 = 87°$, and $T_m = 37°$. This gives $70 = 37 + (87 - 37)e^{-k25}$, or $33 = 50e^{-k25}$ or $e^{-25k} = 0.66$ which is $\left(e^{-k}\right)^{25} = 0.66$, and so $e^{-k} = 0.66^{1/25} = 0.9835$. Using this value of $e^{-k} = 0.9835$ with $T = 45$ you get $45 = 37 + (87 - 37)e^{-kt}$ or $8 = 50e^{-kt}$ or $e^{-kt} = 0.16$, and so $0.9835^t = 0.16$. Thus, $t = \frac{\log 0.16}{\log 0.9835} = 110.26$, and we have $t = 110.26$ or about 1 hr and 50 min.

23. Use $Q = Q_o e^{-t/T}$, where $Q_0 = CV$, and $T = RC$. We have $C = 130 \ \mu\text{F} = 130 \times 10^{-6}\text{F}$, $V = 120$ V, and $R = 4500 \ \Omega$. So, $Q_0 = \left(130 \times 10^{-6}\right)120 = 1.560 \times 10^{-2}$, and $T = 4500 \times 130 \times 10^{-6} = 0.585$. Hence $Q = 1.56 \times 10^{-2} \cdot e^{\frac{-0.5}{0.585}} = 0.0066$ C.

25. A half-life of 5.4 yields $\frac{1}{2}y_0 = y_0 e^{k(5.4)}$ or $\frac{1}{2} = e^{k(5.4)}$. Hence, $5.4k = \ln \frac{1}{2}$ or $k = \dfrac{\ln \frac{1}{2}}{5.4} = -0.12836$. Thus, the amount of material still radioactive after 30 days is $y = 1200e^{-0.12836(30)} \approx 25.5 \ \mu$Ci.

27. Here $C = 500 \times 10^{-6} = 5 \times 10^{-4}$ F, $R = 1 \ \text{k}\Omega = 10^3 \ \Omega$. Using the other given information, we have $i = \frac{V}{R}e^{-t/RC} = \frac{50}{10^3}e^{-t/(10^3 \cdot 5 \times 10^{-4})} = 0.05e^{-2t}$
(a) When $t = 0$, we get $i = 0.05e^{-2 \cdot 0} = 0.05e^0 = 0.05$ A
(b) Here $t = 0.02$, so $i = 0.05e^{-2 \cdot 0.02} = 0.05e^{-0.04} \approx 0.048$ A
(c) Here $t = 0.04$, so $i = 0.05e^{-2 \cdot 0.04} = 0.05e^{-0.08} \approx 0.046$ A

≡ 13.3 LOGARITHMIC FUNCTIONS

1. $\log_6 216 = 3; \ 6^3 = 216$

3. $\log_4 16 = 2; \ 4^2 = 16$

5. $\log_{1/7} \frac{1}{49} = 2; \ \left(\frac{1}{7}\right)^2 = \frac{1}{49}$

7. $\log_2 \frac{1}{32} = -5; \ 2^{-5} = \frac{1}{32}$

9. $\log_9 2187 = \frac{7}{2}; \ 9^{\frac{7}{2}} = 2187$

11. $5^4 = 625; \ \log_5 625 = 4$

13. $2^7 = 128; \ \log_2 128 = 7$

15. $7^3 = 343; \ \log_7 343 = 3$

17. $5^{-3} = \frac{1}{125}; \ \log_5 \frac{1}{125} = -3$

19. $4^{7/2} = 128; \ \log_4 128 = \frac{7}{2}$

21. $\ln 5 = 1.609 \ 437 \ 912$

23. $\ln 4.751 = 1.558 \ 355 \ 122$

25. $\log 4 = 0.602 \ 059 \ 991 \ 3$

27. $\log 12.67 = 1.102 \ 776 \ 615$

29. $\log_5 8 = \frac{\ln 8}{\ln 5} = 1.292 \ 029 \ 674$

31. $\log_{12} 16.4 = \frac{\ln 16.4}{\ln 12} = 1.125 \ 708 \ 821$

35.

33.

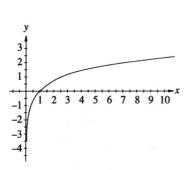

x	0.1	0.5	1	2	4	6	8	10
$\ln x$	−2.30	−0.69	0	0.69	1.39	1.79	2.08	2.30

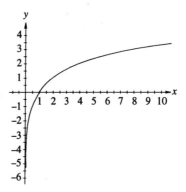

x	0.1	0.5	1	2	4	6	8	10
$\log_2 x$	−3.32	−1	0	1	2	2.58	3	3.32

37.

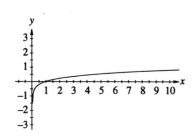

x	0.1	0.5	1	2	4	6	8	10
$\log_{12} x$	−0.93	−0.28	0	0.28	0.56	0.72	0.84	0.93

39.

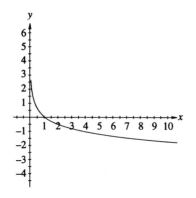

x	0.1	0.5	1	2	4	6	8	10
$\log_{1/4} x$	1.66	0.5	0	−0.5	−1	−1.29	−1.5	−1.66

41. **(a)** $\beta = 10 \log \frac{I}{I_0};$ $\frac{\beta}{10} = \log \frac{I}{I_0}, 10^{\beta/10} = \frac{I}{I_0}; I = I_0 10^{\beta/10}$

(b) $\beta = 10 \log \left(\frac{10^{-7}}{10^{-12}} \right) = 10 \log 10^5 = 10 \cdot 5 = 50$ dB

(c) $\beta = 10 \log \left(\frac{10^{-3}}{10^{-12}} \right) = 10 \log 10^9 = 10 \cdot 9 = 90$ dB

43. Substituting the given values of A, S, and BA, we get

$$\ln Y = 5.36437 - 101.16296 S^{-1} - 22.00048 A^{-1} + 0.97116 \ln BA$$
$$= 5.36437 - 101.16296(110)^{-1} - 22.00048(40)^{-1} + 0.97116 \ln 90$$
$$= 8.26473$$
$$Y = e^{8.26473} \approx 3884$$

The predicted yield is about 3,884 ft³/acre.

≡ 13.4 PROPERTIES OF LOGARITHMS

1. $\log \frac{2}{3} = \log 2 - \log 3$

3. $\log 14 = \log(2 \cdot 7) = \log 2 + \log 7$

5. $\log 12 = \log(2 \cdot 2 \cdot 3) = \log 2 + \log 2 + \log 3$

7. $\log \frac{150}{7} = \log \left(\frac{2 \cdot 3 \cdot 5 \cdot 5}{7} \right) = \log 2 + \log 3 + \log 5 + \log 5 - \log 7$

9. $\log 2x = \log 2 + \log x$

11. $\log \frac{2ax}{3y} = \log 2ax - \log 3y = \log 2 + \log a + \log x - \log 3 - \log y = \log 2 + \log a + \log x - (\log 3 + \log y) = \log 2ax - \log 3y$

13. $\log 2 + \log 11 = \log(2 \cdot 11) = \log 22$

15. $\log 11 - \log 3 = \log \frac{11}{3}$

17. $\log 2 + \log 2 + \log 3 = \log(2 \cdot 2 \cdot 3) = \log 12$

19. $\log 4 + \log x - \log y = \log \frac{4x}{y}$

21. $5 \log 2 + 3 \log 5 = \log 2^5 + \log 5^3 = \log \left(2^5 \cdot 5^3 \right) = \log 4,000$

23. $\log \frac{2}{3} + \log \frac{6}{7} = \log \left(\frac{2}{3} \cdot \frac{6}{7} \right) = \log \frac{4}{7}$

25. $\log 8 = \log 2^3 = 3 \log 2 = 3(0.3010) = 0.9030$

27. $\log 12 = \log 2^2 \cdot 3 = 2 \log 2 + \log 3 = 2(0.3010) + (0.4771) = 0.6020 + 0.4771 = 1.0791$

29. $\log 15 = \log 3 \cdot 5 = \log 3 + \log 5 = 0.4771 + 0.6990 = 1.1761$

31. $\log \frac{75}{2} = \log(3 \cdot 5 \cdot 5/2) = \log 3 + 2 \log 5 - \log 2 = 0.4771 + 2(0.6990) - (0.3010) = 1.5741$

33. $\log 200 = \log 2 \cdot 10^2 = \log 2 + 2 \log 10 = 0.3010 + 2 \cdot 1 = 2.3010$ (Note: $\log 10 = 1$)

35. $\log 5000 = \log 10^3 \cdot 5 = 3 \log 10 + \log 5 = 3 \cdot 1 + 0.6990 = 36990$

37. $\log_b 16 = \log_b 2^4 = 4 \log_b 2 = 4(0.3869) = 1.5476$

39. $\log_b \frac{5}{3} = \log_b 5 - \log_b 3 = 0.8082 - 0.6131 = 0.2851$

41. $\log 2 = \frac{\ln 2}{\ln b} = 0.3869; \ln b = \ln 2/0.3869 = 1.7915;$ Since e^x and $\ln x$ are inverse functions; $b = e^{1.7915} = 5.99$ or $b = 6$; (Note: $\log_b 6 = \log_b 2 + \log_b 3 = 0.3869 + 0.6131 = 1.0000$)

43.
$$\Delta \text{dB} = 20(\log I_2 + \frac{1}{2}\log R_2 - \log I_1 - \frac{1}{2}\log R_1)$$
$$= 20(\log I_2 + \log\sqrt{R_2} - \log I_1 - \log\sqrt{R_1})$$
$$= 20\log\left(\frac{I_2\sqrt{R_2}}{I_1\sqrt{R_1}}\right)$$
$$= \log\left(\frac{I_2\sqrt{R_2}}{I_1\sqrt{R_1}}\right)$$
$$= \log\left(\frac{I_2^{20}R_2^{10}}{I_1^{20}R_1^{10}}\right)$$

☰ 13.5 EXPONENTIAL AND LOGARITHMIC EQUATIONS

1. $5^x = 29; \log 5^x = \log 29; x\log 5 = \log 29; x = \frac{\log 29}{\log 5} = 2.0922$

3. $4^{6x} = 119; \log 4^{6x} = \log 119; 6x\log 4 = \log 119; x = \frac{\log 119}{6\log 4} = 0.5746$

5. $3^{y+5} = 16; \ln 3^{y+5} = \ln 16; (y+5)\ln 3 = \ln 16; y+5 = \frac{\ln 16}{\ln 3}; y = \frac{\ln 16}{\ln 3}; -5 = -2.4763$

7. $e^{2x-3} = 10; \ln e^{2x-3} = \ln 10; 2x-3 = \ln 10; 2x = \ln 10 + 3; x = \frac{\ln 10+3}{2} = 2.6513$

9. $3^{4x+1} = 9^{3x-5}; 3^{4x+1} = (3^2)^{3x-5}; 3^{4x+1} = 3^{6x-10}; \log_3 3^{4x+1} = \log_3 3^{6x-10}; 4x+1 = 6x-10; -2x = -11; x = \frac{11}{2} = 5.5$

11. $e^{3x-1} = 5e^{2x}; \ln e^{3x-1} = \ln(5e^{2x}); 3x-1 = \ln 5 + 2x; x = \ln 5 + 1 = 2.6094$

13. $\log x = 2.3$; rewrite as $x = 10^{2.3} \approx 199.53$

15. $\log(x-5) = 17; x-5 = 10^{17}; x = 10^{17} + 5$

17. $\ln 2x + \ln x = 9; \ln 2x^2 = 9; 2x^2 = e^9; x^2 = \frac{e^9}{2}; x = \sqrt{\frac{1}{2}e^9} \approx 63.6517$

19. $2\ln 3x + \ln 2 = \ln x; \ln 9x^2 + \ln 2 = \ln x; \ln 18x^2 = \ln x; 18x^2 = x; 18x = e^0; 18x = 1; x = \frac{1}{18} \approx 0.056$

21. $\log(x-1) = 2; x-1 = 10^2 = 100; x = 101$

23. $2\log(x-4) = 2\log x - 4; \log(x-4)^2 - \log x^2 = -4; \log\left(\frac{x-4}{x}\right)^2 = -4; \left(\frac{x-4}{x}\right)^2 = 10^{-4} = 0.0001; \frac{x-4}{x} = \sqrt{0.0001} = 0.1; x-4 = .01x; 0.99x = 4; x = \frac{4}{0.99} = 4.040404$

25. $t = \frac{\ln 2}{r} = \frac{\ln 2}{0.06} = 11.55$ years

27. $t = \frac{\ln 2}{0.005} = 138.63$ years

29. $12.5 = \frac{\ln 2}{r}; r = \frac{\ln 2}{12.5} = 0.055$ or 5.5%

31. $8.5 = \frac{\ln 2}{r}; r = \frac{\ln 2}{8.5} = 0.0815$ or 8.15%

33. $-\log(2.5 \times 10^{-6}) = 5.6$

35. $H = (30T+8000)\ln\frac{P_0}{P} = (30(-5)+8000)\ln\frac{76}{43.29} = (7850)\ln\left(\frac{76}{43.29}\right) = 4418.07$ m

37. $Q = Q_0 e^{-t/T}; T = RC = 500 \times 100 \times 10^{-6} = 0.05; 0.05 = 0.40e^{-t/0.05}; 0.125 = e^{-t/0.05}; \ln 0.125 = -t/0.05; t = -0.05\ln 0.125 = 0.104$ s

39. $I = I_0(1 - e^{-Rt/L})L = 0.2, R = 4; I_0 = 1.5; I = 1.4; 1.4 = 1.5(1 - e^{-4t/0.2}); 0.9\overline{3} = 1 - e^{-20t}; -0.06\overline{6} = -e^{-20t}; e^{-20t} = 0.0\overline{6}; -20t = \ln 0.0\overline{6}; t = \frac{\ln 0.0\overline{6}}{-20} = 0.1354$ seconds

41.
$$E = \frac{U^2 C}{2}\left(1 - e^{-2t/RC}\right)$$

$$\frac{2E}{U^2 C} = 1 - e^{-2t/RC}$$

$$e^{-2t/RC} = 1 - \frac{2E}{U^2 C}$$

$$\frac{-2t}{RC} = \ln\left(1 - \frac{2E}{U^2 C}\right)$$

$$t = -\frac{RC}{2}\ln\left(1 - \frac{2E}{U^2 C}\right)$$

13.6 GRAPHS USING SEMILOGARITHMIC AND LOGARITHMIC PAPER

1.

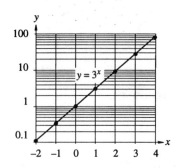

x	-3	-2	-1	0	1	2	3	4
y	0.037	0.11	0.33	1	3	9	27	81

3.

x	-3	-2	-1	0	1	2	3	4
y	0.014	0.083	0.5	3	18	108	648	3888

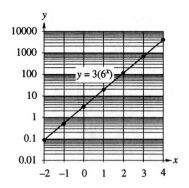

5.

x	-5	-4	-3	-2	-1	0	1	2	3
y	32	16	8	4	2	1	1/2	1/4	1/8

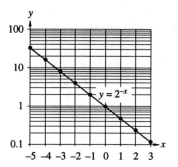

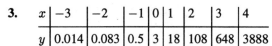

7.

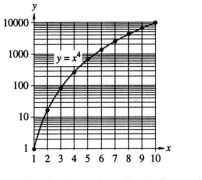

x	1	2	3	4	5	6	7	8	9	10
y	1	16	81	256	625	1296	2401	4096	6561	10000

9.

x	1	2	3	4	5	6	7	8	9	10
y	8	64	216	512	1000	1728	2744	4096	5842	8000

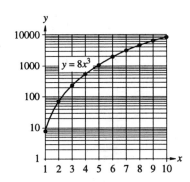

11.

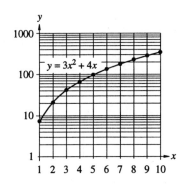

x	1	2	3	4	5	6	7	8
y	7	20	39	64	95	132	175	224

13.

x	.1	.25	1	4	9	16	25	36	49	64	81
y	.316	.5	1	2	3	4	5	6	7	8	9

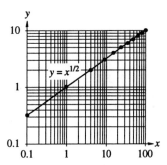

15.

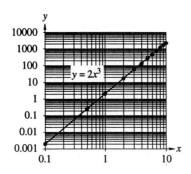

x	.1	.5	1	2	3	4	5	6	7	8
y	.002	.25	2	16	54	128	250	432	686	1024

17.

x	1	2	3	4	5	6	7	9
y	2	5.41	10.7	18	27.24	38.45	51.65	84

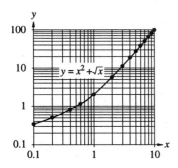

19.

x	1	2	3	4	5	6	7	8	9	10
y	5	2.5	1.67	1.25	1	.83	.71	.625	.56	.5

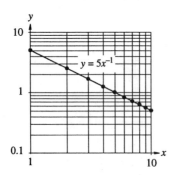

21.

x	1	2	3	4	5	7	9
y	2	5.66	10.39	16	22.36	37.04	54

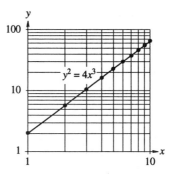

23. Rewrite $x^2 y^3 = 8$, first as $y^3 = 8x^{-2}$, and then, solve for y with the result that $y = (8x^{-2})^{1/3}$ and use this third equation to complete the following table:

x	1	2	3	4	6	8	10
y	1	1.26	0.96	0.79	0.61	0.5	0.43

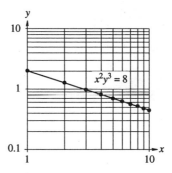

25. $P = e^{(-1.25 \times 10^{-4})h}$:

h	0	10	100	1000	5000	10000
P	1	0.999	0.987	0.882	0.535	0.287

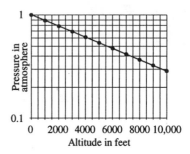

27. Rewrite $RD^2 = 0.9$ as $D^2 = 0.9R^{-1}$ and then, solve for D with the result $D = (0.9R^{-1})^{1/2}$. Now, use this last equation to complete the following table:

R	0.1	0.9	10	100	90
D	3	1	0.3	0.095	0.1

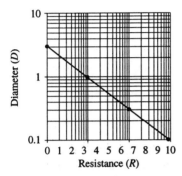

29.

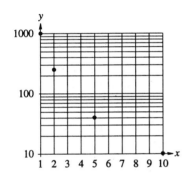

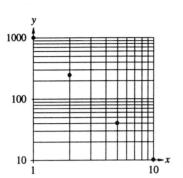

From the graph on the log-log paper we can conclude that the equation is of the form $y = ax^m$ when $x = 1$ $y = 1000$, so we can conclude that $a = 1000$. To find m take

$$m = \frac{\log 250 - \log 1000}{\log 2 - \log 1} = \frac{2.39794 - 3}{0.30103 - 0} - 2$$

hence $y = 1000x^{-2}$

31.

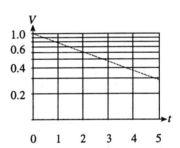

CHAPTER 13 REVIEW

1. $e^5 = 148.41316$

2. $e^{-7} = 9.118819 \times 10^{-4} = 0.00091188$

3. $e^{4.67} = 106.69774$

4. $e^{-3.91} = 0.0200405$

5. $\log 8 = 0.90309$

6. $\log 196.5 = 2.2933626$

7. $\ln 81.3 = 4.398146$

8. $\ln 325.6 = 5.7856696$

9.

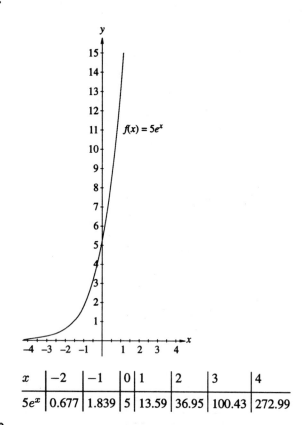

x	-2	-1	0	1	2	3	4
$5e^x$	0.677	1.839	5	13.59	36.95	100.43	272.99

10.

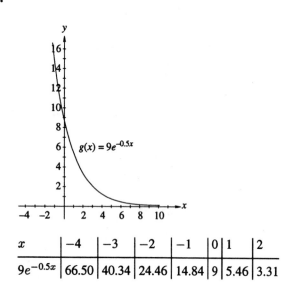

x	-4	-3	-2	-1	0	1	2
$9e^{-0.5x}$	66.50	40.34	24.46	14.84	9	5.46	3.31

11.

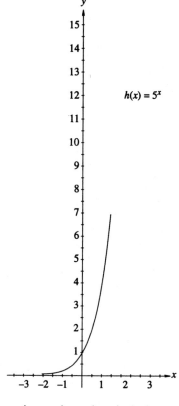

x	-3	-2	-1	0	1	2
5^x	0.008	0.04	0.2	1	5	25

12.

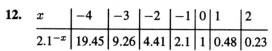

x	-4	-3	-2	-1	0	1	2
2.1^{-x}	19.45	9.26	4.41	2.1	1	0.48	0.23

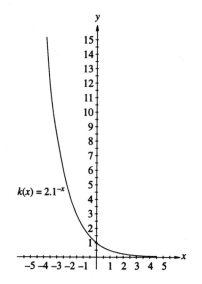

13.

x	0	1	2	3	4	5	6	7	8	9	10
$\log x$	not defined	0	0.301	0.477	0.602	0.699	0.778	0.845	0.903	0.954	1

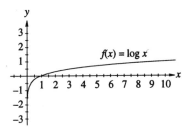

$f(x) = \log x$

14.

x	0	1	2	3	4	5	6	7	8	9	10
$3\log 2x$	not defined	0.9031	1.806	2.334	2.709	3	3.238	3.438	3.612	3.766	3.903

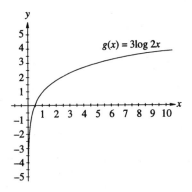

$g(x) = 3\log 2x$

15.

x	$-\frac{1}{2}$	0	1	2	3	4	5	6	7	8	9	10
$\ln(2x+1)$	not defined	0	1.099	1.609	1.946	2.197	2.398	2.565	2.708	2.833	2.944	3.045

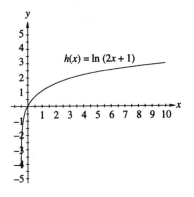

$h(x) = \ln (2x + 1)$

16.

x	0	1	2	3	4	5	6	7	8	9	10
$2 + 3\ln x$	not defined		2	4.079	5.296	6.159	6.828	7.375	7.838	8.238	8.592	8.908

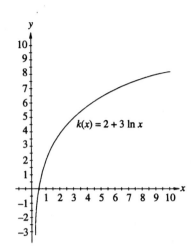

17. $\log \frac{3}{4} = \log 3 - \log 4$

18. $\log 77 = \log(7 \cdot 11) = \log 7 + \log 11$

19. $\log 5x = \log 5 + \log x$

20. $\ln \frac{7x}{3} = \ln 7 + \ln x - \ln 3$

21. $\ln(4x)^3 = 3\ln 4x = 3[\ln 4 + \ln x]$

22. $\ln 4x^3 = \ln 4 + 3\ln x$

23. $\log \sqrt{48} = \log 48^{1/2} = \frac{1}{2}\log 48 = \frac{1}{2}[\log 6 + \log 8]$. Answers may vary.

24. $\log \frac{\sqrt[3]{2x^5}}{4x} = \log(2x^5)^{1/3} - \log 4x = \frac{1}{3}[\log 2 + 5\log x] - \log 4 - \log x$. Answers may vary.

25. $\log 5 + \log 9 = \log(5 \cdot 9) = \log 45$

26. $\log 19 - \log 11 = \log \frac{19}{11}$

27. $4\log x + \log x = \log x^4 + \log x = \log x^5$

28. $\log 3a + \log b - \log a = \log\left(\frac{3ab}{a}\right) = \log 3b$

29. $7\log x + 2\log x - \log x = 8\log x = \log x^8$

30. $\ln \frac{2}{3} + \ln \frac{15}{8} = \ln\left(\frac{2}{3} \cdot \frac{15}{8}\right) = \ln\left(\frac{5}{4}\right)$

31. $4^x = 28; \log 4^x = \log 28; x\log 4 = \log 28; x = \frac{\log 28}{\log 4} = 2.404$

32. $5^{3x} = 20; \ln 5^{3x} = \ln 250; 3x\ln 5 = \ln 250; x = \frac{\ln 250}{3\ln 5} = 1.1436$

33. $e^{5x} = 11; \ln e^{5x} = \ln 11; 5x = \ln 11; x = \frac{\ln 11}{5} = 0.4796$

34. $e^{5x-1} = 2e^x; \ln^{5x-1} = \ln(2e^x) = \ln 2 + \ln e^x; 5x - 1 = \ln 2 + x; 4x = \ln 2 + 1; x = \frac{\ln 2 + 1}{4} = 0.4233$

35. $\ln x = 9.1; e^{\ln x} = e^{9.1}; x = e^{9.1} = 8955.29$

36. $4\log x = 49; \log x = \frac{49}{4}; x = 10^{49/4} = 1.778 \times 10^{12}$

37. $\log(2x + 1) = 50; 2x + 1 = 10^{50}; x = \frac{10^{50}-1}{2}$ or $0.5 \times 10^{50} - 0.5$ (Note: to 50 significant digits this is $0.5 \times 10^{50} = 5 \times 10^{49}$.)

38. $2\ln 4x = 100 + \ln 2x; \ln 16x^2 - \ln 2x = 100; \ln\left(\frac{16x^2}{2x}\right) = 100; \ln 8x = 100; 8x = e^{100}; x = \frac{e^{100}}{8} = 3.36 \times 10^{4}2$

39.

x	1	2	3	4	5	6	7	8	9
y	8	16	32	64	128	256	512	1024	2048

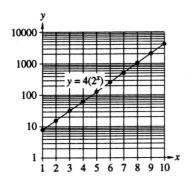

$y = 4(2^x)$

40.

x	1	2	4	6	8	10
y	8	128	2048	10,368	32,768	80,000

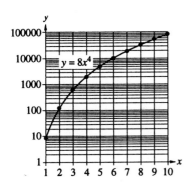

$y = 8x^4$

41.

x	1	2	3	4	5	6
y	32	128	512	2048	8192	32,768

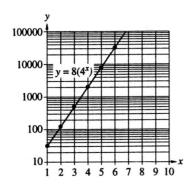

$y = 8(4^x)$

42.

x	1	2	3	4	6	8
y	3	12	48	192	3072	49,152

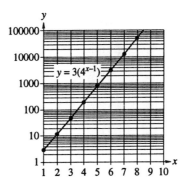

$y = 3(4^{x-1})$

43.

x	1	10	16	81	256	625	2401	10000
y	1	1.78	2	3	4	5	7	10

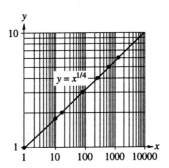

$y = x^{1/4}$

44.

x	1	2	3	4	6	8	10
y	5	40	135	320	1080	2560	5000

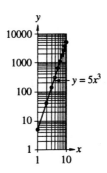

$y = 5x^3$

45.

x	0.001	0.01	0.1	1	10
y	211	37.6	6.69	1.18	0.211

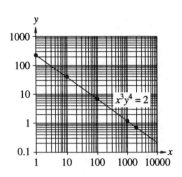

46. $y^3 = 10x^2$ or $y = (10x^2)^{1/3}$.

x	0.1	.1	10	100	1000
y	0.46	2.15	10	46.4	215.4

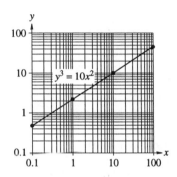

47. First Bank $S = 1000\left(1 + \frac{.075}{2}\right)^{2\times10} 2088.15$, Second Bank $S = 1000\left(1 + \frac{0.75}{12}\right)^{2\times10} = 2112.06$, Third Bank $S = 1000 \times e^{.075\times10} = 2117.00$, Double time $= \frac{\ln 2}{r} = \frac{\ln 2}{.075} = 9.24$ years

48. $e^{1600k} = \frac{1}{2}$; After 100 years $y = 100e^{100k} = 100\left(e^{1600k}\right)^{1/16} = 100(0.5)^{1/16} = 95.76$ mg. After 1000 yr, $y = 100e^{1000k} = 100\left(e^{1600k}\right)^{\frac{10}{16}} = 100(.5)^{5/8} = 64.84$ mg.

49. $y = ce^{kt}$; $25{,}000 = 4000e^{24k}$; $e^{24k} = \frac{25}{4}$; $24k = \ln \frac{25}{4}$; $k = 0.0763571$.
(a) $y = 4000e^{.07635761 \times 12} = 10000$.
(b) $y = 4000e^{0.7635761 \times 48} = 156{,}250$.

50. $R = \log I$ or San Francisco of 1906 $I = 10^{8.3}$;
(a) San Francisco of 1979 $I = 10^6$; $\frac{10^{8.3}}{10^6} = 10^{2.3} = 199.53$ times
(b) San Francisco of 1989 $I = 10^{7.1}$; $\frac{10^{8.3}}{10^{7.1}} = 10^{1.2} = 15.85$ times

51. Since the graph is a straight line on semi log paper the equation must be of the form $y = ab^x$. Hence $15.76 = ab^2$ and $620.84 = ab^4$. This gives $\frac{ab^4}{ab^2} = \frac{620.84}{15.76}$ or $b^2 = 39.3934$ or $b = 6.2764$. Back substituting to find a, $15.76 = a(6.2764)^2 a = \frac{15.76}{6.2764^2} = 0.4$. Answer: $y = 0.4(6.2764)^x$

52. Since the graph is a straight line on log-log paper, the equation must be of the form $y = ax^m$. $m = \frac{\log 2.1875 - \log 8.75}{\log 4 - \log 2} = -2$. Back substituting to find a, $8.75 = a2^{-2}$ or $2^2 \cdot 8.75 = 35$. Answer $y = 35x^{-2}$

53. $M = 6 - 2.5 \log \frac{I}{I_0}$
(a) $M - 6 = -2.5 \log \frac{I}{I_0}$; $\log \frac{I}{I_0} = \frac{6-M}{2.5}$; $\frac{I}{I_0} = 10^{\frac{6-M}{2.5}}$; $I = I_0 \cdot 10^{\frac{6-M}{2.5}}$
(b) $\dfrac{I_0 10^{\frac{6+12.5}{2.5}}}{I_0 10^{\frac{6-2}{2.5}}} = \dfrac{10^{18.5/2.5}}{10^{4/2.5}} = 10^{5.8} = 630957.34$

54. Using $Q = \frac{MQ_0}{Q_0 + (m - Q_0)e^{-kmt}}$ we have $800 = \frac{2000 \cdot 400}{400 + (2000 - 400)e^{-k \cdot 2000 \cdot 1/2}}$ or $800 = \frac{800000}{400 + (1600)e^{-1000k}}$ and so $400 + 1600e^{-1000k} = \frac{800000}{800}$ or $400 + 1600e^{-1000k} = 1000$. This yields $1600e^{-1000k} = 600$ or $e^{1000k} = \frac{600}{1600} = \frac{3}{8} = 0.375$. Taking the natural logarithm of both sides produces $-1000k = \ln 0.375 = -0.9808$ and so $k = \frac{0.9808}{1000} = 0.0009808 = 9.808 \times 10^{-4}$.

CHAPTER 13 TEST

1. $e^{5.3} = 200.33681$

2. $\log 715.3 = 2.854488225$

3. $\ln 72.35 = 4.281515453$

4. $\log \frac{9}{15} = \log 9 - \log 15$ or $\log \frac{9}{15} = \log \frac{3}{5} = \log 3 - \log 5$

5. $\log 65x = \log 65 + \log x = \log 5 + \log 13 + \log x$

6. $\ln(7x)^{-2} = -2\ln 7x = -2[\ln 7 + \ln x] = -2\ln 7 - 2\ln x$

7. $\log \frac{\sqrt[5]{5x^3}}{9x} = \log \sqrt[5]{5x^3} - \log 9x; \frac{1}{5}\log 5x^3 - \log 9x = \frac{1}{5}[\log 5 + 3\log x] - \log 9 - \log x$

8. $\log 11 + \log 3 = \log(11 \cdot 3) = \log 33$

9. $\log 17 - 2\log x = \log 17 - \log x^2 = \log \frac{17}{x^2}$

10. $\ln 4 - \ln 5 + \ln 15 - \ln 8 = \ln = \frac{4 \cdot 15}{5 \cdot 8} = \ln \frac{3}{2}$

11. $\ln x = 17.2; x = e^{17.2} = 29502925.92$

12. $3\log x = 55; \log x = \frac{55}{3}, x = 10^{55/3} = 2.154434 \times 10^{18}$

13. $3^x = 35; \ln 3^x = \ln 35; x \ln 3 = \ln 35; x = \frac{\ln 35}{\ln 3} = 3.23621727$

14. $e^{4x+1} = 2e^x$. Taking the natural logarithm of both sides gives $\ln e^{4x+1} = \ln 2e^x = \ln 2 + \ln e^x$ or $4x + 1 = \ln 2 + x$ and so $3x = \ln 2 - 1$, which means that $x = \frac{\ln 2 - 1}{3} \approx -0.10228$.

15. $\log(2x^2 + 4x) = 5 + \log 2x; \log(2x^2 + 4x) - \log 2x = 5; \log\left(\frac{2x^2 + 4x}{2x}\right) = 5; \log(x + 2) = 5; x + 2 = 10^5; x = 10^5 - 2 = 99,998$

16.

x	-3	-2	-1	0	1	2	3	4
$f(x)$	0.037	0.111	0.333	1	3	9	27	81

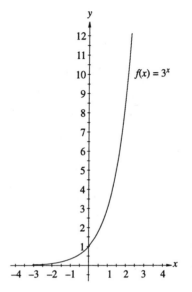

17.

x	1	2	3	4	5	6	7
y	1.2	3.6	10.8	32.4	97.2	291.6	874.8

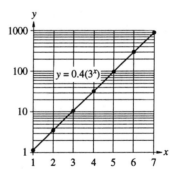

18.

x	0.1	1	2	10	32	243	1024
y	1.26	2	2.30	3.17	4	6	8

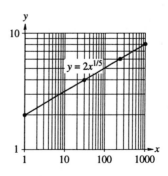

19. $\ln Y = 5.36437 - 101.16296S^{-1} - 22.00048A^{-1} + 0.97116\ln BA$

$= 5.36437 - 101.16296(120)^{-1} - 22.00048(60)^{-1} + 0.97116\ln 130$

$= 8.8818$

$Y = e^{8.8818} \approx 7199.9$

The predicted yield is about 7,200 ft^3/acre.

20. $S = P\left(1 + \dfrac{r}{k}\right)^{kt}$

$= 5,000\left(1 + \dfrac{0.055}{12}\right)^{12\cdot 8}$

$= 7,753.74$

21. We use the formula $y = ce^{kt}$. If 7.5% decays in one year, then $\frac{y}{c} = 1 - 0.75\% = 1 - 0.0075 = 0.9925$. Hence, $0.9925 = e^k$ and so $\ln 0.9925 = k$ or $k \approx -0.007528$. To find the half-life, we need to solve $\frac{y}{c} = \frac{1}{2}$ or $\frac{1}{2} = e^{-0.007528t}$ and so, $\ln(1/2) = \ln 0.5 = -0.007528t$ and $t = \frac{\ln 0.5}{-0.007528} \approx 92.07$. The half-life is about 92.07 yr.

14

Complex Numbers

≡ 14.1 IMAGINARY AND COMPLEX NUMBERS

1. $\sqrt{-25} = \sqrt{25}\sqrt{-1} = 5j$

3. $\sqrt{-0.04} = \sqrt{0.04}\sqrt{-1} = 0.2j$

5. $\sqrt{-75} = \sqrt{25}\sqrt{3}\sqrt{-1} = 5\sqrt{3}j$ or $5j\sqrt{3}$

7. $-3\sqrt{-20} = -3\sqrt{4}\sqrt{5}\sqrt{-1} = -3 \cdot 2 \cdot \sqrt{5}j = -6j\sqrt{5}$

9. $\sqrt{\frac{-9}{16}} = \sqrt{\frac{9}{16}} = \sqrt{-1} = \frac{3}{4}j$

11. $-4\sqrt{\frac{-9}{16}} = -4\sqrt{\frac{9}{16}}\sqrt{-1} = -4 \cdot \frac{3}{4}j = -3j$

13. $\left(\sqrt{-11}\right)^2 = \left(\sqrt{11}j\right)^2 = \sqrt{11}^2 j^2 = 11(-1) = -11$

15. $\left(3\sqrt{-2}\right)^2 = 3^2\sqrt{2}^2 j^2 = 9 \cdot 2(-1) = -18$

17. $\sqrt{-4}\sqrt{-25} = 2j \cdot 5j = 10j^2 = -10$

19. $\left(\sqrt{-49}\right)\left(2\sqrt{-9}\right) = 7j \cdot 2 \cdot 3j = 42j^2 = -42$

21. $\left(\sqrt{-5}\right)\left(-\sqrt{5}\right) = \sqrt{5}j\left(-\sqrt{5}\right) = -5j$

23. $j^7 = j^4 j^3 = 1(-j) = -j$

25. $\sqrt{-\frac{1}{4}}\sqrt{\frac{16}{9}} = \left(\sqrt{\frac{1}{4}}j\right)\sqrt{\frac{16}{9}} = \left(\frac{1}{2}j\right)\left(\frac{4}{3}\right) = \left(\frac{1}{2}j\right)\left(\frac{4}{3}\right) = \frac{2}{3}j$

27. $\sqrt{-\frac{3}{25}}\sqrt{\frac{3}{16}} = \left(\sqrt{\frac{3}{25}}j\right)\left(\sqrt{\frac{3}{16}}\right) = \left(\frac{\sqrt{3}}{5}\right)j\left(\frac{\sqrt{3}}{4}\right) = \frac{3}{20}j$

29. $\sqrt{-\frac{25}{36}}\sqrt{-\frac{9}{16}} = \left(\sqrt{\frac{25}{36}}j\right)\left(\sqrt{\frac{9}{16}}j\right) = \left(\frac{5}{6}j\right)\left(\frac{3}{4}j\right) = \left(\frac{5}{6}\right)\left(\frac{3}{4}\right)j^2 = \frac{5}{8}j^2 = -\frac{5}{8}$

31. $\left(\sqrt{-0.5}\right)\left(\sqrt{0.5}\right) = \sqrt{0.5}j \cdot \sqrt{0.5} = 0.5j$

33. $\left(-2\sqrt{2.7}\right)\left(\sqrt{-3}\right) = \left(-2\sqrt{2.7}\right)\left(\sqrt{3}j\right) = -2\sqrt{8.1} = -2\sqrt{9}\sqrt{0.9}j = -2\cdot3\sqrt{0.9}j = -6j\sqrt{0.9}$

35. $\left(\sqrt{-5}\right)\left(\sqrt{-6}\right)\left(\sqrt{-2}\right) = \sqrt{5}j\sqrt{6}j\sqrt{2}j = \sqrt{60}j^3 = 2\sqrt{15}(-j) = -2\sqrt{15}j = -2j\sqrt{15}$ (Note: $j^3 = -j$.)

37. $\left(-\sqrt{-7}\right)^2\left(\sqrt{-2}\right)^2 j^3 = \left(-\sqrt{7}j\right)^2\left(\sqrt{2}j\right)^2 j^3 = 7j^2 \cdot 2j^2 j^3 = 14j^3 = -14j$ (Note: $j^3 = -j$.)

39. $x + yj = 7 - 2j$ so $x = 7, y = -2$

41. $x + 5 + yj = 15 - 3j$ so $x + 5 = 15$ or $x = 10$ and $y = -3$

43. $x - 5j + 2 = 4 - 3j + yj$ so $x + 2 = 4$ or $x = 2$ and $-5 = -3 + y$ or $y = -2$

45. The real part yields $\frac{1}{2}x = -\frac{1}{4}$, so $x = -\frac{1}{2}$ and the imaginary part gives $\frac{3}{4} = 2+y$, so $y = \frac{3}{4}-2 = -\frac{5}{4}$.

47. Comparing the real parts, we see that $1.2x + 3 = 7.2$ or $1.2x = 4.2$ and so, $x = 4.2 \div 1.2 = 3.5$. Comparing the imaginary parts, we obtain $y = -4.3$

49. $2(4 + 5j) = 8 + 10j$

51. $-5(2 + j) = -10 - 5j$

53. $j(3 - 2j) = 3j - 2j^2 = 3j + 2 = 2 + 3j$

55. $2j(4 + 3j) = 8j + 6j^2 = 8j - 6 = -6 + 8j$

57. $\frac{1}{2}(6 - 8j) = 3 - 4j$

59. $\frac{5-10j}{5} = \frac{5}{5} - \frac{10}{5}j = 1 - 2j$

61. $\frac{6+\sqrt{-18}}{3} = \frac{6+3j\sqrt{2}}{3} = \frac{6}{3} + \frac{3j\sqrt{2}}{3} = 2 + j\sqrt{2}$

63. $\frac{8-\sqrt{-24}}{4} = \frac{8-\sqrt{24}j}{4} = \frac{8-2\sqrt{6}j}{4} = \frac{8}{4} - \frac{2j\sqrt{6}}{4} = 2 - \frac{1}{2}j\sqrt{6}$

65. $\frac{2}{3}\left(\frac{3}{4}-\frac{1}{3}j\right) = \frac{2}{3}\left(\frac{3}{4}\right) - \frac{2}{3}\left(\frac{1}{3}\right)j = \frac{6}{12} - \frac{2}{9}j = \frac{1}{2} - \frac{2}{9}j$

67. $-\frac{5}{3}\left(-\frac{3}{8}+\frac{6}{15}j\right) = -\frac{5}{3}\left(-\frac{3}{8}\right)+\left(-\frac{5}{3}\right)\frac{6}{15}j = \frac{15}{24} - \frac{30}{45}j = \frac{5}{8} - \frac{2}{3}j$

69. $1.5(2.4 - 3j) = 1.5(2.4) - 1.5(3j) = 3.6 - 4.5j$

71. The conjugate of $7 + 2j$ is $7 - 2j$

73. The conjugate of $6 - 5j$ is $6 + 5j$

75. The conjugate of 19 is 19

77. The conjugate of $-8j$ is $8j$

79. $\sqrt{2} - 7.3j$

81. $x^2 + x + 2.5 = 0$. Here $a = 1, b = 1$, and $c = 2.5$. Thus $x = \frac{-1\pm\sqrt{1^2-4\cdot1\cdot2.5}}{2\cdot1} = \frac{-1\pm\sqrt{1-10}}{2} = \frac{-1\pm\sqrt{-9}}{2} = \frac{-1\pm3j}{2}$. Hence $x = -\frac{1}{2} + \frac{3}{2}j$ and $x = -\frac{1}{2} - \frac{3}{2}j$.

83. $x^2 + 9 = 0$. Here $a = 1, b = 0$, and $c = 9$. Thus $x = \frac{-0\pm\sqrt{0^2-4\cdot9}}{2} = \frac{\pm\sqrt{-36}}{2} = \frac{\pm6j}{2}$. Hence $x = 3j$ and $x = -3j$.

85. $2x^2 + 3x + 7 = 0$. Here $a = 2, b = 3$, and $c = 7$. Thus $x = \frac{-3\pm\sqrt{3^2-4\cdot2\cdot7}}{2\cdot2} = \frac{-3\pm\sqrt{9-56}}{4} = \frac{-3\pm\sqrt{-47}}{4} = \frac{-3\pm j\sqrt{47}}{4}$. Hence $x = \frac{-3}{4} + \frac{\sqrt{47}}{4}j$ and $x = \frac{-3}{4} - \frac{\sqrt{47}}{4}j$.

87. $5x^2 + 2x + 5 = 0$. Here $a = 5, b = 2$, and $c = 5$. Thus $x = \frac{-2\pm\sqrt{2^2-4\cdot5\cdot5}}{2\cdot5} = \frac{-2\pm\sqrt{4-100}}{10} = \frac{-2\pm\sqrt{-96}}{10} = \frac{-2\pm4j\sqrt{6}}{10}$. Hence $x = \frac{-1}{5} + \frac{2\sqrt{6}}{5}j$ and $x = \frac{-1}{5} - \frac{2\sqrt{6}}{5}j$.

89. Substituting the given data, we obtain $0.08 = \frac{X}{50^2+X^2}$ and so $0.08(50^2 + X^2) = X$ or $0.08X^2 - X + 200 = 0$. Using the quadratic formula with $a = 0.08, b = -1$, and $c = 200$, we get $6.25 + 49.61j\,\Omega$ or $6.25 - 49.61j\,\Omega$.

91. $R^2 + X^2 = \dfrac{X}{B}$

$R^2 = \dfrac{X}{B} - X^2$

$R = \pm\sqrt{\dfrac{X}{B} - X^2}$

≡ 14.2 OPERATIONS WITH COMPLEX NUMBERS

1. $(5 + 2j) + (-6 + 5j) = -1 + 7j$

3. $(11 - 4j) + (-6 + 2j) = 5 - 2j$

5. $(4 + \sqrt{-9}) + (3 - \sqrt{-16}) = (4 + 3j) + (3 - 4j) = 7 - j$

7. $\left(2 + \sqrt{-9}\right) + \left(8j - \sqrt{5}\right) = (2+3j)+\left(8j - \sqrt{5}\right) = 2 - \sqrt{5} + 11j$

9. $(14 + 3j) - (6 + j) = 8 + 2j$

11. $\left(9 - \sqrt{-4}\right) - \left(\sqrt{-16} + 6\right) = (9 - 2j) - (4j + 6) = 3 - 6j$

13. $(4 + 2j) + j + (3 - 5j) = 7 - 2j$

15. $(2 + j)3j = 6j + 3j^2 = -3 + 6j$

17. $(9 + 2j)(-5j) = -45j - 10j^2 = 10 - 45j$

19. $(2 + j)(5 + 3j) = 10 + 6j + 5j + 3j^2 = 7 + 11j$

21. $(6 - 2j)(5 + 3j) = 30 + 18j - 10j - 6j^2 = 36 + 8j$

23. $(2\sqrt{-9} + 3)(5\sqrt{-16} - 2) = (6j + 3)(20j - 2) = 120j^2 - 12j + 60j - 6 = -126 + 48j$

25. $\left(\sqrt{-3}\right)^4 = \left(\sqrt{3}j\right)^4 = \sqrt{3}^4 j^4 = 9(1) = 9$; (Note: $j^4 = 1$)

27. $(1 + 2j)^2 = 1 + 4j + 4j^2 = -3 + 4j$

29. $(7 - j)^2 = 49 - 14j + j^2 = 48 - 14j$

31. $(5 + 2j)(5 - 2j) = 5^2 + 2^2 = 25 + 4 = 29$ (Note: $(a + bj)(a - bj) = a^2 + b^2$)

33. $\frac{6-4j}{1+j} \cdot \frac{1-1j}{1-j} = \frac{6-6j-4j+4j^2}{1+1} = \frac{2-10j}{2} = 1 - 5j$

35. $\frac{6-3j}{1+2j} \cdot \frac{1-2j}{1-2j} = \frac{6-12j-3j+6j^2}{1+4} = \frac{-15j}{5} = -3j$

37. $\frac{4+2j}{1-2j} \cdot \frac{1+2j}{1+2j} = \frac{4+8j+2j+4j^2}{1+4} = \frac{10j}{5} = 2j$

39. $\frac{2j}{5+j} \cdot \frac{5-j}{5-j} = \frac{10j-2j^2}{25+1} = \frac{2+10j}{26} = \frac{1}{13} + \frac{5}{13}j$

41. $\frac{\sqrt{3}-\sqrt{-6}}{\sqrt{-3}} = \frac{\sqrt{3}-\sqrt{6}j}{\sqrt{3}j} \cdot \frac{-\sqrt{3}j}{-\sqrt{3}j} = \frac{-3j+3\sqrt{2}j^2}{3} = \frac{-3\sqrt{2}-3j}{3} = -\sqrt{2} - j$

43. $\frac{(5+2j)(3-j)}{4+j} = \frac{15-5j+6j-2j^2}{4+j} = \frac{17+j}{4+j} \cdot \frac{4-j}{4-j} =$

$\frac{68-17j+4j-j^2}{16+1} = \frac{69-13j}{17} = \frac{69}{17} - \frac{13}{17}j \approx 4.059 - 0.765j$

45. $(1+j)^4 = [(1+j)(1+j]^2 = (1+2j+j^2)^2 = (2j)^2 = 4j^2 = -4$

47. $\left(\frac{1}{2} + \frac{\sqrt{3}}{2}j\right)^2 = \frac{1}{4} + \frac{\sqrt{3}}{2}j + \frac{3}{4}j^2 = -\frac{1}{2} + \frac{\sqrt{3}}{2}j = -0.5 + \frac{i\sqrt{3}}{2}$

49. $(a+bj)+(a-bj) = (a+a)+(b-b)j = 2a+0j = 2a$ which is a real number

51. $(a+bj)(a-bj) = a^2 - (bj)^2 = a^2 - b^2j^2 = a^2 + b^2$ which is a real number

53. $(6.21-1.37j)+(4.32-2.84j) = 10.53-4.21j$ V

55. $(7.42+1.15j)-(2.34-1.73j) = 5.08+2.88j$ V

57. $(9.13-4.27j)-(3.29-5.43j) = 5.84+1.16j$ Ω

59. $Z = \frac{(20+10j)(10-20j)}{(20+10j)+(10-20j)} = \frac{200-400j+100j-200j^2}{30-10j} = \frac{15000-5000j}{1000} = 15-5j$ Ω

61. $I = \frac{V}{Z} = \frac{5.2+3j}{4-2j} \cdot \frac{4+2j}{4+2j} = \frac{20.8+10.4j+12j+6j^2}{16+4} = \frac{14.8+22.4j}{20} = 0.74+1.12j$ A

63. See *Computer Programs* in main text.

65. See *Computer Programs* in main text.

≡ 14.3 GRAPHING COMPLEX NUMBERS; POLAR FORM OF A COMPLEX NUMBER

1. $4(\cos 30° + j \sin 30°) = 2\sqrt{3}+2j; a = 4\cos 30° = 2\sqrt{3} \approx 3.4641; b = 4\sin 30° = 2$

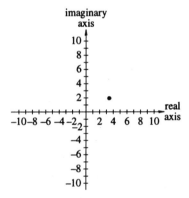

3. $5(\cos 135° + j \sin 135°) = \frac{-5\sqrt{2}}{2} + \frac{5j\sqrt{2}}{2} \approx -3.5355 + 3.5355j$

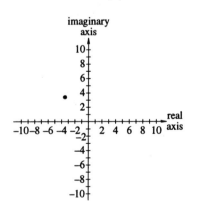

5. $7 \operatorname{cis} 260° = 7\cos 260° + 7j\sin 260° = -1.2155 - 6.8937j$

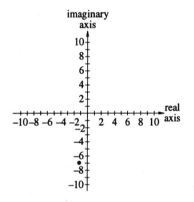

7. $2 \operatorname{cis} 115° = 2\cos 115° + 2j\sin 115° = -0.8452 + 1.8126j$

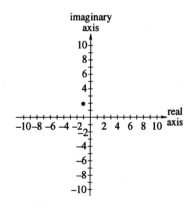

9. $3\underline{/25°} = 3\cos 25° + 3j\sin 25° = 2.7189 + 1.2679j$

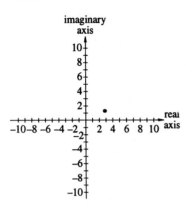

11. $5\underline{/340°} = 5\cos 340° + 5j\sin 340° = 4.6985 - 1.7101j$

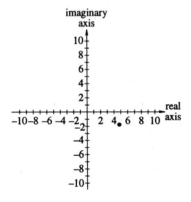

13. $4.5\underline{/245°} = 4.5\cos 245° + 4.5j\sin 245° = -1.9018 - 4.0784j$

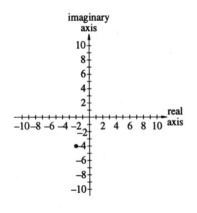

15. $2.5\underline{/180°} = 2.5\cos 180° + 2.5j\sin 180° = -2.5$

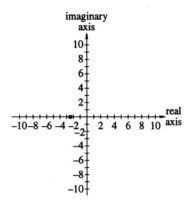

17. $5 + 5j; r = \sqrt{5^2 + 5^2} = \sqrt{50} = 5\sqrt{2} \approx 7.071; \theta = \tan^{-1}\frac{5}{5} = 45°; 7.071\underline{/45°}$ or 7.071 cis $45°$

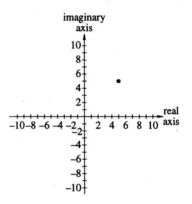

19. $4 - 8j; r = \sqrt{4^2 + (-8)^2} = 8.944; \theta = \tan^{-1}\left(-\frac{8}{4}\right) = -63.4°$ or $296.6°; 8.944\underline{/296.6°}$ or 8.944 cis $296.6°$

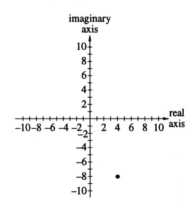

21. $-4 + 7j; r = \sqrt{(-4)^2 + 7^2} = 8.0623; \theta =$
$\tan^{-1} \frac{-7}{4} = 119.7°; 8.0623\underline{/119.7°}$

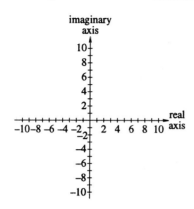

27. $4.2 - 6.3j; r = \sqrt{4.2^2 + (-6.3)^2} = 7.5717; \theta =$
$\tan^{-1} \frac{-6.3}{4.2} = -56.3 \text{ or } 303.7; 7.5717\underline{/303.7°}$

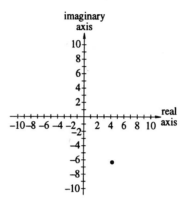

23. $-6 - 2j; r = \sqrt{(-6)^2 + (-2)^2} = 6.3246;$
$\theta = \tan^{-1} \frac{-2}{-6} = 198.4°; 6.3246\underline{/198.4°}$

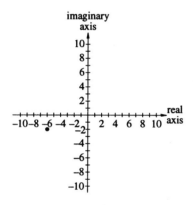

29. $-5.8 + 0.2j; r = \sqrt{(-5.8)^2 + 0.2^2} = 5.8034; \theta =$
$\tan^{-1} \frac{0.2}{-5.8} = 178.0°; 5.8034\underline{/178.0°}$

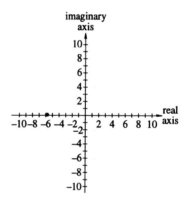

25. $6; r = 6; \theta = 0; 6\underline{/0°}$

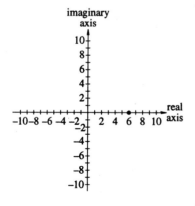

31. $-2.7; r = 2.7; \theta = 180°; 2.7\underline{/180°}$

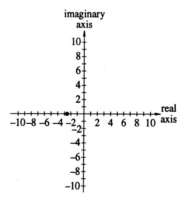

33.

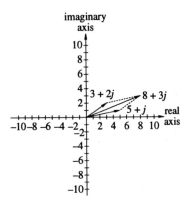

35.

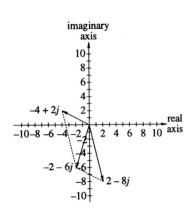

37.

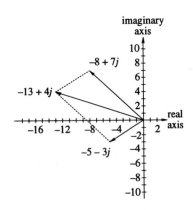

39.

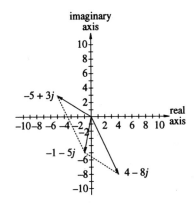

41. First, $r = \sqrt{4.7^2 + (-6.5)^2} = \sqrt{64.34} \approx 8.0$. Then we have $\tan\theta = \frac{-6.5}{4.7}$. Since θ is in the fourth quadrant, $\theta = \tan^{-1}\frac{-6.5}{4.7} \approx -54.1°$ or $-54.1° + 360° = 305.9°$. Thus, $4.7 - 6.5j$ written in polar form is $8.0\underline{/-54.1°} = 8.0\underline{/305.9°}$ A.

43. $r = \sqrt{(-110.4)^2 + 46.1^2} = \sqrt{14,313.37} \approx 119.6$. $\tan\theta = \frac{46.1}{-110.4}$. Since θ is in the second quadrant, $\theta = \tan^{-1}\frac{46.1}{-110.4} + 180° \approx 157.3°$. Thus, $-110.4 + 46.1j$ V is written in polar form as $119.6\underline{/157.3°}$ V.

45. $a = r\cos\theta = 2.5\cos(-50°) \approx 1.61$. $b = r\sin\theta = 2.5\sin(-50°) \approx -1.92$. Thus, $2.5\underline{/-50°}$ A is written in rectangular form as $1.61 - 1.92j$ A.

47. See *Computer Programs* in main text.

≡ 14.4 EXPONENTIAL FORM OF A COMPLEX NUMBER

1. $3(\cos\frac{3\pi}{2} + j\sin\frac{3\pi}{2}); r = 3; \theta = \frac{3\pi}{2}$; Answer: $3e^{3\pi/2 j} \approx 3e^{4.7j}$

3. $2(\cos 60° + j\sin 60°); r = 2, \theta = 60° = \frac{\pi}{3}$; Answer: $2e^{\frac{\pi}{3}j} \approx 2e^{1.05j}$

5. $1.3(\cos 5.7 + \sin 5.7) = 1.3e^{5.7j}$

7. $25° = 0.44; 3.1(\cos 25° + j\sin 25°) = 3.1e^{0.44j}$

9. $8 + 6j = 10\,\text{cis}\,0.64 = 10e^{0.64j}$

11. $-9 + 12j = 15\,\text{cis}\,2.21 = 15e^{2.21j}$

13. $5e^{0.5j} = 5\,\text{cis}\,0.5 = 4.3879 + 2.3971j$

15. $2.3e^{4.2j} = 2.3\,\text{cis}\,4.2 = -1.1276 - 2.0046j$

17. $2e^{3j} \cdot 6e^{2j} = (2 \cdot 6)e^{3j+2j} = 12e^{5j}$

19. $e^{1.3j} \cdot 2.4e^{4.6j} = (1 \cdot 2.4)e^{(1.3+4.6)j} = 2.4e^{5.9j}$

21. $7e^{4.3j} \cdot 4e^{5.7j} = (7 \cdot 4)e^{(4.3+5.7)j} = 28e^{10j} = 28e^{3.72j}$
(Note: $10 - 2\pi \approx 3.72$)

23. $8e^{3j} \div 2e^{j} = (8/2)e^{3j-j} = 4e^{2j}$

25. $17e^{4.3j} \div 4e^{2.8j} = 17/4e^{(4.3-2.8)j} = 4.25e^{1.5j}$

27. $(3e^{2j})^4 = 3^4 e^{2j \cdot 4} = 81e^{8j} \approx 81e^{1.72j}$ (Note: $8 - 2\pi \approx 1.72$.)

29. $(2.5e^{1.5j})^4 = 2.5^4 e^{1.5j \cdot 4} = 39.0625e^{6j}$

31. $(4e^{6j})^{1/2} = 4^{1/2}e^{\frac{6j}{2}} = 2e^{3j}$

33. $(6.25e^{4.2j})^{1/2} = 6.25^{1/2}e^{\frac{4.2j}{2}} = 2.5e^{2.1j}$

35. $V = 56.5 + 24.1j = 61.41 \operatorname{cis} 0.4031 = 61.41e^{0.4031j} \text{ V}$

37. $Z = 135\underline{/-52.5°} \ \Omega = 135e^{5.37j} \ \Omega = 82.2 - 107.1j \ \Omega$

39. $V = IZ = (12.5e^{-0.7256j})(6.4e^{1.4285j}) = (12.5 \times 6.4)e^{(-0.7256+1.4285)j} = 80e^{0.7029j} \text{ V}$

41. $V = IZ \Rightarrow I = \frac{V}{Z} = \frac{115e^{-0.2125j}}{2.5e^{0.5792j}} = \frac{115}{2.5}e^{(-0.2145-0.5792)j} = 46e^{-0.7937j} \text{A} \approx 46e^{5.49j} \text{A}$

43. $Z = \frac{V}{I} = \frac{122.4e^{0.2551j}}{36e^{-0.8189j}} = \frac{122.4}{36}e^{0.2551j-(-0.8189j)} = \frac{122.4}{36}e^{(0.2551+0.8189)j} = 3.4e^{1.074j} \ \Omega$

≡ 14.5 OPERATIONS IN POLAR FORM; DEMOIVRE'S FORMULA

1. $(3 \operatorname{cis} 46°) \cdot (5 \operatorname{cis} 23°) = 3 \cdot 5 \operatorname{cis}(46° + 23°)$
$= 15 \operatorname{cis} 69° = 15(\cos 69° + j \sin 69°)$

3. $(2.5 \operatorname{cis} 1.4°) \cdot (4 \operatorname{cis} 2.67) = 2.5 \cdot \operatorname{cis}(1.43 + 2.67) = 10 \operatorname{cis} 4.1 = 10(\cos 4.1° + j \sin 4.1°)$

5. $\frac{8 \operatorname{cis} 85°}{2 \operatorname{cis} 25°} = \frac{8}{2} \operatorname{cis}(85° - 25°) = 4 \operatorname{cis} 60° = 4(\cos 60° + j \sin 60°)$

7. $\frac{9\underline{/137°}}{2\underline{/26°}} = \frac{9}{2}\underline{/137° - 26°} = 4.5\underline{/111°}$

9. $(3\underline{/2.7})(4\underline{/5.3}) = 3 \cdot 4\underline{/2.7 + 5.3} = 12\underline{/8} = 12\underline{/1.72}$
(Note: $8 - 2\pi \approx 1.72$.)

11. $[5 \operatorname{cis} 84°]^6 = 5^6 \operatorname{cis}(6 \cdot 84) = 5^6 \operatorname{cis}(504) = 15,625 \operatorname{cis} 144°$

13. $[10.4 \operatorname{cis} 3.42]^3 = 10.4^3 \operatorname{cis}(3 \cdot 3.42)$
$= 1124.864 \operatorname{cis} 10.26 = 1124.864 \operatorname{cis} 3.98$

15. $(3.4\underline{/5.3})^4 = 3.4^4\underline{/4 \cdot 5.3} = 133.6\underline{/21.2} \approx 133.6\underline{/2.35}$ (Note: $21.2 \approx 2.35 + 6\pi$.)

17. $(4e^{2.1j})(3e^{1.7j}) = 12e^{(2.1+1.7)j} = 12e^{3.8j}$

19. $(0.5e^{0.3j})^3 = 0.5^3 e^{3 \times 0.3j} = 0.125e^{0.9j}$

21. Cube roots of 1. First the polar form of 1 is $1 \operatorname{cis} 0° = \operatorname{cis} 0°$. $w_0 = \sqrt[3]{1} \operatorname{cis}\left(\frac{0}{3} + \frac{360 \cdot 0}{3}\right)° = 1 \operatorname{cis} 0° = 1; w_1 = \sqrt[3]{1} \operatorname{cis}\left(\frac{0}{3} + \frac{360 \cdot 1}{3}\right)° = 1 \operatorname{cis} 120° = -0.5 + 0.866j; w_2 = \sqrt[3]{1} \operatorname{cis}\left(\frac{0}{3} + \frac{360 \cdot 2}{3}\right)° = 1 \operatorname{cis} 240° = -0.5 - 0.866j°$

23. We want the cube roots of $-8j$. Now, $-8j = 8 \operatorname{cis} 270°$. $w_0 = \sqrt[3]{8} \operatorname{cis}\left(\frac{270}{3} + \frac{360 \cdot 0}{3}\right)° = 2 \operatorname{cis} 90°$

$= 2j; w_1 = \sqrt[3]{8} \operatorname{cis}\left(90 + \frac{360 \cdot 1}{3}\right)° = 2 \operatorname{cis} 210° = -1.732 - j; w_2 = 2 \operatorname{cis}\left(90 + \frac{360 \cdot 2}{3}\right)° = 2 \operatorname{cis} 330° = 1.732 - j$

25. Fourth roots of $-16j$. $-16j = 16 \operatorname{cis} 270° w_0 = \sqrt[4]{16} \operatorname{cis}\left(\frac{270}{4} + \frac{360 \cdot 0}{4}\right)° = 2 \operatorname{cis} 67.5° = 0.7654 + 1.8478j; w_1 = 2 \operatorname{cis}\left(67.5 + \frac{360 \cdot 1}{4}\right)° = 2 \operatorname{cis} 157.5° = -1.8478 + 0.7654j; w_2 = 2 \operatorname{cis}\left(67.5 + \frac{360 \cdot 2}{4}\right)° = 2 \operatorname{cis} 247.5° = -0.7654 - 1.8478j; w_3 = 2 \operatorname{cis}\left(67.5 + \frac{360 \cdot 3}{4}\right)° = 2 \operatorname{cis} 337.5° = 1.8478 - 0.7654j$

27. Fifth roots of $1 + j$. $1 + j = \sqrt{2} \operatorname{cis} 45°$. $w_0 = \sqrt[5]{\sqrt{2}} \operatorname{cis}\left(\frac{45}{5} + \frac{360 \cdot 0}{5}\right)° = 1.0718 \operatorname{cis} 9° = 1.0586 + 0.1677j; w_1 = \sqrt[10]{2} \operatorname{cis}\left(9 + \frac{360 \cdot 1}{5}\right)° = 1.0718 \operatorname{cis} 81° = 0.1677 + 1.0586j; w_2 = \sqrt[10]{2} \operatorname{cis}\left(9 + \frac{360 \cdot 2}{5}\right)° = 1.0718 \operatorname{cis} 153° = -0.9550 + 0.4866j; w_3 = \sqrt[10]{2} \operatorname{cis}\left(9 + \frac{360 \cdot 3}{5}\right)° = 1.0718 \operatorname{cis} 225° = -0.7579 - 0.7579j; w_4 = \sqrt[10]{2} \operatorname{cis}\left(9 + \frac{360 \cdot 4}{5}\right) = 1.0718 \operatorname{cis} 297° = 0.4866 - 0.9550j$

29. $x^3 = -j$: the solutions are the three cube roots of $-j = 1 \operatorname{cis} 270°$. $w_0 = \sqrt[3]{1} \operatorname{cis}(270/3)° = 1 \operatorname{cis} 90° = j; w_1 = 1 \operatorname{cis}(90 + 360/3)° = 1 \operatorname{cis} 210° = -0.8660 - 0.5j; w_2 = 1 \operatorname{cis}(90 + 120 \cdot 2)° = 1 \operatorname{cis} 330° = 0.8660 - 0.5j$

31. The solution of $x^6 - 64j = 0$ are the six sixth roots of $64j = 64 \operatorname{cis} 90°$. $w_0 = \sqrt[6]{64} \operatorname{cis}(90/6)° = 2 \operatorname{cis} 15° = 1.9319 + 0.5176j; w_1 = 2 \operatorname{cis}(15 + \frac{360}{6})° = 2 \operatorname{cis} 75° = 0.5176 + 1.9319j; w_2 =$

$2\operatorname{cis}(15 + 60 \times 2)° = 2\operatorname{cis}135° = -1.4142 + 1.4142j; w_3 = 2\operatorname{cis}(15 + 60 \times 3)° = 2\operatorname{cis}195° = -1.9319 - 0.5176j; w_4 = 2\operatorname{cis}(15 + 60 \times 4)° = 2\operatorname{cis}255° = -0.5176 - 1.9319j; w_5 = 2\operatorname{cis}(15 + 60 \times 5)° 2\operatorname{cis}315° = 1.412 - 1.4142j$

33. $V = IZ; I = 12\underline{/-23°}, Z = 9\underline{/42°}; V = (12\underline{/-23°})(9\underline{/42°}) = (12 \times 9)\underline{/-23° + 42°} = 108\underline{/19°}$

35. We first convert Z_T to polar form. $Z_T = 4\underline{/-90°} + 3\underline{/0°} = -4j + 3 = 5\underline{/-53.13°}\ \Omega$. Then,

$$V_x = \frac{Z_x E}{Z_T} = \frac{(4\underline{/-90°})(100\underline{/0°})}{5\underline{/-53.13°}}$$
$$= \frac{4 \cdot 100}{5}\underline{/-90° + 0° - (-53.13°)}$$
$$= 80\underline{/-36.87°}\ \text{V}$$

37. First, convert each of the expressions to polar form: $1 + j = \sqrt{2}\underline{/45°}; 1 - j = \sqrt{2}\underline{/-45°};$ and $3 - 4j = 5\underline{/-53.13°}$.

Substituting these polar expressions in the given equation, we obtain

$$Z = \frac{(1+j)^2(1-j)^2}{(3-4j)^2}$$
$$= \frac{(\sqrt{2}\underline{/45°})^2(\sqrt{2}\underline{/-45°})^2}{(5\underline{/-53.13°})^2}$$
$$= \frac{(2\underline{/90°})(2\underline{/-90°})}{25\underline{/-106.26°}}$$
$$= \frac{4}{25}\underline{/106.26°} = 0.16\underline{/106.26°}\ \Omega$$

If we change this to rectangular form, we see that $Z = 0.16\underline{/106.26°} = 0.16\cos 106.26° + 0.16j\sin 106.26° = -0.0448 + 0.1536j\ \Omega$.

39. **(a)** $Y = Z^{-1} = \frac{1}{Z} = \frac{1}{4.68\underline{/20.56°}} = \frac{1}{4.68}\underline{/-20.56°} \approx 0.2137\underline{/-20.56°}$ S

(b) Converting $Y = Z^{-1} \approx 0.2137\underline{/-20.56°}$ S to rectangular form, we get $a = 0.2137\cos -20.56° \approx 0.200$ and $b = 0.2137\sin -20.56° \approx -0.075$. Thus, $Y = 0.200 - 0.075j$ S.

41. See *Computer Programs* in main text.

43. See *Computer Programs* in main text.

▤ 14.6 COMPLEX NUMBERS IN AC CIRCUITS

1. **(a)** in Series
$Z = Z_1 + Z_2 = (2 + 3j) + (1 - 5j) = 3 - 2j;$

(b) in Parallel $Z = \frac{Z_1 Z_2}{Z_1 + Z_2} = \frac{(2+3j)(1-5j)}{3-2j} = \frac{2-7j-15j^2}{3-2j} = \frac{17-7j}{3-2j} \cdot \frac{3+2j}{3+2j} = \frac{51+34j-21j-14j^2}{9+4} = \frac{65+13j}{13} = 5 + j$

3. **(a)** in Series $Z = Z_1 + Z_2 = (1 - j) + 3j = 1 + 2j;$

(b) in Parallel

$Z = \frac{Z_1 Z_2}{Z_1 + Z_2} = \frac{(1-j)(3j)}{1+2j} = \frac{3+3j}{1+2j} \cdot \frac{1-2j}{1-2j} = \frac{3-6j+3j-6j^2}{1+4} = \frac{9-3j}{5} = 1.8 - 0.6j$

5. $Z_1 = 2.19\underline{/18.4°} = 2.0780 + 0.6913j;$
$Z_2 = 516\underline{/67.3°} = 1.9913 + 4.7603j;$
(a) $Z = Z_1 + Z_2 = 4.0693 + 5.4516j = 6.803\underline{/53.26°};$

(b) $Z = \frac{Z_1 Z_2}{Z_1 + Z_2} = \frac{(2.19\underline{/18.4°})(5.16\underline{/67.3°})}{6.803\underline{/53.26°}} = \frac{2.19 \cdot 5.16}{6.803}\underline{/18.4 + 67.3 - 53.26} = 1.6611\underline{/32.44°}$

7. $Z_1 = 3\sqrt{5}\underline{/\pi/7} = 6.0439 + 2.9106j;$
$Z_2 = 1.5\underline{/0.45} = 1.3507 + 0.6524j;$
(a) $Z = Z_1 + Z_2 = 7.3946 + 3.5630j = 8.2082\underline{/0.449};$

(b) $Z = \frac{Z_1 Z_2}{Z_1 + Z_2} = \frac{(3\sqrt{5}\underline{/\pi/7})(1.5\underline{/0.45})}{8.2082\underline{/0.449}} = \left(\frac{3\sqrt{5} \cdot 1.5}{8.2082}\right)\underline{/\pi/7 + 0.45 - 0.449} = 1.226\underline{/0.45}$

9. **(a)** $Z = Z_1 + Z_2 + Z_3 = (4 + 3j) + (3 - 2j) + (5 + 4j) = 12 + 5j;$

(b) $Z = \frac{Z_1 Z_2 Z_3}{Z_1 Z_2 + Z_1 Z_3 + Z_2 Z_3} = \frac{(4+3j)(3-2j)(5+4j)}{(4+3j)(3-2j)+(4+3j)(5+4j)+(3-2j)(5+4j)} = \frac{(12-8j+9j-6j^2)(5+4j)}{(12-8j+9j-6j^2)+(20+16j+15j+12j^2)+(15+12j-10j-8j^2)} = \frac{(18+j)(5+4j)}{(18+j)+(8+31j)+(23+2j)} = \frac{86+77j}{49+34j} \cdot \frac{49-34j}{49-34j} = \frac{6832+849j}{2401+1156} = 1.9207 + 0.2387j$

11. **(a)** $Z_1 = 1.64\underline{/38.2°} = 1.2888 + 1.0142j;$
$Z_2 = 2.35\underline{/43.7°} = 1.6990 + 1.6236j;$
$Z_3 = 4.67\underline{/-39.6°} = 3.5983 - 2.9768j; Z = Z_1 + Z_2 + Z_3 = 6.5861 - 0.3390j = 6.595\underline{/-2.95°};$
(b) $Z_1 Z_2 = (1.64\underline{/38.2})(2.35\underline{/43.7°}) = 3.854\underline{/81.9°} = 0.5430 + 3.8156j;$
$Z_1 Z_3 = (1.64\underline{/38.2°})(4.67\underline{/-39.6°}) = 7.6588\underline{/-1.4°} = 7.6565 - 0.1871j;$
$Z_2 Z_3 = (2.35\underline{/43.7°})(4.67\underline{/-39.6°}) = 10.9745\underline{/4.1} = 10.9464 + 0.7846j;$
$Z_1 Z_2 + Z_1 Z_3 + Z_2 Z_3 = 19.1459 + 4.4131j = 19.6479\underline{/12.98°}; Z = \frac{Z_1 Z_2 Z_3}{Z_1 Z_2 + Z_1 Z_3 + Z_2 Z_3} = \frac{(1.64\underline{/38.2°})(2.35\underline{/43.7°})(4.67\underline{/-39.6°})}{19.648\underline{/12.98°}} = 0.916\underline{/29.32°}$

13. $V = IZ = (4 - 3j)(8 - 15j) = 32 - 60j - 24j + 45j^2 = -13 - 84j$ V

15. $V = IZ = (1 - j)(1 + j) = 2$ V

17. $Z = \frac{V}{I} = \frac{7\underline{/36.3°}}{2.5\underline{/12.6°}} = 2.8\underline{/23.7°}$ Ω

19. $X_L = 2\pi fL = 2\pi 60 \cdot 0.2 = 75.40$ Ω;
$X_C = \frac{1}{2\pi fC} = \frac{1}{2\pi 60 \cdot 40 \times 10^{-6}} = 66.31$ Ω;
$X = X_L - X_C = 9.09$ Ω;
$R + Xj = 38 + 9.09j = 39.07\underline{/13.45°}; Z = 39.07$ Ω, $\phi = 13.45°$

21. $X_L = 2\pi fL = 2\pi \times 60 \times 0.4 = 150.80$ Ω;
$X_C = \frac{1}{2\pi fC} = \frac{1}{2\pi 60 \cdot 60 \times 10^{-6}} = 44.21$ Ω;
$X = X_L - X_C = 106.59$ Ω;
$R + Xj = 20 + 106.59j = 108.45\underline{/79.37°}$;
$Z = 108.45$ Ω, $\phi = 79.37°$

23. $X_L = \omega L = 50 \times 0.25 = 12.5$ Ω; $X_C = \frac{1}{\omega C} = \frac{1}{50 \times 200 \times 10^{-6}} = 100$ Ω; $X = X_L - X_C = -87.5$ Ω;

$R + Xj = 28 - 87.5j = 91.87\underline{/-72.26°}; Z = 91.87$ Ω, $\phi = -72.26°$

25. $X = X_L - X_C = 60 - 40 = 20$ Ω;
$R + Xj = 75 + 20j = 77.62\underline{/14.93°}; Z = 77.62$ Ω, $\phi = 14.93°; V = IZ = 77.62 \times 3.50 = 271.67$ V

27. $X = X_L - X_C = 6.0 - 5.0 = 1.0$ Ω;
$R + Xj = 3.0 + 1.0j = 3.162\underline{/18.43°}; Z = 3.162$ Ω, $\phi = 18.43°; V = IZ = 2.85 \times 3.162 = 9.01$ V

29. $I_2 = I_1 + I_3 \Rightarrow I_3 = I_2 - I_1 = (9 - 7j) - (7 + 2j) = 2 - 9j$

31. $\frac{1}{2\pi fC} = 2\pi fL \Rightarrow 1 = 4\pi^2 f^2 LC; f^2 = \frac{1}{4\pi^2 LC}; f = \frac{1}{2\pi\sqrt{LC}}; f = \frac{1}{2\pi\sqrt{2.5 \times 20 \times 10^{-6}}} = 22.51$ Hz

33. $Z = \frac{1}{Y} = \frac{1}{4 + 3j} = \frac{1}{4 + 3j} \cdot \frac{4 - 3j}{4 - 3j} = \frac{4 - 3j}{25} = 0.16 - 0.12j$

35. See *Computer Programs* in main text.

CHAPTER 14 REVIEW

1. $\sqrt{-49} = \sqrt{49}\sqrt{-1} = 7j$

2. $\sqrt{-36} = \sqrt{36}\sqrt{-1} = 6j$

3. $\sqrt{-54} = \sqrt{9}\sqrt{6}\sqrt{-1} = 3\sqrt{6}j$ or $3j\sqrt{6}$

4. $(2j^3)^3 = 2^3 j^9 = 8j$ (Note: $j^8 = 1$)

5. $\sqrt{-2}\sqrt{-18} = (\sqrt{2}j)(3j\sqrt{2}) = 6j^2 = -6$

6. $\sqrt{-9}\sqrt{-27} = (3j)(3j\sqrt{3}) = 9j^2\sqrt{3} = -9\sqrt{3}$

7. $(2 - j) + (7 - 2j) = 9 - 3j$

8. $(9 + j)(4 + 7j) = 36 + 63j + 4j + 7j^2 = 29 + 67j$

9. $(5 + j) - (6 - 3j) = -1 + 4j$

10. $\frac{1}{11 - j} = \frac{1}{11 - j} \cdot \frac{11 + j}{11 + j} = \frac{11 + j}{11^2 + 1^2} = \frac{11}{122} + \frac{j}{122} = 0.0902 + 0.0082j$

11. $(6 + 2j)(-5 + 3j) = -30 + 18j - 10j + 6j^2 = -36 + 8j$

12. $\frac{2 - 5j}{6 + 3j} \cdot \frac{6 - 3j}{6 - 3j} = \frac{12 - 6j - 30j + 15j^2}{36 + 9} = \frac{-3 - 36j}{45} = -0.0667 - 0.8j$

13. $(4 - 3j)(4 + 3j) = 4^2 + 3^2 = 16 + 9 = 25$

14. $\frac{-4}{\sqrt{3} + 2j} \cdot \frac{\sqrt{3} - 2j}{\sqrt{3} - 2j} = \frac{-4\sqrt{3} + 8j}{3 + 4} = \frac{-4\sqrt{3}}{7} + \frac{8}{7}j = -0.9897 + 1.1429j$

15. $9 - 6j; r = \sqrt{9^2 + 6^2} = 10.82; \tan\theta = \frac{-6}{9}; \theta = -33.69$ or 326.31; Answer: $10.82\underline{/326.31°}$

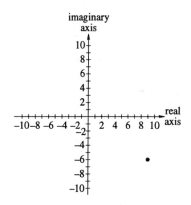

16. $-8 + 2j; r = \sqrt{8^2 + 2^2} = \sqrt{68} = 8.246; \tan\theta = \frac{2}{-8} = 165.96°$; Answer: $8.246\underline{/165.96°}$

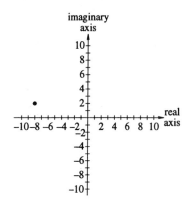

17. $4 - 4j = 4\sqrt{2}\,\underline{/315°}$ or $5.657\,\underline{/315°}$

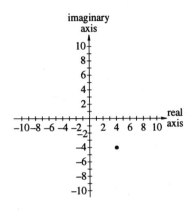

18. $4\,\text{cis}\,60° = 4\cos 60° + 4j\sin 60° = 2 + 3.464j$

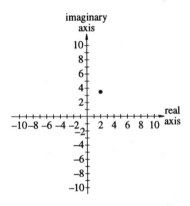

19. $6.5\,\underline{/2.3} = 6.5\cos 2.3 + 6.5j\sin 2.3 = -4.331 + 4.847j$

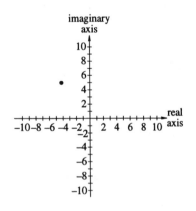

20. $10\,\underline{/20°} = 10\cos 20° + 10j\sin 20° = 9.397 + 3.420j$

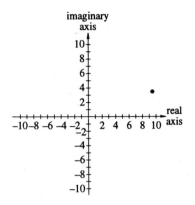

21. $(2\,\text{cis}\,30°)(5\,\text{cis}\,150°) = 10\,\text{cis}\,180° = -10$

22. $\frac{3\,\text{cis}\,20°}{6\,\text{cis}\,80°} = 0.5\,\text{cis}(-60°) = 0.25 - 0.433j$

23. $(3\,\text{cis}\,\frac{5\pi}{4})^{14} = 3^{14}\,\text{cis}\,\frac{14\cdot 5\pi}{4} = 4,782,969\,\text{cis}\,\frac{35\pi}{2} = -4,782,969j$

24. $(324\,\text{cis}\,225°)^{1/5} = \sqrt[5]{324}\,\text{cis}\,\frac{225}{5} = 3.178\,\text{cis}\,45° = 2.247 + 2.247j$

25. $(3\,\underline{/\pi/4})(9\,\underline{/2\pi/3}) = 3\cdot 9\,\underline{/\pi/4 + 2\pi 3} = 27\,\underline{/11\pi/12} = -26.080 + 6.988j$

26. $(44\,\underline{/125°}) \div (4\,\underline{/97°}) = \frac{44}{4}\,\underline{/125 - 97} = 11\,\underline{/28} = 9.7124 + 5.1642j$

27. $(2\,\underline{/\pi/6})^{12} = 2^{12}\,\underline{/\frac{12\pi}{6}} = 4096\,\underline{/2\pi} = 4096$

28. $(2048\,\underline{/330°})^{1/11} = \sqrt[11]{2048}\,\underline{/300/11} = 2\,\underline{/30} = 1.732 + j$

29. Cube roots of $-j = 1\,\text{cis}\,270°$; $w_0 = \sqrt[3]{1}\,\text{cis}\,\frac{270}{3} = 1\,\text{cis}\,90° = j$; $w_1 = 1\,\text{cis}(90 + \frac{360}{3}) = 1\,\text{cis}\,210° = -0.866 - 0.5j$; $w_2 = 1\,\text{cis}(90 + 120\times 2) = 1\,\text{cis}\,330° = 0.866 - 0.5j$

30. Fourth roots of $16 = 16\,\text{cis}\,0°$; $w_0 = \sqrt[4]{16}\,\text{cis}\,\frac{0}{4} = 2$; $w_1 = 2\,\text{cis}(0 + \frac{360}{4}) = 2\,\text{cis}\,90° = 2j$; $w_2 = 2\,\text{cis}\,180° = -2$; $w_3 = 2\,\text{cis}\,270° = -2j$

31. Square roots of $16\,\text{cis}\,120°$; $w_0 = \sqrt[4]{16}\,\text{cis}\,\frac{120}{2} = 4\,\text{cis}\,60° = 2 + 3.464j$; $w_1 = 4\,\text{cis}(60 + 180) - 4\,\text{cis}\,240° = -2 - 3.464j$

32. Cube roots of $27j = 27\,\text{cis}\,90°$; $w_0 = \sqrt[3]{27}\,\text{cis}\,\frac{90}{3} = 3\,\text{cis}\,30° = 2.598 + 1.5j$; $w_1 = 3\,\text{cis}(30 + \frac{360}{3}) = 3\,\text{cis}\,150° = -2.598 + 1.5j$; $w_2 = 3\,\text{cis}(30 + 120\times 2) = 3\,\text{cis}\,270° = -3j$

33. $(3+2j)(5-j) = (3.60555e^{0.588j})(5.099e^{-0.197j}) = 18.385e^{0.391j}$

34. $(4-7j)/(3+j) = (8.062e^{-1.052j})/(3.162e^{0.322j}) = 2.55e^{-1.374j}$

35. $(5+3j)^5 = (5.831e^{0.5404j})^5 = 5.831^5 e^{5 \times 0.5404j} = 6740.6e^{2.702j}$

36. $(-7-2j)^{1/3} = (7.280e^{-2.863j})^{1/3} = \sqrt[3]{7.280}e^{\frac{-2.863}{3}j}$
$= 1.938e^{-0.954j}$

37. $7.3 - 1.4j = 7.433 \operatorname{cis}(-10.86°)$; magnitude is 7.433, direction is $-10.86°$

38. $X = X_L - X_C = 7.0 - 4.5 = 2.5\ \Omega; R + Xj = 3 + 2.5j = 3.905 \underline{/39.81°}; Z = 3.905\ \Omega, \phi = 39.81°; V = IZ = 1.5 \times 3.905 = 5.858$ V

39. $X_L = 2\pi f L = 2\pi(0.3)60 = 113.10\ \Omega; X_C = \frac{1}{2\pi f C} = \frac{1}{2\pi 60 \times 5 \times 10^{-6}} = 53.05\ \Omega; X = X_L - X_C = 60.05\ \Omega; R + Xj = 55 + 60.05j = 81.43 \underline{/47.51°}; Z = 81.43\ \Omega, \phi = 47.51°$

40. **(a)** in series $Z = Z_1 + Z_2 = (3+5j) + (6-3j) = 9 + 2j$;

(b) in parallel $Z = \frac{Z_1 Z_2}{Z_1 + Z_2} = \frac{(3+5j)(6-3j)}{9+2j} = \frac{18-9j+30j-15j^2}{9+2j} = \frac{33+21j}{9+2j} \cdot \frac{9-2j}{9-2j} = \frac{297-66j+189j-42j^2}{81+4} = \frac{339+123j}{85} = 3.988 + 1.447j$

≡ CHAPTER 14 TEST

1. $\sqrt{-80} = \sqrt{80}\sqrt{-1} = \sqrt{16}\sqrt{5}j = 4j\sqrt{5}$

2. $7 - 2j; r = \sqrt{7^2 + 2^2} = \sqrt{53} \approx 7.28; \tan\theta = \frac{-2}{7} = -15.9$ or $344.1°$; Answer: $7.28 \underline{/344.1°}$

3. $8 \operatorname{cis} 150° = 8\cos 150° + 8j \sin 150° = -6.9282 + 4j$ or $-4\sqrt{3} + 4j$

4. $(5 + 2j) + (8 - 6j) = 13 - 4j$

5. $(-5 + 2j) - (8 - 6j) = -13 + 8j$

6. $(2+3j)(4-5j) = 8 - 10j + 12j - 15j^2 = 23 + 2j$

7. $\frac{6+5j}{-3-4j} \cdot \frac{-3+4j}{-3+4j} = \frac{-18+24j-15j+20j^2}{9+16} = \frac{-38+9j}{25} = -1.52 + 0.36j$

8. $(7 \operatorname{cis} 75°)(2 \operatorname{cis} 105°) = 14 \operatorname{cis} 180° = -14$

9. $\frac{4 \operatorname{cis} 115°}{3 \operatorname{cis} 25°} = \frac{4}{3} \operatorname{cis} 90° = \frac{4}{3}j$ or $1.333j$

10. $\left(9 \operatorname{cis} \frac{2\pi}{3}\right)^{5/2} = (9)^{5/2} \operatorname{cis} \frac{5\pi}{3} = 243 \operatorname{cis} \frac{5\pi}{3} = \frac{243}{2} - \frac{243}{2}j\sqrt{3}$

11. $(27 \underline{/129°})^{1/3} = \sqrt[3]{27} \operatorname{cis} \frac{129°}{3} = 3 \operatorname{cis} 43° \approx 2.194 + 2.046j$

12. Since $j = 1 \operatorname{cis} 90°$, then the four fourth roots of j are: $w_0 = \sqrt[4]{1} \operatorname{cis} \frac{90°}{4} = 1 \operatorname{cis} 22.5° = 0.924 + 0.383j; w_1 = 1 \operatorname{cis}(22.5 + 90)° = 1 \operatorname{cis} 112.5° = -0.383 + 0.924j; w_2 = 1 \operatorname{cis}(22.5 + 90 \times 2)° = 1 \operatorname{cis} 202.5° = -0.924 - 0.383j; w_3 = 1 \operatorname{cis}(22.5 + 90 \times 3) = 1 \operatorname{cis} 292.5° = 0.383 - 0.924j$

13. **(a)** in series $Z = Z_1 + Z_2 = (4+2j) + (5-3j) = 9 - j$;

(b) in parallel $Z = \frac{Z_1 Z_2}{Z_1 + Z_2} = \frac{(4+2j)(5-3j)}{9-j} = \frac{20-12j+10j-6j^2}{9-j} = \frac{26-2j}{9-j} \cdot \frac{9+j}{9+j} = \frac{234+26j-18j-2j^2}{81+1} = \frac{236+8j}{82} = \frac{118}{41} + \frac{4}{41}j = 2.878 + 0.098j$

14. $Z^{-1} = \frac{1}{9-3j} \frac{9+3j}{9-3j} = \frac{9+3j}{90} = \frac{1}{10} + \frac{1}{30}j \approx 0.1 + 0.033j$ S.

15. $V = 5 + 2j \approx 5.385 \underline{/21.80°}$ and $I = 3 - 4j \approx 5 \underline{/-53.13°}$. Since $Z = \frac{V}{I} = \frac{5.385 \underline{/21.80°}}{5 \underline{/-53.13°}} = 1.077 \underline{/74.93°} = 0.28 + 1.04j\ \Omega$.

16.   **(a)** $Z = Z_1 + Z_2 = (9 + 3j) + (7 - 2j) = (9 + 7) + (3 - 2)j = 16 + j \, \Omega$

   (b) $\frac{1}{Z_1} = \frac{1}{9+3j} = \frac{1}{9-3j} \cdot = \frac{9-3j}{9-3j} = \frac{9-3j}{90} \approx 0.1 -$

   $0.0333j$ and $\frac{1}{Z_2} = \frac{1}{7-2j} = \frac{1}{7-2j} \cdot \frac{7+2j}{7+2j} = \frac{7-2j}{53} \approx 0.1321 + 0.0377j.$

   Then,

$$Z = \frac{1}{\frac{1}{Z_1} + \frac{1}{Z_2}}$$

$$= \frac{1}{(0.1 - 0.0333j) + (0.1321 + 0.0377j)}$$

$$= \frac{1}{0.2321 + 0.0044j}$$

$$= \frac{1}{0.2321 + 0.0044j} \cdot \frac{0.2321 - 0.0044j}{0.2321 - 0.0044j}$$

$$= \frac{0.2321 - 0.0044j}{0.2321^2 + 0.0044^2}$$

$$= \frac{0.2321 - 0.0044j}{0.0539}$$

$$= \frac{0.2321}{0.0539} - \frac{0.0044}{0.0539}j$$

$$\approx 4.3061 - 0.0816 \, \Omega$$

15

An Introduction to Plane Analytic Geometry

☰ 15.1 BASIC DEFINITIONS AND STRAIGHT LINES

1. $d = \sqrt{(2-7)^2 + (4+9)^2} = \sqrt{(-5)^2 + 13^2} = \sqrt{25 + 169} = \sqrt{194} \approx 13.928$

3. $d = \sqrt{(5+3)^2 + (-6+5)^2} = \sqrt{8^2 + (-1)^2} = \sqrt{64 + 1} = \sqrt{65} \approx 8.062$

5. $d = \sqrt{(12-3)^3 + (1+13)^2} = \sqrt{9^2 + 14^2} = \sqrt{81 + 196} = \sqrt{277} \approx 16.643$

7. $d = \sqrt{(-2-5)^2 + (5-5)^2} = \sqrt{(-7)^2} = 7$

9. midpoint $= \left(\frac{2+7}{2}, \frac{4-9}{2}\right) = \left(\frac{9}{2}, \frac{-5}{2}\right)$

11. midpoint $= \left(\frac{5-3}{2}, \frac{-6-5}{2}\right) = \left(1, \frac{-11}{2}\right)$

13. midpoint $= \left(\frac{12+3}{2}, \frac{1-13}{2}\right) = \left(\frac{15}{2}, -6\right)$

15. midpoint $= \left(\frac{-2+5}{2}, \frac{5+5}{2}\right) = \left(\frac{3}{2}, 5\right)$

17. slope $= \frac{-9-4}{7-2} = \frac{-13}{5} = -13/5 = -2.6$

19. slope $= \frac{-5+6}{-3-5} = \frac{1}{-8} = -1/8$

21. slope $= \frac{-13-1}{3-12} = \frac{-14}{-9} = \frac{14}{9}$

23. slope $= \frac{5-5}{5+2} = \frac{0}{7} = 0$

25. $y - 4 = -2.6(x - 2)$ or $y = -2.6x + 9.2$

27. $y + 6 = -\frac{1}{8}(x - 5)$ or $y + 6 = -\frac{1}{8}x + \frac{5}{8}$ or $y = -\frac{1}{8}x - 5\frac{3}{8}$

29. $y - 1 = \frac{14}{9}(x - 12)$ or $y - 1 = \frac{14}{9}x - \frac{56}{3}$ or $y = \frac{14}{9}x - 17\frac{2}{3}$

31. $y - 5 = 0(x + 2)$ or $y - 5 = 0$ or $y = 5$

33. $m = \tan 45° = 1$

35. $m = \tan 0.15 \approx 0.1511352$

37. $\alpha = \tan^{-1} 2.5 = 68.19859°$

39. $\alpha = 180 + \tan^{-1}(-0.50) = 153.43495°$

41. $m = -\frac{1}{3} = -1/3$

43. $m = -\frac{1}{-1/2} = 2$

45. $y + 5 = 6(x - 2)$ or $y + 5 = 6x - 12$ or $y = 6x - 17$

47. $m = \frac{-2-3}{4-2} = \frac{-5}{2}$; $y + 2 = -\frac{5}{2}(x - 4)$ or $y + 2 = -\frac{5}{2}x + 10$ or $y = \frac{-5}{2}x + 8$

49. $m = \tan 60° = 1.732$; $y + 4 = 1.732(x + 2)$ or $y = 1.732x - 0.536$

51. $m = \tan 20° = 0.364$; $y = 0.364x + 3$

53. $2x - 3y + 4 = 0 \Rightarrow -3y = -2x + 4 \Rightarrow y = \frac{2}{3}x - \frac{4}{3}$; so the slope is $\frac{2}{3}$; $y - 5 = \frac{2}{3}(x - 2)$ or $3y - 15 = 2x - 4$. Combining terms produces $-2x + 3y = 11$ or $-2x + 3y - 11 = 0$ or $2x - 3y + 11 = 0$

55. $2x + 5y = 20 \Rightarrow 5y = -2x + 20 \Rightarrow y = -\frac{2}{5}x + 4$; the slope of this line is $-\frac{2}{5}$ and the slope of the perpendicular is $\frac{5}{2}$. Using the point-slope form, we get $y + 1 = \frac{5}{2}(x - 4)$ or $2y + 2 = 5x - 20$ or $-5x + 2y = -22$ or $5x - 2y = 22$

57. $3x + 2y = 12 \Rightarrow 2y = -3x + 12 \Rightarrow y = -\frac{3}{2}x + 6$; $m = -\frac{3}{2}$, y-intercept $= 6$; when $y = 0, x = 4$ so the x-intercept $= 4$.

59. $x - 3y = 9 \Rightarrow -3y = -x + 9 \Rightarrow y = \frac{1}{3}x - 3; m = \frac{1}{3}$, y- intercept $= -3$, when $y = 0, x = 9$ so x-intercept$= 9$

61. $v = v_0 + at, v_0 = 2.6, v = 5.8$ when $t = 8; 5.8 = 2.6 + a \cdot 8; 3.2 = 8a; a = 0.4; v = 2.6 + 0.4t$

63. $E = IR + Ir; 12 = 2.5 \times 4 + 2.5 \times r; 12 = 10 + 2.5r; 2 = 2.5r; r = \frac{2}{2.5} = 0.8\,\Omega$

65. (a) $C = 1,225 + 1.25n$,
(b) Substituting 20,000 for n, we get \$26,225.

67. (a) Here, $t = 1995 - 1940 = 55$. Thus, $C = 1.44(55) + 282.88 = 362.08$ ppm.
(b) Here, $t = 2050 - 1940 = 110$. Thus, $C = 1.44(110) + 282.88 = 441.28$ ppm.
(c) In this exercise, we have $t =$ current year -1940. Answer varies with year.

15.2 THE CIRCLE

1. $C = (2, 5), r = 3$: The standard equation for the circle is $(x-2)^2 + (y-5)^2 = 3^2$ or $(x-2)^2 + (y-5)^2 = 9$. This gives $x^2 - 4x + 4 + y^2 - 10y + 25 = 9$ or $x^2 + y^2 - 4x - 10y + 20 = 0$ for the general equation.

3. $C = (-2, 0), r = 4$: $(x+2)^2 + y^2 = 16$ is the standard form. $x^2 + y^2 + 4x - 12 = 0$ is the general form.

5. $C = (-5, -1), r = \frac{5}{2}$: $(x+5)^2 + (y+1)^2 = \frac{25}{4}$ is the standard form. $x^2 + 10x + 25 + y^2 + 2y + 1 = \frac{25}{4}$ or $x^2 + y^2 + 10x + 2y + 19.75$ is the general form.

7. $C = (2, -4), r = 1$: $(x-2)^2 + (y+4)^2 = 1$ is the standard form. $x^2 - 4x + 4 + y^2 + 8y + 16 = 1$ or $x^2 + y^2 - 4x + 8y + 19 = 0$ is the general form.

9. $C = (3, 4), r = 3$

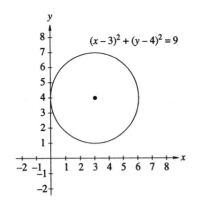

$(x-3)^2 + (y-4)^2 = 9$

11. We first write this in the standard form for the equation of a circle, $(x-h)^2 + (y-k)^2 = r^2$ or $\left(x - \left(-\frac{1}{2}\right)\right)^2 + \left(y - \left(-\frac{13}{4}\right)\right)^2 = \left(\sqrt{7}\right)^2$. Thus, the center is $C = (h, k) = (-\frac{1}{2}, -\frac{13}{4})$, and the radius is $r = \sqrt{7} \approx 2.646$.

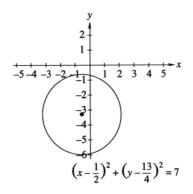

$$\left(x - \frac{1}{2}\right)^2 + \left(y - \frac{13}{4}\right)^2 = 7$$

13. $C = (0, \frac{7}{3}), r = \sqrt{6} \approx 2.449$

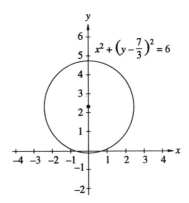

$$x^2 + \left(y - \frac{7}{3}\right)^2 = 6$$

15. $x^2 + y^2 + 4x - 6y + 4 = 0$. Completing the square leads to $x^2 + 4x + 4 + y^2 - 6y + 9 = -4 + 4 + 9$ or $(x+2)^2 + (y-3)^2 = 9$. This is the standard equation for a circle with center $(-2, 3)$ and radius $r = 3$.

17. $x^2 + y^2 + 10x - 6y - 47 = 0$; $x^2 + 10x + 25 + y^2 - 6y + 9 = 47 + 25 + 9$; $(x+5)^2 + (y-3)^3 = 81$; Circle: $C = (-5, 3), r = 9$

19. $x^2 + y^2 - 2x + 2y + 3 = 0$; $x^2 - 2x + 1 + y^2 + 2y + 1 = -3 + 1 + 1$; $(x-1)^2 + (y+1)^2 = -1$; $r^2 = -1$ so it is not a circle.

21. $x^2 + y^2 + 6x - 16 = 0$; $x^2 + 6x + 9 + y^2 = 16 + 9$; $(x+3)^2 + y^2 = 25$; Circle: $C = (-3, 0), r = 5$

23. $x^2 + y^2 + 5x - 9y = 9.5$; $x^2 + 5x + \frac{25}{4} + y^2 - 9y + \frac{81}{4} = \frac{19}{2} + \frac{25}{4} + \frac{81}{4}$; $(x + \frac{5}{2})^2 + (y - \frac{9}{2})^2 = \frac{144}{4} = 36$; Circle: $C = (-\frac{5}{2}, \frac{9}{2}), r = 6$

25. $9x^2 + 9y^2 + 18x - 15y + 27 = 0$; $9(x^2 + 2x) + 9(y^2 - \frac{5}{3}y) = -27$; $9(x^2 + 2x + 1) + 9(y^2 - \frac{5}{3}y + \frac{25}{36}) = -27 + 9 \cdot 1 + 9 \cdot \frac{25}{36}$; $9(x+1)^2 + 9(y - \frac{5}{6})^2 = -11.75$; $(x+1)^2 + (y - \frac{5}{6})^2 = -1.306$; Not a circle since r^2 is negative.

27. $\sigma = \frac{mv}{\beta q} = \frac{(1.673 \times 10^{-27})(1.186 \times 10^7)}{(4 \times 10^{-4})(1.5 \times 10^{-19})} = 330.70$. Choosing the center of the circle to be at the origin of the coordinate system we get $x^2 + y^2 = (330.70)^2$ or $x^2 + y^2 \approx 1.094 \times 10^5$.

29. $(x - 1.495 \times 10^8)^2 + y^2 = (3.844 \times 10^5)^2$ or $(x - 1.495 \times 10^8)^2 + y^2 = 1.478 \times 10^{11}$

31. If the center is 0.9 directly above the origin, the center is (0, 0.9), the radius is 1.1/2 so $x^2 + (y - 0.9)^2 = 0.55^2$ or $x^2 + (y - 0.9)^2 = 0.3025$.

33. Center is at (0, 5) and $r = \frac{26}{2} = 13$
(a) $x^2 + (y - 5)^2 = 13^2 = 169$
(b) Drawing a radius to where the floor would intersect the flywheel and the y-axis, we get a right triangle with hypotenuse 13 and one leg of 5. Using the Pythagorean theorem we find the other leg is 12, hence the chord is 2×12 or 24. Allowing for a 2 cm clearance on both sides, the opening must be $24 + 4 \times 2 = 28$ cm.

≡ 15.3 THE PARABOLA

1. $x^2 = 4y, p = 1$, vertical so $F = (0, 1)$, directrix: $y = -1$, opens upward

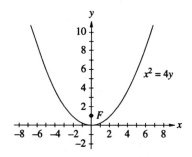

3. $y^2 = -4x, p = -1$, horizontal so $F = (-1, 0)$, directrix: $x = 1$, opens left.

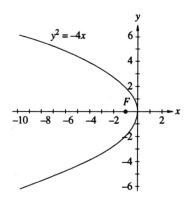

5. $x^2 = -8y, p = -2$, vertical so $F = (0, -2)$, directrix: $y = 2$, opens downward.

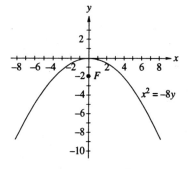

7. $y^2 = 10x, p = \frac{10}{4} = 2.5$, horizontal so $F = (2.5, 0)$, directrix: $x = -2.5$, opens right.

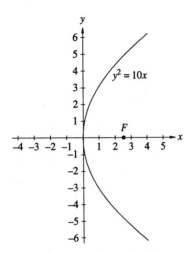

9. $x^2 = 2y, p = \frac{2}{4} = 0.5$, vertical so $F = (0, 0.5)$, directrix: $y = -0.5$, opens upward.

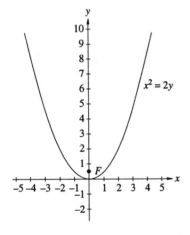

11. $y^2 = -21x, p = \frac{-21}{4} = -5.25$, horizontal so $F = (-\frac{21}{4}, 0)$, directrix: $x = \frac{21}{4}$, opens left.

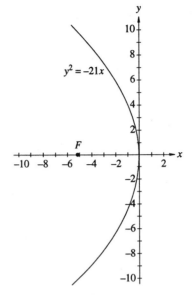

13. Focus $(0, 4)$, directrix: $y = -4, p = 4$ opens upward so $x^2 = 4 \cdot 4y$ or $x^2 = 16y$.

15. Focus $(-6, 0)$, directrix: $x = 6, p = -6$ opens left so $y^2 = 4(-6)x$ or $y^2 = -24x$.

17. Focus $(2, 0)$, vertex $(0, 0)$, $p = 2$, opens right so $y^2 = 4 \cdot 2x$ or $y^2 = 8x$

19. Vertex $(0, 0)$, directrix: $x = -\frac{3}{2}, p = \frac{3}{2}$, opens right so $y^2 = 4 \cdot \frac{3}{2}x$ or $y^2 = 6x$.

21. Choosing the vertex to be at the center of the bridge and the origin, we get $x^2 = 4py$ where $(180, 90)$ is a point on the parabola. Solving for p we get $(180)^2 = 4 \cdot 90p$ or $p = 90$. $x^2 = 360y$. Now setting $x = 100$ and solving for y, $100^2 = 360y$ or $y = \frac{10000}{360} = 27.78$ m.

23. $d = 5 \Rightarrow r = 2.5$. Using $x^2 = 4py$ and the point $(2.5, 1.5)$ we solve for p. $2.5^2 = 4(1.5)p$ or $p = \frac{6.25}{6}$ or 1.042 m from vertex.

25.

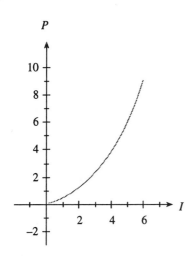

27. Position the "dish" so that the vertex is at (0, 0) and the focus at $(0, f)$. The directrix is $y = -f$. One point on the dish is at (5, 2) and, since the dish is a parabola, we know that distance from the focus to (5, 2) is equal to the distance from the focus to the directrix. This leads to the equation

$$(f + 2)^2 = 5^2 + (f - 2)^2$$
$$f^2 + 4f + 4 = 25 + f^2 - 4f + 4$$
$$8f = 25$$
$$f = \frac{25}{8} = 3.125$$

The focus is 3.125 ft from the vertex.

≡ 15.4 THE ELLIPSE

1. $F(4, 0), V(6, 0)$, so $c = 4, a = 6, b^2 = a^2 - c^2$ or $b^2 = 6^2 - 4^2 = 20$. Major axis is horizontal so the equation is $\frac{x^2}{36} + \frac{y^2}{20} = 1$.

3. $F(0, 2), V = (0, -4)$. Major axis is vertical so $b = 4$, $c = 2, a^2 = b^2 - c^2 = 4^2 - 2^2 = 12$. $\frac{x^2}{12} + \frac{y^2}{16} = 1$.

5. $F(3, 0), V(5, 0)$. Major axis is horizontal so $a = 5, c = 3, b^2 = a^2 - c^2 = 25 - 9 = 16; \frac{x^2}{25} + \frac{y^2}{16} = 1$.

7. Minor axis length 6, vertex (4, 0). Major axis is horizontal so $a = 4, b = \frac{6}{2} = 3$;

$$\frac{x^2}{16} + \frac{y^2}{9} = 1$$

9. $\frac{x^2}{4} + \frac{y^2}{9} = 1$. Major axis vertical $b = 3, a = 2, c^2 = b^2 - a^2 = 9 - 4 = 5, c = \sqrt{5}, V = (0, 3), V' = (0, -3), F = (0, \sqrt{5}), F' = (0, -\sqrt{5}), M = (2, 0), M' = (-2, 0)$

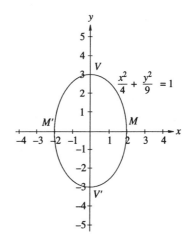

11. $\frac{x^2}{4} + \frac{y^2}{1} = 1, a = 2, b = 1, c^2 = a^2 - b^2 = 3, c = \sqrt{3}, V = (2, 0), V' = (-2, 0), F = (\sqrt{3}, 0), F' = (-\sqrt{3}, 0), M = (0, 1), M' = (0, -1)$.

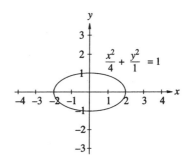

13. $4x^2 + y^2 = 4 \Rightarrow \frac{x^2}{1} + \frac{y^2}{4} = 1, a = 1, b = 2, c = \sqrt{3}, V = (0, \pm 2), F = (-0, \pm\sqrt{3}), M = (\pm 1, 0)$

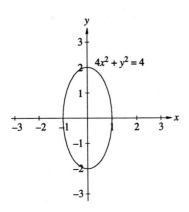

15. $25x^2 + 36y^2 = 900 \Rightarrow \frac{x^2}{36} + \frac{y^2}{25} = 1, a = 6, b = 5, c = \sqrt{11}, V = (\pm 6, 0), F = (\pm\sqrt{11}, 0), M = (0, \pm 5)$

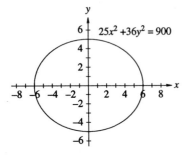

17. $a = 2.992 \times 10^8/2 = 1.496 \times 10^8, c = a/60 = 2.493 \times 10^6, b^2 = a^2 - c^2$ or $b = \sqrt{a^2 - c^2} = 1.49579 \times 10^8$ Minor axis $= 2b = 2.9916 \times 10^8$ km.

19. Minor axis is the diameter of the pipe is 10 in. Major axis is $10\sqrt{2}$ or 14.14 in.

21. $a = 3600/2 = 1800, \ b = 44/2 = 22, \ c = \sqrt{a^2 - b^2} = \sqrt{1800^2 - 22^2} = 1799.865551,$ eccentricity $= \frac{c}{a} = \frac{1,799.865551}{1800} = 0.9999253.$

23. Looking at the cross-section we see that the length of the major axis is the same as the length of the diagonal of a square with sides 8.0 in. Hence, $2a = 8\sqrt{2} \approx 11.3$ in.

≡ 15.5 THE HYPERBOLA

1. $F(6, 0), V(4, 0)$: Transverse axis is on the x-axis so $a = 4, c = 6,$ and $b^2 = c^2 - a^2 = 36 - 16 = 20.$ The equation is $\frac{x^2}{16} - \frac{y^2}{20} = 1.$

3. $F(0, 5), V(0, -3)$: Transverse axis is on the y-axis so $c = 5, a = 3,$ and $b^2 = 5^2 - 3^2 = 16.$ The equation is $\frac{y^2}{9} - \frac{x^2}{16} = 1.$

5. $F(5, 0), V = (3, 0): c = 5, a = 3,$ and $b^2 = 5^2 - 3^2 = 16.$ Transverse axis is on the x-axis so the equation is $\frac{x^2}{9} - \frac{y^2}{16} = 1.$

7. $F(4, 0),$ length of conjugate axis is 6 so $b = \frac{6}{2} = 3, c = 4, a^2 = c^2 - b^2 = 16 - 9 = 7.$ The Transverse axis is on the x-axis so the equation is $\frac{x^2}{7} - \frac{y^2}{9} = 1.$

9. $\frac{x^2}{4} - \frac{y^2}{9} = 1, a = 2, b = 3, c = \sqrt{4+9} = \sqrt{13}. V = (\pm 2, 0), F(\pm\sqrt{13}, 0), M = (0, \pm 3).$

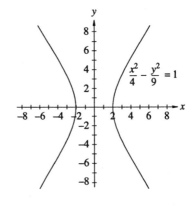

11. $\frac{y^2}{4} - \frac{x^2}{1} = 1, a = 2, b = 1, c = \sqrt{4+1} = \sqrt{5}$. The transverse axis is on the y-axis, so $V = (0, \pm 2), F = (0, \pm\sqrt{5}), M = (\pm 1, 0)$.

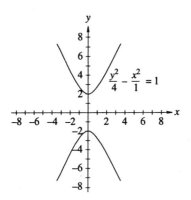

$\frac{y^2}{4} - \frac{x^2}{1} = 1$

13. $4x^2 - y^2 = 4 \Rightarrow \frac{x^2}{1} - \frac{y^2}{4} = 1, a = 1, b = 2, c = \sqrt{1+4} = \sqrt{5}$. The transverse axis is on the x-axis, so $V = (\pm 1, 0), F = (\pm\sqrt{5}, 0), M = (0, \pm 2)$.

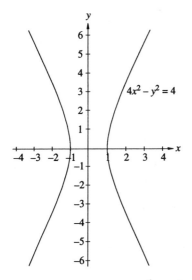

$4x^2 - y^2 = 4$

15. $36y^2 - 25x^2 = 900 \Rightarrow \frac{y^2}{25} - \frac{x^2}{36} = 1, a = 5, b = 6, c = \sqrt{25+36} = \sqrt{61}$. The transverse axis is on the y-axis so $V = (0, \pm 5), F = (0, \pm\sqrt{61}), M = (\pm 6, 0)$.

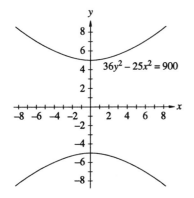

$36y^2 - 25x^2 = 900$

17. If the vertex is $(7, 0)$ then $a = 7, \frac{c}{a} = 1.5$ so $c = 1.5 \times 7 = 10.5, b^2 = c^2 - a^2 = 61.25$. The transverse axis is on the x-axis so the equation is $\frac{x^2}{49} - \frac{y^2}{61.25} = 1$.

19. If $a = b$ then $c = \sqrt{a^2 + b^2} = \sqrt{2a^2} = a\sqrt{2}, e = \frac{c}{a} = \frac{a\sqrt{2}}{a} = \sqrt{2}$

21. The product of the current I and resistance R in a simple circuit as a constant is represented by $IR = k$. Since $i = 3.2$ and $R = 18$, we see that $IR = 3.2 \times 18 = 57.6$. Thus, $k = 57.6$ is the constant. Solving for I, we obtain $I = \frac{57.6}{R}$. Graphing this function produces the following figure.

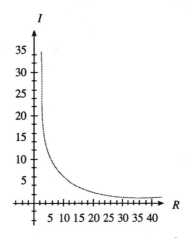

23. Letting $a = 42$, we get $\frac{x^2}{42^2} - \frac{y^2}{b^2} = 1$. Substituting for x and y, we obtain $\frac{(-126)^2}{42^2} - \frac{(30\sqrt{3})^2}{b^2} = 1$ or

$$126^2 b^2 - (30\sqrt{3})^2 42^2 = 42^2 b^2$$
$$b^2(126^2 - 42^2) = 42^2(30\sqrt{3})^2$$
$$14{,}112 b^2 = 4{,}762{,}800$$
$$b^2 = 337.5$$
$$b \approx 18.37$$

The desired equation is $\frac{x^2}{1764} - \frac{y^2}{337.5} = 1$.

25. Substituting, we obtain $2250 = 8.34FC$. Solving for F, we get $F = \frac{2250}{8.34C} = \frac{270}{C}$. The graph is shown below.

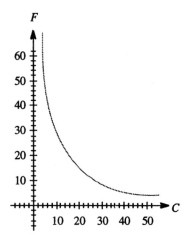

≡ 15.6 TRANSLATION OF AXES

1. $\frac{(x-4)^2}{9} + \frac{(y+3)^2}{4} = 1, x' = x - 4, y' = y + 3$ so $(h, k) = (4, -3); \frac{x'^2}{9} + \frac{y'^2}{4} = 1$.
This is an ellipse with $a = 3, b = 2$, and $c = \sqrt{9-4} = \sqrt{5}$. The major axis is horizontal.

	$x'y'$-system	xy-system
center	$(0', 0')$	$(4, -3)$
vertices	$(\pm 3', 0')$	$(7, -3), (1, -3)$
foci	$(\pm\sqrt{5}', 0')$	$(\sqrt{5}+4, -3),$ $(-\sqrt{5}+4, -3)$
endpoints of minor axis	$(0', \pm 2')$	$(4, -1), (4, -5)$

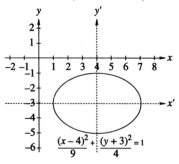

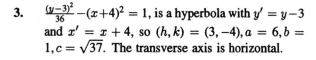

3. $\frac{(y-3)^2}{36} - (x+4)^2 = 1$, is a hyperbola with $y' = y - 3$ and $x' = x + 4$, so $(h, k) = (3, -4), a = 6, b = 1, c = \sqrt{37}$. The transverse axis is horizontal.

	$x'y'$-system	xy-system
center	$(0', 0')$	$(-4, 3)$
vertices	$(0', \pm\sqrt{37'})$	$(-4, 9), (-4, -3)$
foci	$(0', \pm 6)$	$(-4, \sqrt{37}+3)$ $(-4, -\sqrt{37}+3)$
endpoints of conjugate axis	$(\pm 1', 0')$	$(-3, 3), (-5, 3)$

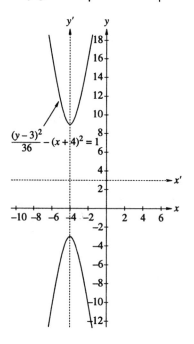

5. $100(x + 5)^2 - 4y^2 = 400 \Rightarrow \frac{(x+5)^2}{4} - \frac{y^2}{100} = 1$;
This is a hyperbola with $x' = x + 5$ and $y' = y$
so $(h, k) = (-5, 0); a = 2, b = 10, c = \sqrt{104} = 2\sqrt{26}$. The transverse axis is vertical.

	$x'y'$-system	xy-system
center	$(0', 0')$	$(-5, 0)$
vertices	$(\pm 2', 0)$	$(-3, 0),\ (-7, 0)$
foci	$(\pm 2\sqrt{26}', 0')$	$(2\sqrt{26} - 5, 0),$ $(-2\sqrt{26} - 5, 0)$
endpoints of conjugate axis	$(0', \pm 10')$	$(-5, 10),\ (-5, -10)$

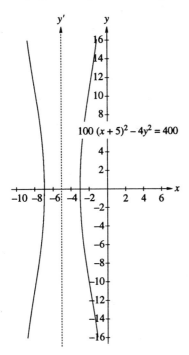

7. $16x^2 + 4y^2 + 64x - 12y + 57 = 0$ or $16(x^2 + 4x) + 4(y^2 - 3y) = -57$. Completing the square
produces $16(x^2 + 4x + 4) + 4(y^2 - 3y + \frac{9}{4}) = -57 + 16 \cdot 4 + 4 \cdot \frac{9}{4}$ or $16(x+2)^2 + 4(y - \frac{3}{2})^2 = 16$. This has

the standard form equation $\frac{(x+2)^2}{1} + \frac{(y - \frac{3}{2})}{4} = 1$
and so it is an ellipse with $x' = x + 2$ and $y' = y - \frac{3}{2}$. Thus, $(h, k) = (-2, \frac{3}{2}), a = 1, b = 2$, and
$c = \sqrt{4 - 1} = \sqrt{3}$. The major axis is vertical.

	$x'y'$-system	xy-system
center	$(0', 0')$	$(-2, \frac{3}{2})$
vertices	$(0', \pm 2')$	$(-2, \frac{7}{2}),\ (-2, -\frac{1}{2})$
foci	$(0', \pm\sqrt{3}')$	$(-2, \sqrt{3} + \frac{3}{2}),$ $(-2, -\sqrt{3} + \frac{3}{2})$
endpoints of minor axis	$(\pm 1', 0')$	$(-1, \frac{3}{2}),\ (-3, \frac{3}{2})$

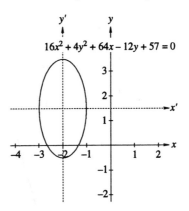

9. $25x^2 + 4y^2 - 250x - 16y + 541 = 0$ or $25(x^2 - 10x) + 4(y^2 - 4y) = -541$. Completing the square
produces $25(x^2 - 10x + 25) + 4(y^2 - 4y + 4) = -541 + 25 \cdot 25 + 4 \cdot 4$ or $25(x - 5)^2 + 4(y - 2)^2 = 100$.
This has the standard form $\frac{(x-5)^2}{4} + \frac{(y-2)^2}{25} = 1$ and
so it is an ellipse with $x' = x - 5, y' = y - 2$. Here
$(h, k) = (5, 2), a = 2, b = 5, c = \sqrt{21}$. The major
axis is vertical.

	$x'y'$-system	xy-system
center	$(0', 0')$	$(5, 2)$
vertices	$(0', \pm 5')$	$(5, 7),\ (5, -3)$
foci	$(0', \pm\sqrt{21}')$	$(5, 2 + \sqrt{21}),$ $(5, 2 - \sqrt{21})$
endpoints of minor axis	$(\pm 2', 0')$	$(7, 2),\ (3, 2)$

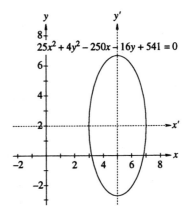

11. $2x^2 - y^2 - 16x + 4y + 24 = 0$ or $2(x^2 - 8x) - (y^2 - 4y) = -24$. Completing the square yields $2(x^2 - 8x + 16) - (y^2 - 4y + 4) = -24 + 2 \cdot 16 - 1 \cdot 4$ or $2(x - 4)^2 - (y - 2)^2 = 4$ which has the standard form equation $\frac{(x-4)^2}{2} - \frac{(y-2)^2}{4} = 1$. This is an hyperbola with $a = \sqrt{2}, b = 2; c = \sqrt{6}, x' = x - 4, y' = y - 2, (h, k) = (4, 2)$. The transverse axis is horizontal.

	$x'y'$-system	xy-system
center	$(0', 0')$	$(4, 2)$
vertices	$(\pm\sqrt{2}, 0')$	$(4 + \sqrt{2}, 2),$ $(4 - \sqrt{2}, 2)$
foci	$(\pm\sqrt{6}, 0')$	$(4 + \sqrt{6}, 2),$ $(4 - \sqrt{6}, 2)$
endpoints of conjugate axis	$(0', \pm2')$	$(4, 0), (4, 4)$

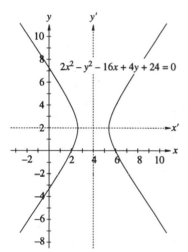

$2x^2 - y^2 - 16x + 4y + 24 = 0$

13. Parabola, Vertex $= (2, -3), p = 8$, axis vertical $x' = x - 2, y' = y + 3, x'^2 = 4py'$ or $(x - 2)^2 = 32(y + 3)$

15. Ellipse, center at $(4, -3)$, focus at $(8, -3)$, vertex at $(10, -3), a = |10 - 4| = 6, c = |8 - 4| = 4, b^2 = a^2 - c^2 = 36 - 16 = 20$. Major axis is horizontal. $x' = x - 4, y' = y + 3$, and so $\frac{x'^2}{36} + \frac{y'^2}{20} = 1$ or $\frac{(x-4)^2}{36} + \frac{(y+3)^2}{20} = 1$.

17. Hyperbola, center at $(-3, 2)$ focus at $(-3, 7)$, transverse axis 6 units. Center and focus vertically displaces so transverse axis is vertical. $c = |7 - 2| =$

$5, a = \frac{6}{2} = 3, b^2 = c^2 - a^2 = 16, x' = x + 3, y' = y - 2$ and so $\frac{y'^2}{9} - \frac{x'^2}{16} = 1$ or $\frac{(y-2)^2}{9} - \frac{(x+3)^2}{16} = 1$.

19. Parabola, vertex at $(-5, 1)$ so $x' = x + 5, y' = y - 1, p = -4$, axis parallel to x axis $y'^2 = 4px'$ or $(y - 1)^2 = -16(x + 5)$.

21. Ellipse, center at $(-4, 1)$ so $x' = x + 4, y' = y - 1$, and the focus is at $(-4, 9)$, so major axis is vertical and $c = |9 - 1| = 8$, minor axis is 12. Hence, $a = \frac{12}{2} = 6, b^2 = a^2 + c^2 = 64 + 36 = 100. \frac{x'^2}{36} + \frac{y'^2}{100} = 1$ or $\frac{(x+4)^2}{36} + \frac{(y-1)^2}{100} = 1$.

23. $s = 29.4t - 4.9t^2$ or $4.9t^2 - 29.4t = -s, 4.9(t^2 - 6t) = -s, 4.9(t^2 - 6t + 9) = -s + 4.9 \times 9, 4.9(t - 3)^2 = -s + 44.1$ or $(t - 3)^2 = -\frac{1}{4.9}(s - 44.1)$. This is a parabola with vertex at $(3, 44.1)$ it opens downward so the maximum height is 44.1 m.

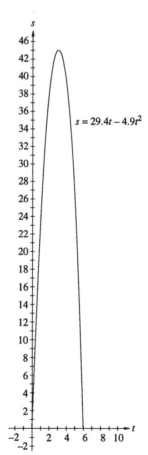

$s = 29.4t - 4.9t^2$

25. Since the signal from A arrives $600\mu s$ after B the common difference on the hyperbola is $300 \times 600 = 180,000$ m $= 180$ km so $2a = 180$ hence $a = 90, b^2 = c^2 - a^2 = 250000 - 8100 = 241,900$. The equation is $\frac{x^2}{8100} - \frac{y^2}{241,900} = 1$. If the transverse axis passes through A and B and the conjugate axis passes through the midpoint of $\overline{AB}$, then A is at $(-500, 0)$ and B at $(500, 0)$. $AP + BP = 300 \times 8,000 = 2,400,000$ m $= 2,400$ km; $AP + BP = 2,400$; $AP - BP = 180$; $AB = 1000$; $2AP = 2580$; $AP = 1290$; $BP = 1110$; $\alpha = \cos^{-1}\left(\frac{1000^2 + 1290^2 - 1110^2}{2 \cdot 1000 \cdot 1290}\right) = 56.2866°$; $y = AP \sin\alpha = 1073.054$ km; $x = AP\cos\alpha - 500 = 216$ km. The plane is at $(216, 1073.054)$.

27. (a) $x^2 + 200y - 4.00 = 0$ or $x^2 = -200y + 4.00$ and so $x^2 = -200(y - 0.02)$

(b) In the no load position $y - 0.02 = 0$ or $y = 0.02$ m $= 2$ cm.

29. We begin by using the distance formula. The distance from the airplane to the transmitter at $(0, 0)$ is $\sqrt{x^2 + y^2}$. Similarly, the distance from the airplane to the transmitter at $(0, 250)$ is $\sqrt{x^2 + (y - 250)^2}$. If the place is 50 mi further from $(0, 0)$ than from $(0, 250)$, we have $\sqrt{x^2 + y^2} = \sqrt{x^2 + (y - 250)^2} + 50$. We solve this equation as follows

$$\sqrt{x^2 + y^2} = \sqrt{x^2 + (y - 250)^2} + 50$$
$$\sqrt{x^2 + y^2} - 50 = \sqrt{x^2 + y^2 - 500y + 250^2}$$
$$x^2 - 100\sqrt{x^2 + y^2} + y^2 + 50^2 = x^2 + y^2 - 500y + 250^2$$
$$-100\sqrt{x^2 + y^2} = -500y + 250^2 - 50^2$$
$$= -500y + 60,000$$
$$100^2(x^2 + y^2) = (-500y + 60,000)^2$$
$$10,000x^2 + 10,000y^2 = 250,000y^2 - 60,000,000y + 3,600,000,000$$
$$24y^2 - 6,000y - x^2 = -360,000$$
$$24(y^2 - 250y) - x^2 = -360,000$$
$$24(y^2 - 250y + 125^2) - x^2 = -360,000 + 24 \cdot 125^2$$
$$24(y - 125)^2 - x^2 = 15,000$$

Thus, the equation of the hyperbola is $24(y - 125)^2 - x^2 = 15,000$. Since it is closer to $(0, 250)$ than to $(0, 0)$, it is on the upper half of the hyperbola.

(b) Since the plane is on the line $y = x$ and the hyperbola $24(y - 125)^2 - x^2 = 15,000$, we can substitute x for y and solve the resulting equation.

$$24(x - 125)^2 - x^2 = 15,000$$
$$24(x^2 - 250x + 15,625) - x^2 = 15,000$$
$$23x^2 - 6,000x + 360,000 = 0$$

Using the quadratic formula, we find that $x = \frac{3000 \pm 600\sqrt{2}}{23}$. Because the plane is on the upper-half of the hyperbola, we have $x = \frac{3000 + 600\sqrt{2}}{23}$ and the plane is located at $\left(\frac{3,000 + 600\sqrt{2}}{23}, \frac{3,000 + 600\sqrt{2}}{23}\right) \approx (167.3, 167.3)$.

31. (a) Solving the equation for the circle produces the equation $y = \pm 2\sqrt{0.0625 + 0.25x^2}$. We will select the positive square root, $y = 2\sqrt{0.0625 - 0.25x^2}$, to get the graph of the upper semicircle as the top part of the cam. Solving the equation for the ellipse we get $y = \pm 2.5\sqrt{0.1 - 0.4x^2}$. Here we take the negative square root, $y = -2.5\sqrt{0.1 - 0.4x^2}$, to get its graph as the bottom part of the cam.

(b)

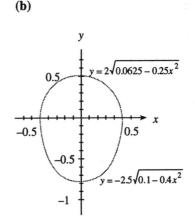

33. **(a)** The cost is 0.9 times the number of items, less 0.01 times the number of items over 24. This yields the function $C = 0.9x - 0.01x(x - 24) = 0.9x - 0.01x^2 + 0.24x = 1.14x - 0.01x^2$.

(b)

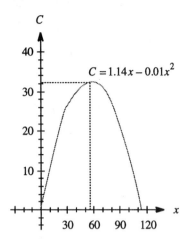

(c) From the graph, we estimate the maximum cost when 57 items are packaged.

(d) The maximum cost is \$32.49.

≡ 15.7 ROTATION OF AXES; THE GENERAL SECOND-DEGREE EQUATION

1. **(a)** $xy = -9$. Here, $A = 0, B = 1$, and $C = 0$, and the discriminant is $B^2 - 4AC = 1 > 0$; so the curve is a hyperbola,

(b) $\cot 2\theta = \frac{A-C}{B} = 0$ or $2\theta = 90°$ so the angle needed to rotate the coordinate axes to remove the xy-term is $\theta = 45°$,

(c) The rotation formulas are $x = x'' \cos\theta - y'' \sin\theta$ and $y = y'' \cos\theta + x'' \sin\theta$ we get $x = 0.7071x'' - 0.7071y''$ and $y = 0.7071y'' + 0.7071x''$,

(d) Substituting the equations from (c) into the original equation $xy = -9$ we get $(0.7071x'' - 0.7071y'')(0.7071y'' + 0.7071x'') = -9$ or $\frac{1}{2}x''^2 - \frac{1}{2}y''^2 = -9$ which is equivalent of $\frac{y''^2}{18} - \frac{x''^2}{18} = 1$.

(b) $\cot 2\theta = \frac{A-C}{B} = \frac{0}{-6}$ so $2\theta = 90°$ and so the angle needed to rotate the coordinate axes to remove the xy-term is $\theta = 45°$,

(c) The rotation formulas are $x = x'' \cos 45° - y'' \sin 45°, y = y'' \cos 45° + x'' \sin 45°$,

(d) Substituting the equations from (c) into the original equation $x^2 - 6xy + y^2 - 8 = 0$ we get $\left(\frac{\sqrt{2}}{2}x'' - \frac{\sqrt{2}}{2}y''\right)^2 - 6\left(\frac{\sqrt{2}}{2}x'' - \frac{\sqrt{2}}{2}y''\right)\left(\frac{\sqrt{2}}{2}y'' + \frac{\sqrt{2}}{2}x''\right) + \left(\frac{\sqrt{2}}{2}y'' + \frac{\sqrt{2}}{2}x''\right)^2 - 8 = 0$; $\frac{1}{2}x''^2 - x''y'' + \frac{1}{2}y''^2 - 6(\frac{1}{2}x''^2 - \frac{1}{2}y''^2) + (\frac{1}{2}y''^2 + x''y'' + \frac{1}{2}x''^2) - 8 = 0$ or $-2x''^2 + 4y''^2 - 8 = 0$.

This is equivalent to $\frac{y''^2}{2} - \frac{x''^2}{4} = 1$.

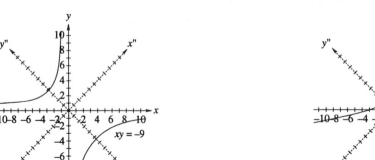

3. **(a)** $x^2 - 6xy + y^2 - 8 = 0$. Here, $A = 1, B = -6$, and $C = 1$. The discriminant is $B^2 - 4AC = 36 - 4 = 32 > 0$, so the curve is a hyperbola

5. **(a)** $52x^2 - 72xy + 73y^2 - 100 = 0$. Here $A = 52, B = -72$, and $C = 73$, so the discriminant is $B^2 - 4AC = -10000 < 0$ so the curve is an ellipse,

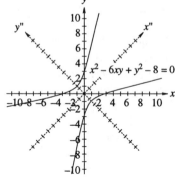

(b) $\cot 2\theta = \frac{A-C}{B} = \frac{52-73}{-72} = 0.29167, 2\theta = $ 73.7398 so the angle needed to rotate the coordinate axes to remove the xy-term is $\theta = 36.870$ with $\cos\theta = 0.8$ and $\sin\theta = 0.6$,

(c) The rotation formulas are $x = 0.8x'' - 0.6y''$ and $y = 0.8y'' + 0.6x''$,

(d) $52(0.8x'' - 0.6y'')^2 - 72(0.8x'' - 0.6y'')(0.8y'' + 0.6x'') + 73(0.8y'' + 0.6x'')^2 - 100 = 0$ which expands to $52(0.64x''^2 - 0.96x''y'' + 0.36y''^2) - 72(0.48x''^2 + 0.28x''y'' - 0.48y''^2) + 73(0.64y''^2 + 0.96x''y'' + 0.36x''^2) - 100 = 0$ or $25x''^2 + 100y''^2 = 100$ which is equivalent to $\frac{x''^2}{4} + y''^2 = 1$.

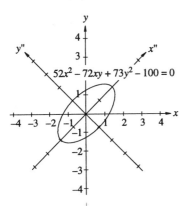

$52x^2 - 72xy + 73y^2 - 100 = 0$

7. **(a)** $x^2 - 2xy + y^2 + x + y = 0$. The discriminant is $B^2 - 4AC = (-2)^2 - 4 \cdot 1 \cdot 1 = 4 - 4 = 0$ so the curve is a parabola,

(b) $\cot 2\theta = \frac{A-C}{B} = \frac{1-1}{-2} = \frac{0}{-2} = 0$ so $2\theta = 90°$ and so $\theta = 45°$,

(c) The rotation formulas are $x = \frac{\sqrt{2}}{2}x'' - \frac{\sqrt{2}}{2}y''$ and $y = \frac{\sqrt{2}}{2}y'' + \frac{\sqrt{2}}{x}x''$,

(d) Substituting the equations from (c) into the original equation $x^2 - 2xy + y^2 + x + y = 0$ we get $\left(\frac{\sqrt{2}}{2}x'' - \frac{\sqrt{2}}{2}y''\right)^2 - 2\left(\frac{\sqrt{2}}{2}x'' - \frac{\sqrt{2}}{2}y''\right)\left(\frac{\sqrt{2}}{2}y'' + \frac{\sqrt{2}}{2}x''\right) + \left(\frac{\sqrt{2}}{2}y'' + \frac{\sqrt{2}}{2}x''\right)^2 + \left(\frac{\sqrt{2}}{2}x'' - \frac{\sqrt{2}}{2}y''\right) + \left(\frac{\sqrt{2}}{2}y'' + \frac{\sqrt{2}}{2}x''\right) = 0$; $\left(\frac{1}{2}x''^2 - x''y'' + \frac{1}{2}y''^2\right) - 2\left(\frac{1}{2}x''^2 - \frac{1}{2}y''^2\right) + \left(\frac{1}{2}y''^2 + x''y'' + \frac{1}{2}x''^2\right) + \sqrt{2}x'' = 0$; $2y''^2 + \sqrt{2}x'' = 0$ or $2y''^2 = -\sqrt{2}x''$ which is equivalent to $y''^2 = \frac{-\sqrt{2}}{2}x''$.

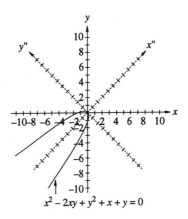

$x^2 - 2xy + y^2 + x + y = 0$

9. **(a)** $2x^2 + 12xy - 3y^2 - 42 = 0$. The discriminant is $B^2 - 4AC = 144 + 24 = 168 > 0$ so the curve is a hyperbola,

(b) $\cot 2\theta = \frac{A-C}{B} = \frac{2+3}{12} = \frac{5}{12}, 2\theta = 67.380$ so the angle needed to rotate the coordinate axes to remove the xy-term is $\theta = 33.690, \cos\theta = 0.8321, \sin\theta = 0.5547$,

(c) The rotation formulas are $x = 0.8321x'' - 0.5547y''$ and $y = 0.8321y'' + 0.5547x''$,

(d) Substituting the equations from (c) into the original equation $2x^2 + 12xy - 3y^2 - 42 = 0$ we get $2(0.8321x'' - 0.5547y'')^2 + 12(0.8321x'' - 0.5547y'')(0.8321y'' + 0.5547x'') - 3(0.8321y'' + 0.5547x'')^2 - 42 = 0$ or $2(0.6923x''^2 - 0.9231x''y'' + 0.3077y''^2) + 12(0.4615x''^2 + 0.3846x''y'' - 0.4615y''^2) - 3(0.6923y''^2 + 0.9231x''y'' + 0.3077x''^2) - 42 = 0$; $1.3846x''^2 - 1.8462x''y'' + 0.6154y''^2 + 5.538x''^2 + 4.6152x''y'' - 5.538y''^2 - 2.0769y''^2 - 2.7693x''y'' - .9231x''^2 - 42 = 0$; $6x''^2 - 7y''^2 = 42$ which is equivalent to $\frac{x''^2}{7} - \frac{y''^2}{6} = 1$.

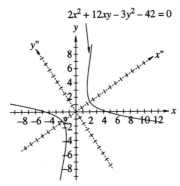

$2x^2 + 12xy - 3y^2 - 42 = 0$

11. **(a)** $6x^2 - 5xy + 6y^2 - 26 = 0$. The discriminant is $B^2 - 4AC = 25 - 144 = -119$ so the curve is an ellipse,

(b) $\cot 2\theta = \frac{A-C}{B} = \frac{6-6}{-5} = 0$ so $2\theta = 90°$ and we see that $\theta = 45°$,

(c) The rotation formulas are $x = \frac{\sqrt{2}}{2}x'' - \frac{\sqrt{2}}{2}y''$ and $y = \frac{\sqrt{2}}{2}y'' + \frac{\sqrt{2}}{2}x''$,

(d) Substituting the equations from (c) into the original equation $6x^2 - 5xy + 6y^2 - 26 = 0$ yields $6\left(\frac{\sqrt{2}}{2}x'' - \frac{\sqrt{2}}{2}y''\right)^2 - 5\left(\frac{\sqrt{2}}{2}x'' - \frac{\sqrt{2}}{2}y''\right)\left(\frac{\sqrt{2}}{2}y'' + \frac{\sqrt{2}}{2}x''\right) + 6\left(\frac{\sqrt{2}}{2}y'' + \frac{\sqrt{2}}{2}x''\right)^2 - 26 = 0$ or $6\left(\frac{1}{2}x''^2 - x''y'' + \frac{1}{2}y''^2\right) - 5\left(\frac{1}{2}x''^2 - \frac{1}{2}y''^2\right) + 6\left(\frac{1}{2}y''^2 + x''y'' + \frac{1}{2}x''^2\right) - 26 = 0$; $3x''^2 - 6x''y'' + 3y''^2 - \frac{5}{2}x''^2 + \frac{5}{2}y''^2 + 3y''^2 + 6x''y'' + 3x''^2 - 26 = 0$ which simplifies to $\frac{7}{2}x''^2 + \frac{17}{2}y''^2 = 26$ or $7x''^2 + 17y''^2 = 52$ or the standard form equation $\dfrac{x^2}{\frac{52}{7}} + \dfrac{y''^2}{\frac{52}{17}} = 1$.

Note: $a = \sqrt{\frac{52}{7}} \approx 2.726, b = \sqrt{\frac{52}{17}} \approx 1.749$.

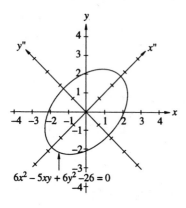

$6x^2 - 5xy + 6y^2 - 26 = 0$

13. $k = PV = 15 \times 400 = 6000, PV = 6000$

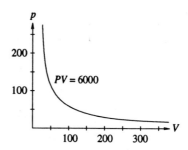

$PV = 6000$

15. $(\overline{v_1} - \overline{v_2})^2 = 19.62 h_L$ is a parabola; If $h_L < 5$ then $(\overline{v_1} - \overline{v_2}) < 9.9$

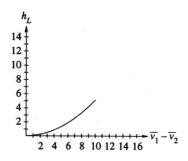

15.8 CONIC SECTIONS IN POLAR COORDINATES

1. $r = \frac{6}{1 + 3\cos\theta}, e = 3 > 1$, so this is a hyperbola.

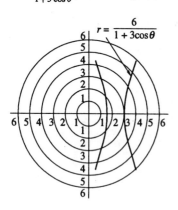

$r = \dfrac{6}{1 + 3\cos\theta}$

3. $r = \frac{12}{3 - \cos\theta} = \dfrac{4}{1 - \frac{1}{3}\cos\theta}, e = \frac{1}{3}$, so this is an ellipse.

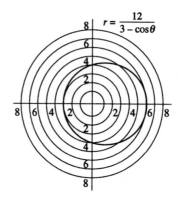

$r = \dfrac{12}{3 - \cos\theta}$

5. $r = \frac{12}{4-4\cos\theta} = \frac{3}{1-1\cos\theta}$, $e = 1$, so this is a parabola.

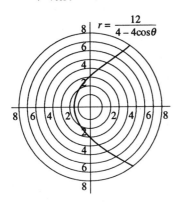

7. $r = \frac{12}{3+2\sin\theta} = \frac{4}{1+\frac{2}{3}\cos\theta}$, $e = \frac{2}{3}$, so this is an ellipse.

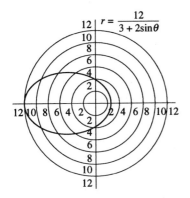

9. $r = \frac{12}{2-4\cos(\theta+\frac{\pi}{3})} = \frac{6}{1-2\cos(\theta+\frac{\pi}{3})}$, $e = 2$,

so this is a hyperbola rotated $\frac{\pi}{3}$ counterclockwise.

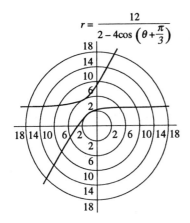

11. $r = \frac{pe}{1-e\cos\theta} = \frac{4\cdot\frac{3}{2}}{1-\frac{3}{2}\cos\theta} = \frac{12}{2-3\cos\theta}$, so this is a hyperbola.

13. $r = \frac{pe}{1-e\sin\theta} = \frac{5}{1-\sin\theta}$, so this is a parabola.

15. $r = \frac{1\cdot\frac{2}{3}}{1-\frac{2}{3}\cos\theta} = \frac{2}{3-2\cos\theta}$, so this is an ellipse.

17. Greatest distance is when $\theta = 0$, so $r = 4.335 \times 10^7$. The shortest distance is when $\theta = \pi$ so $r = 2.854 \times 10^7$.

19.

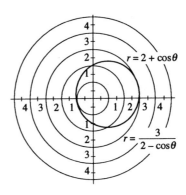

CHAPTER 15 REVIEW

1. **(a)** $d = \sqrt{(2+1)^2 + (5-9)^2} = \sqrt{3^2 + (-4)^2} = 5$

(b) midpoint $= \left(\frac{2-1}{2}, \frac{5+9}{2}\right) = \left(\frac{1}{2}, 7\right)$

(c) $m = \frac{9-5}{-1-2} = \frac{4}{-3} = -\frac{4}{3}$

(d) $y - 5 = -\frac{4}{3}(x-2)$ or $3y - 15 = -4x + 8$ or $4x + 3y = 23$

2. **(a)** $d = \sqrt{12^2 + 5^2} = 13$

(b) midpoint $= \left(\frac{8}{2}, \frac{-15}{2}\right) = \left(4, \frac{-15}{2}\right)$

(c) $m = \frac{-10+5}{10+2} = \frac{-5}{12}$

(d) $y + 5 = -\frac{5}{12}(x+2)$ or $12y + 60 = -5 - 10$ or $5x + 12y = -70$

3. **(a)** $d = \sqrt{2^2 + 10^2} = \sqrt{104} = 2\sqrt{26}$

(b) midpoint $= \left(\frac{4}{2}, \frac{2}{2}\right) = (2, 1)$

(c) $m = \frac{6+4}{3-1} = \frac{10}{2} = 5$

(d) $y + 4 = 5(x-1)$ or $y + 4 = 5x - 5$ or $5x - y = 9$ or $5x - 9$ or $y - 5x = -9$

4. **(a)** $d = \sqrt{8^2 + 8^2} = 8\sqrt{2}$

(b) midpoint $= \left(\frac{2-6}{2}, \frac{-5+3}{2}\right) = \left(\frac{-4}{2}, \frac{-2}{2}\right) = (-2, -1)$

(c) $m = \frac{8}{-8} = -1$

(d) $y + 5 = -1(x-2)$ or $y + 5 = -x + 2$ or $x + y = -3$

5. #1: $\frac{3}{4}$; #2: $\frac{12}{5}$; #3: $-\frac{1}{5}$; #4: 1

6. #1: $y - 7 = \frac{3}{4}\left(x - \frac{1}{2}\right)$ or $4y - 28 = 3x - \frac{3}{2}$ or $8y - 56 = 6x - 3$ or $8y - 6x = 53$;

#2: $y + \frac{15}{2} = \frac{12}{5}(x - 4)$ or $5y + \frac{75}{2} = 12x - 48$ or $10y + 75 = 24x - 96$ or $10y - 24x = -171$;

#3: $y - 1 = -\frac{1}{5}(x - 2)$ or $5y - 5 = -x + 2$ or $x + 5y = 7$;

#4: $y + 1 = 1(x + 2)$ or $y + 1 = x + 2$ or $y = x + 1$ or $y - x = 1$ or $x - y = -1$

7. The slope of $2y + 4x = 9$ is obtained by solving for y: This gives $y = -2x + \frac{9}{2}$ so the slope is -2. $y - 5 = -2(x + 3)$ or $y - 5 = -2x - 6$ or $y + 2x = -1$

8. $y + 7 = 4(x - 2)$ or $y + 7 = 4x - 8$ or $y - 4x = -15$ or $4x - y = 15$ or $y = 4x - 15$

9. $x^2 + y^2 = 16$: circle with center at $(0, 0)$ and $r = 4$

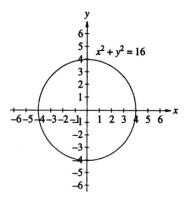

10. $x^2 - y^2 = 16 \Rightarrow \frac{x^2}{16} - \frac{y^2}{16} = 1$, hyperbola, center at $(0, 0)$, $a = 4, b = 4, c = 4\sqrt{2}$

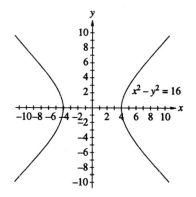

11. $x^2 + 4y^2 = 16 \Rightarrow \frac{x^2}{16} + \frac{y^2}{4} = 1$, ellipse, center at $(0, 0)$, $a = 4, b = 2, c = \sqrt{12} = 2\sqrt{3}$

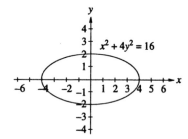

12. $y^2 = 16x$, parabola, vertex at $(0, 0)$ focus at $(4, 0)$, directrix $x = -4$

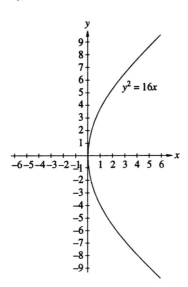

13. $(x - 2)^2 + (y + 4)^2 = 16$, circle center $(2, -4)$, radius 4

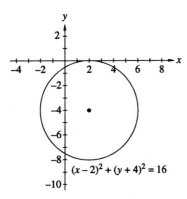

14. $(x - 2)^2 - (y + 4)^2 = 16 \Rightarrow \frac{(x-2)^2}{16} - \frac{(y+4)^2}{16} = 1$ hyperbola, center at $(2, -4), a = 4, b = 4, c = 4\sqrt{2}$

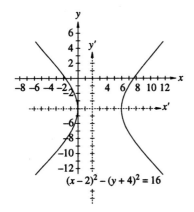

15. $(x - 2)^2 + 4(y + 4)^2 = 16 \Rightarrow \frac{(x-2)^2}{16} + \frac{(y+4)^2}{4} = 1$ ellipse, center $(2, -4), a = 4, b = 2, c = \sqrt{12} = 2\sqrt{3}$

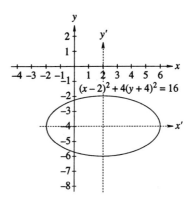

16. $(y+4)^2 = 16(x-2)$, parabola, vertex $(2, -4)$ opens right, focus at $(6, -4)$

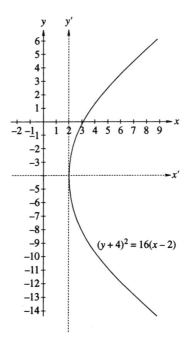

17. $x^2 + y^2 + 6x - 10y + 18 = 0$, $x^2 + 6x + 9 + y^2 - 10y + 25 = -18 + 9 + 25$, $(x+3)^3 + (y-5)^2 = 16$ circle, center $(-3, 5)$ radius 4

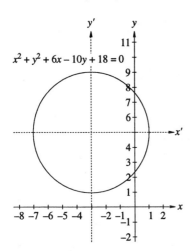

18. $x^2 - 4y^2 + 6x + 40y - 107 = 0$, $x^2 + 6x - 4(y^2 - 10y) = 107$, $(x^2 + 6x + 9) - 4(y^2 - 10y + 25) = 107 + 9 - 4 \cdot 25$, $(x+3)^2 - 4(y-5)^2 = 16$, $\frac{(x+3)^2}{16} - \frac{(y-5)^2}{4} = 1$ Hyperbola, center $(-3, 5)$, $a = 4, b = 2, c = \sqrt{20} = 2\sqrt{5}$

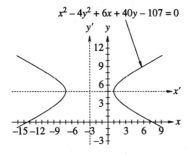

19. $x^2 + 4y^2 + 6x - 40y + 93 = 0$, $(x^2 + 6x) + 4(y^2 - 10y) = -93$, $(x^2 + 6x + 9) + 4(y^2 - 10y + 25) = -93 + 9 + 4 \cdot 25$, $(x+3)^2 + 4(y-5)^2 = 16$, $\frac{(x+3)^2}{16} + \frac{(y-5)^2}{4} = 1$, ellipse, center $(-3, 5)$, $a = 4, b = 2, c = \sqrt{12} = 2\sqrt{3}$

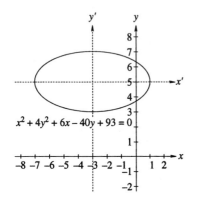

20. $x^2 + 6x - 4y + 29 = 0$, $x^2 + 6x = 4y - 29$, $(x^2 + 6x + 9) = 4y - 29 + 9$, $(x+3)^2 = 4y - 20$, $(x+3)^2 = 4(y - 5)$ parabola, vertex $(-3, 5)$, $p = 1$, opens upward, Focus $(-3, 6)$, directrix, $y = 4$

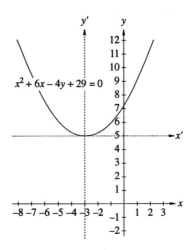

21. $2x^2 + 12xy - 3y^2 - 42 = 0$. The discriminant is $B^2 - 4AC = 144 + 24 = 168$ so the curve is a hyperbola. Thus, $\cot 2\theta = \frac{A-C}{B} = \frac{2+3}{12} = \frac{5}{12}$, and so $2\theta = 67.3801°$. Hence, the angle of rotation is $\theta = 33.69°$ and we obtain $\cos \theta = 0.83205$, $\cos^2 \theta = 0.6923$, $\sin \theta = 0.5547$, $\sin^2 \theta = 0.3077$, and $\sin \theta \cos \theta = 0.4615$. The rotation equations are $x = 0.83205x'' - 0.5547y''$ and $y = 0.83205y'' + 0.5547x''$. When these are substituted into the given equation, we obtain $2(0.83205x'' - 0.5547y'')^2 + 12(0.83205x'' - 0.5547y'') \times (0.83205y'' + 0.5547x'') - 3(0.83205y'' + 0.5547x'')^2 - 42 = 0$ or $2(0.6923x''^2 - 0.9230x''y'' + 0.3077y''^2) + 12(0.4615x''^2 + 0.3846x''y'' - 0.4615y''^2) - 3(0.6923y''^2 + 0.9230x''y'' + 0.3077x''^2) - 42 = 0$. This simplifies to $6x''^2 - 7y''^2 = 42$ or the standard form equation $\frac{x''^2}{7} - \frac{y''^2}{6} = 1$ with $a = \sqrt{7}$, $b = \sqrt{6}$, and $c = \sqrt{13}$.

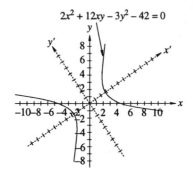

22. $5x^2 - 4xy + 8y^2 - 36 = 0$. The discriminant is $B^2 - 4AC = 16 - 160 = -144$ so the curve is an ellipse. Thus, $\cot 2\theta = \frac{A-C}{B} = \frac{5-8}{-4} = \frac{-3}{-4} = \frac{3}{4}$, so $2\theta = 53.1301$, and the angle of rotation is $\theta = 26.565$. As a result, we have $\cos \theta = 0.8944, \cos^2 \theta = 0.8, \sin \theta = 0.4472, \sin^2 \theta = 0.2$, and $\cos \theta \sin \theta = 0.4$. The rotation equations are $x = 0.8944x'' - 0.4472y''$ and $y = 0.8944y'' + 0.4472x''$. When these are substituted into the given equation, we obtain $5(0.8944x'' - 0.4472y'')^2 - 4(0.8944x'' - 0.4472y'')(0.8944y'' + 0.4472x'') + 8(0.8944y'' + 0.4472x'')^2 - 36 = 0$ or $5(0.8x''^2 - 0.8x''y'' + 0.2y''^2) - 4(0.4x''^2 + 0.6x''y'' - 0.4y''^2) + 8(0.8y''^2 + 0.8x''y'' + 0.2x''^2) - 36 = 0$. This simplifies to $4x''^2 + 9y''^2 = 36$ which has the standard form equation $\frac{x''^2}{9} + \frac{y''^2}{4} = 1$ with $a = 3, b = 2$, and $c = \sqrt{5}$.

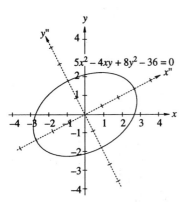

23. $3x^2 + 2\sqrt{3}xy + y^2 + 8x - 8\sqrt{3}y = 32$. The discriminant is $B^2 - 4AC = 12 - 12 = 0$ so the curve is a parabola. Thus, $\cot 2\theta = \frac{3-1}{2\sqrt{3}} = \frac{2}{2\sqrt{3}} = \frac{1}{\sqrt{3}}$ and so $2\theta = 60°$, and the angle of rotation is $\theta = 30°$. As a result, we have $\cos 30° = \frac{\sqrt{3}}{2}, \cos^2 30° = \frac{3}{4}, \sin 30° = \frac{1}{2}$, and $\sin^2 30° = \frac{1}{4}$. The rotation equations are $x = \frac{\sqrt{3}}{2}x'' - \frac{1}{2}y''$ and $y = \frac{\sqrt{3}}{2}y'' + \frac{1}{2}x''$. When these are substituted into the given equation, we obtain $3(\frac{\sqrt{3}}{2}x'' - \frac{1}{2}y'')^2 + 2\sqrt{3}(\frac{\sqrt{3}}{2}x'' - \frac{1}{2}y'')(\frac{\sqrt{3}}{2}y'' + \frac{1}{2}x'') + (\frac{\sqrt{3}}{2}y'' + \frac{1}{2}x'')^2 + 8(\frac{\sqrt{3}}{2}x'' - \frac{1}{2}y'') - 8\sqrt{3}(\frac{\sqrt{3}}{2}y'' + \frac{1}{2}x'') = 32$ or $3(\frac{3}{4}x''^2 - \frac{\sqrt{3}}{2}x''y'' + \frac{1}{4}y''^2) + 2\sqrt{3}(\frac{\sqrt{3}}{2}x''^2 - \frac{\sqrt{3}}{4}y''^2 + \frac{1}{2}x''y'') + (\frac{3}{4}y''^2 + \frac{\sqrt{3}}{2}x''y'' + \frac{1}{4}x''^2) + 4\sqrt{3}x'' - 4y'' - 12y'' - 4\sqrt{3}x'' = 32$. This simplifies to $4x''^2 = 16y'' + 32$ which has the standard form equation for a parabola $x''^2 = 4(y'' + 2)$ with $p = 1$. The parabola opens upward and has its vertex at $(0'', -2'')$.

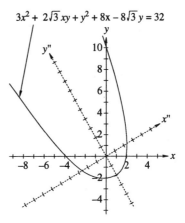

$3x^2 + 2\sqrt{3}\,xy + y^2 + 8x - 8\sqrt{3}\,y = 32$

24. $2x^2 - 4xy - y^2 = 6$. The discriminant is $B^2 - 4AC = 16 + 8 = 24 > 0$ so the curve is a hyperbola. Thus, $\cot 2\theta = \frac{A-C}{B} = \frac{2+1}{-4} = -\frac{3}{4}$ and so $2\theta = 126.8699$ and the angle of rotation is $\theta = 63.4349$. As a result, we have $\cos \theta = 0.447, \cos^2 \theta = 0.2, \sin \theta = 0.8944, \sin^2 \theta = 0.8, \cos \theta \sin \theta = 0.4$. The rotation equations are $x = x'' \cos \theta - y'' \sin \theta = 0.447x'' - 0.8944y''$ and $y = y'' \cos \theta + x'' \sin \theta = 0.8944y'' + 0.447x''$. When these are substituted into the given equation, we obtain $2(x'' \cos \theta - y'' \sin \theta)^2 - 4(x'' \cos \theta - y'' \sin \theta)(y'' \cos \theta + x'' \sin \theta) - (y'' \cos \theta + x'' \sin \theta)^2 = 6$ or $2(0.2x''^2 - 0.8x''y'' + 0.8y''^2) - 4(0.4x''^2 - 0.6x''y'' - 0.4y''^2) - (0.2y''^2 + 0.8x''y'' + 0.8x''^2) = 6$. This simplifies to $-2x''^2 + 3y''^2 = 6$ or the standard form equation $\frac{y''^2}{2} - \frac{x''^2}{3} = 1$ with $a = \sqrt{2}, b = \sqrt{3}$, and $x = \sqrt{5}$.

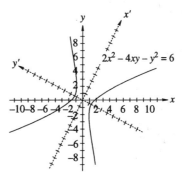

$2x^2 - 4xy - y^2 = 6$

25. $r = \frac{16}{5-3\cos\theta} = \frac{\frac{16}{5}}{1-\frac{3}{5}\cos\theta}$; $e = \frac{3}{5}$ so ellipse

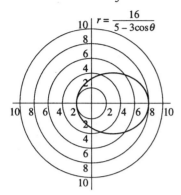

26. $r = \frac{9}{3-5\cos\theta} = \frac{3}{1-\frac{5}{3}\cos\theta}$; $e = \frac{5}{3}$ so hyperbola

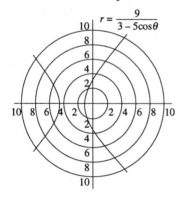

27. $r = \frac{9}{3+3\sin\theta} - \frac{3}{1+\sin\theta}$; $e = 1$ so parabola

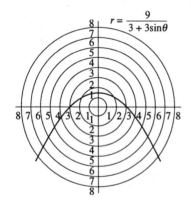

28. $r = \frac{2}{1+\cos\theta}$; $e = 1$, so parabola

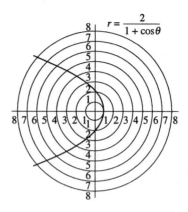

29. **(a)** Completing the square, we obtain $(y+3)^2 = 16x + 4$. We graph $y_1 = \sqrt{16x+4} - 3$ and $y_2 = -\sqrt{16x+4} - 3$. Their graphs produce the figure shown below.

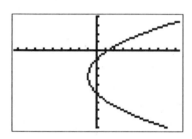

[–9.4, 9.4] × [–9, 4]

(b) This is a parabola, which has the standard form equation $(y+3)^2 = 4(x+1)$.

(c) The parabola has a horizontal axis, opens to the right with vertex $(-1,-3)$, directrix $x = -2$, and focus $(0,-3)$.

30. **(a)**

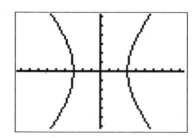

[–9.4, 9.4] × [–6.2, 6.2]

(b) This is a hyperbola with standard form equation $\frac{x^2}{9} - \frac{y^2}{16} = 1$.

(c) From the standard form equation, we see that $a = 3, b = 4$, and so, $c = 5$. Thus, the "important" points are, Center $(0, 0)$, Vertices: $(-3,0)$ and $(3, 0)$, Foci: $(-5,0)$ and $(5,0)$; endpoints of conjugate axis: $(0, 4)$ and $(0, -4)$

31. Since $2a = 100, a = 50$, this gives $\frac{x^2}{50^2} + \frac{y^2}{b^2} = 1$. The clearance constraint gives the point (45, 6.1). Substituting and solving for b^2 results in $\frac{45^2}{50^2} + \frac{6.1^2}{b^2} = 1$ or $b^2 \cdot 45^2 + 6.1^2 \cdot 50^2 = 50^2 b^2$ or $50^2 b^2 - 45^2 b^2 = 50^2 \cdot 6.1^2$. As a result, $b^2 = \frac{50^2 \cdot 6.1^2}{50^2 - 45^2} = 195.84$ and we get the standard form equation $\frac{x^2}{2500} + \frac{y^2}{195.84} = 1$.

32. Taking the road to be the x-axis and the center of the road as the origin, the vertex of the parabola is at (0, 10). This gives an equation of the form $x^2 = 4p(y - 10)$. The top of the tower gives the point (200,120). Substituting and solving for $4p$ we obtain $4p = \frac{x^2}{y - 10} = \frac{200^2}{120 - 10} = \frac{40000}{110} = 363.64$. The equation is $x^2 = 363.64(y - 10)$.

33. The equation is a hyperbola. The common difference is $AP - BP = 320 \times 500 = 160,000\,\text{m} = 160$ km. Thus, $2a = 160, a = 80, 2c = 400$, and so $c = 200$ and $b^2 = c^2 - a^2 = 33600$. The equation is $\frac{x^2}{6400} - \frac{y^2}{33600} = 1$. The y-coordinate for p is -50. Solving for x we get $\frac{x^2}{6400} = 1 + \frac{50^2}{33600} = 1.0744$ or $x^2 = 6876.19$ and so $x = 82.92$. The plane is at $(82.92, -50)$.

34. $x^2 + y^2 - 2x + 4y = 6361, x^2 - 2x + 1 + y^2 + 4y + 4 = 6361 + 1 + 4 = 6366, (x - 1)^2 + (y + 2)^2 = (79.7872)^2$. The radius is Earth's radius plus the number of units the satellite is orbiting above Earth. This is $79.7872 + 0.8 = 80.587$ km and so $r^2 = 6494.3$. Thus, we get the equation for a circle: $(x - 1)^2 + (y + 2)^2 = 6494.3, x^2 - 2x + 1 + y^2 + 4y + 4 = 6494.3, x^2 + y^2 - 2x + 4y - 6489.3 = 0$.

▤ CHAPTER 15 TEST

1. Comparing the given equation with the basic equation $y^2 = 4px$, we see that $4p = -18$ and so $p = -\frac{18}{4} = -\frac{9}{2} = -4.5$. The focus of the parabola is at $(p, 0) = (-4.5, 0)$ and the directrix is the line $x = -p = 4.5$.

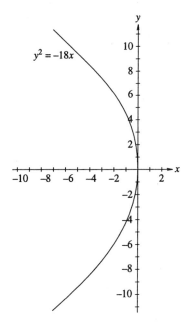

2. To write this in standard form, divide both sides by 36, obtaining $\frac{9x^2}{36} - \frac{4y^2}{36} = 1$ which simplifies to $\frac{x^2}{4} - \frac{y^2}{9} = 1$. From this we see that $a^2 = 4$ and $b^2 = 9$. Since $c^2 = a^2 + b^2 = 4 + 9 = 13$, we have $c = \sqrt{13}$. The foci are at $F' = (-c, 0) =$

$(-\sqrt{13}, 0)$ and $F = (c, 0) = (\sqrt{13}, 0)$; the vertices are $V' = (-a, 0) = (-2, 0)$ and $V = (a, 0) = (2, 0)$; and the endpoints of the conjugate axis are $M' = (0, -b) = (0, -3)$ and $M = (0, b) = (0, 3)$.

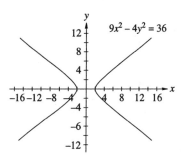

3. (a)

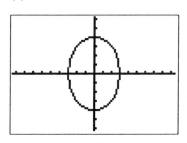

[–9.4, 9.4] x [–6.2, 6.2]

(b) This is an ellipse with standard form equation $\frac{x^2}{9} + \frac{y^2}{16} = 1$.

(c) From the standard form equation, we see that $a = 3, b = 4$, and so, $c = \sqrt{4^2 - 3^2} = \sqrt{7}$. Thus, the "important" points are, Center: (0, 0), Vertices $(-3, 0)$ and $(3, 0)$, Foci: $(-\sqrt{7}, 0)$ and $(\sqrt{7}, 0)$; endpoints of minor axis: (0, 4) and (0, −4)

4. Using the distance formula for the distance between two points P_1 and P_2, we obtain

$d(P_1, P_2) = \sqrt{(x_2 - x_1)^2 + (y_2 - y_1)^2} =$
$\sqrt{(5 - (-2))^2 + (-6 - 4)^2} = \sqrt{7^2 + (-10)^2} =$
$\sqrt{49 + 100} = \sqrt{149}$

5. The midpoint between these points is $\left(\frac{x_1 + x_2}{2}, \frac{y_1 + y_2}{2}\right)$
$= \left(\frac{4+2}{2}, \frac{-5+(-8)}{2}\right) = \left(3, \frac{-13}{2}\right)$. The slope of the line through the two given points is $m = \frac{y_2 - y_1}{x_2 - x_1} =$
$\frac{-8-(-5)}{2-4} = \frac{-3}{-2} = \frac{3}{2}$ and so the slope of a line perpendicular to this line is $-\frac{2}{3}$. Hence the desired line is $y - \left(\frac{-13}{2}\right) = -\frac{2}{3}(x - 3)$ or $y + \frac{13}{2} = -\frac{2}{3}(x - 3)$ or $6y + 4x + 27 = 0$.

6. A line with an angle of inclination 30° has a slope $m = \tan 30° = \frac{\sqrt{3}}{3}$. An x-intercept of 3 means it passes through (3, 0). The equation has the point-slope form $y - 0 = \frac{\sqrt{3}}{3}(x - 3)$. Changing this to the slope-intercept form produces $y = \frac{\sqrt{3}}{3}x - \sqrt{3}$.

7. **(a)**

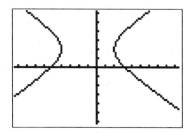

$[-9.4, 9.4] \times [-6.2, 6.2]$

(b) This is a hyperbola with standard form equation $\frac{(x+1)^2}{9} - \frac{(y-2)^2}{4} = 1$.
(c) From the standard form equation, we see that $a = 3, b = 2$, and so, $c = \sqrt{3^2 + 2^2} = \sqrt{13}$. Thus, the "important" points are, Center: $(-1, 2)$, Vertices: $(-4, 2)$ and $(2, 2)$, Foci: $(-1 - \sqrt{13}, 2)$ and $(-1 + \sqrt{13}, 2)$; endpoints of minor axis: $(-1, 4)$ and $(-1, 0)$

8. **(a)** Completing the square we obtain $(3x^2 - 6x) + (4y^2 + 16y) = -7$ or $3(x^2 - 2x + 1) + 4(y^2 + $

$4y + 4) = -7 + 3 + 16$, which can be written as $3(x - 1)^2 + 4(y + 2)^2 = 12$. Dividing both sides by 12, we obtain a standard form equation for a horizontal ellipse $\frac{(x-1)^2}{4} + \frac{(y+2)^2}{3} = 1$;
(b) Ellipse;
(c) center: $(1, -2)$, foci: $(0, -2), (2, -2)$, vertices: $(-1, -2), (3, -2)$, endpoints of minor axis: $(1, -2 - \sqrt{3}), (1, -2 + \sqrt{3})$

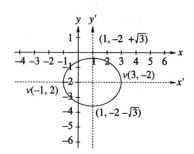

9. **(a)** The angle is determined by the equation $\cot 2\theta = \frac{A - C}{B}$. Here $A = 1, B = 1$, and $C = 1$, and so $\cot 2\theta = \frac{0}{1}$, which means that $\theta = \frac{\pi}{4}$.
(b) Using $B^2 - 4AC = 1^2 - 4(1)(1) = -3$. Since this is negative, the conic is an ellipse.

10. The denominator has a constant term of 1 so $e = 4$. Since $e > 1$, the conic is a hyperbola

$$r = \frac{2}{1 + 4\cos\theta}$$

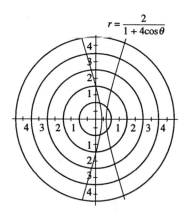

11. Consider this parabola to have its lowest point (vertex) at the origin and its axis the y-axis. Then the equation is of the form $x^2 = 4py$. Then the supports can be represented by the lines $x = -60$ and $x = 60$. The distance of 48 ft in from each support and 3 ft above the lowest point are the points $(-12, 3)$ and $(12, 3)$. Substituting the point $(12, 3)$ into the equation for the parabola we obtain $12^2 = 4p(3)$ and so $p = 12$. The equation is now $x^2 = 48y$. Substituting $x = 60$ in this equation and solving for y, we get $y = 75$.

16

Systems of Equations and Inequalities

≡ 16.1 SOLUTIONS OF NONLINEAR SYSTEMS OF EQUATIONS

1. $\begin{cases} x - 2y = 5 & (1) \\ x^2 - 4y^2 = 45 & (2) \end{cases}$. Solve (1) for x: $x = 2y + 5$. Substitute this result into (2) to get $(2y + 5)^2 - 4y^2 = 45$ or $4y^2 + 20y + 25 - 4y^2 = 45$ which simplifies to $20y = 20$ and so $y = 1$. Back substitution produces $x = 2 \cdot 1 + 5 = 7$. The only solution is $(7, 1)$

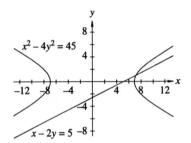

3. $\begin{cases} x^2 + 4y^2 = 32 & (1) \\ x + 2y = 0 & (2) \end{cases}$. Solving (2) for x: $x = -2y$. Substituting into (1) we get $(-2y)^2 + 4y^2 = 32, 4y^2 + 4y^2 = 32, 8y^2 = 32, y^2 = 4; y = \pm 2$. If $y = 2 \Rightarrow x = -2(2) = -4$; if $y = -2 \Rightarrow x = -2(-2) = 4$. Solutions $(-4, 2)$ and $(4, -2)$

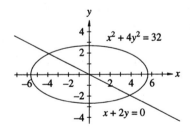

5. $\begin{cases} x^2 + y^2 = 4 & (1) \\ x^2 - 2y = 1 & (2) \end{cases}$. Subtracting (2) from (1) yields $y^2 + 2y = 3$ or $y^2 + 2y - 3 = 0$. This factors as $(y + 3)(y - 1) = 0$ and so $y = -3$ or $y = 1$. If $y = 1$, then, from (2), we obtain $x^2 - 2 = 1$, or $x^2 = 3$ and so $x = \pm\sqrt{3}$. If $y = -3$, then, from (2), we get $x^2 - 2 \cdot (-3) = 1$, or $x^2 + 6 = 1, x^2 = -5, x = \pm j\sqrt{5}$, which is not

real. Solution: $(\sqrt{3}, 1), (-\sqrt{3}, 1), (j\sqrt{5}, -3), (-j\sqrt{5}, -3)$

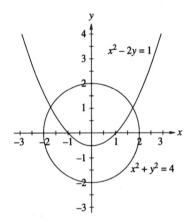

7. $\begin{cases} x^2 + y^2 = 7 & (1) \\ y^2 = 6x & (2) \end{cases}$. Substituting $6x$ for y^2 in (1) yields: $x^2 + 6x = 7, x^2 + 6x - 7 = 0, (x + 7)(x - 1) = 0, x = -7$ or $x = 1, x = -7$ yields $y^2 = 6(-7) = -42, y = \pm j\sqrt{42}$. $x = 1$ yields $y^2 = 6 \cdot 1 = 6, y = \pm\sqrt{6}$. Solutions $(1, \sqrt{6}), (1, -\sqrt{6}), (-7, j\sqrt{42}), (-7, -j\sqrt{42})$. (Note: the last two solutions are not real numbers.)

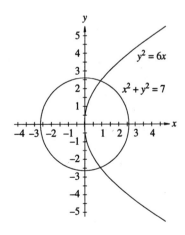

9. $\begin{cases} xy = 3 & (1) \\ 2x^2 - 3y^2 = 15 & (2) \end{cases}$. Solving (1) for x : $x = \frac{3}{y}$. Substituting into (2) $2 \cdot \left(\frac{3}{y}\right)^2 - 3y^2 = 15$, or $\frac{18}{y^2} - 3y^2 = 15$, or $18 - 3y^4 = 15y^2$, or $3y^4 + 15y^2 - 18 = 0$, which simplifies to $y^4 + 5y^2 - 6 = 0$, which factors as $(y^2 + 6)(y^2 - 1) = 0$. Thus, $y^2 = -6$ and $y = \pm j\sqrt{6} = \pm 3j\sqrt{\frac{2}{3}}$ or $y^2 = 1$ and $y = \pm 1$. Now, $y = 1 \Rightarrow x = 3$ which is the solution $(3, 1)$ and $y = -1 \Rightarrow x = -3$ and the solution $(-3, -1)$. Also, $y = j\sqrt{6} \Rightarrow x = \frac{3}{j\sqrt{6}} = -\frac{\sqrt{6}}{2}j = -j\sqrt{\frac{3}{2}}$ which produces the solution $\left(-\frac{\sqrt{6}}{2}j, j\sqrt{6}\right)$ or $\left(-j\sqrt{\frac{3}{2}}, 3j\sqrt{\frac{2}{3}}\right)$. Similarly, $y = -j\sqrt{6} \Rightarrow x = \frac{\sqrt{6}}{2}j = j\sqrt{\frac{3}{2}}$ and we have the solution $\left(\frac{\sqrt{6}}{2}j, -j\sqrt{6}\right)$ or $\left(j\sqrt{\frac{3}{2}}, -3j\sqrt{\frac{2}{3}}\right)$.

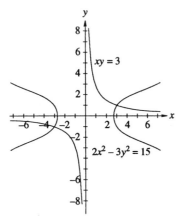

11. $t = \frac{d}{r}, t_1 = \frac{100}{r}, t_2 = \frac{20}{r-10}, t_1 + t_2 = \frac{100}{r} + \frac{20}{r-10} = 3, (r - 10)100 + 20r = 3r(r - 10), 100r - 1000 + 20r = 3r^2 - 30r, 3r^2 - 150r + 1000 = 0$. Using the quadratric formula, $r = 42.08$ mph for the first 100 miles, 32.08 mph for the last 20 miles.

13. When the resistors are connected in series, we get $R_1 + R_2 = 100$ and when they are connected in parallel $\frac{R_1 R_2}{R_1 + R_2} = 24$. Since $R_1 + R_2 = 100$, the second equation is equivalent to $R_1 R_2 = 2400$. Now, $R_1 = 100 - R_2$ and so $(100 - R_2)R_2 = 2400$, or $100R_2 - R_2^2 = 2400$, or $R_2^2 - 100R_2 + 2400 = 0$. This factors as $(R_2 - 60)(R_2 - 40) = 0$ which means that $R_2 = 60$ or 40. The resistances are $40\,\Omega$ and $60\,\Omega$.

15. $40v_1 + 60v_2 = 450 \Rightarrow 4v_1 + 6v_2 = 45, 20v_1^2 + 30v_2^2 = 1087.5 \Rightarrow 2v_1^2 + 3v_2^2 = 108.75$. Solving the first equation for v_1 yields: $v_1 = \frac{45 - 6v_2}{4}$. Substituting into the second equation yields $2\left(\frac{45 - 6v_2}{4}\right)^2 + 3v_2^2 = 108.75$ or $2\left(\frac{2025 - 540v_2 + 36v_2^2}{16}\right) + 3v_2^2 = 108.75$, or $2025 - 540v_2 + 36v_2^2 + 24v_2^2 = 870$, which is equivalent to $60v_2^2 - 540v_2 + 1155 = 0$ or $4v_2^2 - 36v_2 + 77 = 0$. Factoring the last equation produces $(2v_2 - 7)(2v_2 - 11) = 0$, and so $v_2 = \frac{7}{2}$ or $\frac{11}{2}$. If $v_2 = \frac{7}{2} = 3.5$, then $v_1 = 6$. If $v_2 = 5.5$, then $v_1 = 3$. Either $v_1 = 6$ m/s and $v_2 = 3.5$ m/s or $v_1 = 3$ m/s and $v_2 = 5.5$ m/s.

17. Since $F\ell = 200(0.8) = 160 \Rightarrow F = \frac{160}{\ell}$ and so $(F + 50)(\ell - 1) = 160 \Rightarrow \left(\frac{160}{\ell} + 50\right)(\ell - 1) = 160$, or $(160 + 50\ell)(\ell - 1) = 160\ell$, or $50\ell^2 + 110\ell - 160 = 160\ell$, which is equivalent to $50\ell^2 - 50\ell - 160 = 0$ and simplifies as $5\ell^2 - 5\ell - 16 = 0$. Using the quadratic formula $\ell = 2.36$ or -1.36. But, ℓ cannot be negative so 2.36 is the only answer. $F = \frac{160}{2.36} = 67.80$ gives $F = 67.80$ N, $\ell = 2.36$ m.

19. The first equation can be rewritten as $x = y + 5.59$. Subtracting the two areas we obtain the equation $\pi x^2 - y^2 = 303.22$. Substituting 3.14 for π and $y + 5.59$ for x we obtain

$$3.14(y + 5.59)^2 - y^2 = 303.22$$

Expanding and collecting like terms produces the equation

$$2.14y^2 + 35.11y - 205.10 = 0$$

The quadratic formula yields the answers $y = -20.98$ and $y = 4.57$. Since y must be positive, the answers are $x = 10.16$ mm and $y = 4.57$ mm.

21. Substituting $0.334v^2$ for r into the first equation and simplifying we have $L = (0.334v^2)x = 1.336v^2x$.

23. Solving the first equation for n produces $15n = 4291 - 4g^2$ or $n = 286.07 - 0.267g^2$. Substituting into the second equation we get

$$8g^2 + 15g - 5(286.07 - 0.267g^2) = 8607$$
$$\text{or} \quad 9.33g^2 + 15g - 10{,}037.34 = 0$$

The quadratic formula yields $g = 32$ and back substitution gives $n = 13$.

16.2 PROPERTIES OF INEQUALITIES; LINEAR INEQUALITIES

1. $x < 10 \Rightarrow x + 5 < 15$ **3.** $x < 10 \Rightarrow 3x < 30$ **5.** $x < 10 \Rightarrow -\frac{x}{2} > -5$

7. $x < 10 \Rightarrow x^2 < 100$ (Note $x > 0$)

9. Dividing $3x > 6$ by 3 leads to $x > 2$.

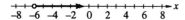

11. Subtracting 5 from both sides of the inequality $2x + 5 > -7$ produces $2x > -12$, which is equivalent to $x > -6$.

13. Adding 5 to both sides of the inequality $2x - 5 < x$ produces $2x < x + 5$ and subtracting x from both sides yields $x < 5$

15. Adding $7 - 2x$ to both sides of the inequality $4x - 7 \geq 2x + 5$ gives $2x \geq 12$, and dividing by 2, produces $x \geq 6$.

17. First we multiply the inequality $\frac{2}{3}x - 4 < \frac{1}{3}x + 2$ by 3, with the result $2x - 12 < x + 6$. Adding $12 - x$ to both sides, yields the result $x < 18$.

19. Multiplying $\frac{x-2}{4} \leq \frac{3}{8}$ by 8 produces $2(x - 2) \leq 3$ or $2x - 4 \leq 3$. Adding 4 to both sides, gives $2x \leq 7$, and then dividing by 2, we get $x \leq \frac{7}{2}$ or $x \leq 3\frac{1}{2}$.

21. Multiplying $\frac{x-3}{4} \leq \frac{2x}{3}$ by 12 gives $3(x-3) \leq 4 \cdot 2x$ which simplifies as $3x - 9 \leq 8x$, or $-5x \leq 9$, and so, after dividing by -5, we get $x \geq -\frac{9}{5}$.

23. The absolute value inequality $|x + 1| < 5$ is equivalent to $-5 < x + 1 < 5$ which simplifies to $-6 < x < 4$.

25. The inequality $|x + 4| > 6$ is equivalent to $x + 4 < -6$ or $x + 4 > 6$. Simplifying each of these inequalities, we get $x < -10$ or $x > 2$.

27. $|x + 5| < -3$. But, an absolute value can never be negative, so there are no real numbers that satisfy this inequality. This is a contradictory inequality and so there is no solution.

29. The inequality $-7 \leq 3x + 5 < 26$ is equivalent to $-12 \leq 3x < 21$, or $-4 \leq x < 7$

31. The inequality $3x + 1 < 5 < 2x - 3$ is equivalent to the two inequalities $3x + 1 < 5$ and $5 < 2x - 3$. Subtracting 1 from the first inequality, gives $3x < 4$ or $x < \frac{4}{3}$. Adding 3 to the second inequality, gives $8 < 2x$ which can be simplified as $4 < x$ or $x > 4$. Thus, the solution is $x < \frac{4}{3}$ and $x > 4$. This is a contradictory statement, there are no real numbers that satisfy the given inequality.

33. $6357 \text{ km} \leq r \leq 6378 \text{ km}$

35. $0.60 - 0.05 \leq c \leq 0.60 + 0.05, 0.10° \leq c \leq 1.10°$

37. $200 < \frac{800R}{800+R} < 500, 200(800 + R) < 800R < 500(800+R), 160000 + 200R < 800R < 400000 + 500R$, so $160000 + 200R < 800R$ and $800R < 400000 + 500R, 160000 < 600R$ and $300R < 400000, 266.67 < R$ and $R < 1333.33, 266.67 \,\Omega < R < 1333.33 \,\Omega$

39. $R - C > 0, 60x - (1,500 + 25x) > 0, 35x - 1,500 > 0, 35x > 1,500, x > 42.86$ or, since we cannot have a fraction, $x \geq 43$

41. (a) $|\epsilon| < 0.6$,
(b) $|P - 9.0| < 0.6$,
(c) Since $|P - 9.0| < 0.6$, then $-0.6 < P - 9.0 < 0.6$. Adding 9.0 to all three terms, we get $8.4 < P < 9.6$ W.

43. Let x represent the number of copies sold. Since profit P is the difference between the revenue R and the cost C, we have $P = R - C = 0.35x + (0.10)(0.35)(x - 10{,}000) - 0.38x$. Solving this equation, we obtain
$$P = 0.35x + (0.10)(0.35)(x - 10{,}000) - 0.38x$$
$$= 0.35x - 350 + 0.35x - 0.38x$$
$$= 0.005x - 350$$

The break-even point is when $P = 0$ or $0.005x - 350 = 0$ or $x = 70{,}000$. A profit is made on the 70,001st copy. So, the company must sell at least 70,001 copies in order for the company to make a profit on this magazine.

$\equiv$ 16.3 NONLINEAR INEQUALITIES

1. $\{x | x < -1 \text{ or } x > 3\}$

3. $\{x | -4 \le x \le 1\}$

5. $x^2 - 1 < 0$ is equivalent to $(x + 1)(x - 1) < 0$. The zero values are -1 and 1, so the solution set is $\{x | -1 < x < 1\}$.

7. $x^2 - 2x - 15 \le 0$ is equivalent to $(x+3)(x-5) \le 0$. The zero values are -3 and 5, so the solution set is $\{x | -3 \le x \le 5\}$.

9. $x^2 - x - 2 \ge 0$ is equivalent to $(x + 1)(x - 2) \ge 0$. The zero values are -1 and 2, so the solution set is $\{x | x \le -1 \text{ or } x \ge 2\}$.

11. $x^2 - 5x > -6$ is equivalent to $x^2 - 5x + 6 > 0$ which factors as $(x - 2)(x - 3) > 0$. The zero values are 2 and 3, so the solution set is $\{x | x < 2 \text{ or } x > 3\}$.

13. $2x^2 + 7x + 3 < 0$ is equivalent to $(2x+1)(x+3) < 0$. The zero values are -3 and $-\frac{1}{2}$, so the solution set is $\{x | -3 < x < -\frac{1}{2}\}$

15. $2x^2 - x < 1$ is equivalent to $2x^2 - x - 1 < 0$, or $(2x + 1)(x - 1) < 0$. The zero values are $-\frac{1}{2}$ and 1, so the solution set is $\{x | -\frac{1}{2} < x < 1\}$.

17. $4x^2 + 2x > x^2 - 1$ is equivalent to $3x^2 + 2x + 1 > 0$. The discriminant of the equation $3x^2 + 2x + 1 = 0$ is $b^2 - 4ac = 4 - 4 \cdot 3 \cdot 1 = -8 < 0$, so this equation has no real solutions. When $x = 0$, we get $1 > 0$ which is true. Thus, the given inequality

is an absolute inequality, all real numbers satisfy this inequality.

19. The zero values are $-1, 2$, and -3, so the solution is $x < -3$ or $-1 < x < 2$.

21. The zero values are $-3, \frac{5}{2}$, and -4, so the solution set is $\{x | -4 < x < -3 \text{ or } x > \frac{5}{2}\}$.

23. $(x - 2)^2(x + 4) < 0$. The zero values are 2 and -4, so the solution set is $\{x | x < -4\}$.

25. $x^3 - 4x > 0$ is equivalent to $x(x + 2)(x - 2) > 0$. The zero values are $0, -2$, and 2, so the solution set is $\{x | -2 < x < 0 \text{ or } x > 2\}$.

27. $x^2 + 2x + 3 \le 0$. The discriminant of the equation $x^2 + 2x + 3 = 0$ is $2^2 - 4 \cdot 3 = -8 < 0$, so there are no real solutions to this equation. When $x = 0, 3 \le 0$ is false so the given inequality is a contradictory inequality and there are no real solutions.

29. The zero values are $2, 5$, and -1, so the solution set is $\{x | x < -1 \text{ or } 2 < x < 5\}$

31. $\frac{x}{(x+1)(x-2)} > 0$. The zero values are $0, -1$, and 2, so the solution set is $\{x | -1 < x < 0 \text{ or } x > 2\}$.

33. $\frac{(3x+1)(x-3)}{2x-1} \leq 0$. The zero values are $-\frac{1}{3}, \frac{1}{2}$, and 3, so the solution set is $\{x | x \leq -\frac{1}{3}$ or $\frac{1}{2} < x \leq 3\}$.

35. The inequality $\frac{4}{x-1} < \frac{5}{x+1}$ is equivalent to $\frac{4}{x-1} - \frac{5}{x+1} < 0$ or $\frac{4(x+1)-5(x-1)}{(x-1)(x+1)} < 0$. This simplifies to $\frac{-x+9}{(x-1)(x+1)} < 0$. The zero values are 9, -1, and 1, so the solution set is $\{x | -1 < x < 1$ or $x > 9\}$.

37. $\frac{x^2+2x+3}{x-1} \geq 0$. The numerator x^2+2x+3 is always > 0, so the original inequality is true when $x > 1$.

39. $|x^2 + 3x + 2| \leq 4$ is equivalent to $-4 \leq x^2 + 3x + 2 \leq 4$. We first examine the right-hand inequality $x^2 + 3x + 2 \leq 4$, which simplifies to $x^2 + 3x - 2 \leq 0$. Using the quadratic formula, we see that the equation $x^2 + 3x - 2 = 0$ is zero when $x = \frac{-3 \pm \sqrt{9+8}}{2} = \frac{-3 \pm \sqrt{17}}{2}$. Next, we examine the left-hand inequality and see that $-4 \leq x^2 + 3x + 2$ is always true. So the solution set is $\left\{x | \frac{-3-\sqrt{17}}{2} \leq x \leq \frac{-3+\sqrt{17}}{2}\right\}$.

41. Since $I = 75d^2$, the solution to $75 < I < 450$ is equivalent to $75 < 75d^2 < 450$ or $1 < d^2 < 6$. Since the context of the problem means that d is positive, and so we see that 1 m $< d < \sqrt{6}$ m.

43. $x^2 - 1.1x + 0.2 > 0.08$ is equivalent to $x^2 - 1.1x + 0.12 > 0$. The equation $x^2 - 1.1x + 0.12 = 0$ has solutions $x = \frac{1.1 \pm \sqrt{1.21-.48}}{2} = \frac{1.1 \pm \sqrt{0.73}}{3}$. So, the

solution set to the given inequality is $x < 0.123$ or $x > 0.977$.

45. **(a)** The height will be greater than 500 ft when $h(t) = -16t^2 + 64t + 452 > 500$. Solve the inequality as follows
$$-16t^2 + 64t + 452 > 500$$
$$-16t^2 + 64t + 452 - 500 > 0$$
$$-16t^2 + 64t - 48 > 0$$
$$t^2 - 4t + 3 < 0$$
$$(t-3)(t-1) < 0$$
A sign chart, or your graphing calculator, will show that $t^2 - 4t + 3 < 0$ between $1 < t < 3$ s and so the ball is more than 500 ft above the ground when 1 s $< t < 3$ s.

(b) The height will be less than 395 ft when $h(t) = -16t^2 + 64t + 452 < 395$ or $-16t^2 + 64t + 452 - 395 < 0$. This is equivalent to $-16t^2 + 64t + 57 < 0$. The quadratic formula shows that $-16t^2 + 64t + 57 = 0$ when $t = -0.75$ or $t = 4.75$. The first result is not possible since that is before the ball is thrown. Examining a sign chart or a graph we see that the solution will be when $t > 4.75$ s.

47.
$$0.79D^2 - 2D - 4 > 926$$
$$0.79D^2 - 2D - 9308 > 0$$
By the quadratic formula, the roots of $0.79D^2 - 2D - 9308 = 0$ are $D = -33.1$ and $D = 35.6$. Hence, by a sign chart or graphing calculator, you can see that $D \geq 35.6$ in. is the only realistic solution.

≡ 16.4 INEQUALITIES IN TWO VARIABLES

1.

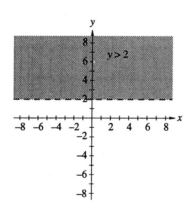

3.

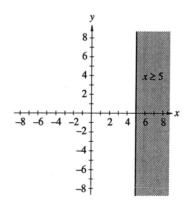

5.

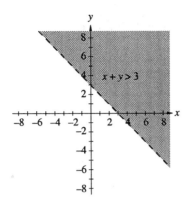

11.

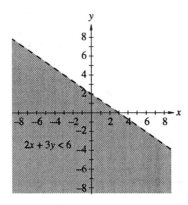

7.

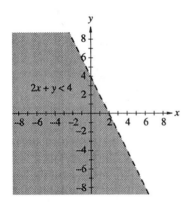

13.

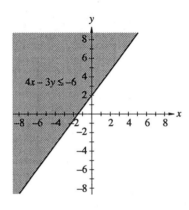

9.

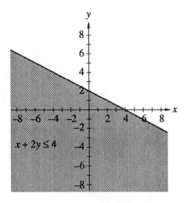

15

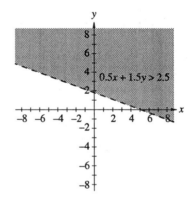

17.

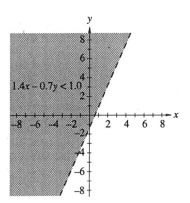

19.

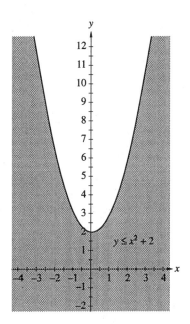

21.

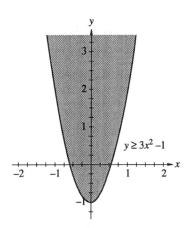

23.

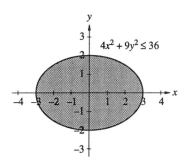

25.

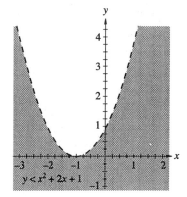

27.

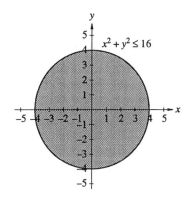

29. $|x+1| < 4, -4 < x+1 < 4, -5 < x < 3$

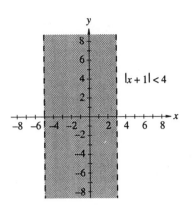

33. The inequality is $0.085x + 0.0675y \geq 3.6$ or $y \geq 53.5 - 1.25925x$.

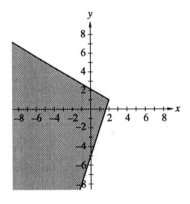

31.

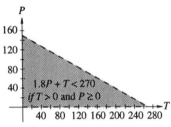

≡ 16.5 SYSTEMS OF INEQUALITIES: LINEAR PROGRAMMING

1.

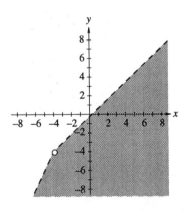

5.

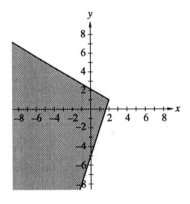

3.

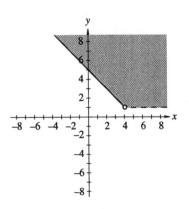

7.

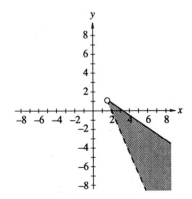

9.

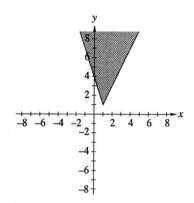

11.

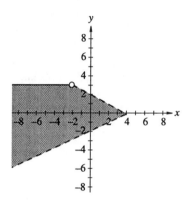

13.

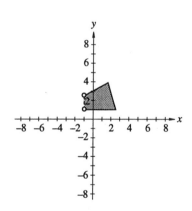

15. First graph the constraints. The vertex points of the feasible region are (2, 3), (2, 4.5) and (4, 3). (Fig. 16.79s). Evaluating $P = 2x + y$ at each of these points (2, 3): $2 \cdot 2 + 3 = 7, (2, 4.5): 2 \cdot 2 + 4.5 = 8.5, (4, 3): 2 \cdot 4 + 3 = 11$. Maximum of 11 at (4, 3).

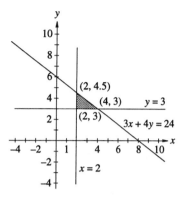

17. After graphing the constraints, the vertex points of the feasible region are $(0, 0), (0, 23\frac{1}{3}), (26\frac{2}{3}, 0)$, and $(20, 10)$. $F = 15x + 10y$, so $F(0, 0) = 15 \cdot 0 + 10 \cdot 0 = 0, F(0, 23\frac{1}{3}) = 10 \times 23\frac{1}{3} = 233.3, F(26\frac{2}{3}, 0) = 15 \times 26\frac{2}{3} = 400, F(20, 10) = 15 \cdot 20 + 10 \cdot 10 = 400$ also. The maximum of 400 occurs at any point on the segment $3x + 2y = 80$ and $20 \leq x \leq 26\frac{2}{3}$.

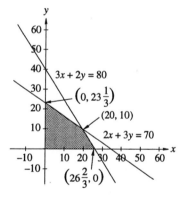

19. After graphing the constraints, the vertex points of the feasible region are at (0,3), (0,18), (12,12), and (16.5,3). Evaluating $P = 8x + 10y$ at each of these vertex points, we obtain $P(0, 3) = 30$, $P(0, 18) = 180$, $P(12, 12) = 216$, and $P(16.5, 3) = 162$. Maximum of 216 at (12, 12).

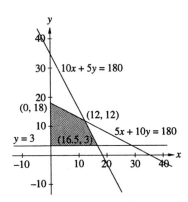

21. The constraints are as follows: $X \geq 0, Y \geq 0, \frac{1}{2}X + \frac{1}{4}Y \leq 100$, or $Y \leq 400 - 2X, \frac{3}{4}X + \frac{1}{6}Y \leq 80$ or $Y \leq 480 - \frac{9}{2}X$. Graphing the constraints gives vertex points at (0, 0), (0, 400), $(106\frac{2}{3}, 0)$, and (32, 336). Evaluating $P(x, y) = 10x + 8y$ at each of these vertex points, we get $P(0,0) = 0, P(0, 400) = 3200, P(106\frac{2}{3}, 0) = 1067$, and $P(32, 336) = 3008$. Maximum profit of \$3200 when you produce 0 of X and 400 of Y. Adding people at A will increase profits.

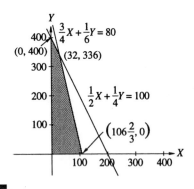

23. $P(A, B) = 8A + 9.50B, P(0,0) = 0, P(0, 120) = 1140, P(30, 90) = 1095, P(97.5, 0) = 780$. The maximum profit will be when the company makes 120 of B and none of A for a profit of \$1,140.00.

25. Let x be the amount in millions of dollars for home loans and y the amount for commercial loans. Then the constant equations are $x \geq 0, y > 0, x + y \leq 50, x \geq 5y$, and $0.085x + 0.0675y \geq 3.6$. The vertices are at $\left(\frac{7200}{197}, \frac{1440}{197}\right) \approx (36.5, 7.3), \left(\frac{125}{3}, \frac{25}{3}\right) \approx (41.7, 8.3), (50, 0)$, and $\left(\frac{720}{17}, 0\right) \approx (42.4, 0)$. The bank can make as few as 37 home loans and 7 commercial loans in order to reach its goal of at least \$3.6 million dollars.

27. Let x represent the number of small tubes and y the number of large ones. The constraints are $x \geq 8,000, y \geq 5,000, x+y \geq 15,000$, and $x \geq 2y$. The only vertex of the feasible region is (10000, 5000). Buy 10,000 small and 5,000 large for a total cost of \$2,100.

CHAPTER 16 REVIEW

1. $3x < -12$ or $x < -4$

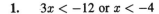

2. $4x + 5 < 6$ or $4x < 1$ and so $x < \frac{1}{4}$

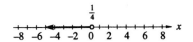

3. $2x - 7 \geq 15$ or $2x \geq 22$ which gives $x \geq 11$

4. $7 - 5x < 2 + 3x$ or $-8 < -5$, and so $x > \frac{5}{8}$

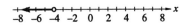

5. $\frac{2x+5}{4} < \frac{4x-1}{3}$ or $6x + 15 < 16x - 4$ which simplifies to $-10x < -19$, and so $x > \frac{19}{10}$.

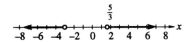

6. $|2x - 1| \le 5$ is equivalent to $-5 \le 2x - 1 \le 5$ or $-4 \le 2x \le 6$, and so $-2 \le x \le 3$.

7. $|3x + 2| > 7$ is equivalent to

$3x + 2 < -7$ or $3x + 2 > 7$
$3x < -9$ or $3x > 5$
$x < -3$ or $x > \frac{5}{3}$

8. $2x + 3 < 13 \le 3x - 9$ is equivalent to
$2x + 3 < 13$ and $13 \le 3x - 9$
$2x < 10$ and $22 \le 3x$
$x < 5$ and $x \ge \frac{22}{3}$
This is a contradictory inequality and so there is no solution.

9. $-2x + y < 5$ or $y < 2x + 5$

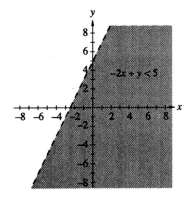

10. $2x + 3y \ge 6$ or $y \ge \frac{-2x}{3} + 2$

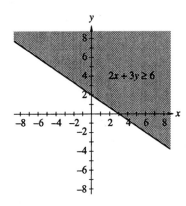

11. $4x - 3y < 3$ or $y > \frac{4}{3}x - 1$

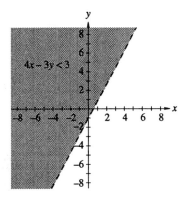

12. $y > 2x^2 - 5$

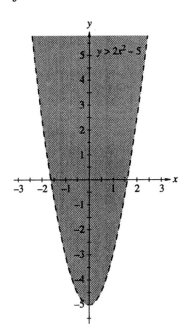

13. $y \le x^2 - 6x + 9$

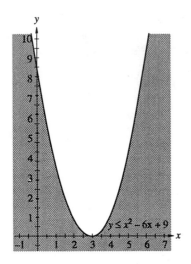

14.

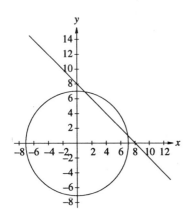

15.

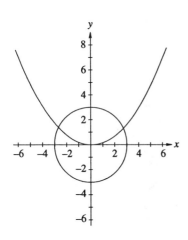

16.

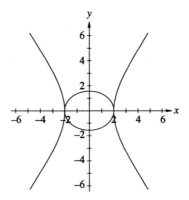

17.

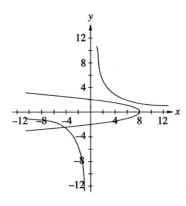

18.

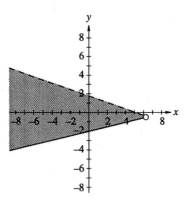

19.

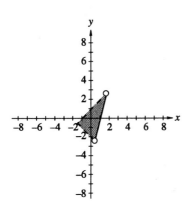

20.

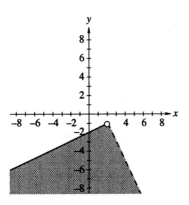

21.

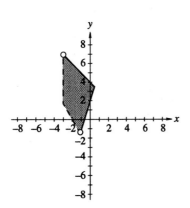

22. Graphing the constraints as shown, we see that the vertices of the feasible region are $(-1, 6), (-1, -3)$, $(3, 1)$ and $(3, 2)$. We have $P(x, y) = 3x + 5y, P(-1, 6) = -3 + 30 = 27, P(-1, -3) = -3 - 15 = -18$, $P(3, 1) = 9 + 5 = 14$, and $P(3, 2) = 9 + 10 = 19$. We can see that there is a maximum of 27 at $(-1, 6)$

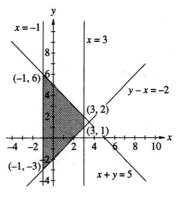

23. After graphing, we see that the vertices of the feasible region are $\left(\frac{-4}{7}, \frac{-23}{7}\right), \left(\frac{18}{5}, -\frac{6}{5}\right)$, and $\left(\frac{7}{6}, \frac{11}{3}\right)$. Checking $F(x, y) = 4x - 3y$ at these points, we see that $F\left(\frac{4}{7}, \frac{23}{7}\right) = 7\frac{4}{7}, F\left(\frac{7}{6}, \frac{11}{3}\right) = -6\frac{1}{3}$, and $F\left(\frac{18}{5}, -\frac{6}{5}\right) = 18$. So, the minimum of $-6\frac{1}{3}$ is at $\left(\frac{7}{6}, \frac{11}{3}\right)$ or $\left(1\frac{1}{6}, 3\frac{2}{3}\right)$.

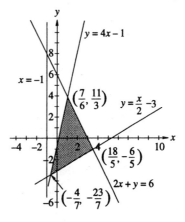

24. $A = lw., 7.0 \times 1.4 < A < 7.5 \times 1.6, 9.8 \text{ mm}^2$
$A < 12 \text{ mm}^2$

25. Since $R_1 R_2 = 20$, we see that $\frac{R_1 R_2}{R_1 + R_2} \geq 4$ is equivalent to $\frac{20}{R_1 + R_2} \geq 4$ or $20 \geq 4(R_1 + R_2)$ and so $R_1 + R_2 \leq 5\,\Omega$.

26.

	r	t	d
normal	r	8	$8r$
increased	$r+4$	7.5	$7.5(r+4)$

Thus, we have the equation $8r = 7.5r + 30$, or $0.5r = 30$, and so $r = 60$. Usual speed is 60 mph, distance is $8 \cdot 60 = 480$. mi.

27. We have the system: $12C + 18S \le 220, 15C + 10S \le 180, C \ge 0, S \ge 0, C$ and S must be integers. We graph C on the horizontal axis, and S on the vertical axis. Solving the equation derived from the second inequality, we get $S = 18 - 1.5C$. Substituting this value of S into the equation from the first inequality yields $12C + 18(18 - 1.5C) = 220$ or $12C + 324 - 27C = 220$ which simplifies to $-15C = -104$. Thus, we find that these two lines intersect when $C = 6.9\overline{3}$. Possible vertex points

are $(0, 0), (0, 12), (6, 8), (7, 7)$ and $(12, 0)$. Evaluating $P(C, S) = 1250C + 1620S$ at each of these vertex points, we find $P(0,0) = 0, P(0, 12) = 19,400, P(6,8) = 20,460, P(7,7) = 20,090$, and $P(12,0) = 15,000$. Maximum profit of \$20,460 at $(6, 8)$ or 6 of C and 8 of S.

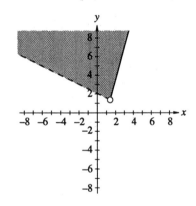

CHAPTER 16 TEST

1. $5x > 30, x > 6$

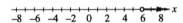

2. $|4x + 5| < 17, -17 < 4x + 5 < 17, -22 < 4x < 12, -\frac{11}{2} < x < 3$

3. $\frac{4x-5}{2} \le \frac{7x-2}{3}, 12x - 15 \le 14x - 4, -2x \le 11, x \ge -\frac{11}{2}$

4. $(x - 2)(x - 5) > 0$ has two cases. In Case I, we have $x - 2 > 0$ or $x > 2$ and that $x - 5 > 0$ or $x > 5$. These both occur when $x > 5$. In Case II, we have $x - 2 < 0$ or $x < 2$ and $x - 5 < 0$ or $x < 5$. These both occur when $x < 2$. Thus the solution set is $\{x | x < 2 \text{ or } x > 5\}$

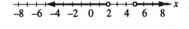

5. $5x^2 + 3x \le 2x^2 - x - 1, 3x^2 + 4x + 1 \le 0, (3x + 1)(x + 1) \le 0$. The solution set is $\{x | -1 \le x \le -\frac{1}{3}\}$

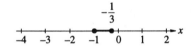

6.

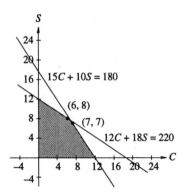

7. Iron constraint gives the inequality $12x + 5y \geq 80$, the zinc constraint gives the inequality $10x + 8y \geq 90$, while $x \geq 0$, and $y \geq 0$. The vertex points on the axes are (0, 16) and (9, 0). The two equations $10x + 8y = 90$ and $12x + 5y = 80$ intersect at $\left(\frac{95}{23}, \frac{140}{23}\right) \approx (4.13, 6.09)$. The points (5, 5) and (4, 7) are the closest to this vertex. The cost is given by the equation $C(x, y) = 4x + 6y$. Evaluating the cost function at each of these four points we get $C(0, 15) = 90\cent$, $C(4, 7) = 58\cent$, $C(5, 5) = 50\cent$, and $C(9, 0) = 36\cent$.

(a) The minimum cost is obtained by taking 9 of brand X and none of Y.

(b) The minimum cost is $36\cent$.

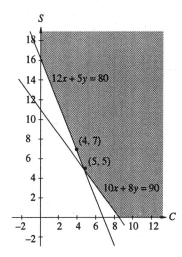

CHAPTER
17
Matrices

17.1 MATRICES

1. 2×3

3. 3×6

5. 4×2

7. $a_{11} = 1; a_{24} = 13; a_{21} = 8; a_{32} = 6$

9. $x = 12, y = -6, z = 2, w = 5$

11. $\begin{bmatrix} -3 & 6 & 6 & -8 \\ 12 & 9 & 17 & 0 \\ 9 & 0 & 5 & 10 \end{bmatrix}$

13. $\begin{bmatrix} 2 & -1 & 4 \\ 5 & 9 & -2 \end{bmatrix}$

15. $A + B = \begin{bmatrix} -1 & 7 & 4 \\ 8 & 1 & 15 \end{bmatrix}$

17. $3A = \begin{bmatrix} 12 & 9 & 6 \\ 15 & 0 & 21 \end{bmatrix}$

19. $2B + C = \begin{bmatrix} -10 & 8 & 4 \\ 6 & 2 & 16 \end{bmatrix} + \begin{bmatrix} 5 & 8 & 4 \\ 6 & -2 & 9 \end{bmatrix} = \begin{bmatrix} -5 & 16 & 8 \\ 12 & 0 & 25 \end{bmatrix}$

21. $2B - 3A = \begin{bmatrix} -10 & 8 & 4 \\ 6 & 2 & 16 \end{bmatrix} - \begin{bmatrix} 12 & 9 & 6 \\ 15 & 0 & 21 \end{bmatrix} = \begin{bmatrix} -22 & -1 & -2 \\ -9 & 2 & -5 \end{bmatrix}$

23. $D - E = \begin{bmatrix} -3 \\ -7 \\ 3 \\ -5 \end{bmatrix}$

25. $3E = \begin{bmatrix} 21 \\ 27 \\ -12 \\ 30 \end{bmatrix}$

27. $3D + 2E = \begin{bmatrix} 12 \\ 6 \\ -3 \\ 15 \end{bmatrix} + \begin{bmatrix} 14 \\ 18 \\ -8 \\ 20 \end{bmatrix} = \begin{bmatrix} 26 \\ 24 \\ -11 \\ 35 \end{bmatrix}$

29. $7D - 2F = \begin{bmatrix} 28 \\ 14 \\ -7 \\ 35 \end{bmatrix} - \begin{bmatrix} 28 \\ 36 \\ -16 \\ 10 \end{bmatrix} = \begin{bmatrix} 0 \\ -22 \\ 9 \\ 25 \end{bmatrix}$

31. $\begin{bmatrix} 18 & 7 & 9 \\ 7 & 11 & 4 \end{bmatrix} + \begin{bmatrix} 9 & 5 & 6 \\ 11 & 4 & 12 \end{bmatrix} = \begin{bmatrix} 27 & 12 & 15 \\ 18 & 15 & 16 \end{bmatrix}$

27 chip A, 12 chip B, 15 EPROMS, 18 keyboards, 15 motherboards and 16 disk drives

33. $\begin{bmatrix} 70 & 40 \\ 30 & 60 \end{bmatrix} + \begin{bmatrix} 65 & 42 \\ 35 & 58 \end{bmatrix} = \begin{bmatrix} 135 & 82 \\ 65 & 118 \end{bmatrix}$

35. The answer is the last sales matrix times 1.1. Note: Answers are rounded to the nearest integer.

Chip Type:	286	386	486	586
Warehouse A	297	2343	3531	292
Warehouse B	1232	4653	3436	83
Warehouse C	352	3439	3017	1113

37. See *Computer Programs* in main text.

39. See *Computer Programs* in main text.

17.2 MULTIPLICATION OF MATRICES

1. $\begin{bmatrix} 2 & 3 \\ 1 & -4 \end{bmatrix} \times \begin{bmatrix} 4 & -3 \\ 6 & 5 \end{bmatrix} = \begin{bmatrix} 8+18 & -6+15 \\ 4-24 & -3-20 \end{bmatrix} = \begin{bmatrix} 26 & 9 \\ -20 & -23 \end{bmatrix}$

3. $\begin{bmatrix} 5 & 1 & 6 & 2 \end{bmatrix} \times \begin{bmatrix} 2 \\ 9 \\ -8 \\ 3 \end{bmatrix} = \begin{bmatrix} 10+9-48+6 \end{bmatrix} = \begin{bmatrix} -23 \end{bmatrix}$

5. $\begin{bmatrix} 1 & 2 & 4 \\ 2 & 3 & 6 \end{bmatrix} \times \begin{bmatrix} 5 & -7 & 4 \\ 4 & 6 & 0 \\ -4 & 2 & 1 \end{bmatrix} = \begin{bmatrix} 5+8-16 & -7+12+8 & 4+0+4 \\ 10+12-24 & -14+18+12 & 8+0+6 \end{bmatrix} = \begin{bmatrix} -3 & 13 & 8 \\ -2 & 16 & 14 \end{bmatrix}$

7. $\begin{bmatrix} 1 & 2 & 3 \\ 6 & 5 & 4 \\ -1 & 0 & 2 \end{bmatrix} \times \begin{bmatrix} 9 & 8 & 7 \\ -2 & 3 & -1 \\ 0 & 1 & 0 \end{bmatrix} =$

$\begin{bmatrix} 9-4+0 & 8+6+3 & 7-2+0 \\ 54-10+0 & 48+15+4 & 42-5+0 \\ -9+0+0 & -8+0+2 & -7+0+0 \end{bmatrix} =$

$\begin{bmatrix} 5 & 17 & 5 \\ 44 & 67 & 37 \\ -9 & -6 & -7 \end{bmatrix}$

9. $\begin{bmatrix} 3 & 4 & -2 \\ 7 & 8 & 10 \end{bmatrix} \times \begin{bmatrix} 0 & 4 & 1 \\ 1 & 0 & 2 \\ 8 & -1 & 0 \end{bmatrix} =$

$\begin{bmatrix} 0+4-16 & 12+0+2 & 3+8+0 \\ 0+8+80 & 28+0-10 & 7+16+0 \end{bmatrix} =$

$\begin{bmatrix} -12 & 14 & 11 \\ 88 & 18 & 23 \end{bmatrix}$

11. $AB = \begin{bmatrix} 1 & 2 \\ 3 & 4 \end{bmatrix} \times \begin{bmatrix} 3 & 6 \\ 2 & 5 \end{bmatrix} = \begin{bmatrix} 3+4 & 6+10 \\ 9+8 & 18+20 \end{bmatrix} =$ $\begin{bmatrix} 7 & 16 \\ 17 & 38 \end{bmatrix}$

13. $AC = \begin{bmatrix} 1 & 2 \\ 3 & 4 \end{bmatrix} \begin{bmatrix} 4 & -2 \\ -6 & 2 \end{bmatrix} =$

$\begin{bmatrix} 4-12 & -2+4 \\ 12-24 & -6+8 \end{bmatrix} = \begin{bmatrix} -8 & 2 \\ -12 & 2 \end{bmatrix}$

15. $A(BC) = A \begin{bmatrix} -24 & 6 \\ -22 & 6 \end{bmatrix} = \begin{bmatrix} 1 & 2 \\ 3 & 4 \end{bmatrix} \begin{bmatrix} -24 & 6 \\ -22 & 6 \end{bmatrix} =$

$\begin{bmatrix} -24-44 & 6+12 \\ -72-88 & 18+24 \end{bmatrix} = \begin{bmatrix} -68 & 18 \\ -160 & 42 \end{bmatrix}$

17. $A(B+C) = \begin{bmatrix} 1 & 2 \\ 3 & 4 \end{bmatrix} \left(\begin{bmatrix} 3 & 6 \\ 2 & 5 \end{bmatrix} + \begin{bmatrix} 4 & -2 \\ -6 & 2 \end{bmatrix} \right) =$

$\begin{bmatrix} 1 & 2 \\ 3 & 4 \end{bmatrix} \begin{bmatrix} 7 & 4 \\ -4 & 7 \end{bmatrix} = \begin{bmatrix} 7-8 & 4+14 \\ 21-16 & 12+28 \end{bmatrix} =$

$\begin{bmatrix} -1 & 18 \\ 5 & 40 \end{bmatrix}$

19. $(2A)(3B)$: Scaler multiplication is commutative so

$(2A)(3B) = 6(AB) = 6 \cdot \begin{bmatrix} 7 & 16 \\ 17 & 38 \end{bmatrix} = \begin{bmatrix} 42 & 96 \\ 102 & 228 \end{bmatrix}$

21. $A = \begin{bmatrix} 2 & 10 \\ 3 & 15 \end{bmatrix}$, $B = \begin{bmatrix} -10 & 25 \\ 2 & -5 \end{bmatrix}$, $AB =$ $\begin{bmatrix} -20+20 & 50-50 \\ -30+30 & 75-75 \end{bmatrix} = \begin{bmatrix} 0 & 0 \\ 0 & 0 \end{bmatrix}$. We cannot *conclude* that, if $AB = 0$, then either $A = 0$ or $B = 0$. Hence, matrix multiplication does *not* have a zero product principal.

23. Here we have the product

$\begin{bmatrix} 270 & 2130 & 3210 & 265 \\ 1120 & 4230 & 3124 & 75 \\ 320 & 3126 & 2743 & 1012 \end{bmatrix} \times \begin{bmatrix} 25 \\ 35 \\ 55 \\ 95 \end{bmatrix} =$

$\begin{bmatrix} 283,025 \\ 354,995 \\ 364,415 \end{bmatrix}$

From this we see that Warehouse A generated sales of $283,025, Warehouse B sales of $354,995, and Warehouse C generated sales of $364,415.

25. $A = \begin{bmatrix} 0 & 1 \\ 1 & 0 \end{bmatrix}$:

$A^2 = \begin{bmatrix} 0 & 1 \\ 1 & 0 \end{bmatrix} \times \begin{bmatrix} 0 & 1 \\ 1 & 0 \end{bmatrix} = \begin{bmatrix} 1 & 0 \\ 0 & 1 \end{bmatrix} = I$,

$B = \begin{bmatrix} 0 & -i \\ i & 0 \end{bmatrix}$: $B^2 = \begin{bmatrix} 0 & -i \\ i & 0 \end{bmatrix} \begin{bmatrix} 0 & -i \\ i & 0 \end{bmatrix} =$

$\begin{bmatrix} -i^2 & 0 \\ 0 & -i^2 \end{bmatrix} = \begin{bmatrix} 1 & 0 \\ 0 & 1 \end{bmatrix} = I; C = \begin{bmatrix} 1 & 0 \\ 0 & -1 \end{bmatrix}$:

$C^2 = \begin{bmatrix} 1 & 0 \\ 0 & -1 \end{bmatrix} \cdot \begin{bmatrix} 1 & 0 \\ 0 & -1 \end{bmatrix} = \begin{bmatrix} 1 & 0 \\ 0 & 1 \end{bmatrix} = I$

27. $AB - BA = \begin{bmatrix} i & 0 \\ 0 & -i \end{bmatrix} - \begin{bmatrix} -i & 0 \\ 0 & i \end{bmatrix} =$

$\begin{bmatrix} 2i & 0 \\ 0 & -2i \end{bmatrix} = 2i \begin{bmatrix} 1 & 0 \\ 0 & -1 \end{bmatrix} = 2iC. AC - CA =$

$$\begin{bmatrix} 0 & -1 \\ 1 & 0 \end{bmatrix} - \begin{bmatrix} 0 & 1 \\ -1 & 0 \end{bmatrix} = \begin{bmatrix} 0 & -2 \\ 2 & 0 \end{bmatrix} = -2iB.$$

$$-2i \begin{bmatrix} 0 & -i \\ i & 0 \end{bmatrix} = \begin{bmatrix} 0 & 2i^2 \\ -2i^2 & 0 \end{bmatrix} = \begin{bmatrix} 0 & -2 \\ 2 & 0 \end{bmatrix}.$$

$$BC - CB = \begin{bmatrix} 0 & i \\ i & 0 \end{bmatrix} - \begin{bmatrix} 0 & -i \\ -i & 0 \end{bmatrix} = \begin{bmatrix} 0 & 2i \\ 2i & 0 \end{bmatrix} =$$

$$2i \begin{bmatrix} 0 & 1 \\ 1 & 0 \end{bmatrix} = 2iA$$

29. See *Computer Programs* in main text.

≡ 17.3 INVERSES OF MATRICES

1. $\begin{vmatrix} 6 & 1 \\ 5 & 1 \end{vmatrix} = 6 - 5 = 1.$ Inverse is $\begin{bmatrix} 1 & -1 \\ -5 & 6 \end{bmatrix}$

3. $\begin{vmatrix} 10 & 4 \\ 8 & 2 \end{vmatrix} = 30 - 32 = -2.$ Inverse is

$$\frac{1}{-2} \begin{bmatrix} 3 & -4 \\ -8 & 10 \end{bmatrix} = \begin{bmatrix} \frac{-3}{2} & 2 \\ 4 & -5 \end{bmatrix} = \begin{bmatrix} -1.5 & 2 \\ 4 & -5 \end{bmatrix}$$

5. $\begin{vmatrix} 8 & -6 \\ -6 & 4 \end{vmatrix} = 32 - 36 = -4.$ Inverse is

$$-\frac{1}{4} \begin{bmatrix} 4 & 6 \\ 6 & 8 \end{bmatrix} = \begin{bmatrix} -1 & -1.5 \\ -1.5 & -2 \end{bmatrix}$$

7. $\begin{vmatrix} 15 & 10 \\ 4 & 3 \end{vmatrix} = 45 - 40 = 5.$ Inverse is

$$\frac{1}{5} \begin{bmatrix} 3 & -10 \\ -4 & 15 \end{bmatrix} = \begin{bmatrix} .6 & -2 \\ -0.8 & 3 \end{bmatrix}$$

9.
$$\begin{bmatrix} 0 & -3 & 0 & \vdots & 1 & 0 & 0 \\ 1 & 0 & 0 & \vdots & 0 & 1 & 0 \\ 0 & 0 & 4 & \vdots & 0 & 0 & 1 \end{bmatrix} \begin{matrix} -\frac{1}{3}R_1 \\ \Rightarrow \\ \frac{1}{4}R_3 \end{matrix}$$

$$\begin{bmatrix} 0 & 1 & 0 & \vdots & -\frac{1}{3} & 0 & 0 \\ 1 & 0 & 0 & \vdots & 0 & 1 & 0 \\ 0 & 0 & 1 & \vdots & 0 & 0 & \frac{1}{4} \end{bmatrix}.$$ Interchange R_1 and

$$R_2 \begin{bmatrix} 1 & 0 & 0 & \vdots & 0 & 1 & 0 \\ 0 & 1 & 0 & \vdots & -\frac{1}{3} & 0 & 0 \\ 0 & 0 & 1 & \vdots & 0 & 0 & \frac{1}{4} \end{bmatrix}.$$ Inverse is

$$\begin{bmatrix} 0 & 1 & 0 \\ -\frac{1}{3} & 0 & 0 \\ 0 & 0 & \frac{1}{4} \end{bmatrix}$$

11. $\begin{bmatrix} 1 & 2 & 6 \\ 0 & 0 & 2 \\ -3 & -6 & -9 \end{bmatrix}$ Since column 2 is twice column 1, this is a singular matrix and it has no inverse.

13. $\begin{bmatrix} 8 & 7 & -1 & \vdots & 1 & 0 & 0 \\ -5 & -5 & 1 & \vdots & 0 & 1 & 0 \\ -4 & -4 & 1 & \vdots & 0 & 0 & 1 \end{bmatrix} 2R_3 + R_1$

$$\begin{bmatrix} 0 & -1 & 1 & \vdots & 1 & 0 & 2 \\ -5 & -5 & 1 & \vdots & 0 & 1 & 0 \\ -4 & -4 & 1 & \vdots & 0 & 0 & 1 \end{bmatrix} -1R_2$$

$$\begin{bmatrix} 0 & -1 & 1 & \vdots & 1 & 0 & 2 \\ 5 & 5 & -1 & \vdots & 0 & -1 & 0 \\ -4 & -4 & 1 & \vdots & 0 & 0 & 1 \end{bmatrix} R_3 + R_2$$

$$\begin{bmatrix} 0 & -1 & 1 & \vdots & 1 & 0 & 2 \\ 1 & 1 & 0 & \vdots & 0 & -1 & 1 \\ -4 & -4 & 1 & \vdots & 0 & 0 & 1 \end{bmatrix} 4R_2 + R_3$$

$$\begin{bmatrix} 0 & -1 & 1 & \vdots & 1 & 0 & 2 \\ 1 & 1 & 0 & \vdots & 0 & -1 & 1 \\ 0 & 0 & 1 & \vdots & 0 & -4 & 5 \end{bmatrix} \text{Swap } R_1 + R_2$$

$$\begin{bmatrix} 1 & 1 & 0 & \vdots & 0 & -1 & 1 \\ 0 & -1 & 1 & \vdots & 1 & 0 & 2 \\ 0 & 0 & 1 & \vdots & 0 & -4 & 5 \end{bmatrix} R_2 + R_1$$

$$\begin{bmatrix} 1 & 0 & 1 & \vdots & 1 & -1 & 3 \\ 0 & -1 & 1 & \vdots & 1 & 0 & 2 \\ 0 & 0 & 1 & \vdots & 0 & -4 & 5 \end{bmatrix} \begin{matrix} -R_3 + R_1 \\ -R_3 + R_2 \end{matrix}$$

$$\begin{bmatrix} 1 & 0 & 0 & \vdots & 1 & 3 & -2 \\ 0 & -1 & 0 & \vdots & 1 & 4 & -3 \\ 0 & 0 & 1 & \vdots & 0 & -4 & 5 \end{bmatrix} -1R_2$$

$$\Rightarrow \begin{bmatrix} 1 & 0 & 0 & \vdots & 1 & 3 & -2 \\ 0 & 1 & 0 & \vdots & -1 & -4 & 3 \\ 0 & 0 & 1 & \vdots & 0 & -4 & 5 \end{bmatrix}.$$ Inverse is

$$\begin{bmatrix} 1 & 3 & -2 \\ -1 & -4 & 3 \\ 0 & -4 & 5 \end{bmatrix}$$

15. $\begin{bmatrix} 3 & -1 & 0 \\ -6 & 2 & 0 \\ 1 & 0 & 5 \end{bmatrix}.$ Since $R_2 = -2R_1$ the matrix is singular, so there is no inverse.

17. $\begin{bmatrix} 2 & 0 & 0 & \vdots & 1 & 0 & 0 \\ 2 & 2 & 0 & \vdots & 0 & 1 & 0 \\ 2 & 2 & 2 & \vdots & 0 & 0 & 1 \end{bmatrix}.$

First multiply the entire matrix by $\frac{1}{2}$ to get

$$\begin{bmatrix} 1 & 0 & 0 & \vdots & \frac{1}{2} & 0 & 0 \\ 1 & 1 & 0 & \vdots & 0 & \frac{1}{2} & 0 \\ 1 & 1 & 1 & \vdots & 0 & 0 & \frac{1}{2} \end{bmatrix} \begin{matrix} -R_1 + R_2 \\ -R_1 + R_3 \end{matrix}$$

$$\begin{bmatrix} 1 & 0 & 0 & \vdots & \frac{1}{2} & 0 & 0 \\ 0 & 1 & 0 & \vdots & -\frac{1}{2} & \frac{1}{2} & 0 \\ 0 & 1 & 1 & \vdots & -\frac{1}{2} & 0 & \frac{1}{2} \end{bmatrix} -R_2 + R_3$$

$$\begin{bmatrix} 1 & 0 & 0 & \vdots & \frac{1}{2} & 0 & 0 \\ 0 & 1 & 0 & \vdots & -\frac{1}{2} & \frac{1}{2} & 0 \\ 0 & 0 & 1 & \vdots & 0 & -\frac{1}{2} & \frac{1}{2} \end{bmatrix}.$$

The inverse is $\begin{bmatrix} \frac{1}{2} & 0 & 0 \\ -\frac{1}{2} & \frac{1}{2} & 0 \\ 0 & -\frac{1}{2} & \frac{1}{2} \end{bmatrix}$ or

$$\begin{bmatrix} 0.5 & 0 & 0 \\ -0.5 & 0.5 & 0 \\ 0 & -0.5 & 0.5 \end{bmatrix}.$$

19. $\begin{bmatrix} 1 & -1 & 1 & \vdots & 1 & 0 & 0 \\ 0 & 2 & -1 & \vdots & 0 & 1 & 0 \\ 2 & 3 & 0 & \vdots & 0 & 0 & 1 \end{bmatrix} \; -2R_1 + R_3$

$\begin{bmatrix} 1 & -1 & 1 & \vdots & 1 & 0 & 0 \\ 0 & 2 & -1 & \vdots & 0 & 1 & 0 \\ 0 & 5 & -2 & \vdots & -2 & 0 & 1 \end{bmatrix} \begin{array}{l} \frac{1}{2}R_2 + R_1 \\ \\ -\frac{5}{2}R_2 + R_3 \end{array}$

$\begin{bmatrix} 1 & 0 & \frac{1}{2} & \vdots & 1 & \frac{1}{2} & 0 \\ 0 & 2 & -1 & \vdots & 0 & 1 & 0 \\ 0 & 0 & \frac{1}{2} & \vdots & -2 & -\frac{5}{2} & 1 \end{bmatrix} \begin{array}{l} -R_3 + R_1 \\ 2R_3 + R_2 \end{array}$

$\begin{bmatrix} 1 & 0 & 0 & \vdots & 3 & 3 & -1 \\ 0 & 2 & 0 & \vdots & -4 & -4 & 2 \\ 0 & 0 & \frac{1}{2} & \vdots & -2 & -\frac{5}{2} & 1 \end{bmatrix} \begin{array}{l} \frac{1}{2}R_2 \\ \\ 2R_3 \end{array}$

$\begin{bmatrix} 1 & 0 & 0 & \vdots & 3 & 3 & -1 \\ 0 & 1 & 0 & \vdots & -2 & -2 & 1 \\ 0 & 0 & 1 & \vdots & -4 & -5 & 2 \end{bmatrix}.$

The inverse is $\begin{bmatrix} 3 & 3 & -1 \\ -2 & -2 & 1 \\ -4 & -5 & 2 \end{bmatrix}.$

21. All problems should check.

23. (a) Not a square matrix
(b) Every element in a row (or column) is 0
(c) Every element in a row (or column) is a constant multiple of every corresponding element in another row (or column)
(d) The determinant is 0.

17.4 MATRICES AND LINEAR EQUATIONS

1. The coefficient matrix of $A = \begin{bmatrix} 6 & 1 \\ 5 & 1 \end{bmatrix}$ has an inverse $A^{-1} = \begin{bmatrix} 1 & -1 \\ -5 & 6 \end{bmatrix}$. Thus, we see that $\begin{bmatrix} x \\ y \end{bmatrix} = \begin{bmatrix} 1 & -1 \\ -5 & 6 \end{bmatrix}\begin{bmatrix} -4 \\ -3 \end{bmatrix} = \begin{bmatrix} -1 \\ 2 \end{bmatrix}$, and so $x = -1$ and $y = 2$

3. Here the coefficient matrix is $A = \begin{bmatrix} 8 & -6 \\ -6 & 4 \end{bmatrix}$, which has an inverse matrix $A^{-1} = \begin{bmatrix} -1 & -1.5 \\ -1.5 & -2 \end{bmatrix}$. Thus, we see that the solution is $\begin{bmatrix} x \\ y \end{bmatrix} = \begin{bmatrix} -1 & -1.5 \\ -1.5 & -2 \end{bmatrix}\begin{bmatrix} -27 \\ 19 \end{bmatrix} = \begin{bmatrix} -1.5 \\ 2.5 \end{bmatrix}$, and so $x = -1.5$ and $y = 2.5$

5. This coefficient matrix $A = \begin{bmatrix} 1 & 2 & 6 \\ 0 & 0 & 2 \\ -3 & -6 & -9 \end{bmatrix}$, is a singular matrix, and so there is no solution.

7. The coefficient matrix is $A = \begin{bmatrix} 1 & 2 & 3 \\ 2 & 5 & 7 \\ 1 & 1 & 1 \end{bmatrix}$; which has the inverse $A^{-1} = \begin{bmatrix} 2 & -1 & 1 \\ -5 & 2 & 1 \\ 3 & -1 & -1 \end{bmatrix}$; and the constant matrix is $K = \begin{bmatrix} 4 \\ 7.5 \\ 1 \end{bmatrix}$. Thus we have

$\begin{bmatrix} x \\ y \\ z \end{bmatrix} = A^{-1}K = \begin{bmatrix} 1.5 \\ -4 \\ 3.5 \end{bmatrix}$; with the result that $x = 1.5, y = -4$, and $z = 3.5$.

9. The coefficient matrix $A = \begin{bmatrix} 2 & 0 & 0 \\ 2 & 2 & 0 \\ 2 & 2 & 2 \end{bmatrix}$ has inverse $A^{-1} = \begin{bmatrix} .5 & 0 & 0 \\ -.5 & .5 & 0 \\ 0 & -.5 & .5 \end{bmatrix}$ and the constant matrix is $K = \begin{bmatrix} 11 \\ 4 \\ -9 \end{bmatrix}$. Thus, we have

$\begin{bmatrix} x \\ y \\ z \end{bmatrix} = A^{-1}K = \begin{bmatrix} 5.5 \\ -3.5 \\ -6.5 \end{bmatrix}$, which produces the solution $x = 5.5, y = -3.5$, and $z = -6.5$.

11. $A = \begin{bmatrix} 2 & 1 \\ -3 & 2 \end{bmatrix}, K = \begin{bmatrix} 1 \\ 16 \end{bmatrix},$

$A^{-1} = \frac{1}{7}\begin{bmatrix} 2 & -1 \\ 3 & 2 \end{bmatrix}, \begin{bmatrix} x \\ y \end{bmatrix} = A^{-1}K = \begin{bmatrix} -2 \\ 5 \end{bmatrix}, x = -2, y = 5.$

13. $A = \begin{bmatrix} 2 & 2 \\ 4 & 3 \end{bmatrix}, A^{-1} = -\frac{1}{2}\begin{bmatrix} 3 & -2 \\ -4 & 2 \end{bmatrix},$

$K = \begin{bmatrix} 4 \\ 1 \end{bmatrix}, \begin{bmatrix} x \\ y \end{bmatrix} = A^{-1}K = \begin{bmatrix} -5 \\ 7 \end{bmatrix}, x = -5, y = 7.$

15. $A = \begin{bmatrix} 1.5 & 2.5 \\ 3.2 & 2.6 \end{bmatrix}, A^{-1} = \frac{1}{-4.1} \begin{bmatrix} 2.6 & -2.5 \\ -3.2 & 1.5 \end{bmatrix},$

$K = \begin{bmatrix} 0.3 \\ 7.2 \end{bmatrix}, \begin{bmatrix} x \\ y \end{bmatrix} = A^{-1}K = \begin{bmatrix} 4.2 \\ -2.4 \end{bmatrix}, x = 4.2,$

$y = -2.4.$

17. $A = \begin{bmatrix} 5 & 2 & 3 \\ -3 & 2 & -8 \\ 4 & -2 & 9 \end{bmatrix}, K = \begin{bmatrix} 1 \\ 6 \\ -7 \end{bmatrix},$

$\begin{bmatrix} x \\ y \\ z \end{bmatrix} = A^{-1}K = \begin{bmatrix} -2 \\ 4 \\ 1 \end{bmatrix}, x = -2, y = 4, z = 1.$

19. $A = \begin{bmatrix} 1 & 2 & 4 \\ 3 & 1 & 4 \\ 2 & 9 & -2 \end{bmatrix}, K = \begin{bmatrix} 7 \\ -2 \\ 10 \end{bmatrix},$

$\begin{bmatrix} x \\ y \\ z \end{bmatrix} = A^{-1}K = \begin{bmatrix} -3.4 \\ 2.2 \\ 1.5 \end{bmatrix}, x = -3.4, y = 2.2,$

$z = 1.5.$

21. $A = \begin{bmatrix} 1 & -1 & -1 \\ 16 & 0 & 4 \\ 0 & 12 & -4 \end{bmatrix}, K = \begin{bmatrix} 0 \\ 100 \\ 60 \end{bmatrix},$

$\begin{bmatrix} x \\ y \\ z \end{bmatrix} = A^{-1}K = \begin{bmatrix} 6.053 \\ 5.263 \\ 0.789 \end{bmatrix}, I_A = 6.05, I_B =$

$5.26, I_C = 0.79$ all in Amperes

23. $A = \begin{bmatrix} 7.18 & -1 & 2.2 \\ -1 & 5.8 & 1.5 \\ 2.2 & 1.5 & 8.4 \end{bmatrix}, K = \begin{bmatrix} 10 \\ 15 \\ 20 \end{bmatrix},$

$\begin{bmatrix} I_1 \\ I_2 \\ I_3 \end{bmatrix} = A^{-1}K = \begin{bmatrix} 1.22 \\ 2.37 \\ 1.64 \end{bmatrix}, I_1 = 1.22, I_2 = 2.37$

and $I_3 = 1.64$ all in amperes.

25. See *Computer Programs* in main text.

27. The problem yields the system of equations

$\begin{cases} A + B + C = 1,900 \\ -A + B = 90 \\ -B + C = 130 \end{cases}$. Using matrices we get

$\begin{bmatrix} 1 & 1 & 1 \\ -1 & 1 & 0 \\ 0 & -1 & 1 \end{bmatrix}^{-1} \begin{bmatrix} 1900 \\ 90 \\ 130 \end{bmatrix} = \begin{bmatrix} 530 \\ 620 \\ 750 \end{bmatrix}$. Hence,

$A = 530$ km, $B = 620$ km, and $C = 750$ km.

29. **(a)** The given information leads to the system of equations

$\begin{cases} 2A + 2B + C = 5 \\ A + 2B + 5C = 6 \\ A + 2B + C = 4 \end{cases}$. Using matrices we get

$\begin{bmatrix} 2 & 2 & 1 \\ 1 & 2 & 5 \\ 1 & 2 & 1 \end{bmatrix}^{-1} \begin{bmatrix} 5 \\ 6 \\ 4 \end{bmatrix} = \begin{bmatrix} 1 \\ 1.25 \\ 0.5 \end{bmatrix}$. Multiplying this

last matrix by the common denominator of 4, we get the ratio $4 : 5 : 2$. Hence, the city should buy from Companies A, B, and C in the proportion $A : B : C = 4 : 5 : 2$ units.

(b) The proportion of $4 : 5 : 2$ units is a total of 11 units. Dividing the number of units the city expects to need for this winter by 11, we get $3,630,000 \div 11 = 330,000$. Multiplying each number in the continued proportion by 330,000, we determine that the city should purchase the following number of units from the three companies. Company A: 1,320,000 units, Company B: 1,650,000 units, Company C: 660,000 units.

31. Using matrices, we have $\begin{bmatrix} 1 & 1 & 1 & 1 \\ 1 & 1 & -1 & 1 \\ -1.64 & 0 & 0 & 1 \\ 3 & 0 & -2 & 2 \end{bmatrix}^{-1}$

$= \begin{bmatrix} 100 \\ 0 \\ 0 \\ -1 \end{bmatrix} = \begin{bmatrix} 12.5 \\ 27.25 \\ 39.75 \\ 20.5 \end{bmatrix}$. Hence, $A = 12.5, B =$

$27.25, C = 39.75,$ and $D = 20.5.$

CHAPTER 17 REVIEW

1. $A + C = \begin{bmatrix} 4 & 3 & 2 & 5 \\ 6 & 7 & -1 & 4 \\ 9 & 10 & -8 & 3 \end{bmatrix} +$

$\begin{bmatrix} 0 & 1 & 0 & 2 \\ 3 & 0 & 4 & 0 \\ 0 & -5 & 0 & -6 \end{bmatrix} = \begin{bmatrix} 4 & 4 & 2 & 7 \\ 9 & 7 & 3 & 4 \\ 9 & 5 & -8 & -3 \end{bmatrix}$

2. $B + C = \begin{bmatrix} 3 & -2 & 1 & 0 \\ 5 & -1 & 2 & 0 \\ 4 & 3 & -2 & 0 \end{bmatrix} +$

$\begin{bmatrix} 0 & 1 & 0 & 2 \\ 3 & 0 & 4 & 0 \\ 0 & -5 & 0 & -6 \end{bmatrix} = \begin{bmatrix} 3 & -1 & 1 & 2 \\ 8 & -1 & 6 & 0 \\ 4 & -2 & -2 & -6 \end{bmatrix}$

3. $A - B = \begin{bmatrix} 4 & 3 & 2 & 5 \\ 6 & 7 & -1 & 4 \\ 9 & 10 & -8 & 3 \end{bmatrix} -$

$\begin{bmatrix} 3 & -2 & 1 & 0 \\ 5 & -1 & 2 & 0 \\ 4 & 3 & -2 & 0 \end{bmatrix} = \begin{bmatrix} 1 & 5 & 1 & 5 \\ 1 & 8 & -3 & 4 \\ 5 & 7 & -6 & 3 \end{bmatrix}$

4. $C - B = \begin{bmatrix} 0 & 1 & 0 & 2 \\ 3 & 0 & 4 & 0 \\ 0 & -5 & 0 & -6 \end{bmatrix} -$

$\begin{bmatrix} 3 & -2 & 1 & 0 \\ 5 & -1 & 2 & 0 \\ 4 & 3 & -2 & 0 \end{bmatrix} = \begin{bmatrix} -3 & 3 & -1 & 2 \\ -2 & 1 & 2 & 0 \\ -4 & -8 & 2 & -6 \end{bmatrix}$

5. $2A - 3C = \begin{bmatrix} 8 & 6 & 4 & 10 \\ 12 & 14 & -2 & 8 \\ 18 & 20 & -16 & 6 \end{bmatrix} -$

$\begin{bmatrix} 0 & 3 & 0 & 6 \\ 9 & 0 & 12 & 0 \\ 0 & -15 & 0 & -18 \end{bmatrix} = \begin{bmatrix} 8 & 3 & 4 & 4 \\ 3 & 14 & -14 & 8 \\ 18 & 35 & -16 & 24 \end{bmatrix}$

6. $A + B - 2C = \begin{bmatrix} 16 & 12 & 8 & 20 \\ 24 & 28 & -4 & 16 \\ 36 & 40 & -32 & 12 \end{bmatrix} +$

$\begin{bmatrix} 3 & -2 & 1 & 0 \\ 5 & -1 & 2 & 0 \\ 4 & 3 & -2 & 0 \end{bmatrix} - \begin{bmatrix} 0 & 2 & 0 & 4 \\ 6 & 0 & 8 & 0 \\ 0 & -10 & 0 & -12 \end{bmatrix} =$

$\begin{bmatrix} 19 & 8 & 9 & 16 \\ 23 & 27 & -10 & 16 \\ 40 & 53 & -34 & 24 \end{bmatrix}$

7. $\begin{bmatrix} 3 & 2 \\ 4 & 5 \\ 1 & 0 \end{bmatrix} \times \begin{bmatrix} 1 & 2 \\ 0 & 1 \end{bmatrix} = \begin{bmatrix} 3+0 & 6+2 \\ 4+0 & 8+5 \\ 1+0 & 2+0 \end{bmatrix} =$

$\begin{bmatrix} 3 & 8 \\ 4 & 13 \\ 1 & 2 \end{bmatrix}$

8. $\begin{bmatrix} 1 & -4 \\ 5 & 1 \end{bmatrix} \begin{bmatrix} 3 & 4 & 1 \\ 2 & 5 & 0 \end{bmatrix} =$

$\begin{bmatrix} 3-8 & 4-20 & 1+0 \\ 15+2 & 20+5 & 5+0 \end{bmatrix} = \begin{bmatrix} -5 & -16 & 1 \\ 17 & 25 & 5 \end{bmatrix}$

9. $\begin{bmatrix} 3 \\ 2 \\ -1 \end{bmatrix} [4 \;\; 5 \;\; 1] = \begin{bmatrix} 12 & 15 & 3 \\ 8 & 10 & 2 \\ -4 & -5 & -1 \end{bmatrix}$

10. $[4 \;\; 5 \;\; 1] \begin{bmatrix} -2 \\ 7 \\ -3 \end{bmatrix} = [-8+35-3] = [24]$

11. $\begin{bmatrix} 2 & 3 \\ -4 & -5 \end{bmatrix}^{-1} = \frac{1}{(-10+12)} \begin{bmatrix} -5 & -3 \\ 4 & 2 \end{bmatrix} =$

$\begin{bmatrix} -\frac{5}{2} & -\frac{3}{2} \\ 2 & 1 \end{bmatrix}$

12. $\begin{bmatrix} 2 & -1 \\ 0 & 4 \end{bmatrix}^{-1} = \frac{1}{8} \begin{bmatrix} 4 & 1 \\ 0 & 2 \end{bmatrix} = \begin{bmatrix} \frac{1}{2} & \frac{1}{8} \\ 0 & \frac{1}{4} \end{bmatrix} =$

$\begin{bmatrix} 0.5 & 0.125 \\ 0 & 0.25 \end{bmatrix}$

13. $\begin{bmatrix} -2 & 1 & 0 & \vdots & 1 & 0 & 0 \\ 0 & 4 & 0 & \vdots & 0 & 1 & 0 \\ 1 & 0 & 1 & \vdots & 0 & 0 & 1 \end{bmatrix} \; 2R_3 + R_1$

$\begin{bmatrix} 0 & 1 & 2 & \vdots & 1 & 0 & 2 \\ 0 & 4 & 0 & \vdots & 0 & 1 & 0 \\ 1 & 0 & 1 & \vdots & 0 & 0 & 1 \end{bmatrix}$ Swap R_1 and R_3

$\begin{bmatrix} 1 & 0 & 1 & \vdots & 0 & 0 & 1 \\ 0 & 4 & 0 & \vdots & 0 & 1 & 0 \\ 0 & 1 & 2 & \vdots & 1 & 0 & 2 \end{bmatrix} \; -\frac{1}{4}R_2 + R_3$

$\begin{bmatrix} 1 & 0 & 1 & \vdots & 0 & 0 & 1 \\ 0 & 4 & 0 & \vdots & 0 & 1 & 0 \\ 0 & 0 & 2 & \vdots & 1 & -\frac{1}{4} & 2 \end{bmatrix} \; -\frac{1}{2}R_3 + R_1$

$\begin{bmatrix} 1 & 0 & 0 & \vdots & -\frac{1}{2} & \frac{1}{8} & 0 \\ 0 & 4 & 0 & \vdots & 0 & 1 & 0 \\ 0 & 0 & 2 & \vdots & 1 & -\frac{1}{4} & 2 \end{bmatrix} \begin{matrix} \\ \frac{1}{4}R_2 \\ \frac{1}{2}R_3 \end{matrix}$

$\begin{bmatrix} 1 & 0 & 0 & \vdots & -\frac{1}{2} & \frac{1}{8} & 0 \\ 0 & 1 & 0 & \vdots & 0 & \frac{1}{4} & 0 \\ 0 & 0 & 1 & \vdots & \frac{1}{2} & -\frac{1}{8} & 1 \end{bmatrix}$. The inverse is

$\begin{bmatrix} -\frac{1}{2} & \frac{1}{8} & 0 \\ 0 & \frac{1}{4} & 0 \\ \frac{1}{2} & -\frac{1}{8} & 1 \end{bmatrix}$ or $\begin{bmatrix} -0.5 & 0.125 & 0 \\ 0 & 0.25 & 0 \\ 0.5 & -0.125 & 1 \end{bmatrix}$.

14. $\begin{bmatrix} 2 & 3 & -1 & \vdots & 1 & 0 & 0 \\ 1 & 2 & 1 & \vdots & 0 & 1 & 0 \\ -1 & -1 & 3 & \vdots & 0 & 0 & 1 \end{bmatrix} \begin{matrix} -R_1 + 2R_2 \\ R_2 + R_3 \end{matrix}$

$\begin{bmatrix} 2 & 3 & -1 & \vdots & 1 & 0 & 0 \\ 0 & 1 & 3 & \vdots & -1 & 2 & 0 \\ 0 & 1 & 4 & \vdots & 0 & 1 & 1 \end{bmatrix} \begin{matrix} -3R_2 + R_1 \\ \\ -R_2 + R_3 \end{matrix}$

$\begin{bmatrix} 2 & 0 & -10 & \vdots & 4 & -6 & 0 \\ 0 & 1 & 3 & \vdots & -1 & 2 & 0 \\ 0 & 0 & 1 & \vdots & 0 & -1 & 1 \end{bmatrix} \begin{matrix} 10R_3 + R_1 \\ -3R_3 + R_2 \end{matrix}$

$\begin{bmatrix} 2 & 0 & 0 & \vdots & 14 & -16 & 10 \\ 0 & 1 & 0 & \vdots & -4 & 5 & -3 \\ 0 & 0 & 1 & \vdots & 1 & -1 & 1 \end{bmatrix} \; \frac{1}{2}R_1$

$\begin{bmatrix} 1 & 0 & 0 & \vdots & 7 & -8 & 5 \\ 0 & 1 & 0 & \vdots & -4 & 5 & -3 \\ 0 & 0 & 1 & \vdots & 1 & -1 & 1 \end{bmatrix}$. The inverse is

$\begin{bmatrix} 7 & -8 & 5 \\ -4 & 5 & -3 \\ 1 & -1 & 1 \end{bmatrix}$

15. $A = \begin{bmatrix} 12 & 5 \\ 3 & 1 \end{bmatrix}, A^{-1} = -\frac{1}{3}\begin{bmatrix} 1 & -5 \\ -3 & 12 \end{bmatrix}, K = \begin{bmatrix} -2 \\ 1.1 \end{bmatrix}, A^{-1}K = \begin{bmatrix} 2.5 \\ -6.4 \end{bmatrix}, x = 2.5, y = -6.4$

16. $A = \begin{bmatrix} 4 & 1 \\ -3 & 2 \end{bmatrix}, A^{-1} = \frac{1}{11}\begin{bmatrix} 2 & -1 \\ 3 & 4 \end{bmatrix}, K = \begin{bmatrix} -4 \\ 14 \end{bmatrix}, A^{-1}K = \begin{bmatrix} -2 \\ 4 \end{bmatrix}, x = -2, y = 4$

17. $A = \begin{bmatrix} 2 & 3 & 5 \\ -2 & 3 & 5 \\ 5 & -3 & -2 \end{bmatrix}, K = \begin{bmatrix} 20 \\ 12 \\ 9 \end{bmatrix},$

$A^{-1}K = \begin{bmatrix} 2 \\ -3 \\ 5 \end{bmatrix}, x = 2, y = -3, z = 5$

18. $A = \begin{bmatrix} 1 & 1 & 6 \\ -2 & 2 & 4 \\ 3 & 2 & 4 \end{bmatrix}, K = \begin{bmatrix} -3 \\ -2 \\ 14 \end{bmatrix},$

$A^{-1}K = \begin{bmatrix} 3.2 \\ 6.4 \\ -2.1 \end{bmatrix}, x = 3.2, y = 6.4, z = -2.1$

19. $\begin{cases} -9a + T = 75.4 \\ -11a - T = 100.0 \end{cases}$ $A = \begin{bmatrix} -9 & 1 \\ -11 & -1 \end{bmatrix},$

$K = \begin{bmatrix} 75.4 \\ -100.0 \end{bmatrix}, A^{-1}K = \begin{bmatrix} 1.23 \\ 86.47 \end{bmatrix}, a = 1.23,$

$T = 86.47$

20. $\begin{cases} 4P + 8B + 12T = 1872 \\ 2P + 3B + 5T = 771 \\ 7P + 6B = 11T = 1770 \end{cases}$ $A = \begin{bmatrix} 4 & 8 & 12 \\ 2 & 3 & 5 \\ 7 & 6 & 11 \end{bmatrix},$

$K = \begin{bmatrix} 1872 \\ 771 \\ 1770 \end{bmatrix}, A^{-1}K = \begin{bmatrix} 45 \\ 72 \\ 93 \end{bmatrix}$, and so they can
make 45 PCs, 72 BCs and 93 TCs.

21. The given systems can be expressed in matrices as
follows: $\begin{bmatrix} \frac{1}{2} & \frac{\sqrt{3}}{2} \\ -\frac{\sqrt{3}}{2} & \frac{1}{2} \end{bmatrix} \begin{bmatrix} x \\ y \end{bmatrix} = \begin{bmatrix} x' \\ y' \end{bmatrix}$ and

$\begin{bmatrix} -\frac{1}{2} & \frac{\sqrt{3}}{2} \\ -\frac{\sqrt{3}}{2} & -\frac{1}{2} \end{bmatrix} \begin{bmatrix} x' \\ y' \end{bmatrix} = \begin{bmatrix} x'' \\ y'' \end{bmatrix}.$ Hence,

$\begin{bmatrix} x'' \\ y'' \end{bmatrix} = \begin{bmatrix} -\frac{1}{2} & \frac{\sqrt{3}}{2} \\ -\frac{\sqrt{3}}{2} & -\frac{1}{2} \end{bmatrix} \begin{bmatrix} \frac{1}{2} & \frac{\sqrt{3}}{2} \\ -\frac{\sqrt{3}}{2} & \frac{1}{2} \end{bmatrix} \begin{bmatrix} x \\ y \end{bmatrix} =$

$\begin{bmatrix} -1 & 0 \\ 0 & -1 \end{bmatrix} \begin{bmatrix} x \\ y \end{bmatrix} = \begin{bmatrix} -x \\ -y \end{bmatrix}.$ Thus, we have shown
that $x'' = -x$ and $y'' = -y$.

22. Since $x'' = -x$ and $y'' = -y$, we have that
$x\cos\theta + y\sin\theta = -x$ or $\cos\theta = -1$ and $\sin\theta = 0$. This is a 180° rotation.

23. $M = \begin{bmatrix} 1 & -\frac{1}{f_2} \\ 0 & 1 \end{bmatrix} \cdot \begin{bmatrix} 1 & 0 \\ d & 1 \end{bmatrix} \cdot \begin{bmatrix} 1 & -\frac{1}{f_1} \\ 0 & 1 \end{bmatrix} =$

$\begin{bmatrix} 1 - \frac{d}{f_2} & -\frac{1}{f_2} \\ d & 1 \end{bmatrix} \begin{bmatrix} 1 & -\frac{1}{f_1} \\ 0 & 1 \end{bmatrix} =$

$\begin{bmatrix} 1 - \frac{d}{f_2} & -\frac{1}{f_1} + \frac{d}{f_1 f_2} - \frac{1}{f_2} \\ d & -\frac{d}{f_1} + 1 \end{bmatrix}$ or

$M = \begin{bmatrix} 1 - \frac{d}{f_2} & \frac{d}{f_1 f_2} - \frac{1}{f_1} - \frac{1}{f_2} \\ d & 1 - \frac{d}{f_1} \end{bmatrix}.$

24. $-\frac{1}{f} = \frac{d}{f_1 f_2} - \frac{1}{f_1} - \frac{1}{f_2}$ so $\frac{1}{f} = \frac{1}{f_1} + \frac{1}{f_2} - \frac{d}{f_1 f_2}$

▤ CHAPTER 17 TEST

1. **(a)** $A+B = \begin{bmatrix} 8 & 0 & -4 \\ 16 & -6 & 2 \end{bmatrix} + \begin{bmatrix} -1 & 5 & -3 \\ 3 & 0 & 4 \end{bmatrix} =$

$\begin{bmatrix} 7 & 5 & -7 \\ 19 & -6 & 6 \end{bmatrix}$

(b) $3A - 2B = \begin{bmatrix} 24 & 0 & -12 \\ 48 & -18 & 6 \end{bmatrix} -$

$\begin{bmatrix} -2 & 10 & -6 \\ 6 & 0 & 8 \end{bmatrix} = \begin{bmatrix} 26 & -10 & -6 \\ 42 & -18 & -2 \end{bmatrix}$

2. $[1 \quad -2 \quad 3] \begin{bmatrix} -4 \\ -6 \\ 8 \end{bmatrix} = [-4 + 12 + 24] = [32]$

3. **(a)** $CD = \begin{bmatrix} 4 & 6 \\ -10 & 4 \end{bmatrix} \cdot \begin{bmatrix} \frac{1}{2} & 0 \\ -\frac{3}{2} & 4 \end{bmatrix} =$

$\begin{bmatrix} 2-9 & 0+24 \\ -5-6 & 0+16 \end{bmatrix} = \begin{bmatrix} -7 & 24 \\ -11 & 16 \end{bmatrix}.$

(b) $DC = \begin{bmatrix} \frac{1}{2} & 0 \\ -\frac{3}{2} & 4 \end{bmatrix} \begin{bmatrix} 4 & 6 \\ -10 & 4 \end{bmatrix} =$

$\begin{bmatrix} 2+0 & 3+0 \\ -6-40 & -9+16 \end{bmatrix} = \begin{bmatrix} 2 & 3 \\ -46 & 7 \end{bmatrix}.$

4. **(a)** $E = \begin{bmatrix} 2 & -3 \\ 7 & 9 \end{bmatrix}, E^{-1} = \frac{1}{18+21} \begin{bmatrix} 9 & 3 \\ -7 & 2 \end{bmatrix} \approx$

$\begin{bmatrix} 0.2308 & 0.0769 \\ -0.1795 & 0.0513 \end{bmatrix}$

(b) $EX = F$ so $X = E^{-1}F = \begin{bmatrix} -5 \\ 7 \end{bmatrix}.$

5. The coefficient matrix is $A = \begin{bmatrix} 1 & 3 & 1 \\ 2 & 5 & 1 \\ 1 & 2 & 3 \end{bmatrix}$ and the

constant matrix is $K = \begin{bmatrix} -2 \\ -5 \\ 6 \end{bmatrix}$. If $X = \begin{bmatrix} x \\ y \\ z \end{bmatrix}$, we

have the system $AX = K$, so $X = A^{-1}K$. You

can determine that $A^{-1} = \begin{bmatrix} -\frac{13}{3} & \frac{7}{3} & \frac{2}{3} \\ \frac{5}{3} & -\frac{2}{3} & -\frac{1}{3} \\ \frac{1}{3} & -\frac{1}{3} & \frac{1}{3} \end{bmatrix} =$

$\frac{1}{3} \begin{bmatrix} -13 & 7 & 2 \\ 5 & -2 & -1 \\ 1 & -1 & 1 \end{bmatrix}$ and so, $A^{-1}K = \begin{bmatrix} 1 \\ -2 \\ 3 \end{bmatrix}.$

Thus, $x = 1, y = -2, z = 3$.

6. Let the costs be $x, y,$ and z. Then we get the system $\begin{cases} x+y+z=60 \\ x=y+z \\ y=2z+3 \end{cases}$ or $\begin{cases} x+y+z=60 \\ x-y-z=0 \\ y-2z=3 \end{cases}$. Using matrices we get $\begin{bmatrix} 1 & 1 & 1 \\ 1 & -1 & -1 \\ 0 & 1 & -2 \end{bmatrix}^{-1} \begin{bmatrix} 60 \\ 0 \\ 3 \end{bmatrix} = \begin{bmatrix} 30 \\ 21 \\ 9 \end{bmatrix}$. Thus, we see that $x=30, y=21,$ and $z=9$ and so, the parts cost \$30, \$21, and \$9.

7. By Kirchhoff's laws, we have the following equations

$$I_1 - I_2 - I_3 = 0 \qquad (1)$$
$$6I_1 + 4I_2 = 20 \qquad (2)$$
$$4I_2 - 2I_3 = 10 \qquad (3)$$

This yields the matrix equation

$$\begin{bmatrix} 1 & -1 & -1 \\ 6 & 4 & 0 \\ 0 & 4 & -2 \end{bmatrix}^{-1} \begin{bmatrix} 0 \\ 20 \\ 10 \end{bmatrix} = \begin{bmatrix} \frac{20}{11} \\ \frac{25}{11} \\ -\frac{5}{11} \end{bmatrix}.$$ Thus, we have determined that $I_1 = \frac{20}{11}$ A, $I_2 = \frac{25}{11}$ A, and $I_3 = -\frac{5}{11}$ A.

18

Higher Degree Equations

≡ 18.1 THE REMAINDER AND FACTOR THEOREM

1. $P(2) = 3 \cdot 2^2 - 2 \cdot 2 + 1 = 12 - 4 + 1 = 9$

3. $P(-1) = (-1)^4 + (-1)^3 + (-1)^2 - (-1) + 1 = 1 - 1 + 1 + 1 + 1 = 3$

5. $P(3) = 5(3)^3 - 4(3) + 7 = 135 - 12 + 7 = 130$

7. $P(-1) = (-1)^5 - (-1)^4 + (-1)^3 + (-1)^2 - (-1) + 1 = -1 - 1 - 1 + 1 + 1 + 1 = 0$

9. $R = P(1) = 1^3 + 2 \cdot 1^2 - 1 - 2 = 1 + 2 - 1 - 2 = 0$

11. $R = P(3) = 3^3 - 3 \cdot 3^2 + 2 \cdot 3 + 5 = 27 - 27 + 6 + 5 = 11$

13. $R = P(-2) = 4(-2)^4 + 13(-2)^3 - 13(-2)^2 - 40(-2) + 12 = 64 - 104 - 52 + 80 + 12 = 0$

15. $R = P(5) = 3 \cdot 5^4 - 12 \cdot 5^3 - 60 \cdot 5 + 4 = 1875 - 1500 - 300 + 4 = 79$

17. $P(3) = 3^3 + 2 \cdot 3^2 - 12 \cdot 3 - 9 = 27 + 18 - 36 - 9 = 0$; yes

19. $P(-1) = 2(-1)^5 - 6(-1)^3 + (-1)^2 + 4(-1) - 1 = -2 + 6 + 1 - 4 - 1 = 0$; yes

21. $P(-3) = 3(-3)^5 + 3(-3)^4 - 14(-3)^3 + 4(-3)^2 - 24(-3) = -729 + 243 + 378 + 36 + 72 = 0$; yes

23. $P\left(\frac{5}{2}\right) = 6 \cdot \left(\frac{5}{2}\right)^4 - 15\left(\frac{5}{2}\right)^3 - 8\left(\frac{5}{2}\right)^2 + 20\left(\frac{5}{2}\right) = 234\frac{3}{8} - 234\frac{3}{8} - 50 + 50 = 0$; yes

25.
$$\begin{array}{r|rrrrrr} 3 & 1 & 0 & -17 & 0 & 75 & 9 \\ & & 3 & 9 & -24 & -72 & 9 \\ \hline & 1 & 3 & -8 & -24 & 3 & 18 \end{array}$$
So, $Q(x) = x^4 + 3x^3 - 8x^2 - 24x + 3$, $R(x) = 18$

27.
$$\begin{array}{r|rrrr} -3 & 5 & 7 & 0 & 9 \\ & & -15 & 24 & -72 \\ \hline & 5 & -8 & 24 & -63 \end{array}$$
So, $Q(x) = 5x^2 - 8x + 24$; $R(x) = -63$

29.
$$\begin{array}{r|rrrrrr} \frac{1}{2} & 8 & 0 & -4 & 7 & -2 & 0 \\ & & 4 & 2 & -1 & 3 & \frac{1}{2} \\ \hline & 8 & 4 & -2 & 6 & 1 & 1\frac{1}{2} \end{array}$$
So, $Q(x) = 8x^4 + 4x^3 - 2x^2 + 6x + 1$ and $R(x) = \frac{1}{2}$

31.
$$\begin{array}{r|rrrrr} \frac{3}{2} & 4 & -12 & 9 & -8 & 12 \\ & & 6 & -9 & 0 & -12 \\ \hline & 4 & -6 & 0 & -8 & 0 \end{array}$$

Since we used $\frac{2x-3}{2} = x - \frac{3}{2}$, we need to divide the depressed polynomial by 2 also. So, $Q(x) = \frac{4x^3 - 6x^2 - 8}{2} = 2x^3 - 3x^2 - 4$; $R(x) = 0$.

≡ 18.2 ROOTS OF AN EQUATION

1. Using synthetic division, we get

$$\begin{array}{r|rrrr} 1 & 5 & 0 & -8 & 3 \\ & & 5 & 5 & -3 \\ \hline & 5 & 5 & -3 & 0 \end{array}$$

We now use the quadratic equation on the depressed equation $5x^2 + 5x - 3 = 0$, with the results $x = \frac{-5 \pm \sqrt{5^2 - 4 \cdot 5 \cdot (-3)}}{2 \cdot 5} = \frac{-5 \pm \sqrt{25 + 60}}{10} = \frac{-5 \pm \sqrt{85}}{10}$. The roots are 1, $\frac{-5 + \sqrt{85}}{10}$ and $\frac{-5 - \sqrt{85}}{10}$.

3. Using synthetic division, we get

$$\begin{array}{r|rrrr} \frac{1}{3} & 9 & -3 & -81 & 27 \\ & & 3 & 0 & -27 \\ \hline & 9 & 0 & -81 & 0 \end{array}$$

Now, $9x^2 - 81 = 9(x^2 - 9) = 9(x + 3)(x - 3)$ and so the roots are $\frac{1}{3}, 3$, and -3.

5. Synthetic division and the complex conjugate of r_1, which is $-j$, yields

$$
\begin{array}{r|rrrrr}
j & 1 & 0 & -3 & 0 & -4 \\
  &   & j & -1 & -4j & 4 \\
\hline
-j & 1 & j & -4 & -4j & 0 \\
   &   & -j & 0 & 4j & \\
\hline
  & 1 & 0 & -4 & 0 & \\
\end{array}
$$

Since $x^2 - 4 = (x + 2)(x - 2)$, we see that the roots are $2, -2, j$, and $-j$.

7. Synthetic division and the complex conjugate of r_1, which is $-1 - 2j$, yields

$$
\begin{array}{r|rrrrr}
-1+2j & 1 & 2 & -4 & -18 & -45 \\
      &   & -1+2j & -5 & 9-18j & 45 \\
\hline
-1-2j & 1 & 1+2j & -9 & -9-18j & 0 \\
      &   & -1-2j & 0 & 9+18j & \\
\hline
      & 1 & 0 & -9 & 0 & \\
\end{array}
$$

Since $x^2 - 9 = (x + 3)(x - 3)$, the roots are $3, -3, -1 + 2j$, and $-1 - 2j$.

9. Repeated use of synthetic division yields

$$
\begin{array}{r|rrrrr}
2 & 1 & -3 & -3 & 7 & 6 \\
  &   & 2 & -2 & -10 & -6 \\
\hline
3 & 1 & -1 & -5 & -3 & 0 \\
  &   & 3 & 6 & 3 & \\
\hline
  & 1 & 2 & 1 & 0 & \\
\end{array}
$$

Since $x^2 + 2x + 1 = (x + 1)^2$, the roots are $-1, -1, 2$, and 3.

11. Repeated use of synthetic division yields

$$
\begin{array}{r|rrrrr}
-1 & 3 & 12 & 6 & -12 & -9 \\
   &   & -3 & -9 & 3 & 9 \\
\hline
-1 & 3 & 9 & -3 & -9 & 0 \\
   &   & -3 & -6 & 9 & \\
\hline
   & 3 & 6 & -9 & 0 & \\
\end{array}
$$

Since $3x^2 + 6x - 9 = 3(x^2 + 2x - 3) = 3(x + 3)(x - 1)$, the roots are $-1, -1, -3$, and 1.

13. Repeated use of synthetic division yields

$$
\begin{array}{r|rrrrr}
-\frac{1}{2} & 6 & 25 & 33 & 1 & -5 \\
   &   & -3 & -11 & -11 & 5 \\
\hline
\frac{1}{3} & 6 & 22 & 22 & -10 & 0 \\
   &   & 2 & 8 & 10 & \\
\hline
   & 6 & 24 & 30 & 0 & \\
\end{array}
$$

Here $6x^2 + 24x + 30 = 6(x^2 + 4x + 5)$, so using the quadratic formula, we get $x = \frac{-4 \pm \sqrt{16-20}}{2} = -2 \pm j$. The roots are $-\frac{1}{2}, \frac{1}{3}, -2 + j$, and $-2 - j$.

15. Repeated use of synthetic division yields

$$
\begin{array}{r|rrrrr}
\frac{2}{3} & 3 & -2 & 0 & -3 & 2 \\
   &   & 2 & 0 & 0 & -2 \\
\hline
1 & 3 & 0 & 0 & -3 & 0 \\
   &   & 3 & 3 & 3 & \\
\hline
   & 3 & 3 & 3 & 0 & \\
\end{array}
$$

Here $3x^2 + 3x + 3 = 3(x^2 + x + 1)$, and by using the quadratic formula, we get $x = \frac{-1 \pm \sqrt{1-4}}{2} = -\frac{1}{2} \pm \frac{\sqrt{3}}{2}j$. The roots are $1, \frac{2}{3}, -\frac{1}{2} + \frac{\sqrt{3}}{2}j$, and $-\frac{1}{2} - \frac{\sqrt{3}}{2}j$.

17. Here we use synthetic division three times, with the following results:

$$
\begin{array}{r|rrrrrr}
1 & 1 & -1 & 1 & -7 & 10 & -4 \\
  &   & 1 & 0 & 1 & -6 & 4 \\
\hline
1 & 1 & 0 & 1 & -6 & 4 & 0 \\
  &   & 1 & 1 & 2 & -4 & \\
\hline
1 & 1 & 1 & 2 & -4 & 0 & \\
  &   & 1 & 2 & 4 & & \\
\hline
  & 1 & 2 & 4 & 0 & & \\
\end{array}
$$

Here $x^2 + 2x + 4$ has roots $x = \frac{-2 \pm \sqrt{4-16}}{2} = -1 \pm j\sqrt{3}$. The roots are $1, 1, 1, -1 + j\sqrt{3}$, and $-1 - j\sqrt{3}$.

19. Here we use synthetic division three times, with the following results:

$$
\begin{array}{r|rrrrrr}
3 & 3 & -2 & -24 & 1 & 28 & -12 \\
  &   & 9 & 21 & -9 & -24 & 12 \\
\hline
-2 & 3 & 7 & -3 & -8 & 4 & 0 \\
   &   & -6 & -2 & 10 & -4 & \\
\hline
\frac{2}{3} & 3 & 1 & -5 & 2 & 0 & \\
   &   & 2 & 2 & -2 & & \\
\hline
   & 3 & 3 & -3 & 0 & & \\
\end{array}
$$

Since $3x^2 + 3x - 3 = 3(x^2 + x - 1)$, we find that $x = \frac{-1 \pm \sqrt{1+4}}{2} = \frac{-1 \pm \sqrt{5}}{2}$. The roots are $3, -2, \frac{2}{3}, \frac{-1+\sqrt{5}}{2}$, and $\frac{-1-\sqrt{5}}{2}$.

21. Use of synthetic division four times yields

$$
\begin{array}{c|cccccc}
j & 1 & -6 & 3 & -60 & -61 & -54 & -63 \\
 & & j & -1-6j & 6+2j & -2-54j & 54-63j & 63 \\
\hline
-j & 1 & -6+j & 2-6j & -54+2j & -63-54j & -63j & 0 \\
 & & -j & 6j & -2j & & 54j & 63j \\
\hline
-3j & 1 & -6 & 2 & -54 & -63 & 0 \\
 & & -3j & -9+18j & 54+21j & 63 \\
\hline
3j & 1 & -6-3j & -7+18j & 21j & 0 \\
 & & 3j & -18j & -21j \\
\hline
 & 1 & -6 & -7 & 0
\end{array}
$$

Since $x^2 - 6x - 7 = (x+1)(x-7)$, the roots are $j, -j, 3j, -3j, -1$, and 7.

18.3 RATIONAL ROOTS

1. $P(x) = x^3 - 3x - 2$ has one sign change, so there is one positive root. $P(-x) = -x^3 + 3x - 2$ has two sign changes so there are either 0 or 2 negative roots. $c = \pm 1, \pm 2; d = 1$, possible rational roots are ± 1 and ± 2.

3. $P(x) = x^3 - 8x^2 - 17x + 6$ has two sign changes so 0 or 2 positive roots. $P(-x) = -x^3 - 8x^2 + 17x + 6$ has one sign change so 1 negative root. $c = \pm 1, \pm 2, \pm 3, \pm 6; d = 1$ so possible rational roots are $\pm 1, \pm 2, \pm 3, \pm 6$.

5. $P(x) = x^3 - 2x^2 - 5x + 6$ has two sign changes so 0 or 2 positive roots. $P(-x) = -x^3 - 2x^2 + 5x + 6$ has one sign change so 1 negative root. $c = \pm 1, \pm 2, \pm 3, \pm 6; d = 1$ so possible rational roots are $\pm 1, \pm 2, \pm 3, \pm 6$.

7. $P(x) = 2x^4 - x^3 - 5x^2 + x + 3$ has two sign changes so 0 or 2 positive roots. $P(-x) = 2x^4 + x^3 - 5x - x + 3$ has 2 sign changes so 0 or 2 negative roots. $c = \pm 1, \pm 3; d = 1, 2$. Possible rational roots are $\pm 1, \pm 3, \pm \frac{1}{2}, \pm \frac{3}{2}$.

9. $P(x) = 6x^4 - 7x^3 - 13x^2 + 4x + 4$ has two sign changes so 0 or 2 positive roots. $P(-x) = 6x^4 + 7x^3 - 13x^2 - 4x + 4$ has two sign changes so 0 or 2 negative roots. $c = \pm 1, \pm 2, \pm 4; d = 1, 2, 3, 6$. Possible rational roots are $\pm 1, \pm 2, \pm 4, \pm \frac{1}{2}, \pm \frac{1}{3}, \pm 23, \pm \frac{4}{3}, \pm \frac{1}{6}$.

11. Possible rational roots are $\pm 1, \pm 3, \pm 5, \pm 15$. $P(-x) = -x^3 - 9x^2 - 23x - 15$ so no negative roots

$$
\begin{array}{c|cccc}
1 & 1 & -9 & 23 & -15 \\
 & & 1 & -8 & 15 \\
\hline
 & 1 & -8 & 15 & 0
\end{array}
$$

Since $x^2 - 8x + 15 = (x-3)(x-5)$, we see that the roots are $1, 3, 5$.

13. Possible rational roots are ± 1 and ± 2. Using synthetic division, we obtain

$$
\begin{array}{c|cccc}
-1 & 1 & 0 & -3 & -2 \\
 & & -1 & 1 & 2 \\
\hline
 & 1 & -1 & -2 & 0
\end{array}
$$

Since $x^2 - x - 2 = (x-2)(x+1)$, the roots are $-1, -1$, and 2.

15. Possible roots are $\pm 1, \pm 2, \pm 3, \pm 4, \pm 6, \pm 12, \pm \frac{1}{2}, \pm \frac{3}{2}, \pm \frac{1}{4}$, and $\pm \frac{3}{4}$. Synthetic division produces

$$
\begin{array}{c|cccc}
-\frac{3}{4} & 4 & -5 & 10 & 12 \\
 & & -3 & 6 & -12 \\
\hline
 & 4 & -8 & 16 & 0
\end{array}
$$

We see that $4x^2 - 8x + 16 = 4(x^2 - 2x + 4)$ and so $x = \frac{+2 \pm \sqrt{4-16}}{2} = 1 \pm j\sqrt{3}$. Roots are $-\frac{3}{4}, 1 + j\sqrt{3}$, and $1 - j\sqrt{3}$.

17. Possible rational roots are $\pm 1, \pm 2, \pm 3, \pm 4, \pm 6, \pm 12$.

$$
\begin{array}{c|ccccc}
2 & 1 & -4 & 7 & -12 & 12 \\
 & & 2 & -4 & 6 & -12 \\
\hline
2 & 1 & -2 & 3 & -6 & 0 \\
 & & 2 & 0 & 6 \\
\hline
 & 1 & 0 & 3 & 0
\end{array}
$$

Thus, $x^2 + 3 = 0$ and so $x^2 = -3$ and $x = \pm j\sqrt{3}$. The roots are $2, 2, j\sqrt{3}$, and $-j\sqrt{3}$.

19. Possible rational roots are $\pm 1, \pm 2, \pm 3, \pm 6, \pm \frac{1}{2}, \pm \frac{3}{2}, \pm \frac{1}{4}, \pm \frac{3}{4}$. There are either 0 or 2 positive roots, and 0 or 2 negative roots.

$$
\begin{array}{c|ccccc}
-\frac{1}{2} & 4 & -20 & 1 & 18 & 6 \\
 & & -2 & 11 & -6 & -6 \\
\hline
-\frac{1}{2} & 4 & -22 & 12 & 12 & 0 \\
 & & -2 & 12 & -12 \\
\hline
 & 4 & -24 & 24 & 0
\end{array}
$$

Since $4x^2 - 24x + 24 = 4(x^2 - 6x + 6)$ we can use the quadratic formula to see that $x = \frac{6 \pm \sqrt{36-24}}{2} = \frac{6 \pm \sqrt{12}}{2} = 3 \pm \sqrt{3}$. Hence, the roots are $-\frac{1}{2}, -\frac{1}{2}, 3 + \sqrt{3}, 3 - \sqrt{3}$.

21. Possible rational roots are $\pm 1, \pm 2, \pm 5, \pm 10$. There are no rational roots.

23. $P(-x) = -2x^5 - 13x^4 - 26x^3 - 22x^2 - 24x - 9$ so no sign changes so no negative roots. Possible rational roots are $1, 3, 9, \frac{1}{2}, \frac{3}{2}$, and $\frac{9}{2}$.

Using synthetic division three times, we get

$$\begin{array}{r|rrrrrr}
3 & 2 & -13 & 26 & -22 & 24 & -9 \\
  &   & 6 & -21 & 15 & -21 & 9 \\
\hline
3 & 2 & -7 & 5 & -7 & 3 & 0 \\
  &   & 6 & -3 & 6 & -3 & \\
\hline
\frac{1}{2} & 2 & -1 & 2 & -1 & 0 & \\
  &   & 1 & 0 & 1 & & \\
\hline
  & 2 & 0 & 2 & 0 & &
\end{array}$$

We have $2x^2 + 2 = 0$ or $2x^2 = -2$ or $x^2 = -1$ and so, $x = \pm j$. The roots are $\frac{1}{2}, 3, 3, j$, and $-j$.

25. Possible rational roots are $\pm 1, \pm 2, \pm 3, \pm 5, \pm 6, \pm 10, \pm 15, \pm 30$. Using synthetic division three times, we get

$$\begin{array}{r|rrrrrr}
3 & 1 & -5 & 6 & 11 & -43 & 30 \\
  &   & 3 & -6 & 0 & 33 & -30 \\
\hline
1 & 1 & -2 & 0 & 11 & -10 & 0 \\
  &   & 1 & -1 & -1 & 10 & \\
\hline
-2 & 1 & -1 & -1 & 10 & 0 & \\
  &   & -2 & 6 & -10 & & \\
\hline
  & 1 & -3 & 5 & 0 & &
\end{array}$$

We have $x^2 - 3x + 5 = 0$ and so, $x = \frac{3 \pm \sqrt{9-20}}{2} = \frac{3 \pm j\sqrt{11}}{2}$. The roots are $1, 3, -2, \frac{3}{2} + \frac{\sqrt{11}}{2}j$, and $\frac{3}{2} - \frac{\sqrt{11}}{2}j$.

27. From the given information, we have the equation $(8 - 2x)(10 - 2x)x = 48$ where $0 < x < 4$. Expanding this equation produces $4x^3 - 36x^2 + 80x - 48 = 0$. Dividing both sides of this equation by 4 we get the equation $x^3 - 9x^2 + 20x - 12 = 0$. Possible rational roots are 1, 2, and 3. (Notice, that because $0 < x < 4$, we can ignore any possible negative roots.) Using synthetic division, we get

$$\begin{array}{r|rrrr}
1 & 1 & -9 & 20 & -12 \\
  &   & 1 & -8 & 12 \\
\hline
  & 1 & -8 & 12 & 0
\end{array}$$

The depressed equation factors: $x^2 - 8x + 12 = (x - 2)(x - 6) = 0$. Solutions to the equation are

1, 2 and 6. Solutions to the problem are 1 in. and 2 in.

29. The volume of a cylinder is $\pi r^2 h$ and the volume of a hemisphere is one-half the volume of a sphere or $\frac{1}{2}\frac{4}{3}\pi r^3 = \frac{2}{3}\pi r^3$. Since the total height of the silo is 34 ft, and the radius or the hemisphere is r ft, the height of the cylinder is $34 - r$ ft. Thus, the total volume of this grain silo is $V = \pi r^2 h + \frac{2}{3}\pi r^3 = \pi r^2(34 - r) + \frac{2}{3}\pi r^3 = 2511\pi$. Expanding, and dividing by π, we obtain $34r^2 - r^3 + \frac{2}{3}r^3 = 2511$. Collecting terms we get $-\frac{1}{3}r^3 + 34r^2 - 2511 = 0$ or $r^3 - 102r^2 + 7533 = 0$. Now, $7533 = 3^5 \cdot 31$ so factors include 1, 3, 9, 27, and 31. Synthetic division shows that 9 is a solution

$$\begin{array}{r|rrrr}
9 & 1 & -102 & 0 & 7533 \\
  &   & 9 & -837 & -7533 \\
\hline
  & 1 & -93 & -837 & 0
\end{array}$$

The radius is 9 ft.

31. Multiplying $\frac{1}{R} + \frac{1}{R+4} + \frac{1}{R+1} = 1$ by the common denominator $R(R + 1)(R + 4)$ yields

$(R + 1)(R + 4) + R(R+1) + R(R + 4)$
$\qquad = R(R + 1)(R + 4);$
$R^2 + 5R + 4 + R^2 + R + R^2 + 4R$
$\qquad = R^3 + 5R^2 + 4R;$
$\qquad 3R^2 + 10R + 4 = R^3 + 5R^2 + 4R;$
$\qquad R^3 + 2R^2 - 6R - 4 = 0$

Using synthetic division, produces

$$\begin{array}{r|rrrr}
2 & 1 & 2 & -6 & -4 \\
  &   & 2 & 8 & 4 \\
\hline
  & 1 & 4 & 2 & 0
\end{array}$$

There are no other positive solutions, and so we have $R = 2, R + 4 = 6$ and $R + 1 = 5$.

33. If the side of the square that is cut from each corner has length x, then the volume of the box is given by

$$(1 - 2x)(19 - 2x)x = 210$$
$$(228 - 62x + 4x^2)x = 210$$
$$4x^3 - 62x^2 + 228x - 210 = 0$$
$$2x^3 - 31x^2 + 114x - 105 = 0$$

Now, x must be positive and the realistic domain is $0 < x < 6$. The possible rational solutions include $1, 5, 7, \frac{1}{2}, \frac{5}{2}$. and $\frac{7}{2}$.

$$\begin{array}{r|rrrr}
\frac{7}{2} & 2 & -31 & 114 & -105 \\
  &   & 7 & -84 & 105 \\
\hline
  & 2 & -24 & 30 & 0
\end{array}$$

Thus, $\frac{7}{2}$ is a root. The depressed polynomial $2x^2 -$ $24x + 30$ has roots at $x = \frac{24 \pm \sqrt{(-24)^2 - 4(2)(30)}}{2(2)} =$ $\frac{24 \pm \sqrt{336}}{4} = \frac{24 \pm 4\sqrt{21}}{4} = 6 \pm \sqrt{21}$. There are no other real solutions. Since $6 + \sqrt{21} \approx 10.583$ is not in the realistic domain, it is not a solution. So the only solutions are $x = \frac{5}{2} = 3.5$ cm or $6 - \sqrt{21} \approx 1.417$ cm.

≡ 18.4 IRRATIONAL ROOTS

1. $P(x) = x^3 + 5x - 3$, $P(0) = -3$, and $P(1) = 3$ so

a	$P(a)$	b	$P(b)$	$c = a - \dfrac{P(a)}{m}$	$P(c)$
0	-3	1	3	0.5	-0.375
0	-3	0.5	-0.375	0.5714	0.0437
0	-3	0.5714	0.0437	0.5632	-0.0052
0	-3	0.5632	-0.0052	0.5642	0.0006

Thus, $c \approx 0.56$

3.   Let $P(x) = x^4 + 2x^3 - 5x^2 + 1$, $P(0) = 1$ and $P(1) = -1$, with the following results:

a	$P(a)$	b	$P(b)$	$c = a - \dfrac{P(a)}{m}$	$P(c)$
0	1	1	-1	0.5	0.0625
0	1	0.5	0.0625	0.5333	-0.379
0	1	0.5333	-0.0379	0.5139	0.0208
0	1	0.5139	0.0208	0.5248	-0.0121
0	1	0.5248	-0.0121	0.5185	0.0068
0	1	0.5185	0.0068	0.5221	-0.0039

Thus, $c \approx 0.52$.

5.   Let $P(x) = x^3 + x^2 - 7x + 3$, $P(0) = 3$, and $P(1) = -2$. We get

a	$P(a)$	b	$P(b)$	$c = a - \dfrac{P(a)}{m}$	$P(c)$
0	3	1	-2	0.6	-0.624
0	3	0.6	-0.624	0.4967	-0.1076
0	3	0.4967	-0.1076	0.4795	-0.0163
0	3	0.4795	-0.0163	0.4769	-0.0024
0	3	0.4769	-0.0024	0.4765	-0.00036

Thus, $c \approx 0.48$.

7. $P(x) = x^4 - x - 2$. Since $P(-1) = 0$, one root is at -1. Using synthetic division we get the depressed equation $D(x) = x^3 - x^2 + x - 2$. Since $D(1) = -1$ and $D(2) = 4$, there is a real root between 1 and 2.

a	$P(a)$	b	$P(b)$	$c = a - \dfrac{P(a)}{m}$	$P(c)$
1	-1	2	4	1.2	-0.512
1.2	-0.512	2	4	1.29	-0.2274
1.29	-0.2274	2	4	1.328	-0.0935
1.328	-0.0935	2	4	1.343	-0.03835
1.343	-0.03835	2	4	1.349	-0.0159
1.349	-0.0159	2	4	1.352	-0.0046

Thus, $c \approx 1.35$ and we have found an approximation of the second real root. The other two roots are complex numbers.

9. Letting $P(x) = x^4 + x^3 - 2x^2 - 7x - 5$, we see that $P(-1) = 0$, so -1 is a root. Using synthetic division, we obtain

$$
\begin{array}{r|rrrrr}
-1 & 1 & 1 & -2 & -7 & -5 \\
   &   & -1 & 0 & 2 & 5 \\
\hline
   & 1 & 0 & -2 & -5 & 0
\end{array}
$$

This gives the depressed equation $D(x) = x^3 - 2x - 5$. Since $D(2) = -1$ and $D(3) = 16$, we see that there is a root between 2 and 3.

a	$P_1(a)$	b	$P_1(b)$	$c = a - \dfrac{P_1(a)}{m}$	$P_1(c)$
2	-1	3	16	2.0588	-0.39105
2.0588	-0.39105	3	16	2.081	-0.1501
2.081	-0.1501	3	16	2.0895	-0.5622
2.0895	-0.05622	3	16	2.0927	-0.02064

Thus, $c \approx 2.09$ is a second root. The other two roots are complex numbers.

11. Here $P(x) = 2x^4 + 3x^3 - x^2 - 2x - 2$. $P(1) = 0$ and synthetic division gives the depressed equation $D(x) = 2x^3 + 5x^2 + 4x + 2$. Since $D(-2) = -2$ and $D(-1) = 1$, we see that there is a root between -2 and -1. Linear interpolation shows that this root is approximately -1.66.

13. Here $P(x) = 2x^5 - 5x^3 + 2x^2 + 4x - 1$. $P(-1) = 0$ and synthetic division gives the depressed equation $D(x) = 2x^4 - 2x^3 - 3x^2 + 5x - 1$. Since $D(-2) = 25$, $D(-1) = -5$, $D(0) = -1$, and $D(1) = 1$, we see that there is a root between -2 and -1 and

another root between 0 and 1. Linear interpolation shows that these roots are approximately -1.43 and 0.24.

15. If we let $P(x) = 8x^4 + 6x^3 - 15x^2 - 12x - 2$, we can see that there are rational roots at $-\frac{1}{2}$ and $-\frac{1}{4}$. Using synthetic division with these factors, we get the depressed equation $D(x) = 8x^2 - 16 = 8(x^2 - 2)$. This depressed equation has roots at $\pm\sqrt{2}$. Thus, this equation has the four real roots $-\frac{1}{2}, -\frac{1}{4}, -1.41$, and 1.41.

17. Here we have $P(t) = 150t - 20t + t^3$. We want to know the values of t when $P(t) = 400$. Thus, we want the solution to $P(t) = 150t - 20t^2 + t^3 - 400 = 0$. Since $P(6) = -4$ and $P(7) = 13$, we see that this will be between 6 and 7. Using linear interpolation, we obtain

a	$P(a)$	b	$P(b)$	$c = a - \dfrac{P(a)}{m}$	$P(c)$
6	-4	7	-13	6.23529	0.13759
6	-4	6.23529	0.13759	6.22747	0.002736
6	-4	6.22747	0.002736	6.22731	0.0000546
6	-4	6.22731	0.0000546	6.22731	0.000001

Thus, $c \approx 6.22731$ and so it will take about 6 years 83 days.

19. **(a)** Since $h = 2r$, we have the following

$$V = \pi(r + 15.5)^2(2r + 31) = 1\,000\,000\pi$$
$$(r^2 + 31r + 240.25)(2r + 31) = 1\,000\,000$$
$$2r^3 + 93r^2 + 1\,441.5r + 7\,447.75 = 1\,000\,000$$
$$2r^3 + 93r^2 + 1\,441.5r - 992\,552.25 = 0$$

Using linear interpolation, we get

Thus, $r \approx 63.870$ cm.

(b) $h = 2r$, $V = \pi(63.87)^2(2 \cdot 63.87) = 521\,100\pi$ cm^3 $1\,637\,000$ cm^3.

a	$V(a)$	b	$V(b)$	$c = a - \dfrac{V(a)}{m}$	$V(c)$
63	-32526	64	4919	63.869	-54.17
63	-32526	63.869	-54.177	63.870	0.5983
63	-32526	63.870	0.5983	63.870	-0.0066

21. R is an even function so all solutions come in pairs. Evaluating R at various points, we see that $R(0.3) = -0.675; R(0.5) = 1.536; R(0.8) = 0.864; R(0.9) = -0.509; R(1.4) = -1.278; R(1.5) = 0.700; R(1.7) = 1.0183; R(1.8) = -3.561; R(2.2) = -13.699;$ and $R(2.3) = 51.874$. By the locator theorem, there are roots between 0.3 and 0.5, between 0.8 and 0.9, between 1.4 and 1.5, between 1.7 and 1.8 as well as between 2.2 and 2.3.

Linear interpolation for the one between 1.7 and 1.8 is as follows

a	$R_1(a)$	b	$R_1(b)$	$c = a - \dfrac{R_1(a)}{m}$	$R_1(c)$
1.7	1.0183	1.8	-3.5614	1.7222	0.3546
1.7	1.0183	1.7222	0.3546	1.7341	-0.0796
1.7	1.0183	1.7341	-0.0796	1.7316	0.01548
1.7	1.0183	1.7316	0.01548	1.7321	-0.0031
1.7	1.0183	1.7321	-0.0031	1.7320	0.00062
1.7	1.0183	1.7320	0.00062	1.7321	-0.0001
1.7	1.0183	1.7321	-0.00012	1.7321	0.000025

The approximate roots are $\pm 0.3536, \pm 0.8654, \pm 1.4639, \pm 1.7321,$ and ± 2.2323.

23. We want to determine the value of t for which $-0.009t^5 + t^3 + 5.5 = 32.6$ or $-0.009t^5 + t^3 - 27.1 = 0$. Using a graphing calculator we estimate one root at 3.095 h or 3 h 5.7 min and the other root is at 10.414 h or about 10 h 24.8 min.

≡ 18.5 RATIONAL FUNCTIONS

1. Vertical asymptote: $x = -2$; Horizontal asymptote: $y = 0$

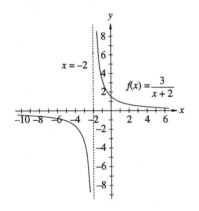

3. Vertical asymptote: $x = -4$; Horizontal asymptote: $y = 2$

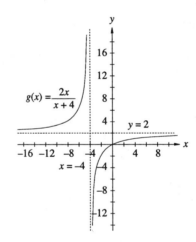

5. $\frac{3x^2}{x^2+6x+8} = \frac{3x^2}{(x+4)(x+2)}$, Vertical asymptotes: $x = -2$ and $x = -4$; Horizontal asymptote: $y = 3$

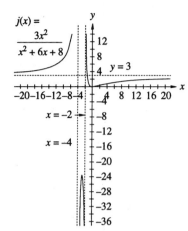

7. $\frac{x(x+1)}{(x+1)(x+2)}$; Vertical asymptote: $x = -2$; Horizontal asymptote: $y = 1$; (Note: hole at $x = -1$)

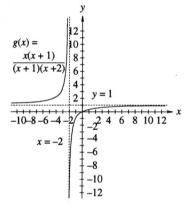

9. $\frac{4x^2-12x-27}{x^2-6x+8} = \frac{(2x-9)(2x+3)}{(x-2)(x-4)}$. Vertical asymptotes: $x = 2$ and $x = 4$; Horizontal asymptote: $y = 4$

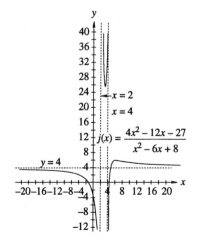

11. $\frac{(x+2)(x-3)}{(x+5)(x+4)} = 0 \Rightarrow (x+2)(x-3) = 0; x = -2$ or $x = 3$

13. $\frac{x^2-1}{x^2+2x+1} = \frac{(x+1)(x-1)}{(x+1)^2} = 0; x \neq -1$ so $x = 1$ is the only solution.

15. The given equation $\frac{4}{x-3} = \frac{2}{x+1}$. Subtracting, we get $\frac{4}{x-3} - \frac{2}{x+1} = 0$. This has a common denominator of $(x-3)(x+1)$. Multiplying the equation by this common denominator produces $\frac{4(x+1)-2(x-3)}{(x-3)(x+1)}$ $= 0$ which simplifies to $\frac{4x+4-2x+6}{(x-3)(x+1)} = 0$ or $\frac{2x+10}{(x-3)(x+1)}$ $= 0$. This has the solution $2x+10 = 0$ or $x = -5$.

17. Subtracting $\frac{2}{3}$ from both sides of the equation produces $\frac{5}{2x} - \frac{3}{4x} - \frac{2}{3} = 0$. Finding a common denominator, we get $\frac{6\cdot5-3\cdot3-2\cdot4x}{12x} = 0$ or $\frac{30-9-8x}{12x} = 0$. Any solutions will be when $30 - 9 - 8x = 0$ or $21 - 8x = 0$ which means that $x = \frac{21}{8}$.

19. Finding that $(x+2)(x-2) = x^2 - 4$ is a common denominator for $\frac{3}{x-2} + \frac{5}{x+2} = \frac{20}{x^2-4}$, we get $\frac{3(x+2)}{x^2-4} + \frac{5(x-2)}{x^2-4} - \frac{20}{x^2-4} = 0$ or $\frac{3x+6+5x-10-20}{x^2-4} = 0$. The numerator simplifies to $8x - 24$ and is 0 only when $x = 3$.

21. If we let $P(t) = 3t^4 + 2t^3 - 300t - 50$, we see that $P(4.4) = -75.20$ and $P(4.5) = 12.4375$. Using linear interpolation, we get

a	$P(a)$	b	$P(b)$	$c = a - \dfrac{P(a)}{m}$	$P(c)$
4.4	−75.20	4.5	12.4375	4.4858	−.469
4.4	−75.20	4.4858	−.469	4.4863	+0.01787
4.4	−75.20	4.4863	0.01787	4.4863	−0.00068

Thus, $t \approx 4.49$ hours.

23. As x gets very large, the fraction $\frac{7,250}{x}$ gets close to 0. Hence, the lowest cost is $73 + 0 = 73$ or \$73/set.

CHAPTER 18 REVIEW

1. $P(2) = 9 \cdot 2^2 - 4 \cdot 2 + 2 = 36 - 8 + 2 = 30$, degree $= 2$

2. $P(3) = 2 \cdot 3^3 - 4 \cdot 3^2 + 7 = 54 - 36 + 7 = 25$, degree $= 3$

3. $P(-1) = 4(-1)^3 - 5(-1)^2 + 7(-1) + 3 = -4 - 5 - 7 + 3 = -13$, degree $= 3$

4. $P(-3) = 5(-3)^2 + 6(-3) - 3 = 45 - 18 - 3 = 24$, degree $= 2$

5.
$$\begin{array}{r|rrr} -3 & 4 & -6 & -3 \\ & & -12 & 54 \\ \hline & 4 & -18 & 51 \end{array}$$
So, $Q(x) = 4x - 18$; $R = 51$

6.
$$\begin{array}{r|rrrr} 1 & 6 & 0 & -2 & 1 \\ & & 6 & 6 & 4 \\ \hline & 6 & 6 & 4 & 5 \end{array}$$
So, $Q(x) = 6x^2 + 6x + 4$; $R = 5$

7.
$$\begin{array}{r|rrrrr} 1 & 9 & -5 & 2 & -7 & 1 \\ & & 9 & 4 & 6 & -1 \\ \hline & 9 & 4 & 6 & -1 & 0 \end{array}$$
So, $Q(x) = 9x^3 + 4x^2 + 6x - 1$; $R = 0$

8.
$$\begin{array}{r|rrrr} -2 & 7 & -3 & 0 & -12 \\ & & -14 & 34 & -68 \\ \hline & 7 & -17 & 34 & -80 \end{array}$$
So, $Q(x) = 7x^2 - 17x + 34$; $R = -80$

9. $x^3 - 3x^2 + 3x - 1 = 0$; possible rational roots are ± 1. Using synthetic division, we find
$$\begin{array}{r|rrrr} 1 & 1 & -3 & 3 & -1 \\ & & 1 & -2 & 1 \\ \hline & 1 & -2 & 1 & 0 \end{array}$$

The quotient is $x^2 - 2x + 1 = (x-1)(x-1)$. Thus, the roots are 1, 1, and 1.

10. $x^4 - 5x^3 + 9x^2 - 7x + 2 = 0$; possible roots $\pm 1, \pm 2$. Using synthetic division with 1, we find
$$\begin{array}{r|rrrrr} 1 & 1 & -5 & 9 & -7 & 2 \\ & & 1 & -4 & 5 & -2 \\ \hline 1 & 1 & -4 & 5 & -2 & 0 \\ & & 1 & -3 & 2 & \\ \hline & 1 & -3 & 2 & 0 & \end{array}$$
Since $x^2 - 3x + 2 = (x-2)(x-1)$. Thus, the roots are 1, 1, 1, and 2.

11. $x^4 - 6x^2 - 8x - 3 = 0$. Possible roots ± 1, and ± 3.

$$\begin{array}{r|rrrrr} -1 & 1 & 0 & -6 & -8 & -3 \\ & & -1 & 1 & 5 & 3 \\ \hline -1 & 1 & -1 & -5 & -3 & 0 \\ & & -1 & 2 & 3 & \\ \hline & 1 & -2 & -3 & 0 & \end{array}$$
Since $x^2 - 2x - 3 = (x+1)(x-3)$, the roots are $-1, -1, -1$, and 3.

12. $x^4 + 10x^3 + 25x^2 - 16 = 0$; Possible roots $\pm 1, \pm 2, \pm 4, \pm 8, \pm 16$

$$\begin{array}{r|rrrrr} -1 & 1 & 10 & 25 & 0 & -16 \\ & & -1 & -9 & -16 & 16 \\ \hline -4 & 1 & 9 & 16 & -16 & 0 \\ & & -4 & -20 & 16 & \\ \hline & 1 & 5 & -4 & 0 & \end{array}$$
Using the quadratic formula on $x^2 + 5x - 4 = 0$, we get $x = \frac{-5 \pm \sqrt{25+16}}{2} = \frac{-5 \pm \sqrt{41}}{2}$. The roots are $-1, -4, \frac{-5 \pm \sqrt{41}}{2}$, and $\frac{-5 - \sqrt{41}}{2}$.

13. $x^4 + x^3 - 6x^2 + x + 1 = 0$. Only possible rational roots are ± 1. Neither works

a	$P(a)$	b	$P(b)$	$c = a - \dfrac{P(a)}{m}$	$P(c)$
−3.2	8.4496	−3	−2	−3.038	−0.2580
−3.2	8.4496	−3.038	−0.2580	−3.043	−0.3169
−3.2	8.4496	−3.043	−0.3169	−3.0436	−.00387
−3.2	8.4496	−3.0436	−0.00387	−3.0437	−.00047
−0.4	−0.3984	−0.3	0.1411	−0.3262	0.01221
−0.4	−0.3984	−0.3262	0.01221	−0.3283	0.00099
−0.4	−0.3984	−0.3283	0.00099	−0.3285	0.00008
−0.4	−3.984	−0.3285	0.00008	−0.3285	0.000006
0.5	0.1875	0.6	−0.2144	0.5467	0.00633
0.5	0.1875	0.5467	0.00633	0.5482	−0.00021
0.5	0.1875	0.5482	−0.00021	0.5482	0.000007
1.8	−0.3104	1.9	1.1311	1.8215	−0.03354
1.8	−0.3104	1.8215	−0.03354	1.8241	0.00121
1.8	−0.3104	1.8241	0.00121	1.8240	−0.00004

The approximate solutions are $-3.044, -0.328, 0.548,$ and 1.824.

14. $2x^5 - 3x^4 - 18x^3 + 75x^2 - 104x + 48 = 0$; possible roots include $\pm 1, \pm 2, \pm 3, \pm 4, \pm 6, \pm 8, \pm \frac{1}{2}$, and $\pm \frac{3}{2}$.

$$
\begin{array}{r|rrrrrr}
1 & 2 & -3 & -18 & 75 & -104 & 48 \\
  &   & 2 & -1 & -19 & 56 & -48 \\
\hline
-4 & 2 & -1 & -19 & 56 & -48 & 0 \\
   &   & -8 & 36 & -68 & 48 & \\
\hline
\frac{3}{2} & 2 & -9 & 17 & -12 & 0 & \\
   &   & 3 & -9 & 12 & & \\
\hline
   & 2 & -6 & 8 & 0 & &
\end{array}
$$

Here $2x^2 - 6x + 8 = 2(x^2 - 3x + 4)$. Using the quadratic formula, we get $x = \frac{3 \pm \sqrt{9-16}}{2} = \frac{3 \pm \sqrt{-7}}{2} = \frac{3}{2} \pm \frac{\sqrt{7}}{2}j$. The roots are $1, -4, \frac{3}{2}, \frac{3}{2} + \frac{\sqrt{7}}{2}j$, and $\frac{3}{2} - \frac{\sqrt{7}}{2}j$.

15. $x^6 + 2x^5 + 3x^4 + 4x^3 + 3x^2 + 2x + 1 = 0$; possible solution is -1.

$$
\begin{array}{r|rrrrrrr}
-1 & 1 & 2 & 3 & 4 & 3 & 2 & 1 \\
   &   & -1 & -1 & -2 & -2 & -1 & -1 \\
\hline
-1 & 1 & 1 & 2 & 2 & 1 & 1 & 0 \\
   &   & -1 & 0 & -2 & 0 & -1 & \\
\hline
   & 1 & 0 & 2 & 0 & 1 & 0 &
\end{array}
$$

Notice that $x^4 + 2x^2 + 1 = (x^2 + 1)^2$ and that $x^2 + 1 = 0 \Rightarrow x^2 = -1$ or $x = \pm j$. This happens twice. The solutions are $-1, -1, j, j, -j$, and $-j$.

16. $x^6 - 2x^5 - 4x^4 + 6x^3 - x^2 + 8x + 4 = 0$. Possible solutions are $\pm 1, \pm 2, \pm 4$

$$
\begin{array}{r|rrrrrrr}
-2 & 1 & -2 & -4 & 6 & -1 & 8 & 4 \\
   &   & -2 & 8 & -8 & 4 & -6 & -4 \\
\hline
2 & 1 & -4 & 4 & -2 & 3 & 2 & 0 \\
   &   & 2 & -4 & 0 & -4 & -2 & \\
\hline
   & 1 & -2 & 0 & -2 & -1 & 0 &
\end{array}
$$

Here $x^4 - 2x^3 - 2x - 1 = (x^2 + 1)(x^2 - 2x - 1)$ and that $x^2 + 1 = 0$ or $x^2 = -1$ and $x = \pm j$. Also, $x^2 - 2x - 1$ and $x = \frac{2 \pm \sqrt{4+4}}{2} = 1 \pm \sqrt{2}$. The roots are $-2, 2, 1 + \sqrt{2}, 1 - \sqrt{2}, j, -j$.

17. $\frac{5x}{x^2+3x-4} = \frac{5x}{(x+4)(x-1)}$; Vertical asymptotes: $x = -4$ and $x = 1$; Horizontal asymptote: $y = 0$. A rational equation is 0 only when the numerator is 0, and so $5x = 0$ or $x = 0$, the solution is $x = 0$.

18. $\frac{7x^3}{x^3+1} = \frac{7x^3}{(x+1)(x^2-x+1)}$; Vertical asymptote: $x = -1$; Horizontal asymptote: $y = 7$. A rational equation is 0 only when the numerator is 0, and since $7x^3 = 0 \Rightarrow x = 0$, the solution is $x = 0$.

19. $\frac{4}{x-2} - \frac{5}{x+2} = \frac{4(x+2)-5(x-2)}{(x-2)(x+2)} = \frac{4x+8-5x+10}{(x-2)(x+2)} = \frac{-x+18}{(x-2)(x+2)}$ Vertical asymptote: $x = 2, x = -2$; Horizontal asymptote: $y = 0$. The solution is when $-x + 18 = 0$ or when $x = 18$.

20. $\frac{3}{x-1} + \frac{2}{x+1} - \frac{4}{x^2-1} = \frac{3(x+1)+2(x-1)-4}{(x-1)(x+1)} = \frac{3x+3+2x-2-4}{(x-1)(x+1)} = \frac{5x-3}{(x-1)(x+1)}$. Vertical asymptotes: $x = 1$ and $x = -1$; Horizontal asymptote: $y = 0$. The solution is when $5x - 3 = 0$ or when $x = \frac{3}{5}$.

21. $P(n) = \frac{0.5+0.8(n-1)}{1.5+0.8(n-1)} = \frac{0.8n-0.3}{0.8n+0.7}$. Horizontal asymptote is $\frac{0.8}{0.8} = 1$ or 100%.

22. $N(t) = \frac{5t^2+6t+15}{t^2}$ Horizontal asymptote is $y = \frac{5}{1} = 5$.

CHAPTER 18 TEST

1. $P(x) = 7x^5 - 2x^4 - 5$. Degree is 5

2.
$$P(x) = 4x^3 - 5x^2 + 2x - 8$$
$$P(-2) = 4(-2)^3 - 5(-2)^2 + 2(-2) - 8$$
$$= -32 - 20 - 4 - 8 = -64$$
or

$$
\begin{array}{r|rrrr}
-2 & 4 & -5 & 2 & -8 \\
   &   & -8 & 26 & -56 \\
\hline
   & 4 & -13 & 28 & -64
\end{array}
$$

and so, $R = -64$.

3.
$$
\begin{array}{r|rrrrr}
6 & 1 & -8 & 3 & 44 & 75 \\
  &   & 6 & -12 & -54 & -60 \\
\hline
  & 1 & -2 & -9 & -10 & 15
\end{array}
$$

and so, $Q(x) = x^3 - 2x^2 - 9x - 10$ and $R = 15$.

4. $c = \pm 1, \pm 2, d = 1, 3$ and so $\frac{c}{d} = \pm 1, \pm 2, \pm \frac{1}{3}, \pm \frac{2}{3}$.

5.
$$
\begin{array}{r|rrrr}
1 & 3 & 5 & -4 & -4 \\
  &   & 3 & 8 & 4 \\
\hline
  & 3 & 8 & 4 & 0
\end{array}
$$

and $3x^2 + 8x + 4 = (3x + 2)(x + 2)$ and the roots are $1, -2, -\frac{2}{3}$.

6. Possible roots include $\pm 1, \pm 2, \pm 3, \pm 4$, and ± 6.

$$
\begin{array}{r|rrrrrr}
-1 & 1 & 2 & -4 & -10 & -41 & -72 & -36 \\
   &   & -1 & -1 & 5 & 5 & 36 & 36 \\
\hline
-1 & 1 & 1 & -5 & -5 & -36 & -36 & 0 \\
   &   & -1 & 0 & 5 & 0 & 36 & \\
\hline
3 & 1 & 0 & -5 & 0 & -36 & 0 & \\
  &   & 3 & 9 & 12 & 36 & & \\
\hline
-3 & 1 & 3 & 4 & 12 & 0 & & \\
   &   & -3 & 0 & -12 & & & \\
\hline
   & 1 & 0 & 4 & 0 & & &
\end{array}
$$

From the quotient we see that

$$x^2 + 4 = 0$$
$$x^2 = -4$$
$$x = \pm 2j$$

and so the roots are $-1, -1, -3, 3, 2j$, and $-2j$.

7. $\frac{3x^2 - 2x - 1}{x^2 + 5x + 6} = \frac{(3x+1)(x-1)}{(x+2)(x+3)}$. Vertical asymptote: $x = -2, x = -3$. Horizontal asymptote: $y = 3$. Solutions: $x = -\frac{1}{3}, x = 1$.

8. The volume of the box is $V = lwh = 175$ or $(12 - 2x)(22 - 2x)x = 175$. Expanding, we get $(4x^2 - 68x + 264)x = 175$ or $4x^3 - 68x^2 + 264x - 175 = 0$. Graphing the left-hand side of this equation, we get roots at about 0.83, 4.50 and 11.66. Since the answer must be less than half of the shortest measurement, this is 6, we obtain 0.8 in. or 4.5 in.

9. We have $t^3 - 22t^2 + 250t = 500$ and so $t^3 - 22t^2 + 250t - 500 = 0$. Graphing, we obtain a root at $t = 2.48$. Thus, after 2.48 yr, the pressure has dropped 500 lb/in.^{2}.

19

Sequences, Series, and the Binomial Formula

≡ 19.1 SEQUENCES

1. $a_1 = \frac{1}{1+1} = \frac{1}{2}$; $a_2 = \frac{1}{2+1} = \frac{1}{3}$; $a_3 = \frac{1}{3+1} = \frac{1}{4}$; $a_4 = \frac{1}{4+1} = \frac{1}{5}$; $a_5 = \frac{1}{5+1} = \frac{1}{6}$; $a_6 = \frac{1}{6+1} = \frac{1}{7}$

3. $a_1 = (1+1)^2 = 4$; $a_2 = (2+1)^2 = 9$; $a_3 = (3+1)^2 = 16$; $a_4 = (4+1)^2 = 25$; $a_5 = (5+1)^2 = 36$; $a_6 = (6+1)^2 = 49$

5. $a_1 = (-1)^1 \cdot 1 = -1$; $a_2 = (-1)^2 \cdot 2 = 2$; $a_3 = (-1)^3 \cdot 3 = -3$; $a_4 = (-1)^4 \cdot 4 = 4$; $a_5 = (-1)^5 \cdot 5 = -5$; $a_6 = (-1)^6 \cdot 6 = 6$

7. $a_1 = \frac{2 \cdot 1 - 1}{2 \cdot 1 + 1} = \frac{1}{3}$; $a_2 = \frac{2 \cdot 2 - 1}{2 \cdot 2 + 1} = \frac{3}{5}$; $a_3 = \frac{2 \cdot 3 - 1}{2 \cdot 3 + 1} = \frac{5}{7}$; $a_4 = \frac{2 \cdot 4 - 1}{2 \cdot 4 + 1} = \frac{7}{9}$; $a_5 = \frac{2 \cdot 5 - 1}{2 \cdot 5 + 1} = \frac{9}{11}$; $a_6 = \frac{2 \cdot 6 - 1}{2 \cdot 6 + 1} = \frac{11}{13}$

9. $a_1 = \left(\frac{2}{3}\right)^0 = 1$; $a_2 = \left(\frac{2}{3}\right)^1 = \frac{2}{3}$; $a_3 = \left(\frac{2}{3}\right)^2 = \frac{4}{9}$; $a_4 = \left(\frac{2}{3}\right)^3 = \frac{8}{27}$; $a_5 = \left(\frac{2}{3}\right)^4 = \frac{16}{81}$; $a_6 = \left(\frac{2}{3}\right)^5 = \frac{32}{243}$

11. $a_1 = \left(\frac{1-1}{1+1}\right)^2 = 0$; $a_2 = \left(\frac{2-1}{2+1}\right)^2 = \frac{1}{9}$; $a_3 = \left(\frac{3-1}{3+1}\right)^2 = \frac{4}{16} = \frac{1}{4}$; $a_4 = \left(\frac{4-1}{4+1}\right)^2 = \frac{9}{25}$; $a_5 = \left(\frac{5-1}{5+1}\right)^2 = \frac{16}{36} = \frac{4}{9}$; $a_6 = \left(\frac{6-1}{6+1}\right)^2 = \frac{25}{49}$

13. $a_1 = 1$; $a_2 = 2 \cdot 1 = 2$; $a_3 = 3 \cdot 2 = 6$; $a_4 = 4 \cdot 6 = 24$; $a_5 = 5 \cdot 24 = 120$; $a_6 = 6 \cdot 120 = 720$

15. $a_1 = 5$; $a_2 = 5 + 3 = 8$; $a_3 = 8 + 3 = 11$; $a_4 = 11 + 3 = 14$; $a_5 = 14 + 3 = 17$; $a_6 = 17 + 3 = 20$

17. $a_1 = 2$; $a_2 = 2^1 = 2$; $a_3 = 2^2 = 4$; $a_4 = 4^3 = 64$; $a_5 = 64^4 = 16,777,216$; $a_6 = (16,777,216)^5 = 1.329228 \times 10^{36}$

19. $a_1 = 1$; $a_2 = 2^1 = 2$; $a_3 = 3^2 = 9$; $a_4 = 4^9 = 262,144$; $a_5 = 5^{262,144}$; $a_6 = 6^{(5^{262,144})}$

21. $a_1 = 1$; $a_2 = \frac{1}{2}$; $a_3 = 1 \cdot \frac{1}{2} = \frac{1}{2}$; $a_4 = \frac{1}{2} \cdot \frac{1}{2} = \frac{1}{4}$; $a_5 = \frac{1}{2} \cdot \frac{1}{4} = \frac{1}{8}$; $a_6 = \frac{1}{4} \cdot \frac{1}{8} = \frac{1}{32}$

23. $a_1 = 0$; $a_2 = 2$; $a_3 = 2 - 0 = 2$; $a_4 = 2 - 2 = 0$; $a_5 = 0 - 2 = -2$; $a_6 = -2 - 0 = -2$

25. $a_1 = 25$ ft-3 in.; $a_{25} = 25$ ft$-3(25)$ in. $= 25$ ft-75 in.; Since 75 in. $= 6'3'' = 6.25$ ft, $a_{25} = 25 - 6.25 = 18.75$ ft

27. $a_1 = 26000$, $a_2 = 26000 + 1100 = 27,100$; $a_n = 26000 + (n-1) \cdot 1100$; $a_5 = 26000 + (5-1)1100 = 26000 + 4400 = \$30,400$; $a_{10} = 26000 + 9(1100) = \$35,900$

≡ 19.2 ARITHMETIC AND GEOMETRIC SEQUENCES

1. Common difference, so arithmetic: $d = 9 - 1 = 8$; $a_9 = 1 + (9-1)8 = 1 + 64 = 65$

3. Common ratio, so geometric: $r = \frac{-4}{8} = -\frac{1}{2}$; $a_{10} = \left(-\frac{1}{2}\right)^9 \cdot 8 = -\frac{1}{512} \cdot 8 = -\frac{1}{64}$

5. Geometric: $r = \frac{4}{-1} = -4$; $a_6 = (-4)^5 \cdot (-1) = 1024$

7. Arithmetic: $d = -2 - 3 = -5$; $a_8 = 3 + 7(-5) = -32$

9. Neither a common difference nor a common ratio so it is neither arithmetic nor geometric

11. Arithmetic: $d = 0.4$; $a_9 = 0.4 + 8(0.4) = 9(0.4) = 3.6$

13. Arithmetic: $d = -\frac{1}{6}; a_8 = \frac{2}{3} + 7\left(-\frac{1}{6}\right) = \frac{4}{6} - \frac{7}{6} = -\frac{3}{6} = -\frac{1}{2}$

15. Geometric: $r = \frac{1}{10}$ or 0.1; $a_6 = (0.1)^5 \cdot 5 = 0.00005$

17. Arithmetic: $d = 2; a_9 = (1+\sqrt{3})+8(2) = 17+\sqrt{3}$

19. Neither. It looks like an arithmetic sequence with alternating signs, so try $[4.3 - (n-1)1.1](-1)^{n-1}$; $a_6 = 1.2$

21. $d = a_6 - a_5 = 24 - 9 = 15; a_5 = a_1 + 4d; 9 = a_1 + 4 \cdot 15; a_1 = 9 - 60 = -51$

23. $r = \frac{a_5}{a_3} = \frac{3}{9} = \frac{1}{3}; a_4 = \left(\frac{1}{3}\right)^3 \cdot a_1; 9 = \frac{1}{27} \cdot a_1; a_1 = 243$

25. $81 = 5+(n-1)4; 76 = 4(n-1); 19 = (n-1); n = 20$

27. $\frac{1}{559,872} = 3\left(\frac{1}{6}\right)^{n-1}; \frac{1}{1,679.616} = \left(\frac{1}{6}\right)^{n-1}; 6^{n-1} = 1,679,619; \ln 6^{n-1} = (n-1)\ln 6 = \ln 1,679,619; n-1 = \frac{\ln 1,679,619}{\ln 6} = 8; n = 9$

29. $u_1 = -5; a_7 = 4; a_7 = a_1 + (n-1)d; 4 = -5 + 6d; 9 = 6d; d = \frac{3}{2}; a_2 = -5 + \frac{3}{2} = -3\frac{1}{2}$ or -3.5

31. $a_{61} = a_1 + 60d; 8 = \frac{1}{2} + 60d; 7\frac{1}{2} = 60d; d = \frac{7.5}{60} = 0.125 = \frac{1}{8}$ mi

33. $a_{10} = a_1 + (10-1)d; a_{10} = 15 + 9(-0.3); a_{10} = 15-2.7 = 12.3$ cm; $0 = 15+(n-1)(-0.3); -15 = -0.3(n-1); n-1 = 50; n = 51$

35. **(a)** Geometric sequence: Here $r = (1 - 0.162) = 0.838$ and $a_1 = 940$, so the concentration is reduced to 100 ppm when $100 = 940(0.838)^{n-1}$. Thus, $a_n = \frac{100}{940} = (0.838)^{n-1}$ and so, $(n-1)\ln(0.838) = \ln\frac{10}{94}$. Dividing, we get $n-1 = \frac{\ln\left(\frac{10}{94}\right)}{\ln(0.838)} \approx 12.678$, and hence $n = 13.678$. Thus, the concentration will be reduced to 100 ppm about 13,678 mi from the first monitor or about 14.678 mi from the spill.
(b) Here $a_n = 1.5 = 940(0.838)^{n-1}$ and so $n - 1 = \frac{\ln\left(\frac{1.5}{940}\right)}{\ln(0.838)} = 36.44$ which is $n = 37.44$ mi from the monitor so about 38.44 mi from the spill.

19.3 SERIES

1. $\sum_{k=1}^{20} 3\left(\frac{1}{2}\right)^k = 3\left(\frac{1}{2}\right)^1 + 3\left(\frac{1}{2}\right)^2 + 3\left(\frac{1}{2}\right)^3 + 3\left(\frac{1}{2}\right)^4 + \cdots = 3\left(\frac{1}{2}\right) + 3\left(\frac{1}{4}\right) + 3\left(\frac{1}{8}\right) + 3\left(\frac{1}{16}\right)\cdots = \frac{3}{2} + \frac{3}{4} + \frac{3}{8} + \frac{3}{16} + \cdots$

3. $\sum_{n=0}^{20} (-1)^n \frac{3^n}{n+1} = 1 - \frac{3}{2} + 3 - \frac{27}{4} + \cdots$

5. $\sum_{k=1}^{5} (k-3) = -2 - 1 + 0 + 1 + 2 = 0$

7. $\sum_{k=1}^{6} k^2 = 1 + 4 + 9 + 16 + 25 + 36 = 91$

9. $\sum_{n=1}^{5} \sum n^3 = 1 + 8 + 27 + 64 + 125 = 225$

11. $\sum_{n=1}^{8} \frac{2^i}{3i+1} = \frac{8}{10} + \frac{16}{13} + \frac{32}{16} + \frac{64}{19} + \frac{128}{22} + \frac{256}{25} = \frac{1,593,342}{67,925} \approx 23.457$

13. Aritimetic: $d = 3; a_{10} = 3 + 9 \cdot 3 = 30; S_{10} = \frac{10(3+30)}{2} = 165$

15. Geometric: $r = 2; S_8 = 1 \cdot \frac{1-2^8}{1-2} = \frac{-255}{-1} = 255$

17. Arithmetic: $d = 4; a_{12} = -6 + 11 \cdot 4 = 38; S_{12} = \frac{12(-6+38)}{2} = 192$

19. Arithmetic: $d = 0.25, a_{20} = 0.5 + 19(.25) = 5.25; S_{20} = \frac{20}{2}(0.5 + 5.25) = 57.5$

21. Geometric: $r = -\frac{1}{3}; S_{16} = \frac{3}{4}\frac{\left(1 - \left(-\frac{1}{3}\right)^{16}\right)}{\left(1 + \frac{1}{3}\right)} = \frac{9}{10}\left(1 - \left(\frac{1}{3}\right)^{16}\right) \approx 0.5625$

23. Arithmetic: $d = 1.2; a_{14} = 0.4+13(1.2) = 16; S_{14} = \frac{14}{2}(0.4 + 16) = 114.8$

25. Geometric: $r = \sqrt{3}; S_{15} = 2\sqrt{3} \cdot \left(\frac{1-\sqrt{3}^{15}}{1-\sqrt{3}}\right) = 17,920.253$

27. Geometric: $r = 3; S_8 = \left(\frac{1-3^8}{1-3}\right) = 9840$

29. First time it hits the ground it has only fallen 80m. After this it rises and falls so $2 \times 80(.8)$. So this is a sequence $80 + 160(.8) + 160(0.8)^2 + 160(0.8)^3 = 160 + 160(0.8) + 160(0.8)^2 + 160(0.8)^3 - 80 = 160\left(\frac{1-0.8^4}{1-0.8}\right) - 80 = 392.32$m when it hits the fourth time. $160\left(\frac{1-0.8^{10}}{1-0.8}\right) - 80 = 634.10$ when it hits the tenth time.

31. In a 16 ft span there will be 12 studs $16''$ apart. The longest is $8'$ and the shortest is $8''$ or $\frac{2}{3}$ ft.

$S_{12} = \frac{12}{2}\left(8 + \frac{2}{3}\right) = 6\left(8\frac{2}{3}\right) = 52$ ft. This is only half of the roof. The whole roof will be twice that or 104 ft.

33. The rate of 8% compounded monthly is $\frac{8}{12} = \frac{2}{3}\%$ per month $\approx 0.00\overline{6}$ per month, 3 years have 36 months $A = 2000(1 + 0.00\overline{6})^{36} = \$2540.47$. This is $3.99 more than exercise #32.

35. (a) Here the sequence is $\$750,000 + \$320,000 + \$240,000 + \$250,000 + \cdots + \$336,000$. Here $a_1 = 240,000$ is the cost of the 2nd floor, so $a_9 = 336,000$ and $S_9 = \frac{9(240,000+336,000)}{2} = 2,592,000$ is the cost of floors 2–10. So, the cost of a 10-story building is $\$750,000 + \$320,000 + \$2,592,000 = \$3,662,000$.

(b) $\$750,000 + \$320,000 + S_{19} = \$7,682,000$. ($S_{19} =$ cost of floors 2–20 $= \frac{19}{2}(240,000 + 456,000) = 6,612,000$.)

≡ 19.4 INFINITE GEOMETRIC SERIES

1. $r = \frac{-1}{2}$ so converges; $S = \dfrac{\frac{1}{2}}{1 - \frac{-1}{2}} = \dfrac{\frac{1}{2}}{\frac{3}{2}} = \frac{1}{3}$

3. $r = -\frac{2}{3}$ so converges; $S = \dfrac{1}{1 - \frac{-2}{3}} = \dfrac{1}{\frac{5}{3}} = \frac{3}{5}$ or 0.6

5. $r = \frac{1}{4}$ so converges; $S = \dfrac{3}{1 - \frac{1}{4}} = \dfrac{3}{\frac{3}{4}} = 4$

7. $r = \frac{1}{10}$ so converges; $S \dfrac{\frac{5}{4}}{1 - \frac{1}{10}} = \dfrac{\frac{5}{4}}{\frac{9}{10}} = \frac{5}{4} \cdot \frac{10}{9} = \frac{50}{36} = \frac{25}{18} = 1.3888\ldots$

9. $r = 0.1$ so converges; $S = \frac{0.03}{1-0.1} = \frac{0.03}{0.9} = \frac{3}{90} = \frac{1}{30} = 0.03333\ldots$

11. $r = (-0.8)$ so converges; $S = \frac{1}{1+0.8} = \frac{1}{1.8} = \frac{10}{18} = \frac{5}{9} = 0.555\ldots$

13. $r = \frac{5}{4}$ so diverges

15. $r = 10$ so diverges

17. $r = \frac{2}{3}$ so converges; $S = \dfrac{\frac{2}{3}}{1 - \frac{2}{3}} = \dfrac{\frac{2}{3}}{\frac{1}{3}} = 2$

19. $r = \frac{1}{\sqrt{2}}$ so converges; $S = \dfrac{\frac{1}{\sqrt{2}}}{1 - \frac{1}{\sqrt{2}}} = \dfrac{\frac{1}{\sqrt{2}}}{\frac{\sqrt{2}-1}{\sqrt{2}}} = \frac{1}{\sqrt{2}-1} = \sqrt{2} + 1 \approx 2.4142$

21. $r = \sqrt{5}$ so diverges

23. $a_1 = 0.4; r = 0.1; S = \frac{0.4}{1-0.1} = \frac{0.4}{0.9} = \frac{4}{9}$

25. $a_1 = 0.57; r = 0.01; S = \frac{0.57}{1-0.01} = \frac{0.57}{0.99} = \frac{57}{99} = \frac{19}{33}$

27. $a_1 = 1.352; r = 0.0001; S = \frac{1.352}{1-0.0001} = \frac{1.352}{0.9999} = \frac{13520}{9999}$

29. $6.3021021\ldots = 6.3 + (0.0021 + 0.0000021 \cdots);$ $a_1 = 0.0021; r = 0.001; S = \frac{0.0021}{.999} = \frac{21}{9990} = \frac{7}{3330}; 6.3 + \frac{7}{3330} = \frac{63}{10} + \frac{7}{3330} = \frac{20986}{3330} = \frac{10493}{1665}$

31. After the first bounce it is a geometric series with $a_1 = 8; r = 0.8; S = \frac{8}{1-0.8} = \frac{8}{0.2} = 40$. The total distance is $5 + 40 = 45$ m.

33. Geometric series with $a_1 = 250; r = \frac{8}{10} = 0.8$ so $S = \frac{250}{1-0.8} = \frac{250}{0.2} = 1250$ rev.

▤ 19.5 THE BINOMIAL THEOREM

1. $(a + 1)^4 = 1a^4 + 4a^3 \cdot 1^1 + 6a^2 1^2 + 4a 1^3 + 1^4 = a^4 + 4a^3 + 6a^2 + 4a + 1$

3. $(3x - 1)^4 = (3x)^4 + 4(3x)^3(-1) + 6(3x)^2(-1)^2 + 4(3x)(-1)^3 + (-1)^4 = 81x^4 - 108x^3 + 54x^2 - 12x + 1$

5. $\left(\frac{x}{2}+d\right)^6 = \left(\frac{x}{2}\right)^6 + 6\left(\frac{x}{2}\right)^5 d + 15\left(\frac{x}{2}\right)^4 d^2 + 20\left(\frac{x}{2}\right)^3 d^3 + 15\left(\frac{x}{2}\right)^2 d^4 + 6\left(\frac{x}{2}\right)d^5 + d^6 = \frac{x^6}{64} + \frac{3x^5 d}{16} + \frac{15x^4 d^2}{16} + \frac{5x^3 d^3}{2} + \frac{15x^2 d^4}{4} + 3xd^5 + d^6$

7. $\left(\frac{a}{2} - \frac{4}{b}\right)^5 = \left(\frac{a}{2}\right)^5 + 5\left(\frac{a}{2}\right)^4\left(-\frac{4}{b}\right) + 10\left(\frac{a}{2}\right)^3\left(-\frac{4}{b}\right)^2 + 10\left(\frac{a}{2}\right)^2\left(-\frac{4}{b}\right)^3 + 5\left(\frac{a}{2}\right)\left(-\frac{4}{b}\right)^4 + \left(-\frac{4}{b}\right)^5 = \frac{a^5}{32} - \frac{5a^4}{4b} + 20\frac{a^3}{b^2} - 160\frac{a^2}{b^3} + 640\frac{a}{b^4} - \frac{1024}{b^5}$

9. $(a + b)^7 = \binom{7}{0}a^7 + \binom{7}{1}a^6 b + \binom{7}{2}a^5 b^2 + \binom{7}{3}a^4 b^3 + \binom{7}{4}a^3 b^4 + \binom{7}{5}a^2 b^5 + \binom{7}{6}ab^6 + \binom{7}{7}b^7 = a^7 + 7a^6 b + 21a^5 b^2 + 35a^4 b^3 + 35a^3 b^4 + 21a^2 b^5 + 7ab^6 + b^7$

11. $(t - a)^8 = \binom{8}{0}t^8 + \binom{8}{1}t^7(-a)^1 + \binom{8}{2}t^6(-a)^2 + \binom{8}{3}t^5(-a)^3 + \binom{8}{4}t^4(-a)^4 + \binom{8}{5}t^3(-a)^5 + \binom{8}{6}t^2(-a)^6 + \binom{8}{7}t(-a)^7 + \binom{8}{8}(-a)^8 = t^8 - 8t^7 a + 28t^6 a^2 - 56t^5 a^3 + 70t^4 a^4 - 56t^3 a^5 + 28t^2 a^6 - 8ta^7 + a^8$

13. $(2a - 1)^6 = \binom{6}{0}(2a)^6 + \binom{6}{1}(2a)^5(-1) + \binom{6}{2}(2a)^4(-1)^2 + \binom{6}{3}2a)^3(-1)^3 + \binom{6}{4}(2a)^2(-1)^4 + \binom{6}{5}(2a)(-1)^5 + \binom{6}{6}(-1)^6 = 64a^6 - 192a^5 + 240a^4 - 160a^3 + 60a^2 - 12a + 1$

15. $\left(x^2 y + \frac{a}{2}\right)^7 = \binom{7}{0}\left(x^2 y\right)^7 + \binom{7}{1}\left(x^2 y\right)^6\left(\frac{a}{2}\right) + \binom{7}{2}\left(x^2 y\right)^5\left(\frac{a}{2}\right)^2 + \binom{7}{3}\left(x^2 y\right)^4\left(\frac{a}{2}\right)^3 \binom{7}{4}\left(x^2 y\right)^3\left(\frac{a}{2}\right)^4 + \binom{7}{5}\left(x^2 y\right)^2\left(\frac{a}{2}\right)^5 + \binom{7}{6}\left(x^2 y\right)\left(\frac{a}{2}\right)^6 + \binom{7}{7}\left(\frac{a}{2}\right)^7 = x^{14}y^7 + \frac{7}{2}x^{12}y^6 a + \frac{21}{4}x^{10}y^5 a^2 + \frac{35}{8}x^8 y^4 a^3 + \frac{35}{16}x^6 y^3 a^4 + \frac{21}{32}x^4 y^2 a^5 + \frac{7}{64}x^2 ya^6 + \frac{a^7}{128}$

17. $(x + y)^{12} = \binom{12}{0}x^{12} + \binom{12}{1}x^{11}y + \binom{12}{2}x^{10}y^2 + \binom{12}{3}x^9 y^3 + \cdots = x^{12} + 12x^{11}y + 66x^{10}y^2 + 220x^9 y^3 + \cdots$

19. $(1 - a)^{-2} = 1 + (-2)(-a) + \frac{(-2)(-3)}{2!}(-a)^2 + \frac{(-2)(-3)(-4)}{3!}(-a)^3 + \cdots = 1 + 2a + 3a^2 + 4a^3 + \cdots$

21. $(1 + b)^{-\frac{1}{3}} = 1 + \left(-\frac{1}{3}\right)b + \frac{\left(-\frac{1}{3}\right)\left(-\frac{4}{3}\right)}{2!}b^2 + \frac{\left(-\frac{1}{3}\right)\left(-\frac{4}{3}\right)\left(-\frac{7}{3}\right)}{3!}b^3 + \cdots = 1 - \frac{b}{3} + \frac{2b^2}{9} + \frac{14b^3}{81} + \cdots$

23. $(1.1)^4 = (1 + 0.1)^4 = 1 + 4(0.1) + \frac{4 \cdot 3}{2!}(0.1)^2 + \frac{4 \cdot 3 \cdot 2}{3!}(0.1)^3 = 1 + 0.4 + 0.06 + 0.004 = 1.464$

25. $\sqrt{1.1} = (1 + 0.1)^{\frac{1}{2}} = 1 + \frac{1}{2}(0.1) + \frac{\frac{1}{2} \cdot \left(-\frac{1}{2}\right)}{2!}(0.1)^2 + \frac{\frac{1}{2}\left(-\frac{1}{2}\right)\left(-\frac{3}{2}\right)}{3!}(0.1)^3 = 1 + 0.05 - 0.00125 + 0.00006 = 1.049$

27. $\sqrt[5]{1.04} = (1 + 0.04)^{\frac{1}{5}} = 1 + \frac{1}{5}(0.04) + \frac{\frac{1}{5}\left(-\frac{4}{5}\right)}{2!}(0.04)^2 = 1 + 0.008 - 0.000128 \approx 1.008$

29. $\sqrt[3]{0.95} = (1 - 0.05)^{\frac{1}{3}} = 1 + \frac{1}{3}(-0.05) = 1 - 0.017 \approx 0.983$

31. Counting from 0, the sixth term is $\binom{15}{5}(x^2)^{10}y^5 = 3003x^{20}y^5$

33. Fifth term is $\binom{12}{4}(2x)^8(-y)^4 = 126{,}720x^8 y^4$

35. Term involving b^4 is the 5th term: $\binom{14}{4}a^{10}b^4 = 1001a^{10}b^4$

37. **(a)** $\left(1 - \frac{v^2}{c^2}\right)^{-\frac{1}{2}} = 1 + \left(-\frac{1}{2}\right)\left(-\frac{v^2}{c^2}\right) + \frac{\left(-\frac{1}{2}\right)\left(-\frac{3}{2}\right)}{2!}\left(-\frac{v^2}{c^2}\right)^2 = 1 + \frac{v^2}{2c^2} + \frac{3v^4}{8c^4}$

(b) $mc^2 + \frac{mv^2}{2} + \frac{3mv^4}{8c^2}$

39. $(r^2)^{3/2}\left(1 + \frac{\ell^2}{r^2}\right)^{3/2}$
$= r^3\left[1 + \frac{3}{2} \cdot \frac{\ell^2}{r^2} + \frac{\frac{3}{2} \cdot \frac{1}{2}}{2!}\left(\frac{\ell^2}{r^2}\right)^2\right]$
$= r^3\left[1 + \frac{3\ell^2}{2r^2} + \frac{3\ell^4}{8r^4}\right]$
$= r^3 + \frac{3}{2}r\ell^2 + \frac{3\ell^4}{8r}$

▤ CHAPTER 19 REVIEW

1. $\frac{1}{1+2} = \frac{1}{3}; \frac{1}{2+2} = \frac{1}{4}; \frac{1}{3+2} = \frac{1}{5}; \frac{1}{4+2} = \frac{1}{6}; \frac{1}{5+2} = \frac{1}{7}; \frac{1}{6+2} = \frac{1}{8}$

2. $\frac{3}{2-1} = 3; \frac{3}{4-1} = 1; \frac{3}{6-1} = \frac{3}{5}; \frac{3}{8-1} = \frac{3}{7}; \frac{3}{10-1} = \frac{1}{3}; \frac{3}{12-1} = \frac{3}{11}$

3. $\frac{-1}{3}; \frac{1}{2\cdot 5} = \frac{1}{10}; \frac{-1}{3(7)} = \frac{-1}{21}; \frac{1}{4(9)} = \frac{1}{36}; \frac{-1}{5(11)} = \frac{-1}{55};$
$\frac{1}{6(13)} = \frac{1}{78}$

4. $\frac{1+1}{3-1} = \frac{2}{2} = 1; \frac{5}{5} = 1; \frac{10}{8} = \frac{5}{4}; \frac{17}{11}; \frac{26}{14} = \frac{13}{7}; \frac{37}{17}$

5. $a_1 = 1; a_2 = \frac{1}{2}; a_3 = \frac{\frac{1}{2}}{3} = \frac{1}{6}; a_4 = \frac{\frac{1}{6}}{4} = \frac{1}{24}; a_5 =$
$\frac{\frac{1}{24}}{5} = \frac{1}{120}; a_6 = \frac{\frac{1}{120}}{6} = \frac{1}{720}$

6. $a_1 = 5; a_2 = 5 - 2 = 3; a_3 = 3 - 3 = 0; a_4 = 0 - 4 = -4; a_5 = -4 - 5 = -9; a_6 = -9 - 6 = -15$

7. $a_1 = 1; a_2 = 3; a_3 = 1 + 3 = 4; a_4 = 3 + 4 = 7; a_5 = 4 + 7 = 11; a_6 = 7 + 11 = 18$

8. $a_1 = 1; a_2 = 2; a_3 = 1 \cdot 2 = 2; a_4 = 2 \cdot 2 = 4; a_5 = 2 \cdot 4 = 8; a_6 = 4 \cdot 8 = 32$

9. Arithmetic; $d = -3, a_{10} = 10 + 9(-3) = -17$

10. Geometric: $r = \frac{1}{4}; a_8 = 8 \cdot \left(\frac{1}{4}\right)^7 = \frac{1}{2048}$

11. Arithmetic: $d = 4; a_7 = -1 + 6(4) = 23$

12. Neither; $7, 4, 0, -5, -11, -18, -26, -35, -45, = a_9$

13. Geometric: $r = -\frac{1}{6}; a_{10} = 3\left(-\frac{1}{6}\right)^9 = \frac{-1}{3,359,232} \approx$
$2.97687 \times 10^{-7} = 0.0000003$

14. Geometric: $r = 10; a_7 = 2.05 \times 10^6 = 2,050,000$

15. $d = 18 - 12 = 6; a_7 = a_1 + 6d; 12 = a_1 + 6 \cdot 6; a_1 = 12 - 36 = -24$

16. $r = \frac{\frac{9}{7}}{9} = \frac{1}{7}; a_8 = a_7 \cdot r = \frac{9}{7} \cdot \frac{1}{7} = \frac{9}{49}$

17. **(a)** $4 \cdot \frac{1}{3} + 4 \cdot \frac{1}{9} + 4 \cdot \frac{1}{27} + 4 \cdot \frac{1}{81} = \frac{4}{3} + \frac{4}{9} + \frac{4}{27} + \frac{4}{81}$

(b) $2 + \frac{3}{2} + \frac{4}{3} + \frac{5}{4}$

18. $2 + 3 + 4 + 5 = 14$

19. $-1 + 2 + 5 + 8 + 11 + 14 = 39$

20. $\frac{-1}{3} + \frac{2}{5} - \frac{3}{9} + \frac{4}{17} = \frac{-8}{255} = -0.03137$

21. Arithmetic: $d = 5; a_{10} = 4 + 9 \cdot 5 = 49; S_{10} = \frac{10}{2}(4 + 49) = 5(53) = 265$

22. Arithmetic: $d = -7; a_{12} = 2 + 11(-7) = -75; S_{12} = \frac{12}{2}(2 - 75) = 6(-73) = -438$

23. Geometric: $r = \frac{1}{3}; S_{14} = 1 \cdot \dfrac{1 - \frac{1}{3}^{14}}{1 - \frac{1}{3}} = 1.4999997$

24. Arithmetic: There are two possible ways to look at the sequence. One way is to let $a_1 = \sqrt{5} + 1$ and $d = \sqrt{5} + 1$. Then $a_8 = \sqrt{5} + 1 + (7)(\sqrt{5} + 1) = 8\sqrt{5} + 8$ and $S_8 = \frac{8}{2}(\sqrt{5} + 1 + 8\sqrt{5} + 8) = 4(9\sqrt{5} + 9) = 36\sqrt{5} + 36$. The other way to look at this sequence is to let $a_1 = \sqrt{5}$ and $d = 1 + \sqrt{5}$. Then, $a_8 = \sqrt{5} + 7(1 + \sqrt{5}) = 7 + 8\sqrt{5}$ and $S_8 = \frac{8}{2}(\sqrt{5} + 7 + 8\sqrt{5}) = 4(7 + 9\sqrt{5}) = 28 + 36\sqrt{5}$.

25. Converges: $r = \frac{1}{2}; S = \dfrac{\frac{1}{3}}{1 - \frac{1}{2}} = \frac{2}{3}$

26. Converges: $r = \frac{3}{4}; S = \dfrac{8}{1 - \frac{3}{4}} = 32$

27. Diverges: $r = 10$

28. Converges: $r = -\frac{1}{10}; S = \frac{1.5}{1 + 0.1} = \frac{1.5}{1.1} = \frac{15}{11} \approx 1.3636$

29. Converges: $r = \frac{1}{\sqrt{3}}; S = \dfrac{\frac{1}{\sqrt{3}}}{1 - \frac{1}{\sqrt{3}}} = \dfrac{\frac{1}{\sqrt{3}}}{\frac{\sqrt{3}-1}{\sqrt{3}}} =$
$\frac{1}{\sqrt{3}-1} = \frac{\sqrt{3}+1}{2} \approx 1.3660$

30. Converges: $r = \frac{1}{\sqrt{2}}; a_1 = 1; S = \dfrac{1}{1 - \frac{1}{\sqrt{2}}} =$
$\dfrac{1}{1 - \frac{\sqrt{2}-1}{\sqrt{2}}} = \frac{\sqrt{2}}{\sqrt{2}-1} = \frac{2+\sqrt{2}}{1} \approx 3.4142$

31. $a_1 = 0.185; r = 0.001; S = \frac{0.185}{0.999} = \frac{185}{999} = \frac{5}{27}$

32. Start with $a_1 = 0.01; r = 0.1; S = \frac{0.01}{.9} = \frac{1}{90}; 0.6 +$
$\frac{1}{90} = \frac{6}{10} + \frac{1}{90} = \frac{54}{90} + \frac{1}{90} = \frac{55}{90} = \frac{11}{18}$

33. $(a + 2)^5 = a^5 + 5a^4 \cdot 2 + 10a^3 \cdot 2^2 + 10a^2 \cdot 2^3 + 5a^1 \cdot 2^4 + 2^5 = a^5 + 10a^4 + 40a^3 + 80a^2 + 80a + 32$

34. $(3x - y)^6 = (3x)^6 + 6(3x)^5(-y) + 15(3x)^4(-y)^2 + 20(3x)^3(-y)^3 + 15(3x)^2(-y)^4 + 6(3x)(-y)^5 + (-y)^6$
$= 729x^6 - 1458x^5y + 1215x^4y^2 - 540x^3y^3 + 135x^2y^4 - 18xy^5 + y^6$

35. $\left(\frac{x}{2} - 3y^2\right)^6 = \left(\frac{x}{2}\right)^6 + 6\left(\frac{x}{2}\right)^5(-3y^2) +$
$15\left(\frac{x}{2}\right)^4(-3y^2)^2 + 20\left(\frac{x}{2}\right)^3(-3y^2)^3 + 15\left(\frac{x}{2}\right)^2(-3y^2)^4$
$+ 6\left(\frac{x}{2}\right)(-3y^2)^5 + (-3y^2)^6 = \frac{x^6}{64} - \frac{9}{16}x^5y^2 + \frac{135}{16}x^4y^4$
$- \frac{135}{2}x^3y^6 + \frac{1215}{4}x^2y^8 - 729xy^{10} + 729y^{12}$

36. $\left(\frac{2a^3}{5} + \frac{5b}{2}\right)^5 = \left(\frac{2a^3}{5}\right)^5 + 5\left(\frac{2a^3}{5}\right)^4\left(\frac{5b}{2}\right) +$
$10\left(\frac{2a^3}{5}\right)^3\left(\frac{5b}{2}\right)^2 + 10\left(\frac{2a^3}{5}\right)^2\left(\frac{5b}{2}\right)^3 + 5\left(\frac{2a^3}{5}\right)\left(\frac{5b}{2}\right)^4 +$
$\left(\frac{5b}{2}\right)^5 = \frac{32a^{15}}{3125} + \frac{8a^{12}b}{25} + 4a^9b^2 + 25a^6b^3 + \frac{625a^3b^4}{8} +$
$\frac{3125b^5}{32}$

37. $(2x + y)^{15} = (2x)^{15} + 15(2x)^{14}y + \binom{15}{2}(2x)^{13}y^2 +$
$\binom{15}{3}(2x)^{12}y^3 + \cdots = 32{,}768x^{15} + 245{,}760x^{14}y +$
$860{,}160x^{13}y^2 + 1{,}863{,}680x^{12}y^3 + \cdots$

38. $(1 + ax^2)^{10} = 1^{10} + 10 \cdot 1^9(ax^2) + \binom{10}{2}1^8(ax^2)^2 +$
$\binom{10}{3}1^7(ax^2)^3 + \cdots = 1 + 10ax^2 + 45a^2x^4 + 120a^3x^6$
$+ \cdots$

39. $(1 - x)^{-5} = 1 - 5(-x) + \frac{(-5)(-6)}{2!}x^2 +$
$\frac{(-5)(-6)(-7)}{3!}(-x)^3 + \cdots = 1 + 5x + 15x^2 + 35x^3 + \cdots$

40. $(1 + b)^{-\frac{1}{4}} = 1 - \frac{1}{4}b + \frac{\left(-\frac{1}{4}\right)\left(-\frac{5}{4}\right)}{2!}b^2 +$
$\frac{\left(-\frac{1}{4}\right)\left(-\frac{5}{4}\right)\left(-\frac{9}{4}\right)}{3!}b^3 + \cdots = 1 - \frac{b}{4} + \frac{5b^2}{32} - \frac{15b^3}{128} + \cdots$

41. $\binom{20}{6}x^{14}(2y)^6 = 2{,}480{,}640x^{14}y^6$

42. $\binom{12}{4}a^8(-3b)^4 = 40{,}095a^8b^4$

43. $\sqrt[5]{1.02} = (1 + 0.02)^{\frac{1}{5}} = 1 + \frac{1}{5}(0.02) + \frac{\frac{1}{5}\left(-\frac{4}{5}\right)}{2!}(0.02)^2$
$= 1 + 0.004 - 0.000032 \approx 1.004$

44. $\sqrt[7]{0.98} = (1 - 0.02)^{\frac{1}{7}} = 1 - \frac{1}{7}(0.02) \approx 1 - 0.003 = 0.997$

45. $\frac{1}{2}\%$ per month $= 0.005$; 10 years $= 120$ months;
$A = 400(1.005)^{120} = \$727.76$

46. Geometric sequence: $a_1 = 20; r = \frac{3}{4}; a_6 = 20 \cdot$
$\left(\frac{3}{4}\right)^5 = 4.75$ cm

47. $S_6 = 20 \cdot \frac{1 - \left(\frac{3}{4}\right)^6}{1 - \frac{3}{4}} = 65.76$ cm

48. $S = \frac{20}{1 - \frac{3}{4}} = \frac{20}{\frac{1}{4}} = 80$ cm

49. First convert 250 ft to inches. Since $250 \times 12 = 3000$, we see that 250 ft $= 3000$ in. $S_n = \frac{n}{2}(a_1 + a_n); a_n = a_1 + (n-1)d$ so $a_n = 10 + (n-1)15 = -5 + 15n$. Substituting, we get $S_n = \frac{n}{2}(10 - 5 + 15n) = \frac{n}{2}(5 + 15n)\frac{5n + 15n^2}{2} = 3000$. Hence, $15n^2 + 5n - 6000 = 0$. Dividing by 5, we have $3n^2 + n - 1200 = 0$. The quadratic formula gives answers of -20.167 and 19.834. Since n must be positive, the answer is 19.834 s

50. $a_{10} = 250{,}000 + 9(40{,}000) + \$610{,}000$. $S_{10} = \frac{10}{2}(250{,}000 + 610{,}000) = \$4{,}300{,}000$

≡ CHAPTER 19 TEST

1. $\frac{5}{-1} = -5; \frac{5}{1} = 5; \frac{5}{3}; \frac{5}{5} = 1; \frac{5}{7}; \frac{5}{9}$

2. $a_1 = -2; a_2 = 3 + 2(-2) = -1; a_3 = 3 + 3(-1) = 0; a_4 = 3 + 4 \cdot 0 = 3; a_5 = 3 + 5 \times 3 = 18; a_6 = 3 + 6 \times 18 = 111$

3. Arithmetic: $d = -2.5; a_{10} = 15 + 9(-2.5) = -7.5$

4. $d = 14 - 10 = 4; a_5 = a_1 + 4d; 10 = a_1 + 4 \cdot 4; a_1 = -6$

5. $\sum_{k=1}^{5}(2k - 1) = 1 + 3 + 5 + 7 + 9 = 25$

6. Converges: $r = -\frac{1}{2}; S = \frac{\frac{1}{5}}{1 + \frac{1}{2}} = \frac{\frac{1}{5}}{\frac{3}{2}} = \frac{2}{15}$

7. $a_1 = 0.435; r = 0.001; S = \frac{0.435}{0.999} = \frac{435}{999} = \frac{145}{333}$

8. $(3x - 2y)^{12} = (3x)^{12} - 12(3x)^{11}(2y) + \binom{12}{2}(3x)^{10}(2y)^2$
$- \binom{12}{3}(3x)^9(2y)^3 + \cdots = 531{,}441x^{12} - 4{,}251{,}528x^{11}$
$y + 15{,}588{,}936x^{10}y^2 - 34{,}642{,}080x^9y^3$

9. Arithmetic Series: $a_1 = 2000; d = 1500; a_{12} = 2000 + 11(1500) = 18{,}500; S_{12} = \frac{12}{2}(2000 + 18{,}500) = 123{,}000$ copies

10. Using $a_1 = 1.4$ and $n = 11$, we have that $a_n = a_1r^{10} = 1.40 \times 1.01^{10} = 1.55$ nA.

20

Trigonometric Formulas, Identities and Equations

≡ 20.1 BASIC IDENTITIES

1. $\sin^2\theta + \cos^2\theta = 1$; Dividing by $\sin^2\theta$ produces
$\frac{\sin^2\theta}{\sin^2\theta} + \frac{\cos^2\theta}{\sin^2\theta} = \frac{1}{\sin^2\theta}$ or $1 + \cot^2\theta = \csc^2\theta$

3. $\sin\theta\sec\theta = \tan\theta$

$$
\begin{array}{c|c}
\sin\theta\frac{1}{\cos\theta} & \tan\theta \\
\frac{\sin\theta}{\cos\theta} & \\
\tan\theta &
\end{array}
$$

5.
$$\frac{\sin\theta}{\cot\theta} = \sec\theta - \cos\theta$$

$$
\begin{array}{c|c}
\sin\theta & \frac{1}{\cos\theta} - \cos\theta \\
\hline
\frac{\cos\theta}{\sin\theta} & \\
\frac{\sin^2\theta}{\cos\theta} & \\
\frac{1-\cos^2\theta}{\cos\theta} & \\
\frac{1}{\cos\theta} - \cos\theta &
\end{array}
$$

7.
$$(1 - \sin^2\theta)(1 + \tan^2\theta) = 1$$

$$
\begin{array}{c|c}
1 + \tan^2\theta - \sin^2\theta - \sin^2\theta\tan^2\theta & 1 \\
1 + \frac{\sin^2\theta}{\cos^2\theta} - \sin^2\theta - \sin^2\theta\cdot\frac{\sin^2\theta}{\cos^2\theta} & \\
\frac{\cos^2\theta}{\cos^2\theta} + \frac{\sin^2\theta}{\cos^2\theta} - \frac{\sin^2\theta\cos^2\theta}{\cos^2\theta} - \frac{\sin^2\theta\sin^2\theta}{\cos^2\theta} & \\
\frac{\cos^2\theta+\sin^2\theta-\sin^2\theta(\cos^2\theta+\sin^2\theta)}{\cos^2\theta} & \\
\frac{1-\sin^2\theta}{\cos^2\theta} & \\
\frac{\cos^2\theta}{\cos^2\theta} & \\
1 &
\end{array}
$$

9.
$$1 - \frac{\sin A}{\csc A} = \cos^2 A$$

$$
\begin{array}{c|c}
1 - \frac{\sin A}{1/\sin A} & \cos^2 A \\
1 - \sin^2 A & \\
\cos^2 A &
\end{array}
$$

11. $(1 + \cos x)(1 - \cos x) = \sin^2 x$

$$
\begin{array}{c|c}
1 - \cos^2 x & \sin^2 x \\
\sin^2 x &
\end{array}
$$

13. $2\csc\theta = \frac{\sin\theta}{1+\cos\theta} + \frac{1+\cos\theta}{\sin\theta}$

$$
\begin{array}{c|c}
2\csc\theta & \frac{\sin^2\theta}{\sin\theta(1+\cos\theta)} + \frac{(1+\cos\theta)(1+\cos\theta)}{\sin\theta(1+\cos\theta)} \\
& \frac{\sin^2\theta+1+2\cos\theta+\cos^2\theta}{\sin\theta(1+\cos\theta)} \\
& \frac{\sin^2\theta+\cos^2\theta+1+2\cos\theta}{\sin\theta(1+\cos\theta)} \\
& \frac{2+2\cos\theta}{\sin\theta(1+\cos\theta)} \\
& \frac{2(1+\cos\theta)}{\sin\theta(1+\cos\theta)} \\
& \frac{2}{\sin\theta} \\
& 2\csc\theta
\end{array}
$$

15.
$$(\sin\theta + \cos\theta)^2 = 1 + 2\sin\theta\cos\theta$$

$$
\begin{array}{c|c}
\sin^2\theta + 2\sin\theta\cos\theta + \cos^2\theta & 1 + 2\sin\theta\cos\theta \\
\sin^2\theta + \cos^2\theta + 2\sin\theta\cos\theta & \\
1 + 2\sin\theta\cos\theta &
\end{array}
$$

17.
$$\frac{\tan\theta+\cot\theta}{\tan\theta-\cot\theta} = \frac{\tan^2\theta+1}{\tan^2\theta-1}$$

$$
\begin{array}{c|c}
\frac{\frac{\sin\theta}{\cos\theta} + \frac{\cos\theta}{\sin\theta}}{\frac{\sin\theta}{\cos\theta} - \frac{\cos\theta}{\sin\theta}} & \frac{\frac{\sin^2\theta}{\cos^2\theta} + \frac{\cos^2\theta}{\cos^2\theta}}{\frac{\sin^2\theta}{\cos^2\theta} - \frac{\cos^2\theta}{\cos^2\theta}} \\
\frac{\frac{\sin^2\theta+\cos^2\theta}{\sin\theta\cos\theta}}{\frac{\sin^2\theta-\cos^2\theta}{\sin\theta\cos\theta}} & \frac{\sin^2\theta+\cos^2\theta}{\sin^2\theta-\cos^2\theta} \\
\frac{\sin^2\theta+\cos^2\theta}{\sin^2\theta-\cos^2\theta} &
\end{array}
$$

19.
$$\frac{\sec\theta-\csc\theta}{\sec\theta+\csc\theta} = \frac{\tan\theta-1}{\tan\theta+1}$$

$$
\begin{array}{c|c}
\frac{\frac{1}{\cos\theta} - \frac{1}{\sin\theta}}{\frac{1}{\cos\theta} + \frac{1}{\sin\theta}} & \frac{\tan\theta-1}{\tan\theta+1} \\
\frac{\left(\frac{1}{\cos\theta} - \frac{1}{\sin\theta}\right)\sin\theta}{\left(\frac{1}{\cos\theta} + \frac{1}{\sin\theta}\right)\sin\theta} & \\
\frac{\frac{\sin\theta}{\cos\theta} - \frac{\sin\theta}{\sin\theta}}{\frac{\sin\theta}{\cos\theta} + \frac{\sin\theta}{\sin\theta}} & \\
\frac{\tan\theta-1}{\tan\theta+1} &
\end{array}
$$

21. $\tan^2 x \cos^2 x + \cot^2 x \sin^2 x = 1$

$$
\begin{array}{c|c}
\frac{\sin^2 x}{\cos^2 x} \cdot \frac{\cos^2 x}{1} + \frac{\cos^2 x}{\sin^2 x} \cdot \frac{\sin^2 x}{1} & 1 \\
\sin^2 x + \cos^2 x & \\
1 &
\end{array}
$$

23. $\sec^4 x - \sec^2 x = \tan^2 x \sec^2 x$

$$
\begin{array}{c|c}
\sec^2 x (\sec^2 x - 1) & \sec^2 x \tan^2 x \\
\sec^2 x \tan^2 x &
\end{array}
$$

25.

$$
\begin{array}{c|c}
\frac{\tan \theta - \sin \theta}{\sin^3 \theta} & = \frac{\sec \theta}{1 + \cos \theta} \\
\frac{\frac{\sin \theta}{\cos \theta} - \sin \theta}{\sin \theta \cdot \sin^2 \theta} & \frac{\sec \theta}{1 + \cos \theta} \\
\frac{\sin \theta \left(\frac{1}{\cos \theta} - 1 \right)}{\sin \theta (1 - \cos^2 \theta)} & \\
\frac{\frac{1 - \cos \theta}{\cos \theta}}{(1 + \cos \theta)(1 - \cos \theta)} & \\
\frac{\frac{1}{\cos \theta}}{1 + \cos \theta} & \\
\frac{\sec \theta}{1 + \cos \theta} &
\end{array}
$$

27. $\tan^2 \theta \csc^2 \theta \cot^2 \theta \sin^2 \theta = 1$

$$
\begin{array}{c|c}
\frac{\sin^2 \theta}{\cos^2 \theta} \cdot \frac{1}{\sin^2 \theta} \cdot \frac{\cos^2 \theta}{\sin^2 \theta} \cdot \frac{\sin^2 \theta}{1} & 1 \\
1 &
\end{array}
$$

29.

$$
\begin{array}{c|c}
\frac{\sec A + \csc A}{\tan A + \cot A} & = \sin A + \cos A \\
\frac{\left(\frac{1}{\cos A} + \frac{1}{\sin A} \right) \sin A \cos A}{\left(\frac{\sin A}{\cos A} + \frac{\cos A}{\sin A} \right) \sin A \cos A} & \sin A + \cos A \\
\frac{\sin A + \cos A}{\sin^2 A + \cos^2 A} & \\
\frac{\sin A + \cos A}{1} & \\
\sin A + \cos A &
\end{array}
$$

31. This identity is true based on the graphs of $y = 2 \csc 2x$ and $y = \sec x \csc x$ shown below.

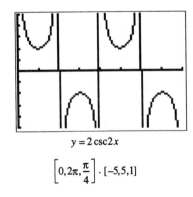

$y = 2\csc 2x$

$\left[0, 2\pi, \frac{\pi}{4} \right] \cdot [-5, 5, 1]$

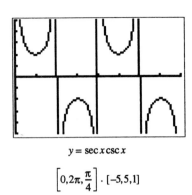

$y = \sec x \csc x$

$\left[0, 2\pi, \frac{\pi}{4} \right] \cdot [-5, 5, 1]$

33. This identity is not true based on the graphs of $y = \sin \frac{1}{2} x$ and $y = \frac{1}{2} \sin x$ shown below.

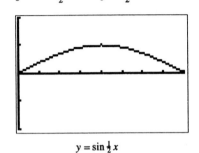

$y = \sin \frac{1}{2} x$

$\left[0, 2\pi, \frac{\pi}{4} \right] \cdot [-2, 2, 1]$

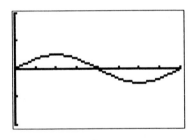

$y = \frac{1}{2} \sin x$

$\left[0, 2\pi, \frac{\pi}{4} \right] \cdot [-2, 2, 1]$

35. For $2 \sin \theta \neq \sin 2\theta$ let $\theta = 90°$. $2 \sin 90° \neq \sin(2 \cdot 90°)$, $2 \times 1 \neq \sin 180°$ and $2 \neq 0$.

37. If $\theta = \pi$, then we have $\cos \pi^2 \neq (\cos \pi)^2$ or $-0.90269 \neq (-1)^2$, which is equivalent to $-0.90269 \neq 1$.

39. Here we let $x = 120°$. Now, $\sin 120° = \frac{\sqrt{3}}{2}$ and $\frac{\tan 120°}{\sqrt{1 + \tan^2 120°}} = \frac{-\sqrt{3}}{\sqrt{1 + (-\sqrt{3})^2}}$, which is negative. Hence, we see that $\sin x \neq \frac{\tan x}{\sqrt{1 + \tan^2 x}}$ when $x = 120°$.

41. $-2\cot x\csc^2 x = -2\frac{\cos x}{\sin x}\cdot\csc^2 x = -2\cos x\cdot\frac{1}{\sin x}\cdot$
$\csc^2 x = -2\cos x\csc x\csc^2 x = -2\cos x\csc^3 x$

43. $\frac{(1.2\sin\omega t-1.6\cos\omega t)^2+(1.6\sin\omega t+1.2\cos\omega t)^2}{2L} =$

$\frac{1.2^2\sin^2\omega t-2(1.2)(1.6)\sin\omega t\cos\omega t+(1.6)^2\cos^2\omega t}{2L} +$

$\frac{(1.6)^2\sin^2\omega t+2(1.6)(1.2)\sin\omega t\cos\omega t+(1.2)^2\cos^2\omega t}{2L} =$

$\frac{(1.2)^2+(1.6)^2}{2L} = \frac{1.44+2.56}{2L} = \frac{4.00}{2L} = \frac{2.0}{L}$

20.2 THE SUM AND DIFFERENCE IDENTITIES

1. $\sin 15° = \sin(45° - 30°) = \sin 45°\cos 30° -$
$\cos 45°\sin 30° = \frac{\sqrt 2}{2}\cdot\frac{\sqrt 3}{2} - \frac{\sqrt 2}{2}\cdot\frac{1}{2} = \frac{\sqrt 6-\sqrt 2}{4}$

3. $\sin 120° = \sin(60° + 60°) = \sin 60°\cos 60° +$
$\cos 60°\sin 60° = \frac{\sqrt 3}{2}\cdot\frac{1}{2}+\frac{1}{2}\cdot\frac{\sqrt 3}{2} = \frac{\sqrt 3}{4}+\frac{\sqrt 3}{4} = \frac{\sqrt 3}{2}$

5. $\tan 15° = \tan(45° - 30°) = \frac{\tan 45° - \tan 30°}{1+\tan 45°\tan 30°} =$

$\frac{1 - \frac{\sqrt 3}{3}}{1+1\cdot\frac{\sqrt 3}{3}} = \frac{\frac{3-\sqrt 3}{3}}{\frac{3+\sqrt 3}{3}} = \frac{3-\sqrt 3}{3+\sqrt 3} = \frac{3-\sqrt 3}{3+\sqrt 3}\cdot\frac{3-\sqrt 3}{3-\sqrt 3} =$

$\frac{9-6\sqrt 3+3}{9-3} = \frac{12-6\sqrt 3}{6} = 2-\sqrt 3$

7. $\sin 150° = \sin(180° - 30°) = \sin 180°\cos 30° -$
$\cos 180°\sin 30° = 0\cdot\frac{\sqrt 3}{2} - (-1)\cdot\frac{1}{2} = \frac{1}{2}$

9. $\sin(x+90°) = \sin x\cos 90°+\cos x\sin 90° = \sin x\cdot$
$0 + \cos x\cdot 1 = \cos x$

11. $\cos\left(x + \frac{\pi}{2}\right) = \cos x\cos\frac{\pi}{2} - \sin x\sin\frac{\pi}{2} = \cos x\cdot$
$(0) - \sin x\cdot(1) = -\sin x$

13. $\cos(\pi - x) = \cos\pi\cos x+\sin\pi\sin x = -1\cos x+$
$0\sin x = -\cos x$

15. $\sin(180° - x) = \sin 180°\cos x - \cos 180°\sin x =$
$0\cos x - (-1)\sin x = \sin x$

17. $\sin(\alpha + \beta) = \sin\alpha\cos\beta + \cos\alpha\sin\beta = \frac{3}{4}\cdot\frac{7}{8} +$
$\frac{\sqrt 7}{4}\cdot\frac{\sqrt{15}}{8} = \frac{21+\sqrt{105}}{32} \approx 0.97647$

19. $\tan(\alpha + \beta) = \frac{\sin(\alpha+\beta)}{\cos(\alpha+\beta)} = \frac{21+\sqrt{105}}{7\sqrt 7-3\sqrt{15}} = 4.52768$

or $\tan(\alpha + \beta) = \frac{\tan\alpha+\tan\beta}{1-\tan\alpha-\tan\beta} = \frac{\frac{3\sqrt 7}{7} + \frac{\sqrt{15}}{7}}{1 - \frac{3\sqrt 7}{7}\cdot\frac{\sqrt{15}}{7}} =$

$\frac{\frac{3\sqrt 7+\sqrt{15}}{7}}{\frac{49-3\sqrt{105}}{49}} = \frac{7(3\sqrt 7+\sqrt{15})}{49-3\sqrt{105}} \approx 4.52768$

21. $\cos(\alpha - \beta) = \cos\alpha\cos\beta + \sin\alpha\sin\beta = \frac{\sqrt 7}{4}\cdot\frac{7}{8} +$
$\frac{3}{4}\cdot\frac{\sqrt{15}}{8} = \frac{7\sqrt 7+3\sqrt{15}}{32} \approx 0.94185$

23. Since both $\sin(\alpha + \beta)$ and $\cos(\alpha + \beta)$ are positive $(\alpha - \beta)$ is in Quadrant I.

25. $\sin(\alpha + \beta) = \sin\alpha\cos\beta + \cos\alpha\sin\beta = \frac{3}{4}\cdot\frac{-7}{8} +$
$\frac{-\sqrt 7}{4}\cdot\frac{-\sqrt{15}}{8} = \frac{-21+\sqrt{105}}{32} = \frac{\sqrt{105}-21}{32} \approx -0.33603$

27. $\tan(\alpha + \beta) = \frac{\sin(\alpha+\beta)}{\cos(\alpha+\beta)} = \frac{\sqrt{105}-21}{7\sqrt 7+3\sqrt{15}} \approx -0.35678$

or $\tan(\alpha + \beta) = \frac{\tan\alpha+\tan\beta}{1-\tan\alpha\tan\beta} = \frac{\frac{-3\sqrt 7}{7} + \frac{\sqrt{15}}{7}}{1 - \frac{-3\sqrt 7}{7}\cdot\frac{\sqrt{15}}{7}} =$

$\frac{7(\sqrt{15}-3\sqrt 7)}{49+3\sqrt{105}} \approx -0.35678$

29. $\sin(\alpha - \beta) = \cos\alpha\cos\beta+\sin\alpha\sin\beta = \frac{-7}{8}\cdot\frac{-\sqrt 7}{4} +$
$\frac{3}{4}\cdot\frac{-\sqrt{15}}{8} = \frac{7\sqrt 7-3\sqrt{15}}{32} \approx 0.21567$

31. Since $\sin(\alpha + \beta)$ is negative and $\cos(\alpha + \beta)$ is positive, $(\alpha + \beta)$ is in Quadrant IV.

33. $\sin 47°\cos 13°+\cos 47°\sin 13° = \sin(47°+13°) =$
$\sin 60° = \frac{\sqrt 3}{2}$

35. $\cos 32°\cos 12°-\sin 32°\sin 12° = \cos(32°+12°) =$
$\cos 44°$

37. $\cos(\alpha + \beta)\cos\beta + \sin(\alpha + \beta)\sin\beta = \cos(\alpha + \beta -$
$\beta) = \cos\alpha$

39. $\cos(x + y)\cos(x - y) - \sin(x + y)\sin(x - y) =$
$\cos(x + y + x - y) = \cos 2x = \cos x\cos x -$
$\sin x\sin x = \cos^2 x - \sin^2 x = \cos 2x$

41.

$\sin(x + y)\sin(x - y) = \sin^2 x - \sin^2 y$	
$(\sin x\cos y + \cos x\sin y)$	$\sin^2 x - \sin^2 y$
$\cdot(\sin x\cos y - \cos x\sin y)$	
$\sin^2 x\cos^2 y - \cos^2 x\sin^2 y$	
$\sin^2 x(1 - \sin^2 y)$	
$-(1 - \sin^2 x)\sin^2 y$	
$\sin^2 x - \sin^2 x\sin^2 y$	
$-\sin^2 y + \sin^2 y\sin^2 x$	
$\sin^2 x - \sin^2 y$	

43.

$\cos\theta = \sin(\theta + 30°) + \cos(\theta + 60°)$	
$\cos\theta$	$\sin\theta\cos 30° + \cos\theta\sin 30°$
	$+ \cos\theta\cos 60° - \sin\theta\sin 60°$
	$\frac{\sqrt 3}{2}\sin\theta + \frac{1}{2}\cos\theta + \frac{1}{2}\cos\theta - \frac{\sqrt 3}{2}\sin\theta$
	$\cos\theta$

45. $\tan x - \tan y = \frac{\sin(x-y)}{\cos x \cos y}$

$$
\begin{array}{c|c}
\tan x - \tan y & \frac{\sin x \cos y - \cos x \sin y}{\cos x \cos y} \\
\hline
& \frac{\sin x \cos y}{\cos x \cos y} - \frac{\cos x \sin y}{\cos x \cos y} \\
\hline
& \tan x - \tan y
\end{array}
$$

47. Both sides equal 0.8386705679

49. Both sides equal 14.10141995

51. $\sin(55° + 37°) = 0.99939$;
$\sin 55° + \sin 37° = 1.420967$; not equal

53. $\cos(40° + 35°) = 0.258819$;
$\cos 40° + \cos 35° = 1.585196$; not equal

55. $[r_1(\cos\theta_1 + j\sin\theta_1)][r_2(\cos\theta_2 + j\sin\theta_2)] = r_1 r_2[\cos\theta_1\cos\theta_2 + j\cos\theta_2\sin\theta_1 + j\sin\theta_2\cos\theta_1 + j^2\sin\theta_1\sin\theta_2] = r_1 r_2[(\cos\theta_1\cos\theta_2 - \sin\theta_1\sin\theta_2) + j(\sin\theta_1\cos\theta_2 + \cos\theta_1\sin\theta_2)] = r_1 r_2[\cos(\theta_1 + \theta_2) + j\sin(\theta_1 + \theta_2)]$

57. $d = \frac{h}{\cos\theta_r} \cdot \sin(\theta_i - \theta_r) =$
$\frac{h}{\cos\theta_r}[\sin\theta_i\cos\theta_r - \cos\theta_i\sin\theta_r] =$
$h\left[\sin\theta_i\cos\theta_r \cdot \frac{1}{\cos\theta_r} - \cos\theta_i\sin\theta_r \cdot \frac{1}{\cos\theta_r}\right] =$
$h[\sin\theta_i - \cos\theta_i\tan\theta_r]$

▤ 20.3 THE DOUBLE- AND HALF-ANGLE IDENTITIES

1. $\cos 15° = \cos\frac{30°}{2} = \sqrt{\frac{1+\cos 30°}{2}} = \sqrt{\frac{1+\sqrt{3}/2}{2}} = \sqrt{\frac{2+\sqrt{3}}{4}} \approx 0.96593$

3. $\sin 15° = \sin\frac{30°}{2} = \sqrt{\frac{1-\cos 30°}{2}} = \sqrt{\frac{1-\sqrt{3}/2}{2}} = \sqrt{\frac{2-\sqrt{3}}{4}} \approx 0.25882$

5. $\sin 105° = \sin\frac{210°}{2} = \sqrt{\frac{1-\cos 210°}{2}} = \sqrt{\frac{1+\sqrt{3}/2}{2}} = \sqrt{\frac{2+\sqrt{3}}{4}} \approx 0.96593$

7. $\cos 7\frac{1}{2}° = \cos\frac{15°}{2} = \sqrt{\frac{1+\cos 15°}{2}} = \sqrt{\frac{1+\sqrt{\frac{2+\sqrt{3}}{4}}}{2}} = \sqrt{\frac{2+\sqrt{2+\sqrt{3}}}{4}} \approx 0.99145$

9. $\tan 22\frac{1}{2}° = \tan\frac{45°}{2} = \sqrt{\frac{1-\cos 45°}{1+\cos 45°}} = \sqrt{\frac{1-\sqrt{2}/2}{1+\sqrt{2}/2}} = \sqrt{\frac{2-\sqrt{2}}{2+\sqrt{2}}} \approx 0.41421$

11. $\cos 75° = \cos\frac{150°}{2} = \sqrt{\frac{1+\cos 150°}{2}} = \sqrt{\frac{1-\frac{\sqrt{3}}{2}}{2}} = \frac{\sqrt{2-\sqrt{3}}}{2} \approx 0.25882$

13. $\sin 127\frac{1}{2}° = \sin(105° + 22.5°)$. From exercise 5, we have $\sin 105° = \frac{\sqrt{2+\sqrt{3}}}{2}$, and from exercise 4, $\cos 105° = \frac{-\sqrt{2+\sqrt{3}}}{2}$. We determine that

$\sin 22\frac{1}{2}° = \sin\frac{45°}{2} = \sqrt{\frac{1-\frac{\sqrt{2}}{2}}{2}} = \frac{\sqrt{2-\sqrt{2}}}{2}$ and
that $\cos 22\frac{1}{2}° = \cos\frac{45°}{2} = \sqrt{\frac{1+\frac{\sqrt{2}}{2}}{2}} = \frac{\sqrt{2+\sqrt{2}}}{2}$.
Using these values, we find the desired result as follows:
$\sin(105° + 22.5°) = \sin 105°\cos 22.5° + \cos 105°\sin 22.5° = \sin\left(\frac{\sqrt{2+\sqrt{3}}}{2}\right)\left(\frac{\sqrt{2+\sqrt{2}}}{2}\right) + \left(\frac{-\sqrt{2-\sqrt{3}}}{2}\right)\left(\frac{\sqrt{2-\sqrt{2}}}{2}\right)$. This "simplifies" as
$\frac{1}{4}\left[\sqrt{4+2\sqrt{3}+2\sqrt{2}+\sqrt{6}} - \sqrt{4-2\sqrt{2}-2\sqrt{3}+\sqrt{6}}\right] \approx 0.79335$

15. $\sin x = \frac{7}{25}$; $\cos x = -\sqrt{1-\left(\frac{7}{25}\right)^2} = -\frac{24}{25}$; $\sin 2x = 2\sin x\cos x = 2\cdot\frac{7}{25}\cdot\frac{-24}{25} = \frac{-336}{625}$; $\cos 2x = \cos^2 x - \sin^2 x = \left(\frac{24}{25}\right)^2 - \left(\frac{7}{25}\right)^2 = \frac{527}{625}$; $\tan 2x = \frac{-336}{527}$;
$\sin\frac{x}{2} = \sqrt{\frac{1-\cos x}{2}} = \sqrt{\frac{1+\frac{24}{25}}{2}} = \sqrt{\frac{49}{50}} = \frac{7\sqrt{2}}{10}$;
$\cos\frac{x}{2} = \sqrt{\frac{1+\cos x}{2}} = \sqrt{\frac{1-\frac{24}{25}}{2}} = \sqrt{\frac{1}{50}} = \frac{\sqrt{2}}{10}$;
$\tan\frac{x}{2} = \frac{7\sqrt{2}}{\sqrt{2}} = 7$

17. $\sec x = \frac{29}{20}$; $\cos x = \frac{20}{29}$; $\sin x = \frac{21}{29}$; $\sin 2x = 2\left(\frac{21}{29}\right)\left(\frac{20}{29}\right) = \frac{840}{841}$; $\cos 2x = \cos^2 x - \sin^2 x =$

$\frac{400-441}{841} = \frac{-41}{841}$; $\tan 2x = -\frac{840}{41}$; $\sin \frac{x}{2} = \sqrt{\frac{1-\frac{20}{29}}{2}}$

$= \sqrt{\frac{9}{58}} = \frac{3}{\sqrt{58}}$; $\cos \frac{x}{2} = \sqrt{\frac{1+\frac{20}{29}}{2}} = \sqrt{\frac{49}{58}} = \frac{7}{\sqrt{58}}$;

$\tan \frac{x}{2} = \frac{3}{7}$

19. $\tan x = \frac{35}{12}$; $\sqrt{u^2+v^2} = \sqrt{35^2+12^2} = 37$; $\sin x = -\frac{35}{37}$; $\cos x = -\frac{12}{37}$; $\sin 2x = 2\left(\frac{-35}{37}\right)\left(\frac{-12}{37}\right) = \frac{840}{1369}$;

$\cos 2x = 1 - 2\sin^2 x = 1 - 2\left(\frac{-35}{37}\right)^2 = \frac{-1081}{1369}$;

$\tan 2x = -\frac{840}{1081}$; $\sin \frac{x}{2} = \sqrt{\frac{1+\frac{12}{37}}{2}} = \sqrt{\frac{49}{74}} = \frac{7}{\sqrt{74}}$; $\cos \frac{x}{2} = -\sqrt{\frac{1-\frac{12}{37}}{2}} = -\sqrt{\frac{25}{74}} = \frac{-5}{\sqrt{74}}$; $\tan \frac{x}{2} = -\frac{7}{5}$

21.

$\cos^2 x = \sin^2 x + \cos 2x$

$\cos^2 x$	$\sin^2 x + 2\cos^2 x - 1$
	$\sin^2 x + \cos^2 x + \cos^2 x - 1$
	$1 + \cos^2 x - 1$
	$\cos^2 x$

23.

$\cos 4x = 1 - 8\sin^2 x \cos^2 x$

$\cos(2\cdot 2x)$	$1 - 8\sin^2 x \cos^2 x$
$1 - 2\sin^2 2x$	
$1 - 2(2\sin x \cos x)^2$	
$1 - 2(4\sin^2 x \cos^2 x)$	
$1 - 8\sin^2 x \cos^2 x$	

25. $\frac{1+\tan^2 \alpha}{1-\tan^2 \alpha} = \sec 2\alpha$

$\frac{1+\tan^2 \alpha}{1-\tan^2 \alpha}$	$\frac{1}{\cos 2\alpha}$
	$\frac{1}{\cos^2 \alpha - \sin^2 \alpha}$
	$\frac{\cos^2 \alpha + \sin^2 \alpha}{\cos^2 \alpha - \sin^2 \alpha}$
	$\frac{\frac{\cos^2 \alpha}{\cos^2 \alpha} + \frac{\sin^2 \alpha}{\cos^2 \alpha}}{\frac{\cos^2 \alpha}{\cos^2 \alpha} - \frac{\sin^2 \alpha}{\cos^2 \alpha}}$
	$\frac{1+\tan^2 \alpha}{1-\tan^2 \alpha}$

27.

$1 - 2\sin^2 3x = \cos 6x$

$1 - 2\sin^2 3x$	$\cos(2\cdot 3x)$
	$1 - 2\sin^2 3x$

29. $\sin^2 x \cos^2 x = \frac{1}{4}\sin^2 2x$

$\sin^2 x \cos^2 x$	$\frac{1}{4}[2\sin x \cos x]^2$
	$\frac{1}{4}(4\sin^2 x \cos^2 x)$
	$\sin^2 x \cos^2 x$

31. $\tan\left(\frac{\alpha+\beta}{2}\right)\cot\left(\frac{\alpha-\beta}{2}\right) = \left(\frac{1-\cos(\alpha+\beta)}{\sin(\alpha+\beta)}\right)\left(\frac{1+\cos(\alpha-\beta)}{\sin(\alpha-\beta)}\right)$

$= \left(\frac{1-\cos\alpha\cos\beta+\sin\alpha\sin\beta}{\sin\alpha\cos\beta+\cos\alpha\sin\beta}\right)\left(\frac{1+\cos\alpha\cos\beta+\sin\alpha\sin\beta}{\sin\alpha\cos\beta-\cos\alpha\sin\beta}\right)$

$= \frac{1+2\sin\alpha\sin\beta+\sin^2\alpha\sin^2\beta-\cos^2\alpha\cos^2\beta}{\sin^2\alpha\cos^2\beta-\cos^2\alpha\sin^2\beta}.$

In the denominator, let $\cos^2\beta = 1 - \sin^2\beta$ and $\cos^2\alpha = 1 - \sin^2\alpha$ to produce

$\frac{1+2\sin\alpha\sin\beta+\sin^2\alpha\sin^2\beta-(1-\sin^2\alpha)(1-\sin^2\beta)}{\sin^2\alpha(1-\sin^2\beta)-(1-\sin^2\alpha)\sin^2\beta}$

$= \frac{1+\alpha\sin\alpha\sin\beta+\sin^2\alpha\sin^2\beta-1+\sin^2\alpha+\sin^2\beta-\sin^2\alpha\sin^2\beta}{\sin^2\alpha-\sin^2\alpha\sin^2\beta-\sin^2\beta+\sin^2\alpha\sin^2\beta}$

$= \frac{\sin^2\alpha+2\sin\alpha\sin\beta+\sin^2\beta}{\sin^2\alpha-\sin^2\beta} = \frac{(\sin\alpha+\sin\beta)^2}{\sin^2\alpha-\sin^2\beta}$

33. $\cos 2\cdot 45° = \cos 90° = 0$; $2\cos 45° = 2\cdot\frac{\sqrt{2}}{2} = \sqrt{2}$; $0 \neq \sqrt{2}$

35. $\cot 2\cdot 150° = \cot 300° \approx -0.5774$; $2\cot 150° \approx -3.4641$; $-0.5774 \neq -3.4641$

37. $n = \dfrac{\sin\frac{\alpha+\phi}{2}}{\sin\frac{\alpha}{2}} = \dfrac{\sqrt{\frac{1-\cos(\alpha+\phi)}{2}}}{\sqrt{\frac{1-\cos\alpha}{2}}} = \sqrt{\dfrac{1-\cos(\alpha+\phi)}{1-\cos\alpha}} = \sqrt{\dfrac{1-\cos\alpha\cos\phi+\sin\alpha\sin\phi}{1-\cos\alpha}}$

39. $P = V_{\max}I_{\max}\cos\omega t\sin\omega t = V_{\max}I_{\max}\frac{1}{2}\cdot 2\cos\omega t\sin\omega t = \frac{V_{\max}I_{\max}}{2}\cdot\sin 2\omega t$

≣ 20.4 TRIGONOMETRIC EQUATIONS

1. $2\cos\theta = 0; \cos\theta = 0; \theta = 90°, 270°$

3. $\sqrt{3}\tan x = 1; \tan x = \frac{1}{\sqrt{3}}; x = 30°, 210°$

5. $4\sin\theta = -3; \sin\theta = \frac{-3}{4}; \theta \approx 228.59°, 311.41°$

7. $4\tan\alpha = 5; \tan\alpha = \frac{5}{4}; \alpha = 51.34°, 231.34°$

9. $\cos 2x = -1; 2x = 180°$ or $540°; x = 90°$ or $270°$

11. $\tan\frac{\theta}{4} = 1; \frac{\theta}{4} = 45°; \theta = 180°$

13. $\sin^2\alpha = \sin\alpha; \sin^2\alpha - \sin\alpha = 0; \sin\alpha(\sin\alpha - 1) = 0$, so $\sin\alpha = 0$ or $\sin\alpha = 1; \alpha = 0°, 90°, 180°$

15. $\sin x\cos x = 0; \sin x = 0\cos x = 0; x = 0°, 90°, 180°, 270°$

17. $3\tan^2 x = 1; \tan^2 x = \frac{1}{3}; \tan x = \pm\sqrt{\frac{1}{3}}; x = 30°, 150°, 210°, 330°$

19. $4\sin\alpha\cos\alpha = 1; 2(2\sin\alpha\cos\alpha) = 2(\sin 2\alpha) = 1; \sin 2\alpha = \frac{1}{2}; 2\alpha = 30°, 150°, 390°, 510°$ and so, $\alpha = 15°, 75°, 195°, 255°$

21. $\sin\theta - \cos\theta = 0; \sin\theta = \cos\theta; \frac{\sin\theta}{\cos\theta} = 1$ or $\tan\theta = 1$, and so $\theta = 45°, 225°$

23. $\sin 6\theta + \sin 3\theta = 0; 2\sin 3\theta\cos 3\theta + \sin 3\theta = 0; \sin 3\theta(2\cos 3\theta+1) = 0; \sin 3\theta = 0$ or $\cos 3\theta = -\frac{1}{2}$. If $\sin 3\theta = 0$, then $3\theta = 0°, 180°, 360°, 540°, 720°, 900°$; and if $\cos 3\theta = -\frac{1}{2}$, then $3\theta = 120°$,

$240°; 480°; 600°; 840°; 960°$. Thus, we see that $\theta = 0°, 60°, 120°, 180°, 240°, 300°, 40°, 80°, 160°, 200°, 280°, 320°$

25. $2\cos^2 x - 3\cos 2x = 1; 2\cos^2 x - 3(2\cos^2 x - 1) = 1; -4\cos^2 x + 3 = 1; \cos^2 x = \frac{1}{2}; \cos x = \pm\frac{\sqrt{2}}{2}; x = 45°, 135°, 225°, 315°$

27. $\sec^2\theta + \tan\theta = 1; (\tan^2\theta + 1) + \tan\theta = 1; \tan^2\theta + \tan\theta = 0; \tan\theta(\tan\theta + 1) = 0; \tan\theta = 0$ or $\tan\theta = -1; \theta = 0°, 180°, 135°, 315°$

29. $\sin 2x = \cos x; 2\sin x\cos x = \cos x; 2\sin x\cos x - \cos x = 0; \cos x(2\sin x - 1) = 0; \cos x = 0$ or $\sin x = \frac{1}{2}; x = 30°, 90°, 150°, 270°$

31. $\frac{\sin 35°}{\sin\theta_r} = 1.61; \sin\theta_r = \frac{\sin 35°}{1.61} = 0.356258; \theta_r = 20.87°$

33. Since $E = 125\cos(\omega t - \phi)$, we have $60 = 125\cos(120\pi t - \frac{\pi}{2})$ so $\cos(120\pi t - \frac{\pi}{2}) = \frac{60}{125}$. Thus, $120\pi t - \frac{\pi}{2} \approx 1.070\ 14$ and so $120\pi t = 2.640\ 938$. Hence, $t = \frac{2.640\ 938}{120\pi} \approx 0.007\ 005$ s.

35. Since $d = \sin\omega t + \frac{1}{2}\sin 2\omega t$ we use a double-angle formula to get $0 = \sin\omega t + \frac{1}{2}\cdot 2\sin\omega t\cos\omega t$ which factors as $0 = \sin\omega t(1+\cos\omega t)$. Thus, $\sin\omega t = 0$ or $\cos\omega t = -1$ and so, $\omega t = 0, \pi$

≣ CHAPTER 20 REVIEW

1.

$$\dfrac{\sin(x+y)}{\cos x\cos y} = \tan x + \tan y$$

$\dfrac{\sin x\cos y+\cos x\sin y}{\cos x\cos y}$	$\tan x + \tan y$
$\dfrac{\sin x\cos y}{\cos x\cos y} + \dfrac{\cos x\sin y}{\cos x\cos y}$	
$\dfrac{\sin x}{\cos x} + \dfrac{\sin y}{\cos y}$	
$\tan x + \tan y$	

2.

$$(\sin x + \cos x)^2 = 1 + \sin 2x$$

$\sin^2 x + 2\sin x\cos x + \cos^2 x$	$1 + \sin 2x$
$1 + 2\sin x\cos x$	

3.

$$\frac{\sin 3x}{\sin x} - \frac{\cos 3x}{\cos x} = 2$$

$$\frac{\sin(x+2x)}{\sin x} - \frac{\cos(x+2x)}{\cos x} \quad \bigg| \quad 2$$

$$\frac{\sin x \cos 2x + \cos x \sin 2x}{\sin x} - \frac{\cos x \cos 2x - \sin x \sin 2x}{\cos x}$$

$$\frac{\sin x(\cos^2 x - \sin^2 x) + \cos x(2 \sin x \cos x)}{\sin x}$$

$$-\frac{\cos x(\cos^2 x - \sin^2 x) - \sin x(2 \sin x \cos x)}{\cos x}$$

$$\cos^2 x + \sin^2 x + 2\cos^2 x - \cos^2 x + \sin^2 x + 2\sin^2 x$$

$$2\cos^2 x + 2\sin^2 x$$

$$2(\cos^2 x + \sin^2 x)$$

$$2$$

4.

$$\cos(\theta + \phi)\cos(\theta - \phi) = \cos^2 \phi - \sin^2 \theta$$

$$(\cos\theta\cos\phi - \sin\theta\sin\phi)(\cos\theta\cos\phi + \sin\theta\sin\phi) \quad \bigg| \quad \cos^2\theta - \sin^2\theta$$

$$\cos^2\theta\cos^2\phi - \sin^2\theta\sin^2\phi$$

$$(1 - \sin^2\theta)\cos^2\phi - \sin^2\theta(1 - \cos^2\phi)$$

$$\cos^2\phi - \sin^2\theta\cos^2\phi - \sin^2\theta + \sin^2\theta\cos^2\phi$$

$$\cos^2\phi - \sin^2\theta$$

5.

$$\sin(\alpha + \beta)\cos\beta - \cos(\alpha + \beta)\sin\beta = \sin\alpha$$

$$(\sin\alpha\cos\beta + \cos\alpha\sin\beta)\cos\beta \quad \bigg| \quad \sin\alpha$$
$$-(\cos\alpha\cos\beta - \sin\alpha\sin\beta)\sin\beta$$
$$\sin\alpha\cos^2\beta + \cos\alpha\sin\beta\cos\beta$$
$$-\cos\alpha\cos\beta\sin\beta + \sin\alpha\sin^2\beta$$
$$\sin\alpha(\cos^2\beta \sin^2\beta)$$
$$\sin\alpha$$

6.

$$\tan 2x = \frac{2\cos x}{\csc x - 2\sin x}$$

$$\frac{2\tan x}{1 - \tan^2 x} \quad \bigg| \quad \frac{2\cos x}{\csc x - 2\sin x}$$

$$\frac{2\frac{\sin x}{\cos x}}{1 - \frac{\sin^2 x}{\cos^2 x}} \cdot \frac{\frac{\cos^2 x}{\sin x}}{\frac{\cos^2 x}{\sin x}}$$

$$\frac{2\cos x}{\frac{\cos^2 x}{\sin x} - \frac{\sin^2 x}{\sin x}}$$

$$\frac{2\cos x}{\frac{1}{\sin x}(\cos^2 x - \sin^2 x)}$$

$$\frac{2\cos x}{\csc x(1 - 2\sin^2 x)}$$

$$\frac{2\cos x}{\csc x(1 - 2\sin x)}$$

7.

$$\cos^4 x - \sin^4 x = \cos 2x$$

$$(\cos^2 x + \sin^2 x)(\cos^2 x - \sin^2 x) \quad \bigg| \quad \cos 2x$$
$$1 \cdot \cos 2x$$
$$\cos 2x$$

8.

$$\frac{\sin 2x - \sin x}{\cos 2x + \cos x} = \tan \frac{x}{2}$$

$$\frac{2\sin x \cos x - \sin x}{2\cos^2 x - 1 + \cos x} \quad \bigg| \quad \frac{\sin x}{\cos x + 1}$$

$$\frac{\sin x(2\cos x - 1)}{(2\cos x - 1)(\cos x + 1)}$$

$$\frac{\sin x}{\cos x + 1}$$

9.

$$\sin 3\theta = 2\sin\theta\cos 2\theta + \sin\theta$$

$$\sin(\theta + 2\theta) \quad \bigg| \quad 2\sin\theta\cos 2\theta + \sin\theta$$
$$\sin\theta\cos 2\theta + \cos\theta\sin 2\theta$$
$$\sin\theta(1 - 2\sin^2\theta) +$$
$$\cos\theta(2\sin\theta\cos\theta)$$
$$\sin\theta - 2\sin^3\theta + 2\sin\theta\cos^2\theta$$
$$\sin\theta + 2\sin\theta(\cos^2\theta - \sin^2\theta)$$
$$2\sin\theta\cos 2\theta + \sin\theta$$

10.

$$\frac{\cos 3x-\cos 5x}{\sin 3x+\sin 5x} = \tan x$$

	$\tan x$
$\frac{\cos(4x-x)-\cos(4x+x)}{\sin(4x-x)+\sin(4x+x)}$	$\tan x$
$(\cos 4x \cos x + \sin 4x \sin x) -$	
$\frac{(\cos 4x \cos x-\sin 4x \sin x)}{(\sin 4x \cos x-\cos 4x \sin x)+} \sin 4x \cos x + \cos 4x \sin x$	
$\frac{2\sin 4x \sin x}{2 \sin 4x \cos x}$	
$\frac{\sin x}{\cos x}$	
$\tan x$	

11. $2\tan x = -\sqrt{3}$ and so $\tan x = -\frac{\sqrt{3}}{2}$ and $x \approx$ 139.11° or 319.11°.

12. $3\sin x = -2; \sin x = \frac{-2}{3}; x = 221.81°, 318.19°$

13. $\cos 2x + \cos x = -1; 2\cos^2 x - 1 + \cos x = -1;$ $2\cos^2 x+\cos x = 0; \cos x(2\cos x+1) = 0; \cos x =$ 0 or $\cos x = \frac{-1}{2}; x = 90°, 120°, 240°, 270°$

14. $\cos x - \sin 2x - \cos 3x = 0; \cos x - 2\sin x \cos x -$ $\cos x \cos 2x+\sin x \sin 2x = 0; \cos x - 2\sin x \cos x$ $-\cos x(1-2\sin^2 x)+\sin x(2\sin x \cos x) = 0; \cos x$ $-2\sin x \cos x-\cos x+2\sin^2 x \cos x+2\sin^2 x \cos x$ $= 0; 4\sin^2 x \cos x - 2\sin x \cos x = 0; 2\sin x \cos x$ $(2\sin x - 1) = 0$. Hence $\sin x = 0$ or $\cos x = 0$ or $\sin x = \frac{1}{2}$ and so $x = 0°, 30°, 90°, 150°, 180°, 270°$

15. $\sin 4x - 2\sin 2x = 0; 2\sin 2x \cos 2x - 2\sin 2x =$ $0; 2\sin 2x(\cos 2x - 1) = 0; \sin 2x = 0$ or $\cos 2x =$ $1; 2x = 0°, 180°, 360°, 540°, x = 0°, 90°, 180°,$ $270°$

16. $\sin(30° +x)-\cos(60° +x) = \frac{-\sqrt{3}}{2}; \sin 30° \cos x+$ $\cos 30° \sin x-\cos 60° \cos x+\sin 60° \sin x = \frac{-\sqrt{3}}{2};$ $\frac{1}{2}\cos x + \frac{\sqrt{3}}{2}\sin x - \frac{1}{2}\cos x + \frac{\sqrt{3}}{2}\sin x = \frac{-\sqrt{3}}{2};$ $\sqrt{3}\sin x = \frac{-\sqrt{3}}{2}; \sin x = -\frac{1}{2}; x = 210°, 330°$

17. $2\sin\theta = \sin 2\theta; 2\sin\theta = 2\sin\theta \cos\theta; \sin\theta -$ $\sin\theta \cos\theta = 0; \sin\theta(1 - \cos\theta) = 0$, so $\sin\theta = 0$ or $\cos\theta = 1$ with the result that $\theta = 0; 180°$.

18. $\sin^2 \alpha + 5\cos^2 \alpha = 3; \sin^2 \alpha + \cos^2 \alpha + 4\cos^2 \alpha =$ $3; 1 + 4\cos^2 \alpha = 3; 4\cos^2 \alpha = 2$, or $\cos^2 \alpha = \frac{1}{2}$.

Thus, we have $\cos\alpha = \pm\frac{\sqrt{2}}{2}$, with the result that $\alpha = 45°, 135°, 225°, 315°$.

19. $\sin^2 x = 1 + \sin x; \sin^2 x - \sin x = -1 = 0; \sin x$ $= \frac{1\pm\sqrt{1+4}}{2} = \frac{1\pm\sqrt{5}}{2} \approx 1.618$ or -0.618. Since 1.618 is not in the range of the sine function, only -0.618 yields answers. They are $x = 218.17°$ and 321.83°.

20. $\sin\theta - 2\csc\theta = -1; \sin\theta-\frac{2}{\sin\theta} = -1; \sin^2\theta-2 =$ $-\sin\theta; \sin^2\theta+\sin\theta-2 = 0; (\sin\theta+2)(\sin\theta-1) =$ $0; \sin\theta = -2$ or $\sin\theta = 1$. But, $\sin\theta$ cannot be -2, so only $\sin\theta = 1$ yields an answer and it is $\theta = 90°$.

21. $\sin 2x = 2\sin x \cos x = 2\left(\frac{5}{13}\right)\left(\frac{-12}{13}\right) = \frac{-120}{169}$

22. $\cos\frac{x}{2} = \sqrt{\frac{1+\cos x}{2}} = \sqrt{\frac{1+\frac{-12}{13}}{2}} = \sqrt{\frac{1}{26}} \approx$ 0.196116

23. $\sin\frac{x}{2} = \sqrt{\frac{1-\cos x}{2}} = \sqrt{\frac{1+\frac{12}{13}}{2}} = \sqrt{\frac{25}{26}} \approx 0.98058$

24. $\tan 2x = \frac{2\tan x}{1-\tan^2 x} = \frac{2\cdot\frac{-5}{12}}{1-\left(\frac{-5}{12}\right)^2} = \frac{\frac{-10}{12}}{\frac{119}{144}} = \frac{-120}{119}$

25. $\sin(x+y) = \sin x \cos y + \cos x \sin y = \frac{-5}{13}\cdot\frac{-15}{17} +$ $\frac{-12}{13}\cdot\frac{8}{17} = \frac{75}{221} - \frac{96}{221} = \frac{-21}{221} \approx -0.09502$

26. $\cos(x-y) = \cos x \cos y + \sin x \sin y = \frac{-12}{13}\cdot\frac{-15}{17} +$ $\frac{-5}{13}\cdot\frac{8}{17} = \frac{180}{221} - \frac{40}{221} = \frac{140}{221} \approx 0.63348$

27. $\cos(x+y) = \frac{-12}{13}\cdot\frac{-15}{17} - \frac{-5}{13}\cdot\frac{8}{17} = \frac{180}{221} + \frac{40}{221} =$ $\frac{220}{221} \approx 0.99548$

28. $\tan(x+y) = \frac{\tan x+\tan y}{1-\tan x \tan y} = \frac{\frac{5}{12} + \frac{-8}{15}}{1 - \frac{5}{12}\cdot\frac{-8}{15}} = \frac{\frac{-21}{180}}{\frac{220}{180}} =$ $\frac{-21}{220} \approx -0.09545$ or $\tan(x+y) = \frac{\sin(x+y)}{\cos(x+y)} =$ $\frac{\frac{-21}{221}}{\frac{220}{221}} = \frac{-21}{220} \approx -0.09545$

29. $\cos(y-x) = \cos y \cos x + \sin y \sin x = \frac{-15}{17}\cdot\frac{-12}{13} +$ $\frac{8}{17}\cdot\frac{-5}{13} = \frac{140}{221} \approx 0.63348$ (Note #26 and #29 are the same answer.)

30. $\sin(x-y) = \sin x \cos y - \cos x \sin y = \frac{-5}{13}\cdot\frac{-15}{17} -$ $\frac{-12}{13}\cdot\frac{8}{17} = \frac{75}{221} + \frac{96}{221} = \frac{171}{221} \approx 0.77376$

31. We begin with the given equation $a = 5.0(\sin \omega t + \cos 2\omega t)$ and substitute $a = 0$ and a double-angle identity for $\cos \omega t$ with the result that $0 = \sin \omega t + 1 - 2\sin^2 \omega t$ which is equivalent to $2\sin^2 \omega t - \sin \omega t - 1 = 0$. This last equation factors as $(2\sin \omega t + 1)(\sin \omega t - 1) = 0$. Thus, by the zero product rule $\sin \omega t = -\frac{1}{2}$ or $\sin \omega t = 1$, and so $\omega t = 90°, 210°, 330°$.

32. $y = y_1 + y_2 = A_1 \cos \omega_1 t + A_2 \cos \omega_2 t$. If $\alpha = \frac{\omega_1 + \omega_2}{2}$ and $\beta = \frac{\omega_2 + \omega_1}{2}$, then $\alpha - \beta = \omega_1$ and $\alpha + \beta = \omega_2$. Substituting produces $A_1 \cos \omega_1 t = A_1 \cos(\alpha - \beta)t$ and $A_2 \cos \omega_2 t = A_2 \cos(\alpha + \beta)t$, and so $y_1 + y_2 = A_1 \cos(\alpha - \beta)t + A_2 \cos(\alpha + \beta)t$.

33. We want to show that $x = 0.01e^{-6t}(\cos 8t + \sin 8t)$ and $x = \frac{\sqrt{2}}{100}e^{-6t}\cos(8t - \frac{\pi}{4})$ are identically equivalent. Now, using a difference formula, we see that the second equation can be rewritten as $\frac{\sqrt{2}}{100}e^{-6t}\cos(8t - \frac{\pi}{4}) = 0.01e^{-6t}\sqrt{2}[\cos 8t \cos \frac{\pi}{4} + \sin 8t \sin \frac{\pi}{4}] = 0.01e^{-6t}\sqrt{2}[\cos 8t \cdot \frac{1}{\sqrt{2}} + \sin 8t \cdot \frac{1}{\sqrt{2}}] = 0.01e^{-6t}\cos 8t + \sin 8t)$, which is the first equation. Hence the two equations are identical.

☰ CHAPTER 20 TEST

1. Since, $\sin \alpha = \frac{4}{5}$ and α is in Quadrant II, we have $\cos \alpha = -\frac{3}{5}$. Also, since $\cos \beta = -\frac{12}{13}$, and β is in Quadrant III, then $\sin \beta = -\frac{5}{13}$. Thus, $\sin(\alpha + \beta) = \frac{4}{5} \cdot \frac{-12}{13} + \frac{-3}{5} \cdot \frac{-5}{13} = \frac{-48}{65} + \frac{15}{65} = \frac{-33}{65}$.

2. Using the values of $\sin \alpha = \frac{4}{5}, \cos \alpha = -\frac{3}{5}, \cos \beta = -\frac{12}{13}$, and $\sin \beta = -\frac{5}{13}$ from Exercise #1, we see that $\cos(\alpha + \beta) = \frac{-3}{5} \cdot \frac{-12}{13} - \frac{4}{5} \cdot \frac{-5}{13} = \frac{36}{65} + \frac{20}{65} = \frac{56}{65}$.

3. Using the values of $\sin \alpha = \frac{4}{5}, \cos \alpha = -\frac{3}{5}, \cos \beta = -\frac{12}{13}$, and $\sin \beta = -\frac{5}{13}$ from Exercise #1, we see that $\sin(\alpha - \beta) = \frac{4}{5} \cdot \frac{-12}{13} - \frac{-3}{5} \cdot \frac{-5}{13} = \frac{-48}{65} - \frac{15}{65} = \frac{-63}{65}$.

4. Using the values of $\sin \alpha = \frac{4}{5}, \cos \alpha = -\frac{3}{5}, \cos \beta = -\frac{12}{13}$, and $\sin \beta = -\frac{5}{13}$ from Exercise #1, we see that $\cos(\alpha - \beta) = \frac{-3}{5} \cdot \frac{-12}{13} + \frac{4}{5} \cdot \frac{-5}{13} = \frac{36}{65} - \frac{20}{65} = \frac{16}{65}$.

5. Using the values of $\sin \alpha = \frac{4}{5}, \cos \alpha = -\frac{3}{5}$ from Exercise #1, we see that $\sin 2\alpha = 2 \cdot \frac{4}{5} \cdot \frac{-3}{5} = \frac{-24}{25}$.

6. Using the values of $\cos \beta = -\frac{12}{13}$, and $\sin \beta = -\frac{5}{13}$ from Exercise #1, we see that $\cos 2\beta = 2\left(\frac{-12}{13}\right)^2 - 1 = \frac{288}{169} - \frac{169}{169} = \frac{119}{169}$.

7. Using the values of $\sin \alpha = \frac{4}{5}, \cos \alpha = -\frac{3}{5}$ from Exercise #1, we see that $\cos \frac{\alpha}{2} = \sqrt{\frac{1 + \cos \alpha}{2}} = \sqrt{\frac{1 - \frac{3}{5}}{2}} = \sqrt{\frac{8}{10}} = \sqrt{\frac{4}{5}} = \frac{2}{\sqrt{5}}$.

8. Using the values of $\cos \beta = -\frac{12}{13}$, and $\sin \beta = -\frac{5}{13}$ from Exercise #1, we see that $\cos \frac{\beta}{2} = -\sqrt{\frac{1 + \cos \beta}{2}} = -\sqrt{\frac{1 - \frac{12}{13}}{2}} = -\sqrt{\frac{1}{26}} = \frac{-1}{\sqrt{26}}$.

9. Using the double angle formula, $2\cos x \sin x = \sin 2x$, we see that can be rewritten as $8\cos 6x \sin 6x = 4 \cdot 2\sin\left(\frac{12x}{2}\right)\cos\left(\frac{12x}{2}\right) = 4\sin 12x$.

10. $\tan x = \frac{\sec x}{\csc x}; \frac{\sec x}{\csc x} = \frac{\frac{1}{\cos x}}{\frac{1}{\sin x}} = \frac{\sin x}{\cos x} = \tan x$

11.
$$\frac{\sin x}{1 - \cos x} + \frac{\sin x}{1 + \cos x} = 2\csc x$$

$$\left(\frac{\sin x}{1 - \cos x}\right)\left(\frac{1 + \cos x}{1 + \cos x}\right) + \left(\frac{\sin x}{1 + \cos x}\right)\left(\frac{1 - \cos x}{1 - \cos x}\right) \quad \Big| \quad 2\csc x$$

$$\frac{\sin x(1 + \cos x)}{1 - \cos^2 x} + \frac{\sin x(1 - \cos x)}{1 - \cos^2 x}$$

$$\frac{\sin x(1 + \cos x)}{\sin^2 x} + \frac{\sin x(1 - \cos x)}{\sin^2 x}$$

$$\frac{1 + \cos x}{\sin x} + \frac{1 - \cos x}{\sin x}$$

$$\frac{2}{\sin x}$$

$$2\csc x$$

12. The equation $6\cos^2 x + \cos x = 2$ is equivalent to $6\cos^2 x + \cos x - 2 = 0$. This expression factors as $(3\cos x + 2)(2\cos x - 1) = 0$ and so by the zero-product rule, $3\cos x + 2 = 0$ or $2\cos x - 1 = 0$. If $3\cos x + 2 = 0$, then $\cos x = -\frac{2}{3}$ and $x \approx 131.81°$ or $228.19°$. If $2\cos x - 1 = 0$, then $\cos x = \frac{1}{2}$ and $x = 60°$ or $300°$. In radians the solutions are $\frac{\pi}{3}, \frac{5\pi}{3}, 2.3$, and 3.98.

13. $x = 4r\sin^2\left(\frac{\theta}{2}\right) = 4r\left(\frac{1 - \cos \theta}{2}\right) = 2r - 2r\cos \theta = 2r(1 - \cos \theta)$

CHAPTER
21
Statistics and Empirical Methods

≡ 21.1 PROBABILITY

1. $\frac{13}{52} = \frac{1}{4}$ **5.** $\frac{12}{52} = \frac{3}{13}$ **9.** $\frac{1}{2} \cdot \frac{26}{51} = \frac{13}{51}$ **13.** $\frac{3}{6} = \frac{1}{2}$

3. $\frac{4}{52} = \frac{1}{13}$ **7.** $\frac{1}{2} \cdot \frac{1}{2} = \frac{1}{4}$ **11.** $\frac{1}{6}$ **15.** $\frac{1}{6} \cdot \frac{1}{6} = \frac{1}{36}$

17. $\frac{5}{6} \cdot \frac{5}{6} = \frac{25}{36}; P = 1 - \frac{25}{36} = \frac{11}{36}$

19.

Die 1 \ Die 2	1	2	3	4	5	6
1	(1, 1)	(1, 2)	(1, 3)	(1, 4)	(1, 5)	(1, 6)
2	(2, 1)	(2, 2)	(2, 3)	(2, 4)	(2, 5)	(2, 6)
3	(3, 1)	(3, 2)	(3, 3)	(3, 4)	(3, 5)	(3, 6)
4	(4, 1)	(4, 2)	(4, 3)	(4, 4)	(4, 5)	(4, 6)
5	(5, 1)	(5, 2)	(5, 3)	(5, 4)	(5, 5)	(5, 6)
6	(6, 1)	(6, 2)	(6, 3)	(6, 4)	(6, 6)	(6, 6)

21. $\frac{1}{36}$

23. 0

25. $\frac{6}{36} = \frac{1}{6}$

27. $\frac{1+2+3}{36} = \frac{6}{36} = \frac{1}{6}$

29. $\frac{1}{36} \cdot \frac{1}{36} = \frac{1}{1,296}$

31. $\frac{6}{21} \cdot \frac{6}{21} = \frac{36}{441} = \frac{4}{49}$

33. $\frac{2+4+6}{21} = \frac{12}{21} = \frac{4}{7}$

35. **(a)** $.75 \times .75 = 0.5625 = \frac{9}{16}$,
(b) $.25 \times .25 = 0.0625 = \frac{1}{16}$,
(c) $1 - 0.5625 - 0.0625 = 0.375 = \frac{3}{8}$

37. **(a)** $0.7 \times 0.8 = 0.56$,
(b) $0.3 \times 0.2 = 0.06$,
(c) $1 - 0.56 - 0.06 = 0.38$

39. If one is defective, then the other is not defective. There are two ways this can happen, so the probability is $2(0.025)(1 - 0.025) = 2(0.025)(0.975) = 0.04875$.

41. The total number of people is 200. Since 20 are blood group B, the probability is $\frac{20}{200} = \frac{1}{10} = 0.10$.

≡ 21.2 MEASURES OF CENTRAL TENDENCY

1.

Number	2	3	4	6	7
Frequency	2	1	4	2	1

3.

Number	73	74	77	80	82	83	84	87	89	92	94	96	100
Frequency	1	1	1	1	1	1	2	1	1	1	3	5	1

5.

Interval	1 − 2	3 − 4	5 − 6	7 − 8
Frequency	2	5	2	1

7.

Interval	71 − 75	76 − 80	81 − 85	86 − 90	91 − 95	96 − 100
Frequency	2	2	4	2	4	6

9.

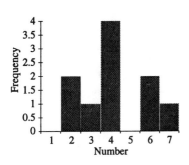

11.

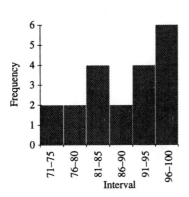

13.

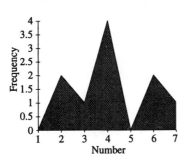

15.

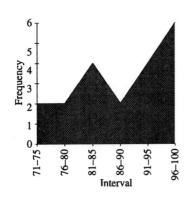

17. Mean $= \frac{2 \times 2 + 3 + 4 \times 4 + 6 \times 2 + 7}{10} = \frac{42}{10} = 4.2$; Median $= 4$; Mode $= 4$; $Q_1 = 3$; $Q_2 = 4$; $Q_3 = 6$;

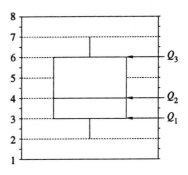

19. Mean $= \frac{\substack{73 + 74 + 77 + 80 + 82 + 83 + 84 \times 2 + 87 \\ + 89 + 92 + 94 \times 3 + 96 \times 5 + 100}}{20} = \frac{1767}{20} = 88.35$;

Median $= \frac{89 + 92}{2} = 90.5$; Mode $= 96$; $Q_1 = 82.5$; $Q_2 = 90.5$; $Q_3 = 96$;

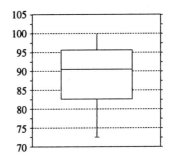

21. See *Computer Programs* in main text.

23.

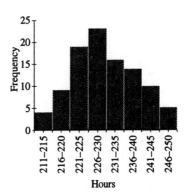

25.

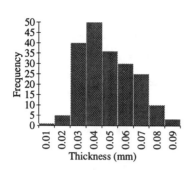

27.

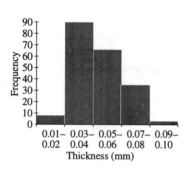

29.

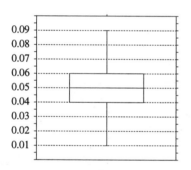

31.

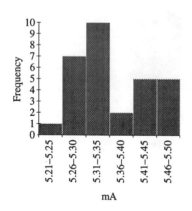

33. $Q_1 = 5.29; Q_2 = 5.345; Q_3 = 5.43$

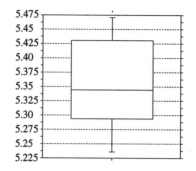

35. **(a)** Mean $= 10.01$, Median $= 10.31$, Mode: None, $Q_1 = 7.44, Q_2 = 10.31, Q_3 = 12.36$
(b)

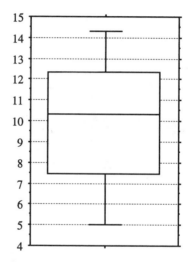

37. Mean $= 0.17$, Median $= 0.16$, Mode $= 0.13$, $Q_1 = 0.12, Q_2 = 0.16, Q_3 = 0.22$

≡ 21.3 MEASURES OF DISPERSION

1. $v = 2.84$

3. $v = 67.71$

5. $\sigma_n = 1.69$

7. $\sigma = 8.23$

9.   The mean is 4.2; $\overline{x} - s = 2.51; \overline{x} + s = 5.89; 5$ of the 10 values are within one standard deviation. $\overline{x} - 2s = 0.82; \overline{x} + 2s = 7.58;$ All 10 values are within two standard deviations.

11.   The mean is $88.35; \overline{x} - 2s = 80.12; \overline{x} + s = 96.58.$ There are 15 values within one standard deviation. $\overline{x} - 2s = 71.89; \overline{x} + 2s = 104.81;$ there are 20 values within two standard deviations.

13.   See *Computer Programs* in main text.

15.   **(a)** The standard deviation is 0.0217.
   (b) The mean is $10.0214; \overline{x} - s = 9.9997; \overline{x} + s 10.0431;$ The number of rods within one standard deviation is 390.
   (c) $\overline{x} - 2s = 9.978; \overline{x} + 2s = 10.0648;$ The number of rods within two standard deviations is 480.

17.   **(a)** The standard deviation is 0.01636;
   (b) $\overline{x} = 0.04865; \overline{x} - s = 0.03229; \overline{x} + s = 0.06501;$ 116 are within one standard deviation.
   (c) $\overline{x} - 2s = 0.01593; \overline{x} + 2s = 0.08137;$ 196 are within two standard deviations.

19.   **(a)** The mean is 60.75 and the standard deviation is 5.85.
   (b) $\overline{x} - s = 60.75 - 5.85 = 54.9$ and $\overline{x} + s = 60.75 + 5.85 = 66.6.$ There are 8 data points within one standard deviation of the mean and this is $\frac{8}{12} \approx 66.7\%.$
   (c) $\overline{x} - 2s = 60.75 - 2(5.85) = 49.05$ and $\overline{x} + 2s = 60.75 + 2(5.85) = 72.45.$ There are 12 data points within two standard deviations of the mean and this is $\frac{12}{12} = 100\%.$

21.   **(a)** $\overline{x} = 659.25, s = 174.67,$
   (b) $\overline{x} - s = 659.25 - 174.67 = 484.58$ and $\overline{x} + s = 659.25 + 174.67 = 833.92.$ There are 14 data points within one standard deviation of the mean and this is $\frac{14}{24} \approx 58.3\%,$
   (c) $\overline{x} - 2s = 659.25 - 2(174.67) = 309.91$ and $\overline{x} + 2s = 659.25 + 2(171.01) = 1008.59.$ There are 23 data points within two standard deviations of the mean and this is $\frac{23}{24} \approx 95.8\%$

23.   **(a)** $\overline{x} = 25,905.65, s = 13,280.32,$
   (b) $\overline{x} - s = 25,905.65 - 13,280.32 = 12,625.33$ and $\overline{x} + s = 25,905.65 - 13,280.32 = 39,185.97.$ There are 23 data points within one standard deviation of the mean and this is $\frac{23}{31} \approx 74.2\%,$
   (c) $\overline{x} - 2s = 25,905.65 - 2(13,280.32) = -654.99$ and $\overline{x} + 2s = 25,905.65 + 2(13,280.32) = 52,466.29.$ There are 29 data points within two standard deviations of the mean and this is $\frac{29}{31} = 93.5\%.$

≡ 21.4 FITTING A LINE TO DATA

1. $y = 0.833x - 1.333, r = 0.943$

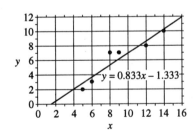

3. $y = 13.55x + 5.1, r = 0.986$

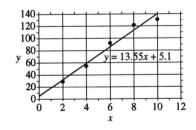

5. $y = 14.955x + 5.379; r = 0.996$

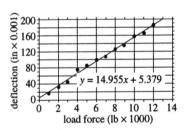

7. $y = 0.000425x + 0.072; r = 0.987$

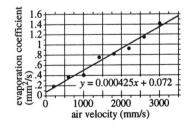

9. See *Computer Programs* in main text.

11. $y = -0.26825x + 74.6865, r = -0.70$. Note that $|r| < 0.75$, so the correlation is poor.

13. $y = 5.1x - 10,121, r = 0.98$. Since $|r| > 0.75$, in fact very close to 1, there is a strong correlation.

≡ 21.5 FITTING NONLINEAR CURVES TO DATA

1. **(a)**

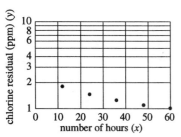

(b) $y = (2.002)(0.988)^x$ Calculator Function

3. **(a)** See the graph below.

(b)

x	5	10	4	10	7	8	8	5	10	5	12	6
y	30	51	26	52	40	43	45	31	52	30	59	36
x^2	25	100	16	100	49	64	64	25	100	25	144	36
x^3	125	1000	64	1000	343	512	512	125	1000	125	1728	216
x^4	625	10000	256	10000	2401	4096	4096	625	10000	625	20736	1296
xy	150	510	104	520	280	344	360	155	520	150	708	216
x^2y	750	5100	416	5200	1960	2752	2880	775	5200	750	8496	1296

$\sum x = 90; \sum x^2 = 748; \sum xy = 4017; \sum y = 495; \sum x^3 = 6750; \sum x^2y = 35,575; \sum x^4 = 64,756.$

Hence:

$$495 = 12a_0 + 90a_1 + 748a_2$$
$$4017 = 90a_0 + 748a_1 + 6750a_2$$
$$35,575 = 748a_0 + 6750a_1 + 64,756a_2$$

Solving yields $a_0 = 4.084; a_1 = 5.848, a_2 = -0.107$. The equation is $y = 4.084 + 5.848x - 0.107x^2$

(c)

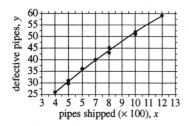

5. **(a)** See the graph below.

(b) To find the least square curve for $y = ae^{-kx}$ we first take natural logarithms giving $\ln y = \ln a - kx$. To use the normal equations of the previous section we need to compute several sums. These are $\sum x = 36; \sum x^2 = 204; \sum \ln y = -4.48183$ and $\sum x \ln y = -25.3578$. Also $n = 8$.

$$\sum \ln y = nb - k \sum x$$
$$\sum x \ln y = b \sum x - k \sum x^2 \quad \text{or}$$
$$-4.48183 = 8b - 36k$$
$$-25.3578 = 36b - 204k$$

Solving we get $b = -0.004208$ and $k = 0.12356$. Since $b = \ln a; a = e^{-0.004208} = 0.9958$. The final equation is $y = (0.9958)e^{-0.12356x}$. Note: Some calculators have built in functions to compute the least square curve of the form $y = ab^x$. This yields $a = 0.9958$ and $b = 0.883768$. To convert to $y = ae^{-kx}$ find $\ln b = \ln(0.883768) = -0.12356$ or $y = (0.9958)e^{-0.12356x}$.

(c)

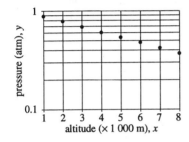

altitude ($\times$ 1 000 m), x

7. **(a)** See the graph below.
(b) As in the previous examples we need sums for the following: $\sum \frac{1}{p}; \sum \frac{1}{p^2}; \sum y$; and $\sum \frac{y}{p}$. These are 2.0290; 1.5498; 15.2; and 8.1939, respectively. Solving the system $15.2 = 10b + 2.929m$; $8.1939 = 2.9292b + 1.5498m$ yields $b = -0.064$ and $m = 5.41$. The final equation is $y = \frac{5.41}{p} - 0.064$.

(c)

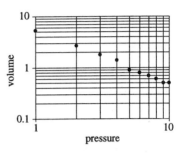

pressure

CHAPTER 21 REVIEW

1. $\frac{1}{2}$

2. $\frac{2}{54} = \frac{1}{27}$

3. $\frac{10}{16} = \frac{5}{8}$

4. $\frac{48}{52} = \frac{12}{13}$

5. $\frac{2}{6} = \frac{1}{3}$

6. $\frac{1}{36} + \frac{2}{36} + \frac{3}{36} = \frac{6}{36} = \frac{1}{6}$

7. $\frac{1}{6} \times \frac{1}{6} = \frac{1}{36}$

8. $\frac{4}{52} \times \frac{3}{51} = \frac{1}{221}$

9. $\frac{46}{50} = \frac{23}{25}$ = probability that a drive is not defective $\left(\frac{23}{25}\right)^3 = \frac{12167}{15625}$

10. 297.8, 298.9, 299.6, 299.8, 300.1, 300.1, 300.3, 300.4, 300.5, 300.7, 301.2. The median is 300.1

11. $Q_1 = \frac{299.6 + 299.8}{2} = 299.7; Q_2 = 300.1; Q_3 = 300.45$

12.

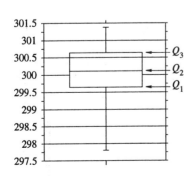

13. 300.1

14. $\bar{x} = 299.945$

15. 0.931

16. $\frac{8}{11} = 72.7\%$

17. $\frac{10}{11} = 90.9\%$

18. 23, 24, 26, 32, 33, 37, 37, 38, 39, 42, 43, 45, 48, 50, 52;
(a) 37.93,
(b) 38,
(c) $Q_1 = 32.5, Q_2 = 38, Q_3 = 44$;
(d) 8.79

19. **(a)** See the figure below.
(b) $y = 43.94x + 18.94$,

(c)

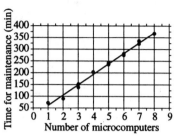

(d) $r = 0.996$

20. **(a)** See the figure below.
(b) $P = 3983.96V^{-1.405}$ (Note: enter data with V first and P second),
(c)

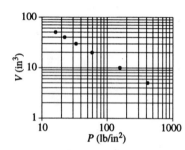

21. **(a)** See the figure below.
(b) $n = -0.686p + 83.714, r = -0.9384$
(c) Shown by curve in the figure below.

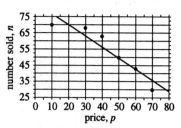

22. To find this equation we need the followings sums $\sum \sqrt{x} = 31.775; \sum x = 110; \sum y = 111.5; \sum \sqrt{x}y = 346.772$. This gives the normal equations $111.5 = 11b + 31.775m, 346.772 = 31.775b + 110m$. Solving this system yields $b = 6.221, m = 1.356$: Thus the equation is $y = 1.356\sqrt{x} + 6.221$.

CHAPTER 21 TEST

1. A deck of 52 cards has 2 red queens, so the probability of drawing a red queen from this deck is $\frac{2}{52} = \frac{1}{26}$.

2. The probability of no defect is $\frac{49}{50}$. To get 5 with no defect, the probability is $\left(\frac{49}{50}\right)^5 \approx 0.904$.

3. 3598.2, 3598.6, 3598.7, 3598.9, 3599.2, 3599.2, 3599.6, 3599.6, 3599.9, 3600.1, 3600.1, 3600.2, 3600.4, 3600.4, 3600.4, 3600.4, 3600.8, 3601.2, 3601.4, 3601.6, median $= 3600.1$

4. $Q_1 = 3599.2, Q_2 = 3600.1, Q_3 = 3600.4$

5.

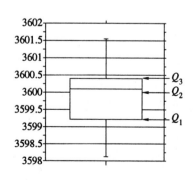

6. mode $= 3600.4$

7. mean $= 3599.945$

8. $\sigma = 0.9206, s = 0.9445$

9. $\frac{13}{20} = 65\%$

10. $\frac{20}{20} = 100\%$

11. **(b)** $y = -39.774x + 1063.76$,
(c) See line in graph;
(d) $r = -0.8997$

CHAPTER

22

An Introduction to Calculus

≣ 22.1 THE TANGENT QUESTION

1. $\frac{f(3)-f(1)}{3-1} = \frac{24-6}{3-1} = \frac{18}{2} = 9$

3. $\frac{h(-1)-h(-3)}{-1--3} = \frac{-6--20}{-1+3} = \frac{14}{2} = 7$

5. $\frac{k(-1-k(-2)}{-1--2} = \frac{-11--39}{-1+2} = \frac{28}{1} = 28$

7. $\frac{f(5)-f(2)}{5-2} = \frac{1-\frac{5}{2}}{5-2} = \frac{\frac{-3}{2}}{3} = -\frac{1}{2}$

9. $\frac{h(0)-h(1)}{0-1} = \frac{0-1}{-1} = 1$

11. $\frac{k(x_1)-k(1)}{x_1-1} = \frac{x_1^2+x_1-7--5}{x_1-1} = \frac{x_1^2+x_1-2}{x_1-1} =$
    $\frac{(x_1+2)(x_1-1)}{x_1-1} = x_1 + 2$

13. $\frac{f(x_1+h)-f(x_1)}{(x_1+h)-x_1} = \frac{[(x_1+h)^2+1]-[x_1^2+1]}{h} =$
    $\frac{x_1^2+2x_1h+h^2+1-x_1^2-1}{h} = \frac{2x_1h+h^2}{h} = \frac{h(2x_1+h)}{h} =$
    $2x_1 + h$

15. $V(t) = 5t^2 + 4t$;
    (a) $\frac{V(3)-V(2)}{3-2} = \frac{57-28}{1} = 29$ L/min

(b) $\frac{V(4)-V(3)}{4-3} = \frac{96-57}{1} = 39$ L/min

(c) $\frac{V(4)-V(2)}{4-2} = \frac{96-28}{2} = \frac{68}{2} = 34$ L/min

17. (a) $\frac{f(6)-f(4)}{6-4} = \frac{103-43}{2} = \frac{60}{2} = 30$

(b) $\frac{f(5)-f(4)}{5-4} = \frac{70-43}{1} = 27$

(c) $\frac{f(4.5)-f(4)}{4.5-4} = \frac{55.75-43}{0.5} = \frac{12.75}{0.5} = 25.5$

(d) $\frac{f(4.25)-f(4)}{4.25-4} = \frac{49.1875-43}{0.25} = \frac{6.1875}{0.25} = 24.75$

(e) $\frac{f(4.1)-f(4)}{4.1-4} = \frac{45.43-43}{0.1} = \frac{2.43}{0.1} = 24.3$

(f) $\frac{f(4.05)-f(4)}{4.05-4} = \frac{44.2075-43}{0.05} = \frac{1.2075}{0.05} = 24.15$

19. See *Computer Programs* in main text.

21. (a) Using the formula $d = rt$, the two distances are $d_1 = (20)\frac{15}{60} = 5$ mi and $d_2 = (54)\frac{35}{60} = 31.5$ mi. The total distance is $d = d_1+d_2 = 5+31.5 = 36.5$ mi.

(b) The total time of the trip was 50 min or $\frac{50}{60} = \frac{5}{6}$ h. $v = \frac{d}{t} = \frac{36.5}{5/6} = 43.8$ mph.

≣ 22.2 THE AREA QUESTION

1. $\Delta x = \frac{b-a}{6} = \frac{7-1}{6} = 1$;

x^*	1.5	2.5	3.5	4.5	5.5	6.5
y^*	6.5	9.5	12.5	15.5	18.5	21.5

$\sum_{i=1}^{6} y_i^* \Delta x = 84$

3. $\Delta x = \frac{3-0}{6} = \frac{1}{2}$ or 0.5;

x^*	0.25	0.75	1.25	1.75	2.25	2.75
y^*	1.0625	1.5625	2.5625	4.0625	6.0625	8.5625

$\sum_{i=1}^{6} y_i^* \Delta x = 11 \cdot 9375$

5. $\Delta x = \frac{5-1}{8} = 0.5$;

x^*	1.25	1.75	2.25	2.75	3.25	3.75	4.25	4.75
y^*	5	12.5	22	33.5	47	62.5	80	99.5

$\sum_{i=1}^{8} y_i^* \Delta x = 362(0.5) = 181$

7. $\Delta x = \frac{4--2}{8} = \frac{6}{8} = \frac{3}{4} = 0.75$;

x^*	−1.625	−0.875	−0.125	0.625	1.375	2.125	2.875	3.625
y^*	13.359375	15.234375	15.984375	15.609375	14.109375	11.484375	7.734375	2.859375

$\sum_{i=1}^{8} y_i^* \Delta x = 96.375(0.75) = 72.28125$

9. $\Delta x = \frac{-1--5}{8} = \frac{1}{2} = 0.5$;

x^*	−4.75	−4.25	−3.75	−3.25	−2.75	−2.25	−1.75	−1.25
y^*	84.1875	69.6875	56.6875	45.1875	35.1875	26.6875	19.6875	14.1875

$\sum y_i^* \Delta x = 351.5(0.5) = 175.75$

11. $\Delta x = 1, \sum y_i^* = 4.05 + 3.85 + 3.9 + 3.75 + 3.45 + 3.65 + 4.0 + 4.2 = 30.85$; $\sum y_i^* \Delta x = 30.85$

13. See *Computer Programs* in main text.

15. Shading the desired area on a graph of $P'(x)$ indicates that the desired area is a trapezoid with bases of length $P'(0) = 78$ and $P'(40) = 46$ and height 40. The area is $\frac{40}{2}(78 + 40) = 2,480$. Thus, the profit for the first 40 coats produced and sold is $2,480.

17. The total length of the interval is $3 - 1 = 2$, so 8 divisions yields $\Delta x = 0.25$. This produces the following table

x^*	1.125	1.375	1.625	1.875	2.125	2.375	2.625	2.875
y^*	0.475	0.614	0.726	0.791	0.790	0.705	0.517	0.207

The total charge on this capacitor is $\sum y^* \Delta x = (4.825)(0.25) \approx 1.206 \ \Omega$.

22.3 LIMITS: AN INTUITIVE APPROACH

1.

x	0.9	0.99	0.999	0.9999	1.0001	1.001	1.01	1.1
$f(x) = 3x$	2.7	2.97	2.997	2.9997	3.0003	3.003	3.03	3.3

$\lim_{x \to 1} 3x = 3$

3.

x	−1.1	−1.01	−1.001	−1.0001	−0.9999	−0.999	−0.99	−0.9
$h(x) = x^2 + 2$	3.21	3.0201	3.002	3.0002	2.9998	2.998	2.9801	2.81

$\lim_{x \to -1} (x^2 + 2) = 3$

5.

x	-0.1	-0.01	-0.001	-0.0001	0.0001	0.001	0.01	0.1
$f(x) = \frac{\tan x}{x}$	1.0033467	1.0000333	1.0000003	1	1	1.0000003	1.0000333	1.0033467

$$\lim_{x \to 0} \frac{\tan x}{x} = 1$$

7.

x	0.9	0.99	0.999	0.9999	1.0001	1.001	1.01	1.1
$h(x) = \frac{x}{x-1}$	-9	-99	-999	-9999	10001	1001	101	11

$$\lim_{x \to 1} \frac{x}{x-1} \text{ Does not exist}$$

9. (a) -1, both sides of graph near $x = 0$ converge on -1.
(b) Does not exist, graph goes toward -2 when $x < 2$ and goes toward 1 for $x > 2$;
(c) 4, both sides of graph near $x = 3$ converge on 4.

11.

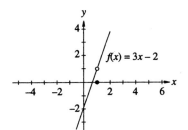

$$\lim_{x \to 1} f(x) = 1$$

15.

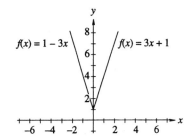

$$\lim_{x \to 0} f(x) = 1$$

13.

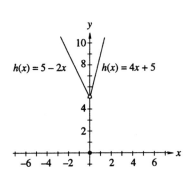

$$\lim_{x \to 0} h(x) = 5$$

17.

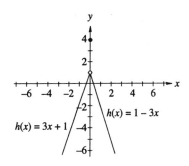

$$\lim_{x \to 0} h(x) = 1$$

19. $\lim_{x \to 2} (-15) = -15$

23. $\lim_{x \to 8} x = 8$

27. 4

21. $\lim_{x \to -4} x = -4$

25. 0

≡ 22.4 ONE-SIDED LIMITS

1. (a) $\lim_{x \to 2^-} f(x) = -2$;
(b) $\lim_{a \to 2^+} f(x) = 1$;
(c) $\lim_{x \to 3^-} f(x) = 4$;
(d) $\lim_{x \to 3^+} f(x) = 4$

3. $\lim\limits_{x \to 5^+} (x - 2) = 3$;

x	6	5.5	5.1
$x - 2$	4	3.5	3.1

5. $\lim\limits_{x \to 4^+} \dfrac{x}{x - 4} = \infty$;

x	5	4.5	4.1	4.01
$\frac{x}{x-4}$	5	9	41	401

7. $\lim\limits_{x \to -2^+} \dfrac{x^2 - 4}{x + 2} = -4$;

x	-1	-1.5	-1.9	-1.99
$\frac{x^2-4}{x+2}$	-3	-3.5	-3.9	-3.99

9. $\lim\limits_{x \to 1^-} \dfrac{|x - 1|}{x - 1} = -1$;

x	0	0.5	0.9
$\frac{|x-1|}{x-1}$	-1	-1	-1

11. $\lim\limits_{x \to 2^-} \dfrac{|x - 2|}{x - 2} = -1$

x	1	1.5	1.9
$\frac{|x-2|}{x-2}$	-1	-1	-1

13. $\lim\limits_{x \to 2^+} \sqrt{x - 2} = 0$;

x	3	2.5	2.1	2.01
$\sqrt{x - 2}$	1	0.707	0.316	0.1

15. $\lim\limits_{x \to \infty} \dfrac{9}{3x} = 0$;

x	10	100	1000
$\frac{9}{3x}$	0.3	0.03	0.003

17. $\lim\limits_{x \to -\infty} \dfrac{1}{2x + 1} = 0$;

x	-10	-100	-1000
$\frac{1}{2x+1}$	-0.0526	-0.00503	-0.0005

19. 2

21. 0

23. $-\infty$

25. **(a)** $C(0.5) = \dfrac{0.25(0.5)}{(0.5)^2 + 5} = 0.02381$

(b) $C(2) = \dfrac{0.25(2)}{(2)^2 + 5} = 0.0556$

(c) $\lim\limits_{t \to \infty} C(t) = \lim\limits_{t \to \infty} \dfrac{0.25t}{t^2 + 5} = \dfrac{\frac{0.25t}{t^2}}{\frac{t^2}{t^2} + \frac{5}{t^2}} =$

$\lim\limits_{t \to \infty} \dfrac{\frac{0.25}{t}}{1 + \frac{5}{t^2}} = \lim\limits_{t \to \infty} \dfrac{0}{1 + 0} = 0$

27. **(a)** $\lim\limits_{t \to 3^-} C(t) = 4.75$ since $t < 3$

(b) $\lim\limits_{x \to 3^+} C(t) = 0.65(3) + 2.80 = 4.75$ since $t > 3$.

(c) $\lim\limits_{x \to 3} C(t) = 4.75$ since $\lim\limits_{t \to 3^-} C(t) = \lim\limits_{x \to 3^+} C(t) = 4.75$.

22.5 ALGEBRAIC TECHNIQUES FOR FINDING LIMITS

1. $\lim\limits_{x \to 5} 8 = 8$

3. $\lim\limits_{x \to 8} (x - 3) = 8 - 3 = 5$

5. $\lim\limits_{x \to -6} (x + 7) = -6 + 7 = 1$

7. $\lim\limits_{x \to 2} 3x = 3 \cdot 2 = 6$

9. $\lim\limits_{x \to 6} \dfrac{2}{3}x + 5 = \dfrac{2}{3} \cdot 6 + 5 = 4 + 5 = 9$

11. $\lim\limits_{p \to 3} \dfrac{p^2 + 6}{p} = \dfrac{3^2 + 6}{3} = \dfrac{15}{3} = 5$

13. $\lim\limits_{s \to 5} \dfrac{3s^2 + 5}{2s} = \dfrac{3 \cdot 5^2 + 5}{2 \cdot 5} = \dfrac{80}{10} = 8$

15. $\lim\limits_{x \to 2} (x + 1)^3 = (2 + 1)^3 = 3^3 = 27$

17. $\lim\limits_{x \to -1} (2x^2 - 1)^2 = (2(-1)^2 - 1)^2 = 1^2 = 1$

19. $\lim\limits_{x \to 3} \sqrt{x + 3} = \sqrt{3 + 3} = \sqrt{6}$

21. $\lim\limits_{x \to 2} \dfrac{x^2 - 2x}{x - 2} = \lim\limits_{x \to 2} \dfrac{x(x - 2)}{x - 2} = \lim\limits_{x \to 2} x = 2$

23. $\lim\limits_{x \to -1} \dfrac{x^2 - 1}{x + 1} = \lim\limits_{x \to -1} \dfrac{(x + 1)(x - 1)}{x + 1} = \lim\limits_{x \to -1} (x - 1) = -2$

25. $\lim\limits_{x \to -4} \dfrac{x^2 + 2x - 8}{x^2 + 5x + 4} = \lim\limits_{x \to -4} \dfrac{(x + 4)(x - 2)}{(x + 4)(x + 1)} =$

$\lim\limits_{x \to -4} \dfrac{x - 2}{x + 1} = \dfrac{-6}{-3} = 2$

27. $\lim\limits_{t \to -7} \dfrac{2t^2 + 15t + 7}{t^2 + 5t - 14} = \lim\limits_{t \to -7} \dfrac{(2t + 1)(t + 7)}{(t - 2)(t + 7)} =$

$\lim\limits_{t \to -7} \dfrac{2t + 1}{t - 2} = \dfrac{-14 + 1}{-7 - 2} = \dfrac{-13}{-9} = \dfrac{13}{9}$

29. $\lim\limits_{h \to 0} \dfrac{(4 + h)^2 - 4^2}{h} = \lim\limits_{h \to 0} \dfrac{16 + 8h + h^2 - 16}{h} =$

$\lim\limits_{h \to 0} (8 + h) = 8$

31. $\lim\limits_{x \to \infty} \dfrac{2x^2 + 4}{5x^2 + 3x + 2} =$

$\lim\limits_{x \to \infty} \dfrac{2 + \frac{4}{x^2}}{5 + \frac{3}{x} + \frac{2}{x^2}} x^2 = \dfrac{2}{5}$

33. $\lim\limits_{x \to \infty} \dfrac{7x^4 + 5x^2}{x^5 + 2} = \lim\limits_{x \to \infty} \dfrac{\frac{7}{x} + \frac{5}{x^3}}{1 + \frac{2}{x^5}} = \dfrac{0}{1} = 0$

35. **(a)** This is true for any two functions f and g where $\lim\limits_{x \to 1} f(x) = 0$ and $\lim\limits_{x \to 1} g(x) = \infty$. For example, $f(x) = x - 1$ and $g(x) = \frac{1}{(x-1)^2}$.

(b) One such example is $f(x) = (x - 1)^3$ and $g(x) = \frac{1}{(x-1)^2}$.

(c) One such example is $f(x) = 5(x - 1)^2$ and $g(x) = \frac{1}{(x-1)^2}$.

≡ 22.6 CONTINUITY

1. $f(x) = \begin{cases} 3x + 1 & \text{if } x \leq 4 \\ x^2 - 3 & \text{if } x > 4 \end{cases} \quad c = 4 =$
Continuous

3. Not continuous, $\lim\limits_{x \to 1} h(x) = 4 \neq 3 = h(1)$

5. Not continuous, $k(2)$ is not defined

13. -1

15. $-3, 2$

21. $f(x) = \frac{x}{x+3}; (-\infty, -3), (-3, \infty)$

23. $h(x) = \frac{x+2}{x^2-4}; (-\infty, -2), (-2, 2), (2, \infty)$

25. $k(x) = \sqrt{x + 3}; [-3, \infty)$

27. $g(x) = \sqrt{x^2 - 9}; (-\infty, -3), [3, \infty)$

29. $j(x) = \frac{1}{x}; (-\infty, 0, (0, \infty)$

31. **(a)** Yes, because when $r = R$, $\frac{GMr}{R^3} = \frac{GMr}{r^3} = \frac{GM}{r^2}$ and the two formulas connect.

(b)

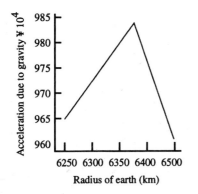

7. Not continuous, $\lim\limits_{x \to -2^-} g(x) = 4$ and $\lim\limits_{x \to -2^+} g(x) = 10$. Since $\lim\limits_{x \to -2^-} g(x) \neq \lim\limits_{x \to -2^+} g(x)$, the limit does not exist.

9. Not continuous, $\lim\limits_{x \to 2} j(x) = 4 \neq 5 = j(2)$

11. None

17. None   **19.** 2

33. **(a)** $\lim\limits_{x \to 23,900^-} T(x) = 0.15(23,900) = \$3,585$

(b) $\lim\limits_{x \to 23,900^+} T(x) = 0.28(23,900) - 3,107 = \$3,585$

(c) $\lim\limits_{x \to 23,900} T(x) = \$3,585$ since $\lim\limits_{x \to 23,900^-} T(x) = \lim\limits_{x \to 23,900^+} T(x)$

(d) Yes

(e) $\lim\limits_{x \to 61,650^-} T(x) = 0.28(61,650) - 3,107 = \$14,155$

(f) $\lim\limits_{x \to 61,650^+} T(x) = 0.33(61,650) - 6,189.50 = \$14,155$

(g) $\lim\limits_{x \to 61,650} T(x) = 14,155$ because $\lim\limits_{x \to 61,650^-} T(x) = \lim\limits_{x \to 61,650^+} T(x)$

(h) Yes

≡ CHAPTER 22 REVIEW

1. $\overline{m} = \frac{f(2) - f(0)}{2 - 0} = \frac{-3 - -7}{2} = \frac{4}{2} = 2$

2. $\overline{m} = \frac{g(-1) - g(-3)}{-1 - -3} = \frac{1 - -7}{-1 + 3} = \frac{8}{2} = 4$

3. $\overline{m} = \frac{h(-2) - h(-3)}{-2 - -3} = \frac{9 - 19}{-2 + 3} = \frac{-10}{1} = -10$

4. $\overline{m} = \frac{j\left(\frac{1}{2}\right) - j(0)}{\frac{1}{2} - 0} = \frac{\frac{8}{9/2} - 2}{\frac{1}{2}} = \frac{\frac{16}{9} - 2}{\frac{1}{2}} = \frac{\frac{-2}{9}}{\frac{1}{2}} = \frac{-4}{9}$

5. $\overline{m} = \frac{k(b) - k(-1)}{b - -1} = \frac{b^2 + 2b - (-1)}{b + 1} = \frac{b^2 + 2b + 1}{b + 1} = b + 1$

6. $\overline{m} = \frac{m(x_1) - m(6)}{x_1 - 6} = \frac{x_1^2 - 5x_1 - 6}{x_1 - 6} = \frac{(x_1 - 6)(x_1 + 1)}{x_1 - 6} = x_1 + 1$

7. $\Delta x = \frac{-2 + 5}{6} = \frac{3}{6} = 0.5;$

x^*	-4.75	-4.25	-3.75	-3.25	-2.75	-2.25
y^*	12	10	8	6	4	2

$\sum y_i^* \Delta x = 42(0.5) = 21$

8. $\Delta x = \frac{4 - 0}{8} = 0.5;$

x^*	0.25	0.75	1.25	1.75	2.25	2.75	3.25	3.75
y^*	5.125	6.125	8.125	11.125	15.125	20.125	26.125	33.125

$\sum y_i^* \Delta x = 125(0.5) = 62.5$

9. $\Delta x = \frac{4 - 2}{8} = \frac{2}{8} = 0.25;$

x^*	2.125	2.375	2.625	2.875	3.125	3.375	3.625	3.875
y^*	16.796875	20.671875	24.921875	29.546875	34.546875	39.921875	45.671875	51.796875

$\sum y_i^* \Delta x = 263.875(0.25) = 65.96875$

10. $\Delta x = \frac{1 - (-1)}{8} = \frac{2}{8} = 0.25;$

x^*	-0.875	-0.625	-0.375	-0.125	0.125	0.375	0.625	0.875
y^*	3.46875	4.21875	4.71875	4.96875	4.96875	4.71875	4.21875	3.46875

$\sum y_i^* \Delta x = 34.75(0.25) = 8.6875$

11. $\lim\limits_{x \to 1}(3x^2 + 2x - 5) = 3 + 2 - 5 = 0$

12. $\lim\limits_{x \to -2} \frac{3x^2 - 12}{x + 2} = \lim\limits_{x \to -2} \frac{3(x + 2)(x - 2)}{x + 2} = \lim\limits_{x \to -2} 3(x - 2) = 3(-2 - 2) = -12$

13. $\lim\limits_{x \to 0} \frac{3x^2 - 12}{x - 2} = \frac{-12}{-2} = 6$

14. $\lim\limits_{x \to -3} \frac{x^2 - 9}{3x + 9} = \lim\limits_{x \to -3} \frac{(x + 3)(x - 3)}{3(x + 3)} = \lim\limits_{x \to -3} \frac{x - 3}{3} = \frac{-6}{3} = -2$

15. $\lim\limits_{x \to \infty} \frac{6x^2 + 3}{2x^2} = \lim\limits_{x \to \infty} \frac{6 + \frac{3}{x^2}}{2} = \frac{6}{2} = 3$

16. $\lim\limits_{x \to \infty} \frac{9x^3 + 2x - 1}{3x^2 + x + 1} = \lim\limits_{x \to \infty} \frac{9x + \frac{2}{x} - \frac{1}{x^2}}{3 + \frac{1}{x} + \frac{1}{x^2}} = \lim\limits_{x \to \infty} \frac{9}{3} x = \infty$ or does not exist

17. $\lim\limits_{x \to 5^+} \sqrt{x - 5} = \sqrt{5 - 5} = \sqrt{0} = 0$

18. $\lim\limits_{x \to 4^-} \frac{x + 4}{x^2 - 16} = \lim\limits_{x \to 4^-} \frac{1}{x - 4} = -\infty$ or does not exist

19. $\displaystyle\lim_{x\to\infty}\frac{x+4}{2x^2+1}=\lim_{x\to\infty}\frac{\frac{1}{x}+\frac{4}{x^2}}{2+\frac{1}{x^2}}=\frac{0}{2}=0$

20. Since both $\displaystyle\lim_{x\to1^-}f(x)=0$ and $\displaystyle\lim_{x\to1^+}f(x)=0$, we have $\displaystyle\lim_{x\to1}f(x)=0$.

21. At $x=5$

22. At $x=0$

23. At $x=-3$

24. At $x=-1$ and 3

25. x cannot equal -7 so $(-\infty,-7),(-7,\infty)$

26. x cannot equal -3 or -2 so $(-\infty,-3),(-3,-2),$ $(-2,\infty)$

27. x^2-25 must be greater than 0 so $x^2>25$; or $x<-5$ or $x>5;(-\infty,-5),(5,\infty)$

28. $3+x>0$ and $3-x\geq0$ so $x>-3$ and $x\leq3$; $(-3,3]$

CHAPTER 22 TEST

1. $\displaystyle\overline{m}=\frac{f(1)-f(0)}{1-0}=\frac{-1-(-3)}{1-0}=2$

2. Here $\Delta x=\frac{5-1}{4}=1$, so the area is $\displaystyle\sum_{i=1}^{4}g(x_i^*)\Delta x=$ $[g(1.5)+g(2.5)+g(3.5)+g(4.5)]\Delta x=(3.75+13.75+29.75+51.75)(1)=99.$

3. $\displaystyle\lim_{x\to2}(5x^2-7)=\left(\lim_{x\to2}5\right)\left(\lim_{x\to2}x^2\right)-\lim_{x\to2}7=5(2^2)-7=5\cdot4-7=20-7=13$

4. $\displaystyle\lim_{x\to2}\left(\frac{x^2-4}{3x-6}\right)=\lim_{x\to2}\left(\frac{(x+2)(x-2)}{3(x-2)}\right)=$ $\displaystyle\lim_{x\to2}\left(\frac{x+2}{3}\right)=\frac{4}{3}$

5. $\displaystyle\lim_{x\to1^+}\begin{cases}x^3-1 & \text{if } x<1\\ 4 & \text{if } x=1\\ 1+x & \text{if } x>1\end{cases}=\lim_{x\to1^+}(1+x)=1+1=2$

6. $\displaystyle\lim_{x\to\infty}\frac{5x^3-4x+1}{7x^3-2}=\lim_{x\to\infty}\frac{\frac{5x^3-4x+1}{x^3}}{\frac{7x^3-2}{x^3}}=$ $\displaystyle\lim_{x\to\infty}\frac{\frac{5x^3}{x^3}-\frac{4x}{x^3}+\frac{1}{x^3}}{\frac{7x^3}{x^3}-\frac{2}{x^3}}=\lim_{x\to\infty}\frac{5-\frac{4}{x^2}+\frac{1}{x^3}}{7-\frac{2}{x^3}}=$ $\displaystyle\frac{5-0+0}{7-0}=\frac{5}{7}$

7. $\displaystyle\lim_{x\to2^-}h(x)=\lim_{x\to2^-}(x+7)=9;\ \lim_{x\to2^+}h(x)=$ $\displaystyle\lim_{x\to2^+}\frac{1}{x^2-9}=-\frac{1}{5}$. Since $\displaystyle\lim_{x\to2^-}h(x)\neq\lim_{x\to2^+}h(x)$, $x=2$ is a point of discontinuity. Also, $\frac{1}{x^2-9}=\frac{1}{(x+3)(x-3)}$, so h is not defined for $x=3$, and 3 is a point of discontinuity.

8. The only possible trouble points are where the denominator is 0. The denominator of $j(x)=\frac{x^2-9}{x^2-x-12}=\frac{x^2-9}{(x+3)(x-4)}$ is 0 when $x^2-x-12=(x+3)(x-4)=0$ or when $x=-3$ or $x=4$. So, these are the only two points of discontinuity and j is continuous over the intervals $(-\infty,-3),(-3,4),$ $(4,\infty)$.

CHAPTER

23

The Derivative

≡ 23.1 THE TANGENT QUESTION AND THE DERIVATIVE

1. Step 1: $f(x + h) = 3(x + h) + 2 = 3x + 3h + 2.$
Step 2: $f(x+h) - f(x) = (3+3h+2) - (3x+2) = 3h.$
Step 3: $\frac{3h}{h} = 3.$
Step 4: $\lim_{h \to 0} 3 = 3 : f'(x) = 3$

3. Step 1: $f(x + h) = 4 - 3(x + h) = 4 - 3x - 3h.$
Step 2: $f(x+h) - f(x) = (4-3x-3h) - (4-3x) = -3h.$
Step 3: $\frac{-3h}{h} = -3.$
Step 4: $\lim_{h \to 0} -3 = -3 : f'(x) = -3$

5. Step 1: $3(x + h)^2 = 3(x^2 + 2xh + h^2) = 3x^2 + 6xh + 3h^2.$
Step 2: $(3x^2 + 6xh + 3h^2) - 3x^2 = 6xh + 3h^2.$
Step 3: $\frac{6xh+3h^2}{h} = 6x + 3h.$
Step 4: $\lim_{h \to 0} (6x + 3h) = 6x : y' = 6x$

7. Step 1: $2(x + h)^2 + 5 = 2(x^2 + 2xh + h^2) + 5 = 2x^2 + 4xh + 2h^2 + 5.$
Step 2: $(2x^2 + 4xh + 2h^2 + 5) - (2x^2 + 5) = 4xh + 2h^2.$
Step 3: $\frac{4xh+2h^2}{h} = 4x + 2h.$
Step 4: $\lim_{h \to 0} (4x + 2h) = 4x : y' = 4x$

9. Step 1: $k(x + h) = 3 - 4(x + h)^2 = 3 - 4(x^2 + 2xh + h^2) = 3 - 4x^2 - 8xh - 4h^2.$
Step 2: $(3-4x^2-8xh-4h^2) - (3-4x^2) = -8xh - 4h^2$
Step 3: $\frac{-8xh-4h^2}{h} = -8x - 4h$
Step 4: $\lim_{h \to 0} (-8x - 4h) = -8x : k'(x) = -8x$

11. Step 1: $-4(x+h)^3 = -4(x^3+3x^2h+3xh^2+h^3) = -4x^3 - 12x^2h - 12xh^2 - 4h^3.$
Step 2: $(-4x^3-12x^2h-12xh^2-4h^3) - (-4x^3) = -12x^2h - 12xh^2 - 4h^3.$
Step 3: $\frac{12x^2h-12xh^2-4h^3}{h} = -12x^2 - 12xh - 4h^2.$
Step 4: $\lim_{h \to 0} -12x^2 - 12xh - 4h^2 = -12x^2 : y' = -12x^2$

13. Step 1: $(t + h) - 3(t + h)^3 = t + h - 3t^3 - 9t^2h - 9th^2 - 3h^3.$
Step 2: $(t+h-3t^3-9t^2h-9th^2-3h^3) - (t-3t^3) = h - 9t^2h - 9th^2 - 3h^3.$
Step 3: $\frac{h-9t^2h-9th^2-3h^3}{h} = 1 - 9t^2 - 9th - 3h^2.$
Step 4: $\lim_{h \to 0} (1 - 9t^2 - 9th - 3h^2) = 1 - 9t^2 : y' = 1 - 9t^2.$

15. Step 1: $f(x + h) = \frac{1}{x+h+1}.$
Step 2: $\frac{1}{x+h+1} - \frac{1}{x+1} = \frac{(x+1)-(x+h+1)}{(x+1)(x+h+1)} = \frac{-h}{(x+1)(x+h+1)}.$
Step 3: $\frac{\frac{-h}{(x+1)(x+h+1)}}{h} = \frac{-1}{(x+1)(x+h+1)}.$
Step 4: $\lim_{h \to 0} \frac{-1}{(x+1)(x+h+1)} = \frac{-1}{(x+1)^2} : f'(x) = \frac{-1}{(x+1)^2}$

17. Step 1: $j(x + h) = \frac{4}{1-(x+h)^2}.$
Step 2: $\frac{4}{1-(x+h)^2} - \frac{4}{1-x^2} = \frac{4(1-x^2)-4[1-(x+h)^2]}{[1-(x+h)^2](1-x^2)} = \frac{4-4x^2-4+4x^2+8xh+4h^2}{[1-(x+h)^2](1-x^2)} = \frac{8xh+4h^2}{[1-(x+h)^2](1-x^2)}.$
Step 3: $\frac{\frac{8xh+4h^2}{[1-(x+h)^2](1-x^2)}}{h} = \frac{8x+4h}{[1-(x+h^2](1-x^2)}.$
Step 4: $\lim_{h \to 0} \frac{8x+4h}{[1-(x+h)^2](1-x^2)} = \frac{8x+4\cdot 0}{[1-(x+0)^2](1-x^2)} = \frac{8x}{(1-x^2)^2} : j'(x) = \frac{8x}{(1-x^2)^2}$

19. Step 1: $16(t + h)^2 - 6(t + h) + 3 = 16t^2 + 32th + 16h^2 - 6t - 6h + 3.$
Step 2: $(16t^2+32th+16h^2-6t-6h+3) - (16t^2-6t+3) = 32th + 16h^2 - 6h$
Step 3: $\frac{32th+16h^2-6h}{h} = 32t + 16h - 6$
Step 4: $\lim_{h \to 0} (32t+16h-6) = 32t-6 : s' = 32t-6$

21. Step 1: $\sqrt{(x + h)^2 + 4}$
Step 2: $\sqrt{(x + h)^2 + 4} - \sqrt{x^2 + 4}.$
Step 3: $\frac{\sqrt{(x+h)^2+4}-\sqrt{x^2+4}}{h} \cdot$
$\frac{\sqrt{(x+h)^2+4}+\sqrt{x^2+4}}{\sqrt{(x+h)^2+4}+\sqrt{x^2+4}} = \frac{((x+h)^2+4)-(x^2+4)}{h[\sqrt{(x+h)^2+4}+\sqrt{x^2+4}]}$

$$= \frac{(x^2+2xh+h^2+4)-(x^2+4)}{h[\sqrt{(x+h)^2+4}+\sqrt{x^2+4}]} =$$

$$\frac{2xh+h^2}{h[\sqrt{(x+h)^2+4}+\sqrt{x^2+4}]} =$$

$$\frac{2x+h}{\sqrt{(x+h)^2+4}+\sqrt{x^2+4}}$$

Step 4: $\lim\limits_{h \to 0} \dfrac{2x+h}{h[\sqrt{(x+h)^2+4}+\sqrt{x^2+4}]} = \dfrac{2x}{2\sqrt{x^2+4}} =$

$$\frac{x}{\sqrt{x^2+4}} : \quad y' = \frac{x}{\sqrt{x^2+4}}$$

23. Step 1: $\sqrt{2(x+h)-(x+h)^2}$
Step 2: $\sqrt{2(x+h)-(x+h)^2} - \sqrt{2x-x^2}$
Step 3: $\dfrac{\sqrt{2(x+h)-(x+h)^2}-\sqrt{2x-x^2}}{h} \cdot$

$$\frac{\sqrt{2(x+h)-(x+h)^2}+\sqrt{2x-x^2}}{\sqrt{2(x+h)-(x+h)^2}+\sqrt{2x-x^2}} =$$

$$\frac{[2(x+h)-(x+h)^2]-[2x-x^2]}{h[\sqrt{2(x+h)-(x+h)^2}+\sqrt{2x-x^2}]} =$$

$$\frac{2h-2xh-h^2}{h[\sqrt{2(x+h)-(x+h)^2}+\sqrt{2x-x^2}]} =$$

$$\frac{2-2x-h}{\sqrt{2(x+h)-(x+h)^2}+\sqrt{2x-x^2}}$$

Step 4: $\lim\limits_{h \to 0} \dfrac{2-2x-h}{\sqrt{2(x+h)-(x+h)^2}+\sqrt{2x-x^2}}$

$$= \frac{2-2x}{2\sqrt{2x-x^2}} = \frac{1-x}{\sqrt{2x-x^2}}$$

$$y' = \frac{1-x}{\sqrt{2x-x^2}}$$

25. $f(x) = 3x + 7; f'(x) = 3; f'(2) = 3; f'(9) = 3$

27. $s(t) = 16t^2+2t; s'(t) = 32t+2; s'(3) = 32\cdot3+2 = 98; s'(2) = 32 \cdot 2 + 2 = 66; s'(0) = 32 \cdot 0 + 2 = 2$

29. $q(t) = t^3 - 4t^2; q'(t) = 3t^2 - 8t; q'(0) = 0; q'(3) = 3 \cdot 3^2 - 8 \cdot 3 = 27 - 24 = 3.$

31. $k(x) = \frac{1}{x^2-4}; k'(x) = \frac{-2x}{(x^2-4)^2}; k'(5) = \frac{-10}{21^2} = \frac{-10}{441}; k'(-5) = \frac{-2(-5)}{((-5)^2-4)^2} = \frac{10}{(25-4)^2} = \frac{10}{21^2} = \frac{10}{441}.$

33. $j'(t) = \sqrt{5t - t^2}; j'(t) = \frac{5-2t}{2\sqrt{5t-t^2}}; j'(1) = \frac{5-2}{2\sqrt{5-1}}$

$$= \frac{3}{2\sqrt{4}} = \frac{3}{4}; \quad j'(4) = \frac{5-8}{2\sqrt{20-16}} = \frac{-3}{2\sqrt{4}} = \frac{-3}{4} \text{ or } -\frac{3}{4}.$$

35. $f(x) = \dfrac{5}{\sqrt{25 - x^2}}$

$$f'(x) = \frac{-5(-2x)}{2\sqrt{25-x^2}^3} = \frac{5x}{\sqrt{25-x^2}^3}$$

$$f(0) = 0; \quad f'(3) = \frac{15}{\sqrt{25-9}^3} = \frac{15}{\sqrt{16}^3}$$

$$= \frac{15}{4^3} = \frac{15}{64};$$

$$f'(-4) = \frac{-20}{\sqrt{25-16}^3} = \frac{-20}{3^3} = \frac{-20}{27}$$

or $-\dfrac{20}{27}$

The four-step process for finding $f'(x)$ is as follows:

Step 1: $f(x + h) = \dfrac{5}{\sqrt{25-(x+h)^2}}$

Step 2: $\dfrac{5}{\sqrt{25-(x+h)^2}} - \dfrac{5}{\sqrt{25-x^2}} =$

$$\frac{5\sqrt{25-x^2}-5\sqrt{25-(x+h)^2}}{\sqrt{25-(x+h)^2}\cdot\sqrt{25-x^2}} \times$$

$$\frac{5\sqrt{25-x^2}+5\sqrt{25-(x+h)^2}}{5\sqrt{25-x^2}+5\sqrt{25-(x+h)^2}} =$$

$$\frac{5(25-x^2-(25-x^2-2xh-h^2)}{\sqrt{E_1}\sqrt{E_2}(\sqrt{E_2}+\sqrt{E_1})} =$$

$$\frac{5(2xh+h^2)}{\sqrt{E_2}\sqrt{E_2}(\sqrt{E_2}+\sqrt{E_1})} =$$

Step 3: $\dfrac{10x+5h}{\sqrt{E_1}\sqrt{E_2}(\sqrt{E_2}+\sqrt{E_1})}$

Step 4: $\lim\limits_{h \to 0} \dfrac{10x+5h}{\sqrt{25-(x+h)^2}\sqrt{25-x^2}\times(\sqrt{25-(x+h)^2}+\sqrt{25-x^2})}$

$$= \frac{10x}{2\sqrt{25-x^2}^3} = \frac{5x}{\sqrt{25-x^2}^3}$$

≡ 23.2 DERIVATIVES OF POLYNOMIALS

1. $f(x) = 19; f'(x) = 0$

3. $h(x) = 7x - 5; h'(x) = 7$

5. $k(x) = 9x^2; k'(x) = 9(2x) = 18x$

7. $f(x) = \frac{1}{3}x^{15}; f'(x) = \frac{1}{3}(15x^{14}) = 5x^{14}$

9. $h(x) = 5x^{-2}; h'(x) = 5(-2x^{-3}) = -10x^{-3}$

11. $k(x) = -\frac{2}{3}x^{3/2}; k'(x) = -\frac{2}{3}(\frac{3}{2}x^{1/2}) = -x^{1/2}$

13. $h(x) = 4\sqrt{3x} = 4\sqrt{3}x^{1/2}; h'(x) = 4\sqrt{3} \cdot \frac{1}{2}x^{-1/2} = 2\sqrt{3}x^{-1/2}$ or $\frac{2\sqrt{3}}{\sqrt{x}}$

15. $g(x) = \frac{3}{\sqrt[3]{x}} = 3x^{-1/3}; g(x) = 3(-\frac{1}{3}x^{-4/3}) = -x^{-4/3}$ or $\frac{-1}{\sqrt[3]{x^4}} = \frac{-1}{x\sqrt[3]{x}}$

17. $g(x) = 9x^2 + 3x - 4; g'(x) = 18x + 3$

19. $h(x) = \frac{1}{3}x^3 + \frac{1}{2}x^2 - 5x + \frac{2}{3}x^{-2} + 5; h'(x) = x^2 + x - 5 - \frac{4}{3}x^{-3}$

21. $f(x) = \frac{4}{5}x^5 - \frac{3}{2}x^3 - 7x^0 + \frac{1}{2}x^{-2} - \frac{2}{3}x^{-3}; f'(x) = 4x^4 - \frac{9}{2}x^2 - x^{-3} + 2x^{-4}$

23. $l(x) = 3\sqrt{3}x^4 + 3\sqrt{5}x + \sqrt{4x^4} = 3\sqrt{3}x^4 + 3\sqrt{5}x^{1/2}$
$+2x^2; l'(x) = 12\sqrt{3}x^3 + \frac{3\sqrt{5}}{2}x^{-1/2} + 4x = 12\sqrt{3}x^3$
$+ \frac{3\sqrt{5}}{2\sqrt{x}} + 4x$

25. $s(t) = 16t^2 - 32t + 5; s'(t) = 32t - 32$

27. $\alpha(t) = 30 - 4.0t^2 + 2t^{1/2}; \alpha'(t) = -8.0t + t^{-1/2}$

29. First, rewrite $q(t) = \sqrt{6t} - 4t\sqrt{3t} + \sqrt[3]{t^2}$ as $q(t) = \sqrt{6}t^{1/2} - 4\sqrt{3}t^{3/2} + t^{2/3}$. Then, differentiating produces $q'(t) = \frac{\sqrt{6}}{2}t^{-1/2} - 6\sqrt{3}t^{1/2} + \frac{2}{3}t^{-1/3}$.

31. Rewrite $g(x) = 5x^2\sqrt[3]{x}$ as $g(x) = 5x^{2\frac{1}{3}}$ or $g(x) = 5x^{7/3}$. Then, differentiating gives $g'(x) = 5 \cdot \frac{7}{3}x^{4/3} = \frac{35}{3}x^{4/3}$ or $\frac{35}{3}x\sqrt[3]{x}$.

33. Here, we rewrite $h(x) = \frac{4}{x^2} = 4x^{-2}$. Then, differentiation produces $h'(x) = -8x^{-3} = \frac{-8}{x^3}$

35. Rewrite $K(x) = \frac{4\sqrt[3]{x}}{\sqrt{x}}$ as $K(x) = 4x^{1/3-1/2} = 4x^{-1/6}$. Then, differentiating produces $K'(x) = -\frac{2}{3}x^{-7/6}$

37. $f(x) = \dfrac{4\sqrt{3x} + 3x^2 - 7\sqrt{x^3} + 2x^{-1}}{\sqrt{x}}$
$= \dfrac{4\sqrt{3x}}{\sqrt{x}} + \dfrac{3x^2}{\sqrt{x}} - \dfrac{7\sqrt{x^3}}{\sqrt{x}} + \dfrac{2x^{-1}}{\sqrt{x}}$
$= 4\sqrt{3} + 3x^{3/2} - 7x + 2x^{-3/2}$
$f'(x) = \dfrac{9}{2}x^{1/2} - 7 - 3x^{-5/2}$

39. Here we have $f(x) = x^3 + 3x^2$, and so $f'(x) = 3x^2 + 6x$. Now, $f'(x) = 0$ means that $3x^2 + 6x = 0$ or $3x(x + 2) = 0$. Thus, by the zero product principle, $x = 0$ or $x = -2$.

41. We are given $g(x) = 2x^3 + x^2 - 4x$ and determine that $g'(x) = 6x^2 + 2x - 4$. Now $g'(x) = 0$ is the same as $6x^2 + 2x - 4 = 0$, or $2(3x^2 + x - 2) = 0$. This factors as $2(3x - 2)(x + 1) = 0$ which, by the zero product principle, has solutions when $x = \frac{2}{3}$ or $x = -1$.

43. Since $j(t) = 4t^3 + \frac{1}{t}$, then $j'(t) = 12t^2 - t^{-2} = 12t^2 - \frac{1}{t^2}$. Now, $j'(t) = 0$, means that $12t^2 - \frac{1}{t^2} = 0$ or $12t^2 = \frac{1}{t^2}$ and so $12t^4 = 1$, or $t^4 = \frac{1}{12}$. Hence, $t = \sqrt[4]{\frac{1}{12}} \approx 0.537285$.

45. **(a)** $P(6) = 146,000 + 64(6) - 0.04(6)^2 = 146,382$
(b) Since the rate of change at time t is given by $P'(t) = 64 - 0.08t$, then the rate of change

in the city's population six months from now is $P'(6) = 64 - 0.08(6) = 63.52$. This represents a 63.52 increase in the population per month.

47. **(a)** The distance the ball has traveled when $t = 4$ s is given by $s(4) = 15(4) + 7(4)^2 = 172$. The ball has traveled 172 ft when $t = 4$ s.
(b) The velocity (rate of change) of the ball is given by $s'(t) = 15 + 14t$. Thus, $s'(4) = 15 + 14(4) = 71$ and the velocity of the ball when $t = 4$ s is 71 ft/s.

49. **(a)** The city's population six years from now will be $P(6) = 125,000 + 1,500(6) - 36\sqrt{6} \approx 133,912$ people.
(b) We first determine $P'(t)$ by rewriting $P(t) = 125,000 + 1,500t - 36\sqrt{t}$ as $P(t) = 125,000 + 1,500t - 36t^{1/2}$ and then differentiating. Thus, we get $P'(t) = 1,500 - 36\frac{1}{2}t^{-1/2} = 1,500 - \frac{18}{\sqrt{t}}$. Six years from now the rate at which the city's population is growing will be $P'(6) = 1,500 - \frac{18}{\sqrt{6}} \approx 1,493$ people per year.
(c) The city's population ten years from now will be $P(10) = 125,000 + 1,500(10) - 36\sqrt{10} \approx 139,886$ people.
(d) The rate at which the city's population is growing ten years from now is $P'(10) = 1,500 - \frac{18}{\sqrt{10}} \approx 1,494$ people per year.

51. Since $P'(i) = 2 \cdot 25i = 50i$, then $P'(2.4) = 50(2.4) = 120$ W/A.

53. **(a)** $P(n) = R(n) - C(n) = (78n - 0.025n^2) - (9,800 + 22.5n) = 55.5n - 0.025n^2 - 9,800$
(b) $P'(n) = 55.5 - 0.05n$
(c) $P'(150) = 55.5 - 0.05(150) = 55.5 - 2.5 = \$53/\text{telephone}$

55. **(a)** When the rocket is first launched $t = 0$, so the rocket's altitude at launch is $h(0) = -0^4 + 12(0)^3 - 16(0)^2 + 605 = 605$ ft above sea level.
(b) The rocket will fall into the ocean when $h(t) = 0$. That is, when $-t^4 + 12t^3 - 16t^2 + 605 = 0$. Using a graphing utility, we obtain $t = 11$ s.
(c) The rocket's altitude at $t = 2$ s is $h(2) = -(2)^4 + 12(2)^3 - 16(2)^2 + 605 + 621$ ft. The rocket's velocity at any time t is given by $h'(t)$. Differentiating h, we see that $v(t) = h'(t) = -4t^3 + 36t^2 - 32t$. Thus, the rocket's velocity at $t = 2$ s is $v(2) = -4(2)^3 + 36(2)^2 - 32(2) = 48$ ft/s.
(d) The rocket's altitude at $t = 4$ s is $h(4) = -(4)^4 + 12(4)^3 - 16(4)^2 + 605 = 861$ ft. The rocket's velocity at $t = 4$ s is $v(4) = -4(4)^3 + 36(4)^2 - 32(4) = 192$ ft/s.

(e) The rocket's altitude at $t = 6$ s is $h(6) = -(6)^4 + 12(6)^3 - 16(6)^2 + 605 = 1,325$ ft. The rocket's velocity at $t = 6$ s is $v(6) = -4(6)^3 + 36(6)^2 - 32(6) = 240$ ft/s.

(f) The rocket's altitude at $t = 10$ s is $v(10) = -(10)^4 + 12(10)^3 - 16(10)^2 + 605 = 1,005$ ft. The

rocket's velocity at $t = 10$ s is $v(10) = -4(10)^3 + 36(10)^2 - 32(10) = -720$ ft/s.

(g) The rocket's velocity when it falls into the ocean, at $t = 11$ s is $v(11) = -4(11)^3 + 36(11)^2 - 32(11) = -1,320$ ft/s.

≡ 23.3 DERIVATIVES OF PRODUCT AND QUOTIENTS

1. $f'(x) = (3x + 1)(2) + (3)(2x - 7) = 6x + 2 + 6x - 21 = 12x - 19$

3. $h'(x) = (2x^2 + x - 1)(3) + (4x + 1)(3x - 5) = 6x^2 + 3x - 3 + 12x^2 - 17x - 5 = 18x^2 - 14x - 8$

5. $j'(x) = (4 - 3x)(12x + 6) + (-3)(6x^2 + 6x - 4)$
$= 48x + 24 - 36x^2 - 18x - 18x^2 - 18x + 12$
$= -54x^2 + 12x + 36$

7. $q(t) = 4(t^2 - 3t)(t^2 - 3t)$
$q'(t) = 4[(t^2 - 3t)(2t - 3) + (2t - 3)(t^2 - 3t)]$
$= 4[2t^3 - 9t^2 + 9t + 2t^3 - 9t^2 + 9t]$
$= 16t^3 - 72t^2 + 72t$

(Note: we learn a better way to do this problem in the next unit).

9. $f'(w) = (3w^3 - 4w^2 + 2w - 5)(2w + w^{-2})$
$\qquad + (9w^2 - 8w + 2)(w^2 - w^{-1})$
$= 6w^4 - 8w^3 + 4w^2 - 10w + 3w - 4$
$\qquad + 2w^{-1} - 5w^{-2} + 9w^4 - 8w^3$
$\qquad + 2w^2 - 9w + 8 - 2w^{-1}$
$= 15w^4 - 16w^3 + 6w^2 - 16w + 4 - 5w^{-2}$

11. $f'(x) = \frac{4(2x+3) - (4x-1)(2)}{(2x+3)^2} = \frac{8x+12-8x+2}{(2x+3)^2} = \frac{14}{(2x+3)^2}$

13. $g'(x) = \frac{(4x - 1)(18x) - (9x^2 + 2)(4)}{(4x - 1)^2}$
$= \frac{72x^2 - 18x - 36x^2 - 8}{(4x - 1)^2}$
$= \frac{36x^2 - 18x - 8}{(4x - 1)^2}$

15. $j'(s) = 12s - \frac{-12s}{36s^4} = 12s + \frac{1}{3s^3}$

17. $f'(t) = 12t^2 - \frac{(t-2)2 - 2t}{(t-2)^2} = 12t^2 + \frac{4}{(t-2)^2}$

19. $H'(x) = \frac{(x - 1)(3x^2) - (x^3 - 1)}{(x - 1)^2}$
$= \frac{3x^3 - 3x^2 - x^3 + 1}{(x - 1)^2}$
$= \frac{2x^3 - 3x^2 + 1}{(x - 1)^2}$
$= \frac{(2x + 1)(x - 1)(x - 1)}{(x - 1)^2}$
$= 2x + 1$

(Note: $H(x) = \frac{x^3 - 1}{x - 1} = \frac{(x-1)(x^2 + x + 1)}{x - 1} = x^2 + x + 1$ and so, $H'(x) = 2x + 1$.)

21. $f'(\phi) = \frac{(3\phi^2 - 1)(2\phi) - (\phi^2)(6\phi)}{(3\phi^2 - 1)^2}$
$= \frac{6\phi^3 - 2\phi - 6\phi^3}{(3\phi^2 - 1)^2}$
$= \frac{-2\phi}{(3\phi^2 - 1)^2}$

23. $h'(x) = \frac{(x^3 - 3x - 1)(9x^2 - 1) - (3x^3 - x + 1)(3x^2 - 3)}{(x^3 - 3x - 1)^2}$
$= \frac{(9x^5 - 28x^3 - 9x^2 + 3x + 1) - (9x^5 - 12x^3 + 3x^2 + 3x - 3)}{(x^3 - 3x - 1)^2}$
$= \frac{-16x^3 - 12x^2 + 4}{(x^3 - 3x - 1)^2}$

25. $f'(t) = \frac{t^{1/3}(6t - 1) - (3t^2 - t - 1)(\frac{1}{3}t^{-2/3})}{t^{2/3}}$
$\cdot \frac{t^{2/3}}{t^{2/3}} = \frac{t(6t - 1) - (3t^2 - t - 1)\frac{1}{3}}{t^{4/3}}$
$= \frac{6t^2 - t - t^2 + \frac{1}{3}t + \frac{1}{3}}{t^{4/3}}$
$= \frac{5t^2 - \frac{2}{3}t + \frac{1}{3}}{t^{4/3}} = \frac{15t^2 - 2t + 1}{3t^{4/3}}$

27. $g'(w) = 24w + \frac{(w+1) - (w-1)}{(w+1)^2} = 24w + \frac{2}{(w+1)^2}$

29. $h'(x) = \frac{(x^2+1)0 - 1(2x)}{(x^2+1)^2} = \frac{-2x}{(x^2+1)^2}$

31.

$$H'(x) = \frac{(2-x^2)(3x-x^3)(-3x^2)-}{[(2-x^2)(3x-x^3)]^2}$$

$$= \frac{(6x-5x^3+x^5)(-3x^2)}{[(2-x^2)(3x-x^3)]^2}$$

$$= \frac{-18x^3+15x^5-3x^7-(4-x^3)[5x^4-15x^2+6]}{(2-x^2)(3x-x^3)^2}$$

$$= \frac{-18x^3+15x^5-3x^7-20x^4+60x^2-24+5x^7-15x^5+6x^3}{(2-x^2)^2(3x-x^3)^2}$$

$$= \frac{2x^7-20x^4-12x^3+60x^2-24}{(2-x^2)^2(3x-x^3)^2}$$

33. $n(s) = \dfrac{2s^3}{s^3-s^2-s+1}$

$$n'(s) = \frac{(s^3-s^2-s+1)(6s^2)-2s^3(3s^2-2s-1)}{(s^3-s^2-s+1)^2}$$

$$= \frac{6s^5-6s^4-6s^3+6s^2-6s^5+4s^4+2s^3}{(s^3-s^2-s+1)^2}$$

$$= \frac{-2s^4-4s^3+6s^2}{(s^2-1)^2(s-1)^2}$$

35. $y = \dfrac{(2x-3)(x^2-4x+1)}{3x^3+1}$

$$= \frac{2x^3-11x^2+14x-3}{3x^3+1}$$

$$y' = \frac{(3x^3+1)(6x^2-22x+14)}{(3x^3+1)^2}$$
$$\quad \frac{-(2x^3-11x^2+14x-3)(9x^2)}{(3x^3+1)^2}$$

$$= \frac{18x^5-66x^4+42x^3+6x^2-}{22x+14-18x^5+99x^4-126x^3+27x^2}{(3x^3+1)^2}$$

$$= \frac{33x^4-84x^3+33x^2-22x+14}{(3x^3+1)^2}$$

37. $y = \dfrac{(t^2-1)(t-1)}{(2t+1)^2} = \dfrac{t^3-t^2-t+1}{4t^2+4t+1}$

$$y' = \frac{(4t^2+4t+1)(3t^2-2t-1)}{(4t^2+4t+1)^2}$$
$$\quad \frac{-(t^3-t^2-t+1)(8t+4)}{(4t^2+4t+1)^2} = \frac{12t^4+4t^3-9t^2-6t-1}{-8t^4+4t^3+12t^2-4t-4}{(4t^2+4t+1)^2}$$

$$= \frac{4t^4+8t^3+3t^2-10t-5}{(2t+1)^4}$$

$$= \frac{(2t+1)(2t^3+3t^2-5)}{(2t+1)^4} = \frac{2t^3+3t^2-5}{(2t+1)^3}$$

39. $y' = (3x^2+2x-1)(3x^2-1)+(6x+2)(x^3-x+1)$.
At $x = 1, y' = (3+2-1)(3-1)+(6+2)(1-1+1) = 4\cdot2+8\cdot1 = 8+8 = 16$. The slope at $(1,4)$ is 16.

41. $y' = \dfrac{-8}{(x+1)^2}$; at $x = 3, y' = \dfrac{-8}{4^2} = -\dfrac{1}{2}$. Slope of the tangent is $-\dfrac{1}{2}$. Equation of the tangent is $y - 2 = -\dfrac{1}{2}(x - 3)$ or $2y + x = 7$. The slope of the normal is 2; the equation of the normal is $y - 2 = 2(x - 3)$ or $2x - 4$.

43. $y' = \dfrac{(2x+1)5-(5x+4)2}{(2x+1)^2}$; at $x = 1, y' = \dfrac{15-18}{9} = -\dfrac{1}{3}$. The equation of the tangent is $y - 3 = -\dfrac{1}{3}(x - 1)$ or $3y + x = 10$. The equation of the normal is $y - 3 = 3(x - 1)$ or $y - 3x = 0$.

45. $y = \dfrac{9x+2}{x^3+4x^2}$. $y' = \dfrac{(x^3+4x^2)9-(9x+2)(3x^2+8x)}{(x^3+4x^2)^2}$ at $x = -2, y' = \dfrac{(-8+16)9-(-16)(12-16)}{(-8+16)^2} = \dfrac{72-64}{64} = \dfrac{8}{64} = \dfrac{1}{8}$.
Tangent: $y + 2 = \dfrac{1}{8}(x + 2)$ or $8y - x = -14$.
Normal: $y + 2 = -8(x + 2)$ or $y + 8x = -18$.

47.   First, find $P'(t) = (0.8t - 6)(0.5) + (0.8)(0.5t + 9) = (0.4t - 3) + (0.4t + 7.2) = 0.8t + 4.2$. Then, $P'(5) = 0.8(5) + 4.2 = 8.2$. Thus, in 5 years the population will be growing at a rate of 8.2 thousand people per year or 8,200 people per year.

49. $V' = \dfrac{(R+60)110-110R(1)}{(R+60)^2} = \dfrac{110R+6600-110R}{(R+60)^2} = \dfrac{6600}{(R+60)^2}$

51.   **(a)** $N'(t) = \dfrac{(3\sqrt{t}+2)(2t-2)-(t^2-2t+1)(\frac{3}{2}t^{-1/2})}{(3\sqrt{t}+2)^2}$. Multiply this result by $\dfrac{2\sqrt{t}}{2\sqrt{t}}$ to get

$$N'(t) = \frac{2\sqrt{t}(3\sqrt{t}+2)(2t-2)-2\sqrt{t}(t^2-2t+1)(\frac{3}{2}t^{-1/2})}{2\sqrt{t}(3\sqrt{t}+2)^2}$$

$$= \frac{(6t+4\sqrt{t})(2t-2)-3(t^2-2t+1)}{2(3\sqrt{t}+2)^2\sqrt{t}}$$

$$= \frac{(12t^2+8t\sqrt{t}-12t-8\sqrt{t})-3t^2+6t-3}{2(3\sqrt{t}+2)^2\sqrt{t}}$$

$$= \frac{9t^2+8t^{3/2}-6t-8\sqrt{t}-3}{2(3\sqrt{t}+2)^2\sqrt{t}}$$

(b) $N'(5) = \dfrac{9(5)^2+8(5)^{3/2}-6(5)-8\sqrt{5}-3}{2(3\sqrt{5}+2)^2\sqrt{5}} \approx 0.777$ thousand bacteria/hr = 777 bacteria/hr.

23.4 DERIVATIVES OF COMPOSITE FUNCTIONS

1. $f'(x) = 4(3x - 6)^3(3) = 12(3x - 6)^3$

3. $h'(x) = -5(5x - 7)^{-6}(5) = -25(5x - 7)^{-6}$

5. $f'(x) = 4(x^2+3x)^3(2x+3) = 4(2x+3)(x^2+3x)^3$

7. $H(x) = (4x^2 + 7)^{-1/2}$, so $H'(x) = -\dfrac{1}{2}(4x^2 + 7)^{-3/2}(8x) = -4x(4x^2 + 7)^{-3/2}$ or $\dfrac{-4x}{\sqrt{4x^2+7}^3}$.

9. $s'(t) = 3(t^4 - t^3 + 2)^2(4t^3 - 3t^2)$

11. $f'(u) = 3(u^3 + 2u^{-4})^2(3u^2 - 8u^{-5})$

13. $g'(x) = 3[(x-2)(3x^2-x)]^2$
$\times [(x-2)(6x-1)+3x^2-x]$
$= 3[(x-2)(3x^2-x)]^2$
$\times [6x^2-13x+2+3x^2-x]$
$= 3[(x-2)(3x^2-x)]^2[9x^2-14x+2]$

15. $h'(v) = (v^2+1)^3[3(2v-5)^2(2)]$
$+ (2v-5)^3[2(v^2+1)2v]$
$= 6(v^2+1)^3(2v-5)^2$
$+ 4v(v^2+1)(2v-5)^3$

17. $h'(x) = 4\big[(x^3-6x^2)(5x-6x^2+x^3)\big]^3\big[x^3-6x^2)(5-12x+3x^2)+(3x^2-12x)(5x-6x^2+x^3)\big] = 4\big[(x^3-6x^2)(5x-6x^2+x^3)\big]^3\big[3x^5-30x^4+77x^3-30x^2+3x^5-30x^4+87x^3-60x^2\big] = 4\big[(x^3-6x^2)(5x-6x^2+x^3)\big]^3(6x^5-60x^4+164x^3-90x^2)$

19. $k'(x) = (3x^2+2)^3 \cdot \frac{1}{2}(x^3-7x)^{-1/2}(3x^2-7)$
$+ 3(3x^2+2)^2(6x)(x^3-7x)^{1/2}$
$= \frac{(3x^2+2)^3(3x^2-7)}{2\sqrt{x^3-7x}}$
$+ 18x(3x^2+2)^2\sqrt{x^3-7x}$
$= \frac{(3x^2+2)^3(3x^2-7)}{2\sqrt{x^3-7x}}$
$+ \frac{36x(3x^2+2)^2(x^3-7x)}{2\cdot\sqrt{x^3-7x}}$
$= \frac{(3x^2+2)^2[(3x^2+2)(3x^2-7)+36x(x^3-7x)]}{2\sqrt{x^3-7x}}$
$= \frac{(3x^2+2)^2[9x^4-15x^2-14+36x^4-252x^2]}{2\sqrt{x^3-7x}}$
$= \frac{(3x^2+2)^2[45x^4-267x^2-14]}{2\sqrt{x^3-7x}}$

21. $g'(v) = \frac{(v^3-9)^2 \cdot 3(v^2-4v)^2(2v-4)-(v^2-4v)^3 2(v^3-9)(3v^2)}{(v^3-9)^4}$
$= \frac{3(v^3-9)(v^2-4v)^2(2v-4)-6v^2(v^2-4v)^3}{(v^3-9)^4}$
$= \frac{(v^2-4v)^2[3(v^3-9)(2v-4)-6v^2(v^2-4v)]}{(v^3-9)^3}$
$= \frac{(v^2-4v)^2[6v^4-12v^3-54v+108-6v^4+24v^3]}{(v^3-9)^3}$
$= \frac{(v^2-4v)^2[12v^3-54v+108]}{(v^3-9)^3}$

23. $f'(u) = 6u^5; \; u' = 5x^4$
$y' = 6u^5 \cdot 5x^4 = 6(x^5+4)^5 \cdot 5x^4$
$= 30x^4(x^5+4)^5$

25. $f'(u) = \frac{1}{2}u^{-1/2}; \; u' = 8x$
$y' = \frac{1}{2}u^{-1/2} \cdot 8x = 4x(4x^2-5)^{-1/2}$

27. $f'(u) = \frac{2}{3}u^{-1/3}; \; u' = 21x^2-9$
$y' = \frac{2}{3}(7x^3-9x)^{-1/3}(21x^2-9)$
$= 2(7x^2-3)(7x^3-9x)^{-1/3}$

29. $f'(u) = 3(u^2+1)^2(2u); \; u' = -(x+3)^{-2}$
$y' = 3(u^2+1)^2(2u)(-(x+3)^{-2})$
$= 3\left[\frac{1}{(x+3)^2}+1\right]^2\left[\frac{2}{x+3}\right]\left[\frac{-1}{(x+3)^2}\right]$
$= \frac{-6}{(x+3)^3}\left[\frac{1}{(x+3)^2}+1\right]^2$

31. $f'(u) = \frac{1}{3}(u^2-2u)^{-2/3}(2u-2); \; u' = 3x^2$
$y' = \frac{1}{3}[(x^3+4)^2-2(x^3+4)]^{-2/3}$
$\times [2(x^3+4)-2] \cdot 3x^2$
$= 2x^2[(x^3+4)^2-2(x^3+4)]^{-2/3}(x^3+3)$

33. $g'(x) = 16x^3-3; \; x' = 6t-4$
$y'(t) = [16x^3-3](6t-4)$
$= [16(3t^2-4t)^3-3](6t-4)$

35. $y' = 6(9x^2+4x)^5(18x+4)$

37. $y' = 10(11x^5-2x+1)^9(55x^4-2)$

39. $y = 7(9x^2-4)^{-8}$
$y' = -56(9x^2-4)^{-9}(18x)$
$= \frac{-56(18x)}{(9x^2-4)^9}$

41. $y = (2x^2-5)^{3/4}$
$y' = \frac{3}{4}(2x^2-5)^{1/4}(4x)$
$= \frac{3x}{\sqrt[4]{2x^2-5}}$

43. $\frac{dy}{dx} = \frac{dy}{du} \cdot \frac{du}{dv} \cdot \frac{dv}{dx}; \frac{dy}{du} = 4u^3; \frac{du}{dv} = 6v^2;$
$\frac{dv}{dx} = -8x^{-3} = \frac{-8}{x^3}$
$\frac{dy}{dx} = (4u^3)(6v^2)\left(\frac{-8}{x^3}\right)$
$= (4(2v^3-1)^3)(6v^2)\left(\frac{-8}{x^3}\right)$
$= 4\left(2 \cdot \left(\frac{4}{x^2}\right)^3-1\right)^3\left(6\left(\frac{4}{x^2}\right)^2\right)\left(\frac{-8}{x^3}\right)$
$= 4\left(\frac{128}{x^6}-1\right)^3\left(\frac{96}{x^4}\right)\left(\frac{-8}{x^3}\right)$
$= \frac{-3072}{x^7}\left(\frac{128}{x^6}-1\right)^3$

45. $\dfrac{dy}{du} = 6u; \quad \dfrac{du}{dv} = -4v^{-2}; \quad \dfrac{dv}{dx} = 5x^4$

$\dfrac{dy}{dx} = (6u)\left(\dfrac{-4}{v^2}\right)(5x^4) = 6\left(\dfrac{4}{v}\right)\left(\dfrac{-4}{v^2}\right)(5x^4)$

$= \dfrac{-96}{v^3}(5x^4) = \dfrac{-96}{(x^5)^3} \cdot 5x^4$

$= \dfrac{-480x^4}{x^{15}} = \dfrac{-480}{x^{11}}$

47. $y' = 3(4x^2 - 18)^2(8x)$; at $x = 2, y' = 3(4 \cdot 2^2 - 18)^2(8 \cdot 2) = 3(-2)^2(16) = 192$. Tangent equation is $y + 8 = 192(x - 2)$ or $y = 192x - 392$ or $192x - y = 392$.

49. **(a)** $P = E' = \dfrac{dE}{dt} = 3 \cdot 6(1 + 4t^2)^2(8t) = 144(1 + 4t^2)^2t$ W

(b) $144(1+4t^2)^2t$ W $= 144(1+4t^2)^2t$ W $\cdot \dfrac{1\ kW}{1000\ W} = \dfrac{144(1+4t^2)^2t}{1000} = 0.144(1 + 4t^2)^2t$ kW

(c) $P(2.5) = 0.144(1+4(2.5)^2)^2(2.5) = 243.36$ kW

51. **(a)** Rewrite $P(t) = \dfrac{250t^2}{\sqrt{t^2+15}} = \dfrac{250t^2}{(t^2+15)^{1/2}}$. Then differentiating, we get

$P'(t) = \dfrac{(t^2+15)^{1/2}(500t) - (250t^2)\frac{1}{2}(t^2+15)^{-1/2}(2t)}{\left((t^2+15)^{1/2}\right)^2}$

$= \dfrac{(t^2+15)^{1/2}(500t) - (250t^2)\frac{1}{2}(t^2+15)^{-1/2}(2t)}{\left((t^2+15)^{1/2}\right)^2}$

$\cdot \dfrac{(t^2+15)^{1/2}}{(t^2+15)^{1/2}}$

$= \dfrac{(t^2 + 15)(500t) - (250t^2)\frac{1}{2}(2t)}{(t^2 + 15)^{3/2}}$

$= \dfrac{500t^3 + 7500t - 250t^3}{(t^2+15)^{3/2}} = \dfrac{250t^3 + 7500t}{(t^2+15)^{3/2}}$

$= \dfrac{250t(t^2 + 30)}{(t^2+15)^{3/2}}$

(b) When $t = 5.0$, we have $P'(5.0) = \dfrac{250(5.0)((5.0)^2+30)}{((5.0)^2+15)^{3/2}}$

$= \dfrac{1250(55)}{(40)^{3/2}} \approx 271.8$ ppm/day.

53. $R' = 2\left(\dfrac{n-1}{n+1}\right)^1 \left[\dfrac{(n+1)-(n-1)}{(n+1)^2}\right] = 2\dfrac{(n-1)(2)}{(n+1)^3} = \dfrac{4n-4}{(n+1)^3}$

55. **(a)** Substituting the given information into the formula for the current produces $I = \dfrac{V}{\sqrt{R^2+X_C^2}} = \dfrac{120}{\sqrt{30^2+X_C^2}} = \dfrac{120}{\sqrt{900+X_C^2}} = 120(900 + X_C^2)^{-1/2}$.

Differentiating, we get $I' = -\dfrac{1}{2}(120)(900+X_C^2)^{-3/2} \times (2X_C) = -\dfrac{120X_C}{(900+X_C^2)^{3/2}}$.

(b) $I'(25) = -\dfrac{120(25)}{(900+(25)^2)^{3/2}} = -0.0504$ A/Ω

≡ 23.5 IMPLICIT DIFFERENTIATION

1. $\frac{d}{dx}(4x + 5y) = \frac{d}{dx}(0)$ gives $4 + 5\frac{dy}{dx} = 0$, and so $\frac{dy}{dx} = -\frac{4}{5}$.

3. $\frac{d}{dx}(x - y^2) = \frac{d}{dx}(4)$ yields $1 - 2y\frac{dy}{dx} = 0$. Thus, $\frac{dy}{dx} = \frac{1}{2y}$.

5. Here we see that $\frac{d}{dx}(x^2 - y^2) = \frac{d}{dx}(16)$ produces $2x - 2yy' = 0$. Solving for y', we get $2yy' = 2x$ or $y' = \frac{2x}{2y} = \frac{x}{y}$.

7. $\frac{d}{dx}(9x^2+16y^2) = \frac{d}{dx}(144)$ means that $18x + 32yy' = 0$. Solving for y', we get $y' = \frac{-18x}{32y} = \frac{-9x}{16y}$.

9. $\frac{d}{dx}(4x^2 - y^2) = \frac{d}{dx}4y$ produces $8x - 2yy' = 4y'$ or $8x = 4y' + 2yy' = y'(4 + 2y)$. Solving for y', we get $y' = \frac{8x}{4+2y} = \frac{4x}{2+y}$.

11. Here we have $\frac{d}{dx}(xy + 5xy^2) = \frac{d}{dx}x^2$ or $xy' + y + 5y^2 + 10xyy' = 2x$ which is the same as $xy' + 10xyy' = 2x - y - 5y^2$. Solving for y', we get $y' = \frac{2x-y-5y^2}{x+10xy}$.

13. Here, $\frac{d}{dx}(x^2 + 4xy + y^2) = \frac{d}{dx}y$ gives $2x + 4y + 4xy' + 2yy' = y'$ or $4xy' + 2yy' - y' = -2x - 4y$, and so we get $y' = \frac{-2x-4y}{4x+2y-1}$ or $\frac{2x+4y}{1-2y-4x}$.

15. Here, $\frac{d}{dx}(x+5x^2 - 10y^2) = \frac{d}{dx}3y$ gives $1 + 10x - 20yy' = 3y'$ or $1 + 10x = 3y' + 20yy'$. Solving for y', we get $y' = \frac{1+10x}{2+20y}$.

17. $\frac{d}{dx}(4x^2 + y^3) = \frac{d}{dx}9$ yields $8x + 3y^2y' = 0$, and so $y' = \frac{-8x}{3y^2}$.

19. Here $\frac{d}{dx}(6y^3 + x^3) = \frac{d}{dx}(xy)$ produces $18y^2y' + 3x^2 = xy' + y$ or $18y^2y' - xy' = y - 3x^2$. Solving for y', we get $y' = \frac{y-3x^2}{18y^2-x}$.

21. $\frac{d}{dx}(x^2y) = \frac{d}{dx}(x + 1)$ yields $2xy + x^2y' = 1$ or $x^2y' = 1 - 2xy$. Solving for y', we get $y' = \frac{1-2xy}{x^2}$.

23. $\frac{d}{dx}(x^{-1} + y^{-1}) = \frac{d}{dx}16$ produces $-x^{-2} - y^{-2}y' = -0$. Solving for y', we get $y' = \frac{x^{-2}}{-y^{-2}} = -\frac{y^2}{x^2}$.

25. $\frac{d}{dx}\left(\frac{3}{x+1} + x^2y\right) = \frac{d}{dx}5$ produces $-3(x + 1)^{-2} + 2xy + x^2y' = 0$. Solving for y', we get $y' = \frac{3(x+1)^{-2}-2xy}{x^2} = \frac{\frac{3}{(x+1)^2} - 2xy}{x^2} = \frac{3 - 2xy(x + 1)^2}{x^2(x + 1)^2}$.

27. $\frac{d}{dx}\left(xy + \frac{y}{x}\right) = \frac{d}{dx}x$ gives $y + xy' + \frac{xy'-y}{x^2} = 1$, or $yx^2 + x^3y' + xy' - y = x^2$ which can be written as $x^3y' + xy' = x^2 + y - yx^2$. Solving for y', we get $y' = \frac{x^2+y-yx^2}{x^3+x}$.

29. Here, $\frac{d}{dx}(x^4 - 6x^2y^2 + y^2) = \frac{d}{dx}(10x)$ yields $4x^3 - 6x^2(2yy') - 12xy^2 + 2yy' = 10$, or $2yy' - 12x^2yy' = 10 + 12xy^2 - 4x^3$, and so $y' = \frac{10+12xy^2-4x^3}{2y-12x^2y} = \frac{5+6xy^2-2x^3}{y-6x^2y}$.

31. $\frac{d}{dx}(x^2+y^2)^{1/2} = \frac{d}{dx}\left(\frac{2y}{x}\right)$ gives $\frac{1}{2}(x^2+y^2)^{-1/2}(2x + 2yy') = \frac{x\cdot 2y'-2y}{x^2}$, or $x^2(2x + 2yy') = 2\sqrt{x^2+y^2}(2xy' - 2y)$, or $2x^3 + 2x^2yy' = 4x\sqrt{x^2+y^2}y' - 4y\sqrt{x^2+y^2}$, and so $2x^2yy' - 4x\sqrt{x^2+y^2}y' = -4y\sqrt{x^2+y^2} - 2x^3$. Solving for y', we get $y' = \frac{4y\sqrt{x^2+y^2}+2x^3}{4x\sqrt{x^2+y^2}-2x^2y} = \frac{2y\sqrt{x^2+y^2}+x^3}{2x\sqrt{x^2+y^2}-x^2y}$.

33. $\frac{d}{dx}(x^2 + y^2)^3 = \frac{d}{dx}y$ produces $3(x^2 + y^2)^2(2x + 2yy') = y'$, or $6x(x^2 + y^2)^2 + 6yy'(x^2 + y^2)^2 = y'$, or $6x(x^2 + y^2)^2 = y' - 6yy'(x^2 + y^2)^2$. Solving for y', we get $y' = \frac{6x(x^2+y^2)^2}{1-6y(x^2+y^2)^2}$.

35. $\frac{d}{dx}(x^2 + y^2) = \frac{d}{dx}25$ give $2x + 2yy' = 0$ and so $y' = \frac{-2x}{2y} = \frac{-x}{y}$. Thus, the slope of the tangent

to the circle at the point $(3, -4)$ is $m = \frac{-3}{-4} = \frac{3}{4}$. Using the point-slope form for the equation of a line, we see that the tangent line is $y + 4 = \frac{3}{4}(x - 3)$ or $4y + 16 = 3x - 9$ or $4y - 3x = -25$.

37. Here, $\frac{d}{dx}(9x^2 + 16y^2 - 100) = \frac{d}{dx}0$ gives $18x + 32yy' = \frac{d}{dx}0$, and so $y' = \frac{-18x}{32y} = \frac{-9x}{16y}$. Thus, the slope of the tangent to the ellipse at the point $(-2, 2)$ is $m = \frac{-9(-2)}{16(2)} = \frac{+18}{32} = \frac{9}{16}$. Hence, the tangent line is $y - 2 = \frac{9}{16}(x + 2)$ or $16y - 32 = 9x + 18$ or $16y - 9x = 50$ or $9x - 16y + 50 = 0$.

39. $\frac{d}{dx}(12x^2 - 16y^2 - 192) = \frac{d}{dx}0$ gives $24x - 32yy' = 0$ and so $y' = \frac{24x}{32y} = \frac{3x}{4y}$. Thus, the slope of the tangent to the hyperbola at the point $(-8, -6)$ is $m = \frac{3(-8)}{4(-6)} = \frac{-24}{-24} = 1$ and the slope of the normal is -1.

41. **(a)** Rewrite the given equation $2s^2 + \sqrt{st} - 3t = 0$ as $2s^2 + (st)^{1/2} - 3t = 0$. Differentiating produces $4s\frac{ds}{dt} + \frac{1}{2}(st)^{-1/2}\left(t\frac{ds}{dt} + s\right) - 3 = 0$ or $4s\frac{ds}{dt} + \frac{t}{2(st)^{1/2}}\frac{ds}{dt} + \frac{s}{2(st)^{1/2}} - 3 = 0$. Collecting terms yields $\left(4s + \frac{t}{2(st)^{1/2}}\right)\frac{ds}{dt} = 3 - \frac{s}{2(st)^{1/2}}$ or $\left(\frac{8s(st)^{1/2}+t}{2(st)^{1/2}}\right)\frac{ds}{dt} = \frac{6(st)^{1/2}-s}{2(st)^{1/2}}$. Dividing produces the answer $\frac{ds}{dt} = \frac{6(st)^{1/2}-s}{8s(st)^{1/2}+t} = \frac{6\sqrt{st}-s}{8s\sqrt{st}+t}$.
(b) When $t = 2$ the automobile's position can be determined by solving $2s^2 + \sqrt{2(2)} - 3(2) = 0$ or $2s^2 + \sqrt{2s} - 6 = 0$ for s. A graph yields $s \approx 1.464$ mi.
(c) Substituting $t = 2$ and $s = 1.464$ into $\frac{ds}{dt} = \frac{6\sqrt{st}-s}{8s\sqrt{st}+t}$ produces $\frac{ds}{dt} = \frac{6\sqrt{1.464(2)}-1.464}{11.712\sqrt{2.928}+2} = \frac{6\sqrt{2.928}-1.464}{11.712\sqrt{2.928}+2} \approx 0.399$ mi/min $= 23.94$ mi/hr.

43. **(a)** Differentiating produces $1.4y\frac{dy}{dx} = 12 + y + x\frac{dy}{dx}$ and so $(1.4y - x)\frac{dy}{dx} = 12 + y$. Dividing by $1.4y - x$ gives the result $\frac{dy}{dx} = \frac{12+y}{1.4y-x}$. When $x = 10$ and $y = 25$, we have $\frac{dy}{dx} = \frac{12+25}{1.4(25)-10} = \frac{37}{25} = 1.48$.
(b) The store should sell 1.48 television sets for each digital satellite that is sold.

23.6 HIGHER ORDER DERIVATIVES

1. $y = 4x^3 - 6x^2 + 3x - 10$
$y' = 12x^2 - 12x + 3$
$y'' = 24x - 12$

3. $y = 7x^5 - 3x^3 + x;$
$y' = 35x^4 - 9x^2 + 1;$
$y'' = 140x^3 - 18x;$
$y''' = 420x^2 - 18;$
$y^{(4)} = 840x$

5. $f(x) = x + x^{-1};$
$f'(x) = 1 - x^{-2};$
$f''(x) = 2x^{-3} = \dfrac{2}{x^3}$

7. $f(t) = t^2 + \dfrac{1}{t+1};$
$f'(t) = 2t - \dfrac{1}{(t+1)^2};$
$f''(t) = 2 + \dfrac{2}{(t+1)^3}$

9. $g(u) = \dfrac{u-1}{u^2};$
$g'(u) = \dfrac{u^2 - (u-1)2u}{u^4} = \dfrac{u^2 - 2u^2 + 2u}{u^4}$
$\quad = \dfrac{2u - u^2}{u^4} = \dfrac{2-u}{u^3}$
$g''(u) = \dfrac{u^3(-1) - (2-u)3u^2}{u^6}$
$\quad = \dfrac{-u^3 - 6u^2 + 3u^3}{u^6}$
$\quad = \dfrac{2u^3 - 6u^2}{u^6} = \dfrac{2u - 6}{u^4}$
$g'''(u) = \dfrac{u^4(2) - (2u-6)4u^3}{u^8} = \dfrac{2u - 8u + 24}{u^5}$
$\quad = \dfrac{-6u + 24}{u^5}$

11. $y = (x+1)^{1/2}$
$\dfrac{dy}{dx} = \dfrac{1}{2}(x+1)^{-1/2}$
$\dfrac{d^2y}{dx^2} = -\dfrac{1}{4}(x+1)^{-3/2} = \dfrac{-1}{4\sqrt{x+1}^3}$

13. $y = (2x+1)^3 = 8x^3 + 12x^2 + 6x + 1$
$\dfrac{dy}{dx} = 24x^2 + 24x + 6;\ \dfrac{d^2y}{dx^2} = 48x + 24$
$\dfrac{d^3y}{dx^3} = 48$

15. $y = x^{-2/3};\ \dfrac{dy}{dx} = -\dfrac{2}{3}x^{-5/3};\ \dfrac{d^2y}{dx^2} = \dfrac{10}{9}x^{-8/3}$

17. $f(x) = x(x+1)^{-1};$
$D_x f(x) = 1(x+1)^{-1} - x(x+1)^{-2}$
$D_x^2 f(x) = -(x+1)^{-2} - (x+1)^{-2}$
$\qquad\qquad + 2x(x+1)^{-3}$
$\qquad = -2(x+1)^{-2} + 2x(x+1)^{-3}$
$D_x^3 f(x) = 4(x+1)^{-3} + 2(x+1)^{-3}$
$\qquad\qquad - 6x(x+1)^{-4}$
$\qquad = 6(x+1)^{-3} - 6x(x+1)^{-4}$
$D_x^4 f(x) = -18(x+1)^{-4} - 6(x+1)^{-4}$
$\qquad\qquad + 24x(x+1)^{-5}$
$\qquad = -24(x+1)^{-4} + 24x(x+1)^{-5}$
$\qquad = -24(x+1)(x+1)^{-5} + 24x(x+1)^{-5}$
$\qquad = -24x(x+1)^{-5} - 24(x+1)^{-5}$
$\qquad\qquad + 24x(x+1)^{-5}$
$\qquad = -24(x+1)^{-5}$

19. $f(w) = (3w^2 + 1)^{-2};$
$D_w f(w) = -2(3w^2 + 1)^{-3}(6w)$
$\qquad = -12w(3w^2 + 1)^{-3}$
$D_w^2 f(w) = -12(3w^2 + 1)^{-3}$
$\qquad\qquad + 36w(3w^2 + 1)^{-4}(6w)$
$\qquad = 216w^2(3w^2 + 1)^{-4} - 12(3w^2 + 1)^{-3}$
$\qquad = 216w^2(3w^2 + 1)^{-4}$
$\qquad\qquad - 12(3w^2 + 1)(3w^2 + 1)^{-4}$
$\qquad = (216w^2 - 36w^2 - 12)(3w^2 + 1)^{-4}$
$\qquad = (180w^2 - 12)(3w^2 + 1)^{-4}$

21. $\dfrac{d}{dx}(x^7 - 4x^5 - x^3) = 7x^6 - 20x^4 - 3x^2;$
$\dfrac{d^2}{dx^2}(x^7 - 4x^5 - x^3) = 42x^5 - 80x^3 - 6x;$
$\dfrac{d^3}{dx^3}(x^7 - 4x^5 - x^3) = 210x^4 - 240x^2 - 6;$
$\dfrac{d^4}{dx^4}(x^7 - 4x^5 - x^3) = 840x^3 - 480x$

23. $\dfrac{d}{dx}(x - x^{-2}) = 1 + 2x^{-3};$
$\dfrac{d^2}{dx^2}(x - x^{-2}) = -6x^{-4};$
$\dfrac{d^3}{dx^3}(x - x^{-2}) = 24x^{-5}$

25. $\dfrac{d}{dx}\left(\dfrac{x^3}{3x-1}\right)$

$\qquad = \dfrac{(3x-1)(3x^2)-x^3(3)}{(3x-1)^2} = \dfrac{9x^3-3x^2-3x^3}{(3x-1)^2}$

$\qquad = \dfrac{6x^3-3x^2}{(3x-1)^2};$

$\dfrac{d}{dx}\left(\dfrac{6x^3-3x^2}{(3x-1)^2}\right)$

$\qquad = \dfrac{(3x-1)^2(18x^2-6x)-(6x^3-3x^2)(2)(3x-1)(3)}{(3x-1)^4}$

$\qquad = \dfrac{(3x-1)(18x^2-6x)-6(6x^3-3x^2)}{(3x-1)^3}$

$\qquad = \dfrac{54x^3-36x^2+6x-36x^3+18x^2}{(3x-1)^3}$

$\qquad = \dfrac{18x^3-18x^2+6x}{(3x-1)^3} = \dfrac{6x(3x^2-3x+1)}{(3x-1)^3}$

27. $\dfrac{d}{dx}(x(x+1)^{-1/2})$

$\qquad = x\left(-\dfrac{1}{2}\right)(x+1)^{-3/2}+(x+1)^{-1/2}$

$\qquad = -\dfrac{x}{2}(x+1)^{-3/2}+(x+1)^{-1/2}$

$\dfrac{d^2}{dx^2}[x(x+1)^{-1/2}]$

$\qquad = -\dfrac{x}{2}(-3/2)(x+1)^{-5/2}-\dfrac{1}{2}(x+1)^{-3/2}$

$\qquad\quad -\dfrac{1}{2}(x+1)^{-3/2}$

$\qquad = \dfrac{3}{4}x(x+1)^{-5/2}-(x+1)^{-3/2}$

$\qquad = \dfrac{3}{4}x(x+1)^{-5/2}-(x+1)(x+1)^{-5/2}$

$\qquad = \dfrac{-x-4}{4(x+1)^{5/2}}$

29. $2x-8yy'=1;$

$\qquad y' = \dfrac{1-2x}{-8y} = \dfrac{2x-1}{8y}$

$\qquad y'' = \dfrac{8y(2)-(2x-1)8y'}{64y^2}$

$\qquad\quad = \dfrac{2y-(2x-1)\left(\frac{2x-1}{8y}\right)}{8y^2}$

$\qquad\quad = \dfrac{\frac{16y^2}{8y}-\frac{(2x-1)^2}{8y}}{8y^2} = \dfrac{16y^2-(2x-1)^2}{64y^3}$

$\qquad\quad = \dfrac{16y^2-4x^2+4x-1}{64y^3}$

$\qquad\quad = \dfrac{4(4y^2-x^2)+4x-1}{64y^3}$

$\qquad\quad = \dfrac{4(-x)+4x-1}{64y^3} = \dfrac{-1}{64y^3}$

(Note: In the second-to-last step, we used the fact that since $x^2-4y^2=x$, then $4y^2-x^2=-x$.)

31. Differentiating, we get $x(2yy')+y^2+4=y'$. Then, solving for y' produces

$\qquad 2xyy'-y'=-y^2-4;$

$\qquad y' = \dfrac{y^2+4}{1-2xy}$

$\qquad y'' = \dfrac{(1-2xy)(2yy')-(y^2+4)[-2xy'-2y]}{(1-2xy)^2}$

$\qquad\quad = \dfrac{(1-2xy)\left(2y\cdot\frac{y^2+4}{1-2xy}\right)+(y^2+4)\left(2x\cdot\frac{y^2+4}{1-2xy}+2y\right)}{(1-2xy)^2}$

$\qquad\quad = \dfrac{(2y^3+8y)+(y^2+4)\left[\frac{2xy^2+8x}{1-2xy}+\frac{2y(1-2xy)}{1-2xy}\right]}{(1-2xy)^2}$

$\qquad\quad = \dfrac{(1-2xy)2y(y^2+4)+(y^2+4)(8x+2y-2xy^2)}{(1-2xy)^3}$

$\qquad\quad = \dfrac{(y^2+4)[2y-4xy^2+8x+2y-2xy^2]}{(1-2xy)^3}$

$\qquad\quad = \dfrac{(y^2+4)(4y-6xy^2+8x)}{(1-2xy)^3}$

33. $x^2y'+2xy=0;$

$\qquad y' = \dfrac{-2xy}{x^2} = -\dfrac{2y}{x}$

$\qquad y'' = \dfrac{x(-2y')+(2y\cdot1)}{x^2}$

$\qquad\quad = \dfrac{-2x\cdot\left(\frac{-2y}{x}\right)+2y}{x^2}$

$\qquad\quad = \dfrac{4y+2y}{x^2} = \dfrac{6y}{x^2}$

35. **(a)** $N'(t)=10t-0.08\left(\frac{5}{2}\right)t^{3/2}=10t-0.2t^{3/2},$

$\qquad$ **(b)** $N''(t)=10-0.2\left(\frac{3}{2}\right)t^{1/2}=10-0.3t^{1/2}$

37. **(a)** $h'(t)=120-24t,$

$\qquad$ **(b)** $h''(t)=-24$

CHAPTER 23 REVIEW

1. Step 1 : $2(x+h)^2 - (x+h) = 2x^2 + 4xh + 2h^2 - x - h$

Step 2 : $[2x^2 + 4xh + 2h^2 - x - h] - [2x^2 - x] = 4xh + 2h^2 - h$

Step 3 : $\frac{4xh + 2h^2 - h}{h} = 4x - 1 + 2h$.

Step 4 : $\lim_{h \to 0}(4x - 1 + 2h) = 4x - 1 = f'(x)$

2. Step 1 : $(x+h)^3 - 1 = x^3 + 3x^2h + 3xh^2 + h^3 - 1$

Step 2 : $[x^3 + 3x^2h + 3xh^2 + h^3 - 1] - [x^3 - 1] = 3x^2h + 3xh^2 + h^3$

Step 3 : $\frac{3x^2h + 3xh^2 + h^3}{h} = 3x^2 + 3xh + h^2$.

Step 4 : $\lim_{h \to 0}(3x^2 + 3xh + h^2) = 3x^2 = g'(x)$

3. Step 1 : $f(x+h) = (x+h) - 2(x+h)^3 = x + h - 2x^3 - 6x^2h - 6xh^2 - 2h^3$

Step 2 : $[x + h - 2x^3 - 6x^2h - 6xh^2 - 2h^3] - (x - 2x^3) = h - 6x^2h - 6xh^2 - 2h^3$

Step 3 : $\frac{h - 6x^2h - 6xh^2 - 2h^3}{h} = 1 - 6x^2 - 6xh - 2h^2$.

Step 4 : $\lim_{h \to 0}(1 - 6x^2 - 6xh - 2h^2) = 1 - 6x^2$

$f'(x) = 1 - 6x^2$

4. Step 1 : $k(x+h) = \frac{4}{(x+h)^2}$

Step 2 : $\frac{4}{(x+h)^2} - \frac{4}{x^2} = \frac{4x^2 - 4(x+h)^2}{x^2(x+h)^2}$

$= \frac{4x^2 - 4x^2 - 8xh - 4h^2}{x^2(x+h)^2} = \frac{-8xh - 4h^2}{x^2(x+h)^2}$

Step 3 : $\frac{-8xh - 4h^2}{x^2(x+h)^2} \cdot \frac{1}{h} = \frac{-8x - 4h}{x^2(x+h)^2}$

Step 4 : $\lim_{h \to 0} \frac{-8x - 4h}{x^2(x+h)^2} = \frac{-8x - 0}{x^2(x+0)^2} = \frac{-8x}{x^4} = \frac{-8}{x^3}$

$k'(x) = \frac{-8}{x^3}$

5. Step 1 : $m(x+h) = \sqrt{x+h}$

Step 2 : $m(x+h) - m(x) = \sqrt{x+h} - \sqrt{x}$.

Step 3 : $\frac{\sqrt{x+h} - \sqrt{x}}{h} \cdot \frac{\sqrt{x+h} + \sqrt{x}}{\sqrt{x+h} + \sqrt{x}} = \frac{x+h-x}{h(\sqrt{x+h} + \sqrt{x})} = \frac{h}{h(\sqrt{x+h} + \sqrt{x})} = \frac{1}{\sqrt{x+h} + \sqrt{x}}$

Step 4 : $\lim_{h \to 0} \frac{1}{\sqrt{x+h} + \sqrt{x}} = \frac{1}{\sqrt{x} + \sqrt{x}} = \frac{1}{2\sqrt{x}}$

$m'(x) = \frac{1}{2\sqrt{x}}$

6. Step 1 : $j(t+h) = \sqrt{t+h+4}$

Step 2 : $\sqrt{t+h+4} - \sqrt{t+4}$

Step 3 : $\frac{\sqrt{t+h+4} - \sqrt{t+4}}{h} \cdot \frac{\sqrt{t+h+4} + \sqrt{t+4}}{\sqrt{t+h+4} + \sqrt{t+4}}$

$= \frac{(t+h+4) - (t+4)}{h(\sqrt{t+h+4} + \sqrt{t+4})} = \frac{1}{\sqrt{t+h+4} + \sqrt{t+4}}$

Step 4 : $\lim_{h \to 0} \frac{1}{\sqrt{t+h+4} + \sqrt{t+4}} = \frac{1}{\sqrt{t+4} + \sqrt{t+4}} = \frac{1}{2\sqrt{t+4}}$

$j'(t) = \frac{1}{2\sqrt{t+4}}$

7. $f'(x) = 6x + 2$

8. $g'(x) = -20x^{-6} + 2$

9. $H'(x) = 15x^4 - 12x^{-4} + \frac{3}{2}x^{-1/2} = 15x^4 - \frac{12}{x^4} + \frac{3}{2\sqrt{x}}$

10. $y' = \frac{(x^2+1)(0) - 1(2x)}{(x^2+1)^2} = \frac{-2x}{(x^2+1)^2}$

11. $y' = \frac{(x+3)3 - 3x \cdot 1}{(x+3)^2} = \frac{3x+9-3x}{(x+3)^2} = \frac{9}{(x+3)^2}$

12. Method 1: Expand the given function as $y = (2x+1)(x^2+3) = 2x^3 + x^2 + 6x + 3$; and the differentiating produces $y' = 6x^2 + 2x + 6$. Method 2: Use the Product rule with the result: $y' = (2x+1)(2x) + 2(x^2+3) = 4x^2 + 2x + 2x^2 + 6 = 6x^2 + 2x + 6$.

13. $y' = (x-4)(3x^2 - 3) + (1)(x^2 - 3x)$
$= 3x^3 - 12x^2 - 3x + 12 + x^3 - 3x$
$= 4x^3 - 12x^2 - 6x + 12$

14. $F'(x) = \dfrac{x \cdot \frac{1}{2}(x+1)^{-1/2} - \sqrt{x+1}}{x^2}$

$= \dfrac{\frac{x}{2\sqrt{x+1}} - \sqrt{x+1}}{x^2} = \dfrac{\frac{x}{2\sqrt{x+1}} - \frac{2(x+1)}{2\sqrt{x+1}}}{x^2}$

$= \dfrac{\frac{-x-2}{2\sqrt{x+1}}}{x^2} = \dfrac{-x-2}{2x^2\sqrt{x+1}}$

15. $g'(x) = \dfrac{\sqrt{x+1}(1) - (x+1) \cdot \frac{1}{2}(x+1)^{-1/2}}{\sqrt{x+1}^2}$

$= \dfrac{\sqrt{x+1} - \frac{(x+1)}{2\sqrt{x+1}}}{x+1} = \dfrac{\frac{2(x+1)}{2\sqrt{x+1}} - \frac{(x+1)}{2\sqrt{x+1}}}{x+1}$

$= \dfrac{\frac{x+1}{2\sqrt{x+1}}}{x+1} = \dfrac{1}{2\sqrt{x+1}}$

16. $G'(x) = \dfrac{\sqrt{x+1}(2x) - x^2 \cdot \frac{1}{2}(x+1)^{-1/2}}{x+1}$

$= \dfrac{\frac{2(x+1)2x}{2\sqrt{x+1}} - \frac{x^2}{2\sqrt{x+1}}}{x+1} = \dfrac{4x^2 + 4x - x^2}{2\sqrt{x+1}(x+1)}$

$= \dfrac{3x^2 + 4x}{2(x+1)^{3/2}}$

17. $h'(x) = (x^2 + 4)[(x - 2)(3x^2) + (x^3 - 2)]$
$\qquad + (2x)(x - 2)(x^3 - 2)$
$\quad = (x^2 + 4)[3x^3 - 6x^2 + x^3 - 2]$
$\qquad + (2x^2 - 4x)(x^3 - 2)$
$\quad = (x^2 + 4)(4x^3 - 6x^2 - 2)$
$\qquad + (2x^2 - 4x)(x^3 - 2)$
$\quad = 4x^5 - 6x^4 + 16x^3 - 26x^2 - 8$
$\qquad + 2x^5 - 4x^4 - 4x^2 + 8x$
$\quad = 6x^5 - 10x^4 + 16x^3 - 30x^2 + 8x - 8$

18. $f'(x) = (3x^2 - 4x)(15x^2 + 2)$
$\qquad + (6x - 4)(5x^3 + 2x - 1)$
$\quad = 45x^4 - 60x^3 + 6x^2 - 8x$
$\qquad + 30x^4 - 20x^3 + 12x^2 - 14x + 4$
$\quad = 75x^4 - 80x^3 + 18x^2 - 22x + 4$

19. $F'(x) = \dfrac{1}{2}[(x^3 + 7)(x + 5)]^{-1/2}$
$\qquad \cdot [(x^3 + 7) + 3x^2(x + 5)]$
$\quad = \dfrac{4x^3 + 15x^2 + 7}{2\sqrt{(x^3 + 7)(x + 5)}}$

20. $g'(x) = 3(x + 5)^2$

21. $y' = (x^2 + 5x)^2(-2)(x^3 + 1)^{-3}(3x^2)$
$\qquad + 2(x^2 + 5x)(2x + 5)(x^3 + 1)^{-2}$
$\quad = \dfrac{-6x^2(x^2 + 5x)^2}{(x^3 + 1)^3} + \dfrac{2(x^2 + 5x)(2x + 5)}{(x^3 + 1)^2}$
$\quad = \dfrac{-6x^2(x^2+5x)^2+2(x^2+5x)(2x+5)(x^3+1)^3}{(x^3 + 1)^3}$
$\quad = \dfrac{(x^2+5x)[(-6x^4-30x^3)+2(2x^4+5x^3+2x+5)]}{(x^3 + 1)^3}$
$\quad = \dfrac{(x^2+5x)(-6x^4-30x^3+4x^4+10x^3+4x+10)}{(x^3 + 1)^3}$
$\quad = \dfrac{(x^2 + 5x)(-2x^4 - 20x^3 + 4x + 10)}{(x^3 + 1)^3}$

22. $h'(x) = 4(x^2 + 7x - 1)^3(2x + 7)$

23. $y' = (x + 1)^2 3(x - 1)^2 + 2(x + 1)(x - 1)^3$
$\quad = (x + 1)(x - 1)^2[3(x + 1) + 2(x - 1)]$
$\quad = (x + 1)(x - 1)^2[5x + 1]$

24. $y' = (x + 4)^3 \dfrac{1}{2}(x - 1)^{-1/2} + 3(x + 4)^2\sqrt{x - 1}$

25. Here $y = 5(x^2 + 1)^{-3}$, and so $y' = -15(x^2 + 1)^{-4}(2x) = \dfrac{-30x}{(x^2+1)^4}$.

26. $y' = \dfrac{(3x + 4)^2 \cdot 4 - 4x \cdot 2(3x + 4) \cdot 3}{(3x + 4)^4}$
$\quad = \dfrac{4(3x + 4) - 24x}{(3x + 4)^3} = \dfrac{-12x + 16}{(3x + 4)^3}$

27. Method 1: Using the Quotient rule produces $y' =$
$$\frac{(x^2 - 2)^{3/2}(2x) - x^2 \cdot \frac{3}{2}(x^2 - 2)^{1/2} \cdot 2x}{\left(\sqrt{(x^2 - 2)^3}\right)^2}$$
$$= \frac{(x^2 - 2)^{3/2}(2x) - x^2 \cdot \frac{3}{2}(x^2 - 2)^{1/2} \cdot 2x}{(x^2 - 2)^3}$$
$$= \frac{(x^2-2)2x-3x^3}{(x^2-2)^{5/2}} = \frac{2x^3-4x-3x^3}{(x^2-2)^{5/2}} = \frac{-x^3-4x}{(x^2-2)^{5/2}}$$
Method 2: Rewriting as a product (using negative exponents), we obtain $y = x^2(x^2 - 2)^{-3/2}$, and then, using the Product rule produces $y' = x^2(-\frac{3}{2})(x^2 - 2)^{-5/2}(2x) + 2x(x^2 - 2)^{-3/2} = (x^2 - 2)^{-5/2}[-3x^3 + 2x(x^2 - 2)] = (x^2 - 2)^{-5/2}[-x^3 - 4x]$.

28. $f'(t) = \dfrac{(t^2 - 2)(12t^2) - (4t^3 - 3)(2t)}{(t^2 - 2)^2}$
$\quad = \dfrac{12t^4 - 24t^2 - 8t^4 + 6t}{(t^2 - 2)^2} = \dfrac{4t^4 - 24t^2 + 6t}{(t^2 - 2)^2}$

29. $g'(u) = -4(u^2 + \frac{1}{u} - \frac{4}{u^2})^{-5}(2u - \frac{1}{u^2} + \frac{8}{u^3})$

30. $R(t) = (t^2 - 2t + 1)^{1/2}(t^2 + 1)^{1/3}$
$R'(t) = \dfrac{1}{3}(t^2 + 1)^{-2/3} \cdot (2t)(t^2 - 2t + 1)^{1/2}$
$\qquad + \dfrac{1}{2}(t^2 - 2t + 1)^{-1/2}(2t - 2)(t^2 + 1)^{1/3}$
$\quad = \dfrac{2t}{3}(t^2 - 2t + 1)^{1/2}(t^2 + 1)^{-2/3}$
$\qquad + (t - 1)(t^2 + 1)^{1/3}(t^2 - 2t + 1)^{-1/2}$

31. Here we have $\frac{dy}{du} = 3u^2 - 1$ and $\frac{du}{dx} = 2x$. Applying the Chain rule, we get
$\dfrac{dy}{dx} = (3u^2 - 1)(2x) = [3(x^2 + 1)^2 - 1](2x)$
$\qquad = (3x^4 + 6x^2 + 3 - 1)(2x)$
$\qquad = (3x^4 + 6x^2 + 2)(2x)$
$\qquad = 6x^5 + 12x^3 + 4x$

32. Here $\frac{dy}{du} = 3(u + 4)^2$ and $\frac{du}{dt} = 2$, so if we apply the Chain rule, we obtain
$\dfrac{dy}{dt} = 3(u + 4)^2 \cdot 2 = 6(2t - 5 + 4)^2$
$\qquad = 6(2t - 1)^2 = 6(4t^2 - 4t + 1)$
$\qquad = 24t^2 - 24t + 6$

33. $\dfrac{d}{dx}(x^2 + 4xy + 4y^2) = \dfrac{d}{dx}16;$
$2x + 4xy' + 4y + 8yy' = 0$
$\qquad\qquad (4x + 8y)y' = -2x - 4y$
$\qquad\qquad\qquad y' = -\dfrac{2x + 4y}{4x + 8y} = -\dfrac{1}{2}$

34. $3x^2 - 2xy' - 2y = 2yy';$
$$(-2x - 2y)y' = 2y - 3x^2;$$
$$y' = \frac{3x^2 - 2y}{2x + 2y}$$

35. $2xy' + 2y - 2x = 2yy'x + y^2;$
$$(2x - 2xy)y' = y^2 + 2x - 2y;$$
$$y' = \frac{y^2 + 2x - 2y}{2x - 2xy} \text{ or } \frac{2y - 2x - y^2}{2xy - 2x}$$

36. $x^2y' + 2xy + y3x^2 + y'x^3 = 0;$
$$(x^2 + x^3)y' = -2xy - 3x^2y;$$
$$y' = -\frac{2xy + 3x^2y}{x^2 + x^3}$$
$$= -\frac{2y + 3xy}{x + x^2}$$

37. $x^2 3y^2 y' + 2xy^3 - \dfrac{y'}{y^2} = 1;$
$$\left(3x^2y^2 - \frac{1}{y^2}\right)y' = 1 - 2xy^3$$
$$(3x^2y^4 - 1)y' = y^2 - 2xy^5$$
$$y' = \frac{y^2 - 2xy^5}{3x^2y^4 - 1}$$

38. $8x + x2yy' + y^2 - y'y^{-2} = 2$
$$8xy^2 + 2xy^3y' + y^4 - y' = 2y^2$$
$$(2xy^3 - 1)y' = 2y^2 - 8xy^2 - y^4$$
$$y' = \frac{2y^2 - 8xy^2 - y^4}{2xy^3 - 1}$$

39. $f(x) = x^7 + 2x^4 + 3x;$
$$f'(x) = 7x^6 + 8x^3 + 3$$
$$f''(x) = 42x^5 + 24x^2$$

40. $h(x) = x^{1/2} - x^{-1/2};$
$$h'(x) = \frac{1}{2}x^{-1/2} + \frac{1}{2}x^{-3/2}$$
$$h''(x) = -\frac{1}{4}x^{-3/2} - \frac{3}{4}x^{-5/2};$$
$$h'''(x) = \frac{3}{8}x^{-5/2} + \frac{15}{8}x^{-7/2}$$

41. $g(x) = 4x^3 - 5x + 2x^{-1};$
$$g'(x) = 12x^2 - 5 - 2x^{-2}$$
$$g''(x) = 24x + 4x^{-3}$$
$$g'''(x) = 24 - 12x^{-4} \text{ or } 24 - \frac{12}{x^4}$$

42. $\dfrac{dy}{dx} = 20x^4 + 6x^{-4}; \dfrac{d^2y}{dx^2} = 80x^3 - 24x^{-5}$

43. $\dfrac{dy}{dx} = 21x^2 - 21x^{-4};$
$$\frac{d^2y}{dx^2} = 42x + 84x^{-5}$$
$$\frac{d^3y}{dx^3} = 42 - 420x^{-6}$$

44. $\dfrac{dy}{dx} = -3x^{-4} - 4x^{-2};$
$$\frac{d^2y}{dx^2} = 12x^{-5} + 8x^{-3}$$
$$\frac{d^3y}{dx^3} = -60x^{-6} - 24x^{-4}$$

45. $D_x f(x) = 15x^2 + \dfrac{1}{2}x^{-1/2};$
$$D_x^2 f(x) = 30x - \frac{1}{4}x^{-3/2}$$
$$D_x^3 f(x) = 30 + \frac{3}{8}x^{-5/2}$$

46. $\dfrac{d}{dx}g(x) = 3(x^2 + 1)^2(2x) = 6x(x^2 + 1)^2$
$$\frac{d^2}{dx^2}g(x) = 12x(x^2 + 1)(2x) + 6(x^2 + 1)^2$$
$$= 24x^4 + 24x^2 + 6x^4 + 12x^2 + 6$$
$$= 30x^4 + 36x^2 + 6$$
$$\frac{d^3}{dx^3}g(x) = 120x^3 + 72x$$

47. Using implicit differentiation, we get $x^2y' + 2xy = y' + 2$ or $(x^2 - 1)y' = 2 - 2xy$. Solving for y', we obtain $y' = \frac{2-2xy}{x^2-1}$. Differentiating again yields the following:
$$y'' = \frac{(x^2 - 1)(-2xy' - 2y) - (2 - 2xy)2x}{(x^2 - 1)^2} =$$
$$\frac{(x^2 - 1)(-2x \cdot \frac{2-2xy}{x^2-1} - 2y) - (2 - 2xy)2x}{(x^2 - 1)^2} =$$
$$\frac{-2x(2-2xy)-(x^2-1)2y-2x)(2-2xy)}{(x^2 - 1)^2} =$$
$$\frac{-4x(2 - 2xy) - (x^2 - 1)2y}{(x^2 - 1)^2} =$$
$$\frac{-8x + 8x^2y - 2x^2y + 2y}{(x^2 - 1)^2} = \frac{6x^2y - 8x + 2y}{(x^2 - 1)^2}$$

48. $2yy' = 2xyy' + y^2 + 3x^2$
$$(2y - 2xy)y' = y^2 + 3x^2$$
$$y' = \frac{y^2 + 3x^2}{2y - 2xy}$$

49. Using implicit differentiation, we get $xy' + y = 2x + 2yy'$ or $(x - 2y)y' = 2x - y$. Solving for y', we obtain $y' = \frac{2x-y}{x-2y}$. Differentiating again yields the following:

$$y'' = \frac{(x-2y)(2-y') - (2x-y)(1-2y')}{(x-2y)^2} =$$

$$\frac{2x - xy' - 4y + 2yy' - 2x + 4xy' + y - 2yy'}{(x-2y)^2} =$$

$$\frac{3xy' - 3y}{(x-2y)^2} = \frac{3x\left(\frac{2x-y}{x-2y}\right) - 3y}{(x-2y)^2} =$$

$$\frac{3x(2x-y) - 3y(x-2y)}{(x-2y)^3} =$$

$$\frac{6x^2 - 3xy - 3xy + 6y^2}{(x-2y)^3} = \frac{6x^2 - 6xy + 6y^2}{(x-2y)^3}$$

50. $y' = 8x - 9$; at $x = 3$, $y' = 8 \cdot 3 - 9 = 24 - 9 = 15$. The slope of the tangent is 15 and the slope of the normal is $-\frac{1}{15}$.

51. $f'(x) = 21x^2 - 3$; at $x = -1$, $f'(-1) = 21 - 3 = 18$. The slope of the tangent is 18.

52. $8x + 18yy' = 0$; $y' = \frac{-8x}{18y} = \frac{-4x}{9y}$; at $(-1, 2)$, $y' = \frac{-4(-1)}{9 \cdot 2} = \frac{4}{18} = \frac{2}{9}$. The slope of the tangent is $\frac{2}{9}$.

53. Finding y' by implicit differentiation, we have
$$8x + 6yy' - 5y - 5xy' - 4 + 10y' = 0$$
$$(6y - 5x + 10)y' = 5y + 4 - 8x$$
$$y' = \frac{5y + 4 - 8x}{6y - 5x + 10}$$

At $(2, -1)$, we see that
$$y' = \frac{5(-1) + 4 - 8(2)}{6(-1) - 5 \cdot 2 + 10}$$
$$= \frac{-5 + 4 - 16}{-6 - 10 + 10} = \frac{-17}{-6} = \frac{17}{6}$$

The slope of the tangent is $\frac{17}{6}$ and the slope of the normal is $-\frac{6}{17}$.

54. $f'(x) = 3x^2 - 3$; $3x^2 - 3 = 0 \Rightarrow x^2 = 1 \Rightarrow x = \pm 1$

55. $g'(x) = \frac{(x+1)2x - x^2 \cdot 1}{(x+1)^2} = \frac{2x^2 + 2x - x^2}{(x+1)^2}$
$$= \frac{x^2 + 2x}{(x+1)^2}$$
$g'(x) = 0$ when $x^2 + 2x = 0$ so
$x(x+2) = 0 \Rightarrow x = 0$ or $x = -2$

56. $y' = 2x$, at $(-1, 5)$, $y' = -2$. Equation of tangent is $y - 5 = -2(x + 1)$ or $y = -2x + 3$ or $2x + y - 3 = 0$

57. Finding y' by implicit differentiation, we have
$8x + 10yy' = 0$;
$$y' = \frac{-4x}{5y}$$
At $(-2, 2)$, we see that $y' = \frac{-4(-2)}{5(2)} = \frac{4}{5}$.

Slope of the normal is $-\frac{5}{4}$. Equation of the normal is $y - 2 = -\frac{5}{4}(x + 2)$ or $4y - 8 = -5x - 10$ or $5x + 4y + 2 = 0$

58. $y' = \frac{1}{2}(x + 4)^{-1/2} = \frac{1}{2\sqrt{x+4}}$. This is never equal to zero so the answer is none.

59. $f(x) = 3x^4 - 6x^2 + 2$; $f'(x) = 12x^3 - 12x$
$f''(x) = 36x^2 - 12$
Now, $36x^2 - 12 = 0 \Rightarrow x^2 = \frac{1}{3}$ and so, $x = \pm\sqrt{\frac{1}{3}} = \frac{\pm\sqrt{3}}{3}$

60. $y' = 3x^2 - 18x$; $y'' = 6x - 18$. Now, $y'' = 0$ means that $6x - 18 = 0 \Rightarrow x - 3 = 0$ or $x = 3$.

CHAPTER 23 TEST

1. $f'(x) = 15x^2$

2. $g'(x) = -12x^{-5} + 10x$

3. Rewrite as $h(x) = (x^6 - 2)^{1/3}$ and, using the Chain Rule, we obtain $h'(x) = \frac{1}{3}(x^6 - 2)^{-2/3}(6x^5) = 2x^5(x^6 - 2)^{-2/3}$

4. Using the Product rule, we get $j'(x) = (2x + 1)(6x - 1) + (3x^2 - x)(2) = 12x^2 + 4x - 1 + 6x^2 - 2x = 18x^2 + 2x - 1$

5. Using the Quotient rule produces
$$k'(x) = \frac{(4x^2+1)^{1/2}(2) - (2x+1)\frac{1}{2}(4x^2+1)^{-1/2}(8x)}{[(4x^2+1)^{1/2}]^2}$$
$$= \frac{(4x^2+1)^{-1/2}[(4x^2+1)(2) - (2x+1)4x]}{4x^2+1}$$
$$= \frac{8x^2 + 2 - 4x(2x+1)}{(4x^2+1)^{3/2}} = \frac{2 - 4x}{(4x^2+1)^{3/2}}$$

6. $f'(x) = 20x^3 + 12x^{-4}$; $f''(x) = 60x^2 - 48x^{-5}$; $f'''(x) = 120x + 240x^{-6}$

7. Using implicit differentiation, we obtain $2y + 2xy' + 3y^2xy' + y^3 = 3x^2$. Rewriting, we get $(2x + 3y^2x)y' = 3x^2 - 2y - y^3$ and so $y' = \frac{3x^2 - 2y - y^3}{2x + 3y^2x}$.

8. Taking the derivative of g we obtain $g'(x) = 12x^2 - 4x$. When $x = 1, g'(1) = 8$, and so the slope of the tangent to the curve at (1, 6) is 8. Using the point-slope form for the equation of a line, we see the equation of the tangent line is $y - 6 = 8(x - 1)$ or $y = 8x - 2$.

9. $v(t) = s'(t) = 4.5(2t) + 27 = 9t + 27$

10. **(a)** Using the quotient rule produces
$$P'(t) = \frac{(2\sqrt{t}+7)(2t-5)-(t^2-5t)(t^{-1/2})}{(2\sqrt{t}+7)^2}$$
Multiplying by $\frac{\sqrt{t}}{\sqrt{t}}$, we obtain
$$P'(t) = \frac{(2t + 7\sqrt{t})(2t - 5) - (t^2 - 5t)}{(2\sqrt{t} + 7)^2\sqrt{t}}$$
$$= \frac{4t^2 + 14t^{3/2} - 10t - 35\sqrt{t} - t^2 + 5t}{(2\sqrt{t} + 7)^2\sqrt{t}}$$
$$= \frac{3t^2 + 14t^{3/2} - 5t - 35\sqrt{t}}{(2\sqrt{t} + 7)^2\sqrt{t}}$$

(b) $P'(5) = \frac{3(5)^2 + 14(5)^{3/2} - 5(5) - 35\sqrt{(5)}}{(2\sqrt{(5)}+7)^2\sqrt{(5)}} \approx 0.4358$ thousand bacteria/ hour = 435.8 bacteria/hour.

24

Applications of Derivatives

≣ 24.1 RATES OF CHANGE

1. $s(t) = 3t^2 - 12t + 5; v(t) = s'(t) = 6t - 12; a(t) = s''(t) = 6$

t	0	1	2	3	4	5
$s(t)$	5	−4	−7	−4	5	20
$v(t)$	−12	−6	0	6	12	18
$a(t)$	6	6	6	6	6	6

3. $s(t) = t^3 - 12t + 2; v(t) = s'(t) = 3t^2 - 12; a(t) = s''(t) = 6t$

t	−4	−3	−2	−1	0	1	2	3	4
$s(t)$	−14	11	18	13	2	−9	−14	−7	18
$v(t)$	36	15	0	−9	−12	−9	0	15	36
$a(t)$	−24	−18	−12	−6	0	6	12	18	24

5. $s(t) = t + 4/t; v(t) = 1 - 4/t^2; a(t) = 8/t^3$

t	1	2	3	4
$s(t)$	5	4	4.3333	5
$v(t)$	−3	0	0.5555	0.75
$a(t)$	8	1	0.2963	0.125

7. $s(t) = 2\sqrt{t} + \frac{1}{\sqrt{t}} = 2t^{1/2} + t^{-1/2}; v(t) = t^{-1/2} - \frac{1}{2}t^{-3/2}; a(t) = -\frac{1}{2}t^{-3/2} + \frac{3}{4}t^{-5/2}$

t	1	2	3	4
$s(t)$	3	3.5355	4.0415	4.5
$v(t)$	0.5	0.5303	0.4811	0.4375
$a(t)$	0.25	−0.0442	−0.0481	−0.0391

9. **(a)** The object hits the ground when $s(t) = 0$. Solving $144t - 16t^2 = 0$ we get $t(144 - 16t) = 0$ so $t = 0$ or $t = 9$. The object is fired at $t = 0$ and hits the ground at $t = 9$ s.
(b) $v(t) = 144 - 32t; v(9) = 144 - 32 \cdot 9 = -144$ ft/s; $a(t) = -32; a(9) = -32$ ft/s^2.
(c) $s(0) = 0$ ft = height when fired.

11. **(a)** $s(t) = 29.4t - 4.9t^2 = 0 \Rightarrow t(29.4 - 4.9t) = 0$. By the zero product property, $t = 0$ or $t = \frac{29.4}{4.9} = 6$. Since t must be greater than 0, it hits the ground at 6 s.

(b) $v(t) = 29.4 - 9.8t; v(6) = -29.4$ m/s; $a(t) = -9.8; a(6) = -9.8$ m/s^2.

(c) $s(0) = 0$ m is the initial height.

13. **(a)** The rate of change of $F(s)$ is $\frac{dF}{ds} = F'(s) = -100/s^3$ dynes/cm,

    (b) $F'(3) = -\frac{100}{27}$ dynes/cm

15. **(a)** $\frac{dF}{dC} = \frac{9}{5}$,

    (b) $\frac{dC}{dF} = \frac{d}{dF}\left(\frac{5}{9}(F-32)\right) = \frac{5}{9}$,

    (c) $\frac{dC}{dF}$ at $C = 20°$ is $\frac{5}{9}$

17. **(a)** The rate of change in the quantity of water is $\frac{dQ}{dt} = Q'(t) = -50 + t$ gal/min,

    (b) $Q'(4) = -50 + 4 = -46$ gal/min,

    (c) $Q'(8) = -50 + 8 = -42$ gal/min.

19. **(a)** The velocity, v, at any time t is given by $v(t) = s'(t) = 8.0 - 4.0t$. The initial velocity is when $t = 0$, and so the initial velocity is $v(0) = 8.0$ m/s.

    (b) When the ball stops rolling its velocity is 0. Thus, we want the solution to $v(t) = 0$ or $8.0 - 4.0t = 0$ or $t = 2$ s. When $t = 2$, the ball's

position is $s(2) = 8.0 \cdot 2 - 2.0 \cdot 2^2 = 16 - 8 = 8$ m;

(c) $s(t) = 0$ means that $8.0t - 2.0t^2 = 0$ and so $t(8 - 2t) = 0$. By the zero product property, this means that $t = 0$ or $t = 4$. It will take 4.0 s to return to its initial position.

21. **(a)** $v = -N\frac{d\phi}{dt} = -50\frac{d(2.4t^{1/2} - 0.4t^2)}{dt} = -50(1.2t^{-1/2} - 0.8t); v(t) = -60t^{-1/2} + 40t;$

    (b) $v(1.0) = \frac{-60}{\sqrt{1}} + 40 \cdot 1 = -60 + 40 = -20$ V

    and $v(9.0) = \frac{-60}{\sqrt{9}} + 40 \cdot 9 = \frac{-60}{3} + 360 = 340$ V.

23. $A = \pi r^2$ and $r = 1.7t$ m so $A = 1.7^2\pi t^2 = 2.89\pi t^2$ m^2. Thus, $\frac{dA}{dt} = 5.78\pi t$ and when $t = 8$, the area is increasing at the rate of 46.24π m^2/s.

25. **(a)** $P = \frac{k}{V}$ or $5.21 = \frac{k}{1.23}$ so $k = 6.4083 \approx 6.41$. Hence, $P = \frac{6.41}{V}$;

    (b) $\frac{dP}{dV} = -6.41V^{-2}$ or $\frac{-6.41}{V^2}$.

27. **(a)** $v(t) = s'(t) = 6t - 0.05t^2$,

    (b) $a(t) = v'(t) = s''(t) = 6 - 0.1t$,

    (c) $|v(t)| = |6t - 0.05t^2|$

≡ 24.2 EXTREMA AND THE FIRST DERIVATIVE TEST

1. $f'(x) = 2x - 8$, so we have $2x - 8 = 0$, and $x = 4$ is the critical value. On the interval $(-\infty, 4)$, $f'(0) = -8$ so decreasing. On the interval $(4, \infty)$, $f'(5) = 10 - 8 = 2$, so f is increasing. No maximums; minimum at $x = 4$, $f(4) = 16 - 32 = -16$.

3. $m'(x) = -2x; -2x = 0 \Rightarrow 0$ is the critical value. On the interval $(-\infty, 0)$, $m'(-1)$ is positive so increasing. On the interval $(0, \infty)$, $m'(1)$ is negative so decreasing. Maximum at $x = 0; m(0) = 16$. No minimum.

5. $j'(x) = 16 - 8x; 16 - 8x = 0 \Rightarrow x = 2$ is the critical value. On $(-\infty, 2)$, $j'(0)$ is positive so increasing. On the interval $(2, \infty)$, $j'(3)$ is negative so decreasing. Maximum at $x = 2; j(2) = 31$. Minimum: none.

7. $f'(x) = 8x - 16; x = 2$ is the critical value. On the interval $(-\infty, 2)$, $f'(0)$ is negative so decreasing. On the interval $(2, \infty)$, $f'(3)$ is positive so increasing. Maximum: none; Minimum: $f(2) = -15$.

9. Here $q'(t) = 3t^2 - 3$ and so, $3t^2 - 3 = 3(t+1)(t-1) = 0$. Thus, by the zero product property, -1 and 1 are the critical values. On $(-\infty, -1)$, $q'(-2)$ is positive so increasing. On the interval $(-1, 1)$,

$q'(0)$ is negative so decreasing. On the interval $(1, \infty), q'(2)$ is positive so increasing. Maximum: $q(-1) = 2$. Minimum: $q(1) = -2$.

11. The derivative is $y' = x^3 - 36x$. By factoring we can determine when the derivative is 0, thus $x^3 - 36x = x(x+6)(x-6) = 0$ and so, the critical values are at $-6, 0$, and 6. On the interval $(-\infty, -6), y'$ is negative so decreasing. On the interval $(-6, 0), y'$ is positive so increasing. On the interval $(0, 6), y'$ is negative so decreasing. On the interval $(6, \infty), y'$ is positive so increasing. Maximum: $y(0) = 16$. Minimum: $y(-6) = y(6) = -308$.

13. $f'(z) = 3(z+4)^2$, so $3(z+4)^2 = 0$ when $x = -4$. Thus, -4 is only critical value. On the interval $(-\infty, -4), f'(z)$ is positive so increasing. On $(-4, \infty), f'(z)$ is positive so increasing. Hence $f(z)$ is always increasing and there are no extrema.

15. $g'(v) = 12v^3 - 12v^2 - 24v = 12v(v^2 - v - 2) = 12v(v-2)(v+1)$, so the critical values are at $-1, 0$, and 2. On the interval $(-\infty, -1), g'(v)$ is negative so decreasing. On the interval $(-1, 0), g'(v)$

is positive so increasing. On $(0,2), g'(v)$ is negative so decreasing. On the interval $(2,\infty), g'(v)$ is positive so increasing. Maximum: $g(0) = 12$. Minima: $g(-1) = 7, g(2) = -20$.

17. $s'(t) = \frac{-6}{(t+1)^2}$. Since s is undefined at $t = -1, -1$ is *not* a critical value. On the interval $(-\infty, -1)$, $s'(t)$ is negative, so s decreasing over this interval. On the interval $(-1, \infty), s'(t)$ is negative so s is decreasing over this interval. $s(t)$ is undefined at -1. There are no extrema.

19. $g(x)$ is not defined at $x = 0$. Thus, $g'(x) = 2x + \frac{2}{x^2} = \frac{2x^3 + 2}{x^2}$ which is 0 when $2x^3 + 2 = 0$ or $x = -1$. The critical value is -1. On the interval $(-\infty, -1), g'(x)$ is negative so g is decreasing. On the interval $(-1, 0), g'(x)$ is positive so g is increasing. On the interval $(0, \infty), g'(x)$ is positive so g is increasing. Maximum: none; Minimum: $g(-1) = 3$.

21. $k(x)$ is undefined at $x = 1$. $k'(x) = \frac{(1-x)2x + (x^2)}{(1-x)^2} = \frac{2x - x^2}{(1-x)^2}$. Critical values are at 0 and 2. On the interval $(-\infty, 0), k'(x)$ is negative so $k(x)$ decreasing. On the interval $(0, 1), k'(x)$ is positive, so $k(x)$ increasing. On the interval $(1, 2), k'(x)$ is positive, so $k(x)$ increasing. On the interval $(2, \infty), k'(x)$ is negative, so $k(x)$ is decreasing. Maximim: $k(2) = -4$; Minimum: $k(0) = 0$.

23. The domain of f is all t when $1 - t^2 \geq 0$ or $[-1, 1]$. The derivative is
$$f'(t) = \sqrt{1 - t^2} - t \cdot \frac{1}{2}(1 - t^2)^{-1/2}(2t)$$
$$= \sqrt{1 - t^2} - \frac{t^2}{\sqrt{1 - t^2}}$$
$$= \frac{1 - t^2 - t^2}{1\sqrt{1 - t^2}}$$
$$= \frac{1 - 2t^2}{\sqrt{1 - t^2}}$$
Critical values are when $1 - 2t^2 = 0$, or $t = \pm\frac{\sqrt{2}}{2}$ and when $1 - t^2 = 0$ or $t = \pm 1$. The domain of $f(t)$ is $[-1, 1]$. On the interval $(-1, -\frac{\sqrt{2}}{2}), f'(t)$ is negative, so $f(t)$ decreasing. On the interval $\left(-\frac{\sqrt{2}}{2}, \frac{\sqrt{2}}{2}\right)$, we see that $f'(t)$ is positive, so $f(t)$ increasing. On the interval $\left(\frac{\sqrt{2}}{2}, 1\right), f'(t)$ is negative, so $f(x)$ decreasing. Maximum: $f(-1) = 0$ and $f\left(\frac{\sqrt{2}}{2}\right) = 0.5$; Minimum: $f\left(-\frac{\sqrt{2}}{2}\right) = -0.5$ and $f(1) = 0$.

25. $y' = (x - 1)^{2/3} + \frac{2}{3}(x - 1)^{-1/3} \cdot x$
$$= (x - 1)^{2/3} + \frac{2x}{3(x - 1)^{1/3}}$$
$$= \frac{3(x - 1) + 2x}{3(x - 1)^{1/3}} = \frac{5x - 3}{3(x - 1)^{2/3}}$$
The critical values are 0.6 and 1. On the interval $(-\infty, 0.6), y'$ is positive so y is increasing. On the interval $(0.6, 1), y'$ is negative so y is decreasing. On the interval $(1, \infty), y'$ is positive so y is increasing. Maximum: $(0.6, 0.326)$; Minimum: $(1, 0)$.

27. **(a)** $s(t) = 250 + 256t - 16t^2; s(t) = 256 - 32t$; Critical value when $256 - 32t = 0$ or $t = 8$. On the interval $[0, 8), s'(t)$ is positive so increasing. On the interval $(8, \infty), s'(t)$ is negative, so s is decreasing. Maximum at 8 s
(b) $s(8) = 1274$ ft.

29. $P(t) = 4.7(t - 2.0)^2; P'(t) = 9.4(t - 2.0)$
(a) The critical value is 2.0 s and this will give the lowest power $P(2) = 0.0$ W
(b) This occurs at 2.0 s.

31. $R = \frac{9R_2}{9 + R_2}$;
$$R' = \frac{(9 + R_2)9 - 9R_2(1)}{(9 + R_2)^2} = \frac{81 + 9R_2 - 9R_2}{(9 + R_2)^2}$$
$$= \frac{81}{(9 + R_2)^2}.$$
(a) the only critical value is -9 but R_2 must be greater than or equal to 0. R' is always positive hence R is always increasing. There is no maximum and the minimum value occurs at the minimum value for R_2 which is 0.
(b) The mimimum value for R is 0.

33. **(a)** Using the product rule, we get $P'(n) = (4 - 0.2n)(2.5) + (-0.2)(2.5n + 7) = 10 - 0.5n - 0.5n - 1.4 = -n + 8.6$
(b) The first derivative is 0 when $n = 8.6$. Since $P'(n) > 0$ for $n < 8.6$ and $P'(n) < 0$ for $n > 8.6$, P has a maximum at $n = 8.6$. Thus, 8.6 items must be sold to make the most profit.
(c) The maximum profit is $P(8.6) = [4 - 0.2(8.6)][2.5(8.6) + 7] = 64.98$ hundreds of dollars or $64.98 \times 100 = \$6,498$.

35. **(a)** $N'(t) = 2t^{-1/2} + 3t^{1/2} - 2t^{3/2}$
(b) We want to solve $2t^{-1/2} + 3t^{1/2} - 2t^{3/2} = 0$. Multiplying both sides by $\sqrt{t} = t^{1/2}$ we obtain $2 + 3t - 2t^2 = 0$. This yields critical values of $t = -0.5$ and $t = 2$. Since, $0 \leq t \leq 6$, the only critical value to consider is when $t = 2$. The first derivative test gives us as a maximum $t = 2$ mo.

≡ 24.3 CONCAVITY AND THE SECOND DERIVATIVE TEST

1. Since $f(x) = -3x^2 + 12x$, $f'(x) = -6 + 12 = 0$ when $x = 2$. $f''(x) = -6$ is always negative, so f is concave down over the interval $(-\infty, \infty)$. Maximum at (2, 12).

3. $g(x) = x^3 - 12x + 12$; $g'(x) = 3x^2 - 12 = 0$ when $x = \pm2$; and $g''(x) = 6x = 0$ when $x = 0$. Concave up: $(0, \infty)$; Concave down: $(-\infty, 0)$. Maximum: $(-2, 28)$; Minimum: $(2, -4)$; Inflection Point: (0, 12).

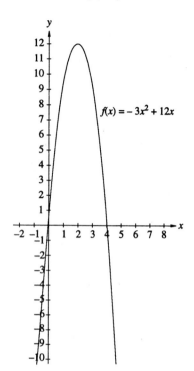

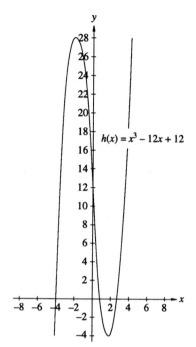

5. $j(x) = x^3 - 6x^2 + 12x - 4$; $j'(x) = 3x^2 - 12x + 12 = 3(x-2)^2 = 0$ when $x = 2$; $j''(x) = 6x - 12 = 0$ when $x = 2$. Inflection point at (2, 4). No extrema. Concave up: $(2, \infty)$; Concave down: $(-\infty, 2)$.

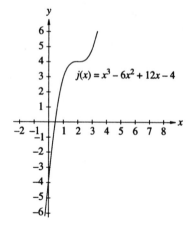

7. $h(x) = -2x^3 + 3x^2 - 12x + 5$; $h'(x) = -6x^2 + 6x - 12 = -6(x^2 - x + 2)$ never equals 0 since the discriminant $= 1 - 8 = -7$ is negative: hence no extrema. $h''(x) = -12x + 6 = 0$ when $x = \frac{1}{2}$.

Concave up: $(-\infty, \frac{1}{2})$; Concave down: $(\frac{1}{2}, \infty)$;

Inflection point: $(\frac{1}{2}, -\frac{1}{2})$.

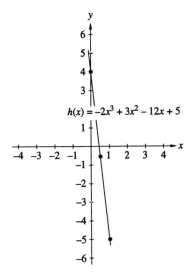

9. $j(x) = 4x^2 - \frac{4}{3}x^3 - x^4$; $j'(x) = 8x - 4x^2 - 4x^3 = -4x(x^2 + x - 2) = -4x(x+2)(x-1) = 0$ when $x = 0, -2$ and 1; $j''(x) = 8 - 8x - 12x^2 = -4(3x^2 + 2x - 2) = 0$ when $x = -1.22$ and 0.55. Concave up: $(-1.22, 0.55)$; Concave down $(-\infty, -1.22)$ and $(0.55, \infty)$; Inflection points: $(-1.22, 6.16)$ and $(0.55, 0.9)$; Maximum: $(-2, 10.67)$ and $(1, 1.67)$; Minimum: $(0,0)$

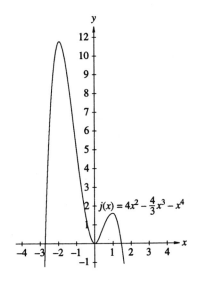

11. $h(x) = (1-x)^3$; $h'(x) = -3(1-x)^2 = 0$ when $x = 1$; $h''(x) = 6(1-x) = 0$ when $x = 1$; Concave down: $(1, \infty)$; Concave up: $(-\infty, 1)$; No extrema. Inflection point: $(1,0)$.

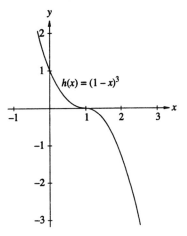

13. Since $h(x) = \sqrt[3]{x} + 2 = x^{1/3} + 2$, then $h'(x) = \frac{1}{3}x^{-2/3}$ and $h''(x) = -\frac{2}{9}x^{-5/3}$. We see that h is defined at $x = 0$, but h' is undefined at $x = 0$, so there is a critical value at $x = 0$; also $h''(x)$ is undefined at $x = 0$. Concave up: $(-\infty, 0)$; Concave down: $(0, \infty)$; Inflection point: $(0, 2)$; No extrema.

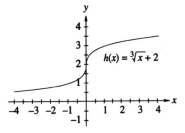

15. $k(x) = x + \frac{4}{x}$ has the derivative $k'(x) = 1 - \frac{4}{x^2} = \frac{x^2-4}{x^2}$. Now, $k'(x) = 0$ when $x = \pm 2$ and k' undefined at $x = 0$, so critical values are at $x = -2$ and $x = 2$. $k''(x) = \frac{8}{x^3}$ which is undefined at $x = 0$; Concave up: $(0, \infty)$; Concave down: $(-\infty, 0)$; Maximum: $(-2, -4)$; Minimum: $(2, 4)$; Vertical Asymptote at $x = 0$; No inflection point

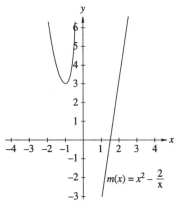

17. $n(x) = x^2 - \frac{2}{x}$ has the derivative $n'(x) = 2x + \frac{2}{x^2} = \frac{2x^3+2}{x^2}$. Critical value at -1. The second derivative is $n''(x) = 2 - \frac{4}{x^3} = \frac{2x^3-4}{x^3}$; this is undefined at 0 and is zero at $x = \sqrt[3]{2}$. Concave up $(-\infty, 0)$ and $(\sqrt[3]{2}, \infty)$, Concave down $(0, \sqrt[3]{2})$. Maximum: none; Minimum: $(-1, 3)$; Vertical Asymptote: $x = 0$; Inflection point: $(\sqrt[3]{2}, 0)$.

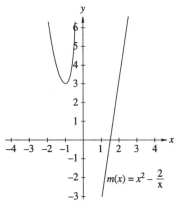

19. $g(x) = \frac{x^3}{3} - \frac{x^2}{2} - 6x$ has the derivative $g'(x) = x^2 - x - 6 = (x-3)(x+2)$, and so $g'(x) = 0$ when $x = -2$ or $x = 3$. The second derivative $g''(x) = 2x - 1 = 0$ when $x = \frac{1}{2}$.

Concave up: $(\frac{1}{2}, \infty)$; Concave down: $(-\infty, \frac{1}{2})$. Maximum: $(-2, 7\frac{1}{3})$; Minimum: $(3, -13\frac{1}{2})$; Inflection point: $(\frac{1}{2}, -3.0833)$.

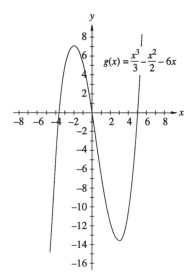

21. $j(x) = 2x^{1/2} - x$ has the derivative $j'(x) = x^{-1/2} - 1$, and $j'(x) = 0$ when $x = 1$ and j' is undefined at $x = 0$. The second derivative is $j''(x) = -\frac{1}{2}x^{-3/2}$ which is undefined at $x = 0$; Concave up: nowhere; Concave down: $(0, \infty)$. Maximum: $(1, 1)$; Minimum: $(0, 0)$; Inflection points: none.

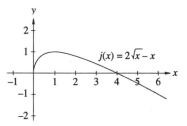

23. $g(x) = 2 + x^{2/3}$, so $g'(x) = \frac{2}{3}x^{-1/3}$. Since g is defined at $x = 0$ and g' is undefined at $x = 0$, 0 is a critical value. $g''(x) = \frac{-2}{9}x^{-4/3}$ is undefined at 0 but negative everywhere else, so concave down over $(-\infty, 0)$ and $(0, \infty)$ and $g'(x) > 0$ on $(0, \infty)$ and $g'(x) < 0$ on $(-\infty, 0)$, so Minimum value: $(0, 2)$. No inflection points.

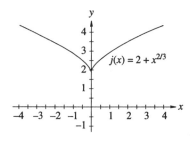

25. $f(x) = x^{2/3}(x - 10) = x^{5/3} - 10x^{2/3}$

$f'(x) = \frac{5}{3}x^{2/3} - \frac{20}{3}x^{-1/3}$

$= \frac{5x^{2/3}}{3} - \frac{20}{3x^{1/3}}$

$= \frac{5x - 20}{3x^{1/3}}$

Critical values are at 0 and 4. $f''(x) = \frac{10}{9}x^{-1/3} + \frac{20}{9}x^{-4/3} = \frac{10x+20}{9x^{4/3}} = 0$ when $x = -2$, undefined at $x = 0$. Concave down $(-\infty, -2)$. Concave up $(-2, 0)$ and $(0, \infty)$. Maximum: $(0, 0)$; Minimum: $(4, -15.119)$. Inflection point: $(-2, -19.049)$.

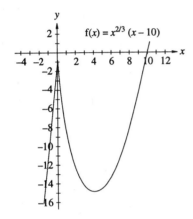

27. $h(x) = x(x^2 + 1)^{-1}$

$h'(x) = (x^2 + 1)^{-1} - 2x^2(x^2 + 1)^{-2}$

$= \frac{1}{x^2 + 1} - \frac{2x^2}{(x^2 + 1)^2} = \frac{x^2 + 1 - 2x^2}{(x^2 + 1)^2}$

$= \frac{1 - x^2}{(x^2 + 1)^2}$

Critical values at $x = \pm 1$.

$h''(x) = \frac{(x^2 + 1)^2(-2x) - (1 - x^2)(x^2 + 1)(2x)2}{(x^2 + 1)^4}$

$= \frac{-2x^3 - 2x - 4x + 4x^3}{(x^2 + 1)^3}$

$= \frac{2x^3 - 6x}{(x^2 + 1)^3}$

You can determine that $h''(x) = 0$ when $x = 0$ and $x = \pm\sqrt{3}$. Thus h is concave up over the intervals $(-\sqrt{3}, 0)$ and $(\sqrt{3}, \infty)$ and h is concave down over the intervals $(-\infty, -\sqrt{3})$ and $(0, \sqrt{3})$. Maximum: $(1, \frac{1}{2})$; Minimum: $(-1, -\frac{1}{2})$; Inflection point: $\left(-\sqrt{3}, \frac{-\sqrt{3}}{4}\right), (0, 0)$, and $\left(\sqrt{3}, \frac{\sqrt{3}}{4}\right)$.

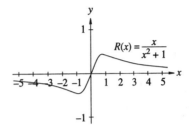

29. To find the potential inflection points, we find $F''(t)$ and then solve $F''(t) = 0$. Differentiating produces $F'(t) = 1.242t - 0.027t^2$ and $F''(t) = 1.242 - 0.054t$. Solving $1.242 - 0.054t = 0$ yields $t = 23$. Since $F''(t)$ changes from concave upward to concave downward at $t = 23$, this is an inflection point. Hence, the best time to harvest is during the 23rd day.

31. **(a)** To find the maximum of $N(t)$, we set $N'(t) = 0$ to determine the critical values of N. Differentiating produces $N'(t) = 6t - 2t^{3/2}$ and $6t - 2t^{3/2} = 2t(3 - t^{1/2}) = 0$ yields critical values at $t = 0$ and $t = 9$. Since $N''(t) = 6 - 3t^{1/2}$ is positive at $t = 0$, N has a relative minimum at $t = 0$. Since $N''(9) < 0$, we see that N has a relative maximum at $t = 9$. Evaluating $N(9)$ we see that $N(9) = 116$. The maximum is 116 crimes in the ninth month.
(b) Solving $N''(t) = 0$ we get $3\sqrt{t} = 6$ or $\sqrt{t} = 2$ and so, $t = 4$. The maximum rate of increase is $N'(4)6(4) - 2(4^{3/2}) = 8$ crimes/mo.

☰ 24.4 APPLIED EXTREMA PROBLEMS

1. $A = \ell \cdot w; \ell + 2w = 10,000 \Rightarrow \ell = 10,000 - 2w$. Substituting into the area formula yields $A = (10,000 - 2w)w = 10,000w - 2w^2$, so $A' = 10,000 - 4w = 0 \Rightarrow w = 2,500$. $A'' = -4 < 0$ so $w = 2,500$ yields a maximum area, and $\ell = 10,000 - 5000 = 5000; A = 5,000 \times 2,500 = 12,500,000$ ft^2.

3. **(a)** Let x be the length of the side along the river, as well as the side opposite. Then the length of the other two sides is $6,000,000/x$. The cost $C(x) = 2x + x + 2(6,000,000/x$, so $C(x) = 3x + \frac{12,000,000}{x}$.
$C'(x) = 3 - \frac{12,000,000}{x^2} = 0 \Rightarrow 3x^2 = 12,000,000 \Rightarrow x^2 = 4,000,000 \Rightarrow x = 2,000$. The other sides are 3,000 ft. The dimensions are 2,000ft. $\times$ 3,000 ft.
(b) The cost is $3(2000) + 2(3,000) = \$12,000$.

5. After cutting out the squares, the resulting length and width of the box will both be $(24 - 2x)$. The height will be x; $V = \ell \cdot w \cdot h = (24 - 2x)^2 x = 4x^3 - 96x^2 + 576x$; $V' = 12x^2 - 192x + 576 = 12(x^2 - 16x + 48) = 12(x-4)(x-12)$. Now, $V' = 0$ when $x = 4$ or 12. $V'' = 24x - 192$; $V''(4) < 0$ so $V(4)$ is maximum. $V''(12) > 0$ so $V(12)$ is minimum. $V(4) = 4 \cdot 4^3 - 96 \cdot 4^2 + 576 \cdot 4 = 1024$ cm^3. The length of the side of a removed square should be 4 cm.

7. First solve $9x^2 + 16y^2 = 144$ for y. This yields $y = \sqrt{9 - \frac{9}{16}x^2}$. The area of the rectangle is $A = xy$, or after substituting for y, can be written as $A = x\sqrt{9 - \frac{9}{16}x^2}$. $A' = \left(9 - \frac{9}{16}x^2\right)^{1/2} + x \cdot \frac{1}{2}\left(9 - \frac{9}{16}x^2\right)^{-1/2}\left(\frac{-18}{16}x\right) = \frac{9 - \frac{9}{16}x^2 - \frac{9}{16}x^2}{\left(9 - \frac{9}{16}x^2\right)^{1/2}} =$

$\frac{9 - \frac{9}{8}x^2}{\left(9 - \frac{9}{16}x^2\right)^{1/2}}$. Critical values are when $9 - \frac{9}{8}x^2 = 0$, or $x^2 = 8$, or $x = 2\sqrt{2}$, and when $9 - \frac{9}{16}x^2 = 0$ or $x = 4$. When $x = 4$, $y = 0$ so $A = 0$ which is the minimum. When $x = 2\sqrt{2}$, we see that $y = \sqrt{9 - \frac{9}{2}} = \sqrt{\frac{9}{2}} = \frac{3}{\sqrt{2}} = \frac{3\sqrt{2}}{2}$ and the area is $A = 2\sqrt{2} \cdot \frac{3\sqrt{2}}{2} = 6$, which is a maximum. The dimensions are $2\sqrt{2} \approx 2.83$ units horizontally and $\frac{3}{2}\sqrt{2} \approx 2.12$ units vertically.

9. Let x be the length and width of the box and $y = \frac{32}{x^2}$ be the height. The material used is proportional to the area of the 4 sides plus the base. $A = x^2 + 4 \cdot x \cdot \frac{32}{x^2} = x^2 + \frac{128}{x}$, and so $A' = 2x - \frac{128}{x^2} = \frac{2x^3 - 128}{x^2}$. Solving $2x^3 - 128 = 0$ yields $x^3 = 64$ or $x = 4$. $A'' = 2 + \frac{256}{x^3}$ is positive at $x = 4$, so this yields a minimum. The height is $\frac{32}{16} = 2$. The dimensions are $4'' \times 4'' \times 2''$.

11. $A = 2\pi r^2 + 2\pi rh = 480 \Rightarrow h = \frac{480 - 2\pi r^2}{2\pi r} = \frac{240 - \pi r^2}{\pi r}$, so the volume is $V = \pi r^2 h = \pi r^2 \left(\frac{240 - \pi r^2}{\pi r}\right) = 240r - \pi r^3$, which has the derivative $V' = 240 - 3\pi r^2$. Setting $240 - 3\pi r^2 = 0$ and solving for r yields $\pi r^2 = 80$ or $r = \sqrt{\frac{80}{\pi}} \approx 5.05$. $h = 10.09$. Radius = 5.05 cm; height = 10.09 cm.

13. $A = \ell \cdot w = 580$, and solving for ℓ, we get $\ell = \frac{580}{w}$. The printing area of the page is $P = (\ell - 5)(w -$

$6) = \left(\frac{580}{w} - 5\right)(w - 6) = 580 - \frac{3480}{w} - 5w + 30 = 610 - 5w - \frac{3480}{w}$. In order to determine when the printed area of the page is the largest, we take the derivative, with the result $P' = -5 + \frac{3480}{w^2}$. Setting the derivative equal to 0, we have $-5 + \frac{3480}{w^2} = 0$ or $w^2 = \frac{3480}{5} = 696$ and so $w = \sqrt{696} = 26.38$ and $\ell = \frac{580}{w} = 21.98$. The largest printed area will have a length of about 21.98 cm and a width of about 26.38 cm.

15. $r = 15$. Placing the center of the log at the origin yields the equation $x^2 + y^2 = 225$ and so, $y = \sqrt{225 - x^2}, d = 2y$, and $w = 2x$. Thus, the strength is $S = wd^2 = 2x\left(2\sqrt{225 - x^2}\right)^2 = 8x(225 - x^2) = 1800x - 8x^3$. To find when the strength is the most, we take the derivative, getting $S' = 1800 - 24x^2$, then determine the critical values by finding when the derivative is 0. Thus, $1800 - 24x^2 = 0 \Rightarrow 24x^2 = 1800 \Rightarrow x^2 = 75$, and so $x = 5\sqrt{3}$. As a result, the width is $w = 2x = 10\sqrt{3} \approx 17.32$ in. and the depth is $d = 2y = 2\sqrt{225 - 75} = 2\sqrt{150} = 24.49$ in.
(b) This time our equation is $x^2 + y^2 = r^2$ so $y = \sqrt{r^2 - x^2}$ and so the strength is $S = 2x\left(2\sqrt{r^2 - x^2}\right)^2 = 8x(r^2 - x^2) = 8xr^2 - 8x^3$. Taking the derivative, we obtain $S' = 8r^2 - 24x^2$, and to find when the derivative is 0, we get $8r^2 - 24x^2 = 0 \Rightarrow 24x^2 = 8r^2 \Rightarrow x^2 = \frac{1}{3}r^2$.

Thus, $x = \sqrt{\frac{r^2}{3}} = \frac{r}{\sqrt{3}} = \frac{r\sqrt{3}}{3}$. For the width we have $w = 2x = 2r\sqrt{3}/3$ and the depth is $d = 2y = 2\sqrt{r^2 - \frac{r^2}{3}} = 2\sqrt{\frac{2r^2}{3}} = \frac{2r\sqrt{6}}{3}$.

17. $A = \ell \cdot w; \ell = 320 - 2w$
$A = (320 - 2w)w = 320w - 2w^2$
$A' = 320 - 4w = 0 \Rightarrow w = 80\,\text{mm}$
$\ell = 320 - 2 \cdot 80 = 320 - 160 = 160\,\text{mm}$
The gutter is 80 mm deep and 160 mm wide.

19. $P = VI - RI^2 = 10I - 4I^2$ and so $P' = 10 - 8I$. Critical value is $\frac{5}{4} = 1.25$. $P'' = -8$ so critical value yields a maximum. Maximum power given by current of 1.25 A.

21. $C(A) = bA + \frac{c}{A}; C' = b - \frac{c}{A^2} = 0$ when $A = \sqrt{c/b}; C'' = 2c/A^3 > 0$, so minimum cost at $A = \sqrt{c/b}$.

23. ℓ = height; $w = d$ = width = $2r$. Thus,

$$P = 8 = 2\ell + w + \frac{\pi w}{2} = 2\ell + \frac{2+\pi}{2}w$$

$$2\ell = 8 - \frac{2+\pi}{2}w; \ell = 4 - \frac{2+\pi}{4}w$$

$$A = \ell \cdot w + \frac{1}{2}\pi\left(\frac{w}{2}\right)^2$$

$$= \left(4 - \frac{2+\pi}{4} \cdot w\right)w + \frac{\pi}{8}w^2$$

$$= 4w - \frac{2+\pi}{4}w^2 + \frac{\pi}{8}w^2 = 4w - \frac{4+\pi}{8}w^2$$

$$A' = 4 - \frac{4+\pi}{4}w = 0; (4+\pi)w = 16;$$

$$w = \frac{16}{4+\pi} = 2.24 \text{ m}.$$

The diameter is 2.24 m.

25. **(a)** $P = I^2R = (4.5-t)^2(25.4)$
$P' = 2(4.5-t)(25.4)(-1) = 0$
when $t = 4.5s$. P'' is always positive so this is a minimum.
(b) $P = (4.5-4.5)^2(25.4) = 0$ W.

27. Let x = distance from less intense light and $8 - x$ = distance from more intense light. Light intensity $= \frac{1}{x^2} + \frac{4}{(8-x)^2}$. Taking the derivative, we get $L' = \frac{-2}{x^3} + \frac{8}{(8-x)^3}$. Critical values will be when $\frac{(8-x)^3}{8} = \frac{x^3}{2}$, or $(8-x)^3 = 4x^3$, or $512 - 192x + 24x^2 - x^3 = 4x^3$, or when $512 - 192x + 24x^2 - 5x^3 = 0$. This has one solution between 0 and 8. It is about 3.1 m. Thus, the desired answer is that the light is the least about 3.1 m from the less intense light.

29. Here, we have $Y = k(2x^4 - 5Lx^3 + 3L^2x^2)$, and so $Y' = k(8x^3 - 15Lx^2 + 6L^2x) = 0$ or $x(8x^2 - 15Lx + 6L^2) = 0$. This yields critical values when $x = 0, 0.5785L$, and $1.2965L$. Now, $Y'' = k(24x^2 - 30Lx + 6L^2)$ and, since $L > 0$, we find that $Y''(0) > 0, Y''(0.5785L) < 0; Y''(1.2965L) > 0$. Thus, the maximum deflection occurs at $x = 0.5785L$.

31. Let d represent the diameter (or width) and h the height of the vertical walls of the tunnel. Then the area of the opening is $A = hd + \frac{1}{2}\pi\left(\frac{d}{2}\right)^2 = hd + \frac{\pi d^2}{8}$, or $hd = A - \frac{\pi d^2}{8}$ and so $h = \frac{A}{d} - \frac{\pi d}{8}$ where A is the constant area. Now, the cost, C, is given by $C = 2h + d + \frac{1}{2}\pi d \cdot 4$. We begin by substituting the above value of h to get the cost as a function of just one variable, d. Thus, we have

$$C = 2h + d + 2\pi d = 2\left(\frac{A}{d} - \frac{\pi d}{8}\right) + (2\pi + 1)d$$

$$= \frac{2A}{d} - \frac{\pi d}{4} + (2\pi + 1)d$$

To determine when the cost is a minimum, we take the derivative, getting

$$C' = \frac{-2A}{d^2} - \frac{\pi}{4} + 2\pi + 1$$

$$= \frac{-2A}{d^2} + \frac{7\pi}{4} + 1$$

Setting this equal to 0, we get $\frac{2A}{d^2} = \frac{7\pi+4}{4}$. Substituting for A, we have

$$\frac{2\left(hd + \frac{\pi d^2}{8}\right)}{d^2} = \frac{7\pi+4}{4}$$

or $\frac{hd}{d^2} + \frac{\pi d^2}{8d^2} = \frac{7\pi+4}{8}$ which produces $\frac{h}{d} = \frac{7\pi+4}{8} - \frac{\pi}{8} = \frac{6\pi+4}{8} = \frac{3\pi+2}{4}$.

33. Draw a line from the oil rig perpendicular to the shore at point P. This line represents the 5 mi from the rig to the shore. Let x be the distance from the connection to P. Then the pipe must go a distance of $\sqrt{5^2+x^2}$ mi underwater and $12 - x$ mi along the shore. The total cost is $C(x) = 1200\sqrt{25+x^2} + 600(12-x) = 600\left(2\sqrt{25+x^2} + 12 - x\right)$.

Finding the derivative, we get $C'(x) = 600\left[\frac{2x}{\sqrt{25+x^2}} - 1\right]$. Setting $C'(x) = 0$, we obtain $\frac{2x}{\sqrt{25+x^2}} - 1 = 0$ or $\frac{2x}{\sqrt{25+x^2}} = 1$ or $2x = \sqrt{25+x^2}$. Squaring both sides, we get $4x^2 = 25 + x^2$ or $3x^2 = 25$. Hence $x = +\frac{5}{\sqrt{3}}$. From the given information, we know $x \geq 0$, and so, $x = \pm\frac{5}{\sqrt{3}} = \frac{5}{3}\sqrt{3}$. The connection should be $\frac{5}{3}\sqrt{3}$ mi from P or $12 - \frac{5}{3}\sqrt{3} \approx 9.11$ mi from the storage tank.

35. **(a)** Let x represent the length of the interior partition wall. Then, there are two exterior walls of length x and two of length $4200/x$. The total cost is $C(x) = 2(177)x + 2(175)(4200/x) + 115x$. This simplifies to $C(x) = 465x + \frac{1,470,000}{x}$. The derivative is $C'(x) = 465 - \frac{1,470,000}{x^2}$. Setting $C'(x) = 0$ we obtain $465 - \frac{1,470,000}{x^2} = 0$ or $465 = \frac{1,470,000}{x^2}$. Hence $x^2 = \frac{1,470,000}{465}$ and $x \approx 56.225$ ft. The other length is $4200/x \approx 74.700$. The dimensions that will minimize costs are about 56.2×74.7 ft.
(b) The minimum cost is $C(56.225) = 465(56.225) + \frac{1,470,000}{56.225} \approx \$52,290$.

≡ 24.5 RELATED RATES

1. As in example 24.19 we have variables t, r, S, and V with formulas $S = 4\pi r^2$ and $V = \frac{4}{3}\pi r^3$. $\frac{dS}{dt} = 8\pi r \frac{dr}{dt}$; $\frac{dV}{dt} = 4\pi r^2 \frac{dr}{dt}$; and $\frac{dV}{dt} = 20$ L/min $= 20{,}000$ cm³/min; $20{,}000 = 4\pi r^2 \frac{dr}{dt} \Rightarrow \frac{dr}{dt} = \frac{5{,}000}{\pi r^2}$. This yields $\frac{dS}{dt} = 8\pi r \cdot \frac{5{,}000}{\pi r^2} = \frac{40{,}000}{r}$. When $r = 300$ cm, then $\frac{dS}{dt} = \frac{40{,}000}{300} = 133\frac{1}{3}$ cm²/min.

3. In this problem we are not concerned about the volume, only the radius and height. These are related by the equation $(24 - h)^2 + r^2 = 24^2$ or $(24^2 - 48h + h^2) + r^2 = 24^2$ or $-48h + h^2 = -r^2$. Taking the derivatives with respect to time we get

$$-48\frac{dh}{dt} + 2h\frac{dh}{dt} = -2r\frac{dr}{dt}$$

and thus

$$\frac{dr}{dt} = \frac{24-h}{r}\frac{dh}{dt}$$

When $h = 8$, then $r^2 = 24^2 - (24 - 8)^2 = 24^2 - 16^2 = 576 - 256$, so $r^2 = 320$ and $r = \sqrt{320} = 8\sqrt{5}$ ft. We were given $\frac{dh}{dt} = 2$ ft/min . Substituting we get
$$\frac{dr}{dt} = \frac{24-8}{8\sqrt{5}}(2)$$
$$= \frac{32}{8\sqrt{5}} = \frac{4}{\sqrt{5}} = \frac{4\sqrt{5}}{5} \approx 1.789 \text{ ft/min}$$

5. Let x be the distance the helicopter flies after passing over you and y the distance from you to the helicopter. x and y are related by the equation $x^2 + (0.75)^2 = y^2$. Taking derivatives with respect to time yields $2x\frac{dx}{dt} + 0 = 2y\frac{dy}{dt}$ or $\frac{dy}{dt} = \frac{x}{y}\frac{dx}{dt}$. After 1 minute, $x = 3$ km and $y = 3.0923$ km. Substituting yields $\frac{dy}{dt} = \frac{3}{3.0923} \cdot 3 = 2.9104$ km/min $= 174.6$ km/h.

7. Using the formula $V = \frac{1}{3}\pi h^2(3R - h)$ with $R = 5$ we get
$$V = \frac{1}{3}\pi h^2(15 - h) = 5\pi h^2 - \frac{1}{3}\pi h^3$$
$$\frac{dV}{dt} = 10\pi h\frac{dh}{dt} - \pi h^2\frac{dh}{dt} = (10\pi h - \pi h^2)\frac{dh}{dt}$$
$$\frac{dh}{dt} = \frac{dV/dt}{10\pi h - \pi h^2}$$
We are given $\frac{dV}{dt} = 10$ ft³/min, so $\frac{dh}{dt} = \frac{10}{10\pi h - \pi h^2}$
$= \frac{10}{(10h - h^2)\pi}$.

(a) When $h = 2$, we get $\frac{dh}{dt} = \frac{10}{(10\cdot2 - 2^2)\pi} = \frac{10}{(20-4)\pi} = \frac{10}{16\pi} = \frac{5}{8\pi} \approx 0.1989$ ft/min

(b) When $h = 3$, then $\frac{dh}{dt} = \frac{10}{(30-9)\pi} = \frac{10}{21\pi} \approx 0.1516$ ft/min

(c) When $h = 4$, we get $\frac{dh}{dt} = \frac{10}{(40-16)\pi} = \frac{10}{24\pi} = \frac{5}{12\pi} \approx 0.1326$ ft/min

9. Let h be the height of the surface of the water. Since the trough is an equilateral triangle we know that the width of the surface and the height are related by the equation $w = \frac{2\sqrt{3}h}{3}$. The volume of the trough is $V = \frac{1}{2}wh \cdot 10 = \frac{10\sqrt{3}h^2}{3}$ and $\frac{dV}{dt} = 25.0$ L/min $= 25$ L/min $\cdot \frac{1\,m^3}{1000\,L} = 0.025$ m³/min; $\frac{dV}{dt} = \frac{20\sqrt{3}}{3}h\frac{dh}{dt} \Rightarrow \frac{dh}{dt} = \frac{3dV/dt}{20\sqrt{3}h}$. When $h = 0.175$, we have $\frac{dh}{dt} = \frac{0.025}{\frac{20\sqrt{3}}{3}(0.175)} \approx 0.0124$ m/min $= 12.4$ mm/min.

11. We start with the formulas $V = \frac{4}{3}\pi r^3$ and $S = 4\pi r^2$. Differentiating V and setting it equal to 60, we get $\frac{dV}{dt} = 4\pi r^2\frac{dr}{dt} = 60 \Rightarrow \frac{dr}{dt} = \frac{60}{4\pi r^2} = \frac{15}{\pi r^2}$. Now, differentiating S, we find $\frac{dS}{dt} = 8\pi r\frac{dr}{dt} = 8\pi r \cdot \frac{15}{\pi r^2} = \frac{120}{r}$. When $r = 8$ cm, $\frac{dS}{dt} = \frac{120}{8} = 15$ cm²/s.

13. We have $V = \frac{4}{3}\pi r^3 + \pi r^2 h$ and $h = 4r$, so $V = \frac{4}{3}\pi r^3 + 4\pi r^3 = \frac{16}{3}\pi r^3$. Differentiating, we obtain $\frac{dV}{dt} = 16\pi r^2\frac{dr}{dt} = 60 \Rightarrow \frac{dr}{dt} = \frac{60}{16\pi r^2} = \frac{15}{4\pi r^2}$. The formula for the surface area is $S = 4\pi r^2 + 2\pi r h = 4\pi r^2 + 2\pi r(4r) = 4\pi r^2 + 8\pi r^2 = 12\pi r^2$. Differentiating this, we get $\frac{dS}{dt} = 24\pi r\frac{dr}{dt} = \frac{24\pi r \cdot 15}{4\pi r^2} = \frac{90}{r} = \frac{90}{8} = 11.25$ cm²/s.

15. $A = s^2$; $\frac{dA}{dt} = 2s\frac{ds}{dt} = 2(40) \cdot (0.3) = 24$ mm²/s.

17. Here $PV^{1.2} = 400$, so $P = 400V^{-1.2}$. $\frac{dP}{dt} = -1.2(400)V^{-2.2}\frac{dV}{dt} = \frac{-480}{V^{2.2}}\frac{dV}{dt}$. We are given that $\frac{dP}{dt} = 0.1P = 0.1(400V^{-1.2}) = 40V^{-1.2}$. Thus, $40V^{-1.2} = \frac{-480}{V^{2.2}}\frac{dV}{dt}$. Solving for dV/dt, we get $\frac{dV}{dt} = \frac{40V^{-1.2} \times V^{2.2}}{-480} = -0.0833V = -8.3\%$ of the volume per hour.

19. Let $a =$ altitude of the rocket and s be the distance from the radar station to the rocket. Then $a^2 + 2.430^2 = s^2$. Taking the derivatives, we obtain

$$2a\frac{da}{dt} = 2s\frac{ds}{dt} \Rightarrow \frac{da}{dt} = \frac{s}{a}\frac{ds}{dt}$$

When $a = 3.750$, $s = \sqrt{3.750^2 + 2.430^2} \approx 4.468$, and so $\frac{da}{dt} = \frac{4.468}{3.750}(325.0 \text{ m/s}) = 387.2$ m/s or 1394 km/hr.

21. **(a)** The derivative of $v^2 = 1200 - 36.0s$ is $2v\frac{dv}{dt} = -36.0\frac{ds}{dt}$. Now, $\frac{dv}{dt} = a$ and $\frac{ds}{dt} = v$, so we get $2va = -36.0v$ or $a = -18.0$ m/s^2.

(b) The acceleration is constant so when $s = 2.45$ m, then a is -18.0 m/s^2.

23. **(a)** We let A be at the origin and B at $(8, 0)$. The location of the ship at 8 mi from A means that the ship is on the circle $x^2 + y^2 = 64$; and the fact that the ship is 6 mi from B means that it is on the circle $(x - 8)^2 + y^2 = 36$. Solving the first equation for $y^2 = 64 - x^2$ and substituting into the second gives $(x - 8)^2 + (64 - x^2) = 36$ or $x^2 - 16x + 64 + 64 - x^2 = 36$, or $-16x = -92$, or $x = 5.75$ and so, $y = \pm\sqrt{64 - 5.75^2} = \pm 5.56$. The ship is 5.75 mi east of A and 5.56 mi north or south of A.

(b) Using the same setup as (a), the velocity vector from A has magnitude 14 and direction $\tan^{-1}\left(\frac{5.56}{5.75}\right) = 44.0°$. The velocity vector from B has magnitude 2 and direction $\tan^{-1}\left(\frac{5.56}{2.25}\right) = 112.0°$. Using component vectors to add gives

$A_x =$	$14\cos 44° =$	10.07	$A_y =$	$14\sin 44° =$	9.73
$B_x =$	$2\cos 112° =$	-0.75	$B_y =$	$2\sin 112° =$	1.85
$R_x =$		9.32	$R_y =$		11.58

Thus, $|R| = \sqrt{9.32^2 + 11.58^2} = 14.86$ mph.

(c) $\theta = 51.2°$ north of east or 38.8° east of north.

25. **(a)** Let x be the distance from the person to the base of the light and y be the distance from the light to the top of the shadow. Then we have similar triangles so $\frac{y}{20} = \frac{y-x}{6}$. Simplifying: $6y = 20y - 20x$ or $-14y = -20x$ or $y = \frac{10}{7}x$, we get $\frac{dy}{dt} = \frac{10}{7}dx/dt = \frac{10}{7}(5) = \frac{50}{7} \approx 7.143$ ft/s.

(b) Same as (a) $\frac{50}{7} = 7.143$ ft/s.

(c) Same as (a) 7.143 ft/s.

27. Here $R = 6 + 0.008T^2$, so $\frac{dR}{dt} = 0.016T\frac{dt}{dt} = 0.016(40)(0.01) = 0.0064$ Ω/s.

29. $Z = \frac{RX}{R+X}$. If R is constant 3 then $Z = \frac{3X}{3+X}$ and $Z' = \frac{(3+X)3X' - 3X(X')}{(3+X)^2} = \frac{9X'}{(3+X)^2}$, or $Z' = \frac{9(1.45)}{(3+1.05)^2} = 0.80$ Ω/min.

31. $R = \frac{1}{R_1} + \frac{1}{R_2} = R_1^{-1} + R_2^{-1}$; $R' = \frac{-R_1'}{R_1^2} + \frac{-R_2'}{R_2^2} = \frac{-0.5}{4^2} + \frac{-0.4}{5^2} = -0.04725$ Ω/s. Thus, R is decreasing at the rate of 0.04725 Ω/s.

33. $R = 35.0 + 0.0174T^2$, so $R' = 0.0348TT'$, and after substituting the given values, we have $R' = 0.0348(47.0)(-1.25) = -2.04$ Ω/min.

35. **(a)** $f = f_s\frac{v_L}{v_L - v_s}$; $f_s = 200$; $v_L = 343$, $v_s = 29$; $f = 200\left(\frac{343}{343 - 29}\right) = 218.47$ Hz.

(b) Here f_s and v_L are constants. We want to find f' and, in this case, we will write f as $f = f_s v_L(v_L - v_s)^{-1}$. Here v_s is the only variable, so $f' = f_s v_L(v_L - v_s)^{-2} = \frac{f_s v_L}{(v_L - v_s)^2}$. Substituting the given values, we obtain $f' = \frac{200(343)}{(343-29)^2} \approx 20.17$ Hz/s.

37. Substituting $k = 637.5$ in $V = k\left(R^2 - r^2\right)$ produces $V = 637.5\left(R^2 - r^2\right) = 637.5R^2 - 637.5r^2$. Then, $\frac{dV}{dt} = 2 \cdot 637.5R\frac{dR}{dt} = 1275R\frac{dR}{dt}$. Evaluating this derivative using the given data we have $\frac{dV}{dt} = 1275(0.0250)(0.002) = 0.06375$ mm/min.

39. Substituting the given values of $R = 7.5$ Ω into $Z = \frac{RX}{R+X}$ we obtain $Z = \frac{7.5X}{7.5+X}$. Differentiating produces $\frac{dZ}{dt} = \frac{(7.5+X)7.5 - 7.5X}{(7.5+X)^2}\frac{dX}{dt} = \frac{7.5^2}{(7.5+X)^2}\frac{dX}{dt}$. Evaluating this when $X = 4.0$ Ω and $\frac{dX}{dt} = -2.5$ Ω/s, we obtain $\frac{dZ}{dt} = \frac{7.5^2}{(7.5+4.0)^2}(-2.5) = -1.06$ Ω/s.

41. As in Exercise 39, $\frac{dZ}{dt} = \frac{7.5^2}{(7.5+X)^2}\frac{dX}{dt}$, so $\frac{dX}{dt} = \frac{dZ}{dt}\Big/\frac{7.5^2}{(7.5+X)^2}$. Substituting, we get $\frac{dX}{dt} = 2.0\Big/\frac{7.5^2}{(7.5+3.5)^2} = 4.30$ Ω/s.

43. Since the cross-section of the pipe is circular its area is given by $A = \pi r^2$ and so, $\frac{dA}{dt} = 2\pi r\frac{dr}{dt}$. Evaluating this derivative at the given data produces $\frac{dA}{dt} = 2\pi(0.5)(-0.1) = -0.1\pi$ in.^{2}/yr. Thus, the cross-sectional area of the pipe's opening is decreasing at the rate of 0.1π in.^{2}/yr ≈ 0.314 in.^{2}/yr.

≡ 24.6 NEWTON'S METHOD

1.   See *Computer Programs* in main text.

3. $x^3 - 4x + 2 = 0$; $P(x) = x^3 - 4x + 2$; $P'(x) = 3x^2 - 4$. There are three solutions; one each near -2; 0.5, and 1.5. Newton's method produces the following results for each solution.

i	x_i	$P(x_i)$	$P'(x_i)$	$x_{i+1} = x_i - \dfrac{P(x_i)}{P'(x_i)}$	$P(x_{i+1})$
1	-2	2	8	-2.25	-0.3906
2	-2.25	-0.3906	11.1875	-2.2151	-0.0084
3	-2.2151	-0.0084	10.7200	-2.2143	2.11×10^{-4}

Hence, one solution is $x \approx -2.2143$.

i	x_i	$P(x_i)$	$P'(x_i)$	$x_{i+1} = x_i - \dfrac{P(x_i)}{P'(x_i)}$	$P(x_{i+1})$
1	0.5	0.125	-3.25	0.5385	0.0022
2	0.5385	0.0022	-3.1301	0.5392	-3.48×10^{-5}

Another solution is $x \approx 0.5392$.

i	x_i	$P(x_i)$	$P'(x_i)$	$x_{i+1} = x_i - \dfrac{P(x_i)}{P'(x_i)}$	$P(x_{i+1})$
1	1.5	-0.625	2.75	1.7273	0.2443
2	1.7273	.2443	4.9507	1.6780	0.0127
3	1.6780	0.127	4.4471	1.6751	-1.36×10^{-4}

And the final solution is $x \approx 1.6751$.

5. $P(x) = x^4 - 2x^3 - 3x + 2$ and $P'(x) = 4x^3 - 6x^2 - 3$. Solutions are near 0.6 and 2.3. Newton's method produces the following results for each solution.

i	x_i	$P(x_i)$	$P'(x_i)$	$x_{i+1} = x_i - \dfrac{P(x_i)}{P'(x_i)}$	$P(x_{i+1})$
1	0.6	-0.1024	-4.2960	0.5762	-9.76×10^{-4}
2	0.5762	-9.76×10^{-4}	-4.2268	0.5760	-1.3×10^{-4}

Hence, one solution is $x \approx 0.5760$.

i	x_i	$P(x_i)$	$P'(x_i)$	$x_{i+1} = x_i - \dfrac{P(x_i)}{P'(x_i)}$	$P(x_{i+1})$
1	2.3	-1.2499	13.9280	2.3897	0.1490
2	2.3897	0.1490	17.3231	2.3811	0.0015
3	2.3811	0.0015	16.9821	2.3810	-1.6×10^{-4}

And the other solution is $x \approx 2.3810$.

7. $x^3 = 91 \Rightarrow x^3 - 91 = 0$; $P(x) = x^3 - 91$; $P'(x) = 3x^2$. A solution is near 4.5. Using Newton's method, you can show that $x \approx 4.4979$.

9. **(a)** The volume is given by $V = \pi r^2 h$ and the total surface area is $A = 2\pi rh + \pi r^2$. Then $2V = 2\pi r^2 h$ and so, $2V - 2\pi r^2 h = 0$.

From the area formula, we see that $A - \pi r^2 = 2\pi rh$ and so, $Ar - \pi r^3 = 2\pi r^2 h$. Substituting this for $2\pi r^2 h$ in the displayed equation produces the desired result

$$2V - (Ar - \pi r^3) = 2V - Ar + \pi r^3 = 0$$
$$2V + \pi r^3 - Ar = 0.$$

(b) Substituting 10 000 for V and 2 400 for A, we get

$$2(10000) + \pi r^3 - (2400)r = 0$$
$$\pi r^3 - 2400r + 20000 = 0.$$

Using Newton's method we obtain solution at $r_1 = 9.4315$ cm, $r_2 = 21.6893$ cm

≡ 24.7 DIFFERENTIALS

1. $dy = f'(x)dx = (4x^3 - 2x)dx$

3. $dy = (15x^2 - 2x + 1)dx$

5. $dy = \frac{1}{3}(4 - 2x)^{-2/3}(-2)dx = \frac{-2}{3}(4 - 2x)^{-2/3}dx$

7. $\Delta y = ((3.2)^2 - 3.2) - (3^2 - 3) = 7.04 - 6 = 1.04, dy = (2x - 1)dx = 5(0.2) = 1.0$

9. $\Delta y = f(5.15) - f(5) = 59.9675 - 56 = 3.9675$, $dy = f'(5)(0.15) = 26(0.15) = 3.9$

11. $\Delta y = h(3.2) - h(3) = 0.0305175781 - 0.037037037 = -0.00651946, dy = h'(3)(0.2) = -0.037037037(0.2) = -0.00740741$

13. **(a)** $V = \frac{1}{3}\pi r^2 h = \frac{1}{3}\pi(2)^2 \cdot 4 = \frac{16\pi}{3} = 16.755$ m^3

(b) $dV = \frac{2}{3}\pi rh\,dr = \frac{2}{3}(\pi \cdot 2 \cdot 4)(0.01) = \frac{0.16\pi}{3} = \frac{16}{300}\pi \approx 0.16755$ m^3

(c) Relative error $= \frac{dV}{V} = \frac{\frac{16\pi}{300}}{16\frac{\pi}{3}} = \frac{1}{100}; \frac{1}{100} = 0.01 = 1\%$.

15. **(a)** $A = s^2; dA = 2s\,ds = 2(24)(0.02) = 0.96$ in^2

(b) $\frac{dA}{A} = \frac{0.96}{24^2} = 0.00167 = 0.167\%$

17. $S = 2\pi r^2; dS = 4\pi r\,dr = 4\pi(100)(0.01) = 4\pi$ m^2. (Note: 10 mm = 0.01 m.)

19. **(a)** The volume of a sphere is $V = \frac{4}{3}\pi r^3$. Since we are given $d = 12.4$ m, we see that $r = 6.2$ m, and $V = \frac{4}{3}\pi(6.2)^3 \approx 317.8\pi$. Taking the derivative of the formula for the volume, we obtain $dV = 4\pi r^2\,dr = 4\pi(6.2)^2(\pm 0.05) = \pm 7.69\pi$ m$^3 \approx \pm 24.2$ m^3

(b) $\frac{dV}{V} = \frac{7.69\pi}{317.8\pi} = 0.024 = \pm 2.4\%$.

21. $R = \sigma T^4$, so $dR = 4\sigma T^3\,dT$ and $\frac{dT}{T} = 0.02; \frac{dR}{R} = \frac{4\sigma T^3 dT}{\sigma T^4} = 4\frac{dT}{T} = 4(0.02) = 0.08$.

23. $V = IR; 30 = IR; I = 30R^{-1}; dI = -30R^{-2}dR = -30(10)^{-2}(0.1) = -0.03$ A

25. $R = 35.0 + 0.0174T^2; dR = 0.0348T \cdot dT = 0.0348(125)(\pm 0.5) = \pm 2.175\,\Omega$.

27. $S = 4\pi r^2; \frac{dS}{S} = \frac{8\pi r\,dr}{4\pi r^2} = 2\frac{dr}{r} = 2(0.5\%) = 1.0\%$

29. A ball bearing is spherical and its volume is given by $V = \frac{4}{3}\pi r^3$. Differentiating produces $\frac{dV}{dr} = 4\pi r^2$ or $dV = 4\pi r^2\,dr$. Substituting the given values yields $dV = 4\pi(0.6)^2(\pm 0.015) = \pm 0.0678$ or ± 0.07 mm^3.

31. The volume of a sphere is given by $V = \frac{4}{3}\pi r^3$. Differentiating produces $\frac{dV}{dr} = 4\pi r^2$ or $dV = 4\pi r^2\,dr$. Substituting the given values yields $dV = 4\pi(15)^2(0.5) = 450\pi \approx 1414$ mm^3.

33. Here $dR = (0.04n + 0.00045n^2)dn$. Using $n = 500$ and $dn = 1$ produces $dR = (0.04(500) + 0.00045(500)^2)(1) = 132.5$. Revenue will increase by \$132.50 from the sale of one more computer when 500 have been sold.

35. The volume of a right circular cylinder is given by $V = \pi r^2 h$. Differentiating with respect to r produces $\frac{dV}{dr} = 2\pi rh$ or $dV = 2\pi rh\,dr$. We are given $h = 3.2$, $r = \frac{1}{2}(30.0) = 15.0$, and $dr = 0.2$. Substituting these values produces $dV = 2\pi(15.0)(3.2)(0.2) = 19.2\pi \approx 60.3$ cm^3.

24.8 ANTIDERIVATIVES

1. $f(x) = 7; F(x) = 7x + C$

3. $h(x) = 2 - 3x^2; H(x) = 2x - 3(\frac{1}{3})x^3 + C = 2x - x^3 + C$

5. $k(x) = 4x^3 - 3x^2 + 2x + 9; K(x) = 4 \cdot \frac{1}{4}x^4 - 3 \cdot \frac{1}{3}x^3 + 2 \cdot \frac{1}{2}x^2 + 9x + C = x^4 - x^3 + x^2 + 9x + C$

7. $f(x) = x^{-4} + 2x^{-3} - x^{-2} + 5; F(x) = \frac{1}{-3}x^{-3} + 2 \cdot \frac{1}{-2}x^{-2} - \frac{1}{-1}x^{-1} + 5x + C = \frac{-1}{3}x^{-3} - x^{-2} + x^{-1} + 5x + C$

9. $h(x) = \sqrt{x} + x - \frac{1}{x^2} = x^{1/2} + x - x^{-2}; H(x) = \frac{1}{3/2}x^{3/2} + \frac{1}{2}x^2 - \frac{1}{-1}x^{-1} + C = \frac{2}{3}x^{3/2} + \frac{1}{2}x^2 + x^{-1} + C$

11. $k(x) = 2x^{-1.4} + 3.5x^{-6} + 1.2x^{-1.3}; K(x) = 2 \cdot \frac{1}{-.4}x^{-.4} + 3.5 \cdot \frac{1}{-5}x^{-5} + 1.2 \cdot \frac{1}{-0.3}x^{-.3} + C = -5x^{-0.4} - 0.7x^{-5} - 4x^{-0.3} + C$

13. $s(t) = t^2 + 2t; S(t) = \frac{1}{3}t^3 + t^2 + C$

15. $v(t) = 42t - 5; V(t) = \frac{42}{2}t^2 - 5t + C = 21t^2 - 5t + C$

17. $a(t) = -32$ ft/s^2; $v(t) = A(t) = -32 \cdot t + C_1$. When the ball is dropped at $t = 0$, we have $v(0) = 0$, so we can see that $C_1 = 0$. Hence, $v(t) = -32t$. Now, $s(t) = V(t) = -16t^2 + C_2$. When $t = 0$, we are given that $s(0) = 600$, so $C_2 = 600$ and we have $s(t) = -16t^2 + 600$.
(a) When the ball hits the ground, $s(t) = 0$. Solving $s(t) = 0$ for t, we have $16t^2 = 600$ or $t^2 = \frac{600}{16}$ and so the ball hits the ground after $t = \sqrt{\frac{600}{16}} \approx 6.12$ s
(b) $v(6.12) = -32(6.12) = -195.96$ ft/s

19. $a(t) = -32; v(t) = -32t + C$. Since the ball was thrown down at $t = 0$ at 25 ft/s, we have $C = -25$ and thus, $v(t) = -32t - 25$.
(a) $v(t) = -185 = -32t - 25 \Rightarrow 32t = 160 \Rightarrow t = 5$ s.
(b) $s(t) = -16t^2 - 25t + C, s(5) = 0 = -16 \cdot 5^2 - 25 \cdot 5 + C, C = 16 \cdot 5^2 + 25 \cdot 5 = 525$. The building is 525 feet high.

21. $a(t) = -32$ and $v(t) = -32t + 160$.
(a) Since $v(t) = -32t + 160 = -384$, then $-32t = -544$ and $t = 17$ s.
(b) $s(t) = -16t^2 + 160t + C$, so $s(17) = -16 \cdot 17^2 + 160 \cdot 17 + C = 0$, thus $C = 4.9(16.837)^2 - 45(16.837) = 631.41$ m = 1904 ft.

23. $a(t) = 3.2; v(t) = 3.2t + 0; 3.2t = 32 \Rightarrow t = 10$ s, $s(t) = 1.6t^2, s(10) = 1.6(10)^2 = 160$ m

25. $v(t) = a \cdot t + 50 \Rightarrow a = \frac{-50}{t}, s(t) = \frac{a}{2}t^2 + 50t + 500 = 0, \frac{-50}{2t}t^2 + 50t = -500$. This simplifies to $-25t + 50t = 500$, or $25t = 500$, or $t = 20$ s. Finally, $a = \frac{-50}{20} = -2.5$ m/s^2.

27. (a) $\theta(t) = 12.4t - \frac{3.4}{2}t^2 + \frac{0.30}{3}t^3 = $ antiderivative of $\omega(t) = 12.4t - 1.7t^2 + 0.10t^3; \theta(3) = 24.6$ rad $= \frac{24.6}{2\pi} = 3.9$ rev.
(b) $\alpha(t) = \omega'(t) = -3.4 + 0.6t = -1, 0.6t = 2.4, t = 4$ s, $\theta(4) = 28.8$ rad $= \frac{28.8}{2\pi} = 4.6$ rev.

29. (a) $i = 4.4t - 2.1t^2; q = 2.2t^2 - 0.7t^3 + C = 2.2t^2 - 0.7t^3 + 5$
(b) $q(3.2) \approx 4.59 \approx 4.6$ C

31. $v(t) = -N\phi'(t)$
(a) $-200\phi'(t) = 2t - 4t^{1/3}, \phi'(t) = -\frac{t}{100} + \frac{1}{50}t^{1/3}, \phi(t) = \frac{-t^2}{200} + \frac{3}{200}t^{4/3} + C, C = 0.02; \phi(t) = -0.005t^2 + 0.015t^{4/3} + 0.02$, or $-\frac{1}{200}(t^2 - 3t^{4/3} - 4)$
(b) $\phi(0.729) = -\frac{1}{200}(0.729^2 - 3(0.729)^{4/3} - 4) = 0.027$ Wb

33. $m = \sqrt{x}$ so $y' = \frac{-1}{\sqrt{x}} = -x^{-1/2}, y = -\frac{1}{1/2}x^{1/2} = -2x^{1/2} + C$, and $-3 = -2(4)^{1/2} + C$ gives us that $C = -3 + 4 = 1$. Hence, $f(x) = -2x^{1/2} + 1 = -2\sqrt{x} + 1$.

35. Rewriting $\frac{dT}{dx} = -\frac{5750}{(x+1)^3}$ as $\frac{dT}{dx} = -5750(x+1)^{-3}$. The antiderivative is $T(x) = -\frac{5750}{-2}(x+1)^{-2} + C = 2875(x+1)^{-2} + C$. Since $T(0) = 2875(0+1)^{-2} + C = 2875 + C = 2900$, we have $C = 25$. Thus, $T = 2875(x+1)^{-2} + 25$.

CHAPTER 24 REVIEW

1. $f(x) = x^4 - x^2; f'(x) = 4x^3 - 2x = 2x(2x^2 - 1) = 0$ when $x = 0$ and $\pm\frac{\sqrt{2}}{2}$. These are the critical values. $f''(x) = 12x^2 - 2; f''(0) < 0$, so there is a maximum at $(0, 0)$. $f''\left(\frac{\sqrt{2}}{2}\right) = f''\left(-\frac{\sqrt{2}}{2}\right) > 0$, so there are minimums at $\left(\pm\frac{\sqrt{2}}{2}, -\frac{1}{4}\right)$. $f''(x) = 12x^2 - 2 = 0$ when $x = \pm\frac{1}{\sqrt{6}} = \pm\frac{\sqrt{6}}{6}$. Inflection points are at $\left(\pm\frac{\sqrt{6}}{6}, -0.1389\right)$. Concave up over the intervals $\left(-\infty, \frac{-\sqrt{6}}{6}\right)$ and $\left(\frac{\sqrt{6}}{6}, \infty\right)$. Concave down over $\left(\frac{-\sqrt{6}}{6}, \frac{\sqrt{6}}{6}\right)$.

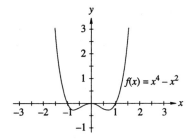

2. $g(x) = x^4 - 32x; g'(x) = 4x^3 - 32 = 4(x^3 - 8)$. Critical value is 2. $g''(x) = 12x^2$ is always ≥ 0 so there is a minimum at $(2, -48)$. There are no maxima. $g''(x) = 0$ when $x = 0$. Since $g''(x)$ is never negative there is no inflection point and $g(x)$ is concave up over the interval $(-\infty, \infty)$.

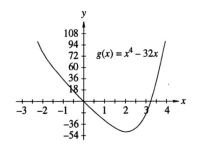

3. $h(x) = 18x^2 - x^4$, so $h'(x) = 36x - 4x^3 = 4x(9 - x^2)$. Critical values are at 0 and ± 3. $h''(x) = 36 - 12x^2 = 12(3 - x^2)$. Now, $h''(x) > 0$, so there is a minimum at $(0, 0)$; and $h''(\pm 3) < 0$, so there is a maximum at $(\pm 3, 81)$. Inflection points at $\left(\pm\frac{\sqrt{3}}{3}, 5.889\right)$. Concave up $\left(\frac{-\sqrt{3}}{3}, \frac{\sqrt{3}}{3}\right)$. Concave down $\left(-\infty, \frac{-\sqrt{3}}{3}\right)$, and $\left(\frac{\sqrt{3}}{3}, \infty\right)$.

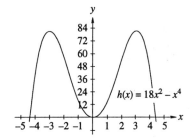

4. $j(x) = 3x^4 - 4x^3 + 1; j'(x) = 12x^3 - 12x^2 = 12x^2(x - 1)$. Critical values at 0, 1. $j''(x) = 36x^2 - 24x; j''(0) = 0$ so the Second Derivative Test fails. $j''(1) > 0$ so minimum at $(1, 0)$. Use the First Derivative Test to determine if the critical value 0 yields a maximum or minimum. On the interval $(-\infty, 0); j' < 0$. Likewise, on $(0, 1), j' < 0$ so also decreasing. No maxima. $j''(x)36x^2 - 24x = 12x(3x - 2)$. Inflection points at $(0, 1)$ and $\left(\frac{2}{3}, 0.4074\right)$. Concave up $(-\infty, 0)$ and $\left(\frac{2}{3}, \infty\right)$. Concave down $\left(0, \frac{2}{3}\right)$.

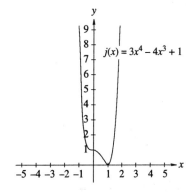

5. $k(x) = \sqrt[5]{x} = x^{1/5}; k'(x) = \frac{1}{5}x^{-4/5} = \frac{1}{5} \cdot \frac{1}{\sqrt[5]{x^4}}$.
Critical value at 0, but since $k'(x)$ is always positive or undefined there are no extrema. $k''(x) = -\frac{4}{25}x^{-8/5}$. This is undefined at 0. On the interval $(-\infty, 0), k'' > 0$ so Concave up. On the interval $(0, \infty), k'' < 0$ so concave down. Inflection point: $(0, 0)$.

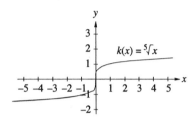

6. Here $f(x) = x\sqrt[3]{4-x}$, so $f'(x) = \sqrt[3]{4-x} + x \cdot \frac{1}{3}(4-x)^{-2/3}(-1) = \sqrt[3]{4-x} - \frac{x}{3\left(\sqrt[3]{4-x}\right)^2} =$

$\frac{3(4-x)-x}{3\left(\sqrt[3]{4-x}\right)^2} = \frac{12-4x}{3\left(\sqrt[3]{4-x}\right)^2}$. Critical values are at 3 and 4. $f''(x) = \frac{3\sqrt[3]{4-x}^2(-4)-(12-4x)2(4-x)^{-1/3}(-1)}{9\left(\sqrt[3]{4-x}\right)^4} =$

$\frac{3(4-x)(-4)+(12-4x)2}{9\left(\sqrt[3]{4-x}\right)^5} = \frac{-48+12x+24-8x}{9\left(\sqrt[3]{4-x}\right)^5} = \frac{4x-24}{9\left(\sqrt[3]{4-x}\right)^5}$.
The second derivative is undefined at 4 and zero at 6. Concave down over $(-\infty, 4)$ and $(6, \infty)$; concave up over $(4, 6)$. Inflection points at $(4, 0)$ and $(6, -7.560)$. Maximum at $(3, 3)$, no minimum.

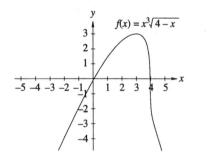

7. $g(x) = \frac{x^2+1}{x^2-4}; g'(x) = \frac{(x^2-4)2x-(x^2+1)(2x)}{(x^2-4)^2} = \frac{-10x}{(x^2-4)^2}$.
Critical values are at 0 and ± 2. Vertical asymptotes at $x = 2$ and $x = -2$. Horizontal asymptote at $y = 1$. $g''(x) = \frac{(x^2-4)^2(-10)+10x(2)(x^2-4)(2x)}{(x^2-4)^4} = \frac{(x^2-4)(-10)+40x^2}{(x^2-4)^3} = \frac{30x^2+40}{(x^2-4)^3}$. $g''(x)$ is undefined at $x = \pm 2$. Concave up over the intervals $(-\infty, -2)$ and $(2, \infty)$. Concave down over $(-2, 2)$. Maximum at the point $(0, -\frac{1}{4})$, no minimum. No inflection points.

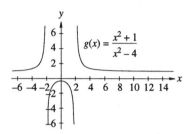

8. $h(x) = \frac{x-1}{x+2}; h'(x) = \frac{(x+2)-(x-1)}{(x+2)^2} = \frac{3}{(x+2)^2}$. Critical value $x = -2$; Vertical asymptote: $x = -2$, horizontal asymptote : $y = 1$. No extrema. $h''(x) = -6(x+2)^{-3}$ undefined at $x = -2$. Concave up: $(-\infty, -2)$; Concave down $(-2, \infty)$; No inflection points.

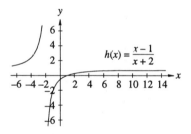

9. $j(x) = \frac{x}{x^2+x-2}; j'(x) = \frac{(x^2+x-2)-x(2x+1)}{(x^2+x-2)^2} = \frac{x^2+x-2-2x^2-x}{(x^2+x-2)^2} = \frac{-x^2-2}{(x^2+x-2)^2}$. Since $j(x)$ is undefined at $x = -2$ and 1, there are no critical values. No extrema. Vertical asymptotes: $x = -2$ and $x = 1$; Horizontal asymptote: $y = 0$.

$j''(x) = \frac{(x^2+x-2)^2(-2x)+(x^2+2)(2)(x^2+x-2)(2x+1)}{(x^2+x-2)^4}$

$= \frac{(x^2+x-2)(-2x)+2(x^2+2)(2x+1)}{(x^2+x-2)^3}$

$= \frac{(-2x^3-2x^2+4x)+(4x^3+2x^2+8x+4)}{(x^2+x-2)^3}$

$= \frac{2x^3+12x+4}{(x^2+x-2)^3}$

Using Newton's method to solve $2x^3 + 12x + 4 = 0$ we get an inflection point when $x = -0.3275$. Concave up: $(-2, -0.3275)$ and $(1, \infty)$, Concave down: $(-\infty, -2)$ and $(-0.3275, 1)$.

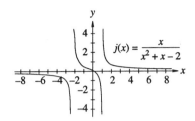

10. $k(x) = \frac{x^3}{x^2 - 9}; k'(x) = \frac{(x^2-9)3x^2 - x^3(2x)}{(x^2-9)^2} = \frac{3x^4 - 27x^2 - 2x^4}{(x^2-9)^2} = \frac{x^4 - 27x^2}{(x^2-9)^2}$. Critical values at 0 and $\pm 3\sqrt{3}$. Vertical asymptotes at $x = \pm 3$. No horizontal asymptote but there is a slant asymptote $y = x$.

$$k''(x) = \frac{(x^2-9)(4x^3-54x) - (x^4-27x^2)(2)(x^2-9)(2x)}{(x^2-9)^4}$$
$$= \frac{(x^2-9)(4x^3-54x) - 4x(x^4-27x^2)}{(x^2-9)^3}$$
$$= \frac{4x^5 - 90x^3 + 486x - 4x^5 + 108x^3}{(x^2-9)^3}$$
$$= \frac{18x^3 + 486x}{(x^2-9)^3}$$

This is undefined at $x = \pm 3$ and zero at $x = 0$. Inflection point: $(0, 0)$; Maximum: $(-3\sqrt{3}, -7.794)$; Minimum: $(3\sqrt{3}, 7.794)$; Concave up: $(-3, 0)$; $(3, \infty)$; Concave down: $(-\infty, -3)$; $(0, 3)$

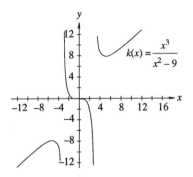

11. $s(t) = t^3 - 9t^2 + t, v(t) = s'(t) = 3t^2 - 18t + 1; a(t) = v'(t) = 6t - 18, a'(t) = 6; v(t) = 0 = 3t^2 - 18t + 1 = 0$ when $t = 0.056$ and 5.944. Distance: maximum at $(0.056, 0.028)$; minimum $(-4, -212)$ and $(4, -76); a(t) = 0$ at $t = 3$. Velocity: minimum $(3, -26)$; maxima at $(-4, 121)$ and $(4, -23)$. Acceleration is always increasing. Minimum: $(-4, -42)$, Maximum: $(4, 6)$.

12. $s(t) = t + 4t^{-1}; v(t) = 1 - 4t^{-2}; a(t) = 8t^{-3}; v(t) = 0 = 1 - \frac{4}{t^2} \Rightarrow \frac{4}{t^2} = 1; t^2 = 4$ or $t = 2$; distance: Maximum: $(4, 5)$; Minimum: $(1, 5), (2, 4); a(t) =$

$0 = 8t^{-3}$ is never 0. $v(1) = -3; v(4) = 0.75$. Velocity: Maximum: $(4, 0.75)$; Minimum: $(1, -3)$. Acceleration is always decreasing on $[1, 4]$, so maximum $(1, 8)$; minimum $(4, 0.125)$.

13. $f(x) = 3x^2 - 4x; F(x) = 3 \cdot \frac{1}{3}x^3 - 4 \cdot \frac{1}{2}x^2 + C; F(x) = x^3 - 2x^2 + C$

14. $g(x) = 2x^{-3} + 5x^6; G(x) = 2 \cdot \frac{1}{-2}x^{-2} + 5 \cdot \frac{1}{7}x^7 + C; G(x) = -x^{-2} + \frac{5}{7}x^7 + C$

15. $h(x) = x^{1/2} + x^2 - 3x^{-2}; H(x) = \frac{1}{3/2}x^{3/2} + \frac{1}{3}x^3 - 3 \cdot \frac{1}{-1}x^{-1} + C; H(x) = \frac{2}{3}x^{3/2} + \frac{1}{3}x^3 + 3x^{-1} + C$

16. $s(t) = 4.9t^2 - 3.6t + 14; S(t) = 4.9 \cdot \frac{1}{3}t^3 - 3.6 \cdot \frac{1}{2}t^2 + 14t + C; S(t) = 1.63t^3 - 1.8t^2 + 14t + C$

17. **(a)** $s(t) = 288t - 16t^2; v(t) = s'(t) = 288 - 32t$ ft/s; $a(t) = v'(t) = -32$ ft/s^2
(b) $s(4) = 896$ ft; $v(4) = 160$ ft/s; $a(4) = -32$ft/s^2
(c) Maximum height is when $v(t) = 0; 288 - 32t = 0 \Rightarrow 32t = 288 \Rightarrow t = 9; s(9) = 1296$ ft.
(d) at $t = 9$ s.
(e) $s(t) = 0 = 288t - 16t^2 = t(288 - 16t); t = 288/16 = 18$ s.
(f) $v(18) = -288$ ft/s

18. **(a)** $s(t) = 15t^2 + 5; v(t) = s'(t) = 30t$ cm/s
(b) $v(2) = 60$ cm/s
(c) $v(t) = 30t = 105 \Rightarrow t = 3.5$ s.

19. The total cost is $C = (1 + 0.00058v^{3/2}) \times 1000 + 25\left(\frac{1000}{v}\right) = (1000 + 0.58v^{3/2}) + \left(\frac{25,000}{v}\right)$. To find the minimum cost, we find the derivative and set it equal to 0. Thus, $C' = \frac{3}{2}(0.58)v^{1/2} - \frac{25,000}{v^2} = 0$. Hence, $0.87v^{5/2} - 25,000 = 0$ or $v^{5/2} = \frac{25,000}{0.87} \approx 28,735$ and $v = (28,735)^{2/5} \approx 60.7$. Checking the second derivative, we have $C'' = 0.435v^{-1/2} + 50,000v^{-3}$, and we see that $C''(60.7) > 0$, so, by the second derivative test, $v \approx 60.7$ is a minimum. Thus, the trucker should drive at 60.7 mph in order to minimize the cost of the trip.

20. $V = \pi r^2 h = 246\pi \Rightarrow h = \frac{246}{r^2}$. Area of top and bottom $= 2\pi r^2$. Area of side $= 2\pi rh = 2\pi r \cdot \frac{246}{r^2} = 492\frac{\pi}{r}$. Cost is $C(r) = 2\pi r^2 + 3 \cdot 492\frac{\pi}{r} = 2\pi r^2 + 1476\pi r^{-1}$ and has the derivative $C'(r) = 4\pi r - 1476\pi r^{-2} = 0 \Rightarrow r - \frac{369}{r^2} = 0$, or $r = \frac{369}{r^2}$, or $r^3 = 369$. Thus, $r = \sqrt[3]{369} \approx 7.17258$ ft; $h = \frac{246}{7.17258^2} = 4.78$ ft.

21. $A = \ell \cdot w; 2\ell + 2w = 1200$ m; $\ell + w = 600; w = 600 - \ell; A = \ell(600 - \ell) = 600\ell - \ell^2; A' = 600 - 2\ell \Rightarrow \ell = 300$ m; $w = 600 - 300 = 300$ m; $A = (300 \times 300) = 90000$ m^2

22. Since the radius of the tank is 8 ft and the height is 11 ft, the tank is more than half full. The height left is $16 - 11 = 5$ ft and the volume remaining is given by the formula $V = \frac{1}{3}\pi h^2(3R - h)$ where h is the height left and $R = 8$ ft. $V = \frac{1}{3}\pi h^2(24) - \frac{1}{3}\pi h^3; \frac{dV}{dt} = (16\pi h - \pi h^2)dh/dt$ and $\frac{dV}{dt} = 20$ gal/min $= 20(0.13358)$ ft^3/min; $\frac{dh}{dt} = \frac{20(0.13358)}{[16\pi \cdot 5 - \pi \cdot 5^2]} = \frac{2.6716}{55\pi} = 0.01546$ ft/min.

23. $V = \frac{4}{3}\pi r^3; dV = 4\pi r^2 dr; dV = 4\pi(1.5)^2(0.01) = 0.2827$ m^3; $A = \pi r^2; dA = 8\pi r dr = 8\pi(1.5)(0.01) \approx 0.3770$ m^2

24. The largest capacity is when the cross-sectional area is the largest. $A = \ell \cdot w; \ell + 2w = 380$ mm; $\ell = 380 - 2w; A = (380 - 2w)w = 380w - 2w^2; A' = 380 - 4w;$ maximum when $A' = 0; 4w = 380; w = 95; \ell = 380 - 2 \cdot 95 = 190$. The gutter should measure 95 mm $\times$ 190 mm.

25. $V = 4.5T + 0.0003T^3; \Delta v \approx dv = (4.5 + 0.0009T^2)dT = (4.5 + 0.0009 \cdot 100^2)(1) = 13.5$ V

26. The distance between the ships is related by the equation $d^2 = x^2 + y^2$ so $2dd' = 2xx' + 2yy'$ or $d' = \frac{xx' + yy'}{d}$. After 3 hours, we have $x = 12 \cdot 3 = 36, y = 5 \cdot 3 = 15, d = \sqrt{36^2 + 15^2} = 39$, and $d' = \frac{36 \cdot 12 + 15 \cdot 5}{39} = 13$ km/h.

27. $V = \pi r^2 h = 20$ kL $= 20\,000$ L$= 20\,000$ dm$^3 = 20$ m$^3; h = \frac{20}{\pi r^2}; A = 2\pi r^2 + 2\pi rh = 2\pi r^2 + 2\pi r\left(\frac{20}{\pi r^2}\right). \frac{dA}{dr} = 4\pi r - 40r^{-2};$ minimized when $\frac{dA}{dr} = 0$ or $40r^{-2} = 4\pi r; \frac{10}{\pi} = r^3; r = \sqrt[3]{\frac{10}{\pi}} \approx 1.4710$ m; $h = \frac{20}{\pi r^2} = 2.942$ m.

28. Distance in the desert is $\sqrt{8^2 + x^2} = \sqrt{64 + x^2}$. Distance on the road is $10 - x$. Total time $= \frac{\sqrt{64 + x^2}}{15} + \frac{10 - x}{55}; \frac{dt}{dx} = \frac{\frac{1}{2}(64 + x^2)^{-1/2}(2x)}{15} - \frac{1}{55}$. Set $\frac{dt}{dx} = 0$ and solve $\frac{x}{15\sqrt{64 + x^2}} = \frac{1}{55}$, or $55x = 15\sqrt{64 + x^2}$, and so $\frac{11}{3}x = \sqrt{64 + x^2}$, or $\frac{121}{9}x^2 = 64 + x^2; \frac{112}{9}x^2 = 64; 112x^2 = 576; x^2 = 5.14287$ and we get $x = 2.2678; 2.2678$ mi from A.

29. **(a)** $a = -32$ ft/s$^2; v = -32t; s(t) = -16t^2 + C; s(7) = 0 = -16 \cdot 7^2 + C; C = 16 \cdot 7^2 = 16 \cdot 49 = 784$ ft. The building is 784 ft tall.
(b) $v(7) = -32 \cdot 7 = -224$ ft/s; Speed $= |v| = |-224| = 224$ ft/s.

CHAPTER 24 TEST

1. $f(x) = x^4 - 8x^2; f'(x) = 4x^3 - 16x = 4x(x^2 - 4)$
(a) Critical values are $0, \pm 2$.

(b) $f''(x) = 12x^2 - 16; x = \pm\sqrt{\frac{16}{12}} = \pm\sqrt{4/3} = \frac{\pm 2\sqrt{3}}{3}$ are inflection points. Maximum: $(0, 0)$; Minima: $(-2, -16), (2, 16)$

(c) Concave up: $\left(-\infty, \frac{-2\sqrt{3}}{3}\right), \left(\frac{2\sqrt{3}}{3}, \infty\right)$

(d) Concave down $\left(\frac{-2\sqrt{3}}{3}, \frac{2\sqrt{3}}{3}\right)$

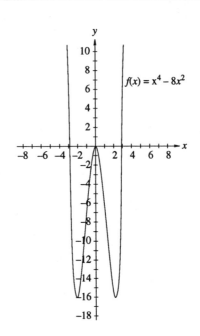

$f(x) = x^4 - 8x^2$

2. $g(x) = \frac{x^2}{9-x^2}; g'(x) = \frac{(9-x^2)(2x)-x^2(-2x)}{(9-x^2)^2} =$

$\frac{18x-2x^3+2x^3}{(9-x^2)^2} = \frac{18x}{(9-x^2)^2}.$

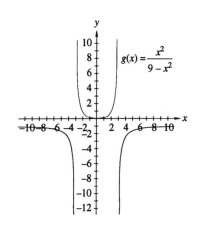

(a) Critical value: $x = 0$. Vertical asymptote at $x = 3, x = -3$

(b) $g''(x) = \frac{(9-x^2)^2(18)-18x(2)(9-x^2)(-2x)}{(9-x^2)^4} =$

$\frac{18(9-x^2)+18\cdot 4x^2}{(9-x^2)^3} = \frac{54x^2+162}{(9-x^2)^3}$. The second derivative is undefined at ± 3. Maximum: none; Minimum: $(0, 0)$,

(c) Concave up: $(-3, 3)$,

(d) Concave down: $(-\infty, -3), (3, \infty)$,

(e) Vertical asymptotes: $x = \pm 3$. Horizontal asymptote: $y = -1$

3. $x^2 - 7 = 0; P(x) = x^2 - 7; P'(x) = 2x$

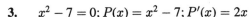

i	x_i	$P(x_i)$	$P'(x_i)$	$x_{i+1} = x_i - \dfrac{P(x_i)}{P'(x_i)}$	$P(x_{i+1})$
1	3	2	6	2.6667	0.1113
2	2.6667	0.1113	5.3333	2.6458	2.58×10^{-4}
3	2.6458	2.58×10^{-4}	5.2916	2.64575	-6.9×10^{-6}

Hence, $x \approx 2.646$

4. $f(x) = 5x^4 - 7x; F(x) = 5 \cdot \frac{1}{5}x^5 - 7 \cdot \frac{1}{2}x^2 + C; F(x) = x^5 - \frac{7}{2}x^2 + C$

5. $s(t) = t^3 - 12t^2 + 5$

(a) $v(t) = s'(t) = 3t^2 - 24t; a(t) = v'(t) = 6t - 24$

(b) Extrema for $s(t)$ are when $v(t) = 0$ that is at 0 and 8; $s(-2) = -51; s(0) = 5; s(4) = -123$. Position: Maximum at $(0, 5)$; Minima at $(-2, -51)$ and $(4, -123)$; Extrema for velocity when $a(t) = 0$ or $t = 4; v(-2) = 60; v(4) = -48$; Velocity: Maximum at $(-2, 60)$, Minimum at $(4, -48)$. Acceleration is always increasing on the interval $[-2, 4]$, so Maximum: $(4, 0)$, Minimum: $(-2, -36)$

6. Let ℓ and w be the dimensions of the printed part. Then $\ell \cdot w = 24$ so $\ell = \frac{24}{w}$. The dimensions of the page are $\ell + 3$ and $w + 2$. Area of the page is $(\ell+3)(w+2) = (\frac{24}{w}+3)(w+2) = 24+\frac{48}{w}+3w+6 = 3w + \frac{48}{w} + 30; A' = 3 - 48/w^2 = 0; 48/w^2 = 3; 16 = w^2; w = 4$. The width is 4 in. and the length is 6 in. for the printed portion. The page is 6 in. wide by 9 in. long.

7. Making the necessary substitutions, we get $800 = \pi r^2(20) + \frac{2}{3}\pi r^3$ or $\frac{2}{3}\pi r^3 + \pi r^2(20) - 800 = 0$. Using Newton's method produces $r \approx 3.38$ m.

CHAPTER

25

Integration

≡ 25.1 THE AREA QUESTION AND THE INTEGRAL

1. $\int_1^3 x\,dx$. The interval $[1,3]$ has length 2 so each subinterval is length $\frac{2}{n}$. Using the right end point of each subinterval, $x_i = 1 + \frac{2i}{n} = \frac{n+2i}{h}$. Since $f(x) = x; f(x_i) = x_i = \frac{n+2i}{n}$

$$\sum_{i=1}^n f(x_i)nx = \sum_{i=1}^n \left(\frac{n+2i}{n}\right)\cdot\frac{2}{n}$$
$$= \sum_{i=1}^n \left(\frac{2}{n} + \frac{4i}{n^2}\right)$$
$$= \frac{2}{n}\sum_{i=1}^n 1 + \frac{4}{n^2}\sum_{i=1}^n i$$
$$= \frac{2}{n}\cdot n + \frac{4}{n^2}\left(\frac{n^2+n}{2}\right)$$
$$= 2 + 2 + \frac{2}{n} = 4 + \frac{2}{n}$$
$$\lim_{n\to\infty}\left(4 + \frac{2}{n}\right) = 4$$

3. $\int_0^4 (3x+1)dx; \Delta x = \frac{4}{n}; x_i = \frac{4i}{n}; f(x_i) = 3\left(\frac{4i}{n}\right) + 1 = \frac{12i}{n} + 1;$

$$\sum_{i=1}^n \left(\frac{12i}{n}+1\right)\cdot\frac{4}{n} = \frac{48}{n^2}\sum_{i=1}^n i + \frac{4}{n}\sum_{i=1}^n 1$$
$$= \frac{48}{n^2}\left(\frac{n^2+n}{2}\right) + \frac{4}{n}\cdot n$$
$$= 24 + \frac{18}{n} + 4;$$
$$\lim_{n\to\infty}\left(28 + \frac{18}{n}\right) = 28$$

5. $\int_1^3 (2x-2)\,dx; \Delta x = \frac{2}{n}; x_i = 1 + \frac{2i}{n}; f(x_i) = 2\left(1 + \frac{2i}{n}\right) - 2 = \frac{4i}{n};$

$$\sum_{i=1}^n \frac{4i}{n}\cdot\frac{2}{n} = \frac{8}{n^2}\sum_{i=1}^n = \frac{8}{n^2}\left(\frac{n^2+n}{2}\right) = 4 + \frac{4}{n};$$
$$\lim_{n\to\infty}\left(4 + \frac{4}{n}\right) = 4$$

7. $\int_1^4 (3-2x)\,dx; \Delta x = \frac{3}{n}; x_i = 1 + \frac{3i}{n}; f(x_i) = 3 - 2\left(1 + \frac{3i}{n}\right) = 3 - 2 - \frac{6i}{n} = 1 - \frac{6i}{n};$

$$\sum_{i=1}^n \left(1 - \frac{6i}{n}\right)\frac{3}{n} = \sum_{i=1}^n \left(\frac{3}{n} - \frac{18i}{n^2}\right)$$
$$= \frac{3}{n}\sum_{i=1}^n 1 - \frac{18}{n^2}\sum_{i=1}^n i$$
$$= \frac{3}{n}\cdot n - \frac{18}{n^2}\left(\frac{n^2+n}{2}\right)$$
$$= 3 - 9 - \frac{9}{n} = -6 - \frac{9}{n};$$
$$\lim_{n\to\infty}\left(-6 - \frac{9}{n}\right) = -6$$

9. $\int_0^2 (x^2-1)\,dx; \Delta x = \frac{2}{n}; x_i = \frac{2i}{n}; f(x_i) = \left(\frac{2i}{n}\right)^2 - 1 = \frac{4i^2}{n^2} - 1;$

$$\sum_{i=1}^n \left(\frac{4i^2}{n^2} - 1\right)\cdot\frac{2}{n} = \sum_{i=1}^n \frac{8i^2}{n^3} - \sum_{i=1}^n \frac{2}{n}$$
$$= \frac{8}{n^3}\sum_{i=1}^n i^2 - \frac{2}{n}\sum_{i=1}^n 1$$
$$= \frac{8}{n^3}\left(\frac{n^3}{3} + \frac{n^2}{2} + \frac{n}{6}\right) - \frac{2}{n}\cdot n$$
$$= \frac{8}{3} + \frac{4}{n} + \frac{4}{3n^2} - 2$$
$$= \frac{2}{3} + \frac{4}{n} + \frac{4}{3n^2};$$
$$\lim_{n\to\infty}\left(\frac{2}{3} + \frac{4}{n} = \frac{4}{3n^2}\right) = \frac{2}{3}$$

11. $\int_0^2 (2x^2 + 1)dx$; $\Delta x = \frac{2}{n}$; $x_i = \frac{2i}{n}$; $f(x_i) = 2 \cdot$

$\left(\frac{2i}{n}\right)^2 + 1 = \frac{8i^2}{n} + 1$;

$$\sum_{i=1}^{n}\left(\frac{8i^2}{n^2} + 1\right) \cdot \frac{2}{n} = \sum_{i=1}^{n}\left(\frac{16i^2}{n^3} + \frac{2}{n}\right)$$
$$= \frac{16i}{n^3}\sum_{i=1}^{n}i^2 + \frac{2}{n}\sum_{i=1}^{n}1$$
$$= \frac{16}{n^3}\left(\frac{n^3}{3} + \frac{n^3}{2} + \frac{n}{6}\right)$$
$$+ \frac{2}{n}(n)$$
$$= \frac{16}{3} + \frac{8}{n} + \frac{8}{3n^2} + 2$$
$$= \frac{22}{3} + \frac{8}{n} + \frac{8}{3n^2};$$
$$\lim_{n\to\infty}\left(\frac{22}{3} + \frac{8}{n} + \frac{8}{3n^2}\right) = \frac{22}{3}.$$

13. $\int_1^3 x^3 dx$; $\Delta x = \frac{2}{n}$; $x_i = 1 + \frac{2i}{n}$; $f(x_i) = \left(1 + \frac{2i}{n}\right)^3 = 1 + \frac{6i}{n} + \frac{12i^2}{n^2} + \frac{8i^3}{n^3}$;

$$\sum_{i=1}^{n}\left(1 + \frac{6i}{n} + \frac{12i^2}{n^2} + \frac{8i^3}{n^3}\right)\frac{2}{n}$$
$$= \sum_{i=1}^{n}\left(\frac{2}{n} + \frac{12i^2}{n^2} + \frac{24i^2}{n^3} + \frac{16i^3}{n^4}\right)$$
$$= \frac{2}{n}\cdot n + \frac{12}{n^2}\left(\frac{n^2 + n}{2}\right) + \frac{24}{n^3}$$
$$\times \left(\frac{n^3}{3} + \frac{n^2}{2} + \frac{n}{6}\right)$$
$$+ \frac{16}{n^4}\left(\frac{n^4}{4} + \frac{n^3}{2} + \frac{n^2}{4}\right)$$
$$= 2 + 6 + 8 + 4 + \frac{6}{n} + \frac{12}{n} + \frac{4}{n^2} + \frac{8}{n} + \frac{4}{n^2};$$
$$\lim_{n\to\infty}\left(20 + \frac{26}{n} + \frac{8}{n^2}\right) = 20$$

15. $\int_0^4 (x^3 + 2)dx$; $\Delta x = \frac{4}{n}$; $x_i = \frac{4i}{n}$; $f(x_i) = \left(\frac{4i}{n}\right)^3 + 2 = \frac{64i^3}{n^3} + 2$;

$$\sum_{i=1}^{n}\left(\frac{64i^3}{n^3} + 2\right)\frac{4}{n} = \sum_{i=1}^{n}\left(\frac{256i^3}{n^4} + \frac{8}{n}\right)$$
$$= \frac{256}{n^4}\left[\frac{n^4}{4} + \frac{n^3}{2} + \frac{n^2}{2}\right]$$
$$+ \frac{8}{n}\cdot n$$
$$\lim_{n\to\infty}\left[\frac{256}{n^4}\left[\frac{n^4}{4} + \frac{n^3}{2} + \frac{n^2}{2}\right] + \frac{8}{n}\cdot n\right] = 64 + 8 = 72$$

17. $\int_0^2 (x^2 + x)dx$; $\Delta x = \frac{2}{n}$; $x_i = \frac{2i}{n}$; $f(x_i) = \left(\frac{2i}{n}\right)^2 + \frac{2i}{n} = \frac{4i^2}{n^2} + \frac{2i}{n}$;

$$\sum_{i=1}^{n}\left(\frac{4i^2}{n^2} + \frac{2i}{n}\right)\frac{2}{n} = \sum_{i=1}^{n}\left(\frac{8i^2}{n^3} + \frac{4i}{n^2}\right)$$
$$= \frac{8}{n^3}\left(\frac{n^3}{2} + \frac{n^2}{2} + \frac{n}{6}\right)$$
$$+ \frac{4}{n^2}\left(\frac{n^2}{2} + \frac{n}{2}\right)$$
$$\lim_{n\to\infty}\left[\frac{8}{n^3}\left(\frac{n^3}{3} + \frac{n^2}{2} + \frac{n}{6}\right) + \frac{4}{n^2}\left(\frac{n^2}{2} + \frac{n}{2}\right)\right] = \frac{8}{2} + 2 = \frac{14}{3}$$

25.2 THE FUNDAMENTAL THEOREM OF CALCULUS

1. $\int_0^5 6\,dx = 6x\big|_0^5 = 6\cdot 5 - 6\cdot 0 = 30$

3. $\int_1^5 2x\,dx = x^2\big|_1^5 = 5^2 - 1^2 = 25 - 1 = 24$

5. $\int_{-1}^3 \frac{5}{3}x^3 dx = \frac{5}{12}x^4\big|_{-1}^3 = \frac{5}{12}\cdot 3^4 - \frac{51}{2}\cdot(-1)^4 = \frac{5\cdot 81}{12} - \frac{5}{12} = \frac{400}{12} = \frac{100}{3}$

7. $\int_1^2 \frac{8}{5}x^3 dx = \frac{2}{5}x^4\big|_1^2 = \frac{2}{5}\cdot 2^4 - \frac{2}{5}\cdot 1^4 = \frac{32}{5} - \frac{2}{5} = \frac{30}{5} = 6$

9. $\int_0^2 (4x + 5)dx = 2x^2 + 5x\big|_0^2 = (2\cdot 2^2 + 5\cdot 2) - (2\cdot 0^2 + 5\cdot 0) = 8 + 10 = 18$

11. $\int_2^0 (4x + 5)dx = 2x^2 + 5x\big|_2^0 = (2\cdot 0^2 + 5\cdot 0) - (2\cdot 2^2 + 5\cdot 2) = 0 - 18 = -18$

13. $\int_1^3 (-4x)dx = -2x^2\big|_1^3 = -2\cdot 3^2 - (-2\cdot 1^2) = -2\cdot 9 - (-2\cdot 1) = -18 + 2 = -16$

15. $\int_0^3 (t^2 + 2t)dt = \frac{1}{3}t^3 + t^2\big|_0^3 = \frac{1}{3}\cdot 3^3 + 3^2 - 0 = 18$

17. $\int_{-1}^3 (3y^2 - 2y)dy = y^3 - y^2\big|_{-1}^3 = (3^3 - 3^2) - ((-1)^3 - (-1)^2) = 18 + 2 = 20$

19. $\int_1^4 (4 - 2w^2)dw = 4w - \frac{2}{3}w^3\big|_1^4 = \left(4\cdot 4 - \frac{2}{3}\cdot 4^3\right) - \left(4\cdot 1 - \frac{2}{3}\cdot 1^3\right) = \left(16 - \frac{128}{3}\right) - \left(4 - \frac{2}{3}\right) = 12 - \frac{126}{3} = 12 - 42 = -30$

21. $\int_2^3 (x^2 + 2x + 1)dx = \frac{x^3}{3} + x^2 + x\big|_2^3 = \left(\frac{3^3}{3} + 3^2 + 3\right) - \left(\frac{2^3}{3} + 2^2 + 2\right) = 21 - \frac{26}{3} = \frac{37}{3}$

23. $\int_3^2 (x^2 + 2x + 1)dx = \frac{x^3}{3} + x^2 + x\big|_3^2 = \left(\frac{2^3}{3} + 2^2 + 2\right) - \left(\frac{3^3}{3} + 3^2 + 3\right) = \frac{26}{3} - 21 = -\frac{37}{3}$

25. $\int_6^8 dx = x\big|_6^8 = 8 - 6 = 2$

27. $\int_{-4}^{-2}(3y^2 - y - 1)dy = y^3 - \frac{1}{2}y^2 - y\big|_{-4}^{-2} = \left((-2)^3 - \frac{1}{2}(-2)^2 - (-2)\right) - \left((-4)^3 - \frac{1}{2}(-4)^2 - (-4)\right) = (-8 - 2 + 2) - (-64 - 8 + 4) = -8 - (-68) = 60$

29. $\int_2^4 (x^2 - 2x + 1)dx = \frac{x^3}{3} - x^2 + x\big|_2^4 = \left(\frac{4^3}{3} - 4^2 + 4\right) - \left(\frac{2^3}{3} - 2^2 + 2\right) = \left(\frac{64}{3} - 16 + 4\right) - \left(\frac{8}{3} - 4 + 2\right) = \frac{64}{3} - 16 + 4 - \frac{8}{3} + 4 - 2 = \frac{56}{3} - 10 = \frac{26}{3}$

31. $\int_1^2 -3x^{-4}dx = x^{-3}\big|_1^2 = \frac{1}{8} - 1 = -\frac{7}{8}$

33. $\int_{-2}^{-1}\frac{t^{-2}}{3}dx = -\frac{1}{3}t^{-1}\big|_{-2}^{-1} = \frac{1}{3} - \frac{1}{6} = \frac{1}{6}$

35. $\int_0^2 \sqrt{x}\,dx = \int_0^2 x^{1/2}dx = \frac{2}{3}x^{3/2}\big|_0^2 = \frac{2}{3}\cdot 2^{3/2} - 0 = \frac{4}{3}\sqrt{2} \approx 1.8856$

37. $\int_{1/2}^3 \left(\frac{1}{\sqrt{x}} - 2\right)dx = \int_{1/2}^3 (x^{-1/2} - 2)dx = 2x^{1/2} - 2x\big|_{1/2}^3 = (2\sqrt{3} - 2\cdot 3) - (2\cdot\sqrt{1/2} - 1) = 2\sqrt{3} - 6 - \sqrt{2} + 1 = 2\sqrt{3} - \sqrt{2} - 5 \approx -2.9501$

39. Since velocity is the derivative of displacement, the displacement is the integral of the velocity. $\int_0^4 (4.00t - 1.00t^2)dt = 2.00t^2 - \frac{1}{3}t^3\big|_0^4 = 2\cdot 4^2 - \frac{1}{3}\cdot 4^3 - 0 = 32 - \frac{64}{3} = 10.7$ m

41. **(a)** $a(t) = -9.8$ m/s^2; $v(t) =$ antiderivative of $a(t)$; $v(t) = -9.8t + C_1$; at $t = 0$; $v(t) = -10$ so $v(t) = -9.8t - 10$; $v(3) = -9.8\cdot 3 - 10 = -39.4$ m/s.
The negative means its velocity is downward. Its speed is $|-39.4| = 39.4$ m/s
(b) Displacement is the integeral of the velocity. Hence $\int_0^3 (-9.8t - 10)dt = -4.9t^2 - 10t\big|_0^3 = -4.9(3^2) - 10.3 - 0 = -74.1$. The negative indicates a downward movement. Distance is $|-74.1| = 74.1$ m.
(c) $4.9t^2 - 10t = 180$; $4.9t^2 + 10t - 180 = 0$. Using the quadratic formula, $t = -7.167$ or $t = 5.1258$. Since t must be positive, $t = 5.1258$ s.

43. **(a)** Acceleration is change in velocity divided by time so $a = (10,500 - 500)/0.01 = 10,000/0.01 = 1\,000\,000$ m/s^2.
(b) $v = \int a(t)dt = 1\,000\,000t + 500$; $s = \int_0^{0.01} v(t) = 500\,000t^2 + 500t\big|_0^{0.01} = 50 + 5 = 55$ m.

45. $\int_0^9 N'(t)\,dt = \int_0^9 25 - 6t^{1/2}\,dt = 25t - 4t^{3/2}\big|_0^9 = 25(9) - 4(9)^{3/2} = 225 - 108 = 117$. There will be an increase of 117 moose.

47. **(a)** $\int_1^2 \left(0.5 + 4t^{-3}\right)dt = 0.5t - 2t^{-2}\big|_1^2 = \left[0.5(2) - 2(2)^{-2}\right] - \left[0.5(1) - 2(1)^{-2}\right] = [1.0 - 0.5] - [0.5 - 2] = 2$. The tree will grow 2 ft in the second year after it is 20 years old.
(b) $\int_4^5 \left(0.5 + 4t^{-3}\right)dt = 0.5t - 2t^{-2}\big|_4^5 = \left[0.5(5) - 2(5)^{-2}\right] - \left[0.5(4) - 2(4)^{-2}\right] = [2.5 - 0.8] - [2.0 - 0.125] = 0.545$. The tree will grow 0.545 ft in the fifth year after it is 20 years old.
(c) $\int_1^5 \left(0.5 + 4t^{-3}\right)dt = 0.5t - 2t^{-2}\big|_1^5 = \left[0.5(5) - 2(5)^{-2}\right] - \left[0.5(1) - 2(1)^{-2}\right] = [2.5 - 0.8] - [0.5 - 2] = 3.92$. The tree will grow 3.92 ft in years 1 through 5 after it is 20 years old.

25.3 THE INDEFINITE INTEGRAL

1. $\int 9\,dx + 9x + C$

3. $\int 6x^2 dx = 6\cdot\frac{1}{3}x^3 + C = 2x^3 + C$

5. $\int x^2\sqrt{x}\,dx = \int x^{5/2}dx = \frac{2}{7}x^{7/2} + C$

7. $\int (t^3 + 1)dt = \frac{1}{4}t^4 + t + C$

9. $\int (y^2 + 4y - 3)dy = \frac{1}{3}y^3 + 2y^2 - 3y + C.$

11. $\int \left(\sqrt[3]{x} - \frac{3}{\sqrt[3]{x^2}}\right)dx = \int \left(x^{1/3} - 3x^{-2/3}\right)dx = \frac{3}{4}x^{4/3} - 9x^{1/3} + C = \frac{3x\sqrt[3]{x}}{4} - 9\sqrt[3]{x} + C.$

13. $\int (x^2 + 3)2x\,dx.$ Let $u = x^2 + 3$, then $du = 2x\,dx.$ Then, substituting we get $\int u\,du = \frac{1}{2}u^2 + C = \frac{1}{2}(x^2 + 3)^2 + C.$

15. $\int (4 - 2x^2)4x\,dx$. Let $u = 4 - 2x^2$, then $du = -4x\,dx$. Substituting, we can rewrite the given integral as $\int (4-2x^2)4x\,dx = -\int u\,du = -\frac{1}{2}u^2 + C = -\frac{1}{2}(4 - 2x^2)^2 + C$.

17. $\int (3 - x^2)^3 2x\,dx$. Let $u = 3 - x^2$, then $du = -2x\,dx$ or $-du = 2x\,dx$. Substituting, we have $-\int u^3 du = -\frac{1}{4}u^4 + C = -\frac{1}{4}(3 - x^2)^4 + C$.

19. $\int (x^2 + 4)^{1/2} x\,dx$. Let $u = x^2 + 4$, then $du = 2x\,dx$ or $x\,dx = \frac{1}{2}du$. Substituting, we obtain $\frac{1}{2}\int u^{1/2}du = \frac{1}{2} \cdot \frac{2}{3}u^{3/2} + C = \frac{1}{3}(x^2 + 4)^{3/2} + C$.

21. $\int \frac{(\sqrt{x}-1)^3}{\sqrt{x}}dx = \int (\sqrt{x} - 1)^3 \cdot x^{-1/2}dx$. Let $u = \sqrt{x} - 1$, then $du = \frac{1}{2}x^{-1/2}$ or $2\,du = x^{-1/2}dx$. Substituting, we have $2\int u^3 du = 2 \cdot \frac{1}{4}u^4 + C = \frac{1}{2}(\sqrt{x} - 1)^4 + C$.

23. $\int (3x^3 + 1)^4 x^2 dx$. Let $u = 3x^3 + 1$, then $du = 9x^2 dx$ or $x^2 dx = \frac{1}{9}du$. Substituting, we obtain $\frac{1}{9}\int u^4 du = \frac{1}{9} \cdot \frac{1}{5}u^5 + C = \frac{1}{45}(3x^3 + 1)^5 + C$.

25. $\int \frac{x\,dx}{\sqrt{x^2+3}} = \int (x^2 + 3)^{-1/2} x\,dx$. Let $u = x^2 + 3$, then $du = 2x\,dx$ or $x\,dx = \frac{1}{2}du$. Substituting produces $\frac{1}{2}\int u^{-1/2}du = \frac{1}{2} \cdot 2u^{1/2} + C = (x^2 + 3)^{1/2} + C = \sqrt{x^2 + 3} + C$.

27. $\int \frac{x\,dx}{(x^2+3)^3} = \int (x^2 + 3)^{-3} x\,dx$. Let $u = x^2 + 3$, then $du = 2x\,dx$, or $x\,du = \frac{1}{2}du$. Substitution yields $\frac{1}{2}\int u^{-3}du = \frac{1}{2} \cdot \frac{-1}{2}u^{-2} + C = \frac{-1}{4}(x^2 + 3)^{-2} + C = \frac{-1}{4(x^2+3)^2} + C = -\frac{1}{4}(x^2 + 3)^{-2} + C$.

29. $\int (3x^2 - 1)^2 dx$. We cannot use substitution so we expand to get $\int (9x^4 - 6x^2 + 1)\,dx = \frac{9}{5}x^5 - 2x^3 + x + C$.

31. $\int \frac{x^2-x-6}{x+1}dx = \int \frac{(x-3)(x+2)}{x+2}dx = \int (x - 3)dx = \frac{1}{2}x^2 - 3x + C$, $x \neq -2$.

33. $\int \frac{5x\,dx}{(x^2-1)^{1/3}}$. Let $u = x^2 - 1$, then $du = 2x\,dx$ or $5x\,dx = \frac{5}{2}du$. Substitution produces $\frac{5}{2}\int u^{-1/3}du = \frac{5}{2} \cdot \frac{3}{2}u^{2/3} + C = \frac{15}{4}(x^2 - 1)^{2/3} + C$.

35. $\int (1+3x^2)^2 x\,dx$. Let $u = 1+3x^2$, and then $du = 6x\,dx$, or $x\,dx = \frac{1}{6}du$. Substitution produces $\frac{1}{6}\int u^2 du = \frac{1}{6} \cdot \frac{1}{3}u^3 + C = \frac{1}{18}(1 + 3x^2)^3 + C$.

37. $\int 3x^2(4x^3 - 5)^{2/3}dx$. Let $u = 4x^3 - 5$, then $du = 12x^2 dx$, and we have $\frac{1}{4}du = 3x^2 dx$. Thus, $\frac{1}{4}\int u^{2/3}du = \frac{1}{4} \cdot \frac{3}{5}u^{5/3} + C = \frac{3}{20}(4x^3 - 5)^{5/3} + C$.

39. $\int (1+3x^2)^2 dx$. We cannot use substitution so we expand to get $\int (1 + 6x^2 + 9x^4)dx = x + 2x^3 + \frac{9}{5}x^5 + C$.

41. $\int \sqrt{4x^2 + 2x}(4x + 1)dx$. Let $du = 4x^2 + 2x$, and then $du = (8x+2)dx$, or $\frac{1}{2}du = (4x+1)dx$. Thus, substituting results in $\frac{1}{2}\int u^{1/2}du = \frac{1}{2} \cdot \frac{2}{3}u^{3/2} + C = \frac{1}{3}(4x^2 + 2x)^{3/2} + C$.

43. $\int \frac{x-1}{(x-1)^3}dx = \int (x - 1)^{-2}dx = -1(x-1)^{-1} + C = \frac{-1}{x-1} + C$

45. $\int (2x^4 - 3)^2 8x\,dx$. Let $u = 2x^4 - 3$ and $du = 8x^3$. We do not have x^3 in the original integrand, so substitution will not work. Expand the indicated binomial to get $\int (4x^8 - 12x^4 + 9)8x\,dx = \int (32x^9 - 96x^5 + 72x)dx = \frac{32}{10}x^{10} - \frac{96}{6}x^6 + \frac{72}{2}x^2 + C = \frac{16}{5}x^{10} - 16x^6 + 36x^2 + C$.

47. $\int (x^3 - 3x)^{2/3}(x^2 - 1)dx$. Let $u = x^3 - 3x$, then $du = (3x^2 - 3)dx$, or $\frac{1}{3}du = (x^2 - 1)dx$; $\frac{1}{3}\int u^{2/3}du = \frac{1}{3} \cdot \frac{3}{5}u^{5/3} + C = \frac{1}{5}(x^3 - 3x)^{5/3} + C$

49. $\int (3x + 1)^3 dx$. Let $u = 3x + 1$, then $du = 3\,dx$, or $\frac{1}{3}du = dx$. Hence, we obtain $\frac{1}{3}\int u^3 du = \frac{1}{3} \cdot \frac{1}{4}u^4 + C = \frac{1}{12}(3x + 1)^4 + C$.

51. $\int (2x^3 - 1)^4 \sqrt{x^4 - 2x}\,dx$. Let $u = x^4 - 2x$, then $du = (4x^3 - 2)dx$, or $\frac{1}{2}du = (2x^3 - 1)dx$. Substitution results in $\frac{1}{2}\int u^{1/4}du = \frac{1}{2} \cdot \frac{4}{5}u^{5/4} + C = \frac{2}{5}(x^4 - 2x)^{5/4} + C$.

53. $\int (t + 7)^{1/2}dt = \frac{2}{3}(t + 7)^{3/2} + C$

55. $\int \frac{y^3-8}{y-2}dy = \int \frac{(y-2)(y^2+2y+4)}{y-2}dy = \int (y^2 + 2y + 4)dy = \frac{1}{3}y^3 + y^2 + 4y + C$, $y \neq 2$.

57. $\int_0^2 (x^2 - 4) 2x \, dx$. Let $u = x^2 - 4$, then $du = 2x \, dx$. For the limits of integration, we have $x = 0 \Rightarrow u = -4$ and $x = 2 \Rightarrow u = 0$. Thus, $\int_{-4}^0 u \, du = \frac{u^2}{2} \Big|_{-4}^0 = 0 - \frac{16}{2} = -8$.

59. $\int_1^2 6x(3x^2 - 7) \, dx$. Letting $u = 3x^2 - 7$, we obtain $du = 6x \, dx$ and $x = 1 \Rightarrow u = -4$, and $x = 2 \Rightarrow u = 5$. Substitution produces $\int_{-4}^5 u \, du = \frac{1}{2} u^2 \Big|_{-4}^5 = \frac{25}{2} - \frac{16}{2} = \frac{9}{2}$.

61. $\int_0^2 2x(20 - 3x^2) \, dx$. Let $u = 20 - 3x^2$, then $du = -6x \, dx$, and so $-\frac{1}{3} du = 2x \, dx$. Here, we see that the limits of integration can be changed as $x = 0 \Rightarrow u = 20$ and $x = 2 \Rightarrow u = 8$. Substitution produces $-\frac{1}{3} \int_{20}^8 u \, du = -\frac{1}{3} \cdot \frac{1}{2} u^2 \Big|_{20}^8 = -\frac{1}{6}(8^2 - 20^2) = 56$.

63. $\int_0^3 (x^3 + 2) x^2 \, dx$. Let $u = x^3 + 2$, then $du = 3x^2 \, dx$, or $\frac{1}{3} du = x^2 \, dx$. Here, we see that $x = 0 \Rightarrow u = 2$ and $x = 3 \Rightarrow u = 29$. Substitution yields $\frac{1}{3} \int_2^{29} u \, du = \frac{1}{3} \cdot \frac{1}{2} u^2 \Big|_2^{29} = \frac{1}{6}(29^2 - 2^2) = 139.5$.

65. $\int_1^4 (3x^2 - 2)^4 x \, dx$. Let $u = 3x^2 - 2$, then $du = 6x \, dx$, or $\frac{1}{6} du = x \, dx$ and we see that $x = 1 \Rightarrow u = 1$ and $x = 4 \Rightarrow 4 = 46$. Substitution produces $\frac{1}{6} \int_1^{46} u^4 \, du = \frac{1}{6} \cdot \frac{1}{5} u^5 \Big|_1^{46} = \frac{1}{30}(46^5 - 1^5) = 6{,}865{,}432.5$

67. $L(x) = \int L'(x) \, dx = \int (0.3 + 0.002x) \, dx = 0.3x + 0.001x^2 + C$. When the population is 20,000 people, $x = 20$ and we are given $L(20) = 7.2$ or $0.3(20) + 0.001(20)^2 + C = 6.4 + C = 7.2$. Thus, $C = 7.2 - 6.4 = 0.8$ and $L(x) = 0.3x + 0.001x^2 + 0.8$

69. $P(x) = \int P'(x) \, dx = \int (4x + 35) \, dx = 2x^2 + 35x + C$. We are given $P(0) = -150$ and so, $C = -150$. Thus, $P(x) = 2x^2 + 35x - 150$

71.
$$T(x) = \int \left(-\frac{5750}{(x+1)^3} \right) dx$$
$$= \int -5750(x+1)^{-3} \, dx$$
$$= \frac{-5750}{-2}(x+1)^{-2} + K$$
$$= 2875(x+1)^{-2} + K$$
Since $T(0) = 2900°C$, then $2875 + K = 2900$ and $K = 25$. Hence, $T(x) = 2875(x+1)^{-2} + 25$.

≡ 25.4 THE AREA BETWEEN TWO CURVES

1.

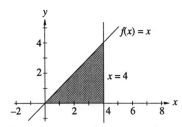

From the graph we can determine that all the area is above the x-axis so we only need to integrate from 0 to 4. $\int_0^4 x \, dx = \frac{x^2}{2} \Big|_0^4 = \frac{4^2}{2} - \frac{0^2}{2} = 8$

3.

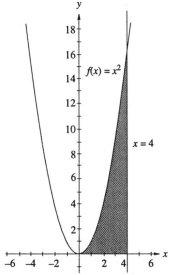

All the area is above the x-axis $\int_0^4 x^2 = \frac{x^3}{3} \Big|_0^4 = \frac{4^3}{3} - \frac{0^3}{3} = \frac{64}{3}$

5.

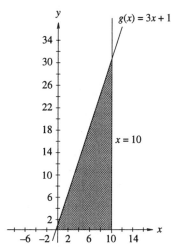

All the area is above the x-axis. We can determine that $3x + 1$ intersects the x-axis at $-\frac{1}{3}$. Hence, the desired area is

$$\int_{-1/3}^{10} (3x + 1)dx = \frac{3}{2}x^2 + x\big|_{-1/3}^{10}$$

$$= \left(\frac{300}{2} + 10\right) - \left(\frac{1}{6} - \frac{1}{3}\right)$$

$$= 160\frac{1}{6}$$

7.

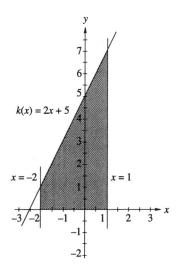

Between -2 and 1, $k(x)$ is above the x-axis. Thus, the desired area is $\int_{-2}^{1}(2x+5)dx = x^2 + 5x\big|_{-2}^{1} = (1 + 5) - (4 - 10) = 6 + 6 = 12$.

9.

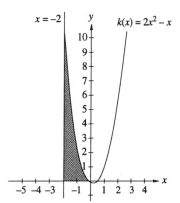

Between -2 and 0, we can see that k is above the x-axis. Thus, the desired area is $\int_{-2}^{0} (2x^2 - x)dx = \frac{2x^3}{3} - \frac{x^2}{2}\big|_{-2}^{0} = 0 - \left(\frac{-16}{3} - \frac{4}{2}\right) = 7\frac{1}{3} = \frac{22}{3}$.

11.

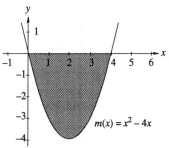

From the graph we can determine that m intersects the x-axis at 0 and 4 and is below the axis. Hence we take the absolute value of the integral.

$$\left|\int_0^4 (x^2 - 4x)dx\right| = \left|\left[\frac{x^3}{3} - 2x^2\right]_0^4\right| = \left|\frac{64}{3} - 32\right|$$

$$= \left|-10\frac{2}{3}\right| = 10\frac{2}{3} \text{ or } \frac{32}{3}$$

13.

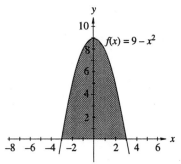

The area is above the x-axis and f intersects the x-axis at -3 and 3. Thus, we have

$$\int_{-3}^{3}\left(9-x^2\right)dx = 9x - \frac{x^3}{3}\Bigg|_{-3}^{3}$$
$$= (27-9)-(-27+9) = 36$$

15.

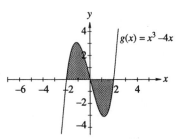

The graph is above the x-axis from -2 to 0 and below the x-axis from 0 to 2.

$$\int_{-2}^{0}\left(x^3-4x\right)dx - \int_{0}^{2}\left(x^3-4x\right)dx$$
$$= \left(\frac{x^4}{4}-2x^2\right)\Bigg|_{-2}^{0} - \left(\frac{x^4}{4}-2x^2\right)\Bigg|_{0}^{2}$$
$$= 0 - \left(\frac{16}{4}-8\right) - \left(\frac{16}{4}-8\right) + 0 = 8$$

17.

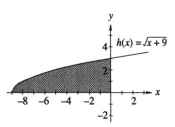

The graph is above the x-axis and intersects at -9. Thus, the desired area is

$$\int_{-9}^{0}\sqrt{x+9}\,dx = \frac{2}{3}(x+9)^{3/2}\Bigg|_{-9}^{0}$$
$$= \frac{2}{3}\left[(0+9)^{3/2}-(-9+9)^{3/2}\right]$$
$$= \frac{2}{3}\cdot 27 = 18.$$

19.

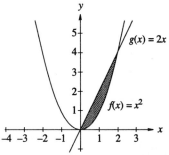

From the graph we determine that $g(x) \geq f(x)$ on $[0, 2]$. Hence, the desired area is

$$\int_{0}^{2}\left(2x-x^2\right)dx = x^2 - \frac{x^3}{3}\Bigg|_{0}^{2} = 4 - \frac{8}{3} = \frac{4}{3}.$$

21.

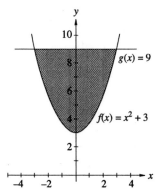

f and g intersect at $(\pm\sqrt{6},\ 9)$. On the interval $[-\sqrt{6},\ \sqrt{6}]$, we see that $g(x) \geq f(x)$. Thus, the desired area is

$$\int_{-\sqrt{6}}^{\sqrt{6}}\left(9-(x^2+3)\right)dx$$
$$= \int_{-\sqrt{6}}^{\sqrt{6}}\left(6-x^2\right)dx = 6x - \frac{x^3}{3}\Bigg|_{-\sqrt{6}}^{\sqrt{6}}$$
$$= \left(6\sqrt{6}-\frac{\sqrt{6}^3}{3}\right) - \left(-6\sqrt{6}+\frac{\sqrt{6}^3}{3}\right)$$
$$= 12\sqrt{6} - 4\sqrt{6} = 8\sqrt{6} \approx 19.596$$

23.

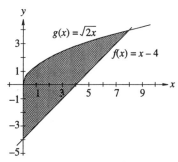

f and g intersect at $(8,4)$. g is defined on $[0,\infty)$. On $[0, 8]$, we see that $g(x) \geq f(x)$.

$$\int_0^8 (g(x) - f(x))dx$$

$$= \int_0^8 (\sqrt{2x} - x + 4)dx$$

$$= \int_0^8 \sqrt{2x}\, dx + \int_0^8 (4-x)dx$$

For the first integral let $u = 2x$, then $du = 2\, dx$, so $dx = \frac{1}{2}du$ and the limits of integration become $x = 0 \Rightarrow u = 0$ and $x = 8 \Rightarrow u = 16$. Thus, by substitution into the first integral, we get

$$\frac{1}{2}\int_0^{16} \sqrt{u}\, du + \int_0^8 (4-x)dx$$

$$= \frac{1}{2}\frac{2}{3}u^{3/2}\Big|_0^{16} + \left(4x - \frac{x^2}{2}\right)\Big|_0^8$$

$$= \left(\frac{64}{3} - 0\right) + \left(32 - \frac{64}{2}\right) = 21\frac{1}{3} = 21.333$$

25.

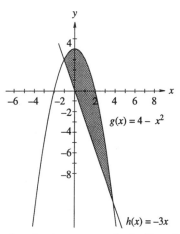

f and g intersect at $(-1,\ 3)$ and $(4,\ -12)$

$$\int_{-1}^4 \left((4 - x^2) - (-3x)\right)dx$$

$$= \int_{-1}^4 (4 + 3x - x^2)dx$$

$$= 4x + \frac{3x^2}{2} - \frac{x^3}{3}\Big|_{-1}^4$$

$$= \left(16 + 24 - \frac{64}{3}\right) - \left(-4 + \frac{3}{2} + \frac{1}{3}\right)$$

$$= 44 - \frac{3}{2} - \frac{65}{3} = 20.8333$$

27.

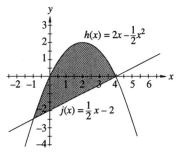

h and j intersect at $\left(-1,\ -2\frac{1}{3}\right)$ and $(4, 0)$

$$\int_{-1}^4 (h(x) - j(x))\, dx$$

$$= \int_{-1}^4 \left(\frac{3}{2}x - \frac{1}{2}x^2 + 2\right)dx$$

$$= \frac{3}{4}x^2 - \frac{1}{6}x^3 + 2x\Big|_{-1}^4$$

$$= \left(12 - \frac{64}{6} + 8\right) - \left(\frac{3}{4} + \frac{1}{6} - 2\right)$$

$$= 10.417$$

29.

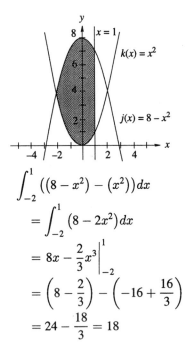

$$\int_{-2}^{1} \left((8 - x^2) - (x^2)\right) dx$$

$$= \int_{-2}^{1} (8 - 2x^2) dx$$

$$= 8x - \frac{2}{3}x^3 \Big|_{-2}^{1}$$

$$= \left(8 - \frac{2}{3}\right) - \left(-16 + \frac{16}{3}\right)$$

$$= 24 - \frac{18}{3} = 18$$

31.

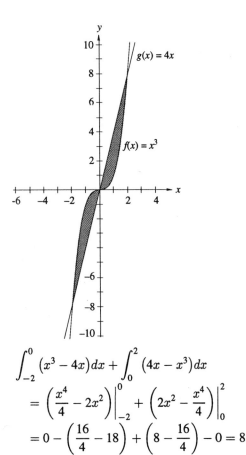

$$\int_{-2}^{0} (x^3 - 4x) dx + \int_{0}^{2} (4x - x^3) dx$$

$$= \left(\frac{x^4}{4} - 2x^2\right)\Big|_{-2}^{0} + \left(2x^2 - \frac{x^4}{4}\right)\Big|_{0}^{2}$$

$$= 0 - \left(\frac{16}{4} - 18\right) + \left(8 - \frac{16}{4}\right) - 0 = 8$$

33.

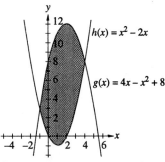

h and g intersect at $(-1,\ 3)$ and $(4,\ 8)$

$$\int_{-1}^{4} (g(x) - h(x)) dx$$

$$= \int_{-1}^{4} (8 + 6x - 2x^2) dx$$

$$= 8x + 3x^2 - \frac{2}{3}x^3 \Big|_{-1}^{4}$$

$$= \left(32 + 48 - \frac{128}{3}\right) - \left(-8 + 3 + \frac{2}{3}\right)$$

$$= 80 - \frac{128}{3} + 8 - 3 - \frac{2}{3}$$

$$= 85 - \frac{130}{3} = \frac{125}{3} = 41\frac{2}{3}$$

35.

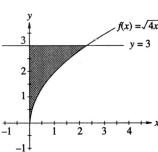

Solving for the point of intersection, we get $\sqrt{4x} = 2\sqrt{x} = 3$; $\sqrt{x} = \frac{3}{2}$, and $x = \frac{9}{4}$. Now integrating, we have

$$\int_{0}^{9/4} (3 - 2\sqrt{x}) dx$$

$$= 3x - \frac{4}{3}x^{3/2} \Big|_{0}^{9/4}$$

$$= \frac{27}{4} - \frac{4}{3} \cdot \frac{27}{8} = \frac{27}{4} - \frac{27}{6} = 2.25$$

37.

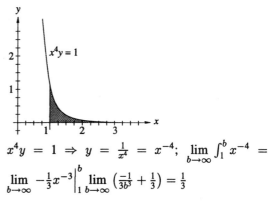

$\int_0^4 (\sqrt{16-2x} - \sqrt{16-4x}\, dx + \int_4^8 \sqrt{16-2x}\, dx =$
$\int_0^4 \sqrt{16-2x}\, dx - \int_0^4 \sqrt{16-4x}\, dx +$
$\int_4^8 \sqrt{16-2x}\, dx.$

For the first and third integrals let $u = 16 - 2x$, then $du = -2\, dx$ or $dx = -\frac{1}{2} du$. For the limits of integration, we obtain $x = 0 \Rightarrow u = 16$; $x = 4 \Rightarrow u = 8$ and $x = 8 \Rightarrow u = 0$.

$-\frac{1}{2} \int_{16}^8 u^{1/2} du = -\frac{1}{2} \cdot \frac{2}{3} u^{3/2} \Big|_{16}^8$
$= -\frac{1}{3}(8^{3/2} - 64)$

$-\frac{1}{2} \int_8^0 u^{1/2} dx = \frac{1}{2} \cdot \frac{2}{3} u^{3/2} \Big|_8^0$
$= -\frac{1}{3}(0 - 8^{3/2})$

The sum of these two integrals is $\frac{64}{3}$. For the second integral let $u = 16 - 4x$, and then $du = -4\, dx$ or $dx = -\frac{1}{4} du$. Here the limits of integration are $x = 0 \Rightarrow u = 16$ and $x = 4 \Rightarrow u = 0$.

$-\frac{1}{4} \int_{16}^0 u^{1/2} du = -\frac{1}{4} \cdot \frac{2}{3} u^{3/2} \Big|_{16}^0$
$= 0 + \frac{1}{6} \cdot 64 = \frac{64}{6}$

For the final answer, we subtract this last answer from the earlier result, and obtain $\frac{64}{3} - \frac{64}{6} = \frac{64}{6} = \frac{32}{3} = 10\frac{2}{3} = 10.667$.

39.

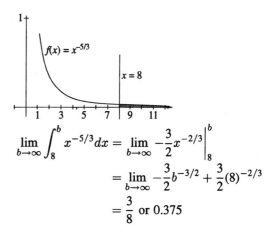

$\lim_{b \to \infty} \int_8^b x^{-5/3} dx = \lim_{b \to \infty} -\frac{3}{2} x^{-2/3} \Big|_8^b$
$= \lim_{b \to \infty} -\frac{3}{2} b^{-3/2} + \frac{3}{2}(8)^{-2/3}$
$= \frac{3}{8}$ or 0.375

41.

$x^4 y = 1 \Rightarrow y = \frac{1}{x^4} = x^{-4};\ \lim_{b \to \infty} \int_1^b x^{-4} =$
$\lim_{b \to \infty} -\frac{1}{3} x^{-3} \Big|_1^b \lim_{b \to \infty} \left(\frac{-1}{3b^3} + \frac{1}{3}\right) = \frac{1}{3}$

43. The area of the blue is $\int_0^1 (\sqrt{x} - x^3) dx = \frac{2}{3} x^{3/2} - \frac{x^4}{4} \Big|_0^1 = \frac{2}{3} - \frac{1}{4} = \frac{5}{12}$. The total area is 1 so the area of the white is $1 - \frac{5}{12} = \frac{7}{12}$. More white enamel is needed.

45. The graphs intersect at $(-2, 9)$ and $(2, 9)$. For the cross-sectional area of the trough, we use $\int_{-2}^2 (9 + 2x^2 - x^4) dx = 2 \int_0^2 (9 + 2x^2 - x^4) dx = 2 \cdot 9x + \frac{2}{3} x^3 - \frac{1}{5} x^5 \Big|_0^2 = \frac{508}{15} \approx 33.87$ cm². The volume is $\frac{508}{15} \times 1\,000 = 33\,866\frac{2}{3} \approx 33\,866.67$ cm³.

47. $\int_{0.6}^{2.4} \left(\frac{3000}{V^{1.33}} - \frac{1000}{V^{1.33}}\right) dV = \int_{0.6}^{2.4} 2000 V^{-1.33} dV = \frac{2000}{-0.33} V^{-0.33} \Big|_{0.6}^{2.4} = 2633.5$ or about 2630 Btu.

49. $\omega = \int_0^{10} \alpha\, dt = \int_0^{10} \frac{100}{\sqrt{(t+1)^3}} dt = \frac{-200}{\sqrt{t+1}} \Big|_0^{10} = -200\left(\frac{1}{\sqrt{11}} - 1\right) \approx 139.7$ rad/s.

25.5 NUMERICAL INTEGRATION

1. $\int_1^2 x^4 dx$; $n = 6$; $\Delta x = \frac{1}{6}$

(a) $\frac{1}{2}\left[1^4 + 2 \cdot \left(\frac{7}{6}\right)^4 + 2\left(\frac{8}{6}\right)^4 + 2\left(\frac{9}{6}\right)^4 + 2\left(\frac{10}{6}\right)^4 + 2\left(\frac{11}{6}\right)^4 + 2^4\right] \cdot \frac{1}{6} = \frac{1}{12}(75.17746914) = 6.2647891$

(b) $\frac{1}{3}\left[1^4 + 4 \cdot \left(\frac{7}{6}\right)^4 + 2\left(\frac{8}{6}\right)^4 + 4\left(\frac{9}{6}\right)^4 + 2\left(\frac{10}{6}\right)^4 + 4\left(\frac{11}{6}\right)^4 + 2^4\right] \cdot \frac{1}{6} = \frac{1}{18}(111.6018519) = 6.20010286$

(c) $\int_1^2 x^4 dx = \frac{x^4}{5}\Big|_1^2 = \frac{32}{5} - \frac{1}{5} = \frac{31}{5} = 6.2$

3. $\int_0^1 \sqrt{x}\, dx$; $n = 4$; $\Delta x = \frac{1}{4}$

(a) $\frac{1}{2}\left(0 + 2\sqrt{\frac{1}{4}} + 2\sqrt{1/2} + 2\sqrt{\frac{3}{4}} + 1\right) \cdot \frac{1}{4} = \frac{1}{8}(5.14626437) = 0.643283$

(b) $\frac{1}{3}\left(0 + 4\sqrt{\frac{1}{4}} + 2\sqrt{1/2_4}\sqrt{\frac{3}{4}} + 1\right) \cdot \frac{1}{4} = \frac{1}{12}(7.878315178) = 0.6565263$

(c) $\int_0^1 x^{1/2} dx = \frac{2}{3} x^{3/2}\Big|_0^1 = \frac{2}{3} = 0.6667$

5. $\int_1^5 \frac{dx}{x}$, $n = 8$
(a) 1.628968,
(b) 1.610847

7. $\int_1^5 \frac{dx}{1+x}$, $n = 8$
(a) 1.1032107,
(b) 1.098726

9. $\int_0^1 \frac{4}{1+x^2} dx$, $n = 10$
(a) 3.139926;
(b) 3.1415926 (Note: The actual value is π.)

11. $\int_0^2 \sqrt{1 + x^3} dx$; $n = 8$
(a) 3.251744,
(b) 3.241238

13. $\int_{-2}^2 \sqrt{x + 4}\, dx$, $n = 8$
(a) 7.909233,
(b) 7.912321

15. $\int_0^4 \frac{\sqrt{x}}{1+\sqrt{x}} dx$; $n = 8$
(a) 2.141030,
(b) 2.170342

17. (a) $\Delta x = 1$; $\frac{1}{2}(4.2 + 2(3.9) + 2(3.8) + 2(4.0) + 2(3.5) + 2(3.4) + 2(3.9) + 2(4.1) + 4.3) = \frac{1}{2}(61.7) = 30.85$

(b) $\frac{1}{3}(4.2 + 4(3.9) + 2(3.8) + 4(4.0) + 2(3.5) + 4(3.4) + 2(3.9) + 4(4.1) + 4.3) = \frac{1}{3}(92.5) = 30.83333$

19. (a) $\Delta x = 0.25$; $\frac{1}{2}(16.32 + 2(16.48) + 2(16.73) + 2(16.42) + 2(16.38) + 2(16.29) + 16.25)(0.25) = \frac{1}{2}(197.17)(0.25) = 24.64625$

(b) $\frac{1}{3}(16.32 + 4(16.48) + 2(16.73) + 4(16.42) + 2(16.38) + 4(16.29) + (16.25)(0.25) = \frac{1}{3}(295.55)(0.25) = 24.629167$

21. $\int_1^2 4.0e^{-10t} - 4.0e^{-20t} dt$; $\Delta t = \frac{1}{4}$
(a) $2.676 \times 10^{-5} C = 26.76 \ \mu C$
(b) $20.34 \ \mu C$

23. See *Computer Programs* in main text.

25. Trapezoidal:
$$A = \frac{100}{2}[300 + 2(300) + 2(350) + 2(400) + 2(500) + 2(600) + 2(500) + 2(600) + 2(550) + 2(650) + 600]$$
$$= 490{,}000 \text{ ft}^2 \approx 11.264 \text{ acres}$$
Simpson:
$$A = \frac{100}{3}[300 + 4(300) + 2(350) + 4(400) + 2(500) + 4(600) + 2(500) + 4(600) + 2(550) + 4(650) + 600]$$
$$= 496{,}667 \text{ ft}^2 \approx 11.417 \text{ acres}$$

CHAPTER 25 REVIEW

1. $\int x^5 dx = \frac{1}{6} x^6 + C$

2. $\int(\sqrt[3]{x} + 4) dx = \int (x^{1/3} + 4)) dx = \frac{3}{4} x^{4/3} + 4x + C$

3. $\int_0^3 (x^2 + 4x - 6) dx = \frac{1}{3} x^3 + 2x^2 - 6x\Big|_0^3 = 9 + 18 - 18 = 9$

4. $\int_1^4 (x + 4)^2 dx = \int_1^4 (x^2 + 8x + 16) dx = \frac{1}{3} x^3 + 4x^2 + 16x\Big|_1^4 = \left(\frac{64}{3} + 64 + 64\right) - \left(\frac{1}{3} + 4 + 16\right) = 129$

5. $\int_1^8 \frac{dt}{\sqrt[3]{t}} = \int_1^8 t^{-1/3} dt = \frac{3}{2} t^{2/3}\Big|_1^8 = 6 - \frac{3}{2} = \frac{9}{2}$

6. $\int_1^4 (\sqrt{x} + 4) dx = \frac{2}{3} x^{3/2} + 4x\Big|_1^4 = \left(\frac{16}{3} + 16\right) - \left(\frac{2}{3} + 4\right) = \frac{14}{3} + 12 = 16\frac{2}{3} = \frac{50}{3}$

7. $\int \frac{dt}{(t+5)^4} = \int (t+5)^{-4} dt = -\frac{1}{3}(t+5)^{-3} + C$

8. $\int \frac{x\,dx}{\sqrt[3]{x^2+4}} = \int (x^2+4)^{-1/3} x\,dx$. Let $u = x^2+4$, and then $du = 2x\,dx$, or $\frac{1}{2}du = x\,dx$. Substitution produces $\frac{1}{2}\int u^{-1/3}du = \frac{1}{2}\cdot\frac{3}{2}u^{2/3} + C = \frac{3}{4}(x^2+4)^{2/3} + C$.

9. $\int_1^8 x\sqrt[3]{x}\,dx = \int_1^8 x^{4/3}dx = \frac{3}{7}x^{7/3}\Big|_1^8 = \frac{3}{7}\cdot 2^7 - \frac{3}{7} = \frac{381}{7} \approx 54.429$.

10. $\int_1^4 \frac{4\,dx}{3\sqrt{x^3}} = \frac{4}{3}\int_1^4 x^{-3/2}dx = \frac{4}{3}\cdot(-2)x^{-1/2}\Big|_1^4 = -\frac{8}{3}(\frac{1}{2}-1) = -\frac{8}{3}\cdot\frac{-1}{2} = \frac{4}{3}$

11. $\int (u^4 - u^{-4})du = \frac{1}{5}u^5 + \frac{1}{3}u^{-3} + C$

12. $\int \frac{x^4-1}{x^3}dx = \int (x - x^{-3})dx = \frac{x^2}{2} + \frac{1}{2}x^{-2} + C = \frac{x^2}{2} + \frac{1}{2x^2} + C$

13. $\int 2x\sqrt{x^2-5}\,dx$. Let $u = x^2-5$, then $du = 2x\,dx$. Thus, $\int u^{1/2}du = \frac{2}{3}u^{3/2} + C = \frac{2}{3}(x^2-5)^{3/2} + C$

14. $\int_0^4 x\sqrt{x^2+9}\,dx$. Let $u = x^2+9$, then $du = 2x\,dx$, or $\frac{1}{2}du = x\,dx$. For the limits of integration, we have $x = 0 \Rightarrow u = 9$ and $x = 4 \Rightarrow u = 25$. Thus $\frac{1}{2}\int_9^{25} u^{1/2}du = \frac{1}{2}\cdot\frac{2}{3}u^{3/2}\Big|_9^{25} = \frac{1}{3}(125-27) = \frac{1}{3}(98) = \frac{98}{3} = 32\frac{2}{3}$.

15. $\int x\sqrt[3]{6-x^2}\,dx$. Let $u = 6-x^2$, then $du = -2x\,dx$, or $-\frac{1}{2}du = x\,dx$. Substitution yields $-\frac{1}{2}\int u^{1/3}du = -\frac{1}{2}\cdot\frac{3}{4}u^{4/3} + C = \frac{-3}{8}(6-x^2)^{4/3} + C$.

16. $\int 3x(4-x^2)^3 dx$. Let $u = 4-x^2$, then $du = -2x\,dx$, or $3x\,dx = -\frac{3}{2}du$. Substituting, we have $-\frac{3}{2}\int u^3 du = -\frac{3}{2}\cdot\frac{1}{4}u^4 + C = \frac{-3}{8}(4-x^2)^4 + C$.

17. $\int_{-1}^2 (x^2+4)(x^3-3)dx = \int_{-1}^2 (x^5 + 4x^3 - 3x^2 - 12)dx = \frac{x^6}{6} + x^4 - x^3 - 12x\Big|_{-1}^2 = (\frac{64}{6}+16-8-24) - (\frac{1}{6}+1+1+12) = -\frac{39}{2}$

18. $\int_1^2 x(x+2)^2 dx = \int_1^2 (x^3 + 4x^2 + 4x)dx = \frac{x^4}{4} + \frac{4}{3}x^3 + 2x^2\Big|_1^2 = (4+\frac{32}{3}+8) - (\frac{1}{4}+\frac{4}{3}+2) = 19\frac{1}{12}$

19. $\int \frac{y^2 dy}{(y^3-5)^{2/3}}$; Let $u = y^3 - 5$, then $du = 3y^2 dy$, or $\frac{1}{3}du = y^3 dy$. Substitution produces $\frac{1}{3}\int u^{-2/3}du = \frac{1}{3}\cdot\frac{3}{1}u^{1/3} + C = (y^3-5)^{1/3} + C$.

20. $\int \frac{5-6x+x^2}{1-x}dx = \int \frac{(5-x)(1-x)}{1-x}dx = \int (5-x)dx = 5x - \frac{x^2}{2} + C$, $x \neq 1$

21. $\int_0^1 (1-x^2)^3 dx = \int_0^1 (1 - 3x^2 + 3x^4 - x^6)dx = (x - x^3 + \frac{3}{5}x^5 - \frac{1}{7}x^7)\Big|_0^1 = 1 - 1 + \frac{3}{5} - \frac{1}{7} = \frac{21}{35} - \frac{5}{35} = \frac{16}{35}$

22. $\int \frac{x+2}{(x^2+4x)^2}dx$. Let $u = x^2 + 4x$, then $du = (2x+4)dx$, or $\frac{1}{2}du = (x+2)dx$. Thus, $\frac{1}{2}\int u^{-2}du = \frac{1}{2}\cdot(-1)u^{-1} + C = -\frac{1}{2}(x^2+4x)^{-1} + C$ or $\frac{-1}{2(x^2+4x)} + C$.

23. $\int (\sqrt{2x} - \frac{1}{\sqrt{2x}})dx = \sqrt{2}\cdot\frac{2}{3}x^{3/2} - \frac{1}{\sqrt{2}}\cdot\frac{2}{1}x^{1/2} + C = \frac{2\sqrt{2}}{3}x^{3/2} - \sqrt{2}x^{1/2} + C = \frac{1}{3}(2x)^{3/2} - (2x)^{1/2} + C$

24. $\int (x^3-4x)(3x^2-4)dx$. Let $u = x^3-4x$, then $du = (3x^2-4)dx$. Thus, $\int u\,du = \frac{u^2}{2} + C = \frac{1}{2}(x^3-4x)^2 + C$.

25. $\int_1^4 \frac{9-t^3}{3t^2}dt = \int_1^4 (3t^{-2} - \frac{1}{3}t)dt = (-3t^{-1} - \frac{1}{3}\cdot\frac{1}{2}t^2)\Big|_1^4 = (\frac{-3}{4} - \frac{8}{3}) - (-3 - \frac{1}{6}) = -\frac{1}{4}$

26. $\int_0^1 \sqrt{x}(3\sqrt{x}-5)dx = \int_0^1 (3x - 5\sqrt{x})dx = \frac{3}{2}x^2 - 5\cdot\frac{2}{3}x^{3/2}\Big|_0^1 = \frac{3}{2} - \frac{10}{3} = -\frac{11}{6}$

27. $\int_{-1}^8 \frac{4\sqrt[3]{u}\,du}{(1+u^{4/3})^3}$. Let $v = 1 + u^{4/3}$, then $dv = \frac{4}{3}u^{1/3}du$, or $3dv = 4u^{1/3}du$. The limits of integration become $u = -1 \Rightarrow v = 2$ and $u = 8 \Rightarrow v = 17$. Thus, we have $3\int_2^{17} v^{-3}dv = 3\cdot\frac{-1}{2}v^{-2}\Big|_2^{17} = \frac{-3}{2\cdot 17^2} + \frac{3}{2\cdot 2^2} \approx -0.00519 + 0.375 = 0.36981$.

28. $\int_1^2 \frac{3x^3-2x^{-3}}{x^2}dx = \int_1^2 (3x - 2x^{-5})dx = (\frac{3}{2}x^2 + \frac{1}{2}x^{-4})\Big|_1^2 = (6 + \frac{1}{32}) - (\frac{3}{2} + \frac{1}{2}) = 6\frac{1}{32} - 2 = 4\frac{1}{32}$

29. $\int_1^\infty \frac{dx}{x^4} = \lim_{b\to\infty}\int_1^b x^{-4}dx = \lim_{b\to\infty}(\frac{-1}{3}x^{-3})\Big|_1^b = \lim_{b\to\infty}(\frac{1}{3b^3} + \frac{1}{3}) = \frac{1}{3}$

30. $\int_0^\infty \frac{4x\,dx}{(3x^2+1)^5}$; Let $u = 3x^2 + 1$, then $du = 6x\,dx$, or $4x\,dx = \frac{2}{3}du$. For the limits of integration, we obtain $x = 0 \Rightarrow u = 1$ and $x \to \infty \Rightarrow u \to \infty$. Thus, $\frac{2}{3}\int_1^\infty u^{-5}du = \frac{2}{3}\left(-\frac{1}{4}\right)u^{-4}\Big|_1^\infty = \lim_{b\to\infty}\left(\frac{-1}{6b^4} + \frac{1}{6}\right) = \frac{1}{6}$.

31. $f(x) = 4x - 6 = 0$; $4x = 6$; $x = \frac{3}{2}$. Thus, $-\int_{-3}^{3/2}(4x - 6)dx + \int_{3/2}^4(4x - 6)dx = -\left(2x^2 - 6x\right)\Big|_{-3}^{3/2} + \left(2x^2 - 6x\right)\Big|_{3/2}^4 = -\left(\frac{9}{2} - 9\right) + (18 + 18) + (32 - 24) - \left(\frac{9}{2} - 9\right) = \frac{9}{2} + 36 + 8 + \frac{9}{2} = 53$.

32. $g(x) = 2 - x - x^2 = (2 + x)(1 - x)$. This crosses the x-axis at -2 and 1. $\int_{-2}^1\left(2 - x - x^2\right)dx = 2x - \frac{x^2}{2} - \frac{x^3}{3}\Big|_{-2}^1 = \left(2 - \frac{1}{2} - \frac{1}{3}\right) - \left(-4 - 2 + \frac{8}{3}\right) = 2 - \frac{5}{6} + 6 - \frac{8}{3} = 8 - \frac{21}{6} = 8 - \frac{7}{2} = \frac{9}{2} = 4\frac{1}{2}$ or 4.5.

33.

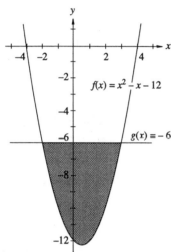

The graphs will intersect when $f(x) = g(x)$ or when $x^2 - x - 12 = -6$. This can be rewritten as $x^2 - x - 6 = 0$ which factors as $(x - 3)(x + 2) = 0$. The graphs intersect at $x = -2$ and $x = 3$. Thus, the desired area is $\left|\int_{-2}^3\left(x^2 - x - 6\right)dx\right| = \left|\frac{x^3}{3} - \frac{x^2}{2} - 6x\Big|_{-2}^3\right| = \left|(9 - \frac{9}{2} - 18) - (\frac{-8}{3} - 2 + 12)\right| = \left|-\frac{125}{6}\right| = \frac{125}{6}$.

34. $h(x) = \sqrt{x + 2} = 0$ at $x = -2$. Thus, $\int_{-2}^7\sqrt{x + 2}\,dx = \frac{2}{3}(x + 2)^{3/2}\Big|_{-2}^7 = \frac{2\cdot 27}{3} - 0 = 18$.

35. $4x^3 = x \Rightarrow 4x^3 - x = 0 \Rightarrow x(4x^2 - 1) = 0 \Rightarrow x = 0, -\frac{1}{2}, \frac{1}{2}$.

$\int_{-1/2}^0(4x^2 - x)dx + \int_0^{1/2}(x - 4x^3)dx$
$= x^4 - \frac{x^2}{2}\Big|_{-1/2}^0 + \frac{x^2}{2} - x^4\Big|_0^{1/2}$
$= -\left(\frac{1}{16} - \frac{1}{8}\right) + \left(\frac{1}{8} - \frac{1}{16}\right) = \frac{1}{8}$

36.

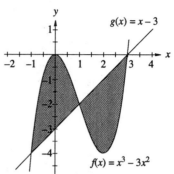

The graphs intersect when $f(x) = g(x)$ or $x^3 - 3x^2 = x - 3$. This can be rewritten as $x^2(x - 3) = (x - 3)$, or $x^2(x - 3) - (x - 3) = 0$, or $(x^2 - 1)(x - 3) = 0$. Thus, we see that the graphs intersect at -1, 1 and 3. The desired area is

$\int_{-1}^1\left(x^3 - 3x^2 - x + 3\right)dx$
$+ \int_1^3\left(x - 3 - x^3 + 3x^2\right)dx$
$= \left(\frac{x^4}{4} - x^3 - \frac{x^2}{2} + 3x\right)\Big|_{-1}^1$
$+ \left(\frac{x^2}{2} - 3x - \frac{x^4}{4} + x^3\right)\Big|_1^3$
$= \left(\frac{1}{4} - 1 - \frac{1}{2} + 3\right) - \left(\frac{1}{4} + 1 - \frac{1}{2} - 3\right)$
$+ \left(\frac{9}{2} - 9 - \frac{81}{4} + 27\right) - \left(\frac{1}{2} - 3 - \frac{1}{4} + 1\right)$
$= \left(1\frac{3}{4}\right) - \left(-2\frac{1}{4}\right) + \left(2 + \frac{1}{4}\right) - \left(-1\frac{3}{4}\right)$
$= 8$

37. $y^2 + x = 0$; $y = 2x + 1$; $(2x + 1)^2 + x = 0$; $4x^2 + 4x + 1 + x = 0$; $4x^2 + 5x + 1 = 0$; $(4x + 1)(x + 1)$. The graphs intersect at $(-1, -2)$ and $\left(-\frac{1}{4}, \frac{1}{2}\right)$.

$$\int_{-1}^{-1/4} \left(2x + 1 + \sqrt{-x}\right) dx$$

$$+ \int_{-1/4}^{0} 2\sqrt{-x}\, dx$$

$$= \left(x^2 + x - \frac{2}{3}(-x)^{3/2}\right)\Big|_{-1}^{-1/4}$$

$$+ \left(-2 \cdot \frac{2}{3}(-x)^{3/2}\right)\Big|_{-1/4}^{0}$$

$$= \left(\frac{1}{16} - \frac{1}{4} - \frac{1}{12}\right) - \left(1 - 1 - \frac{2}{4}\right) + 0 + \frac{1}{6}$$

$$= \frac{9}{16} = 0.5625$$

38. $2x^3 = x^4$; $x^4 - 2x^3 = 0$; $x^3(x-2) = 0$ they intersect at $(0, 0)$ and $(2, 16)$ $\int_0^2 \left(2x^3 - x^4\right) dx =$
$\frac{1}{2}x^4 - \frac{x^5}{5}\Big|_0^2 = 8 - \frac{32}{5} = \frac{8}{5} = 1\frac{3}{5} = 1.6$

39. $\int_0^1 \sqrt{1 - x^2}\, dx$; $n = 10$
(a) 0.776130,
(b) 0.781752

40. $\int_1^4 \frac{1}{3x} dx$; $n = 6$
(a) 0.468452,
(b) 0.462566

41. $\int_2^{2.5} \sqrt{x^3 - 1}\, dx$; $n = 10$
(a) 1.613695,
(b) 1.613656

42. $\int_{-1}^4 \sqrt{1 + x^5}\, dx$; $n = 10$
(a) 38.795198,
(b) 38.464568

43. $\int_0^1 \left(e^x - e^{-x}\right) dx$; $n = 4$

(a) 1.091812,
(b) 1.086185

44. (a) $\frac{1}{2}(4.25 + 2(5.72) + 2(5.13) + 2(3.19) + 2(2.10) + 2(0.15) + 2(1.65) + 2(3.10) + 3.70)(0.25) = \frac{1}{8}(50.03) = 6.25375$
(b) $\frac{1}{3}(4.25 + 4(5.72) + 2(5.13) + 4(3.19) + 2(2.10) + 4(0.15) + 2(1.65) + 4(3.10) + 3.70)(0.25) = \frac{1}{12}(74.35) = 6.195833$

45. (a) $\int_0^{15} \left(t + \frac{1}{\sqrt{1+t}}\right) dt = \frac{t^2}{2} + 2\sqrt{1+t}\Big|_0^{15} = \frac{225}{2} + 2\sqrt{16} - 2 = 112.5 + 8 - 2 = 118.5$ m
(b) The acceleration is given by $a(t) = v'(t) = 1 - \frac{1}{2}(1+t)^{-3/2}$. Thus, the acceleration at $t = 15$ is $a(15) = 1 - \frac{1}{2}(16)^{-3/2} = 1 - \frac{1}{128} = \frac{127}{128} \approx 0.992$ m/s^2.

46. $\omega = \int_3^5 10t^3 dt = \frac{10}{4}t^4\Big|_3^5 = \frac{5}{2}(625 - 81) = \frac{5}{2}(544) = 1360$ J

47. $2.0 - 1.0t^2 = -2\frac{di}{dt} \Rightarrow \frac{di}{dt} = \frac{1}{2}t^2 - 1$. Hence, $i = \int_1^6 \left(\frac{1}{2}t^2 - 1\right) dt = \left(\frac{1}{6}t^3 - t\right)\Big|_1^6 = (36 - 6) - \left(\frac{1}{6} - 1\right) = 30 + \frac{5}{6} = 30\frac{5}{6} \approx 30.8$ A.

48. (a) When the car stops its velocity is 0. Hence, we want to know when $v(t) = 88 - 4t = 0$, which is the same as $4t = 88$, and so $t = 22$ s
(b) The distance the car travels is given by $s(t) = \int v(t) = 88t - 2t^2$, and so, the distance it travels in those 22 s is $s(22) = 88 \cdot 22 - 2 \cdot 22^2 = 968$ ft.

CHAPTER 25 TEST

1. $\int_0^1 2x\, dx = x^2\Big|_0^1 = 1 - 0 = 1$.

2. $\int_1^4 (x+1)(x^2+1) dx = \int_1^4 \left(x^3 + x^2 + x + 1\right) dx = \left(\frac{x^4}{4} + \frac{x^3}{3} + \frac{x^2}{2} + x\right)\Big|_1^4 = \left(64 + \frac{64}{3} + 8 + 4\right) - \left(\frac{1}{4} + \frac{1}{3} + \frac{1}{2} + 1\right) = 95.25$.

3. $\int \left(x^2 + 3\right)^2 dx = \int \left(x^4 + 6x^2 + 9\right) dx = \frac{1}{5}x^5 + 2x^3 + 9x + C$.

4. $\int 4\left(x^4 + 7\right)^5 x^3 dx$; Let $u = x^4 + 7$, then $du = 4x^3 dx$. Thus, $\int u^5 du = \frac{1}{6}u^6 + C = \frac{1}{6}\left(x^4 + 7\right)^6 + C$.

5. $\int \frac{x^2+2}{x^2} dx = \int \left(1 + 2x^{-2}\right) dx = x - 2x^{-1} + C$.

6. $\int \left(x^2 + 8x\right)^3 (x + 4) dx$. Let $u = x^2 + 8x$, then $du = (2x+8) dx$, or $\frac{1}{2}du = (x+4) dx$. Substitution yields $\frac{1}{2}\int u^3 du = \frac{1}{2} \cdot \frac{1}{4}u^4 + C = \frac{1}{8}\left(x^2 + 8x\right)^4 + C$.

7. $\int_1^2 \left(3x^2\sqrt{x^3+x} + \sqrt{x^3+x}\right) dx = \int_1^2 \sqrt{x^3+x} \times \left(3x^2 + 1\right) dx$. Let $u = x^3 + x$, then $du = \left(3x^2 + 1\right) dx$. When $x = 1 \Rightarrow u = 2$ and when $x = 2 \Rightarrow u = 10$. Thus, with these substitutions we have $\int_2^{10} \sqrt{u}\, du = \frac{2}{3}u^{3/2}\Big|_1^{10} = \frac{2}{3}\left(10^{3/2} - 2^{3/2}\right) \approx 19.196$.

8. $4x^3 - 4 = 4(x^3 - 1) = 4(x-1)(x^2 + x + 1)$. This crosses the x-axis at $x = 1$. $\int_{-1}^{1} (4 - 4x^3) dx + \int_{1}^{2} (4x^3 - 4) dx = (4x - x^4)\big|_{-1}^{1} + (x^4 - 4x)\big|_{1}^{2} = (4-1)-(-4-1)+(16-8)-(1-4) = 3+5+8+3 = 19$.

9. The graphs intersect at $(-3, \; 7)$ and $(2, \; 2)$. Thus, the desired area is $\int_{-3}^{2} \left[(4 - x) - (x^2 - 2) \right] dx = \int_{-3}^{2} (6 - x - x^2) dx = \left(6x - \frac{x^2}{2} - \frac{x^3}{3} \right)\Big|_{-3}^{2} = (12 - 2 - \frac{8}{3}) - (-18 - \frac{9}{2} + 9) = \frac{125}{6} \approx 20.833$.

10. $\int_{2}^{4} \frac{1}{x} dx$; $n = 8$; $\Delta x = \frac{2}{8} = \frac{1}{4}$. The trapezoidal rule yields $\frac{1}{2} \left(\frac{1}{2} + \frac{2}{2.25} + \frac{2}{2.5} + \frac{2}{2.75} + \frac{2}{3} + \frac{2}{3.25} + \frac{2}{3.5} + \frac{2}{3.75} + \frac{1}{4} \right) \cdot \frac{1}{4} = \frac{1}{8}(5.55297) = 0.694122$. Simpson's rule produces $\frac{1}{3} \left(\frac{1}{2} + \frac{4}{2.25} + \frac{2}{2.5} + \frac{4}{2.75} + \frac{2}{3} + \frac{4}{3.25} + \frac{2}{3.5} + \frac{4}{3.75} + \frac{1}{4} \right) \cdot \frac{1}{4} = \frac{1}{12}(8.31785) = 0.693154$

11. $\int_{0}^{1} 0.0004 \left(t^{-1/2} + t^{1/2} \right) dt = 0.0004 \left[2t^{1/2} + \frac{2}{3} t^{3/2} \right]_{0}^{1} = 0.0004 \left[\left(2 + \frac{2}{3} \right) - (0 + 0) \right] \approx 0.00107 \text{ C}$

CHAPTER

26

Applications of Integration

1. $\bar{y} = \frac{1}{1}\int_0^1 x^2 dx = \frac{1}{3}x^3 \big|_0^1 = \frac{1}{3}$

   $f_{\text{rms}} = \sqrt{\frac{1}{1}\int_0^1 x^4 dx} = \sqrt{\frac{1}{5}x^5 \big|_0^1} = \sqrt{\frac{1}{5}} = \frac{\sqrt{5}}{5} \approx$ 0.4472

3. $\bar{y} = \frac{1}{3-1}\int_1^3 (x^2 - 1)dx = \frac{1}{2}\left[\frac{x^3}{3} - x\right]_1^3 = \frac{1}{2}\big[(9 -$

   $3) - (\frac{1}{3} - 1)\big] = \frac{1}{2}\big[\frac{20}{3}\big] = \frac{10}{3}$

   $h_{\text{rms}} = \sqrt{\frac{1}{2}\int_1^3 (x^2 - 1)^2 dx} =$

   $\sqrt{\frac{1}{2}\int_1^3 (x^4 - 2x^2 + 1)dx} = \sqrt{\frac{1}{2}\left[\frac{x^5}{5} - \frac{2x^3}{3} + x\right]_1^3}$

   $= \sqrt{\frac{1}{2}\left[\left(\frac{243}{5} - \frac{54}{3} + 3\right) - \left(\frac{1}{5} - \frac{2}{3} + 1\right)\right]} = \sqrt{\frac{248}{15}}$

   $= \sqrt{16.53} \approx 4.0661$

5. $\bar{y} = \int_0^5 \frac{1}{5}\sqrt{x + 4}\,dx = \frac{1}{5}\left[\frac{2}{3}(x + 4)^{3/2}\right]_0^5$

   $= \frac{2}{15}[(27 - 8)] = \frac{38}{15}.$

   $f_{\text{rms}} = \sqrt{\frac{1}{5}\int_0^5 (x + 4)\,dx} =$

   $\sqrt{\frac{1}{5}\left[\frac{x^2}{2} + 4x\right]_0^5} = \sqrt{\frac{1}{5}\left[\frac{25}{2} + 20\right]} = \sqrt{\frac{5}{2} + 4} =$

   $\sqrt{\frac{13}{2}} = 2.5495$

7. $\bar{y} = \frac{1}{4}\int_0^4 (t^2 + 4)dt = \frac{1}{4}\left[\frac{t^3}{3} + 4t\right]_0^4 = \frac{1}{4}\left[\frac{64}{3} + 16\right] =$

   $\frac{16}{3} + 4 = 9\frac{1}{3} = \frac{28}{3}$

   $f_{\text{rms}} = \sqrt{\frac{1}{4}\int_0^4 (t^4 + 8t^2 + 16)\,dt} =$

   $\sqrt{\frac{1}{4}\left[\frac{t^5}{5} + \frac{8t^3}{3} + 16t\right]_0^4} =$

   $\sqrt{\frac{1}{4}\left[\frac{1024}{5} + \frac{512}{3} + 64\right]} = \sqrt{\frac{1648}{15}} = \sqrt{109.8667} \approx$ 10.4817

9. $\bar{y} = \frac{1}{1}\int_1^2 (6x - x^2)dx = \left[3x^2 - \frac{x^3}{3}\right]_1^2$

   $= \left(12 - \frac{8}{3}\right) - \left(3 - \frac{1}{3}\right) = \frac{20}{3}$

$f_{\text{rms}} = \sqrt{\int_1^2 (36x^2 - 12x^3 + x^4)dx} =$

$\sqrt{\left[12x^3 - 3x^4 + \frac{x^5}{5}\right]_1^2} = \sqrt{45.2} \approx 6.7231$

11. $\bar{y} = \frac{1}{2.5}\int_0^{2.5} (4.9t^2 - 2.8t + 4)dt = \frac{1}{2.5}\left[\frac{4.9t^3}{3} -\right.$

    $\left. 1.4t^2 + 4t\right]_0^{2.5} = \frac{1}{2.5}(26.7708) = 10.7083$

    $j_{\text{rms}} = \sqrt{\frac{1}{2.5}\int_0^{2.5} (4.9t^2 - 2.8t + 4)^2 dt}$

    $= \sqrt{\frac{1}{2.5}(415.9765)} = 12.8992$

13. $\frac{1}{90}\int_0^{90} 0.0002(4991 + 366x - x^2)dx = \frac{0.0002}{90}\big[4991x$

    $+ 183x^2 - \frac{x^3}{3}\big]_0^{90} = \frac{0.0002}{90}[1688490] = 3.7522$ cm

    for 90 days. $\frac{0.0002}{365}\int_0^{365} (4991 + 366x - x^2)dx =$

    $\frac{0.0002}{365}\left[4991x + 183x^2 - \frac{x^3}{3}\right]_0^{365} =$

    $\frac{0.0002}{365}(9992848) = 5.4755$ cm for the year.

15. $a(t) = -32;\ v(t) = -32t;\ s(t) = -16t^2 + 360.$
    Setting $s(t) = 0$, and solving for t, we obtain
    $-16t^2 + 360 = 0$, or $16t^2 = 360$, and so $t =$
    4.743416. Thus, $\bar{s} = \frac{1}{4.74}\int_0^{4.74} (-16t^2 + 360)dt =$

    $\frac{1}{4.74}\left[-\frac{16}{3}t^3 + 360t\right]_0^{4.74} = 240$ ft. The average ve-

    locity is $\bar{v} = \frac{1}{4.74}\int_0^{4.74} (-32t)dt = \left[\frac{-16t^2}{4.74}\right]_0^{4.74} =$
    -75.8947 ft/s.

17. (a) $\frac{1}{6}\int_0^6 (60 + 4t - t^{2/3})dt = \frac{1}{6}\left[60t + 2t^2 - \frac{3}{5}t^{5/3}\right]_0^6 =$

    $60 + 2 \cdot 6 - \frac{3}{5}6^{2/3} = 70.018$ or $70°$F

    (b) $\frac{1}{6}\int_3^9 (60 + 4t - t^{2/3})dt = \frac{1}{6}\left[60t + 2t^2 - \frac{3}{5}t^{5/3}\right]_3^9 =$

    $\frac{1}{6}[484.3797] = 80.7°$F

19. $\frac{1}{3}\int_0^3 (1.0 + \sqrt{t})dt = \frac{1}{3}\left[\frac{t^2}{2} + \frac{2}{3}t^{3/2}\right]_0^3 = \frac{3}{2} + \frac{2}{3}\cdot\sqrt{3} \approx$ 2.6547 A.

21. $\frac{1}{4}\int_0^4 4t\sqrt{t^2 + 1} = \int_0^4 t\sqrt{t^2 + 1}.$ Let $u = t^2 + 1$
    and then, $du = 2t\,dt.$ When $t = 0 \Rightarrow u = 1,$

and when $t = 4 \Rightarrow u = 17$. Making these substitutions, we have $\frac{1}{2}\int_1^{17}\sqrt{u}\,du = \frac{1}{2}\cdot\frac{2}{3}u^{3/2}\big|_1^{17} = \frac{1}{3}(69.092796) = 23.0309$ A.

23. $P = \left(i_{\text{eff}}\right)^2 \cdot R = (30.0755)^2\cdot 30 = 27136$ W

25. We have $v = 4\sqrt{t} - 2t = 2(2\sqrt{t} - t)$.

(a) So, $v_{\text{eff}} = \sqrt{\frac{1}{6}\int_0^6 \left[2(2\sqrt{t}-t)\right]^2 dt} =$

$\sqrt{\frac{2}{3}\int_0^6 \left(4t - 4t^{3/2} + t^2\right)dt} =$

$\sqrt{\frac{1}{3}\left[2t^2 - \frac{8}{5}t^{5/2} + \frac{1}{3}t^3\right]_0^6} =$

$\sqrt{\frac{2}{3}\left(72 - \frac{288}{5}\sqrt{6} + 72\right)} = \sqrt{96 - \frac{192}{5}\sqrt{6}}$

$\approx \sqrt{1.939593877} \approx 1.3927$. So, the effective voltage is about 1.3927 V.

26.2 VOLUMES OF REVOLUTION: DISK AND WASHER METHODS

1. $\pi\int_1^5 (4x)^2 dx = \frac{16\pi}{3}r^3\big|_1^5 = \frac{16\pi}{3}(124) = 661.33\pi \approx 2077.64$.

3. $\pi\int_0^4 x^2 dx = \frac{\pi}{3}x^3\big|_0^4 = \frac{64\pi}{3} \approx 67.02$

5. $\pi\int_1^7 (x+1)dx = \pi\left[\frac{x^2}{2}+x\right]_1^7 = \pi\left[\frac{49}{2}+7-\frac{1}{2}-1\right] = 30\pi \approx 94.25$

7. $y = x^2 \Rightarrow x = \sqrt{y}$; $\pi\int_0^4 \sqrt{y}^2 dy = \pi\frac{y^2}{2}\big|_0^4 = 8\pi$

9. The two curves intersect at (2, 2). Thus, $V = \pi\int_0^2 \left((4-x)^2 - x^2\right)dx = \pi\int_0^2 \left(16 - 8x + x^2 - x^2\right)dx = \pi\left[16x - 4x^2\right]_0^2 = 16\pi$

11. $y = \sqrt[3]{x} \Rightarrow x = y^3$ and so $V\pi\int_0^2 \left(8^2 - y^{3\cdot2}\right)dy = \pi\left[64y - \frac{y^7}{7}\right]_0^2 = \pi\left[128 - \frac{128}{7}\right] = \frac{768\pi}{7}$.

13. They intersect at $(\pm2, 0)$ with $8 - 2x^2$ on the outside. Thus, we have
$\pi\int_{-2}^2 \left((8-2x^2)^2 - (4-x^2)^2\right)dx = \pi\int_{-2}^2 \left[(64 - 32x^2 + 4x^4) - (16 - 8x^2 + x^4)\right]dx = \pi\int_{-2}^2 (48 - 24x^2 + 3x^4)dx = \pi\left[48x - 8x^3 + \frac{3}{5}x^5\right]_{-2}^2 = 102.4\pi$ or $\frac{512\pi}{5}$

15. $\pi\int_0^1 \left(x^4 - x^6\right)dx = \pi\left[\frac{x^5}{5} - \frac{x^7}{7}\right]_0^1 = \frac{2\pi}{35}$

17. $x = \sqrt{y}$; $x = \frac{y}{4}$. Hence, we obtain $\pi\int_0^{16}\left(\sqrt{y}^2 - \left(\frac{y}{4}\right)^2\right)dy = \pi\left[\frac{y^2}{2} - \frac{y^3}{48}\right]_0^{16} = \pi\left(\frac{256}{2} - \frac{256}{3}\right) = \pi\left(\frac{256}{6}\right) = \frac{128\pi}{3}$

19. $y = \sqrt{8-x^2}$; $y^2 = 8 - x^2$; $x^2 = 8 - y^2$. So, $x = \sqrt{8-y^2}$. Now, we obtain $\pi\int_0^{\sqrt{8}}\left(8 - y^2\right)dt = \pi\left[8y - \frac{y^3}{3}\right]_0^{\sqrt{8}} = \pi\left[8\sqrt{8} - \frac{8\sqrt{8}}{3}\right] = \pi\left[16\sqrt{2} - \frac{16\sqrt{2}}{3}\right] = \frac{32\sqrt{2}\pi}{3}$

21. Use the equation for the top of a semicircle $y = \sqrt{r^2 - x^2}$ rotated about the x-axis.
$V = \pi\int_{-r}^r \sqrt{r^2-x^2}\,dx = \pi\int_{-r}^r \left(r^2 - x^2\right)dx$
$= \pi\left[r^2 x - \frac{x^3}{3}\right]_{-r}^r = \pi\left[r^3 - \frac{r^3}{3} + r^3 - \frac{r^3}{3}\right]$
$= \frac{4}{3}\pi r^3$

23. (a) The equation must be of the form $y = mx^2$ assuming it passes through the origin. It must also contain the point (1.5, 1) so $1 = m(1.5)^2, 1 = m(2.25)$ or $m = \frac{1}{2.25} = \frac{4}{9}$. The equation is $y = \frac{4}{9}x^2$ of $x = \sqrt{\frac{9}{4}y} = \frac{3}{2}\sqrt{y}$.
(b) To find its volume rotate about the y-axis we evaluate $\pi\int_0^1 \sqrt{\frac{9}{4}y}^2 dy = \pi\int_0^1 \frac{9}{4}y\,dy = \pi\left[\frac{9}{4}\cdot\frac{y^2}{2}\right]_0^1 = \frac{9\pi}{8}$ m^2 = 1.125π m^2.

25. $\pi\int_{-442}^{123}\sqrt{(147^2 + 0.16y^2)^2}\,dy = \pi\int_{-422}^{123}\left(147^2 + 0.16y^2\right)dy = \pi\left[21609y + \frac{0.16y^3}{3}\right]_{-442}^{123} = 16913711\pi$ ft$^2 \approx 53135993$ ft^3.

27. Use the equation for a circle $x^2 + y^2 = 18^2$ or $x = \sqrt{18^2 - y^2}$

(b) The average power is $P = i_{\text{eff}}^2 \cdot R = \left(\sqrt{1.939593877}\right)^2(10) \approx 19.40$. Hence, the average power is about 19.40 W.

27. $q = \int_0^{0.1} 2t\,dt = t^2\big|_0^{0.1} = 0.01$ C is the increase. The new charge is $0.01 + 0.01 = 0.02$ C.

29. $V = 2t + 1$, $i = 0.03t$ W$= \int P\,dt = \int Vi\,dt = \int(2t+1)(0.03t)dt = \int_0^{50}\left(0.06t^2 + 0.03t\right)dt = \left[0.02t^2 + \frac{0.03}{2}t^2\right]_0^{50} = 2537.5$ J

31. $V = \frac{1}{C}\int 0.04t\,dt = \frac{1}{C}0.2t^2 + k$, so $V_c = \frac{0.2t^2}{7.5\times10^{-6}}$. At $t = 0.005$, $V = \frac{0.2(0.005)^2}{7.5\times10^{-6}} = 0.6667$ V.

(a) $\pi \int_{-18}^{-14} (18^2 - y^2)\,dy = \left[18^2 y - \frac{y^3}{3}\right]_{-18}^{-14} =$
$226\frac{2}{3}\pi$ or 837.758 m^3

(b) $\pi \int_{-18}^{6} (18^2 - y^2)\,dy = \pi\left[18^2 y - \frac{y^3}{3}\right]_{-18}^{6} =$
$5760\pi \approx 18\,095.574$ m^3

$\equiv$ 26.3 VOLUMES OF REVOLUTION: SHELL METHOD

1. $2\pi \int_0^2 x(x^2 - 0)\,dx = 2\pi \int_0^2 x^3\,dx = 2\pi \left[\frac{x^4}{4}\right]_0^2 =$
8π

3. When $y = 8$ we see that $x = \pm 3$, we will rotate the portion in the first quadrant. Thus, $2\pi \int_1^3 x(8 -$
$x^2 + 1)\,dx = 2\pi \int_1^3 (9x - x^3)\,dx = 2\pi \left[\frac{9x^2}{2} - \frac{x^4}{4}\right]_1^3$
$= 2\pi\left[\frac{81}{4} - \frac{17}{4}\right] = 32\pi$

5. $2\pi \int_0^5 x(25 - x^2)\,dx = 2\pi \int_0^5 (25x - x^3)\,dx =$
$2\pi \left[\frac{25x^2}{2} - \frac{x^4}{4}\right]_0^5 = \frac{625\pi}{2}$

7. $y = x^3 \Rightarrow x = \sqrt[3]{y} = y^{1/3}$. Hence, we have
$2\pi \int_0^8 y(y^{1/3} - 0)\,dy = 2\pi \int_0^8 y^{4/3}\,dy = 2\pi \left[\frac{3}{7}y^{7/3}\right]_0^8$
$= \frac{768\pi}{7}$

9. $y = \sqrt[3]{x} \Rightarrow x = y^3$; $2\pi \int_0^2 y(y^3)\,dy = 2\pi \int_0^2 y^4\,dy$
$= 2\pi \left[\frac{y^5}{5}\right]^2 = \frac{64\pi}{5}$

11. $2\pi \int_0^1 x(x^{1/2} - x^{3/2})\,dx = 2\pi \int_0^1 (x^{3/2} - x^{5/2})\,dx =$
$2\pi\left[\frac{2}{5}x^{5/2} - \frac{2}{7}x^{7/2}\right]_0^1 = \frac{8\pi}{35}$

13. $y = 2x^3 \Rightarrow x = \sqrt[3]{\frac{y}{2}}$; $y^2 = 4x \Rightarrow x = \frac{1}{4}y^2$.
These graphs intersect at $(0, 0)$ and $(1, 2)$.
$2\pi \int_0^2 y\left[\left(\frac{y}{2}\right)^{1/3} - \frac{y^2}{4}\right]\,dy$
$= 2\pi \int_0^2 \left[\frac{1}{\sqrt[3]{2}}y^{4/3} - \frac{y^3}{4}\right]\,dy$
$= 2\pi \left[\frac{1}{\sqrt[3]{2}}\frac{3}{7}y^{7/3} - \frac{y^4}{16}\right]_0^2$
$= 2\pi \left[\frac{12}{7} - 1\right] = \frac{10\pi}{7}$

15. They intersect at $(0, 0)$ and $(2, 4)$. Rotating about the line $y = 6$ means we must integrate with respect to dy. $y = x^2 \Rightarrow x = \sqrt{y}$ and $y = 2x \Rightarrow$
$x = \frac{1}{2}y$
$2\pi \int_0^4 (6 - y)\left(y^{1/2} - \frac{y}{2}\right)\,dy$
$= 2\pi \int_0^4 \left(6y^{1/2} - 3y - y^{3/2} + \frac{y^2}{2}\right)\,dy$

$= 2\pi \left[4y^{3/2} - \frac{3y^2}{2} - \frac{2}{5}y^{5/2} + \frac{y^3}{6}\right]_0^4$
$= 2\pi \left[32 - 24 - \frac{64}{5} + \frac{32}{3}\right]$
$= \frac{176}{15}\pi \approx 11.7333\pi$

17. $2\pi \int_{-2}^1 (1 - x)(2 - x - x^2)\,dx$
$= 2\pi \int_{-2}^1 (2 - 3x + x^3)\,dx$
$= 2\pi \left[2x - \frac{3}{2}x^2 + \frac{x^4}{4}\right]_{-2}^1$
$= 2\pi \left[2 - \frac{3}{2} + \frac{1}{4} + 4 + 6 - 4\right]$
$= 2\pi \left(6\frac{3}{4}\right) = 2\pi \left(\frac{27}{4}\right) = \frac{27\pi}{2}$

19. Start with the equation $x = \pm\sqrt{25 = y^2}$ rotated about the x axis. We will take the shells starting with $y = 2$ to $y = 5$. The height of a shell is $2\sqrt{25 - y^2}$. $2\pi \int_2^5 y(2\sqrt{25 - y^2})\,dy$; Let $u = 25 - y^2$; $du = -2y\,dy$; $y = 2 \Rightarrow u = 21$; $y = 5 \Rightarrow u = 0$. Substituting, we have
$2\pi \int_{21}^0 -u^{1/2}\,du = 2\pi \left[-\frac{2}{3}u^{3/2}\right]_{21}^0$
$= 2\pi \left(\frac{2}{3} \cdot 21^{3/2}\right) = 28\pi\sqrt{21}$
$= 403.1$
By the washer method:
$\pi \int_{-\sqrt{21}}^{\sqrt{21}} ((25 - y^2) - 4)\,dy$
$= \pi \int_{-\sqrt{21}}^{\sqrt{21}} (21 - y^2)\,dy$
$= \pi \left[(21y - \frac{y^3}{3}\right]_{-\sqrt{21}}^{\sqrt{21}}$
$= \pi(21\sqrt{21} - 7\sqrt{21}) \times 2 = 28\pi\sqrt{21}$

21. The graph intersects the x-axis at 2 and 4.5.

$$V = 2\pi \int_{2}^{4.5} x \left[-(x - 3.25)^4 + \left(\frac{5}{4} \right)^4 \right] dx$$

$$= 2\pi \int_{2}^{4.5} \left(-x^5 + 13x^4 - 63.375x^3 + 137.3125x^2 - 109.125x \right) dx$$

$$= 2\pi \left[-\frac{1}{6}x^6 + 2.6x^5 - 15.84375x^4 + \frac{2197}{48}x^3 - \frac{873}{16}x^2 \right]_{2}^{4.5}$$

$$\approx 2\pi \left[-17.18 - (-33.05) \right]$$

$$\approx 31.74\pi \approx 99.7$$

The volume of the pan is 99.7 in.^{3}.

26.4 ARC LENGTH AND SURFACE AREA

1. $y = \frac{1}{3}(x^2 + 2)^{3/2}$; $y' = \frac{1}{2}(x^2 + 2)^{1/2} \cdot 2x = x(x^2 + 2)^{1/2}$

$$\int_{0}^{3} \sqrt{1 + \left[x(x^2 + 2)^{1/2} \right]} dx$$

$$= \int_{0}^{3} \sqrt{1 + x^2(x^2 + 2)} dx$$

$$= \int_{0}^{3} \sqrt{1 + x^4 + 2x^2} dx$$

$$= \int_{0}^{3} \sqrt{x^4 + 2x^2 + 1} dx$$

$$= \int_{0}^{3} (x^2 + 1) = \frac{x^3}{3} + x \Big|_{0}^{3} = 9 + 3 = 12$$

3. $9x^2 = 4y^3$, or $x^2 = \frac{4}{9}y^3$, or $x = \frac{2}{3}y^{3/2}$, and so $x' = y^{1/2}$

$$\int_{0}^{3} \sqrt{1 + \left(y^{1/2} \right)^2} \, dy = \int_{0}^{3} \sqrt{1 + y} \, dy$$

$$= \frac{2}{3}(1 + y)^{3/2} \Big|_{0}^{3}$$

$$= \frac{2}{3}[4^{3/2} - 1^{3/2}]$$

$$= \frac{2}{3}[8 - 1] = \frac{14}{3}$$

5. $y = \frac{x^3}{6} + \frac{1}{2x}$, so $y' = \frac{x^2}{2} - \frac{1}{2x^2}$

$$\int_{1}^{3} \sqrt{1 + \left[\frac{x^2}{2} - \frac{1}{2x^2} \right]^2} dx$$

$$= \int_{1}^{3} \sqrt{1 + \left(\frac{x^4}{4} - \frac{1}{2} + \frac{1}{4x^4} \right)} dx$$

$$= \int_{1}^{3} \sqrt{\frac{x^4}{4} + \frac{1}{2} + \frac{1}{4x^4}} dx$$

$$= \int_{1}^{3} \left(\frac{x^2}{2} + \frac{1}{2x^2} \right) dx$$

$$= \frac{x^3}{6} - \frac{1}{2x} \Big|_{1}^{3} = \frac{9}{2} - \frac{1}{6} - \frac{1}{6} + \frac{1}{2}$$

$$= 5 - \frac{1}{3} = 4\frac{2}{3} = \frac{14}{3}$$

7. $2\pi \int_{0}^{2} 4x\sqrt{1 + 4^2} \, dx = 8\pi\sqrt{17} \int_{0}^{2} x \, dx = 8\pi\sqrt{17} \left[\frac{x^2}{2} \right]_{0}^{2} = 16\pi\sqrt{17}$

9. $y = x$; $y' = 1$; $2\pi \int_{0}^{2} x\sqrt{1 + 1^2} \, dx = 2\sqrt{2}\pi \frac{x^2}{2} \Big|_{0}^{2} = 4\pi\sqrt{2}$

11. $x = 4y \Rightarrow y = \frac{x}{4}$, $y = 1 \Rightarrow x = 4$, and $y = 3 \Rightarrow x = 12$. We have $y' = \frac{1}{4}$, and so, the surface area is

$$2\pi \int_{4}^{12} \frac{x}{4}\sqrt{1 + \frac{1}{16}} \, dx = \frac{\pi\sqrt{17}}{8} \int_{4}^{12} x \, dx$$

$$= \frac{\pi\sqrt{17}}{8} \left[\frac{x^2}{2} \right]_{4}^{12}$$

$$= \frac{\pi\sqrt{17}}{8} [72 - 8]$$

$$= 8\pi\sqrt{17} \approx 103.62$$

13. Since $y = \frac{2}{75}x^{3/2}$, then $y' = \frac{1}{25}x^{1/2}$. Place the origin at the point where the cable meets the ground, then the length of the cable is given by $L = \int_{0}^{225} \sqrt{1 + \left(\frac{1}{25}x^{1/2} \right)^2} \, dy = \int_{0}^{225} \sqrt{1 + \left(\frac{1}{25} \right)^2 x} \, dy = \int_{0}^{225} \left(1 + \frac{1}{625}x \right)^{1/2} dy = 625 \cdot \frac{2}{3} \left(1 + \frac{1}{625}x \right)^{3/2} \Big|_{0}^{225} \approx 660.841 \ 2147 - 416.666 \ 6667 = 244.174 \ 548$ or approximately 244.17 m.

≡ 26.5 CENTROIDS

1. $\overline{x} = \frac{4 \cdot 5 + 6(-3)}{4+6} = \frac{20-18}{10} = \frac{1}{5}$; $\overline{y} = \frac{4 \cdot 2 + 6 \cdot 7}{4+6} = \frac{8+42}{10} = 5$. The centroid is at $(\overline{x}, \overline{y}) = (0.2, 5)$.

3. $\overline{x} = \frac{2 \cdot 1 + 3(-1) + 5(6)}{2+3+5} = \frac{2-3+30}{10} = 2.9$; $\overline{y} = \frac{2 \cdot 5 + 3(4) + 5(-4)}{2+3+5} = \frac{2}{10} = 0.2$. The centroid is at $(2.9, 0.2)$.

5. The left rectangle has center at $\left(-\frac{1}{2}, 2\right)$ and area 4. The right rectangle has center at $(2, 1)$ and area 8. $\overline{x} = \frac{4\left(-\frac{1}{2}\right) + 8(2)}{4+8} = \frac{-2+16}{12} = \frac{14}{12} = \frac{7}{6}$; $\overline{y} = \frac{4 \cdot 2 + 8 \cdot 1}{12} = \frac{8+8}{12} = \frac{16}{12} = \frac{4}{3}$. The centroid is at $\left(\frac{7}{6}, \frac{4}{3}\right)$.

7. The upper left rectangle has center $(-3.5, 1)$ and area 6. Lower rectangle has center $(-1, -1)$ and area 16. The upper right rectangle has center $(2.5, 1.5)$ and area 15. $\overline{x} = \frac{6(-3.5) + 16(-1) + 15(2.5)}{6+16+15} = \frac{0.5}{37} = \frac{1}{74}$; $\overline{y} = \frac{6 - 16 + 15 \cdot (1.5)}{37} = \frac{12.5}{37} = \frac{25}{74}$. The centroid is at $\left(\frac{1}{74}, \frac{25}{74}\right)$.

9. $y = 2x + 3 [0, 3]$

$M_y = \int_0^3 x(2x+3)dx = \int_0^3 (2x^2 + 3x)dx$

$= \frac{2x^3}{3} + \frac{3x^2}{2}\Big|_0^3 = 18 + \frac{27}{2} = \frac{63}{2}$

$M_x = \frac{1}{2}\int_0^3 (2x+3)^2 dx$

$= \frac{1}{2}\int_0^3 (4x^2 + 12x + 9)dx$

$= \frac{1}{2}\left[\frac{4x^3}{3} + 6x^2 + 9x\right]_0^3 = \frac{1}{2}[36 + 54 + 27]$

$= \frac{117}{2}$

$m = \int_0^3 (2x+3)dx = x^2 + 3x\Big|_0^3 = 9 + 9 = 18$

$\overline{x} = \frac{63}{36} = \frac{7}{4} = 1.75$

$\overline{y} = \frac{117}{36} = \frac{13}{4} = 3.25$

The centroid is $(1.75, 3.25)$.

11. $y = x^{1/3} [0, 8]$

$M_y = \int_0^8 x \cdot x^{2/3} dx = \int_0^8 x^{4/3} dx$

$= \frac{3}{7}x^{7/3}\Big|_0^8 = \frac{384}{7}$

$M_x = \frac{1}{2}\int_0^8 x^{2/3} dx = \frac{1}{2} \cdot \frac{3}{5}x^{5/3}\Big|_0^8 = \frac{48}{5}$

$m = \int_0^8 x^{1/3} dx = \frac{3}{4} \cdot x^{4/3}\Big|_0^8 = 12$

$\overline{x} = \frac{384}{7 \cdot 12} = \frac{32}{7}$

$\overline{y} = \frac{48}{5 \cdot 12} = \frac{4}{5}$.

The centroid is $\left(\frac{32}{7}, \frac{4}{5}\right)$.

13. $y = \sqrt{x+4}$ on $[0, 5]$; $M_y = \int_0^5 x\sqrt{x+4}\,dx$; $u = x + 4$; $du = dx$; $x = u - 4$ $x = 0 \Rightarrow u = 4$ and $x = 5 \Rightarrow u = 9$. Substituting, we obtain

$\int_4^9 (u-4)\sqrt{u}\,du$

$= \int_4^9 (u^{3/2} - 4u^{1/2})du$

$= \left[\frac{2}{5}u^{5/2} - 4 \cdot \frac{2}{3}u^{3/2}\right]_4^9$

$= \frac{2 \cdot 243}{5} - 72 - \frac{64}{5} + \frac{64}{3}$

$= \frac{1458 - 1080 - 192 + 320}{15} = \frac{506}{15}$

$M_x = \frac{1}{2}\int_0^5 (x+4)dx = \frac{1}{2}\left[\frac{x^2}{2} + 4x\right]_0^5$

$= \frac{1}{2}\left[\frac{25}{2} + 20\right] = \frac{65}{4}$

$m = \int_0^5 \sqrt{x+4}\,dx = \frac{2}{3}(x+4)^{3/2}\Big|_0^5$

$= \frac{2}{3}[27 - 8] = \frac{38}{3}$

$\overline{x} = \frac{506}{15} \cdot \frac{3}{38} = \frac{253}{95}$

$\overline{y} = \frac{65}{4} \cdot \frac{3}{38} = \frac{195}{152}$.

The centroid is $\left(\frac{253}{95}, \frac{195}{152}\right)$.

15. $M_y = \displaystyle\int_0^2 x(4x - x^2)dx = \int_0^2 (4x^2 - x^3)dx$

$= \left[\dfrac{4x^3}{3} - \dfrac{x^4}{4}\right]_0^2 = \dfrac{32}{3} - 4 = \dfrac{20}{3}$

$M_x = \dfrac{1}{2}\displaystyle\int_0^2 \left[(4x)^2 - (x^2)^2\right]dx$

$= \dfrac{1}{2}\displaystyle\int_0^2 (16x^2 - x^4)dx$

$= \dfrac{1}{2}\left[\dfrac{16x^3}{3} - \dfrac{x^5}{5}\right]_0^2 = \dfrac{1}{2}\left[\dfrac{128}{3} - \dfrac{32}{5}\right]$

$= \dfrac{64}{3} - \dfrac{16}{5} = \dfrac{272}{15}$

$m = \displaystyle\int_0^2 (4x - x^2)dx = \left[2x^2 - \dfrac{x^3}{3}\right]_0^2$

$= 8 - \dfrac{8}{3} = \dfrac{16}{3}$

$\overline{x} = \dfrac{20}{3} \cdot \dfrac{3}{16} = \dfrac{5}{4}$

$\overline{y} = \dfrac{272}{15} \cdot \dfrac{3}{16} = \dfrac{17}{5}.$

The centroid is $\left(\dfrac{5}{4},\ \dfrac{17}{5}\right)$.

17. The graphs intersect at $(-2,\ 4)$ and $(1,\ 1)$

$M_y = \displaystyle\int_{-2}^1 x(2 - x - x^2)dx$

$= \displaystyle\int_{-2}^1 (2x - x^2 - x^3)dx$

$= \left[x^2 - \dfrac{x^3}{3} - \dfrac{x^4}{4}\right]_{-2}^1$

$= 1 - \dfrac{1}{3} - \dfrac{1}{4} - 4 - \dfrac{8}{4} + 4 = -\dfrac{9}{4}$

$M_x = \dfrac{1}{2}\displaystyle\int_{-2}^1 \left((2-x)^2 - (x^2)^2\right)dx$

$= \dfrac{1}{2}\displaystyle\int_{-2}^1 (4 - 4x + x^2 - x^4)dx$

$= \dfrac{1}{2}\left[4x - 2x^2 + \dfrac{4^3}{3} - \dfrac{x^5}{5}\right]_{-2}^1$

$= \dfrac{1}{2}\left[4 - 2 + \dfrac{1}{3} - \dfrac{1}{5} + 8 + 8 + \dfrac{8}{4} - \dfrac{32}{5}\right]$

$= \dfrac{1}{2}\left[18 + 3 - \dfrac{33}{5}\right] = \dfrac{36}{5}$

$m = \displaystyle\int_{-2}^1 (2 - x - x^2)dx = \left[2x - \dfrac{x^2}{2} - \dfrac{x^3}{3}\right]_{-2}^1$

$= 2 - \dfrac{1}{2} - \dfrac{1}{3} + 4 + 2 - \dfrac{8}{3} = 8 - \dfrac{1}{2} - 3$

$= 4\dfrac{1}{2} = \dfrac{9}{2}$

$\overline{x} = -\dfrac{9}{4} \cdot \dfrac{2}{9} = -\dfrac{1}{2}$

$\overline{y} = \dfrac{36}{5} \cdot \dfrac{2}{9} = \dfrac{8}{5}.$

The centroid is $\left(-\dfrac{1}{2},\ \dfrac{8}{5}\right)$.

19. The graphs intersect at $(\pm 3,\ 9)$

$M_y = \displaystyle\int_{-3}^3 x(18 - x^2 - x^2)dx$

$= \displaystyle\int_{-3}^3 (18x - 2x^3)dx = 0$

$M_x = \dfrac{1}{2}\displaystyle\int_{-3}^3 \left((18 - x^2)^2 - (x^2)^2\right)dx$

$= \dfrac{1}{2}\displaystyle\int_{-3}^3 (324 - 36x^2 + x^4 - x^4)dx$

$= \dfrac{1}{2}\left[324x - 12x^3\right]_{-3}^3$

$= \dfrac{1}{2}[972 - 324 + 972 - 324] = 648$

$m = \displaystyle\int_{-3}^3 (18 - x^2 - x^2)dx = \left[18x - \dfrac{2x^3}{3}\right]_{-3}^3$

$= (54 - 18 + 54 - 18) = 72$

$\overline{x} = 0$

$\overline{y} = \dfrac{648}{72} = 9$

The centroid is $(0,\ 9)$.

21. They intersect at $(-4,\ -4)$ and $(3,\ 3)$

$M_y = \displaystyle\int_{-4}^3 x(12 - x^2 - x)dx$

$= \displaystyle\int_{-4}^3 (12x - x^3 - x^2)dx$

$= \left[6x^2 - \dfrac{x^4}{4} - \dfrac{x^3}{3}\right]_{-4}^3$

$= 54 - \dfrac{81}{4} - 9 - 96 + 64 - \dfrac{64}{3}$

$= 13 - \dfrac{81}{4} - \dfrac{64}{3} = -\dfrac{343}{12}$

$$M_x = \frac{1}{2}\int_{-4}^{3}(12-x^2)^2 - (x)^2 dx$$

$$= \frac{1}{2}\int_{-4}^{3}(144 - 24x^2 + x^4 - x^3)dx$$

$$\times \frac{1}{2}\left[144x - \frac{25x^3}{3} + \frac{x^5}{5}\right]_{-4}^{3}$$

$$= \frac{1}{2}\left[432 - 225 + \frac{243}{5} + 576\right.$$

$$\left. - \frac{1600}{3} + \frac{1024}{5}\right] = \frac{3773}{15}$$

$$m = \int_{-4}^{3}(12 - x^2 - x)dx$$

$$= \left[12x - \frac{x^3}{3} - \frac{x^3}{2}\right]_{-4}^{3}$$

$$= 36 - 9 - \frac{9}{2} + 48 - \frac{64}{3} + 8 = 57\frac{1}{6} = \frac{343}{6}$$

$$\bar{x} = -\frac{343}{12} \times \frac{6}{343} = -\frac{1}{2} = -0.5$$

$$\bar{y} = \frac{3773}{15} \cdot \frac{6}{343} = 4.4$$

The centroid is $(-0.5,\ 4.4)$.

23. $y = x^3;$

$$M_y = \pi\int_{0}^{2}x(x^2)dx = \pi\frac{x^3}{8}\Big|_{0}^{2} = 32\pi$$

$$m = \pi\int_{0}^{2}(x^3)^2 = \pi \cdot \frac{x^7}{7}\Big|_{0}^{2} = \frac{128\pi}{7}.$$

Thus, we find that $\bar{x} = \dfrac{32\pi}{\frac{128\pi}{7}} = 32\pi\left(\frac{7}{128\pi}\right) = \frac{7}{4}$

and $\bar{y} = 0$ since it's rotated about the x-axis. The centroid is $\left(\frac{7}{4},\ 0\right)$.

25. $y = x^4 \Rightarrow x = y^{1/4}$

$$M_x = \pi\int_{0}^{1}y \cdot (y^{1/4})^2 dy = \pi\frac{2}{5} \cdot y^{5/2}\Big|_{0}^{1} = \frac{2\pi}{5}$$

$$m = \pi\int_{0}^{1}(y^{1/4})^2 dy = \pi \cdot \frac{2}{3}y^{3/2}\Big|_{0}^{1} = \frac{2\pi}{3}$$

$$\bar{x} = 0$$

$$\bar{y} = \frac{\frac{2\pi}{5}}{\frac{2\pi}{3}} = \frac{2\pi}{5} \cdot \frac{3}{2\pi} = \frac{3}{5}$$

The centroid is $\left(0,\ \frac{3}{5}\right)$.

27. $x + y = 6 \Rightarrow y = 6 - x;\ x = 3$

$$M_y = \pi\int_{0}^{3}x\left[(6-x)^2\right]dx$$

$$= \pi\int_{0}^{3}(36x - 12x^2 + x^3)dx$$

$$= \pi\left[18x^2 - 4x^3 + \frac{x^4}{4}\right]_{0}^{3} = 74.25\pi = \frac{297}{4}\pi$$

$$m = \pi\int_{0}^{3}(6-x)^2 dx = \pi\int_{0}^{3}(36 - 12x + x^2)dx$$

$$= \pi\left(36x - 6x^2 + \frac{x^3}{3}\right)\Big|_{0}^{3} = 63\pi$$

$$\bar{x} = \frac{297}{4} \cdot \frac{1}{63} = \frac{33}{28}$$

$$\bar{y} = 0$$

The centroid is $\left(\frac{33}{28},\ 0\right)$.

29.
$$M_x = \pi\int_{0}^{4}y \cdot 4^2 dy + \pi\int_{4}^{8}y(8-y)^2 dy$$

$$= \pi\left[128 + 106\frac{2}{3}\right] = 234\frac{2}{3}\pi$$

$$m = \pi\left[\int_{0}^{4}16\,dy + \int_{4}^{8}(8-y)^2 dy\right]$$

$$= \pi\left[64 + 21\frac{1}{3}\right] = 85\frac{1}{3}\pi$$

$$\bar{x} = 0$$

$$\bar{y} = \frac{234\frac{2}{3}}{85\frac{1}{3}} = 2.75$$

The centroid is $(0,\ 2.75)$.

31. $x^2 - y^2 = 1;\ -y^2 = 1 - x^2;\ y^2 = x^2 - 1;\ y = \sqrt{x^2 - 1}$

$$M_y = \pi\int_{1}^{3}x(x^2 - 1)dx = \pi\left[\frac{x^4}{4} - \frac{x^2}{2}\right]_{1}^{3}$$

$$= \pi\left[\frac{81}{4} - \frac{9}{2} - \frac{1}{4} + \frac{1}{2}\right] = \pi[20 - 4] = 16\pi$$

$$m = \pi\int_{1}^{3}(x^2 - 1)dx = \pi\left[\frac{x^3}{3} - x\right]_{1}^{3}$$

$$= \pi\left(9 - 3 - \frac{1}{3} + 1\right) = \pi6\frac{2}{3} \text{ or } \frac{20}{3}\pi$$

$$\bar{x} = \frac{16}{\frac{20}{3}} = \frac{4 \cdot 3}{5} = \frac{12}{5}.$$

The centroid is $\left(\frac{12}{5},\ 0\right)$ or $(2.4,\ 0)$.

33. $y = x^2;\ y = \sqrt{x}$

$$M_y = \pi \int_0^1 x\left[x - x^4\right] dx = \pi \left[\frac{x^3}{3} - \frac{x^6}{6}\right]_0^1 = \frac{\pi}{6}$$

$$m = \pi \int_0^1 x - x^4 dx = \pi \left[\frac{x^2}{2} - \frac{x^5}{5}\right] = \frac{3\pi}{10}$$

$$\bar{x} = \frac{\frac{\pi}{6}}{\frac{3\pi}{10}} = \frac{10}{18} = \frac{5}{9}.$$

The centroid is $\left(\frac{5}{9},\ 0\right)$.

35. **(a)** Make two rectangles by cutting off the top piece. Then the bottom has center at $\left(50,\ \frac{1}{2}\right)$ and area 100 cm^2. The left piece has center at $\left(\frac{1}{2},\ \frac{101}{2}\right)$ and area 99 cm^2.

$$\bar{x} = \frac{100 \times 50 + 99 \cdot \frac{1}{2}}{100 + 99} = \frac{5049.5}{199} \approx 25.374$$

$$\bar{y} = 100 \times \frac{1}{2} + 101\left(\frac{99}{2}\right) = \frac{5049.5}{199} \approx 25.374$$

The centroid of the frame is (25.374, 25.374).

(b) The bottom area is $h \cdot w$ with center $\left(\frac{h}{2},\ \frac{w}{2}\right)$ and the left area is $(h - w)$, center $\left(\frac{w}{2},\ \frac{h+w}{2}\right)$

$$\bar{x} = \frac{hw\left(\frac{h}{2}\right) + w(h-w) \cdot \frac{w}{2}}{hw + (h-w)w} = \frac{\frac{h^2 w}{2} + \frac{hw^2}{2} - \frac{w^3}{2}}{2wh - w^2}$$

$$= \frac{h^2 + hw - w^2}{2(2h - w)}.$$

$$\bar{y} = \frac{\left(\frac{hw \cdot w}{2}\right) + (h-w)w\left(\frac{h+w}{2}\right)}{w(2h - w)}$$

$$= \frac{hw + h^2 - w^2}{2(2h - w)}$$

The centroid is $\left(\frac{h^2 + hw - w^2}{2(2h-w)},\ \frac{h^2 + hw - w^2}{2(2h-w)}\right)$.

37. **(a)** $m = \int_0^{40}(5x + 1)dx = \left[\frac{5x^2}{2} + x\right]_0^{40} = 4040$ g

(b) $M_y = \int_0^{40} x(5x + 1)dx = \left[\frac{5x^3}{3} + \frac{x^2}{2}\right]_0^{40} =$ 107,467; $\bar{x} = \frac{107,467}{4040} = 26.6$ cm from the less dense end.

≡ 26.6 MOMENTS OF INERTIA

1. $I_y = (3 \cdot 4^2) + 5(-3)^2 = 48 + 45 = 93$
$I_x = 3(0)^2 + 5(0)^2 = 0$
$r_y = \sqrt{\frac{93}{3}} = 3.41;\ r_x = 0;\ r_0 = \sqrt{r_x^2 + r_y^2} = 3.4095$

3. $I_y = 4 \cdot 2^2 + 3 \cdot 1^2 = 19;\ r_y = \sqrt{\frac{19}{7}} = 1.6475$
$I_x = 4 \cdot 1^2 + 3 \cdot 4^2 = 52;\ r_x = \sqrt{\frac{52}{7}} = 2.7255;\ r_0 = \sqrt{\frac{19+52}{7}} = 3.1847$

5. $I_y = \rho \int_0^{\sqrt{2}} x^2 \cdot (2 - x^2)\,dx = \rho\left[\frac{2x^3}{3} - \frac{x^5}{5}\right]_0^{\sqrt{2}}$

$$= \rho\left[\frac{4\sqrt{2}}{3} - \frac{4\sqrt{2}}{5}\right] = \rho \cdot \frac{8\sqrt{2}}{15} \approx 0.7542\rho$$

$$m = \rho \int_0^{\sqrt{2}}(2 - x^2)\,dx = \rho\left[2x - \frac{x^3}{3}\right]_0^{\sqrt{2}}$$

$$= \rho\left(2\sqrt{2} - \frac{2\sqrt{2}}{3}\right)$$

$$= \rho \cdot \frac{4\sqrt{2}}{3};\ r_y = \sqrt{\frac{\frac{8}{15}}{4/3}} = \sqrt{\frac{2}{5}} = 0.6325$$

7. $I_x = \rho \int_0^5 y^2 \cdot 3\,dy = \rho y^3 \big|_0^5 = 125\rho$

$$m = \rho \int_0^5 3\,dx = 15$$

$$r_x = \sqrt{\frac{125}{15}} = \sqrt{\frac{25}{3}} \approx 2.8868$$

9. $I_x = \rho \int_0^1 y^2\left(y^{1/3} - y^{1/2}\right)dy$

$$= \rho\left[\frac{3}{10}y^{10/3} - \frac{2}{7}y^{7/2}\right]_0^1$$

$$= \frac{\rho}{70} \approx 0.014286\rho$$

$$m = \rho \int_0^1 (x^2 - x^3)\,dx = \rho\left[\frac{x^3}{3} - \frac{x^4}{4}\right]_0^1 = \frac{1}{12}\rho$$

$$r_x = \sqrt{\frac{\frac{1}{70}}{\frac{1}{12}}} = \sqrt{\frac{12}{70}} \approx 0.4140$$

11.
$$I_y = 5 \int_0^4 x^2\left(4x - x^2\right)dx = 5\left[x^4 - \frac{x^5}{5}\right]_0^4$$
$$= 5\left[256 - \frac{1024}{5}\right] = 256 \text{ g} \cdot \text{cm}^2$$
$$m = 5 \int_0^4 \left(4x - x^2\right)dx = 5\left[2x^2 - \frac{x^3}{3}\right]_0^4$$
$$= 5\left[32 - \frac{64}{3}\right]$$
$$= 53.333; \ r_y = \sqrt{\frac{256}{53.333}} \approx 2.1909 \text{ cm}$$

13.
$$I_y = 8 \int_1^2 x^2\left(x^2 - \frac{1}{x^2}\right)dx = 8\left[\frac{x^5}{5} - x\right]_1^2$$
$$= 8\left[\frac{32}{5} - 2 - \frac{1}{5} + 1\right] = 8\left[\frac{26}{5}\right] = 41.6 \text{ g} \cdot \text{cm}^2$$
$$m = 8 \int_1^2 \left(x^2 - \frac{1}{x^2}\right)dx = 8\left[\frac{x^3}{3} + \frac{1}{x}\right]_1^2$$
$$= 8\left[\frac{8}{3} + \frac{1}{2} - \frac{1}{3} - 1\right] = 14\frac{2}{3} \text{ g}$$
$$r_y = \sqrt{\frac{41.6}{14.6667}} \approx 1.6842 \text{ cm}$$

15.
$$I_y = 2\pi\rho \int_0^{\sqrt{2}} x^3\left[2 - x^2\right]dx$$
$$= 2\pi\rho\left[\frac{2x^4}{4} - \frac{x^6}{6}\right]_0^{\sqrt{2}}$$
$$= 2\pi\rho\left[2 - \frac{8}{6}\right] = \frac{4\pi\rho}{3}$$
$$m = 2\pi\rho \int_0^{\sqrt{2}} x\left(2 - x^2\right)dx = 2\pi\rho\left[x^2 - \frac{x^4}{4}\right]_0^{\sqrt{2}}$$
$$= 2\pi\rho$$
$$r_y = \sqrt{\frac{4/3}{2}} = \sqrt{2/3} \approx 0.8165$$

17.
$$I_y = 2\pi\rho \int_0^4 x^3\left(4x - x^2\right)dx = 2\pi\rho\left[\frac{4x^5}{5} - \frac{x^6}{6}\right]_0^4$$
$$= 2\pi\rho\left(\frac{4096}{30}\right) \approx 857.86\rho \text{ or } 273.07\pi\rho$$
$$m = 2\pi\rho \int_0^4 x\left(4x - x^2\right)dx = 2\pi\rho\left[\frac{4x^3}{3} - \frac{x^4}{4}\right]_0^4$$
$$= \frac{128\pi\rho}{3}$$
$$\approx 134.04\rho \text{ or } 42.667\pi\rho$$
$$r_y = \sqrt{\frac{273.07}{42.667}} = 2.5298$$

19.
$$I_x = 2\pi\rho \int_0^6 y^3 \cdot 4 \, dy = 2\pi\rho y^4 \big|_0^6 = 2592\pi\rho$$
$$m = 2\pi\rho \int_0^6 y \cdot 4 \, dx = 2\pi\rho\left[2y^2\right]_0^6 = 144\pi\rho$$
$$r_x = \sqrt{\frac{2592}{144}} \approx 4.2426$$

21.
$$I_y = 2\pi \cdot 3 \int_1^4 x^3\left(4x - x^2(4 - x)\right)dx$$
$$= 6\pi \int_1^4 x^3\left(5x - x^2 - 4\right)dx$$
$$= 6\pi \int_1^4 \left(5x^4 - x^5 - 4x^3\right)dx$$
$$= 6\pi\left[x^5 - \frac{x^6}{6} - x^4\right]_1^4$$
$$= 6\pi(85.5) = 513\pi \text{ g} \cdot \text{cm}$$
$$m = 6\pi \int_1^4 x\left(5x - x^2 - 4\right)dx$$
$$= 6\pi \int_1^4 \left(5x^2 - x^3 - 4x\right)dx$$
$$= 6\pi\left[\frac{5x^3}{3} - \frac{x^3}{4} - 2x^2\right]_1^4 = 67.5\pi \text{ g}$$
$$r_y = \sqrt{\frac{513}{67.5}} = 2.7568 \text{ cm}$$

23.
$$I_y = 2\pi \cdot 5 \int_1^2 x^3\left(x^3 - x^{-1}\right)dx$$
$$= 10\pi \int_1^2 \left(x^6 - x^2\right)dx = 10\pi\left[\frac{x^7}{7} - \frac{x^3}{3}\right]_1^2$$
$$= 158.0952\pi \text{ g} \cdot \text{cm}^2$$
$$m = 10\pi \int_1^2 x\left(x^3 - x^{-1}\right)dx$$
$$= 10\pi \int_1^2 \left(x^4 - 1\right)dx$$
$$= 10\pi\left[\frac{x^5}{5} - x\right]_1^2 = 52\pi \text{ g}$$
$$r_y = \sqrt{\frac{158.0952}{52}} = 1.7436 \text{ cm}$$

26.7 WORK AND FLUID PRESSURE

1. $F = kx$; $3 = k(1/3)^1$ yields $k = 9$. Hence, $F = 9x$; 10 in. $= \frac{5}{6}$ ft. Work $= \int_0^{5/6} 9x\, dx = \frac{9}{2}x^2\big|_0^{5/6} = \frac{9}{2} \cdot \frac{25}{36} = \frac{25}{8} = 3.125$ ft · lb

3. 20 cm $= 0.2$ m; $6 = k(0.2)$; so $k = 30$ and we get $F = 30x$. Thus, work $= \int_0^1 30x = \frac{30x^2}{2}\big|_0^1 = 15$ J

5. $\int_0^{0.1} kx\, dx = \frac{k}{2}x^2\big|_0^{0.1} = \frac{k(0.01)}{2} = 0.20$. Hence, $k = \frac{0.4}{0.01} = 40$; $F = kx = 40 \cdot (0.1) = 4$N

7. $\int_0^b 9x\, dx = \frac{9}{2}x^2\big|_0^b = \frac{9}{2}b^2$; $\frac{9}{2}b^2 = 105$; $b^2 = 23.333$; $b = 4.8305$ ft

9. $\int_0^{30} 2.5(30-y)dy = 2.5\big[30y - \frac{y^2}{2}\big]_0^{30} = 2.5(450) = 1125$ J

11. For the first 12 feet; each section will have a weight of 0.75 dy. $\int_0^{12} 0.75(12-y)dy = 0.75\big(12y - \frac{y^2}{2}\big)_0^{12} = \frac{3}{4}(72) = 54$ ft · lb. The total weight of the chain is $12 \cdot \frac{3}{4} = 9$ lb. The work in lifting the last 8 feet is 72 ft·lb. The total work is $54 + 72 = 126$ ft · lb.

13. When the tank is x meters above the ground it has taken $2x$ minutes to get there and has used $20 \cdot 2x = 40x$ L. Hence the volume of the tank is $(1000-40x)$ L. Water has a density of 1 kg/L so its mass is $(1000-40x)$kg. The total mass is $(1047-40x)$ kg so its weight is $9.8(1047 - 40x)$. $W = \int_0^{20} 9.8(1047 - 40x)dx = 9.8\big[1047x - 20x^2\big]_0^{20} = 9.8(12940) = 126{,}812$ J or 127 kJ.

15. $W = \int_{0.05}^{10 \times 10^{-9}} 8.988 \times 4 \times 10^{-19} \times 5 \times 10^{-8} \frac{1}{r^2} dr$
$= 1.7976 \times 10^{-16} \Big[\frac{-1}{r}\Big]_{0.05}^{10^{-8}}$
$= 1.7976 \times 10^{-16} \times 10^8$
$= 1.7976 \times 10^{-8}$ J $= 179.76\ \mu$J

17. $W = \int F(r)dr = \int_{0.01}^1 \frac{k}{r^2} = -k\big[\frac{1}{r}\big]_{0.01}^1 = -k[1 - 100] = 99k$ J.

19. $W = \int_{4000}^{5000} \frac{k}{r^2} = -k\big[\frac{1}{r}\big]_{4000}^{5000} = k\big[5 \times 10^{-5}\big]$. Since $k = 16 \times 10^9$, we get $W = 16 \times 10^9 \cdot 5 \times 10^{-5} = 80{,}000$ mi · lb $= 4.224 \times 10^9$ ft · lb.

21. $\int_0^8 880(4\pi)x\, dx = 1.264\pi \times 10^5$ kg · m. Multiplying by the gravitational constant to get joules, we have $1.264\pi \times 10^5 \times 9.8 = 1.23872\pi \times 10^6$ J $= 3.89155 \times 10^6$ J

23. $\int_0^{1.5} 6.86 \times 10^5(2-x)dx = 6.86 \times 10^5\big[2x - \frac{x^2}{2}\big]_0^{1.5} = 6.86 \times 10^5 \times 1875 = 1.28625$ MJ $= 1{,}286{,}250$ J

25. Position the tank so that the bottom is at $(0, 0)$. Then the radius will be $\frac{2}{3}y$ where y is the height. The volume will be $\pi\big(\frac{2}{3}y\big)^2 dy$; $\rho = 847$ kg/m$^3 = 8300.6$ N/m^3. $W = 8300.6\pi \int_0^{3.6} \big(\frac{2}{3}y\big)^2(18.6 - y)dy = 8300.6\pi \cdot \big[\frac{4}{9}(18.6)\frac{y^3}{3} - \frac{4}{9} \cdot \frac{y^4}{4}\big]_0^{3.6} = 8300.6\pi[109.9] = 912\,236\pi$ J $= 2\,866\,000$ J $= 2.866$ MJ

27. $W = 9800\pi \int_{-2}^0 (16-y^2)(-y)dy = 9800\pi\big[-8y^2 + \frac{y^4}{4}\big]_{-2}^0 = 9800\pi[32 - 4] = 9800\pi[28] = 274400\pi$ J

29. (a) $P = \rho g h = 1000 \cdot 9.8 \cdot 2 = 19{,}600$ N/m^2
(b) $F = PA = 19{,}600 \times 10 \times 7 = 1372000$ N

31. The area of the short side is 7 m$\times$ 2m. So, we get $= 9800 \int_0^2 h(y) \cdot L(y)dy = 9800 \int_0^2 (2 - y) \cdot 7 dy = 9800\big[14y - \frac{7y^2}{2}\big]_0^0 = 9800 \cdot 14 = 137200$ N.

33. $9800 \int_0^2 (2-y)\big(3 - \frac{3}{2}y\big)dy = 9800 \int_0^2 \big(6 - 6y + \frac{3}{2}y^2\big)dy = 9800\big[6y - 3y^2 + \frac{y^3}{2}\big]_0^2 = 9800[4] = 39\,200$ N

35. $r = 1.6$ m; $\rho = 680$ kg/m^3; $\rho g = 680 \times 9.8 = 6664$ N/m^3. Thus, we have $\rho g \int_{-1.6}^0 (1 \cdot 6 + y)(2\sqrt{2.56 - y^2})dy = \rho g \cdot 3.2 \int_{-1.6}^0 \sqrt{2.56 - y^2} - \rho g \int_{-1.6}^0 -2y\sqrt{2.56 - y^2}dy$. As in example 26.43. The left integral will be $\frac{1}{4}$ the area of a circle so $\frac{1}{4} \cdot \pi(1.6)^2 = 2.0106$. The right integral is $\frac{2}{3}(2.56 - y^2)^{3/2}\big|_{-1.6}^0 = 4.096 \cdot \frac{2}{3} = 2.7307$; $\rho g[3.2 \times 2.0106 - 2.7307] = \rho g[3.7032] = 6664(3.7032) = 24\,578$ N.

1. $\bar{y} = \dfrac{1}{5}\displaystyle\int_1^6 4x\,dx = \dfrac{1}{5}2x^2\Big|_1^6 = \dfrac{1}{5}(72-2)$

$\quad = \dfrac{20}{5} = 14$

$f_{\text{rms}} = \sqrt{\dfrac{1}{5}\displaystyle\int_1^6 16x\,dx} = \sqrt{\dfrac{1}{5}\left[\dfrac{16x^3}{3}\right]_1^6}$

$\quad = \sqrt{229.\overline{3}} = 15.1438$

2. $\bar{y} = \dfrac{1}{2}\displaystyle\int_0^2 x^3\,dx = \dfrac{1}{2}\dfrac{x^4}{4}\Big|_0^2 = \dfrac{16}{8} = 2$

$y_{\text{rms}} = \sqrt{\dfrac{1}{2}\displaystyle\int_0^2 x^6\,dx} = \sqrt{\dfrac{1}{2}\dfrac{x^7}{7}\Big|_0^7}$

$\quad = \sqrt{9.142857} = 3.0237$

3. $\bar{y} = \dfrac{1}{2}\displaystyle\int_2^4 (x^2-4)\,dx = \dfrac{1}{2}\left[\dfrac{x^3}{3}-4x\right]_2^4$

$\quad = \dfrac{1}{2}\cdot\dfrac{32}{3} = \dfrac{16}{3}$

$h_{\text{rms}} = \sqrt{\dfrac{1}{2}\displaystyle\int_2^4 (x^4-8x^2+16)\,dx}$

$\quad = \sqrt{\dfrac{1}{2}\left[\dfrac{x^5}{5}-\dfrac{8x^3}{3}+16x\right]_2^4}$

$\quad = \sqrt{40.5\overline{3}} = 6.3666$

4. $\bar{y} = \dfrac{1}{2.5}\displaystyle\int_0^{2.5}(4.9t^2-2.8t-4)\,dt = \dfrac{6.7708\overline{3}}{2.5}$

$\quad = 2.7083$

$k_{\text{rms}} = \sqrt{\dfrac{1}{2.5}\displaystyle\int_0^{2.5}(4.9t^2-2.8t-4)^2\,dt}$

$\quad = \sqrt{\dfrac{147.6432}{2.5}}$

$\quad = \sqrt{59.0573} = 7.6849$

5. $V = \pi\displaystyle\int_1^5 (6x)^2\,dx = 36\pi\cdot\dfrac{x^3}{3}\Big|_1^5$

$\quad = 12\pi[125-1] = 1488\pi$

$M_y = \pi\displaystyle\int_1^5 x(6x)^2 = 36\pi\int_1^5 x^3\,dx = 36\pi\left[\dfrac{x^4}{4}\right]_1^5$

$\quad = 9\pi[625-1] = 5616\pi$

$\bar{x} = \dfrac{5676}{1488} = 3.7742$

The centroid is (3.7742, 0).

6. $y = -x \Rightarrow x = -y.$ Hence,

$V = \pi\displaystyle\int_1^4 (-y)^2\,dy = \pi\dfrac{y^3}{3}\Big|_1^4 = \dfrac{\pi}{3}(64-1) = 2\pi$

$M_x = \pi\displaystyle\int_1^4 \pi y^3\,dy = \pi\dfrac{y^4}{4}\Big|_1^4 = \pi\left(64-\dfrac{1}{4}\right)$

$\quad = 63.75\pi$

$\bar{y} = \dfrac{63.75}{21} = 3.0357$

The centroid is (0, 3.0357).

7. $V = \pi\displaystyle\int_1^2 (x^4)^2\,dx = \pi\dfrac{x^9}{9}\Big|_1^2 = \dfrac{\pi}{9}(512-1)$

$\quad = \dfrac{511\pi}{9}$

$M_y = \pi\displaystyle\int_1^2 x^9\,dx = \pi\dfrac{x^{10}}{10}\Big|_1^2 = \dfrac{\pi}{10}(1024-1)$

$\quad = \dfrac{1023\pi}{10}$

$\bar{x} = \dfrac{1023}{10}\times\dfrac{9}{511} = 1.8018$

The centroid is (1.8018, 0).

8. $V = \pi\displaystyle\int_0^1 (x^{2/3})^2\,dx = \pi\dfrac{3}{7}x^{7/3}\Big|_0^1 = \dfrac{3\pi}{7}$

$M_y = \pi\displaystyle\int_0^1 x(x^{4/3})\,dx = \pi\dfrac{3}{10}x^{10/3} = \dfrac{3\pi}{10}$

$\bar{x} = \dfrac{3}{10}\cdot\dfrac{7}{3} = \dfrac{7}{10}$ or 0.7

The centroid is (0.7, 0).

9. $V = \pi\displaystyle\int_0^1 (y^{2/3}-y^6)\,dy = \pi\left[\dfrac{3}{5}y^{5/3}-\dfrac{y^7}{7}\right]_0^1$

$\quad = \dfrac{16\pi}{35} \approx 0.4571\pi$

$M_x = \pi\displaystyle\int_0^1 (y^{5/3}-y^7)\,dy = \pi\left[\dfrac{3}{8}y^{8/3}-\dfrac{y^8}{3}\right]_0^1$

$\quad = \dfrac{\pi}{4}$

$\bar{y} = \dfrac{1}{4}\cdot\dfrac{35}{16} = \dfrac{35}{64} \approx 0.5469$

The centroid is (0, 0.5469).

10. $V = \pi\displaystyle\int_0^1 (x^{2/3}-x^6)\,dx = \dfrac{16\pi}{35}$

$M_y = \pi\displaystyle\int_0^1 x(x^{2/3}-x^6) = \dfrac{\pi}{4}$

$\bar{x} = \dfrac{1}{4}\cdot\dfrac{35}{16} = \dfrac{35}{64} \approx 0.5469$

The centroid is (0.5469, 0).

11. Here $y = x^2 \Rightarrow x = y^{1/2}$, $y = 9 - x^2 \Rightarrow x = \sqrt{9 - y}$, and $x^2 = 9 - x^2 \Rightarrow 2x^2 = 9$. Hence $x = \pm\frac{3}{\sqrt{2}}$ and $y = \frac{9}{2} = 4.5$.

$$V = \pi \int_0^{4.5} y \, dy + \pi \int_{4.5}^9 (9 - y) dy$$

$$= \pi \frac{y^2}{2} \Big|_0^{4.5} + \pi \left[9y - \frac{y^2}{2} \right]_{4.5}^9 = 20.25\pi$$

$$M_x = \pi \int_0^{4.5} y^2 dy + \pi \int_{4.5}^9 \left(9y - y^2 \right) dy$$

$$= \pi \left(\left[\frac{y^3}{3} \right]_0^{4.5} + \left[\frac{9y^2}{2} - \frac{y^3}{3} \right]_{4.5}^9 \right) = 91.125\pi$$

$$\bar{y} = \frac{91.125}{20.25} = 4.5$$

The centroid is (0, 4.5). (Note: this is the answer you would expect due to the symmetry of the curves.)

12. Curves intersect at (0, 0) and (2, 4). Here $x = y^{1/3}$ and $x = \frac{y}{4}$.

$$V = \pi \int_0^4 \left(5 - \frac{y}{4} \right)^2 - \left(5 - y^{1/3} \right)^2 dy$$

$$= \pi \int_0^4 \left[\left(25 - \frac{5}{2}y + \frac{y^2}{16} \right) \right.$$
$$\left. - \left(25 - 10y^{1/3} + y^{2/3} \right) \right] dy$$

$$= \pi \int_0^4 \left(-\frac{5}{2}y + \frac{y^2}{16} + 10y^{1/3} - y^{2/3} \right) dy$$

$$= \pi \left[\frac{-5y^2}{4} + \frac{y^3}{48} + \frac{15}{2}y^{4/3} - \frac{3}{5}y^{5/3} \right]_0^4$$

$$= \pi[22.9077]$$

$$M_x = \pi \int_0^4 \left(-\frac{5}{2}y^2 + \frac{y^3}{16} + 10y^{4/3} - y^{5/3} \right) dy$$

$$= \pi \left[-\frac{5y^3}{6} + \frac{y^3}{64} + \frac{30}{7}y^{7/3} - \frac{3}{8}y^{8/3} \right]_0^4$$

$$= 44.3980\pi$$

$$\bar{y} = \frac{44.3980}{22.9077} = 1.93812$$

Since it's rotated about $x = 5$, then $\bar{x} = 5$. The centroid is (5, 1.0169).

13. $x = y^{3/2}$; $x' = \frac{3}{2}y^{1/2}$; $x'^2 = \frac{9}{4}y$

$$L = \int_0^4 \sqrt{1 + x'^2} \, dy = \int_0^4 \sqrt{1 + \frac{9}{4}y} \, dy$$

$$= \frac{4}{9} \int_0^4 \frac{9}{4}\sqrt{1 + \frac{9}{4}y} \, dy$$

$$= \frac{4}{9} \cdot \frac{2}{3} \left(1 + \frac{9}{4}y \right)^{3/2} \Big|_0^4 = \frac{8}{27}\left(10^{3/2} - 1 \right)$$

$$= 9.0734$$

14. $y = \left(8x^3 \right)^{1/2}$;

$$y' = \frac{1}{2}\left(8x^3 \right)^{-1/2} \cdot 24x^2 = \frac{24x^{1/2}}{4\sqrt{2}} \cdot 24x^2$$

$$= \frac{6x^{3/2}}{\sqrt{2}} = 3\sqrt{2}x^{3/2}$$

$$L = \int_0^2 \sqrt{1 + \left(3\sqrt{2}x^{1/2} \right)^2} \, dx$$

$$= \int_0^2 \sqrt{1 + 18x} \, dx$$

$$= \frac{1}{18} \cdot \frac{2}{3}(1 + 18x)^{3/2} \Big|_0^2 = 8.2986$$

15. **(a)** Here $y = \frac{x^4}{4} + \frac{1}{8x^2}$ and $y' = x^3 - \frac{1}{4x^3}$.

$$L = \int_1^3 \sqrt{1 + \left(x^3 - \frac{1}{4x^3} \right)^2} \, dx$$

$$= \int_1^3 \sqrt{1 + \left(x^6 - \frac{1}{2} + \frac{1}{16x^6} \right)} \, dx$$

$$= \int_1^3 \sqrt{x^6 + \frac{1}{2} + \frac{1}{16x^6}} \, dx$$

$$= \int_1^3 \left(x^3 + \frac{1}{4x^3} \right) dx$$

$$= x^4 - \frac{1}{4} \cdot \frac{1}{2}x^{-2} \Big|_1^3 = \frac{81}{4}$$

$$- \frac{1}{72} - \frac{1}{4} + \frac{1}{8} = 20.111$$

(b) $S = 2\pi \int_1^3 \left(\frac{x^4}{4} + \frac{1}{8x^2} \right)$

$$\times \sqrt{1 + \left(x^6 - \frac{1}{2} + \frac{1}{16x^2} \right)} \, dx$$

$$= 2\pi \int_1^3 \left(\frac{x^4}{4} + \frac{1}{8x^3} \right)\left(x^3 + \frac{1}{4x^3} \right) dx$$

$$= 2\pi \int_1^3 \left(\frac{x^7}{4} + \frac{x}{6} + \frac{1}{8} + \frac{1}{32x^6} \right) dx$$

$$= 2\pi \left[\frac{x^8}{32} + \frac{x^2}{32} + \frac{x}{8} - \frac{1}{32}\frac{1}{5}x^{-5} \right]_1^3$$

$$= 2\pi[205.5062] = 411.0124\pi$$

16. **(a)** $x = \frac{1}{6}y^6 + \frac{1}{16}y^{-4}$; $x' = y^5 - \frac{1}{4}y^{-5}$; $x'^2 = \left(y^{10} - \frac{1}{2} + \frac{1}{16}y^{-10} \right)$

$$L = \int_1^2 \sqrt{1 + \left(y^{10} - \frac{1}{2} + \frac{1}{16}y^{-10} \right)} \, dy$$

$$= \int_1^2 \left(y^5 + \frac{1}{4}y^{-5} \right) dy$$

$$= \left(\frac{y^6}{6} - \frac{1}{16}y^{-4} \right) \Big|_1^2 = 10.5586$$

$$S = 2\pi \int_1^2 \left(\frac{y^6}{6} + \frac{1}{16y^4}\right)\left(y^5 + \frac{1}{4y^5}\right) dy$$

$$= 2\pi \int_1^2 \left(\frac{y^{11}}{6} + \frac{y}{24} + \frac{y}{16} + \frac{1}{64y^9}\right) dy$$

$$= 2\pi \left[\frac{y^{12}}{72} + \frac{y^2}{48} + \frac{y^2}{32} - \frac{1}{512y^8}\right]_1^2$$

$$= 2\pi[57.08332] = 114.0664\pi$$

17. $F = kx$; $5 = k(0.2)$; $k = 25$; $W = \int_0^{0.6} 25x\, dx = \frac{25x^2}{2}\big|_0^{0.6} = 4.5\ \text{N}\cdot\text{m} = 4.5\ \text{J}$

18. $V = \frac{1}{60\times 10^{-6}} \int_0^{0.001} 0.40\, dt = \frac{1}{60\times10^{-6}} 0.40t\,\big|_0^{0.001} = \frac{1}{60\times10^{-6}}(0.004) = 6.6667\ \text{V}$

19. $I_{avg} = \frac{1}{2}\int_0^2 \left(4t - t^3\right) dt = \frac{1}{2}\left[2t^2 - \frac{t^4}{4}\right]_0^2 = \frac{1}{2}[8 - 4] = 2\ \text{A}$

20. $a(t) = -32$; $v(t) = -32t$; $s(t) = -16t^2 + 555$; $s(t) = 0 \Rightarrow 16t^2 = 555 \Rightarrow t = 5.8896$.

$$\overline{s} = \frac{1}{5.8896}\int_0^{5.8896}\left(-16t^2 + 555\right) dt$$

$$= \frac{1}{5.8896}\left[\frac{-16t^3}{3} + 555t\right]_0^{5.8896}$$

$$\overline{s} = 370\ \text{ft}$$

$$\overline{v} = \frac{1}{5.8896}\int_0^{5.8896}(-32t)dt$$

$$= \frac{1}{5.8896}\left[-16t^2\right]_0^{5.8896} = -94.2338\text{ft/s}$$

21. $W = \int_0^{30}[1000 + 5(50 - x)]dx$

$$= \int_0^{30}(1250 - 5x)dx$$

$$= \left[1250x - \frac{5x^2}{2}\right]_0^{30} = 35250\ \text{ft}\cdot\text{lb}$$

22. $h(y) = (100 - y)$; $L(y) = 200 + 2y$

$$P = 62.4\int_0^{100}(100 - y)(200 + 2y)dy$$

$$= 62.4\int_0^{100}\left(20000 - 2y^2\right)dy$$

$$= 64.2\left[20000y - \frac{2}{3}y^3\right]_0^{100} = 62.4\left(1.333\times10^6\right)$$

$$= 8.32\times10^7\ \text{lb} = 4.16\times10^4\ \text{tons}$$

23. The top of the tank is at $x = 0$ and the bottom at $x = 16$ $W = 880\int_0^{16}36\pi\cdot x\, dx = 31680\pi\frac{x^2}{2}\big|_0^{16} = 4\,055\,040\pi\ \text{kg}\cdot\text{m} = 39\,739\,392\pi\ \text{J}$

24. $62.4\pi\int_0^{16}\left(6 - \frac{3x}{8}\right)^2 x\, dx$

$$= 62.4\pi\int_0^{10}\left(36 - \frac{9}{2}x + \frac{9}{64}x^2\right)x\, dx$$

$$= 62.4\pi\int_0^{10}\left(36x - \frac{9}{2}x^2 + \frac{9}{64}x^3\right) dx$$

$$= 62.4\pi\left[18x^2 - \frac{3}{2}x^3 + \frac{9}{256}x^4\right]_0^{16}$$

$$= 62.4\pi(768) = 4.79232\times10^4\pi\ \text{ft}\cdot\text{lb}$$

CHAPTER 26 TEST

1. $\overline{y} = \frac{1}{5-1}\int_1^5 \left(3x^2 + 1\right) dx = \frac{1}{4}\left(x^3 + x\right)\big|_1^5 = \frac{1}{4}(130 - 2) = 32$

2. $g_{rms} = \sqrt{\frac{1}{3-1}\int_1^3\left(x^3 - 1\right)^2 dx} = \sqrt{\frac{1}{2}\int_1^3\left(x^6 - 2x^3 + 1\right)dx} = \sqrt{\frac{1}{2}\left[\frac{1}{7}x^7 - \frac{1}{2}x^4 + x\right]_1^3} \approx \sqrt{137.1429} \approx 11.71$

3. $V = \pi\int_0^7\left[f(x)\right]^2 dx = \pi\int_0^7\sqrt{9x + 1}^2 dx = \pi\int_0^7(9x + 1)dx = \pi\left[\frac{9}{2}x^2 + x\right]_0^7 = 227.5\pi$

4. We begin by determining the moment: $M_y = \pi\int_0^4 \rho x\left(x^2 + 1\right)^2 dx = \rho\pi\int_0^4\left(x^5 + 2x^3 + x\right)dx = \rho\pi\left[\frac{1}{6}x^6 + \frac{1}{2}x^4 + \frac{1}{2}x^2\right]_0^4 = \frac{2456}{3}\rho\pi$. Now we find the mass, $m = \rho V$, from $V == \pi\int_0^4\left(x^2 + 1\right)^2 dx = \pi\int_0^4\left(x^4 + 2x^2 + 1\right)dx = \pi\left[\frac{1}{5}x^5 + \frac{2}{3}x^3 + x\right]_0^4 = \frac{7544}{30}\pi$. so, $m = \frac{7544}{30}\rho\pi$ and $\overline{x} = \frac{M_y}{m} = $

$\left(\frac{2456}{3}\rho\pi\right)/\left(\frac{7544}{30}\rho\pi\right) = \frac{24560}{7544} \approx 3.26$. Thus, the centroid is at $(3.26, 0)$

5. Since $y' = \frac{3}{2}x^{1/2}$, the arc length

$L = \int_1^6\sqrt{1 + \left(\frac{3}{2}x^{1/2}\right)^2}\, dx = \int_1^6\sqrt{1 + \frac{9}{4}x}\, dx = \frac{1}{2}\int_1^6\sqrt{4 + 9x}\, dx$. Let $u = 4 + 9x$ and then $du = 9dx$, and if $x = 1$, $u = 13$ and if $x = 6$, $u = 58$, so $L = \frac{1}{18}\int_{13}^{58}u^{1/2}du = \frac{1}{27}u^{3/2}]_{13}^{58} \approx 14.6238$.

6. Here $y' = \frac{1}{2}x^2$ and so, the surface area $S = \int_1^3 2\pi\left(\frac{1}{6}x^3\right)\sqrt{1 + \left(\frac{1}{2}x^2\right)^2}\, dx = \frac{\pi}{3}\int_1^3 x^3\sqrt{1 + \frac{1}{4}x^4}dx = \frac{4\pi}{9}\left(1 + \frac{1}{4}x^4\right)^{3/2}\big|_1^3 \approx 42.9156\pi \approx 134.82$

7. $W = \int_0^{10} 25\pi\rho(y+4)dy = 25\pi\rho \int_0^{10}(y+4)dy =$
$25\pi\rho\left[\frac{1}{2}y^2 + 4y\right]_0^{10} = 25\pi\rho(50+40) = 2250\pi\rho =$
$2250\ 000\pi$ J

8. $F = \rho\int_5^8 y2(8-y)dy = \rho\int_5^8\left(16y - 2y^2\right)dy =$
$\rho\left[8y^2 - \frac{2}{3}y^3\right]_5^8 \approx 54\rho.$ Since the density of water
is $\rho = 62.5$ lb, we have a force of $F = 54(62.5) =$
$3375.$

CHAPTER
27
Derivatives of Transcendental Functions

≡ 27.1 DERIVATIVES OF THE SINE AND COSINE FUNCTIONS

1. $y = \sin 3x$; $y' = 3\cos 3x$

3. $y = 3\cos 2x$; $y' = 3(-\sin 2x) \cdot 2 = -6\sin 2x$

5. $y = \sin(x^2 + 1)$; $y' = 2x\cos(x^2 + 1)$

7. $y = 4\sin^2 3x = 4(\sin 3x)^2$; $y' = 4 \cdot 2(\sin 3x) \times (\cos 3x)(3) = 24\sin 3x\cos 3x$

9. $y = \cos(3x^2 - 2)$; $y' = -6x\sin(3x^2 - 2)$

11. $y = \sin\sqrt{x} = \sin x^{1/2}$; $y' = \cos x^{1/2} \cdot \frac{1}{2}x^{-1/2} = \frac{\cos\sqrt{x}}{2\sqrt{x}}$

13. $y = \cos\sqrt{2x^3 - 4}$; $y' = -\sin\sqrt{2x^3 - 4} \times \left(\frac{1}{2}(2x^3 - 4)^{-1/2}(6x^2)\right) = \frac{-3x^2\sin\sqrt{2x^3-4}}{\sqrt{2x^3-4}}$

15. $y = x^2 + \sin^2 x$; $y' = 2x + 2\sin x\cos x$

17. $y = \sin x\cos x$; $y' = \sin x(-\sin x) + \cos x\cos x = \cos^2 x - \sin^2 x = \cos 2x$

19. $y = \dfrac{2\cos x}{\sin 2x}$;

$y' = \dfrac{\sin 2x(-2\sin x) - 2\cos x\cos 2x \cdot 2}{\sin^2 2x}$

$= \dfrac{-2\sin x\sin 2x - 4\cos x\cos 2x}{\sin^2 2x}$

$= \dfrac{-2\sin x(2\sin x\cos x) - 4\cos x(2\cos^2 x - 1)}{(2\sin x\cos x)^2}$

$= \dfrac{-4\cos x(\sin^2 x + 2\cos^2 x - 1)}{4\sin^2 x\cos^2 x}$

$= \dfrac{-4\cos x(\sin^2 x + \cos^2 x + \cos^2 x - 1)}{4\sin^2 x\cos^2 x}$

$= \dfrac{-4\cos x(\cos^2 x)}{4\sin^2 x\cos^2 x} = \dfrac{-\cos x}{\sin^2 x}$

21. $y = x^2\sin x$; $y' = 2x\sin x + x^2\cos x$

23. $y = \sqrt{x}\sin x$; $y' = \frac{1}{2}x^{-1/2}\sin x + \sqrt{x}\cos x$

25. $y = (\sin 2x)(\cos 3x)$; $y' = \sin 2x(-3\sin 3x) + 2\cos 2x\cos 3x = 2\cos 2x\cos 3x - 3\sin 2x\sin 3x$

27. $y = \sin^3(x^4)$; $y' = 3\sin^2(x^4)\cos(x^4) \cdot (4x^3) = 12x^3\sin^2(x^4)\cos(x^4)$

29. $y = \sin^2 x + \cos^2 x = 1$; $y' = 0$

31. $f(x) = \sin x$; $f'(x) = \cos x$; $f''(x) = -\sin x$

33. $y = \sin x$; $y' = \cos x$; $y'' = -\sin x$; $y''' = -\cos x$

35. $y = \sin x$; $y' = \cos x$; $y'' = -\sin x$; and $y''' = -\cos x$; $y^{(4)} = \sin x$, so $\frac{d^4}{dx^4}(\sin x) = \sin x$

37. (a) When $x = \frac{\pi}{4}$, then $f(x) = \sqrt{2}\cos\left(\frac{\pi}{4}\right) = \sqrt{2}\left(\frac{\sqrt{2}}{2}\right) = 1$. The line is tangent at the point $\left(\frac{\pi}{4}, 1\right)$. The slope of the tangent line at this point is $f'\left(\frac{\pi}{4}\right)$. Since $f'(x) = -\sqrt{2}\sin x$, the slope of the tangent line is $-\sqrt{2}\sin\left(\frac{\pi}{4}\right) = -1$. Using the slope-intercept form for the equation of a line, the equation of the tangent line is
$y - y_0 = m(x - x_0)$
$y - 1 = -1\left(x - \frac{\pi}{4}\right)$
$\qquad = -x + \frac{\pi}{4}$
$\qquad y = -x + \frac{\pi}{4}$

(b)

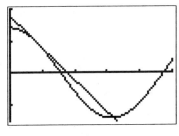

$[0, 5, 1] \times [-1.5, 2, 1]$

39. **(a)** The slope of the tangent line to h at this point is $h'\left(\frac{\pi}{2}\right)$. Since $h'(x) = 4\cos(4x)$, we see that $h'\left(\frac{\pi}{2}\right) = 4\cos(2\pi) = 4$. Thus, the slope of the normal line to h at this point is $-\frac{1}{4}$ and its equation is $y - y_0 = m(x - x_0)$

$$y - 0 = -\frac{1}{4}\left(x - \frac{\pi}{2}\right)$$
$$y = -\frac{1}{4}x + \frac{\pi}{8}$$

(b)

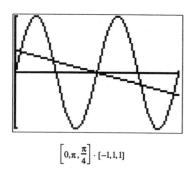

$$\left[0, \pi, \frac{\pi}{4}\right] \cdot [-1, 1, 1]$$

41. As in Example 27.7, $V' = 45{,}240\cos\left[377(1.571)\right]$ $= -3470.332$ or about -3470 V.

43. The current to the capacitor is given by $I = q'(t) = -0.25\sin(t - 1.45)$. When $t = 7.6$ s, the current is $I = q'(7.6) = -0.25\sin(7.6 - 1.45) = -0.25\sin(6.15) = 0.033$ A.

45. $I'(\theta) = -2M\cos(\theta)\sin(\theta)$; $I'\left(\frac{\pi}{3}\right) = -\frac{\sqrt{3}}{2}M \approx -0.866M$ cd/rad

47. To find the maximum and minimum, we first find the critical values. $I'(t) = 2\cos t - \sin t = 0$ or $2\cos t = \sin t$ and so, $\frac{\sin t}{\cos t} = \tan t = 2$. This occurs when $t \approx 1.107$ and $t \approx 4.249$. The maximum is at $I(1.107) = 2\sin(1.107) + \cos(1.107) \approx 2.24$ A. The minimum is $I(4.249) = 2\sin(4.249) + \cos(4.249) \approx -2.24$ A.

≡ 27.2 DERIVATIVES OF THE OTHER TRIGONOMETRIC FUNCTIONS

1. $y = \tan^2\sqrt{x}$; $y' = 2\tan\sqrt{x} \cdot \sec^2\sqrt{x} \cdot \frac{1}{2}x^{-1/2} = \frac{\tan\sqrt{x}\sec^2\sqrt{x}}{\sqrt{x}}$

3. $y = \sec 5x$; $y' = 5\sec 5x\tan 5x$

5. $y = \csc(2x - 1)$; $y' = -2\csc(2x - 1)\cot(2x - 1)$

7. $y = \sin x\tan x$; $y' = \sin x\sec^2 x + \tan x\cos x = \sin x\sec^2 x + \sin x = \sin x\left(\sec^2 x + 1\right)$

9. $y = \tan^3 x$; $y' = 3\tan^2 x \cdot \sec^2 x$

11. $y = \sec^4\left(x^2\right)$; $y' = 4\sec^3\left(x^2\right)\left(\sec\left(x^2\right)\tan\left(x^2\right)\right)(2x) = 8x\sec^4\left(x^2\right)\tan\left(x^2\right)$

13. $y = \tan x\cot x$; $y' = \tan x\left(-\csc^2 x\right) + \cot x \times \left(\sec^2 x\right) = \sec^2 x\cot x - \csc^2 x\tan x$

15. $y = \sin^2 x\cot x$; $y' = 2\sin x\cos x\cot x + \sin^2 x \times \left(-\csc^2 x\right) = 2\sin x\cos x \cdot \frac{\cos x}{\sin x} - \sin^2 x \cdot \frac{1}{\sin^2 x} = 2\cos^2 x - 1 = \cos 2x$

17. $y = \tan\frac{1}{x}$; $y' = \sec^2\frac{1}{x} \cdot \left(-x^{-2}\right) = -\frac{1}{x^2}\sec^2\frac{1}{x}$

19. $y = \sqrt{1 + \tan x^2} = \left(1 + \tan x^2\right)^{1/2}$; $y' = \frac{1}{2}\left(1 + \tan x^2\right)^{-1/2}\left(\sec^2 x^2\right)(2x) = \frac{x\sec^2 x^2}{\sqrt{1 + \tan x^2}}$

21. $y = \frac{\tan x}{1 + \sec x}$;

$$y' = \frac{(1 + \sec x)\sec^2 x - \tan x(\sec x\tan x)}{(1 + \sec x)^2}$$
$$= \frac{\sec^2 x + \sec^3 x - \sec x\tan^2 x}{(1 + \sec x)^2}$$
$$= \frac{\sec x\left[\sec x + \sec^2 x - \tan^2 x\right]}{(1 + \sec x)^2}$$
$$= \frac{\sec x[\sec x + 1]}{(1 + \sec x)^2} = \frac{\sec x}{1 + \sec x}$$

23. Here $y = \frac{\cot x}{1 - \csc x}$, so $y' =$

$$\frac{(1 - \csc x)\left(-\csc^2 x\right) - \cot x(\csc x\cot x)}{(1 - \csc x)^2} =$$
$$\frac{-\csc^2 x + \csc^3 x - \csc x\cot^2 x}{(1 - \csc x)^2} = \frac{\csc x\left(\csc^2 x - \cot^2 x - \csc x\right)}{(1 - \csc x)^2} =$$
$$\frac{\csc x(1 - \csc x)}{(1 - \csc x)^2} = \frac{\csc x}{1 - \csc x}$$

25. $y = (\csc x + 2\tan x)^3$; $y' = 3(\csc x + 2\tan x)^2 \times \left(-\csc x\cot x + 2\sec^2 x\right) = 3(\csc x + 2\tan x)^2 \times \left(2\sec^2 x - \csc x\cot x\right)$

27. $y = (\tan 2x)^{3/5}$; $y' = \frac{3}{5}(\tan 2x)^{-2/5}\left(\sec^2 2x\right)(2) = \frac{6}{5}\sec^2 2x(\tan 2x)^{-2/5}$

29. Here $y = \frac{\sec x}{1+\tan x}$, and so $y' =$

$\frac{(1+\tan x)(\sec x \tan x)-\sec^3 x}{(1+\tan x)^2} = \frac{\sec x \left(\tan x + \tan^2 x - \sec^2 x \right)}{(1+\tan x)^2} =$

$\frac{\sec x(\tan x - 1)}{(1+\tan x)^2}$

31. $\frac{d}{dx} \cot u = \frac{d}{dx}\left(\frac{\cos u}{\sin u} \right) =$

$\frac{\sin u \frac{d}{dx}\cos u - \cos u \frac{d}{dx}\sin u}{\sin^2 u} =$

$\frac{\sin u(-\sin u)\frac{du}{dx} - \cos u \cos u \frac{du}{dx}}{\sin^2 u} =$

$\frac{\left(-\sin^2 u - \cos^2 u \right)\frac{du}{dx}}{\sin^2 u} = -\csc^2 u \frac{du}{dx}$

33. $y = \tan 2x$; $y' = 2\sec^2 2x = 2(\sec 2x)^2$; $y'' = 4(\sec 2x)(\sec 2x \tan 2x)2 = 8 \sec^2 2x \tan 2x$

35. $y = x \tan x$; $y' = x\left(\sec^2 x \right) + \tan x$; $y'' = x(2 \sec x \cdot \sec x \tan x) + \sec^2 x + \sec^2 x = 2x \sec^2 x \tan x + 2 \sec^2 x = 2 \sec^2 x(x \tan x + 1)$

37. Since $y = x \sin y$, then $y' = x(\cos y)y' + \sin y$, or $y' - x(\cos y)y' = \sin y$, or $y'(1 - x \cos y) = \sin y$ and so $y' = \frac{\sin y}{1-x \cos y}$.

39. Here $x + y = \sin(x+y)$, and differentiating, we obtain $1 + y' = \cos(x+y)[1+y'] = \cos(x+y) + y' \cos(x+y)$, or $y' - y' \cos(x+y) = \cos(x+y) - 1$, and so $y' = \frac{\cos(x+y)-1}{1-\cos(x+y)}$ or $\frac{1-\cos(x+y)}{\cos(x+y)-1}$.

41. $n' = -\csc \theta \cot \theta = -\csc \left(\frac{\pi}{8} \right) \cot \left(\frac{\pi}{8} \right) = -6.31$

43. Differentiating $I = 16 \cot^2(2\theta)$, produces
$I' = 2\left[16 \cot(2\theta) \right]\left[-\csc^2(2\theta) \right](2)$
$= -64 \cot(2\theta) \csc^2(2\theta)$
When $\theta = \frac{7\pi}{16}$, then $I'\left(\frac{7\pi}{16} \right) =$
$-64 \cot \left[2\left(\frac{7\pi}{16} \right) \right] \csc^2 \left[2\left(\frac{7\pi}{16} \right) \right] =$
$-64 \cot \left(\frac{7\pi}{8} \right) \csc^2 \left(\frac{7\pi}{8} \right) \approx 1055.05$ mV/rad.

≡ 27.3 DERIVATIVES OF INVERSE TRIGONOMETRIC FUNCTIONS

1. $y = \sin^{-1} 2x$, so $y' = \frac{2}{\sqrt{1-4x^2}}$.

3. $y = \tan^{-1}\frac{x}{2}$, so $y' = \frac{\frac{1}{2}}{1 + \frac{x^2}{4}} = \frac{2}{4+x^2}$.

5. $y = \cos^{-1}\left(1 - x^2 \right)$ and $y' = \frac{2x}{\sqrt{1-\left(1-x^2\right)^2}} = \frac{2x}{\sqrt{2x^2-x^4}}$

7. $y = \sin^{-1}(1-2x)$, so $y' = \frac{-2}{\sqrt{1-(1-2x)^2}} = \frac{-2}{\sqrt{4x-4x^2}} = \frac{-1}{\sqrt{x-x^2}}$

9. $y = \cos^{-1}\left(x^3 - x \right)$; $y' = \frac{1-3x^2}{\sqrt{1-\left(x^3-x\right)^2}} = \frac{1-3x^2}{\sqrt{1-x^2+6x^4-x^6}}$

11. $y = \sec^{-1}(4x+2)$; $y' = \frac{4}{(4x+2)\sqrt{(4x+2)^2-1}} = \frac{2}{(2x+1)\sqrt{(4x+2)^2-1}}$

13. $y = x \sin^{-1} x$; $y' = x \cdot \frac{1}{\sqrt{1-x^2}} + \sin^{-1} x$

15. $y = x^2 \cos^{-1} x$; $y' = 2x \cos^{-1} x - \frac{x^2}{\sqrt{1-x^2}}$

17. Since $y = \tan^{-1}\left(\frac{2x-1}{2x} \right)$, then $y' = \frac{1}{1 + \left(\frac{2x-1}{2x} \right)^2} \times$

$\left(\frac{2x \cdot 2 - (2x-1)2}{4x^2} \right) = \frac{1}{\frac{4x^2+4x^2-4x+1}{4x^2}}\left(\frac{4x-4x+2}{4x^2} \right) =$

$\frac{2}{8x^2-4x+1}$

19. $y = x \tan^{-1}(x+1)$; $y' = \tan^{-1}(x+1) + \frac{x}{1+(x+1)^2}$

21. Here we are given $y = \sqrt{\sin^{-1}\left(1 - x^2 \right)}$, and so

$y' = \frac{1}{2}\left(\sin^{-1}\left(1 - x^2 \right) \right)^{-1/2}\left(\frac{-2x}{\sqrt{1-\left(1-x^2\right)^2}} \right) =$

$\frac{-x}{\sqrt{\sin^{-1}\left(1-x^2 \right)}\sqrt{1-\left(1-x^2\right)^2}} = \frac{-x}{\sqrt{\sin^{-1}\left(1-x^2 \right)}\sqrt{2x^2-x^4}}$

$= \frac{-1}{\sqrt{\sin^{-1}\left(1-x^2 \right)}\sqrt{2-x^2}}$.

23. $y = x \cot^{-1}\left(1 + x^2 \right)$; $y' = \cot^{-1}\left(1 + x^2 \right) + x \frac{-1}{1+\left(1+x^2\right)^2} \cdot 2x = \cot^{-1}\left(1 + x^2 \right) - \frac{2x^2}{2+2x^2+x^4}$

25. Since $y = \frac{\sin^{-1} x}{x}$, then $y' = \frac{x\left(\frac{1}{\sqrt{1-x^2}} \right) - \sin^{-1} x}{x^2}$

$= \frac{1}{x\sqrt{1-x^2}} - \frac{\sin^{-1} x}{x^2} = \frac{1}{x}\left(\frac{1}{\sqrt{1-x^2}} - \frac{\sin^{-1} x}{x} \right)$.

27. $y = (\arcsin x)^{1/3}$; $y' = \frac{1}{3}(\arcsin x)^{-2/3}\left(\frac{1}{\sqrt{1-x^2}}\right)$

$= \dfrac{1}{3\sqrt{1-x^2}(\arcsin x)^{2/3}}$

29. Let $y = \cos^{-1} u$ or $u = \cos y$. Hence $\frac{du}{dx} = \frac{d}{dx}\cos y = -\sin y \frac{dy}{dx}$ so $\frac{dy}{dx} = \frac{-1}{\sin y}\frac{du}{dx}$. Since $\sin^2 y + \cos^2 y = 1$ we have $\sin y = \sqrt{1-\cos^2 y} = \sqrt{1-u^2}$, and so $\frac{dy}{dx} = \frac{d}{dx}\cos^{-1} u = \frac{-1}{\sqrt{1-u^2}}\frac{du}{dx}$.

31. In this problem, let P be a point on the track directly in front of the camera and x the distance

the car has traveled past P, and θ the angle from P to the camera to x. Then $\theta = \tan^{-1}\left(\frac{x}{42}\right)$ and $\frac{d\theta}{dt} = \dfrac{1}{1+\left(\frac{x}{42}\right)^2}\frac{1}{42}\frac{dx}{dt} = \frac{42}{42^2+x^2}\frac{dx}{dt}$. When $\theta = 15°$, then $\theta = \tan^{-1}\left(\frac{x}{42}\right)$ or $\tan\theta = \frac{x}{42}$ and $x = 42\tan 15° \approx 11.25387$. Substituting this value of x and the given information that $\frac{dx}{dt} = 320$ ft/s in the formula for $\frac{d\theta}{dt}$ produces $\frac{d\theta}{dt} = \frac{42}{42^2+(11.25387)^2}(320) \approx 7.1087$ rad/s.

≡ 27.4 APPLICATIONS

1. If $y = \sin x$, then $y' = \cos x$. Since $\cos x > 0$ on $\left(-\frac{\pi}{2}, \frac{\pi}{2}\right)$, $\sin x$ is increasing on $\left[-\frac{\pi}{2}, \frac{\pi}{2}\right]$.

3. If $y = \arccos x$, then $y' = \frac{-1}{\sqrt{1-x^2}}$ when $|x| < 1$. Hence $y' < 0$ which means that the arccos x is decreasing for $|x| \le 1$.

5. $f(x) = 2\sin x + \cos 2x$; $f'(x) = 2\cos x - 2\sin 2x$. The critical values occur when $f'(x) = 0$. Hence, we have $2\cos x - 2\sin 2x = 0$; $2\cos x - 4\sin x\cos x = 0$. Factoring, we obtain $2\cos x(1 - 2\sin x) = 0$. If $\cos x = 0 \Rightarrow x = \frac{\pi}{2}, \frac{3\pi}{2}$ and if $1 - 2\sin x = 0 \Rightarrow \sin x = \frac{1}{2} \Rightarrow x = \frac{\pi}{6}, \frac{5\pi}{6}$. Differentiating again, we get $f''(x) = -2\sin x - 4\cos 2x = 0$, or $-2(\sin x + 2(1 - 2\sin^2 x)) = 0$, or $-2(\sin x + 2 - 4\sin^2 x) = 0$, and so $2(4\sin^2 x - \sin x - 2) = 0$. The quadratic formula yields $\sin x = -0.59307$ or $\sin x = 0.84307$. Hence, $x = 3.7765$, $x = 5.6483$, $x = 1.0030$, or $x = 2.1386$. Maxima are at $\left(\frac{\pi}{6}, 1.5\right)$, $\left(\frac{5\pi}{6}, 1.5\right)$, minima are at $\left(\frac{\pi}{2}, 1\right)$ and $\left(\frac{3\pi}{2}, -3\right)$. Inflection points are at $(1.0030, 1.2646)$, $(2.1386, 1.2646)$, $(3.7765, -0.8896)$ and $(5.6483, -0.8896)$.

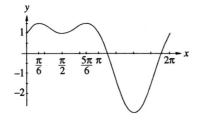

7. $h(x) = \cos x - \sin x$. To find the critical values, we take the derivative and set it equal to 0. Thus, $h'(x) = -\sin x - \cos x = 0$ and so, $-\sin x = \cos x$, or $\tan x = -1$, which means that $x = \frac{3\pi}{4}$ or $\frac{7\pi}{4}$. The second derivative is $h''(x) = -\cos x + \sin x$ and setting it equal to 0 produces $\sin x = \cos x$ or $\tan x = 1$. Thus, $x = \frac{\pi}{4}$ or $x = \frac{5\pi}{4}$. As a result, the maxima are $\left(\frac{7\pi}{4}, \sqrt{3}\right)$; the minimum is $\left(\frac{3\pi}{4}, -\sqrt{3}\right)$, and the inflection points are $\left(\frac{\pi}{4}, 0\right)$ and $\left(\frac{5\pi}{4}, 0\right)$.

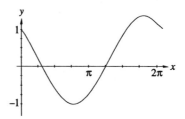

9. $k(x) = \cos^2 x + \sin x$. To find the critical values, we take the derivative and set it equal to 0. $k'(x) = -2\cos x\sin x + \cos x = \cos x(-2\sin x + 1) = 0$. This yields $\cos = 0$ which means that there are critical values at $x = \frac{\pi}{2}$ and $x = \frac{3\pi}{2}$. We also get $\sin x = \frac{1}{2}$, and so there are additional critical values at $x = \frac{\pi}{6}$ and $x = \frac{5\pi}{6}$. For inflection points, we take the second derivatives and set them equal to 0: $k''(x) = \cos x(-2\cos x) - \sin x(-2\sin x + 1) = -2\cos^2 x + 2\sin^2 x - \sin x = -2(1 - 2\sin^2 x) + 2\sin^2 x - \sin x = 6\sin^2 - \sin x - 2 = (3\sin x - 2)(2\sin x + 1) = 0$. Hence, $\sin x = \frac{2}{3} \Rightarrow x = 0.7297$ or $x = 2.4119$, or $\sin x = -\frac{1}{2} \Rightarrow x = \frac{7\pi}{6}$

or $\frac{11\pi}{6}$. Thus, we see that this function has the following maxima: $\left(\frac{\pi}{6},\ 1.25\right)$ and $\left(\frac{5\pi}{6},\ 1.25\right)$; minima: $\left(\frac{\pi}{2},\ 1\right)$ and $\left(\frac{3\pi}{2},\ -1\right)$; and inflection points: $(0.7297,\ 1.2222)$, $(2.4119,\ 1.2222)$, $\left(\frac{7\pi}{6},\ 0\right)$ and $\left(\frac{11\pi}{6},\ 0\right)$.

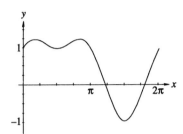

11. $y = \sin x$, at $x = \frac{\pi}{6}$, we get $y = \sin\frac{\pi}{6} = 0.5$. Thus, we want to find the line tangent at this point $\left(\frac{\pi}{6},\ 0.5\right)$. To find the slope of the tangent line we take the derivative $y' = \cos x$ and evaluate it when $x = \frac{\pi}{6}$ to get $\cos\frac{\pi}{6} = \frac{\sqrt{3}}{2}$. Using the point-slope form for the equation of the line, we obtain the desired equation: $y - 0.5 = \frac{\sqrt{3}}{2}\left(x - \frac{\pi}{6}\right)$.

13. $y = \tan^3 x$; at $x = \frac{\pi}{4}$, we get $y = 1$, so point is $\left(\frac{\pi}{4},\ 1\right)$. To find the slope, we take the derivative $y' = 3\tan x\left(\sec^2 x\right)$ and at $\frac{\pi}{4}$, we get $y' = 3 \cdot 1^2(\sqrt{2})^2 = 6$. Hence, the equation of the tangent line is $y - 1 = 6\left(x - \frac{\pi}{4}\right)$.

15. The horizontal distance from the light to the plane is 8000 ft. The velocity is $\frac{dx}{dt} = 400$ mi/h $= 400 \cdot \frac{\text{mile}}{\text{hr}} \cdot \frac{5280\ \text{ft}}{1\ \text{mi}} \cdot \frac{1\ \text{hr}}{3600\ \text{s}} = 586.7$ ft/s; $\tan\theta = \frac{x}{6000}$. Taking derivatives we have $\sec^2\theta\frac{d\theta}{dt} = \frac{1}{6000}\frac{dx}{dt}$. Also, $\tan\theta = \frac{8000}{6000} = \frac{4}{3}$, and so $\theta = 0.9273$ rad. Hence, $\frac{d\theta}{dt} = \frac{dx/dt}{6000\sec^2\theta} = \frac{(dx/dt)\cos^2\theta}{6000} = \frac{(586.7)\cos^2(0.9273)}{6000} = 0.0352$ rad/s.

17. $\frac{d\theta}{dt} = 1$ rpm $= 2\pi$ rad/min $= 0.10472$ rad/s . $\tan\theta = \frac{x}{500}$. Taking derivatives we have $\sec^2\theta\frac{d\theta}{dt} = \frac{1}{500}\frac{dx}{dt}$ or $\frac{dx}{dt} = \sec^2\theta \cdot \frac{d\theta}{dt} \cdot 500 = \left(\sec^2\frac{\pi}{4}\right)(0.10472)(500) = 104.72$ m/s.

19. $\frac{d\theta}{dt} - 30$ rpm $= 3.1416$ rad/s. We know that $\tan\theta = \frac{x}{8}$, and taking derivatives of both sides we have $\sec^2\theta\frac{d\theta}{dt} = \frac{1}{8}\frac{dx}{dt}$. Hence, $\frac{dx}{dt} = \left(8\sec^2 40°\right) \times (3.1416) = 42.8284$ mi/s.

21. (a) $\overline{Q} = 1.33(2)^{5/2}\tan\frac{\theta}{2} = 7.523\tan\frac{\theta}{2}$; $\overline{Q}' = \frac{1}{2}(7.5236)\sec^2\frac{\theta}{2}$
(b) $Q'(30) = 3.7618\sec^2 15° = 3.7518(1.0353)^2 = 4.0319$ m^3/s
(c) $\overline{Q}(40) = 3.7168\sec^2 20° = 4.2602$
(d) $\overline{Q}' = k\sec^2\frac{\theta}{2}$; critical values at $\frac{\theta}{2} = 90°$ or $\theta = 180°$. $\overline{Q}$ is increasing so maximum at $\theta = 180°$.

23. $I = \sin^2 3t$; $I' = 6\sin 3t\cos 3t$; $I'(1.5) = 6(\sin 4.5)(\cos 4.5) = 1.2364$ A/s

25. $x(t) = \frac{1}{4}\cos 2t$; $v(t) = -\frac{1}{2}\sin 2t$; $a(t) = v(t) = -\cos 2t$; $a(t) = 0$ when $2t = \frac{\pi}{2}+2k\pi$ or $\frac{3\pi}{2}+2k\pi$. By the first derivative test the maximums occur at $2t+\frac{3\pi}{2}+2k\pi$ or $t = \frac{3\pi}{4} + k\pi$ or $t = \frac{3\pi+4k\pi}{4}$, $k = 0,\ 1,\ 2,\ 3,\dots$. Thus, $v\left(\frac{3\pi}{4}\right) = -\frac{1}{2}\sin\left(\frac{3\pi}{2}\right) = -\frac{1}{2}(-1) = \frac{1}{2}$ m/s $= 0.5$ m/s.

27. $f(x) = \cos x + x - 2$; $f'(x) = -\sin x + 1$. Use first $x = 3$.

n	x_n	$f(x_n)$	x_{n+1}	$f(x_{n+1})$
1	3	0.0100	2.98835	6.716×10^{-5}
2	2.98835	6.716×10^{-5}	2.9883	3.105×10^{-9}

29. $g(x) = \sin x + \frac{x^2}{4} - 2x$; $g'(x) = \cos x + \frac{3}{4}x^2 - 2$; $x = 0$ is one answer. By Newton's method $x = 2.71995214$. Since $g(x)$ is odd, $x = -2.71995214$ is also an answer.

≡ 27.5 DERIVATIVES OF LOGARITHMIC FUNCTIONS

1. Since $y = \log 5x$, its derivative is $y' = \frac{1}{5x} \cdot \log e \cdot 5 = \frac{1}{x} \cdot \frac{1}{\ln 10}$.

3. $y = \ln\left(x^2 + 4x\right)$; $y' = \frac{2x+4}{x^2+4x}$

5. $y = \sqrt{\ln\left(x^2 + 4x\right)}$; $y' = \frac{1}{2}\left(\ln\left(x^2 + 4x\right)\right)^{-1/2} \cdot \frac{2x+4}{x^2+4x} = \frac{x+2}{\sqrt{\ln\left(x^2+4x\right)} \cdot \left(x^2+4x\right)}$

7. Here $y = \ln\frac{1}{x}$ and its derivative is $y' = \frac{1}{1/x} \cdot$
$\left(-1x^{-2}\right) = -\frac{1}{x}$. By an alternate method: $y = \ln x^{-1} = \ln x$, so $y' = -1\frac{1}{x} = \frac{-1}{x}$.

9. $y = \ln \tan x$; $y' = \frac{1}{\tan x}\left(\sec^2 x\right) = \sec x \csc x$

11. $y = \frac{\ln x}{x}$; $y' = \frac{x \cdot \frac{1}{x} - \ln x}{x^2} = \frac{1 - \ln x}{x^2}$

13. $y = \left[\ln\left(\frac{1+x}{1-x}\right)\right]^{-1/2}$; $y' = \frac{1}{2}\left[\ln\left(\frac{1+x}{1-x}\right)\right]^{-1/2} \cdot \left(\frac{1-x}{1+x}\right)$
$\cdot \left(\frac{(1-x)+(1+x)}{(1-x)^2}\right) = \frac{1}{2}\left[\ln\left(\frac{1+x}{1-x}\right)\right]^{-1/2}\left(\frac{1-x}{1+x}\right)\left(\frac{2}{(1-x)^2}\right)$
$= \frac{1}{(1+x)(1-x)} \cdot \left[\ln\left(\frac{1+x}{1-x}\right)\right]^{-1/2} = $
$\frac{1}{1-x^2}\left[\ln\left(\frac{1+x}{1-x}\right)\right]^{-1/2}$

15. $y = \ln\frac{4x^3}{\sqrt{x^2+4}}$; $y' = \frac{\sqrt{x^2+4}}{4x^3} \cdot$
$\left(\dfrac{\sqrt{x^2+4} \cdot 12x^2 - 4x^3 \cdot \frac{1}{2}\left(x^2+4\right)^{-1/2} \cdot 2x}{x^2+4}\right) = $
$\frac{\sqrt{x^2+4}}{4x^3} \cdot \frac{\left(x^2+4\right) \cdot 12x^2 - 4x^4}{\left(x^2+4\right)^{3/2}} = \frac{1}{4x^3} \cdot \frac{4x^2\left[\left(x^2+4\right) \cdot 3 - x^2\right]}{x^2+4} = $
$\frac{2x^2+12}{x\left(x^2+4\right)}$

17. $y = (\ln x)^2$; $y' = 2\ln x \cdot \frac{1}{x} = \frac{2\ln x}{x}$

19. $y = \ln(\ln x)$; $y' = \frac{1}{\ln x} \cdot \frac{1}{x} = \frac{1}{x\ln x}$

21. Since $y = \ln x$, then $y' = \frac{1}{x} = x^{-1}$, and $y'' = -1x^{-2} = \frac{-1}{x^2}$.

23. $y = \frac{1}{x}\ln x$; $y' = \frac{1}{x} \cdot \frac{1}{x} + \ln x \cdot \left(-1x^{-2}\right) = $
$\frac{1}{x^2} - \frac{\ln x}{x^2} = \frac{1-\ln x}{x^2}$; $y'' = \dfrac{x^2\left(-\frac{1}{x}\right) - (1-\ln x)2x}{x^4}$
$= \frac{-x - 2x(1-\ln x)}{x^4} = \frac{-1-2+2\ln x}{x^3} = \frac{2\ln x - 3}{x^3} = \frac{\ln x^2 - 3}{x^3}$

25. $y = \ln\sqrt{\frac{\sin^2 x}{x^3}} = \frac{1}{2}\ln\frac{\sin^2 x}{x^3}$; $y' = \frac{1}{2} \cdot \frac{x^3}{\sin^2 x} \cdot$
$\frac{2x^3 \sin x \cos x - 3x^2 \sin^2 x}{x^6} = \frac{2x\cos x - 3\sin x}{2\sin x \cdot x} = \cot x - $
$\frac{3}{2x}$; $y'' = -\csc^2 x - \frac{3}{2} \cdot \frac{(-1)}{x^2} = \frac{3}{2x^2} - \csc^2 x$

27. $y = \sqrt{\ln x^2} = \left(\ln x^2\right)^{1/2}$; $y' = \frac{1}{2}\left(\ln x^2\right)^{-1/2} \cdot \frac{1}{x^2} \cdot$
$2x = \frac{1}{x\left(\ln x^2\right)^{1/2}}$; $y'' = \frac{-1}{x^2\ln x^2}\left(x \cdot \frac{1}{2}\left(\ln x^2\right)^{-1/2} \cdot \right.$
$\left.\frac{2}{x} + \left(\ln x^2\right)^{1/2}\right) = \frac{-1}{x^2\ln x^2}\left(\frac{1}{\left(\ln x^2\right)^{1/2}} + \left(\ln x^2\right)^{1/2}\right)$
$= -\frac{1+\ln x^2}{x^2\left(\ln x^2\right)^{3/2}}$

29. Since $y = x^2\ln(\sin 2x)$, then $y' = 2x\ln(\sin 2x) + \frac{2x^2}{\sin 2x}\cos 2x = 2x\ln(\sin 2x) + 2x^2\cot 2x$, and so
$y'' = 2\ln(\sin 2x) + 2x \cdot \frac{1}{\sin 2x} \cdot \cos 2x \cdot 2 - 2x^2\csc^2 2x \cdot$
$2 + 4x\cot 2x = 2\ln(\sin 2x) + 4x\cot 2x - 4x^2\csc^2 2x$
$+ 4x\cot 2x = 2\ln(\sin 2x) + 8x\cot 2x - 4x^2\csc^2 2x.$

31. (a) $N'(t) = -5\ln(0.045t) - \frac{0.225t}{0.045t} - 0.5 = $
$-5\ln(0.045t) - 5.5$. Solving $-5\ln(0.045t) - 5.5 = $
0, or $\ln(0.045t) = -1.1$, we have $0.045t = e^{-1.1}$
and so, $t = \frac{1}{0.045}e^{-1.1} \approx 7.397$ h. The maximum number of medflies will be $N(7.397) \approx 108.96 \approx$
109.
(b) Graphing $N(t)$, we see that it is 0 at about 31.67 h, so the medflies will be eradicated within the 36 h period.

33. (a) In order to find the critical values, we determine when $P'(t) = -0.008\ln(0.025t) - 0.008 = 0$.
The maximum value is when $\ln(0.025t) + 1 = 0$
or when $t = 40e^{-1} \approx 14.72$ h
(b) $P(14.72) = 0.014 - 0.008(14.72)\ln \times$
$[0.025(14.72)] = 0.014 - 0.11776\ln(0.368) \approx$
0.132 ppm.

≡ 27.6 DERIVATIVES OF EXPONENTIAL FUNCTIONS

1. $y = 4^x$; $y' = 4^x \cdot 1 \cdot \ln 4 = 4^x\ln 4$

3. $y = 5^{\sqrt{x}}$; $y' = 5^{\sqrt{x}} \cdot \frac{1}{2\sqrt{x}} \cdot \ln 5 = \frac{5^{\sqrt{x}}}{2\sqrt{x}}\ln 5$

5. $y = 2^{\sin x}$; $y' = 2^{\sin x} \cdot \cos x \cdot \ln 2 = \cos x \cdot 2^{\sin x} \cdot \ln 2$

7. $y = e^{x^2+x}$; $y' = (2x+1)e^{x^2+x}$

9. $y = 4^{x^4}$; $y' = 4^{x^4} \cdot 4x^3 \cdot \ln 4 = 4x^3 4^{x^4}\ln 4$

11. $y = \frac{e^x}{x^2}$; $y' = \frac{x^2 e^x - e^x(2x)}{x^4} = \frac{e^x(x-2)}{x^3}$

13. $y = \frac{1+e^x}{x^2}$; $y' = \frac{x^2\left(e^x\right) - \left(1+e^x\right)2x}{x^4} = \frac{xe^x - 2 - 2e^x}{x^3} = $
$\frac{e^x(x-2)-2}{x^3}$

15. $y = e^{\tan x}$; $y' = e^{\tan x} \cdot \sec^2 x = \sec^2 x \cdot e^{\tan x}$

17. $y = \sin e^{x^2}$; $y' = \cos e^{x^2} \cdot e^{x^2} \cdot 2x = 2xee^{x^2}\cos e^{x^2}$

19. $y = 3^x\left(x^3 - 1\right)$; $y' = 3^x\left(3x^2\right) + \left(x^3-1\right)3^x \cdot \ln 3 = $
$3x^2 3^x + \left(x^3 - 1\right)3^x\ln 3$

21. $y = x^{\cos x}$. Using logarithmic differentiation, we obtain $\ln y = \ln x^{\cos x}$ or $\ln y = \cos x \ln x$, or $\frac{1}{y} y' = \cos x \cdot \frac{1}{x} + (-\sin x) \ln x$, and so $y' = y\left[\frac{1}{x} \cos x - \sin x \cdot \ln x\right]$, or $y' = x^{\cos x} \cdot \left[\frac{\cos x}{x} - (\sin x)(\ln x)\right]$.

23. $y = (\sin x)^x$; Use logarithmic differentiation. Then, $\ln y = x \ln(\sin x)$, or $\frac{1}{y} y' = x \cdot \frac{1}{\sin x} \cdot \cos x + \ln(\sin x)$, and so $y' = y[x \cot x + \ln(\sin x)]$ or $y' = (\sin x)^x [x \cot x + \ln(\sin x)]$.

25. $y = \ln \sin e^{3x}$; $y' = \frac{1}{\sin e^{3x}} \cos e^{3x} \cdot e^{3x} \cdot 3 = \frac{3e^{3x} \cos e^{3x}}{\sin e^{3x}} = 3e^{3x} \cot e^{3x}$

27. $y = e^y + y + x$; this is equivalent to $0 = e^y + x$. By implicit differentiation we obtain $0 = e^y y' + 1$ or $e^y y' = -1$ so $y' = \frac{-1}{e^y}$ or $y' = -e^{-y}$

29. $\frac{d}{dx}(\cosh x) = \frac{d}{dx}\left(\frac{e^x + e^{-x}}{2}\right) = \frac{1}{2}\left(e^x + e^{-x}(-1)\right) = \frac{e^x - e^{-x}}{2} = \sinh x$

31. $\cosh x + \sin hx = \left(\frac{e^x + e^{-x}}{2}\right) + \left(\frac{e^x - e^{-x}}{2}\right) = \frac{2e^x}{2} = e^x$

33. $y = e^{\sinh 3x}$; $y' = e^{\sinh 3x} \cdot \cosh 3x \cdot 3 = 3(\cosh 3x)e^{\sinh 3x}$

35. $g(x) = \tanh \sqrt{x^3 + 4}$; $g'(x) = \operatorname{sech}^2 \sqrt{x^3 + 4} \cdot \frac{1}{2}\left(x^3 + 4\right)^{-1/2} \cdot 3x^2 = \frac{3x^2}{2\sqrt{x^3 + 4}} \cdot \operatorname{sech}^2 \sqrt{x^3 + 4}$

37. $j(x) = \cosh^3\left(x^4 + \sin x\right)$; $j'(x) = 3\cosh^2\left(x^4 + \sin x\right) \cdot \sinh\left(x^4 + \sin x\right)\left(4x^3 + \cos x\right) = 3\left(4x^3 + \cos x\right)\cosh^2\left(x^4 + \sin x\right)\sinh\left(x^4 + \sin x\right)$

39. Here $f(x) = \sinh^{-1} 7x$, so $f'(x) = \frac{1}{\sqrt{(7x)^2 + 1}} \cdot 7 = \frac{7}{\sqrt{49x^2 + 1}}$.

41. $j(x) = x \sinh^{-1} \frac{1}{x}$. Using the product rule with the chain rule we obtain $j'(x) = \sinh^{-1} \frac{1}{x} + x \cdot \frac{1}{\sqrt{\left(\frac{1}{x}\right)^2 + 1}} \cdot \frac{-1}{x^2} = \sinh^{-1} \frac{1}{x} - \frac{1}{x\sqrt{\frac{1}{x^2} + 1}} = \sinh^{-1} \frac{1}{x} - \frac{|x|}{x\sqrt{\frac{1+x^2}{x^2}}} = \sinh^{-1} \frac{1}{x} - \frac{|x|}{x\sqrt{1+x^2}}$. Recall that $\sqrt{x^2} = |x|$.

43. The rate at which N is growing is $N'(t) = 615\left(-e^{0.02t}\right)(-0.02) = 12.3e^{-0.02t}$. After 7 days, N is growing at the rate of $N'(7) = 12.3e^{-0.14} = 10.69$ people/day.

45. **(a)** Using the product rule, we obtain $I = q'(t) = -5e^{-t} \cos 2.5t - 12.5e^{-t} \sin 2.5t$,
(b) $q'(0.45) \approx -8.57$ A

47. **(a)** $N'(t) = 10^6(0.1) \operatorname{sech}^2(0.1t) = 10,000 \operatorname{sech}^2(0.1t)$,
(b) $N'(6.93) \approx 6401$ bacteria/h,
(c) $N(6.93) = 699,906 \approx 7 \times 10^5$ bacteria

≡ 27.7 APPLICATIONS

1. $y = x \ln x$. The domain is $(0, \infty)$. $y' = x \cdot \frac{1}{x} + 1 \ln x = 1 + \ln x$. Setting $y' = 0$ we obtain the critical values $1 + \ln x = 0$; $\ln x = -1$, $x = e^{-1} = 0.3679$; $y'' = \frac{1}{x}$; since $x > 0$, y'' is always positive so y is always concave up. This also means that the critical value yields a minimum. Summary: maxima: none, minima: $\left(e^{-1}, -e^{-1}\right)$ or $(0.3679 - 0.3679)$; inflection points: none.

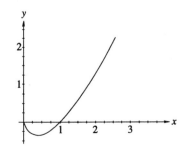

3. $y = xe^x$; $y' = xe^x + e^x = e^x(x + 1)$. The critical values occur when $y' = 0$. Since e^x cannot be 0, the only one is when $x + 1 = 0$ or $x = -1$. $y'' = e^x(1) + (x + 1)e^x = 2e^x + xe^x = e^x(2 + x)$. Inflection point is when $x = -2$; $y''(-1) > 0$ so at $x = -1$ there is a minimum. Summary: Maxima; none; minimum: $\left(-1, -e^{-1}\right)$; inflection point: $\left(-2, -2e^{-2}\right)$

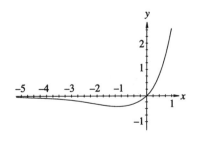

5. Here $y = \ln \frac{1}{x^2+1}$, and so $y' = \frac{x^2+1}{1} \cdot \frac{-2x}{(x^2+1)^2} = \frac{-2x}{x^2+1}$. Since the denominator cannot be 0, the only critical value is when $x = 0$. $y'' = \frac{(x^2+1)(-2)+2x(2x)}{(x^2+1)^2} = \frac{-2x^2-2+4x^2}{(x^2+1)^2} = \frac{2x^2-2}{(x^2+1)^2}$. Inflection points when $x^2 - 1 = 0$ or at $x = \pm 1$. At $x = 0$; $y'' < 0$ so y is a maximum. Summary: maxima; (0, 0); minima: none; inflection points: $\left(-1, \ln \frac{1}{2}\right)$; $\left(1, \ln \frac{1}{2}\right)$

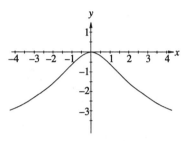

7. $y = \sinh x$; $y' = \cosh x = \frac{e^x+e^{-x}}{2}$, which is never 0. $y'' = \sinh x = \frac{e^x-e^{-x}}{2}$, which is 0 when $e^x = e^{-x}$ or when $x = 0$. Inflection point at $x = 0$. Summary: Maxima: none, minima: none, inflection point (0, 0)

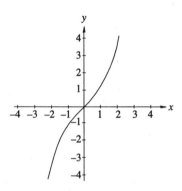

9. $y = e^{-x} \cos x$; $-2\pi \le x \le 2\pi$; $y' = -e^{-x} \sin x - \cos x e^{-x} = -e^{-x}(\sin x + \cos x)$. Critical values occur when $\sin x + \cos x = 0$ or $\sin x = -\cos x$ or $\tan x = -1$. This is at $-\frac{5\pi}{4}$, $-\frac{\pi}{4}$, $\frac{3\pi}{4}$, and $\frac{7\pi}{4}$. $y'' = -e^{-x}(\cos x - \sin x) + e^{-x}(\sin x + \cos x) = e^{-x}(-\cos x + \sin x + \sin x + \cos x) = 2e^{-x} \sin x$. Inflection points occur when $\sin x = 0$ or $x = -2\pi, -\pi, 0, \pi, 2\pi$. At $x = -\frac{5\pi}{4}$, $y'' > 0$ so y is minima. At $x = -\frac{\pi}{4}$, $y'' < 0$ so y is maxima. At $x = \frac{3\pi}{4}$, $y'' > 0$ so y is minima. At $x = \frac{7\pi}{4}$, $y'' < 0$ so y is maxima. Summary: maxima $\left(\frac{-\pi}{4}, 1.5509\right)$, $\left(\frac{7\pi}{4}, 0.002896\right)$, minima at $\left(\frac{-5\pi}{4}, -35.8885\right)$ and $\left(\frac{3\pi}{4}, 0.06702\right)$; Inflection

points at $(-2\pi, 535.49)$, $(-\pi, -23.1407)$, $(0, 1)$, $(\pi, -0.0432)$ and $(2\pi, 0.001867)$.

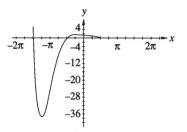

11. $y = x^3 \ln x$; $y' = 3x^2 \ln x + x^2$; at $x = 1$, $y' = 3 \cdot 1^2 \cdot \ln 1 + 1^2 = 1^2 = 1$ so the slope is 1. $y - 0 = 1(x - 1)$ or $y = x - 1$

13. $y = x^2 e^x$; $y' = 2xe^x + x^2 e^x$; at $x = 1$; $y' = 2e + e = 3e$. The equation of the line is $y - e = 3e(x - 1)$

15. (a) $s(t) = e^{-3t}$; $v(t) = s'(t) = -3e^{-3t}$,
(b) $v'(t) = 9e^{-3t}$ which is never undefined or 0 so velocity is never maximum.

17. $s(t) = \sin e^t$;
(a) $v(t) = s'(t) = \cos e^t \cdot e^t$; $v(t) = e^t \cos e^t$; $a(t) = v'(t) = e^t \cos e^t - \sin e^t (e^{2t})$ or $a(t) = e^t \left[\cos e^t - e^t \sin e^t\right]$
(b) $v(t) = 0$ when $\cos e^t = 0$ or $e^t = \frac{\pi}{2}$; this gives $t = \ln \frac{\pi}{2} \approx 0.4516$ s; $a\left(\ln \frac{\pi}{2}\right) = \frac{\pi}{2}\left[0 - \frac{\pi}{2} \sin \frac{\pi}{2}\right] = -\left(\frac{\pi}{2}\right)^2$; $-\left(\frac{\pi}{2}\right)^2 \approx -2.4674$ units/ss.

19. $\rho(x) = xe^{-x^{2/3}}$; this will have a maximum when $\rho'(x) = 0$. $\rho'(x) = x \cdot e^{x^{2/3}} \cdot \left(-\frac{2}{3}x^{-1/3}\right) + e^{-x^{2/3}} = e^{-x^{2/3}}\left(1 - \frac{2}{3}x^{2/3}\right) = 0$ when $\frac{2}{3}x^{\frac{2}{3}} = 1$ or $x^{2/3} = \frac{3}{2}$ or $x = \left(\frac{3}{2}\right)^{3/2} \approx 1.8371$ m. We can use the first derivative test to verify that this is a maximum.

21. $i = 1 - e^{-t^2/2L}$; $i = \frac{t}{L}e^{-t^2/2L}$; The critical value is when $t = 0$. When $t < 0$, $i' < 0$ and when $t > 0$, $i' > 0$ so this yields a maximum.

23. $P(t) = 100e^{-0.015t}$; $P'(t) = -1.5e^{-0.015t}$; $P'(50) = -1.5e^{-0.015 \cdot 50} = -0.7085$ W/day

25. $f(x) = x - \ln x - 2$; $f'(x) = 1 - \frac{1}{x}$. The roots are 0.1586; 3.1462.

27. $h(x) = e^x \cdot \cos x$; $h'(x) = e^x \cos x - e^x \sin x = e^x(\cos x - \sin x)$. Newton's method yields 1.570796327 and several other answers. Reexamining $h(x)$ we can see that e^x is never 0 and $\cos x = 0$ when $x = \frac{\pi}{2} + k\pi$ where k is an integer.

CHAPTER 27 REVIEW

1. $y = \sin 2x + \cos 3x$; $y' = \cos 2x \cdot 2 - \sin 3x \cdot 3 = 2\cos 2x - 3\sin 3x$

2. $y = \tan 3x^2$; $y' = \sec^2 3x^2 \cdot 6x = 6x\sec^2 3x^2$

3. $y = \tan^2 3x$; $y' = 2\tan 3x \cdot \sec^2 3x \cdot 3 = 6\tan 3x \sec^2 3x$

4. $y = \sqrt{\sin 2x} = (\sin 2x)^{1/2}$; $y' = \frac{1}{2}(\sin 2x)^{-1/2} \cdot (\cos 2x) \cdot = \frac{\cos 2x}{\sqrt{\sin 2x}} = \frac{\cos 2x \sqrt{\sin 2x}}{\sin 2x} = \cot 2x\sqrt{\sin 2x}$

5. $y = x^2 \sin x$. Using the product rule we obtain $y' = x^2\cos x + 2x\sin x = x(x\cos x + 2\sin x)$

6. $y = \frac{\cos 3x}{3x}$. Using the quotient rule we obtain $y' = \frac{3x(-\sin 3x)3 - \cos 3x \cdot 3}{9x^2} = \frac{-3x\sin x - \cos 3x}{3x^2}$

7. $y = \sin^{-1}(3x - 2)$; $y' = \frac{1}{\sqrt{1-(3x-2)^2}} \cdot 3 = \frac{3}{\sqrt{12x-9x^2-3}}$

8. $y = \cos^{-1} x^2$; $y' = \frac{-1}{\sqrt{1-x^4}} \cdot 2x = \frac{-2x}{\sqrt{1-x^4}}$

9. $y = \arctan 3x^2$; $y' = \frac{1}{1+9x^4} \cdot 6x = \frac{6x}{1+9x^4}$

10. $y = \arctan\left(\frac{1+x}{1-x}\right)$; $y' = \frac{1}{1+\left(\frac{1+x}{1-x}\right)^2} \cdot \frac{(1-x)1+(1+x)}{(1-x)^2}$ $= \frac{2}{(1-x)^2+(1+x)^2} = \frac{1}{1+x^2}$

11. Here $y = \arcsin\sqrt{1-x^2}$, and so $y' = \frac{1}{\sqrt{1-(1-x)^2}} \cdot$ $\frac{1}{2}\left(1-x^2\right)^{-1/2} \cdot (-2x) = \frac{-x}{\sqrt{x^2}\sqrt{1-x^2}} = \frac{-x}{|x|\sqrt{1-x^2}}$. If $x > 0$ then $y' = \frac{-1}{\sqrt{1-x^2}}$, $x < 0$, $y' = \frac{1}{\sqrt{1-x^2}}$

12. Here $y = \arccos\sqrt{-3+4x-x^2}$, and so $y' = \frac{-1}{\sqrt{1-(-3+4x-x^2)}} \cdot \frac{1}{2}\left(-3+4x-x^2\right)^{-1/2}(4-2x) =$ $\frac{-1}{\sqrt{x^2-4x+4}} \cdot \frac{2-x}{\sqrt{-3+4x-x^2}} = \frac{1}{|x-2|} \cdot \frac{2-x}{\sqrt{-3+4x-x^2}}$. If $x - 2 > 0$ then $y' = \frac{1}{-3+4x-x^2}$.

13. $y = \log_4\left(3x^2 - 5\right)$; $y' = \frac{1}{3x^2-5}(6x)\cdot\frac{1}{\ln 4} = \frac{6x}{3x^2-5} \cdot \frac{1}{\ln 4}$

14. $y = \ln(x+4)^3 = 3\ln(x+4)$; $y' = \frac{3}{x+4}$

15. $y = \ln^2\left(x^3+4\right)$; $y' = \frac{2\ln\left(x^3+4\right)}{x^3+4} \cdot 3x^2 = \frac{6x^2\ln\left(x^3+4\right)}{x^3+4}$

16. $y = \ln\cos x$; $y' = \frac{1}{\cos x} \cdot (-\sin x) = -\tan x$

17. $y = \ln\frac{x^3}{(4x-3)^2}$; $y' = \frac{(4x-3)^2}{x^3} \cdot \frac{(4x-3)^2 \cdot 3x^2 - x^3 \cdot 2 \cdot (4x-3) \cdot 4}{(4x-3)^4}$ $= \frac{(4x-3)^2}{x^3} \cdot \frac{3x^2(4x-3)^2 - 8x^3(4x-3)}{(4x-3)^4} = \frac{3(4x-3)-8x}{x(4x-3)} =$ $\frac{4x-9}{x(4x-3)}$

18. $y = \ln\arctan x$; $y' = \frac{1}{\arctan x} \cdot \frac{1}{1+x^2} = \frac{1}{\left(1+x^2\right)\arctan x}$

19. $y = e^{4x^2}$; $y' = 8xe^{4x^2}$

20. $y = e^{\ln x^2} = x^2$; $y' = 2x$

21. $y = e^{\sin x^2}$; $y' = 2x\cos x^2 e^{\sin x^2}$

22. $y = ex^{x^2} + \sin x$; $y' = xe^{x^2} \cdot 2x + e^{x^2} + \cos x = e^{x^2}\left(2x^2 + 1\right) + \cos x$

23. $y = e^{-x}\sin 5x$; $y' = e^{-x}\cos 5x \cdot 5 + (-1)e^{-x}\sin 5x = e^{-x}(5\cos 5x - \sin 5x)$

24. $y = e^{-x}\ln x$; $y' = e^{-x} \cdot \frac{1}{x} + e^{-x}(-1)\ln x = e^{-x}\left(\frac{1}{x} - \ln x\right)$

25. $y = x^2e^x$; $y' = 2xe^x + x^2e^2 = e^x\left(x^2 + 2x\right)$. The critical values occur when $x^2 + 2x = 0$ or $x = 0$ and $x = -2$. For the inflection points, we find $y'' = 2e^x + 2xe^x + x^2e^x + 2xe^x = e^x\left(x^2 + 4x + 2\right)$. Inflection points occur when $x^2 + 4x + 2 = 0$. The quadratic formula yields $x = -2 \pm \sqrt{2}$. When $x = -2$, then $y'' < 0$, and so y is a maximum. When $x = 0$, then $y'' > 0$, and so y is a minimum. Summary: Maxima: $(-2, 0.5413)$; minima: $(0, 0)$; inflection points: $(-2-\sqrt{2}, 0.3835)$ and $(-2+\sqrt{2}, 0.1910)$.

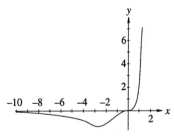

26. $y = 4e^{-9x^2}$. Finding y', we have $y' = (-18x) \times 4e^{-9x^2} = -72xe^{-9x^2}$. The only critical value is 0. Next, we find $y'' = -72e^{-9x^2} + 72x(18x)e^{-9x^2} = -72e^{-9x^2} + 1296x^2e^{-9x^2}$. Setting this equal to 0 we find $1296x^2 - 72 = 0$, or $18x^2 = 1$, or $x = \pm\frac{1}{\sqrt{18}} = \frac{\pm\sqrt{2}}{6}$.

Summary: maxima: (0, 4); minima: none; inflection points: $\left(\pm\frac{\sqrt{2}}{6}, 2.4261\right)$

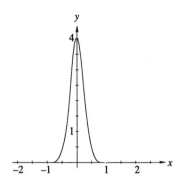

27. $y = e^{-x/2}\sin 2x$ on $[0, 2\pi]$; $y' = -\frac{1}{2}e^{-x/2}\sin 2x$ $+2\cos 2x e^{-x/2} = e^{-x/2}\left(2\cos 2x - \frac{1}{2}\sin 2x\right)$. Since $e^{-\frac{1}{2}x}$ never equals 0, the critical values occur when $2\cos 2x - \frac{1}{2}\sin 2x = 0$. Solving we get $2\cos 2x = \frac{1}{2}\sin 2x$ or $\tan 2x = 4$; so $2x = 1.325817, 4.45741, 7.6090,$ and 10.75060, and as a result $x = -.6629, 2.2337, 2.38045, 5.3753$. Taking the second derivative, we obtain $y'' = -\frac{1}{2}e^{-x/2}\left(2\cos 2x - \frac{1}{2}\sin 2x\right) + e^{-x/2}(-4\sin 2x - \cos 2x) = e^{-x/2} \times \left(-\cos 2x + \frac{1}{4}\sin 2x - 4\sin 2x - \cos 2x\right) = e^{-x/2} \times \left(-2\cos 2x - \frac{15}{4}\sin 2x\right)$. Solving $2\cos 2x + \frac{15}{4}\sin 2x = 0$, we obtain $\frac{15}{4}\sin 2x = -2\cos 2x$, or $\frac{\sin 2x}{\cos 2x} = \frac{-2}{\frac{15}{4}}$, or $\tan 2x = -\frac{8}{14}$ which means that $2x = 2.6516, 5.7932, 8.9348,$ and 12.0764, and as a result, that $x = 1.3258, 2.8966, 4.4675, 6.0382$. Summary: maxima: (0.6629, 0.6964), (3.8045, 0.1448); minima: (2.2337, −0.3175), (5.3753, −0.0660); inflection points: (1.3258, 0.2425), (2.8966, −0.1106), (4.4674, 0.0504), and (6.0382, −0.0230)

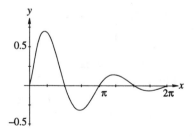

28. $y = e\sin 3x - e^{-1}\cos x$. Noting the e and e^{-1} are constants, we find $y' = e\cos x + e^{-1}\sin x$. Setting y' equal to 0, we have $e\cos x + e^{-1}\sin x = 0$, or $e^{-1}\sin x = -e\cos x$, or $\tan x = -e^2$. Hence, $x = 1.7053$ or 4.8469 are the critical values. Now, $y'' = -e\sin x + e^{-1}\cos x$. Setting this equal to 0 we have $\tan x = e^{-2}$ or $x = 0.1345$ or 3.2761. Thus, we have a maxima at (1.7053, 2.7431), minima at (4.8469, −2.7431), and inflection points at (1.4363, 0) and (3.2761, 0).

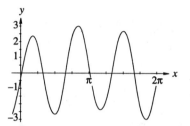

29. With $y = x^3\ln x$, we get $y' = 3x^2\ln x + x^2 = x^2(3\ln + 1)$. Solving $3\ln x + 1 = 0$, we obtain $\ln x = -\frac{1}{3}$ or $e^{-1/3} \approx 0.7165$. For the second derivative, we have $y'' = 6x\ln x + 3x + 2x = 6x\ln x + 5x = x(6\ln x + 5)$. Solving $6\ln x + 5 = 0$, we find $\ln x = -\frac{5}{6}$ or $x = e^{-\frac{5}{6}} \approx 0.4346$. At $x = -0.7165$, y'' is positive so y has a minimum at this point. Summary: Maxima: none; minima; (0.7165, −0.1226); inflection point (0.4346, 0.0684)

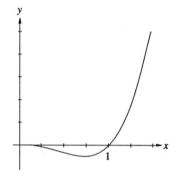

30. $y = 2 \arccos \frac{x}{4}$; $y' = \frac{1}{2} \cdot \dfrac{-1}{\sqrt{1 - \frac{x^2}{16}}} = \frac{1}{2} \cdot \dfrac{-1}{\sqrt{\frac{16-x^2}{16}}} =$

$\dfrac{-2}{\sqrt{16-x^2}}$. This is undefined at -4 and 4 which are the ends of the domain. $y(-4) = 2\pi$, $y(4) = 0$, $y' = -2(16-x)^{-1/2}$; $y'' = (16-x^2)^{-3/2}(-2x)$. This is 0 when $x = 0$ Summary: Maxima: $(-4, 2\pi)$; minima: $(4, 0)$; inflection point $(0, \pi)$

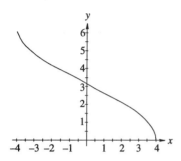

31. $y = \arctan x$; $y' = \frac{1}{1+x^2}$; $y'(1) = \frac{1}{1+1} = \frac{1}{2}$
   (a) The slope of the tangent is $\frac{1}{2}$ and its equation
      is $y = \frac{\pi}{4} = \frac{1}{2}(x-1)$,
   (b) The slope of the normal is -2 and its equation
      is $y = \frac{\pi}{4} = -2(x-1)$

32. $y = \sin^3 x$. The first derivative is $y' = 3\sin^2 x \cos x$ and so $y'\left(\frac{\pi}{4}\right) = 3 \cdot \left(\frac{\sqrt{2}}{2}\right)^2 \cdot \frac{\sqrt{2}}{2} = \frac{3\sqrt{2}}{4}$.
   (a) The slope of the tangent line is $\frac{3\sqrt{2}}{4}$ and its
      equation is $y = \frac{\sqrt{2}}{4} = \frac{3\sqrt{2}}{4}\left(x - \frac{\pi}{4}\right)$.
   (b) The slope of the normal is $-\frac{4}{3\sqrt{2}} = \frac{-2\sqrt{2}}{3}$ and
      its equation is $y - \frac{\sqrt{2}}{4} = -\frac{2\sqrt{2}}{3}\left(x - \frac{\pi}{4}\right)$.

33. $y = \ln 3x$ has derivative $y' = \frac{1}{x}$ and, when $x = 1$, the derivative is $y'(1) = 1$.
   (a) The slope of the tangent at the point $(1, 0)$ is 1 and its equation is $y = x - 1$.
   (b) The slope of the normal line at the point $(1, 0)$ is -1 and its equation is $y = -x + 1$ or $y = 1 - x$.

34. $y = \frac{d^x}{x}$ has the derivative $y' = \frac{xe^x - e^x}{x^2}$ and, when $x = 1$, the derivative is $y'(1) = \frac{1e - e}{1^2} = 0$.
   (a) The slope of the tangent at $(1, e)$ is 0 and its equation is $y = e$.

(b) The slope of the normal at $(1, e)$ is undefined, which means that the normal line is a vertical line. Its equation is $x = 1$.

35. $y = \ln(\sin x)$ has the derivative $y' = \frac{1}{\sin x} \cdot \cos x = \tan x$ and, when $x = \frac{\pi}{4}$, the derivative is $y'\left(\frac{\pi}{4}\right) = \tan\left(\frac{\pi}{4}\right) = 1$.
   (a) The slope of the tangent is 1 and its equation
      is $y + \ln\sqrt{2} = x - \frac{\pi}{4}$.
   (b) The slope of the normal is -1 and its equation
      is $y + \ln\sqrt{2} = \frac{\pi}{4} - x$.

36. $y = \sin(\ln x)$ has the derivative $y' = \cos(\ln x) \cdot \frac{1}{x}$ and, when $x = 1$, the derivative is $y'(1) = \cos(\ln 1) \cdot \frac{1}{1} = \cos 0 = 1$.
   (a) The slope of the tangent is 1 and its equation
      is $y = x - 1$.
   (b) The slope of the normal is -1 and its equation
      is $y = 1 - x$.

37. $T = 75e^{-0.05109t} + 29$
   (a) $T(10) = 75e^{-0.5109} + 29 = 64.9967°C$,
   (b) $T' = (-0.05109)75e^{-0.05109t}$ and $T'(10) = (-0.05109)75e^{-0.5109} = -2.2989°C$

38. $s(t) = 10{,}000 - \frac{400}{3}\ln(\cosh 1.6t)$
   (a) $v(t) = s'(t) = \sqrt{\frac{400}{3} \cdot \frac{1}{\cosh 1.6t}} \cdot \sinh 1.6t \cdot 1.6 = \frac{-640}{3}\tanh 1.6t$
   (b) $v(60) = \frac{-640}{3}\tan 1.6(60) = -213.33$ ft/s

39. Using the same set-up as Example 27.20 we have $\tan(\theta + \phi) = \frac{48}{x}$ and $\tan\phi = \frac{8}{x}$. Writing $\tan\theta = \tan\left[(\theta + \phi) - \phi\right]$, we obtain $\tan\theta = \tan\left[(\theta + \phi) - \phi\right] = \frac{\tan(\theta+\phi)-\tan\phi}{1+\tan(\theta+\phi)\tan\phi} = \dfrac{\frac{48}{x} - \frac{8}{x}}{1 + \frac{48}{x} \cdot \frac{8}{x}} = \dfrac{\frac{40}{x}}{1 + \frac{384}{x^2}} = \frac{40x}{x^2+384}$. Hence, $\theta = \arctan\left(\frac{40x}{x^2+384}\right)$ and $\theta' = \dfrac{1}{1 + \left(\frac{40x}{x^2+384}\right)^2} \cdot \dfrac{\left(x^2+384\right)40 - 40x(2x)}{\left(x^2+384\right)^2} = \dfrac{\left(x^2+384\right)^2}{\left(x^2+384\right)^2 + (40x)^2} \cdot \dfrac{40x^2 + 15360 - 80x^2}{\left(x^2+384\right)^2} = \dfrac{15360 - 40x^2}{\left(x^2+384\right)^2 + (40x)^2}$. The derivative is 0 when $15360 - 40x^2 = 0$ or $40x^2 = 15360$, or $x^2 = 384$, which is $x = 19.5959$. By the first derivative test this is a maximum. Answer 19.5959 ft.

40. $v(t) = 2$ m/s. Letting y represent the height of the balloon above the ground, we have $\tan\theta = \frac{y}{100}$. Taking the derivative of $\tan\theta = \frac{y}{100}$, we obtain $\sec^2\theta\frac{d\theta}{dt} = \frac{1}{100}\frac{dy}{dt}$ and so, $\frac{d\theta}{dt} = \frac{1}{dt}\cdot\frac{dy}{dt}\cdot\frac{1}{\sec^2\theta} = \frac{\cos^2\theta}{100}\frac{dy}{dt}$. Since the balloon is rising at the rate of 2 m/s, we have $\frac{dy}{dt} = 2$ m/s. Thus, when the balloon is 30 m above the ground, $y = 30$, and we see that we obtain $\cos\theta = \frac{100}{\sqrt{100^2+30^2}} = 0.957826$ and the rate of increase of the angle of inclination of the observer's line of sight when the balloon is 30 m high is $\frac{d\theta}{dt} = \frac{(0.957826)^2\cdot 2}{100} = 0.0183$ rad/s.

▰ CHAPTER 27 TEST

1. $f'(x) = \cos 7x \frac{d}{dx}(7x) = 7\cos 7x$

2. $g'(x) = \sec^2\left(3x^2 + 2x\right)\left[\frac{d}{dx}\left(3x^2 + 2x\right)\right] = \left[\sec^2\left(3x^2 + 2x\right)\right](6x + 2) = (6x + 2)\sec^2\left(3x^2 + 2x\right)$

3. $h'(x) = e^{2x}\frac{d}{dx}(2x) = 2e^{2x}$

4. $j'(x) = \frac{1}{5x^2}\frac{d}{dx}\left(5x^2\right) = \frac{1}{5x^2}(10x) = \frac{10x}{5x^2} = \frac{2}{x}$

5. $k'(x) = \frac{1}{\sqrt{1-\left(e^{x^2}\right)^2}}\frac{d}{dx}\left(e^{x^2}\right) = \frac{1}{\sqrt{1-e^{2x^2}}}\left(e^{x^2}\right)\frac{d}{dx}\times\left(x^2\right) = \frac{2xe^{x^2}}{\sqrt{1-e^{2x^2}}}$

6. Using the quotient rule: $f'(x) = \frac{\left(1+e^{2x}\right)4\cos 4x - \sin 4x\left(2e^{2x}\right)}{\left(1+e^{2x}\right)^2}$

7. Using the product rule: $g'(x) = \left(7x^2 + 3x\right)\frac{d}{dx}\times\left(\tan^2 5x\right)\frac{d}{dx}\left(7x^2 + 3x\right)$. Then, by using the chain rule we obtain $g'(x) = \left(7x^2 + 3x\right)(2\tan 5x)\times\left(\sec^2 5x\right)5 + \left(\tan^2 5x\right)(14x+3) = (\tan 5x)\left[10\left(7x^2 + 3x\right)\left(\sec^2 5x\right) + (14x + 3)(\tan 5x)\right]$.

8. Using the product rule: $h'(x) = e^{\sin x}\frac{d}{dx}(\ln\sqrt{x}) + (\ln\sqrt{x})\frac{d}{dx}\left(e^{\sin x}\right)$. We simplify this by using the chain rule: $h'(x) = e^{\sin x}\frac{1}{\sqrt{x}}\times\frac{d}{dx}\sqrt{x} + (\ln\sqrt{x})\times e^{\sin x}\frac{d}{dx}(\sin x) = e^{\sin x}\frac{1}{\sqrt{x}}\frac{1}{2\sqrt{x}} + (\ln\sqrt{x})e^{\sin x}(\cos x) = e^{\sin x}\left(\frac{1}{2x} + \cos x\ln\sqrt{x}\right)$.

9. The current, i, at a particular time is found by taking the derivative of the charge. So, $i(t) = q'(t) = -3e^{-t}\cos 2.5t - 7.5e^{-t}\sin 2.5t$. At $t = 0.65$ sec, $i(0.65) \approx -3.82$ A.

10. The velocity at time t, $v(t)$, is the derivative of the position function. So, $v(t) = x'(t) = 1.6\times(\tan 5.0t)^{-1/2}\left(5\sec^2 5.0t\right) = 8\frac{\sec^2 5.0t}{\sqrt{\tan 5.0t}}$. When $t = 1.5$, we see that $v(1.5) = 8\frac{\sec^2 7.5}{\sqrt{\tan 7.5}} \approx 40.4743$. The particle is at 40.4743 m.

CHAPTER
28
Techniques of Integration

≡ 28.1 THE GENERAL POWER FORMULA

1. $\int \sin^3 x \cos x \, dx$. Let $u = \sin x$ and $du = \cos x \, dx$, then $\int u^3 du = \frac{u^4}{4} + C = \frac{1}{4} \sin^4 x + C$.

3. $\int \sqrt{\sin^3 x} \cos x \, dx$. Let $u = \sin x$ and $du = \cos x \, dx$, so we have $\int u^{3/2} du = \frac{2}{5} u^{5/2} + C = \frac{2}{5} \sin^{5/2} x + C$.

5. $\int x \sec^2 x^2 dx$. If you let $u = x^2$ and $du = 2x \, dx$, or $\frac{1}{2} du = x \, dx$, then you get $\frac{1}{2} \int \sec^2 u \, du = \frac{1}{2} \tan u + C = \frac{1}{2} \tan x^2 + C$.

7. $\int \sec^2 x \tan x \, dx$. If you let $u = \tan x$ and $du = \sec^2 x \, dx$, then you get $\int u \, du = \frac{1}{2} u^2 + C = \frac{1}{2} \tan^2 x + C$ or rewrite the original integral as $\int \sec^2 x \tan x \, dx = \int \sec x (\sec x \tan x \, dx)$ and let $u = \sec x$ and $du = \sec x \tan x \, dx$. Thus, $\int u \, du = \frac{1}{2} u^2 + C = \frac{1}{2} \sec^2 x + C$.

9. $\int \frac{\arccos x}{\sqrt{1-x^2}} dx$. Let $u = \arccos x$ and then $du = \frac{-1}{\sqrt{1-x^2}} dx$, with the result $-\int u \, du = \frac{-1}{2} u^2 + C = \frac{-1}{2} (\arccos x)^2 + C$.

11. $\int \frac{1}{x} \sqrt{\frac{\text{arccsc } x}{x^2-1}} dx$. Let $u = \text{arccsc } x$ and $du = \frac{-1}{x\sqrt{x^2-1}} dx$, with the result $-\int u^{1/2} du = -\frac{2}{3} u^{3/2} + C = -\frac{2}{3} (\text{arccsc } x)^{3/2} + C$.

13. $\int \frac{(\arcsin 2x)^3}{\sqrt{1-4x^2}} dx$. Let $u = \arcsin 2x$ and $du = 2 \cdot \frac{1}{\sqrt{1-4x^2}} dx$, so $\frac{1}{2} du = \frac{dx}{\sqrt{1-4x^2}}$. Substitution yields $\frac{1}{2} \int u^3 du = \frac{1}{2} \cdot \frac{1}{4} u^4 + C = \frac{1}{8} (\arcsin 2x)^4 + C$.

15. $\int \frac{[\ln(x+4)]^4}{x+4} dx$. Let $u = \ln(x+4)$ and $du = \frac{dx}{x+4}$, then $\int u^4 du = \frac{1}{5} u^5 + C = \frac{1}{5} [\ln(x+4)]^5 + C$.

17. $\int \frac{e^x dx}{(e^x+4)^2}$. Let $u = e^x + 4$, and $du = e^x \, dx$, then you have $\int u^{-2} du = -1u^{-1} + C = -(e^x+4)^{-1} + C$.

19. $\int (e^{2x} - 1)^3 e^{2x} dx$. Let $u = e^{2x} - 1$ and $du = 2e^{2x} dx$ so $\frac{1}{2} du = e^{2x} dx$. Then, $\frac{1}{2} \int u^3 du = \frac{1}{2} \cdot \frac{1}{4} u^4 + C = \frac{1}{8} (e^{2x} - 1)^4 + C$.

21. $\int_0^{\pi/4} \sin^3 x \cos x \, dx$. If you let $u = \sin x$ and $du = \cos x \, dx$, then you have $\int u^3 du = \frac{1}{4} u^4$. Substituting for u produces $\frac{1}{4} \sin^4 x \Big|_0^{\pi/4} = \frac{1}{4} \Big[\sin^4 \frac{\pi}{4} - \sin^4 0 \Big] = \frac{1}{4} \Big[\left(\frac{\sqrt{2}}{2} \right)^4 = 0^4 \Big] = \frac{1}{4} \cdot \frac{1}{4} = \frac{1}{16}$.

23. $\int_1^2 \frac{[\ln(x^2+1)]^3}{x^2+1} dx$. Let $u = \ln(x^2 + 1)$ and $du = \frac{2x}{x^2+1} dx$ and so $\frac{1}{2} du = \frac{x}{x^2+1} dx$. Thus, we get $\frac{1}{2} \int u^3 du = \frac{1}{8} u^4$ or $\frac{1}{8} \ln^4(x^2 + 1) \Big|_1^2 = \frac{1}{8} [\ln^4 5 - \ln^4 2] \approx 0.8098$.

25. $\int_1^2 \frac{e^{2x} dx}{(e^{2x}-1)^2}$. Let $u = e^{2x} - 1$ and $du = 2e^{2x} dx$ so $\frac{1}{2} du = e^{2x} dx$. Then, $\frac{1}{2} \int u^{-2} du = -\frac{1}{2} u^{-1}$ or $-\frac{1}{2} (e^{2x} - 1)^{-1} = \Big|_1^2 = -\frac{1}{2} \left((e^4 - 1)^{-1} - (e^2 - 1)^{-1} \right) \approx 0.0689$.

27. $\int_0^\pi \sin^2 x \cos x \, dx$. Let $u = \sin x$ and $du = \cos x \times dx$, and so $\int u^2 du = \frac{1}{3} u^3$. The curve is above the x axis from 0 to $\frac{\pi}{2}$ and below from $\frac{\pi}{2}$ to π. Thus, the area is $\frac{1}{3}\sin^3 x\big|_0^{\pi/2} - \frac{1}{3}\sin^3 x\big|_{\pi/2}^{\pi} = \frac{1}{3}\big(\sin^3\frac{\pi}{2} - \sin^3 0 - \sin^3\pi + \sin^3\frac{\pi}{2}\big) = \frac{1}{3}(1-0-0+1) = \frac{2}{3}$.

29. $\int_{0.7}^{3} e^{-t}(e^{-t}-2)^2 dt$. Let $u = e^{-t}-2$ and $du = -e^{-t}dt$, and so $-\int u^2 du = -\frac{1}{3}u^3$. Substituting for u we obtain $\int_{0.7}^{3} e^{-t}(e^{-t}-2)^2 dt = -\frac{1}{3}\big[(e^{-t}-2)^3\big]_{0.7}^{3} = -\frac{1}{3}\big[(e^{-3}-2)^3 - (e^{-0.7}-2)^3\big]$
≈ 1.3397 C.

31. **(a)** $N(t) = \int -\frac{1020}{t^2} - 340\,dt = \frac{1020}{t} - 340t + C$. Since $N(1) = 1020 - 340 + C = 9750$ and so, $C = 9070$. Thus, $N(t) = \frac{1020}{t} - 340t + 9070$.

(b) $N(10) = \frac{1020}{10} - 340(10) + 9070 = 102 - 3400 + 9070 = 5772$.

≡ 28.2 BASIC LOGARITHMIC AND EXPONENTIAL INTEGRALS

1. $\int \frac{dx}{4x+1}$. Let $u = 4x+1$ and $du = 4\,dx$, so $\frac{1}{4}du = dx$. Then, $\frac{1}{4}\int \frac{du}{u} = \frac{1}{4}|u| + C = \frac{1}{4}\ln|4x+1| + C$.

3. $\int e^{-3x}dx$. Let $u = -3x$ and $du = -3\,dx$, or $-\frac{1}{3}du = dx$, which leads to $-\frac{1}{3}\int e^u du = -\frac{1}{3}e^u + C = -\frac{1}{3}e^{-3x} + C$.

5. $\int \frac{\sin x}{\cos x}dx$. Let $u = \cos x$ and $du = -\sin x\,dx$, then you have $-\int \frac{du}{u} = -\ln|u| + C = -\ln|\cos x| + C$.

7. $\int \frac{2\sec^2 x}{\tan x}$. If you let $u = \tan x$ and $du = \sec^2 x\,dx$, then $2\,du = 2\sec^2 x\,dx$. Substituting we get $2\int \frac{du}{u} = 2\ln|u| + C = 2\ln|\tan x| + C$.

9. $\int 4^x dx = \frac{4^x}{\ln 4} + C$.

11. $\int (e^x + e^{-x})dx = \int e^x dx - \int -e^{-x}dx = e^x - e^{-x} + C$. (Note: for the second integral let $u = -x$, then $du = -dx$)

13. $\int \frac{e^{2x}}{1+e^{2x}}dx$. Let $u = 1 + e^{2x}$, so $du = e^{2x}2\,dx$ or $\frac{1}{2}du = e^{2x}dx$. Substituting we get $\frac{1}{2}\int \frac{du}{u} = \frac{1}{2}\ln|u| + C = \frac{1}{2}\ln(1+e^{2x}) + C = \ln\sqrt{1+e^{2x}} + C$.

15. $\int \frac{\ln(1/x)}{x}dx$. Let $u = \ln\frac{1}{x} = \ln(x)^{-1} = -\ln x$, and then $du = -\frac{1}{x}dx$. Substituting we get $-\int u\,du = -\frac{1}{2}u^2 + C = -\frac{1}{2}\big[\ln\big(\frac{1}{x}\big)\big]^2 + C$.

17. $\int \frac{e^{\sqrt[3]{x}}}{x^{2/3}}dx$. Let $u = \sqrt[3]{x} = x^{1/3}$; $du = \frac{1}{3}x^{-2/3}dx$ so $3\,du = \frac{1}{x^{2/3}}dx$. Substituting we get $3\int e^u du = 3e^u + C = 3e^{\sqrt[3]{x}} + C$.

19. $\int_0^4 \frac{x}{x^2+1}dx$. Let $u = x^2+1$ and $du = 2x\,dx$ or $\frac{1}{2}du = x\,dx$. Substituting we get $\frac{1}{2}\int \frac{du}{u} = \frac{1}{2}\ln|u| = \frac{1}{2}\ln(x^2+1)$. This leads to $\frac{1}{2}\ln(x^2+1)\big|_0^4 = \frac{1}{2}\big[\ln(17) - \ln(1)\big] = \frac{1}{2}\ln 17 \approx 1.4166$.

21. $\int_1^3 \frac{1}{\ln e^x}dx$. Recall that $\ln e^x = x$ so $\int_1^3 \frac{1}{x}dx = \ln|x|\big|_1^3 = \ln 3 - \ln 1 = \ln 3 \approx 1.0986$.

23. $\int_0^1 x^3 e^{x^4} dx$. Let $u = x^4$ and $du = 4x^3 dx$ or $\frac{1}{4}du = x^3 dx$. Substituting produces $\frac{1}{4}\int e^u du = \frac{1}{4}e^u = \frac{1}{4}e^{x^4}$ which leads to $\frac{1}{4}e^{x^4}\big|_0^1 = \frac{1}{4}(e^1 - e^0) = \frac{1}{4}(e-1) \approx 0.4296$.

25. $\int_2^4 \big(\frac{1}{5}\big)^{x/2} dx$. Let $u = \frac{x}{2}$ and $du = \frac{1}{2}dx$. Substituting, we get $2\int \big(\frac{1}{5}\big)^u du = 2\cdot\frac{\big(\frac{1}{5}\big)^u}{\ln\frac{1}{5}} = \frac{2\big(\frac{1}{5}\big)^{x/2}}{\ln\frac{1}{5}}$.

Evaluating, we have $\frac{2\big(\frac{1}{5}\big)^{x/2}}{\ln\frac{1}{5}}\bigg|_2^4 = \frac{2}{\ln\frac{1}{5}}\Big[\big(\frac{1}{5}\big)^2 - \frac{1}{5}\Big]$
$= 0.1998$.

27. Since $e^{x+1} > x^2$ on $[0, 1]$, the area we want is $\int_0^1 (e^{x+1} - x^2)dx = \int_0^1 e^{x+1} - \int_0^1 x^2 dx = e^{x+1}\big|_0^1 - \frac{x^3}{3}\big|_0^1 = e^2 - e^1 - \frac{1^3}{3} = e^2 - e - \frac{1}{3} \approx 4.3374$.

29. **(a)** $s(t) = \int (394e^{-0.025t} - 384)dt = -15760e^{-0.025t} - 384t + C$. At $t = 0$, the object is on the ground, so $s(0) = 0$ and we see that, $C = 15{,}760$. Thus, the position function is $s(t) = -15{,}760e^{-0.025t} - 384t + 15{,}760$;

(b) The object reaches its maximum height when $v(t) = 0$ or when $e^{-0.025t} = \frac{384}{394}$. Taking the natural logarithm of both sides we get $-0.025t = \ln\big(\frac{384}{394}\big) = \ln 384 - \ln 394$. So, $t = \frac{\ln 384 - \ln 394}{-0.025} \approx 1.028$;

(c) $s(1.028) = 5.1196$.

31. Using the disc method this volume is $\pi \int_0^2 (e^{-x})^2 dx$
$= \pi \int_0^2 e^{-2x} dx$. Let $u = -2x$, then $du = -2\,dx$
or $-\frac{1}{2} du = dx$. Thus, the integral becomes $\pi \cdot$
$-\frac{1}{2} \int e^u du = \frac{-\pi}{2} e^u = -\frac{\pi}{2} e^{-2x}$. Evaluating, we
get $\frac{-\pi}{2} e^{-2x} \big|_0^2 = \frac{-\pi}{2}(e^{-4} - e^0) = \frac{\pi}{2}(1 - e^{-4}) \approx$
1.5420.

33. Since $\frac{dV}{ds} = -0.45 e^{-0.15s}$, we see that $V =$
$\int -0.45 e^{-0.15s} ds = \frac{-0.45}{-0.15} e^{-0.15s} + C = 3e^{-0.15s} +$

C. Hence, $V(0) = 3 = 3e^0 + C = 3 + C$ so
$C = 0$ and $V(s) = 3e^{-0.15s}$. Also, $3e^{-0.15s} = 1.5$
yields $e^{-0.15s} = \frac{1}{2}$. Taking the natural logarithm of
both sides we get $-0.15s = \ln\frac{1}{2}$ or $s = \ln(0.5) \div$
$(-0.15) = 4.6210$ mi.

35. First we find $k = PV = 20 \times 0.4 = 8$. Then,
the work done by the gas is $W = \int_{0.400}^{1.500} \frac{8}{V} dN =$
$8(\ln 1.5 - \ln 0.4) \approx 10.574$ ft · lb.

37. $F(x) = \int_0^x 12te^{-3t^2} dt = -2\int_0^x (-6t)e^{-3t^2} dt =$
$-2e^{-3t^2}\big|_0^x = -2e^{-3x^2} - (-2e^{-3\cdot 0^2}) = -2e^{-3x^2} +$
$2 = 2 - 2e^{-3x^2}$

≡ 28.3 BASIC TRIGONOMETRIC AND HYPERBOLIC INTEGRALS

1. $\int \sin\frac{x}{2} dx$. Let $u = \frac{x}{2}$ and $du = \frac{1}{2} dx$ gives
$2\int \sin u\, du = -2\cos u + C = -2\cos\frac{x}{2} + C$.

3. $\int \tan 5x\, dx$. Let $u = 5x$, then $du = 5\, dx$ and
$\frac{1}{5} du = dx$. This gives $\frac{1}{5}\int \tan u\, du = -\frac{1}{5}\ln|\cos u|$
$+ C = -\frac{1}{5}\ln|\cos 5x| + C$ or $\frac{1}{5}\ln|\sec 5x| + C$.

5. $\int \frac{1}{\cos x} dx = \int \sec x\, dx = \ln|\sec x + \tan x| + C$

7. $\int \frac{1}{\cos^2 x} dx = \int \sec^2 x\, dx = \tan x + C$

9. $\int \frac{\tan\sqrt{x}}{\sqrt{x}} dx$. Let $u = \sqrt{x}$ then $du = \frac{1}{2\sqrt{x}} dx$ and
$2du = \frac{1}{\sqrt{x}} dx$. Substituting produces $2\int \tan u\, du =$
$-2\ln|\cos u| + C = -2\ln|\cos\sqrt{x}| + C$ or $2\ln$
$|\sec\sqrt{x}| + C$.

11. $\int \sin(4x - 1) dx = -\frac{1}{4}\cos(4x - 1) + C$

13. $\int \frac{\sin x}{\cos^2 x} dx$. Let $u = \cos x$ and $du = -\sin x\, dx$.
Substituting, we obtain $-\int u^{-2} du = u^{-1} + C =$
$(\cos x)^{-1} + C = \sec x + C$.

15. $\int \frac{\sec 3x \tan 3x}{5 + 2\sec 3x} dx$. Let $u = 5 + 2\sec 3x$, and then
$du = 6\sec 3x \tan 3x\, dx$ or $\frac{1}{6} du = \sec 3x \tan 3x\, dx$.
This yields $\frac{1}{6}\int \frac{1}{u} du = \frac{1}{6}\ln|u| + C = \frac{1}{6}\ln|5 +$
$2\sec 3x| + C$.

17. $\int \sec^4 3x \tan 3x\, dx = \int \sec^3 3x \sec 3x \tan 3x\, dx$.
If you let $u = \sec 3x$, then $du = 3\sec 3x \tan 3x\, dx$.
This leads to $\frac{1}{3}\int u^3 du = \frac{1}{12} u^4 + C = \frac{1}{12}\sec^4 3x +$
C.

19. $\int \frac{3\, dx}{\tan 3x} = \int 3\cot(3x) dx = \ln|\sin 3x| + C$

21. $\int(\sec x + 2)^2 dx = \int(\sec^2 x + 4\sec x + 4) dx =$
$\tan x + 4\ln|\sec x + \tan x| + 4x + C$

23. $\int_0^{\pi/2}(\sec 0.5x + 5)^2 dx = \int_0^{\pi/2}$
$\times (\sec^2 0.5x + 10\sec 0.5x + 25) dx$
$= [2\tan 0.5x + 20\ln|\sec 0.5x$
$+ \tan 0.5x| + 25x]_0^{\pi/2}$
$= \left[2\tan\left(\frac{\pi}{4}\right) + 20\ln\left|\sec\left(\frac{\pi}{4}\right)\right.\right.$
$\left.\left. + \tan\left(\frac{\pi}{4}\right)\right| + 25\left(\frac{\pi}{2}\right)\right]$
$- [2\tan 0 + 20\ln|\sec 0 + \tan 0| + 25(0)]$
$= \left[2(1) + 20\ln\left|\sqrt{2} + 1\right| + \frac{25\pi}{2}\right]$
$- [2(0) + 20\ln|1 + 0| + 25(0)]$
$= 2 + 20\ln\left|\sqrt{2} + 1\right| + \frac{25\pi}{2} \approx 58.897$

25. $\int \frac{1 + \cos 4x}{\sin^2 4x} dx = \int\left(\frac{1}{\sin^2 4x} + \frac{\cos 4x}{\sin^2 4x}\right) dx = \int \csc^2 4x \times$
$dx + \int \frac{\cos 4x}{\sin^2 4x} dx$. Now, the first integral is $\int \csc^2 4x$
$= \frac{-1}{4}\cot^2 4x + C$. For the second integral, let
$u = \sin 4x$ then $du = 4\cos 4x\, dx$ or $\frac{1}{4} du =$
$\cos 4x\, dx$. Substitution yields $\frac{1}{4}\int u^{-2} du = \frac{1}{4} \cdot$
$-1u^{-1} + C = -\frac{1}{4}(\sin 4x)^{-1} + C = -\frac{1}{4}\csc 4x +$
C. Putting these together we get $-\frac{1}{4}\cot^2 4x -$
$\frac{1}{4}\csc 4x + C = -\frac{1}{4}[\cot^2 4x + \csc 4x] + C$ or
$-\frac{1}{4}\left(\frac{\cos 4x + 1}{\sin 4x}\right) + C$.

27. $\int \tan\frac{x}{4} dx = 4\ln|\sec\frac{x}{4}| + C$

29. Using trig identities, we get $\int_{\pi/4}^{\pi/2} \frac{1+\cot^2 x}{\csc^2 x}dx =$ $\int_{\pi/4}^{\pi/2} \left(\frac{1}{\csc^2 x}+\cos^2 x\right)dx = \int_{\pi/4}^{\pi/2}\left(\sin^2 x+\cos^2 x\right)dx$ $= \int_{\pi/4}^{\pi/2} dx = x\big|_{\pi/4}^{\pi/2} = \frac{\pi}{2} - \frac{\pi}{4} = \frac{\pi}{4}.$

31. $\int_{\pi/4}^{\pi/2} \frac{\csc\sqrt{x}\cot\sqrt{x}}{\sqrt{x}}dx$. Let $u = \sqrt{x}$ and $du = \frac{1}{2\sqrt{x}}$. Then, $2\int \csc u \cot u\, du = -2\csc u$. Evaluating, we obtain $-2\csc\sqrt{x}\big|_{\pi/4}^{\pi/2} = 0.4765.$

33. Using a Pythagorean trigonometric substitution gives the result $\int_{\pi/6}^{\pi/2} \frac{\cos^2 x}{\sin x}dx = \int_{\pi/6}^{\pi/2} \frac{1-\sin^2 x}{\sin x}dx = \int_{\pi/6}^{\pi/2} \times$ $\left(\frac{1}{\sin x} - \sin x\right)dx = \int_{\pi/6}^{\pi/2}(\csc x - \sin x)dx =$ $\left[\ln|\csc x - \cot x| + \cos x\right]_{\pi/6}^{\pi/2} = \ln|1 - 0| + 0 -$ $\ln|2 - \sqrt{3}| - \frac{\sqrt{3}}{2} = 0.4509.$

35. $\int x\sinh x^2\, dx$. Let $u = x^2$ and $du = 2x\, dx$ or $\frac{1}{2}du = x\, dx$. Substituting we get $\frac{1}{2}\int \sinh u\, du = \frac{1}{2}\cosh u + C = \frac{1}{2}\cosh x^2 + C$

37. $\int 3x\cosh x^2\sqrt{\sinh x^2}dx$. Let $u = \sinh x^2$, then $du = 2x\cosh x^2 dx$ or $\frac{3}{2}du = 3x\cosh x^2 dx$. Substituting we get $\frac{3}{2}\int \sqrt{u}\, du = \frac{3}{2}\cdot\frac{2}{3}u^{3/2} + C = \left(\sinh x^2\right)^{3/2} + C$

39. $\int \sinh^3 x\cdot\cosh^2 x\, dx$. Recall that $\sinh^3 x = \sinh x\sinh^2 x = \sinh x\left(\cosh^2 x - 1\right)$, and then we have $\int \sinh x\left(\cosh^2 x - 1\right)\cosh^2 x\, dx =$ $\int \cosh^4 x\sinh x\, dx - \int \cosh^2 x\sinh x\, dx =$ $\frac{1}{5}\cosh^5 x - \frac{1}{3}\cosh^3 x + C.$

41. We begin by rewriting $\cot u$ as $\frac{\cos u}{\sin u}$. Here, if we let $v = \sin u$, then $dv = \cos u\, du$, and the integral is $\int \cot u\, du = \int \frac{dv}{v} = \ln|v| + C$. Back substitution gives the desired result: $\int \cot u\, du = \ln|\sin u| + C.$

43. For $0 \leq x \leq \frac{\pi}{2}$, we see that $\sin 2x \geq 0$, so the area is the integral $\int_0^{\pi/2}\sin 2x\, dx = \frac{-1}{2}\cos 2x\big|_0^{\pi/2} =$ $-\frac{1}{2}(\cos\pi - \cos 0) = \frac{-1}{2}(-1 - 1) = 1.$

45. The average value is $\frac{1}{\pi/4}\int_0^{\pi/4}\tan x\, dx =$ $\frac{4}{\pi}\ln|\sec x|_0^{\pi/4} = \frac{4}{\pi}(\ln\sqrt{2} - \ln 1) = \frac{4\ln\sqrt{2}}{\pi} \approx$ $0.4413.$

47. $\bar{E} = \frac{1}{1-0}\int_0^1 5\sin 4t\, dt = -\frac{5}{4}\cos 4t\big|_0^1 =$ $-\frac{5}{4}(-0.6536 - 1) \approx 2.067\text{ V.}$

49. Here $y = 12\cosh\frac{x}{12}$ and so $y' = \sinh\frac{x}{12}$. The length of the cable is the arc length $=$ $\int_{-18}^{18}\sqrt{1 + \sinh^2\frac{x}{12}}\, dx = \int_{-18}^{18}\cosh\frac{x}{12}dx =$ $12\sinh\frac{x}{12}\big|_{-18}^{18} = 12\sinh\frac{3}{2} - 12\sinh\frac{-3}{2} =$ 51.1027 m.

51. The center of the towers is at 0 and the towers are at -150 and 150. We are given $y = 80\cosh\frac{x}{80}$ and differentiating, we obtain $y' = \sinh\frac{x}{80}$. The arc length is $\int_{-150}^{150}\sqrt{1 + \sinh^2\frac{x}{80}}dx = \int_{-150}^{150}\cosh\frac{x}{80}dx$ $= 80\left[\sinh\frac{x}{80}\right]_{-150}^{150} = 509.397\text{ ft.}$

≡ 28.4 MORE TRIGONOMETRIC INTEGRALS

1. $\int \sin^2 3x\cos 3x\, dx$. Legt $u = \sin 3x$ and $du = 3\cos 3x\, dx$. Then, $\frac{1}{3}\int u^2 du = \frac{1}{9}u^3 + C = \frac{1}{9}\sin^3 3x + C.$

3. $\int \cos x\sin^3 x\, dx = \frac{1}{4}\sin^4 x + C.$

5. $\int \cos 5x\sin^4 5x\, dx = \frac{1}{25}\sin^5 5x + C$

7. $\int \sin^2 x\cos^4 x\, dx = \int\left(\frac{1-\cos 2x}{2}\right)\left(\frac{1+\cos 2x}{2}\right)^2 dx = \frac{1}{8}\int (1 - \cos 2x)(1 + 2\cos 2x + \cos^2 2x)dx$

$= \frac{1}{8}\int\left(1 + \cos 2x - \cos^2 2x - \cos^3 2x\right)dx$

$= \frac{1}{8}\int\left(1 + \cos 2x - \cos^3 2x\right) - \frac{1}{8}\int \cos^2 2x\, dx$

$= \frac{1}{8}x - \frac{1}{16}\sin 2x - \frac{1}{8}\int\left(1 - \sin^2 2x\right)\cos 2x\, dx - \frac{1}{8}\int\frac{1 + \cos 4x}{2}dx$

$= \frac{1}{8}x - \frac{1}{16}\sin 2x + \frac{1}{16}\sin 2x + \frac{1}{48}\sin^3 2x - \frac{1}{16}x + \frac{1}{16}\frac{1}{4}\sin 4x + C$

$= \frac{1}{16}x + \frac{1}{48}\sin^3 2x + \frac{1}{64}\sin 4x + C$

9. $\int \cos^6 3x \, dx = \int \left(\frac{1+\cos 6x}{2}\right)^3 dx = \frac{1}{8} \int (1 + 3\cos 6x + 3\cos^2 6x + \cos^3 6x) dx$

$= \frac{1}{8} \int \left[1 + 3\cos 6x + \frac{3}{2}(1 + \cos 12x) + (1 - \sin^2 6x)\cos 6x\right] dx$

$= \frac{1}{8} \int \left[1 + 3\cos 6x + \frac{3}{2} + \frac{3}{2}\cos 12x + \cos 6x - \sin^2 6x \cos 6x\right] dx$

$= \frac{1}{8} \int \left[\frac{5}{2} + 4\cos 6x + \frac{3}{2}\cos 12x - \sin^2 6x \cos 6x\right] dx$

$= \frac{1}{8} \left[\frac{5x}{2} + \frac{2}{3}\sin 6x + \frac{1}{8}\sin 12x - \frac{1}{18}\sin^3 6x\right] + C$

$= \frac{5x}{16} + \frac{1}{12}\sin 6x + \frac{1}{64}\sin 12x - \frac{1}{144}\sin^3 6x + C$

11. $\int \sin^2 2\theta \cos^4 2\theta \, d\theta = \int \left(\frac{1-\cos 4\theta}{2}\right)\left(\frac{1+\cos 2\theta}{2}\right)^2 d\theta = \frac{1}{8} \int (1 + \cos 4\theta - \cos^2 4\theta - \cos^3 4\theta) d\theta$

$= \frac{1}{8} \int \left[1 + \cos 4\theta - \frac{1}{2}(1 + \cos 8\theta) - (1 - \sin^2 4\theta)\cos 4\theta\right] d\theta$

$= \frac{1}{8} \int \left[1 + \cos 4\theta - \frac{1}{2} - \frac{1}{2}\cos 8\theta - \cos 4\theta + \sin^2 4\theta \cos 4\theta\right] d\theta$

$= \frac{1}{8} \int \left[\frac{1}{2} - \frac{1}{2}\cos 8\theta + \sin^2 4\theta \cos 4\theta\right] d\theta = \frac{1}{8} \left[\frac{\theta}{2} - \frac{1}{16}\sin 8\theta + \frac{1}{12}\sin^3 4\theta\right] + C$

$= \frac{\theta}{16} - \frac{1}{128}\sin 8\theta + \frac{1}{96}\sin^3 4\theta + C$

13. $\int \sec^2 x \tan^2 x \, dx$. Let $u = \tan x$ and then $du = \sec^2 x \, dx$. Hence, we get $\int u^2 du = \frac{1}{3}u^3 + C = \frac{1}{3}\tan^3 x + C$.

15. First factor $\int \csc^4 x \cot x \, dx = \int \csc^3 x \times (\csc x \cot x \, dx)$. Then let $u = \csc x$ and $du = -\csc x \cot x \, dx$. These produce $-\int u^3 du = -\frac{1}{4}u^4 + C = -\frac{1}{4}\csc^4 x + C_1$ or $\int \csc^4 x \cot x \, dx = \int \csc^2 x(1 + \cot^2 x) \cot x \, dx = \int (\cot^3 x + \cot x) \times \csc^2 x \, dx$. Let $u = \cot x$ and $du = -\csc^2 x \, dx$, and then you have $-\int (u^3 + u) du = -\frac{1}{4}u^4 - \frac{1}{2}u^2 + C = -\frac{1}{4}\cot^4 x - \frac{1}{2}\cot^2 x + C_2$ where $C_2 = C_1 - 0.25$.

17. We first factor the integrand and use a Pythagorean identity: $\int \sin^{1/2} 3\theta \cos^3 3\theta \, d\theta = \int \sin^{1/2} 3\theta (1 - \sin^2 3\theta) \cos 3\theta \, d\theta$. Then multiply and integrate, with the result $\int \left(\sin^{1/2} 3\theta - \sin^{5/2} 3\theta\right) \cos 3\theta \, d\theta = \frac{1}{3} \cdot \frac{2}{3}\sin^{3/2} 3\theta - \frac{1}{3} \cdot \frac{2}{7}\sin^{7/2} 3\theta + C = \frac{2}{9}\sin^{3/2} 3\theta - \frac{2}{21}\sin^{7/2} 3\theta + C$.

19. $\int \csc x \cot^3 x \, dx = \int (\csc^2 x - 1) \csc x \cot x \, dx = -\frac{1}{3}\csc^3 x + \csc x + C$.

21. $\int \tan^3 x \sec^2 x \, dx = \frac{1}{4}\tan^4 x + C$.

23. $\int \tan^6 x \sec^2 x \, dx = \frac{1}{7}\tan^7 x + C$

25. $\int \cot^6 x \, dx = \int \cot^4 x (\csc^2 x - 1) dx = \int \cot^4 x \csc^2 x \, dx - \int \cot^2 x (\csc^2 x - 1) dx = \int \cot^4 x \csc^2 x \, dx - \int \cot^2 x \csc^2 x \, dx + \int \cot^2 x \, dx = -\frac{1}{5}\cot^5 x + \frac{1}{3}\cot^3 x + \int (\csc^2 x - 1) dx = -\frac{1}{5}\cot^5 x + \frac{1}{3}\cot^3 x - \cot x - x + C$. Recall: $\cot^2 x = \csc^2 x - 1$.

27. $\int_0^{\pi/4} \tan^2 x \, dx = \int_0^{\pi/4} (\sec^2 x - 1) dx = \tan x - x\big|_0^{\pi/4} = 1 - \frac{\pi}{4} - 0 = 1 - \frac{\pi}{4}$

29. $\int_0^{\pi/2} \sin^4 x \, dx = \int_0^{\pi/2} \left(\frac{1-\cos 2x}{2}\right)^2 dx$ by a half-angle identity. This expands as $\frac{1}{4} \int_0^{\pi/2} (1 - 2\cos 2x + \cos^2 2x) dx$ which becomes, by another half-angle identity, $\frac{1}{4} \int_0^{\pi/2} \left(1 - 2\cos 2x + \frac{1+\cos 4x}{2}\right) dx$ or $\frac{1}{4} \int_0^{\pi/2} \left(\frac{3}{2} - 2\cos 2x + \frac{\cos 4x}{2}\right) dx = \frac{1}{4}\left[\frac{3x}{2} - \sin 2x + \frac{1}{8}\sin 4x\right]_0^{\pi/2} = \frac{1}{8}\left[\frac{3\pi}{4} - \sin \pi + \frac{1}{8}\sin 2\pi\right] = \frac{1}{4}\left[\frac{3\pi}{4} - 0 + 0\right] = \frac{3\pi}{16}$.

31. Using the disc method, you get $\pi \int_0^{\pi/3} \cos^2 x \, dx = \pi \int_0^{\pi/3} \frac{1+\cos 2x}{2} dx = \frac{\pi}{2}\left[x + \frac{1}{2}\sin 2x\right]_0^{\pi/3} = \frac{\pi}{2}\left[\frac{\pi}{3} + \frac{\sqrt{3}}{4}\right] \approx 2.3251$.

33. **(a)** $I = \dfrac{mr^2}{4} \displaystyle\int_0^{2\pi} \sin^2 \theta \, d\theta$

$= \dfrac{mr^2}{4} \cdot \dfrac{1}{2} \displaystyle\int_0^{2\pi} (1 - \cos 2\theta) d\theta$

$= \dfrac{mr^2}{8} \left[\theta - \dfrac{1}{2} \sin 2\theta \right]_0^{2\pi}$

$= \dfrac{mr^2}{8} \left[(2\pi - 0) - (0 - 0) \right] = \dfrac{mr^2 \pi}{4}$

(b) Substituting $r = 0.1$ m and $m = 12.4$ kg into
$I = \frac{mr^2 \pi}{4}$ produces $I = \frac{(12.4)(0.1)^2 \pi}{4} \approx 0.0974 \text{ kg} \cdot \text{m}^2$.

35. $i_{\text{eff}} = \sqrt{\dfrac{1}{0.5 - 0} \displaystyle\int_0^{0.5} (2 \sin t \sqrt{\cos t})^2 dt}$

$= \sqrt{8 \displaystyle\int_0^{0.5} \left(\sin^2 t \cos t \right) dt}$

$= \sqrt{\dfrac{8}{3} \sin^3 t \Big|_0^{0.5}}$

$= \sqrt{\dfrac{8}{3} \sin^3 0.5} \approx 0.5421$

The effective current from $t = 0$ to $t = 0.5$ s is about 0.5421 A.

≡ 28.5 INTEGRALS RELATED TO INVERSE TRIGONOMETRIC AND INVERSE HYPERBOLIC FUNCTIONS

1. $\int \dfrac{dx}{\sqrt{4 - x^2}} = \arcsin \dfrac{x}{2} + C$

3. $\int \dfrac{dx}{\sqrt{4 - 9x^2}}$. Let $u = 3x$ and $du = 3 \, dx$ so $\frac{1}{3} du = dx$. Substituting, we have $\frac{1}{3} \int \dfrac{du}{\sqrt{4 - u^2}} = \frac{1}{3} \arcsin \dfrac{u}{2} + C = \frac{1}{3} \arcsin \dfrac{3x}{2} + C$.

5. $\int \dfrac{-x \, dx}{\sqrt{9 - x^2}}$. Let $u = 9 - x^2$ and $du = -2x \, dx$ so $\frac{1}{2} du = -x \, dx$. Then, $\frac{1}{2} \int u^{-1/2} du = \frac{1}{2} \cdot \frac{2}{1} u^{1/2} + C = \sqrt{9 - x^2} + C$.

7. $\int \dfrac{dx}{3x \sqrt{9x^2 - 4}}$. Let $u = 3x$, then $du = 3 \, dx$, so $\frac{1}{3} du = dx$. Thus, we can rewrite the integral as $\frac{1}{3} \int \dfrac{du}{u \sqrt{u^2 - 4}} = \frac{1}{3} \cdot \frac{1}{2} \operatorname{arcsec} \dfrac{u}{2} + C = \frac{1}{6} \operatorname{arcsec} \dfrac{3x}{2} + C$.

9. $\int \dfrac{dx}{1 + (3 - x)^2}$. Here $u = 3 - x$ and $du = -dx$ and so we get $-\int \dfrac{du}{1 + u^2} = -\arctan u + C = -\arctan(3 - x) + C$.

11. $\int \dfrac{4x - 6}{4x^2 + 25} dx = \int \dfrac{4x}{4x^2 + 25} dx - \int \dfrac{6}{4x^2 + 25} dx$. In the first integral let $u = 4x^2 + 25$; $du = 8x \, dx$ and we get by substitution $2 \int \dfrac{du}{u} = 2 \ln |u| = 2 \ln (4x^2 + 25)$. The second integral is $-6 \int \dfrac{dx}{4x^2 + 25} = -6 \cdot \frac{1}{10} \arctan \dfrac{2x}{5}$. Putting these together we have $2 \ln \times (4x^2 + 25) - \frac{3}{5} \arctan \dfrac{2x}{5} + C$

13. Completing the square, we obtain $\int \dfrac{dx}{x^2 + 6x + 10} = \int \dfrac{dx}{(x + 3)^2 + 1} = \arctan(x + 3) + C$.

15. $\int \dfrac{x \, dx}{1 + x^4}$; Let $u = x^2$ and then $du = 2x \, dx$. Substituting, we obtain $\frac{1}{2} \int \dfrac{du}{1 + u^2} = \frac{1}{2} \cdot \arctan u + C = \frac{1}{2} \arctan x^2 + C$.

17. $\int \dfrac{\sin x \, dx}{\sqrt{1 - \cos^2 x}}$. Let $u = \cos x$ and $du = -\sin x \, dx$. Substituting produces $-\int \dfrac{du}{\sqrt{1 - u^2}} = -\arcsin u + C = -\arcsin(\cos x) + C = -\left(\frac{\pi}{2} - x\right) + C = x - \frac{\pi}{2} + C$.

19. $\int_0^{\pi/4} \dfrac{\cos x \, dx}{1 + \sin^2 x} = \arctan(\sin x) \Big|_0^{\pi/4} = \arctan \left(\sin \frac{\pi}{4} \right) - \arctan(\sin 0) = \arctan \frac{\sqrt{2}}{2} - \arctan 0 = 0.6155 - 0 = 0.6155$.

21. $\int \dfrac{dx}{\sqrt{x^2 - 25}} = \cosh^{-1} \dfrac{x}{5} + C$.

23. $\int \dfrac{dx}{\sqrt{25 + 9x^2}}$. Let $u = 3x$ and $du = 3 \, dx$ and so $\frac{1}{3} du = dx$. Substituting produces $\int \dfrac{du}{\sqrt{25 + u^2}} = \frac{1}{3} \sinh^{-1} \dfrac{u}{5} + C = \frac{1}{3} \sinh^{-1} \dfrac{3x}{5} + C$.

25. $\int_0^{0.5} \dfrac{dx}{\sqrt{1 - x^2}} = \arcsin \Big|_0^{0.5} = \arcsin 0.5 - \arcsin 0 = \frac{\pi}{6} - 0 = \frac{\pi}{6} \approx 0.5236$

27. $\int_1^5 \dfrac{dx}{x^2 - 4x + 13} = \int_1^5 \dfrac{dx}{(x^2 - 4x + 4) + 9} = \int_1^5 \dfrac{dx}{3^2 + (x - 2)^2} = \frac{1}{3} \arctan \dfrac{(x - 2)}{3} \Big|_1^5 = \frac{1}{3} \left[\arctan 1 - \arctan \left(\frac{-1}{3} \right) \right] = \frac{1}{3} \left(\frac{\pi}{4} + 0.32175 \right) \approx 0.36905$

29. Let $u = 4 \tan x$ and $du = 4 \sec^2 x \, dx$. Then, $\int_0^{\pi/6} \dfrac{\sec^2 x \, dx}{1 + 16 \tan^2 x} = \frac{1}{4} \int_0^{\pi/6} \dfrac{4 \sec^2 x \, dx}{1 + (4 \tan x)^2} = \frac{1}{4} \int_0^{4\sqrt{3}/3} \dfrac{du}{1 + u^2}$

$= \frac{1}{4} \arctan u \big|_0^{4\sqrt{3}/3} = \frac{1}{4} \left[\arctan \left(\frac{4\sqrt{3}}{3} \right) - \arctan 0 \right] =$

$\frac{1}{4} \arctan \left(\frac{4\sqrt{3}}{3} \right) \approx 0.2905$

31. $\int_0^1 \frac{1}{1+x^2} dx = \arctan x \big|_0^1 = \arctan 1 - \arctan 0 =$
$\frac{\pi}{4} - 0 = \frac{\pi}{4}.$

33. These two graphs intersect at (0, 1) and, from 0 to 1, e^x is always larger. Using the washer method

we get $\pi \int_0^1 \left[(e^x)^2 - \left(\frac{1}{\sqrt{x^2-1}} \right)^2 \right] dx$

$= \pi \int_0^1 \left(e^{2x} - \frac{1}{x^2-1} \right) dx$

$= \pi \int_0^1 e^{2x} dx - \pi \int_0^1 \frac{1}{x^2+1} dx$

$= \frac{\pi}{2} e^{2x} \big|_0^1 - \pi \arctan x \big|_0^1$

$= \frac{\pi}{2} (e^2 - 1) - \pi (\arctan 1 - \arctan 0)$

$= \frac{\pi}{2} (e^2 - 1) - \pi \left(\frac{\pi}{4} - 0 \right)$

$= \frac{\pi}{2} \left(e^2 - 1 - \frac{\pi}{2} \right) \approx 7.5685$

35.

$y = \sqrt{4 - x^2}$

$y' = \frac{1}{2} (4 - x)^{-1/2} \cdot (-2x)$

$= -\frac{x}{\sqrt{4-x^2}}$

$y'^2 = \frac{x^2}{4 - x^2}$

$\int_{-1}^1 \sqrt{1 + (y')^2} = \int_{-1}^1 \sqrt{1 + \frac{x^2}{4-x^2}} \, dx$

$= \int_{-1}^1 \sqrt{\frac{4 - x^2 + x^2}{4 - x^2}} \, dx$

$= \int_{-1}^1 \sqrt{\frac{4}{4 - x^2}} \, dx$

$= \int_{-1}^1 \frac{2 \, dx}{\sqrt{4 - x^2}}$

$= 2 \arcsin \frac{x}{2} \Big|_{-1}^1$

$= 2 \left[\arcsin \frac{1}{2} - \arcsin \frac{-1}{2} \right]$

$= 2 \left[\frac{\pi}{6} - \frac{-\pi}{6} \right] = \frac{2\pi}{3}$

37.
$$f'(x) = -\frac{1}{x\sqrt{1-x^2}} + \frac{x}{\sqrt{1-x^2}}$$

$f(x) = \int f'(x) dx = \int \frac{-1}{x\sqrt{1-x^2}} dx$

$+ \int \frac{x}{\sqrt{1-x^2}} dx$

$= -\text{sech}^{-1} x + \int \frac{x \, dx}{\sqrt{1-x^2}}.$

For the second integral let $u = 1 - x^2$, then $du = -2x \, dx$ or $-\frac{1}{2} du = x \, dx$. These substitutions make $\int \frac{x \, dx}{\sqrt{1-x^2}} = -\frac{1}{2} \int \frac{du}{\sqrt{u}} = -\frac{1}{2} \frac{2}{1} u^{1/2} = -\sqrt{u} = -\sqrt{1 - x^2}$. Putting these together we get $f(x) = -\text{sech}^{-1} x - \sqrt{1 - x^2} + C$. Since $f(1) = 0$ we have $C = \text{sech}^{-1} 1 + \sqrt{1 - 1^2} = \text{sech}^{-1} 1$. Thus, the desired equation is $y = -\text{sech}^{-1} x - \sqrt{1 - x^2} + \text{sech}^{-1} 1$.

≡ 28.6 TRIGONOMETRIC SUBSTITUTION

1. $\int \frac{x}{\sqrt{9-x^2}} dx$. This can be done with trigonometric substitutions but it is much easier to do the following. Let $u = 9 - x^2$; $du = -2x \, dx$ so $x \, dx = -\frac{1}{2} du$. Substituting yields $-\frac{1}{2} \int \frac{du}{\sqrt{u}} = -\frac{1}{2} \frac{2}{1} u^{1/2} + C = -\sqrt{9 - x^2} + C.$

3. $\int \frac{x^2}{\sqrt{9-x^2}} dx$. Let $x = 3 \sin \theta$ and $dx = 3 \cos \theta \, d\theta$.

Substituting produces

$\int \frac{(3 \sin \theta)^2 3 \cos \theta \, d\theta}{\sqrt{9 - 9\sin^2 \theta}} = \int 9 \sin^2 \theta \, d\theta$

$= 9 \int \frac{1 - \cos 2\theta}{2}$

$= \frac{9}{2} \int 1 \, d\theta - \frac{9}{2} \int \cos 2\theta \, d\theta$

$= \frac{9}{2} \theta - \frac{9}{2} \cdot \frac{1}{2} \sin 2\theta + C.$

Since $x = 3 \sin \theta$, we have $\sin \theta = \frac{x}{3}$ and $\theta = \sin^{-1} \frac{x}{3}$. Hence $\sin 2\theta = 2 \sin \theta \cos \theta$, which yields $\sin 2\theta = 2 \cdot \frac{x}{3} \cdot \frac{\sqrt{9-x^2}}{3} = \frac{2}{9} x \sqrt{9 - x^2}$. Thus, the final answer is $\frac{9}{2} \arcsin \frac{x}{3} - \frac{1}{2} x \sqrt{9 - x^2} + C.$

5. $\int x^3\sqrt{9-x^2}\,dx$. Again, this can be done using trig substitutions but the following method is easier. Let $u = 9-x^2$ and $du = -2x\,dx$, and so $x^2 = 9-u$ and $-\frac{1}{2}du = x\,dx$. Thus, $\int x^3\sqrt{9-x^2} = \int x^2\sqrt{9-x^2}x\,dx$. Now substitute

$$\int (9-u)\sqrt{u}\left(-\frac{1}{2}du\right)$$

$$= -\frac{1}{2}\int \left(9\sqrt{u}-u^{3/2}\right)du$$

$$= -\frac{1}{2}\cdot 9\cdot\frac{2}{3}u^{3/2}+\frac{1}{2}\frac{2}{5}u^{5/2}+C$$

$$= -3\left(9-x^2\right)^{3/2}+\frac{1}{5}\left(9-x^2\right)^{5/2}+C$$

$$= -3\left(9-x^2\right)^{3/2}+\frac{1}{5}\left(9-x^2\right)\left(9-x^2\right)^{3/2}+C$$

$$= -\frac{15}{5}\left(9-x^2\right)^{3/2}$$

$$\quad +\frac{1}{5}\left(9-x^2\right)\left(9-x^2\right)^{3/2}+C$$

$$= -\frac{1}{5}\left(9-x^2\right)^{3/2}\left(6+x^2\right)+C.$$

7. $\int \frac{dx}{\sqrt{x^2-9}}$. Let $x = 3\sec\theta$ and then $dx = 3\sec\theta\times\tan\theta\,d\theta$. Substitution into the given integral yields $\int\frac{3\sec\theta\tan\theta}{\sqrt{9\sec^2\theta-9}}d\theta = \int\frac{3\sec\theta\tan\theta d\theta}{3\tan\theta} = \int\sec\theta d\theta = \ln|\sec\theta+\tan\theta|+C$. Since $\sec\theta = \frac{x}{3}$ and $\tan\theta = \frac{\sqrt{x^2-9}}{3}$, the answer is $\ln\left|\frac{x}{3}+\frac{\sqrt{x^2-9}}{3}\right|+C = \ln\frac{1}{3}\left|x+\sqrt{x^2-9}\right|+C = \ln\frac{1}{3}+\ln\left|x+\sqrt{x^2-9}\right|+C$ or $\ln\left|x+\sqrt{x^2-9}\right|+k$ where $k = C+\ln\frac{1}{3} = C-\ln 3$.

9. $\int \left(x^2-9\right)^{1/2}dx = \int\sqrt{x^2-9}dx$. Let $x = 3\sec\theta$, then $dx = 3\sec\theta\tan\theta\,d\theta$. Substituting yields $\int\sqrt{9\sec^2\theta-9}\cdot 3\sec\theta\tan\theta\,d\theta = \int 9\tan^2\theta\sec\times\theta\,d\theta = 9\int\left(\sec^2\theta-1\right)\sec\theta\,d\theta = 9\int\left(\sec^3\theta-\sec\theta\right)d\theta = 9\int\sec^2\theta d\theta - 9\int\sec\theta d\theta$. The first integral is example 20.50 and the second is a formula. They yield $9\left[\frac{1}{2}\sec\theta\tan\theta+\frac{1}{2}\ln|\sec\theta+\tan\theta|-\ln|\sec\theta+\tan\theta|\right] = \frac{9}{2}\sec\theta\tan\theta-\frac{9}{2}\ln|\sec\theta+\tan\theta|$. Since $\sec\theta = \frac{x}{3}$ and $\tan\theta = \frac{\sqrt{x^2-9}}{3}$, we get $\int\left(x^2-9\right)^{1/2}dx = \frac{9}{2}\cdot\frac{x}{3}\cdot\frac{\sqrt{x^2-9}}{3}-\frac{9}{2}\ln\left|\frac{x}{3}+\frac{\sqrt{x^2-9}}{3}\right|+K = \frac{1}{2}\left[x\sqrt{x^2-9}-9\ln\left|x+\sqrt{x^2-9}\right|\right]+C$ where $C = K-\frac{9}{2}\ln 3$.

11. $\int \frac{x}{\sqrt{x^2+9}}dx$. Letting $u = x^2+9$ and $du = 2x\,dx$, produces $\frac{1}{2}\int\frac{du}{\sqrt{u}} = \frac{1}{2}\cdot\frac{2}{1}u^{1/2}+C = \sqrt{x^2+9}+C$.

13. $\int \frac{dx}{x^2+9} = \frac{1}{3}\arctan\frac{x}{3}+C$.

15. $\int x^3\sqrt{9+x^2}dx$. Let $u = 9+x^2$ and $du = 2x\,dx$ or $\frac{1}{2}du = x\,dx$. Thus, $x^2 = u-9$. Substituting these in the given integral yields $\frac{1}{2}\int(u-9)\sqrt{u}\,du = \frac{1}{2}\int\left(u^{3/2}-9u^{1/2}\right)du = \frac{1}{2}\cdot\frac{2}{5}u^{5/2}-\frac{9}{2}\cdot\frac{2}{3}u^{3/2}+C = \frac{1}{5}u^{5/2}-3u^{3/2}+C = \frac{1}{5}uu^{3/2}-3u^{3/2} = \frac{1}{5}\left(9+x^2\right)\left(9+x^2\right)^{3/2}-3\left(9+x^2\right)^{3/2} = \left(\frac{x^2-6}{5}\right)\left(9+x^2\right)^{3/2}$.

17. $\int \frac{\sqrt{4-3x^2}}{x^4}dx$. Let $\sqrt{3}x = 2\sin\theta$, so $x = \frac{2}{\sqrt{3}}\sin\theta$ and $dx = \frac{2}{\sqrt{3}}\cos\theta\,d\theta$. Substitution produces

$$\int \frac{\sqrt{4-4\sin^2\theta}}{\left(\frac{2}{\sqrt{3}}\sin\theta\right)^4}\frac{2}{\sqrt{3}}\cos\theta\,d\theta$$

$$= \int \frac{2\cos\theta\frac{2}{\sqrt{3}}\cos\theta\,d\theta}{\frac{16}{9}\sin^4\theta}$$

$$= \frac{4}{\sqrt{3}}\cdot\frac{9}{16}\int\frac{\cos^2\theta}{\sin^4\theta}d\theta$$

$$= \frac{3\sqrt{3}}{4}\int\csc^2\theta\cot^2\theta d\theta$$

$$= \frac{-3\sqrt{3}}{4}\cdot\frac{1}{3}\cot^3\theta+C$$

$$= \frac{-\sqrt{3}}{4}\left(\frac{\sqrt{4-3x^2}}{\sqrt{3}x}\right)^3+C$$

$$= -\frac{1}{12}\frac{(\sqrt{4-3x^2})^3}{x^3}+C$$

$$= -\frac{1}{12}\frac{\left(4-3x^2\right)^{3/2}}{x^3}+C$$

19. $\int \frac{x^3}{\sqrt{9x^2+4}}dx$; $u = 3x$; $3x = 2\tan\theta$ and so $x = \frac{2}{3}\tan\theta$; and $dx = \frac{2}{3}\sec^2\theta\,d\theta$. Substituting these values in the given integral produces

$$\int \frac{\left(\frac{2}{3}\tan\theta\right)^3}{\sqrt{(2\tan\theta)^2+4}}\frac{2}{3}\sec^2\theta\,d\theta$$

$$= \int \frac{\frac{16}{81}\tan^3\theta\sec^2\theta}{2\sec\theta}d\theta$$

$$= \frac{8}{81}\int\tan^3\theta\sec\theta\,d\theta$$

$$= \frac{8}{81}\int\tan\theta\left(\sec^2\theta-1\right)\sec\theta\,d\theta$$

$$= \frac{8}{81}\int\left(\tan\sec^3\theta-\tan\theta\sec\theta\right)d\theta$$

$$= \frac{8}{81}\left(\frac{1}{3}\sec^3\theta - \sec\theta\right) + C$$

$$= \frac{8}{81}\left[\frac{1}{3}\left(\frac{\sqrt{9x^2+4}}{2}\right)^3 - \left(\frac{\sqrt{9x^2+4}}{2}\right)\right] + C$$

$$= \frac{1}{243}\sqrt{(9x^2+4)^3} - \frac{4}{81}\sqrt{9x^2+4} + C$$

21. $\int \frac{\sqrt{x^2+1}}{x^2}dx$. Let $x = \tan\theta$ and $dx = \sec^2\theta$. Then

$$\int \frac{\sqrt{\tan^2\theta+1}}{\tan^2\theta}\sec^2\theta\, d\theta = \int \frac{\sec\theta}{\tan^2\theta}\sec^2\theta\, d\theta$$

$$= \int \frac{\sec^3\theta}{\tan^2\theta}d\theta = \int \frac{(1+\tan^2\theta)\sec\theta}{\tan^2\theta}d\theta$$

$$= \int \frac{\sec\theta}{\tan^2\theta} + \int \sec\theta\, d\theta$$

$$= \int \frac{\frac{1}{\cos\theta}}{\frac{\sin^2\theta}{\cos^2\theta}}d\theta + \int \sec\theta\, d\theta$$

$$= \int \frac{\cos\theta}{\sin^2\theta}d\theta + \int \sec\theta\, d\theta$$

$$= -(\sin\theta)^{-1} + \ln|\sec\theta + \tan\theta| + C$$

$$= -\frac{\sqrt{x^2+1}}{x} + \ln\left(\sqrt{x^2+1}+x\right) + C$$

23. $\int \sqrt{4-(x-1)^2}dx$. Here we have $a = 2$ and $u = x-1$. Let $x-1 = 2\sin\theta$, then $x = 2\sin\theta + 1$ and $dx = 2\cos\theta\, d\theta$. Substitution yields $\int \sqrt{4-4\sin^2\theta}(2\cos\theta)\, d\theta = \int(2\cos\theta)(2\cos\theta)\times$

$d\theta = 4\int\cos^2\theta\, d\theta = 4\int \frac{1+\cos 2\theta}{2}d\theta = 2\int(1 + \cos 2\theta)\, d\theta + 2\theta + \sin 2\theta + C$. Thus $\theta = \sin^{-1}\frac{x-1}{2}$ and using the identity $\sin 2\theta = 2\sin\theta\cos\theta = 2\cdot\frac{x-1}{2}\cdot\frac{\sqrt{4-(x-1)^2}}{2}$ so the answer is $2\sin^{-1}\left(\frac{x-1}{2}\right) + \frac{(x-1)\sqrt{4-(x-1)^2}}{2} + C$

25. $\int \frac{dx}{\sqrt{(x-1)^2-4}}dx$. Let $x-1 = 2\sec\theta$ and $dx = 2\sec\theta\tan\theta\, d\theta$. Then, you get

$$\int \frac{\sec\theta\tan\theta\, d\theta}{\sqrt{(2\sec\theta)^2-4}} = \int \frac{2\sec\theta\tan\theta\, d\theta}{2\tan\theta}$$

$$= \int \sec\theta\, d\theta$$

$$= \ln|\sec\theta + \tan\theta| + C$$

$$= \ln\left|\frac{x-1}{2}\right.$$

$$\left. + \frac{\sqrt{(x-1)^2-4}}{2}\right| + C$$

$$= \ln\frac{1}{2}\left|x-1\right.$$

$$\left. + \sqrt{(x-1)^2-4}\right| + C$$

$$= \ln\left|x-1\right.$$

$$\left. + \sqrt{(x-1)^2-4}\right| + K$$

where $K = C - \ln 2$.

27. Completing the square produces $x^2-2x-3 = x^2-2x+1-4 = (x-1)^2-4$. Thus $\int \sqrt{x^2-2x-3}\, dx = \int \sqrt{(x-1)^2-4}\, dx$. Let $x-1 = 2\sec\theta$, then $dx = 2\sec\theta\tan\theta\, d\theta$. Substitution gives

$$\int \sqrt{(2\sec\theta)^2-4}(2\sec\theta\tan\theta)\, d\theta$$

$$= \int 2\tan\theta \cdot 2\sec\theta\tan\theta\, d\theta$$

$$= 4\int \tan^2\theta\sec\theta\, d\theta$$

$$= 4\int \left(\sec^2\theta - 1\right)\sec\theta\, d\theta$$

$$= 4\int \sec^3\theta\, d\theta - 4\int \sec\theta\, d\theta$$

$$= 4\cdot\left[\frac{1}{2}\sec\theta\tan\theta + \frac{1}{2}\ln|\sec\theta+\tan|\right.$$

$$\left. - 4\ln|\sec\theta+\tan\theta|\right] + C$$

$$= 4\left[\frac{1}{2}\sec\theta\tan\theta\right.$$

$$\left. - \frac{1}{2}\ln|\sec\theta+\tan\theta|\right] + C$$

$$= 2\cdot\frac{x-1}{2}\cdot\frac{\sqrt{x^2-2x-3}}{2}$$

$$- 2\ln\left|\frac{x-1}{2} + \frac{\sqrt{x^2-2x-3}}{2}\right| + C$$

$$= \frac{1}{2}(x-1)\sqrt{x^2-2x-3}$$

$$- 2\ln\left|x-1 + \sqrt{x^2-2x-3}\right| + K$$

where $K = C + \ln 2$.

29. $\int \frac{dx}{(x-3)\sqrt{x^2-6x+25}} = \int \frac{dx}{(x-3)\sqrt{(x-3)^2+16}}$. Here $u = x-3$ and $a = 4$. Let $x-3 = 4\tan\theta$, so $dx = 4\sec^2\theta\, d\theta$. Then, $\int \frac{4\sec^2\theta\, d\theta}{4\tan\theta\sqrt{(4\tan\theta)^2+16}} = $

$\int \frac{4\sec^2\theta\, d\theta}{4\tan\theta 4\sec\theta} = \frac{1}{4}\int \frac{\sec\theta}{\tan\theta}d\theta = \frac{1}{4}\int \frac{1}{\sin\theta} = \frac{1}{4}\cdot$

$\int \csc\theta\, d\theta = \frac{1}{4}\ln|\csc\theta - \cot\theta| + C =$

$\frac{1}{4}\ln\left|\frac{\sqrt{x^2-6x+25}}{x-3} - \frac{4}{x-3}\right| + C =$

$\frac{1}{4}\ln\left|\frac{\sqrt{x^2-6x+25}-4}{x-3}\right| + C.$

31. Using the trigonometric substitution $x = 3\sin\theta$, produces $\sqrt{9-x^2} = 3\cos\theta$ and $dx = 3\cos\theta\, d\theta$. Substituting these values, we obtain

$$\int_1^3 \frac{\sqrt{9-x^2}}{x^2}dx = \int_{x=1}^{x=3} \frac{3\cos\theta}{(3\sin\theta)^2}3\cos\theta\, d\theta$$

$$= \int_{x=1}^{x=3} \cot^2\theta\, d\theta$$

$$= \int_{x=1}^{x=3} \left(\csc^2\theta - 1\right) d\theta$$

$$= [-\cot\theta - \theta]_{x=1}^{x=3}$$

$$= \left[-\frac{\sqrt{9-x^2}}{x} - \arcsin\left(\frac{x}{3}\right)\right]_1^3$$

$$= \left(-0 - \frac{\pi}{2}\right)$$

$$\quad - (-2.82843 - 0.33984)$$

$$\approx 1.59747$$

33. Let $x = 2\sec\theta$, $\theta = \sec^{1-}\left(\frac{\pi}{2}\right)$, and $dx = 2\sec\theta\tan\theta\, d\theta$ so that $\sqrt{x^2-4} = 2\tan\theta$. Then

$$\int_4^8 \frac{dx}{x\sqrt{x^2-4}} = \int_{x=4}^{x=8} \frac{2\sec\theta\tan\theta\, d\theta}{2\sec\theta 2\tan\theta}$$

$$= \frac{1}{2}\int_{x=4}^{x=8} d\theta = \frac{1}{2}\theta\Big|_{x=4}^{x=8}$$

$$= \frac{1}{2}\sec^{-1}\left(\frac{x}{2}\right)\Big|_4^8$$

$$= \frac{1}{2}\left(\sec^{-1}4 - \sec^{-1}2\right)$$

$$\approx 0.1355$$

35. Let $x = 6\sin\theta$, $dx = 6\cos\theta\, d\theta$, $\sqrt{36-x^2} = 6\cos\theta$, and $\theta = \arcsin\left(\frac{\pi}{6}\right)$. Thus, when $x = 0$, $\theta = 0$ and when $x = 6$, $\theta = \frac{\pi}{2}$. Then

$$\int_0^6 x^2\sqrt{36-x^2}dx = \int_0^{\pi/2} (6\sin\theta)^2$$

$$\times x(6\cos\theta)(6\cos\theta)\, d\theta$$

$$= 1296\int_0^{\pi/2} \sin^2\theta\cos^2\theta\, d\theta$$

$$= 324\int_0^{\pi/2} \sin^2 2\theta\, d\theta$$

$$= 162\int_0^{\pi/2} (1 - \cos 2\theta)\, d\theta$$

$$= 162\left[\theta - \frac{1}{2}\sin 2\theta\right]_0^{\pi/2}$$

$$= 81\pi \approx 254.4690$$

37. The area under $\frac{x^3}{\sqrt{16-x^2}}$ from 0 to 3 is $\int_0^3 \frac{x^2 dx}{\sqrt{16-x^2}}$. Let $x = 4\sin\theta$, then $dx = 4\cos\theta\, d\theta$. Substituting gives

$$\int \frac{(4\sin\theta)^3(4\cos\theta)\, d\theta}{\sqrt{16-(4\sin\theta)^2}}$$

$$= \int \frac{64\sin^3\theta 4\cos\theta}{4\cos\theta}$$

$$= 64\int \sin^3\theta\, d\theta = 64\int \left(1 - \cos^2\theta\right)\sin\theta\, d\theta$$

$$= 64\left[-\cos\theta + \frac{1}{3}\cos^3\theta\right]$$

$$= 64\left[\frac{1}{3}\left(\frac{\sqrt{16-x^2}}{4}\right)^3 - \frac{\sqrt{16-x^2}}{4}\right]_0^3$$

$$= \frac{1}{3}\sqrt{7}^3 - 16\sqrt{7} - \frac{1}{3}\cdot 4^3 + 16\cdot 4$$

$$= \frac{1}{3}(7\sqrt{7} - 48\sqrt{7} - 64 + 192)$$

$$= \frac{1}{3}(128 - 41\sqrt{7}) \approx 6.5080654$$

39. The area described is $\int_0^4 \sqrt{9+x^2}dx$. Let $x = 3\tan\theta$ and $dx = 3\sec^2\theta\, d\theta$. Then, $\int \sqrt{9+9\tan^2\theta} \times (3\sec^2\theta)\, d\theta = \int(3\sec\theta)(3\sec^2\theta)\, d\theta = 9\int \sec^3\theta\, d\theta = 9\left[\frac{1}{2}\sec\theta\tan\theta + \frac{1}{2}\ln|\sec\theta + \tan\theta|\right.$

$$= \frac{9}{2}\left[\frac{\sqrt{9+x^2}}{3}\cdot\frac{x}{3}\right] + \ln\left[\frac{\sqrt{9+x^2}}{3} + \frac{x}{3}\right]\Big|_0^4 \text{ or, we get,}$$

$\frac{9}{2}\left[\frac{5}{3}\cdot\frac{4}{3} + \ln|\frac{5}{3} + \frac{4}{3}| - \frac{3}{3}\cdot\frac{0}{3} - \ln|\frac{3}{3}|\right] = \frac{9}{2}\left[\frac{20}{9} + \ln|3| - 0 - 0\right] = 5 + \frac{9}{2}\ln 3$ or $5 + \ln 3^{9/2} \approx 9.9438$.

41. To find the desired area we first solve $\frac{x^2}{25} + \frac{y^2}{4} = 1$ for $y : \frac{y^2}{4} = 1 - \frac{x^2}{25}$ so $y^2 = 4 - \frac{4x^2}{25}$ or $y^2 = \frac{100-4x^2}{25}$ and we get $y = \sqrt{\frac{100-4x^2}{25}} = \frac{2}{5}\sqrt{25 - x^2}$. We also need $y' = \frac{2}{5}(25 - x^2)^{-1/2}\cdot\frac{1}{2}\cdot 2x = \frac{2}{5}x(25 - x^2)^{-1/2} = \frac{2x}{5\sqrt{25-x^2}}$. Thus, $y'^2 = \frac{4x^2}{25(25-x^2)}$ and so

$$S = \int_{-5}^5 2\pi y\sqrt{1 + y'^2}dx$$

$$= \int_{-5}^5 2\pi\left(\frac{2}{5}\sqrt{25 - x^2}\right)\sqrt{1 + \frac{4x^2}{25(25-x^2)}}dx$$

$$= \int_{-5}^5 2\pi\cdot\frac{2}{5}\sqrt{25 - x^2}\sqrt{\frac{625 - 21x^2}{25(25 - x^2)}}dx$$

$$= \int_{-5}^5 2\pi\cdot\frac{2}{5}\sqrt{25 - x^2}\cdot\frac{1}{5}\frac{\sqrt{625 - 21x^2}}{\sqrt{25 - x^2}}dx$$

$$= \frac{4\pi}{25}\int_{-5}^5 \sqrt{625 - 21x^2}dx.$$

If you let $\sqrt{21}x = 25\sin\theta$, then $dx = \frac{25}{\sqrt{21}}\cos\theta\, d\theta$. Substitution gives

$$\frac{4\pi}{25} \int \sqrt{625 - 625\sin^2\theta} \cdot \frac{25}{\sqrt{21}}\cos\theta\, d\theta$$

$$= \frac{4\pi}{25} \int 25\cos\theta \cdot \frac{25}{\sqrt{21}}\cos\theta\, d\theta$$

$$= \frac{100\pi}{25} \int \cos^2\theta\, d\theta = \frac{100\pi}{\sqrt{21}} \int \frac{1 + \cos 2\theta}{2}\, d\theta$$

$$= \frac{50\pi}{\sqrt{21}}\left[\theta + \frac{1}{2}\sin 2\theta\right] = \frac{50\pi}{\sqrt{21}}[\theta + \sin\theta\cos\theta]$$

$$= \frac{50\pi}{\sqrt{21}}\left[\sin^{-1}\frac{\sqrt{21}x}{25} + \frac{\sqrt{21}x}{25}\right.$$

$$\left.\times \frac{\sqrt{625 - 21x^2}}{25}\right]_{-5}^{5}$$

$$= \frac{50\pi}{\sqrt{21}}\left[\sin^{-1}\frac{\sqrt{21}}{5}\right.$$

$$\left. + \frac{\sqrt{21}\cdot 5\sqrt{625 - 21\cdot 25}}{625}\right] (2)$$

$$= \frac{100\pi}{\sqrt{21}}\left[\sin^{-1}\frac{\sqrt{21}}{5} + \frac{\sqrt{21}\cdot 50}{625}\right]$$

$$= \frac{100\pi \cdot 50}{625} + \frac{100\pi}{\sqrt{21}}\sin^{-1}\frac{\sqrt{21}}{5}$$

$$= 8\pi + \frac{100\pi}{\sqrt{21}}\sin^{-1}\frac{\sqrt{21}}{5} \approx 104.6073.$$

43. Let $u = t = \tan\theta$ and $du = dt = \sec^2\theta\, d\theta$. Then

$$\int \frac{\sqrt{t^2 + 1}}{9t^2}dt = \int \frac{\sqrt{\tan^2\theta + 1}}{9\tan^2\theta}\sec^2\theta\, d\theta$$

$$= \frac{1}{9}\int \frac{\sec^3\theta}{\tan^2\theta}d\theta = \frac{1}{9}\int \csc^2\sec\theta\, d\theta$$

$$= \frac{1}{9}\int \left(1 + \cot^2\theta\right)\sec\theta\, d\theta$$

$$= \frac{1}{9}\int \left(\sec\theta + \frac{\cos\theta}{\sin^2\theta}\right)d\theta$$

$$= \frac{1}{9}\left[\ln|\sec\theta + \tan\theta| - (\sin\theta)^{-1}\right]$$

$$= \frac{1}{9}\left[\ln\left(\sqrt{1 + t^2} + t\right) - \frac{\sqrt{1 + t^2}}{t}\right]$$

So from $t = 0.5$ to $t = 1$, we have $i = \frac{1}{9}\times$

$$\left[\ln\left(\sqrt{1 + t^2} + t\right) - \frac{\sqrt{1+t^2}}{t}\right]_{0.5}^{1} \approx 0.1829\text{ A}$$

≡ 28.7 INTEGRATION BY PARTS

1. $\int x\ln x\, dx$. Completing the table, we get

$u = \ln x$	$v = \frac{1}{2}x^2$
$du = \frac{1}{x}dx$	$dv = x\, dx$

Thus, using the table, we obtain $\int x\ln x\, dx = \frac{1}{2}x^2\ln x - \int \frac{1}{2}x^2 \cdot \frac{1}{x}dx = \frac{1}{2}x^2\ln x - \frac{1}{2}\int x\, dx = \frac{1}{2}x^2\ln x - \frac{1}{4}x^2 + C$ or $\frac{1}{2}x^2(\ln x - \frac{1}{2}) + C$.

3. $\int x\sin 2x\, dx$

$u = x$	$v = -\frac{1}{2}\cos 2x$
$du = dx$	$dv = \sin 2x\, dx$

Using the table, we see that $\int x\sin 2x\, dx = -\frac{1}{2}x\cos 2x - \int -\frac{1}{2}\cos 2x\, dx = -\frac{1}{2}x\cos 2x + \frac{1}{4}\sin 2x + C$.

5. $\int x^2\ln x\, dx$

$u = \ln x$	$v = \frac{1}{3}x^3$
$du = \frac{1}{x}dx$	$du = x^2\, dx$

Thus, from the table, we see that $\int x^2\ln x\, dx = \frac{1}{3}x^3\ln x - \int \frac{1}{3}x^3 \cdot \frac{1}{x}dx = \frac{1}{3}x^3\ln x - \frac{1}{9}x^3 + C$.

7. $\int x\sin^{-1}x^2\, dx$.

$u = \sin^{-1}x^2$	$v = \frac{1}{2}x^2$
$du = \frac{2x}{\sqrt{1-x^2}}dx$	$dv = x\, dx$

Hence, $\int x\sin^{-1}x^2\, dx = \frac{1}{2}x^2\sin^{-1}x^2 - \int \frac{x^3}{\sqrt{1-x^2}}dx$. For the second integral let $w = 1 - x^4$, $dw = -4x^3dx$ or $\frac{1}{4}dw = -x^3dx$. Substituting, we get

$\frac{1}{2}x^2\sin^{-1}x^2 - \int \frac{x^3}{1-x^2}dx = \frac{1}{2}x^2\sin^{-1}x^2 - \frac{1}{4}\int \frac{dw}{\sqrt{w}} = \frac{1}{2}x^2\sin^{-1}x^2 - \frac{1}{4}\int w^{-1/2}dw = \frac{1}{2}x^2\sin^{-1}x^2 + \frac{1}{2}w^{1/2} + C = \frac{1}{2}x^2\sin^{-1}x^2 + \frac{1}{2}(1-x^4)^{1/2} + C$. The final answer is $\frac{1}{2}x^2\sin^{-1}x^2 + \frac{1}{2}(1-x^4)^{1/2} + C = \frac{1}{2}\left[x^2\sin^{-1}x^2 + \sqrt{1-x^4}\right] + C$.

9. $\int x^3e^{2x}dx$

$u = x^3$	$v = \frac{1}{2}e^{2x}$
$du = 3x^2dx$	$dv = e^{2x}dx$

So, $\int x^3e^{2x}\, dx = \frac{1}{2}x^3e^{2x} - \frac{3}{2}\int x^2e^{2x}dx$. Using a second table produces

$u = x^2$ | $v = \frac{1}{2}e^{2x}$

$du = 2x\,dx$ | $dv = e^{2x}dx$

Combined with the result from the first table, this produces $\frac{1}{2}x^3e^{2x} - \frac{3}{2}\int x^2e^{2x}dx = \frac{1}{2}x^3e^{2x} - \frac{3}{2}\left[\frac{1}{2}x^2e^{2x} - \int xe^{2x}dx\right]$. A third table produces

$u = x$ | $v = \frac{1}{2}e^{2x}$

$du = dx$ | $dv = e^{2x}dx$

When combined with the previous result, this yields $\frac{1}{2}x^3e^{2x} - \frac{3}{2}\left[\frac{1}{2}x^2e^{2x} - \int xe^{2x}dx\right] = \frac{1}{2}x^3e^{2x} - \frac{3}{4}x^2e^{2x} + \frac{3}{4}xe^{2x} - \int \frac{1}{2}e^{2x}dx = \frac{1}{2}x^3e^{2x} - \frac{3}{4}x^2e^{2x} + \frac{3}{4}xe^{2x} - \frac{1}{4}e^{2x} + C = e^{2x}\left(\frac{1}{2}x^3 - \frac{3}{4}x^2 + \frac{3}{4}x - \frac{1}{4}\right) + C$

11. $\int x\sqrt{4x+1}\,dx$. Let $u = 4x+1$ and $du = 4\,dx$. So, $\frac{u-1}{4} = x$ and $\frac{1}{4}du = dx$. Thus, the given integral is $\int \frac{u-1}{4}\cdot\sqrt{u}\cdot\frac{1}{4}dx = \frac{1}{16}\int u^{3/2} - u^{1/2}du = \frac{1}{16}\left[\frac{2}{5}u^{5/2} - \frac{2}{3}u^{3/2}\right] + C = \frac{1}{16}\left[\frac{2}{5}(4x+1)^{5/2} - \frac{2}{3}(4x+1)^{3/2}\right] + C = \frac{1}{16}\left[\frac{2}{5}(4x+1)(4x+1)^{3/2} - \frac{2}{3}(4x+1)^{3/2}\right] + C = \frac{1}{16}\left[\left(\frac{8x}{5} - \frac{4}{15}\right)(4x+1)^{3/2}\right] + C = \frac{1}{16}\left[\frac{24x-4}{15}(4x+1)^{3/2}\right] + C = \frac{1}{60}(6x-1)(4x+1)^{3/2} + C$.

13. $\int x^2 e^{x/4}dx$

$u = x^2$ | $v = 4e^{x/4}$

$du = 2x\,dx$ | $dv = e^{x/4}dx$

Thus, $\int x^2 e^{x/4}dx = 4x^2 e^{x/4} - \int 8xe^{x/4}$. Using integration by parts a second time, we get the following table:

$u = 8x$ | $v = 4e^{x/4}$

$du = 8\,dx$ | $dv = e^{x/4}dx$

Hence, we get $4x^2 e^{x/4} - \int 8xe^{x/4} = 4x^2 e^{x/4} - 32xe^{x/4} + \int 32e^{x/4}\,dx = 4x^2 e^{x/4} - 32xe^{x/4} + 128e^{x/4} + C = e^{x/4}(4x^2 - 32x + 128) + C$.

15. $\int e^x \cos x\,dx$

$u = e^x$ | $v = \sin x$

$du = e^x\,dx$ | $dv = \cos x\,dx$

This results in $\int e^x \cos x\,dx = e^x \sin x - \int e^x \sin x\,dx$. Using integration by parts again produces

$u = e^x$ | $v = -\cos x$

$du = -e^x dx$ | $dv = \sin x\,dx$

Combining this with the previous result yields $e^x \sin x - \int e^x \sin x\,dx = e^x \sin x + e^x \cos x - \int e^x \cos x\,dx$. Hence $2\int e^x \cos x\,dx = e^x \sin x + e^x \cos x$ or $\int e^x \cos x\,dx = \frac{1}{2}e^x(\sin x + \cos x) + C$.

17. Rewrite the given integral as $\int x^3 \cos x^2 dx = \int \frac{x^2}{2}\cdot 2x\cos x^2 dx$, and then use integration by parts.

$u = \frac{x^2}{2}$ | $v = \sin x^2$

$du = x\,dx$ | $dv = 2x\cos x^2 dx$

So the integral equals $\frac{x^2}{2}\sin x^2 - \int x\sin x^2 = \frac{x^2}{2}\sin x^2 + \frac{1}{2}\cos x^2 + C$

19. $\int x^3 e^{-x}dx$

$u = x^3$ | $v = -e^{-x}$

$du = 3x^2 dx$ | $dv = e^{-x}dx$

Thus, we obtain $\int x^3 e^{-x}dx = -x^3 e^{-x} + \int 3x^2 e^{-x}$. Using integration by parts a second time, we have the following table.

$u = 3x^2$ | $v = -e^{-x}$

$du = 6x\,dx$ | $dv = e^{-x}dx$

Combined with the earlier result, this produces $-x^3 e^{-x} + \int 3x^2 e^{-x} = -x^3 e^{-x} - 3x^2 e^{-x} + \int 6xe^{-x}$. We now use integration by parts a third time.

$u = 6x$ | $v = -e^{-x}$

$du = 6\,dx$ | $dv = e^{-x}dx$

Hence, $-x^3 e^{-x} - 3x^2 e^{-x} + \int 6xe^{-x} = -x^3 e^{-x} - 3x^2 e^{-x} - 6xe^{-x} + \int 6e^{-x} = -x^3 e^{-x} - 3x^2 e^{-x} - 6xe^{-x} - 6e^{-x} + C = -e^{-x}(x^3 + 3x^2 + 6x + 6) + C$

21. $\int e^{-x}\cos x\,dx$

$u = e^{-x}$ | $v = \sin x$

$du = -e^{-x}dx$ | $dv = \cos x\,dx$

Thus, $\int e^{-x}\cos x\,dx = e^{-x}\sin x + \int \sin xe^{-x}dx$. Applying integration by parts again, we have the following table.

$u = e^{-x}$ | $v = -\cos x$

$du = -e^{-x}dx$ | $dv = \sin x\,dx$

Hence, $e^{-x}\sin x + \int \sin xe^{-x}dx = e^{-x}\sin x - e^{-x}\cos x - \int e^{-x}\cos x\,dx$ which produces $2\int e^{-x}\cos x\,dx = e^{-x}\sin x - e^{-x}\cos x$ and means that $\int e^{-x}\cos x = \frac{1}{2}e^{-x}(\sin x - \cos x) + C$.

23. $\int x \cos 4x \, dx$

$u = x$	$v = \frac{1}{4} \sin 4x$
$du = dx$	$dv = \cos 4x \, dx$

Hence, $\int x \cos 4x \, dx = \frac{x}{4} \sin 4x - \frac{1}{4} \int \sin 4x \, dx = \frac{x}{4} \sin 4x + \frac{1}{16} \cos 4x + C$.

25. Using the disc method, we see that $V = \pi \int_0^{\pi/2} \times \cos^2 x \, dx = \pi \int_0^{\pi/2} \frac{1 + \cos 2x}{2} = \pi \left[\frac{1}{2} x + \frac{1}{4} \sin 2x \right]_0^{\pi/2} = \pi \left(\frac{\pi}{4} + 0 - 0 - 0 \right) = \frac{\pi^2}{4}$.

27. Charge is the integral of current. Hence we must find $\int 4t \sin 2t \, dt$. Use the following integration by parts table.

$u = 4t$	$v = -\frac{1}{2} \cos 2t$
$du = 4 \, dt$	$dv = \sin 2t \, dt$

Thus, $\int 4t \sin 2t \, dt = -2t \cos 2t + \int 2 \cos 2t \, dt = -2t \cos 2t + \sin 2t$. Evaluated from 0 to 4 we have $[-8 \cos 8 + \sin 8] = 2.1534$ C.

29. $M_y = \int_{-1}^{2} x \cdot x^2 e^x dx = \int_{-1}^{2} x^3 e^x dx$. From Exercise #9, we have $\int x^3 e^{2x} dx = e^{2x} \left(\frac{1}{2} x^3 - \frac{3}{4} x^2 + \frac{3}{4} x - \frac{3}{8} \right) + C$. Hence, we see that

$$M_y = \int_{-1}^{2} x \cdot x^2 e^x dx = e^x \left[x^3 - 3x^2 + 6x - 6 \right]_{-1}^{2}$$
$$= 14.7781 + 5.8861 = 20.6642$$

$M_x = \frac{1}{2} \int_{-1}^{2} \left(x^2 e^x \right)^2 dx = \frac{1}{2} \int_{-1}^{2} x^4 e^{2x} dx$. From Exercise #16, we know

$$\int x^4 e^{2x} dx$$
$$= e^{2x} \left(\frac{1}{2} x^4 - x^3 + \frac{3}{2} x^2 - \frac{3}{2} x + \frac{3}{4} \right) + C$$

Thus,

$$M_x = \frac{1}{2} \int_{-1}^{2} \left(x^2 e^x \right)^2 dx = \frac{1}{2} \int_{-1}^{2} x^4 e^{2x} dx$$
$$= \frac{1}{8} e^{2x} \left[2x^4 - 4x^2 + 6x^2 - 6x + 3 \right]_{-1}^{2}$$
$$= \frac{15}{8} e^4 - \frac{21}{8} e^{-2} = 102.0163$$

The mass is $m = \int_{-1}^{2} x^2 e^x dx$. Once again, we need to use integration by parts.

$u = x^2$	$v = e^x$
$du = 2x \, dx$	$dv = e^x dx$

So, $\int x^2 e^x dx = x^2 e^x - \int 2x e^x dx$. A second application of integration by parts produces

$u = 2x$	$v = e^x$
$du = 2 \, dx$	$dv = e^x dx$

With the result that $\int x^2 e^x dx = x^2 e^x - 2xe^x + \int 2e^x dx = x^2 e^x - 2xe^x + 2e^x$. Thus, $m = \int_{-1}^{2} x^2 e^x dx = \left[x^2 e^x - 2xe^x + 2e^x \right]_{-1}^{2} = \left[e^x \left(x^2 - 2x + 2 \right) \right]_{-1}^{2} = 12.9378$. Using the above result, we see that $\bar{x} = \frac{M_y}{m} = \frac{20.6642}{12.9387} = 1.5971$ and $\bar{y} = \frac{M_x}{m} = \frac{102.0163}{12.9387} = 7.8846$.

31. Work is the integral of force so we get $W = \int x^3 \cos x \, dx$. Using integration by parts, we have

$u = x^3$	$v = \sin x$
$du = 3x^2 dx$	$dv = \cos x \, dx$

Thus, $\int x^3 \cos x \, dx = x^3 \sin x - \int 3x^2 \sin x \, dx$. Using integration by parts again yields

$u = -3x^2$	$v = -\cos x$
$du = -6x \, dx$	$dv = \sin x \, dx$

Hence, $\int x^3 \cos x \, dx = x^3 \sin x + 3x^2 \cos x - \int 6x \cos x$. We need integration by parts one more time.

$u = -6x$	$v = \sin x$
$du = -6 \, dx$	$dv = \cos x \, dx$

$$\int x^3 \cos x \, dx$$
$$= x^3 \sin x + 3x^2 \cos x - 6x \sin x$$
$$+ \int 6 \sin x \, dx$$
$$= x^3 \sin x + 3x^2 \cos x - 6x \sin x + 6 \cos x \Big|_0^{\pi/2}$$
$$= \left(\frac{\pi}{2} \right)^3 + 0 - 6 \frac{\pi}{2} - 0 - 0 + 0 - 6$$
$$= \left(\frac{\pi}{2} \right)^3 - 3\pi + 6 \approx 0.4510$$

33. As in example 28.54 set up the coordinate axis so that $(0, 0)$ is at the center of the road between the towers. Since it is a parabola the equation of the main cable fits the form $y = 4px^2$ and contains the point $(500, 200)$. Thus, $200 = 4p \cdot (500)$ or $p = \frac{2}{10000}$. The equation of the cable is thus $y = \frac{1}{1250} x^2$. Differentiating we get $y' = \frac{1}{625} x$. The length of the cable is $\int_{-500}^{500} \sqrt{1 + \left(\frac{x}{625} \right)^2} dx$. Evaluate this integral as follows: $\int \sqrt{1 + \left(\frac{x}{625} \right)^2} = \frac{1}{625} \int \sqrt{625^2 + x^2}$. Let $x = 625 \tan \theta$ and $dx =$

$625 \sec^2 \theta \, d\theta$. This leads to $\frac{1}{625} \int \sqrt{625^2 + 25^2 \tan^2 \theta} \cdot$

$625 \sec \theta \, d\theta = \frac{1}{625} \int 625 \sec \theta \cdot 625 \sec^2 \theta \, d\theta =$

$625 \int \sec^3 \theta \, d\theta$.   Using Example 28.50, we get

$\frac{625}{2} \left[\sec \theta \tan \theta + \ln|\sec \theta + \tan \theta| \right] =$

$\frac{625}{2} \left[\frac{\sqrt{625+x^2}}{625} + \frac{x}{625} + \ln\left| \frac{\sqrt{625+x^2}}{625} + \frac{x}{625} \right| \right]_{-500}^{500} =$

$625(1.02445 + \ln(2.08062) = 625(1.02445 + 0.73267)$.
Thus, the length of the main cable is $625(1.75712) = 1098$ ft.

35. **(a)** $F = 2\pi \int_0^R r \cdot P(r) dr = 2\pi \int_0^R \cdot P_0 e^{-kr} dr$.
Integrating by parts, we have

$u = P_o r$	$v = -\frac{1}{k} e^{-kr}$
$du = P_0 dr$	$dv = e^{-kr} dr$

and so,

$F = 2\pi \int_0^R r \cdot P_0 e^{-kr} dr$

$= -2\pi \frac{1}{k} P_0 r e^{-kr} \Big|_0^R + 2\pi \int_0^R \frac{P_0}{k} e^{-kr} dr$

$= \left[2\pi \frac{1}{k} P_0 r e^{-kr} - 2\pi \frac{P_0}{k^2} d^{-kr} \right]_0^R$

$= \frac{-2\pi P_0 R k e^{-kR} - 2\pi P_0 e^{-kR}}{k^2} + \frac{2\pi P_0}{k^2}$

$= -\frac{e^{-kR} 2\pi (kR + 1) P_0}{k^2} + \frac{2\pi P_0}{k^2}$

(b) $T = 2\pi\mu \int_0^R r^2 \cdot P(r) dr = 2\pi\mu \int_0^R r^2 \times P_0 e^{-kr} dr = 2\pi\mu P_0 \int_0^R r^2 \cdot e^{-kr} dr$. Integrating $\int_0^R r^2 \cdot e^{-kr} dr$ we will use integration by parts twice. The first time produces

$u = r^2$	$v = -\frac{1}{k} e^{-kr}$
$du = 2r \, dr$	$dv = e^{-kr} dr$

and we get $\int r^2 \cdot e^{-kr} dr = -\frac{r^2}{k} e^{-kr} + \int \frac{2}{k} r e^{-kr} dr$.
The second time we use integration by parts, we have

$u = \frac{2}{k} r$	$v = -\frac{1}{k} e^{-kr}$
$du = \frac{2}{k} dr$	$dv = e^{-kr} dr$

and we get

$\int r^2 \cdot e^{-kr} dr = -\frac{r^2}{k} e^{-kr} - \frac{2r}{k^2} + \int \frac{2}{k^2} e^{-kr} dr$

$= -\frac{r^2}{k} e^{-kr} - \frac{2r}{k^2} - \frac{2}{k^3} e^{-kr} + C$

Putting the constant multiples and the limits of integration back in, the result is

$T = 2\pi\mu \int_0^R r^2 \cdot P(r) dr$

$= 2\pi\mu \left[-\frac{r^2}{k} e^{-kr} - \frac{2r}{k^2} - \frac{2}{k^3} e^{-kr} \right]_0^R$

$= -2\pi\mu \left[\frac{1}{k^3} e^{-kr} (k^2 r^2 + 2kr + 2) \right]_0^R$

$= -2\pi\mu \left[\frac{1}{k^3} e^{-kR} (k^2 R^2 + 2kR + 2) \right]$

$\quad + 2\pi\mu \left[\frac{1}{k^3} e^{-0} (k^2 \cdot 0^2 + 2k \cdot 0 + 2) \right]$

$= -2\pi\mu \left[\frac{1}{k^3} e^{-kR} (k^2 R^2 + 2kR + 2) \right]$

$\quad + 4\pi\mu \left(\frac{1}{k^3} \right)$

37. $\overline{C} = \int_0^6 \frac{40 \ln(t+2)}{(t+2)^2} dt = \frac{20}{3} \int_0^6 \frac{\ln(t+2)}{(t+2)^2} dt$. To determine the integral, we use integration by parts

$u = \ln(t+2)$	$v = -(t+2)^{-1}$
$du = \frac{1}{t+2} dt$	$dv = (t+2)^{-2} dt$

and so, $\int \frac{\ln(t+2)}{(t+2)^2} dt = \frac{-\ln(t+2)}{t+2} \int \frac{1}{(t+2)^2} dt = \frac{-\ln(t+2)}{t+2}$

$- \frac{1}{t+2}$. Combining the pieces, we get $\overline{C} = -\frac{20}{3} \times$

$\left[\frac{\ln(t+2)+1}{t+2} \right]_0^6 = -\frac{20}{3} \left[\frac{\ln 8+1}{8} - \frac{\ln 2+1}{2} \right] \approx 3.0776$ ppm

▤ 28.8 USING INTEGRATION TABLES

1. $\int (1 + \tan 3x)^2 dx = \int \left(1 + 2\tan 3x + \tan^2 3x\right) dx$ by formulas #12 and #58 we get $x + 2 \cdot \frac{1}{3} \ln|\sec 3x| + \frac{1}{3}(\tan 3x - 3x) + C = x + \frac{2}{3} \ln|\sec 3x| + \frac{1}{3} \tan 3x - x + C = \frac{1}{3}(2\ln|\sec 3x| + \tan 3x) + C$

3. $\int \frac{dx}{x(4x-3)} = \int \frac{dx}{x(-3+4x)}$ Let $u = x$, $a = -3$, and $du = dx$. By formula #48 this is $\frac{1}{a} \ln\left|\frac{u}{a+bu}\right| + C = \frac{-1}{3} \ln\left|\frac{x}{4x-3}\right| + C$.

5. $\int \frac{x^2 dx}{x^6 \sqrt{16+x^6}}$. Let $u = x^3$ and $du = 3x^2 dx$ or $\frac{1}{3} du = 3x^2 dx$. Substituting yields $\frac{1}{3} \int \frac{du}{u^2 \sqrt{4^2+u^2}}$

By formula #39 this is $\frac{1}{3}\left(\frac{-\sqrt{16+u^2}}{16u}\right)+C=$
$\frac{1}{3}\left(\frac{-\sqrt{16+x^6}}{16x^3}\right)+C=-\frac{\sqrt{16+x^6}}{48x^3}+C.$

7. $\int\frac{x^2}{\sqrt{9-x^2}}dx.$ By formula #25 this is $-\frac{x}{2}\sqrt{9-x^2}+$
$\frac{9}{2}\sin^{-1}\frac{x}{3}+C.$

9. $\int\cos^3 x\, dx.$ By formula #65 this is $\frac{1}{3}\cos^2 x\sin x+$
$\frac{2}{3}\int\cos x\, dx = \frac{1}{3}\cos^2 x\sin x + \frac{2}{3}\sin x + C =$
$\frac{1}{3}(1-\sin^2 x)\sin x+\frac{2}{3}\sin x+C = \sin x-\frac{1}{3}\sin^3 x+$
$C.$ Alternate solution: $\int\cos^3 x\, dx = \int\left(1-\sin^2 x\right)\cos x\, dx = \int\cos x\, dx-\int\sin^2 x\cos x\, dx =$
$\sin x - \frac{1}{3}\sin^3 x + C.$

11. $\int\sin^6 3x\, dx.$ Let $u = 3x$, then $du = 3\, dx$ or
$\frac{1}{3}du = dx.$ Substituting we get $\frac{1}{3}\int\sin^6 u\, du.$ By
formula #64 this is $\frac{1}{3}\left[-\frac{1}{6}\sin^5 u\cos u+\frac{5}{6}\int\sin^4 u\, du\right].$
Using formula #64 once more, you obtain
$\frac{1}{3}\left[-\frac{1}{6}\sin^5 u\cos u+\frac{5}{6}\left(-\frac{1}{4}\sin^3 u\cos u+\frac{3}{4}\int\sin^2 \times u\, du\right)\right].$ By formula #56 we get $\frac{1}{3}\left[-\frac{1}{6}\sin^5 u\cos u+\right.$
$\left.\frac{5}{6}\left(-\frac{1}{4}\sin^3 u\cos u+\frac{3}{4}\left\{\frac{1}{2}u-\frac{1}{2}\sin u\cos u\right\}\right)\right]+C=$
$-\frac{1}{18}\sin^5 u\cos u-\frac{5}{72}\sin^3 u\cos u-\frac{5}{48}\sin u\cos u+$
$\frac{5}{48}u+C = -\frac{1}{18}\sin^5 3x\cos 3x-\frac{5}{72}\sin^3 3x\cos 3x-$
$\frac{5}{48}\sin 3x\cos 3x+\frac{5}{48}x+C.$

13. $\int e^{10x}\cos 6x\, dx$: By formula #88 we have this is
$\frac{e^{10x}}{10^2+6^2}(10\cos 6x+6\sin 6x)+C = \frac{1}{136}e^{10x}(10\cos 6x$
$+6\sin 6x)+C.$

15. $\int x^7\ln x\, dx.$ By formula #90 we obtain $x^8\left(\frac{1}{8}\ln x-\right.$
$\left.\frac{1}{64}\right)+C$

17. $\int\arcsin 4x\, dx.$ Letting $u = 4x$ and $du = 4\, dx$
and then substituting we get $\frac{1}{4}\int\arcsin u\, du.$ By
formula #78, $\int\arcsin 4x\, dx = \frac{1}{4}\left[(4x\sin^{-1}4x+\right.$
$\left.\sqrt{1-(4x)^2}\right]+C = \frac{1}{4}\left[4x\sin^{-1}4x+\sqrt{1-16x^2}\right]$
$+C.$

19. $\int\frac{\sqrt{9+x^2}}{x}dx.$ By formula #32 this is $\sqrt{9+x^2}-$
$3\ln\left|\frac{3+\sqrt{9+x^2}}{x}\right|+C.$

21. $\int x^3 e^{2x}dx.$ By formula #86 we get $\frac{1}{2}x^3 e^{2x}-$
$\frac{3}{2}\int x^2 e^{2x}dx.$ Repeating this process, we get $\frac{1}{2}x^3 e^{2x}$
$-\frac{3}{2}\cdot\frac{1}{2}x^2 e^{2x}+\frac{3}{2}\cdot\frac{2}{2}\int xe^{2x}dx$ or $\frac{1}{2}x^3 e^{2x}-\frac{3}{4}x^2 e^{2x}+$
$\frac{3}{2}\int xe^{2x}dx.$ Repeating again produces $\frac{1}{2}x^3 e^{2x}-$
$\frac{3}{4}x^2 e^{2x}+\frac{3}{4}xe^{2x}-\frac{3}{2}\frac{1}{2}\int e^{2x}dx = \frac{1}{2}x^3 e^{2x}-\frac{3}{4}x^2 e^{2x}+$
$\frac{3}{4}xe^{2x}-\frac{3}{8}e^{2x}+C = \frac{1}{8}e^{2x}\left(4x^3-6x^2+6x-3\right)+C.$

23. $\int x^3\sin 2x\, dx.$ Let $u = 2x$, then $u^3 = 8x^3$ or
$x^3 = \frac{1}{8}u^3$, and also, $du = 2dx$ or $dx = \frac{1}{2}du.$ Substitution produces $\frac{1}{16}\int u^3\sin u\, du.$ By formula
#75 this is $\frac{1}{16}\left[-u^3\cos u+3\int u^2\cos u\, du\right].$ Now
using formula #76, on the integral in this result, we
obtain $\frac{1}{16}\left[-u^3\cos u+3\left(u^2\sin u-2\int u\sin u\, du\right)\right].$
Finally, using formula #75 one more time produces
the result $\frac{1}{16}\left[-u^3\cos u+3u^2\sin u+6u\cos u-\right.$
$\left.6\int\cos u\right]$ or $\frac{1}{16}\left[-u^3\cos u+3u^2\sin u+6u\cos u-\right.$
$\left.6\sin u\right]+C.$ Back substituting for u produces
$\int x^3\sin 2x\, dx = \frac{1}{16}\left[-8x^3\cos 2x+12x^2\sin 2x+\right.$
$\left.12x\cos 2x-6\sin 2x\right]+C = -\frac{1}{2}x^3\cos 2x+\frac{3}{4}x^2\times$
$\sin 2x+\frac{3}{4}x\cos 2x-\frac{3}{8}\sin 2x+C.$

25. $\overline{V} = \frac{1}{2.5}\int_0^{2.5}t^2 e^{5t}dt.$ Using integration form #85
produces
$$\overline{V} = \frac{1}{2.5}\left[\frac{e^{5t}}{125}\left(25t^2-10t+2\right)\right]_0^{2.5}$$
$$= \frac{1}{2.5}(286047.5315) \approx 114419.0126$$

27. $W = \int F\, dx = \int_0^2\frac{1}{49-9x^2}dx = \frac{1}{2.7}\ln\left|\frac{3x+7}{3x-7}\right|_0^2$ by
integration form #19. This evaluates as $\frac{1}{14}(\ln 13-$
$\ln 1) = \frac{1}{14}\ln 13 \approx 0.1832$ N $\cdot$ m.

CHAPTER 28 REVIEW

1. $\int xe^{3x}dx$. Using integration by parts with the table

$u = x$	$v = \frac{1}{3}e^{3x}$
$du = dx$	$dv = e^{3x}dx$

We have $\int xe^{3x}dx = \frac{1}{3}xe^{3x} - \frac{1}{3}\int e^{3x}dx = \frac{1}{3}xe^{3x} - \frac{1}{9}e^{3x} + C = \frac{1}{9}e^{3x}(3x - 1) + C$.

2. $\int \sin^3 x \, \cos^2 x \, dx = \int \sin x (1 - \cos^2 x) \cos^2 x \, dx = \int \sin x (\cos^2 x - \cos^4) dx = -\frac{1}{3}\cos^3 x + \frac{1}{5}\cos^5 x + C$.

3. $\int \frac{x}{\sqrt{25-x^2}}dx$. Let $x = 5\sin\theta$, then $dx = 5\cos\theta \, d\theta$. Thus, substitution produces $\int \frac{5\sin\theta}{\sqrt{25-5\sin^2\theta}} \cdot 5\cos\theta \, d\theta = \int \frac{5\sin\theta \cdot 5\cos\theta \, d\theta}{5\cos\theta} = 5\int \sin\theta \, d\theta = -5\cos\theta + C = 5 \cdot \frac{\sqrt{25-x^2}}{5} + C = -\sqrt{25 - x^2} + C$. Alternate solution: Let $u = 25 - x^2$, then $du = -2x \, dx$ or $-\frac{1}{2}du = x \, dx$. Substituting, we obtain
$$-\frac{1}{2}\int \frac{du}{\sqrt{u}} = -\frac{1}{2}\int u^{-1/2}du = -\frac{1}{2} \cdot \frac{2}{1}u^{1/2} + C = -\sqrt{25 - x^2} + C$$

4. $\int x^3\sqrt{25 - x^2}dx$ Let $u = 25 - x^2$ or $x^2 = 25 - u$, and $du = -2x \, dx$ or $-\frac{1}{2}du = x \, dx$. Substituting we get $-\frac{1}{2}\int (25 - u)\sqrt{u} \cdot du = -\frac{1}{2}\int 25\sqrt{u} \, du + \frac{1}{2}\int u^{3/2}du = -\frac{25}{2} \cdot \frac{2}{3}u^{3/2} + \frac{1}{2}\frac{2}{5}u^{5/2} + C = -\frac{25}{3}(25 - x^2)^{3/2} + \frac{1}{5}(25 - x^2)^{5/2} + C = -\frac{25}{3}(25 - x^2)^{3/2} + \frac{1}{5}(25 - x^2)(25 - x^2)^{3/2} + C = (5 - \frac{25}{3} - \frac{x^2}{5})(25 - x^2)^{3/2} + C = (-\frac{1}{5}x^2 - \frac{10}{3})(25 - x^2) + C$.

5. $\int \sin^4 2x \cos 2x \, dx$. Let $u = \sin 2x$ then $du = 2\cos 2x \, dx$ or $\frac{1}{2}du = \cos 2x \, dx$. Substitution produces $\frac{1}{2}\int u^4 du = \frac{1}{2} \cdot \frac{1}{5}u^5 + C = \frac{1}{10}\sin^5 2x + C$.

6. $\int 2(e^x - e^{-x})dx = 2\int e^x dx - 2\int e^{-x}dx = 2e^x + 2e^{-x} + C$.

7. $\int \frac{dx}{9x+5}$. Let $u = 9x + 5$ and $du = 9 \, dx$ or $\frac{1}{9}du = dx$. Then $\frac{1}{9}\int \frac{du}{u} = \frac{1}{9}\ln|u| = \frac{1}{9}\ln|9x + 5| + C$.

8. $\int \tan^2 8x \, dx = \int \frac{\sin^2 8x}{\cos^2 8x}dx = \int \frac{1 - \cos^2 8x}{\cos^2 8x}dx = \int \frac{1}{\cos^2 8x}dx - \int 1 \, dx = \frac{1}{8}\tan 8x - x + C$.

9. $\int \sin(7x + 2)dx$. Let $u = 7x + 2$ and $du = 7 \, dx$ or $\frac{1}{7}du = dx$. Then, substitution produces $\frac{1}{7}\int \sin u \, du = \frac{1}{7}(-\cos u) + C = -\frac{1}{7}\cos(7x+2) + C$.

10. $\int \sin^3 2x \cos 2x \, dx$. Let $u = \sin 2x$ and then $du = 2\cos 2x \, dx$. Substitution produces $\frac{1}{2}\int u^3 du = \frac{1}{2} \cdot \frac{1}{4}u^4 + C = \frac{1}{8}\sin^4 2x + C$.

11. $\int \sin^5 3x \cos^2 3x \, dx = \int \sin 3x(\sin^4 3x)\cos^2 3x \, dx = \int \sin 3x(1 - \cos^2 3x)^2\cos^2 3x \, dx$. This expands as $\int (\cos^2 3x - 2\cos^4 3x + \cos^6 3x)\sin 3x \, dx$. Let $u = \cos 3x$ and $du = -3\sin 3x \, dx$ or $-\frac{1}{3}du = \sin 3x \, dx$. Substituting we get $-\frac{1}{3}\int (u^2 - 2u^4 + u^6)du = -\frac{1}{3}[\frac{1}{3}u^3 - \frac{2}{5}u^5 + \frac{1}{7}u^7] + C = -\frac{1}{9}\cos^3 3x + \frac{2}{15}\cos^5 3x - \frac{1}{21}\cos^7 3x + C$.

12. $\int \frac{dx}{x^2+4x+20} = \int \frac{dx}{(x+2)^2+16}$. Let $u = x + 2$, $a = 4$, and $du = dx$. Substituting, we get $\int \frac{du}{u^2+4^2} = \frac{1}{4}\tan^{-1}(\frac{u}{4}) + C = \frac{1}{4}\tan^{-1}(\frac{x+2}{4}) + C$.

13. $\int \frac{dx}{(4x^2+49)^{3/2}}$. Let $2x = 7\tan\theta$, or $x = \frac{7}{2}\tan\theta$ and then $dx = \frac{7}{2}\sec^2\theta \, d\theta$. Substituting yields $\frac{7}{2}\int \frac{\sec^2\theta \, d\theta}{((7\tan\theta)^2+49)^{3/2}} = \frac{7}{2} \cdot \int \frac{\sec^2\theta \, d\theta}{7^3\sec^3\theta} = \frac{1}{98}\int \frac{1}{\sec\theta}d\theta = \frac{1}{98}\int \cos\theta \, d\theta = \frac{1}{98} \cdot \sin\theta + C = \frac{1}{98} \cdot \frac{2x}{\sqrt{4x^2+49}} + C = \frac{x}{49\sqrt{4x^2+49}} + C$.

14. $\int x^3 \ln x \, dx$. Using integration by parts with the table

$u = \ln x$	$v = \frac{1}{4}x^4$
$du = \frac{1}{x}dx$	$dv = x^3 dx$

We obtain $\int x^3 \ln x \, dx = \frac{1}{4}x^4 \ln x - \int \frac{1}{4}x^3 dx = \frac{1}{4}x^4 \ln x - \frac{1}{16}x^4 + C = x^4(\frac{1}{4}\ln x - \frac{1}{16}) + C$.

15. $\int x^2 e^{x^3}dx$. Let $u = x^3$ and $du = 3x^2 dx$ or $\frac{1}{3}du = x^2 dx$. Substituting, we get $\frac{1}{3}\int e^u du = \frac{1}{3}e^u + C = \frac{1}{3}e^{x^3} + C$.

16. $\int \frac{dx}{\sqrt{4x^2+49}}$. Let $2x = 7\tan\theta$ or $x = \frac{7}{2}\tan\theta$ and then $dx = \frac{7}{2}\sec^2\theta \, d\theta$. Substituting, we get

$\frac{7}{2}\int \frac{\sec^2\theta\, d\theta}{\sqrt{40\tan^2\theta+49}} = \frac{1}{2}\int \frac{\sec^2\theta\, d\theta}{\sec\theta} = \frac{1}{2}\int \sec\theta\, d\theta =$

$\frac{1}{2}\ln|\sec\theta+\tan\theta| + K = \frac{1}{2}\ln\left|\frac{\sqrt{4x^2+49}}{7} + \frac{2x}{7}\right| +$

$K = \frac{1}{2}\ln\left|\sqrt{4x^2+49}+2x\right| - \frac{1}{2}\ln 7 + K =$

$\frac{1}{2}\ln\left|\sqrt{4x^2+49}+2x\right| + C$ where $C = K - \frac{1}{2}\ln 7$.

17. $\int \frac{x\, dx}{\sqrt{4x^2+49}}$. Let $u = 4x^2 + 49$ and $du = 8x\, dx$ or $\frac{1}{8}du = x\, dx$. Substituting we get $\frac{1}{8}\int \frac{du}{\sqrt{u}} =$

$\frac{1}{8}\int u^{-\frac{1}{2}}du = \frac{1}{8}\cdot\frac{2}{1}u^{1/2} + C = \frac{1}{4}\sqrt{4x^2+49} + C$.

18. $\int \cot 5x \csc^4 5x\, dx = \int \cot 5x \csc 5x \csc^3 5x\, dx$. Let $u = \csc 5x$ and $du = -5\cot 5x \csc 5x\, dx$. Substituting we get $-\frac{1}{5}\int u^3 du = -\frac{1}{5}\cdot\frac{1}{4}u^4 + C = -\frac{1}{20}\csc^4 5x + C$ or $-\frac{1}{20}\cdot\sin^{-4}5x + C$.

19. $\int \frac{\sec 4x \tan 4x}{9+2\sec 4x}dx$. Let $u = 9 + 2\sec 4x$ and $du = 8\sec 4x\tan 4x\, dx$. Substituting we get $\frac{1}{8}\int \frac{du}{u} = \frac{1}{8}\ln|u| + C = \frac{1}{8}\ln|9 + 2\sec 4x| + C$.

20. $\int \sin^6 \frac{3x}{2}dx$. We begin by rewriting this as a cubic expression, then using a half-angle trig identity, and finally expanding the rewritten expression.

$\int \sin^6 \frac{3x}{2}dx$

$= \int \left(\sin^2 \frac{3x}{2}\right)^3 dx$

$= \int \left(\frac{1-\cos 3x}{2}\right)^3 dx$

$= \frac{1}{8}\int \left(1 - 3\cos 3x + 3\cos^2 3x - \cos^3 3x\right)dx$

$= \frac{1}{8}x - \frac{1}{8}\sin 3x$

$\quad + \frac{3}{8}\int \cos^2 3x - \frac{1}{8}\int \cos^3 3x\, dx$

Now $= \frac{3}{8}\int \cos^2 3x\, dx = \frac{3}{8}\int \left(\frac{1+\cos 6x}{2}\right)dx =$

$\frac{3}{16}\int (1 + \cos 6x)dx = \frac{3}{16}x + \frac{3}{96}\sin 6x = \frac{3}{16}x +$

$\frac{1}{32}\sin 6x$. And also, $-\frac{1}{8}\int \cos^3 3x\, dx = -\frac{1}{8}\int \left(1 - \sin^2 3x\right)\cos 3x\, dx = -\frac{1}{8}\cdot\frac{1}{3}\sin 3x + \frac{1}{8}\cdot\frac{1}{9}\sin^3 3x + C$. Putting this all together we get $\frac{1}{8}x - \frac{1}{8}\sin 3x +$

$\frac{3}{16}x + \frac{1}{32}\sin 6x - \frac{1}{24}\sin 3x + \frac{1}{72}\sin^3 3x + C = \frac{5}{16}x - \frac{5}{24}\sin 3x + \frac{1}{32}\sin 6x + \frac{1}{72}\sin^3 3x + C$.

21. $\int \frac{[\ln(2x+1)]^5}{2x+1}dx$. Let $u = \ln(2x+1)$ and then $du = \frac{2dx}{2x+1}$. Substituting yields $\frac{1}{2}\int u^5 du = \frac{1}{12}u^6 + C = \frac{1}{12}[\ln(2x+1)]^6 + C$.

22. $\int \frac{\arctan 7x}{1+49x^2}dx$. Let $u = \arctan 7x$ and $du = \frac{7dx}{1+49x^2}$. Substitution produces $\frac{1}{7}\int u\, du = \frac{1}{7}\cdot\frac{1}{2}u^2 + C = \frac{1}{14}(\arctan 7x)^2 + C$.

23. $\int \frac{x^2}{x^3+4}dx$. Let $u = x^3 + 4$ and $du = 3x^2 dx$ or $\frac{1}{3}du = x^2 dx$. Now substituting we get $\frac{1}{3}\int \frac{du}{u} = \frac{1}{3}\ln|u| + C = \frac{1}{3}\ln|x^3 + 4| + C$.

24. $\int 4x^3 e^{x^4}dx$. Let $u = x^4$ and $du = 4x^3 dx$. Substituting yields $\int e^u du = e^u + C = e^{x^4} + C$.

25. $\int \tan \frac{x}{5}dx = 5\ln\left|\sec \frac{x}{5}\right| + C$ or $-5\ln\left|\cos \frac{x}{5}\right| + C$.

26. $\int x\sin^3 x\, dx = \int x(1 - \cos^2 x)\sin x\, dx = \int x\sin x\, dx - \int x\cos^2 x \sin x\, dx$. For the first integral we use integration by parts with

$u = x$	$v = -\cos x$
$du = dx$	$dv = \sin x\, dx$

Then, $\int x\sin x\, dx = -x\cos x + \int \cos x\, dx = -x\cos x + \sin x + C$. For the second integral by parts

$u = x$	$v = \frac{1}{3}\cos^3 x$
$du = dx$	$dv = -\cos^2 x \sin x\, dx$

Thus, $\int x\left(-\cos^2 x \sin x\right)dx = \frac{x}{3}\cos^3 x - \frac{1}{3}\int \times \cos^3 x\, dx = \frac{x}{3}\cos^3 x - \frac{1}{3}\int \left(1 - \sin^2 x\right)\cos x\, dx = \frac{x}{3}\cos^3 x - \frac{1}{3}\sin x + \frac{1}{9}\sin^3 x + C$. Putting these together we have $\int x\sin^3 x\, dx = -x\cos x + \sin x - \left(\frac{x}{3}\cos^3 x - \frac{1}{3}\sin x + \frac{1}{9}\sin^3 x\right) + C = -x\cos x + \sin x - \frac{x}{3}\cos^3 x + \frac{1}{3}\sin x - \frac{1}{9}\sin^3 x + C = -x\cos x + \frac{x}{3}\cos^3 x + \frac{4}{3}\sin x - \frac{1}{9}\sin^3 x + C$.

27. $\int e^{\cos x}\sin x\, dx = -e^{\cos x} + C$

28. $\int x^4 e^{-x}dx$. We use integration by parts with

$u = x^4$	$v = -e^{-x}$
$du = 4x^3 dx$	$dv = e^{-x}dx$

Hence, $\int x^4 e^{-x}dx = x^4\left(-e^{-x}\right) + \int 4x^3 e^{-x}dx$. Using integration by parts a second time produces

$u = 4x^3$ | $v = -e^{-x}$

$du = 12x^2 dx$ | $dv = e^{-x}dx$

Thus, $x^4\left(-e^{-x}\right) + \int 4x^3 e^{-x}dx = -x^4 e^{-x} - 4x^3 e^{-x} + \int 12x^2 e^{-x}$. Using integration by parts a third time, we have

$u = 12x^2$ | $v = -e^{-x}$

$du = 24x \, dx$ | $dv = e^{-x}dx$

This produces $-x^4 e^{-x} - 4x^3 e^{-x} - 12x^2 e^{-x} - 24xe^{-x} + \int 24xe^{-x}dx$. Finally, using integration by parts a fourth time, we obtain

$u = 24x$ | $v = -e^{-x}$

$du = 24 \, dx$ | $dv = e^{-x}dx$

$-x^4 - e^{-x} - 4x^3 e^{-x} - 12x^2 e^{-x} - 24xe^{-x} + \int 24e^{-x}dx = -x^4 e^{-x} - 4x^3 e^{-x} - 12x^2 e^{-x} - 24xe^{-x} - 24e^{-x} + C = e^{-x}\left(-x^4 - 4x^3 - 12x^2 - 24x - 24\right) + C$.

29. $\int \frac{\cos x}{\sin^2 x + 9} dx$. Let $u = \sin x$ and $du = \cos x \, dx$. Substituting we get $\int \frac{du}{u^2 + 9} = \frac{1}{3} \tan^{-1} \frac{u}{3} + C = \frac{1}{3} \tan^{-1}\left(\frac{\sin x}{3}\right) + C$.

30. $\int \frac{e^x}{d^x + 16} dx$. Let $u = e^x + 16$ and $du = e^x dx$. Substituting yields $\int \frac{du}{u} = \ln|u| + C = \ln|e^x + 16| + C = \ln\left(e^x + 16\right) + C$.

31. Let $u = x^2$ and $du = 2x \, dx$. Then $\int_1^e x^3 \ln x^2 dx = \frac{1}{2}\int_1^e x^2 \ln x^2 2x \, dx$ and by integration formula #90,

$$\frac{1}{2}\int_1^e x^2 \ln x^2 2x \, dx = \frac{1}{2}\left(x^2\right)^2\left(\frac{\ln x^2}{2} - \frac{1}{4}\right)\Big|_1^e$$

$$= \frac{e^4}{2}\left(\frac{\ln e^2}{2} - \frac{1}{4}\right)$$

$$- \frac{1}{2}\left(0 - \frac{1}{4}\right)$$

$$= \frac{e^4}{2}\left(1 - \frac{1}{4}\right) + \frac{1}{8}$$

$$= \frac{3e^4}{8} + \frac{1}{8} \approx 20.5993$$

32. $\int e^{8x} \cos 2x \, dx$. Use integration by parts with $u = e^{8x}$ and $dv = \cos 2x \, dx$. Then $v = \frac{1}{2}\sin 2x$ and $du = 8e^{8x}dx$. This produces $\int e^{8x} \cos 2x \, dx = \frac{1}{2}e^{8x}\sin 2x - \int 4e^{8x}\sin 2x \, dx$. Use integration by parts again, with

$u = 2e^{8x}$ | $v = -\cos 2x$

$du = 16e^{8x}dx$ | $dv = 2\sin 2x \, dx$

Thus we have $\int e^{8x}\cos 2x \, dx = \frac{1}{2}e^{8x}\sin 2x - \int 4e^{8x}\sin 2x = \frac{1}{2}e^{8x}\sin 2x + 2e^{8x}\cos 2x - 16 \times \int e^{8x}\cos 2x$. Adding $16\int e^{8x}\cos 2x$ to both sides of this last equation, we obtain $17\int e^{8x}\cos 2x \, dx = \frac{1}{2}e^{8x}\sin 2x + 2e^{8x}\cos 2x$, and, dividing by 17 produces the desired integral $\int e^{8x}\cos 2x \, dx = \frac{1}{34}e^{8x}\sin 2x + \frac{2}{17}e^{8x}\cos 2x + C = \frac{1}{34}e^{8x}(\sin 2x + 4\cos 2x) + C$.

33. $\int \sin^{1/3} 4x \cos^5 4x \, dx = \int \sin^{1/3} 4x\left(\cos^2 4x\right)^2 \times \cos 4x \, dx = \int \sin^{1/3} 4x\left(1 - \sin^2 4x\right)^2 \cos 4x \, dx = \int \left(\sin^{1/3} 4x - 2\sin^{7/3} 4x + \sin^{13/3} 4x\right)\cos 4x \, dx = \frac{1}{4}\left(\frac{3}{4}\sin^{4/3} 4x - 2 \cdot \frac{1}{10}\sin^{10/3} 4x + \frac{3}{16}\sin^{16/3} 4x\right) + C = \frac{3}{320}\sin^{4/3} 4x\left(20 - 16\sin^2 4x + 5\sin^4 4x\right) + C$.

34. $\int \frac{\sin^3 x \, dx}{\cos^4 x} = \int \frac{\sin^3 x}{\cos^3 x} \cdot \frac{dx}{\cos x} = \int \tan^3 x \sec x \, dx = \int \tan x\left(\sec^2 x - 1\right)\sec x \, dx = \int \sec^2 x \times (\tan x \sec x \, dx) - \int \tan x \sec x \, dx = \frac{1}{3}\sec^3 x - \sec x + C$.

35. $\int \frac{\cos x \, dx}{\sqrt{16 - 4\sin^2 x}}$. Let $u = 2\sin x$ and $du = 2\cos x \, dx$ or $\frac{1}{2}du = \cos x \, dx$. Substituting we get $\frac{1}{2}\int \frac{du}{\sqrt{4^2 - u^2}} = \frac{1}{2}\sin \frac{u}{4} + C = \frac{1}{2}\sin^{-1}\left(\frac{2\sin x}{4}\right) + C = \frac{1}{2}\sin^{-1}\left(\frac{\sin x}{2}\right) + C$.

36. $\int \frac{e^{5x}}{4 - e^{5x}} dx$. Let $u = 4 - e^{5x}$ and $du = -5e^{5x}dx$ or $-\frac{1}{5}du = e^{5x}dx$. Substituting we get $-\frac{1}{5}\int \frac{du}{u} = -\frac{1}{5}\ln|u| + C = -\frac{1}{5}\ln|4 - e^{5x}| + C$.

37. $\int x^5 e^{x^2} dx$. Using integration by parts, we have the following table.

$u = x^4$ | $v = \frac{1}{2}e^{x^2}$

$du = 4x^3 dx$ | $dv = xe^{x^2}dx$

This produces $\int x^5 e^{x^2} dx = \frac{1}{2}e^{x^2}x^4 - 2\int x^3 e^{x^2}dx$. Using integration by parts again with $u = -2x^2$ and $dv = xe^{x^2}dx$, we get $v = \frac{1}{2}e^{x^2}$ and $du = -4x \, dx$. Thus, $\frac{1}{2}e^{x^2}x^4 - 2\int x^3 e^{x^2}dx = \frac{1}{2}e^{x^2}x^4 - e^{x^2} \cdot x^2 + 2\int xe^{x^2} = \frac{1}{2}e^{x^2}x^4 - e^{x^2}x^2 + e^{x^2} + C = e^{x^2}\left(\frac{x^4}{2} - x^2 - 1\right) + C = \frac{e^{x^2}}{2}\left(x^4 - 2x^2 + 2\right) + C$.

38. $\int \frac{e^{3x}}{\left(e^{3x}-1\right)^2}dx$. Let $u = e^{3x} - 1$ and $du = 3e^{3x}dx$

so $\frac{1}{3}du = e^{3x}dx$. Substituting we get

$$\frac{1}{3}\int \frac{du}{u^2} = -\frac{1}{3}u^{-1} + C = -\frac{1}{3}(e^{3x} - 1)^{-1} + C$$

39. $\int \frac{(\arctan 2x)^4}{1+4x^2}dx$. Let $u = \arctan 2x$, then $du = \frac{2}{1+4x^2}dx$ and substituting gives $\frac{1}{2}\int u^4 du = \frac{1}{10}u^5 + C = \frac{1}{10}(\arctan 2x)^5 + C$.

40. $\int \frac{\sec^2 5x\ dx}{2\tan 5x+9}$. Let $u = 2\tan 5x + 9$, and then $du = 10\sec^2 5x\ dx$ or $\frac{1}{10}du = \sec^2 5x\ dx$. Substituting we get

$$\frac{1}{10}\int \frac{du}{u} = \frac{1}{10}\ln|u| + C$$
$$= \frac{1}{10}\ln|2\tan 5x + 9| + C$$

41. Rewrite $\int_0^{\ln 49} \sqrt{9 + e^x}dx = 2\int_0^{\ln 49} \frac{\sqrt{9+e^x}}{e^{x/2}}\frac{e^{x/2}}{2}dx$.

Let $u = e^{x/2}, du = \frac{1}{2}e^{x/2}dx$, and $a = 3$ and this fits integration formula #32. Remember that $\ln 49 = \ln 7^2 = 2\ln 7$, and so, evaluating $e^{x/2}$ when $x = \ln 49$ produces $e^{(\ln 49)/2} = e^{(2\ln 7)/2} = e^{\ln 7} = 7$.

$$\int_0^{\ln 49} \sqrt{9 + e^x}dx$$
$$= 2\left[\sqrt{9 + e^{x/2}}\right.$$
$$\left. - 3\ln\left|\frac{3 + \sqrt{9 + e^{x/2}}}{e^{x/2}}\right|\right]\Big|_0^{\ln 49}$$
$$= 2\left[\sqrt{9 + 7} - 3\ln\left|\frac{3 + \sqrt{9 + 7}}{7}\right|\right]$$
$$- \left[\sqrt{9 + 1} - 3\ln\left|\frac{3 + \sqrt{9 + 1}}{1}\right|\right]$$
$$= 2[4 - 3\ln|1|]$$
$$- 2\left[\sqrt{10} - 3\ln\left|3 + \sqrt{10}\right|\right]$$
$$= 8 - 2\sqrt{10} + 6\ln\left|3 + \sqrt{10}\right| \approx 12.5861$$

42. Using integration by parts, with

$u = \arcsin\left(\frac{x}{2}\right)$	$v = x$
$du = \frac{1}{\sqrt{4-x^2}}dx$	$dv = dx$

Then, we obtain

$$\int_0^1 \arcsin\left(\frac{x}{2}\right)dx$$
$$= x\arcsin\left(\frac{x}{2}\right) - \int \frac{x}{\sqrt{4 - x^2}}dx$$
$$= x\arcsin\left(\frac{x}{2}\right) + \frac{1}{2}\int \frac{2x}{\sqrt{4 - x^2}}dx$$
$$= \left[x\arcsin\left(\frac{x}{2}\right) + \sqrt{4 - x^2}\right]_0^1$$
$$= \arcsin\left(\frac{1}{2}\right) + \sqrt{3} - \arcsin 0 - \sqrt{4}$$
$$= \frac{\pi}{6} + \sqrt{3} - 2 \approx 0.2556$$

43. Using integration by parts, with

$u = x^2$	$v = \sqrt{9 + x^2}$
$du = 2x\ dx$	$dv = \frac{x}{\sqrt{9+x^2}}dx$

Then, we obtain

$$\int_0^3 \frac{x^3 dx}{\sqrt{9 + x^2}}$$
$$= x^2\sqrt{9 + x^2}\Big]_0^3 - \int_0^3 2x\sqrt{9 + x^2}dx$$
$$= \left[x^2\sqrt{9 + x^2} - \frac{2}{3}(9 + x^2)^{3/2}\right]_0^3$$
$$= 9\sqrt{18} - \frac{2}{3}(18)^{3/2} - 0 + \frac{2}{3}(9)^{3/2}$$
$$= 27\sqrt{2} - \frac{2}{3}(3\sqrt{2})^3 + \frac{2}{3}(27)$$
$$= 27\sqrt{2} - 36\sqrt{2} + 18$$
$$= 18 - 9\sqrt{2} \approx 5.2721$$

44. $\int_0^{\pi/2} \sin^3 x \cos^3 x\ dx$

$$= \int_0^{\pi/2} \sin^3 x(1 - \sin^2 x)\cos x\ dx$$
$$= \int_0^{\pi/2} (\sin^3 x - \sin^5 x)\cos x\ dx$$
$$= \left[\frac{1}{4}\sin^4 x - \frac{1}{6}\sin^6 x\right]_0^{\pi/2}$$
$$= \frac{1}{4} - \frac{1}{6} = \frac{1}{12} \approx 0.0833$$

CHAPTER 28 TEST

1. Let $u = 9 - e^x$, then $du = -e^x\, dx$ and $\int \frac{e^x\, dx}{\sqrt{9-e^x}} = -\int u^{-1/2} du = -2u^{1/2} + C = -2\sqrt{9 - e^x} + C$

2. $\int \tan^2 4x \cos^2 4x\ dx = \int \frac{\sin^2 4x}{\cos^2 4x} \cdot \cos^4 4x\ dx = \int \sin^2 4x \cos^2 4x\ dx = \int \sin^2 4x(1 - \sin^2 4x) dx = \int \sin^2 4x\ dx - \int \sin^4 4x\ dx = \int \frac{1-\cos 8x}{2} dx - \int \left(\frac{1-\cos 8x}{2}\right)^2 dx = \frac{1}{2}x - \frac{1}{16}\sin 8x - \frac{1}{4}\int (1 - 2\cos 8x + \cos^2 8x) dx = \frac{1}{2}x - \frac{1}{10}\sin 8x - \frac{1}{4}x + \frac{1}{16}\sin 8x - \frac{1}{4}\int \left(\frac{1+\cos 16x}{2}\right) dx = \frac{1}{4}x - \frac{1}{8}x - \frac{1}{128}\sin 16x + C = \frac{1}{8}x - \frac{1}{128}\sin 16x + C.$

3. $\int \frac{5\, dx}{x^2+1} = 5\tan^{-1} x + C$

4. Let $u = x^2 + 1$, then $du = 2x\, dx$ and so $\int \frac{4x\, dx}{(x^2+1)^3} = 2\int u^{-3} du = -u^{-2} + C = \frac{-1}{(x^2+1)^2} + C.$

5. Use integration by parts with $u = x$ and $dv = e^{4x} dx$. Then $du = dx$ and $v = \frac{1}{4}e^{4x}$ and so $\int xe^{4x} dx = uv - \int v\, du = \frac{1}{4}xe^{4x} - \int \frac{1}{4}e^{4x} dx = \frac{1}{4}xe^{4x} - \frac{1}{16}e^{4x} + C = \frac{1}{16}e^{4x}(4x - 1) + C.$

6. Let $u = \tan x$ and then $du = \sec^2 x\ dx = \frac{1}{\cos^2 x} dx$ and $\int \frac{e^{\tan x}}{\cos^2 x} dx = \int e^u du = e^u + C = e^{\tan x} + C.$

7. Using the disk method, we have $V = \pi \times \int_0^1 \left(\sqrt{x}e^x\right)^2 dx = \pi \int_0^1 xe^{2x} dx = \frac{\pi}{4}e^{2x}(2x-1)\big]_0^1 = \frac{\pi}{4}(e^2 + 1).$

8. The arc length L uses $(y')^2 = (2x)^2$ and so, $L = \int_0^2 \sqrt{1 + (2x)^2} dx = \int_0^2 \sqrt{1 + 4x^2} dx$. From Formula 30 in Appendix C, we see that $L \times \int_0^2 \sqrt{1 + 4x^2} dx = \frac{2x}{2}\sqrt{1 + 4x^2} + \frac{1}{2}\ln\left|2x + \sqrt{1 + 4x^2}\right|\big]_0^2 = 2\sqrt{17} + \frac{1}{2}\ln|4 + \sqrt{17}| \approx 9.2936.$

CHAPTER
29
Parametric Equations and Polar Coordinates

≡ 29.1 DERIVATIVES OF PARAMETRIC EQUATIONS

1. $x = t^2 + t$, $y = t + 1$. Thus, $\frac{dx}{dt} = 2t + 1$ and $\frac{dy}{dt} = 1$, which means that $\frac{dy}{dx} = \frac{1}{2t+1}$. Next, $\frac{dy}{dt}\left(\frac{d}{dt}\right) = \frac{d}{dt}(2t+1)^{-1} = \frac{-2}{(2t+1)^2}$ and we get $\frac{d^2y}{dx^2} = \frac{-2}{(2t+1)^3}$. Critical values occur when $\frac{dx}{dy} = 0$ or is undefined. The derivative $\frac{dy}{dx}$ is undefined at $t = -\frac{1}{2}$. The second derivative is also undefined. There are no extrema. Inflection point at $t = -\frac{1}{2}$: $\left(-\frac{1}{4}, \frac{1}{2}\right)$.

3. Here $x = t^2 - 6t + 12$ and $y = t + 4$, and so, $\frac{dy}{dx} = \frac{dy/dt}{dx/dt} = \frac{1}{2t-6}$. Critical value at $t = 3$. The second derivative is $\frac{d^2y}{dx^2} = \frac{\frac{d}{dt}\left(\frac{dy}{dx}\right)}{\frac{dx}{dt}} = \frac{\frac{-2}{(2t-6)^2}}{2t-6} = \frac{-2}{(2t-6)^3}$. Inflection point at $t = 3$: $(3, 7)$. No extrema.

5. Here $x = t^2 + t$ and $y = t^2 - t$ which means that the first derivative is $\frac{dy}{dx} = \frac{dy/dt}{dx/dt} = \frac{2t-1}{2t+1}$. Critical values are at $t = \frac{1}{2}, -\frac{1}{2}$. The second derivative is $\frac{d^2y}{dx^2} = \frac{\frac{d}{dt}\left(\frac{dy}{dx}\right)}{\frac{dx}{dt}} = \frac{\frac{(2t+1)2-(2t-1)2}{(2t+1)^2}}{2t+1} = \frac{4}{(2t+1)^3}$. Inflection point at $t = -\frac{1}{2}$: $\left(-\frac{1}{4}, \frac{3}{4}\right)$. Minimum at $t = \frac{1}{2}$: $\left(\frac{3}{4}, -\frac{1}{4}\right)$. It is a minimum since the second derivative is positive.

7. Here $x = 3 + 4\cos t$ and $y = -1 + \cos t$ which means that the first derivative is $\frac{dy}{dx} = \frac{dy/dt}{dx/dt} = \frac{-\sin t}{-4\sin t} = \frac{1}{4}$. The second derivative is $\frac{d^2y}{dx^2} = 0$ since $\frac{dy}{dx}$ is a constant. This is segment where the point (x, y) moves between a minimum

when t is an odd multiple of π giving $(-1, -2)$, and a maximum when t is an even multiple of π giving $(7, 0)$.

9. Since $x = 2 + \sin t$ and $y = -1 + \cos t$, the first derivative is $\frac{dy}{dx} = \frac{dy/dt}{dx/dt} = \frac{-\sin t}{\cos t} = -\tan t$. The second derivative is $\frac{d^2y}{dx^2} = \frac{\frac{d}{dt}\left(\frac{dy}{dx}\right)}{\frac{dx}{dt}} = \frac{-\sec^2 t}{\cos t} = -\sec^3 t$. Critical values occur when $\sin t = 0$. At $t = 2k\pi$, the second derivative is negative so maximum $(2, 0)$. At $t = \pi + 2k\pi$, y'' is positive so we have a minimum $(2, -2)$. Inflection points are at $t = \frac{\pi}{2} + 2k\pi$ and $-\frac{\pi}{2} + 2k\pi$. These are $(3, -1)$ and $(1, -1)$. (Note: this is a circle with vertical diameter between $(2, 0)$ and $(2, -2)$ and horizontal diameter between $(3, -1)$ and $(1, -1)$).

11. Here $x = 2t - 1$ and $y = 4t^2 - 2t$ and the first derivative is $\frac{dy}{dx} = \frac{dy/dt}{dx/dt} = \frac{8t-2}{2}$. At $t = 1$, the slope is $m = \frac{dy}{dx} = \frac{8-2}{2} = 3$ and the desired point is $(x, y) = (1, 2)$. Thus, the equation of the tangent is $y - 2 = 3(x - 1) \Rightarrow y = 3x - 1$ and the equation of the normal line is $(y - 2) = -\frac{1}{3}(x - 1) \Rightarrow y = -\frac{1}{3}x + 2\frac{1}{3}$ or $3y + x = 7$.

13. $x = t^3$, $y = t^2$, $t = -3$. Thus, $(x, y) = (-27, 9)$. The derivative is $\frac{dy}{dx} = \frac{2t}{3t^2} = \frac{2}{3t}$; at $t = -3$, $m = -\frac{2}{9}$. Tangent: $y - 9 = -\frac{2}{9}(x + 27)$ or $9y + 2x = 27$; Normal: $y - 9 = \frac{9}{2}(x + 27)$ or $2y - 9x = 261$.

15. $x = 2 + \cos t$, $y = 2\sin t$, $t = \frac{\pi}{2}$, $(x, y) = (2, 2)$. Thus, $\frac{dy}{dx} = \frac{2\cos t}{-\sin t} = -2\cot t$. At $t = \frac{\pi}{2}$, $m = 0$ which means that the desired lines are Tangent: $y = 2$ and Normal: $x = 2$.

337

17. $x = t + 3$, $y = t^2 - 4t$. The derivative is $\frac{dy}{dx} = \frac{dy/dt}{dx/dt} = \frac{2t-4}{2}$. Horizontal tangent when $2t - 4 = 0$ or $t = 2$: $(5, -4)$. Vertical tangents: none.

19. $x = 3\cos t$, $y = 5\sin t$. To find any horizontal tangents, we have $\frac{dy}{dt} = 5\cos t$ and $5\cos t = 0$, when $t = \frac{\pi}{2} + 2k\pi$: $(0, 5)$ and $t = -\frac{\pi}{2} + 2k\pi$: $(0, -5)$. To find vertical tangents, we take $\frac{dx}{dt} = -3\sin t$; $-3\sin t = 0$ when $t = 2k\pi$: $(3, 0)$ and $t = \pi + 2k\pi$: $(-3, 0)$.

21. $s_x(t) = 3780t\cos\frac{4\pi}{15}$; $s_y(t) = 3780t\sin\frac{4\pi}{15} - 16t^2$.

(a) $s_x(5) = 3780 \cdot 5 \cdot \cos\frac{4\pi}{15} \approx 12{,}646.57$ ft;

$s_y(5) = 3780 \cdot 5 \cdot \sin\frac{4\pi}{15} - 16 \cdot 5^2$
$\approx 13{,}645.44$ ft.

(b) $v_x(5) = 3780\cos\frac{4\pi}{15} = 2529.31$ ft/sec, and

$v_y(5) = 3780\sin\frac{4\pi}{15} - 32 \cdot 5 = 2649.09$ ft/sec

(c) $a_x(t) = 0$;

$a_y(t) = -32$ ft/sec^2

23. (a) When $t = 3.0$, then $x = \frac{100}{\sqrt{3^2+1}} = \frac{100}{\sqrt{10}} = 10\sqrt{10}$ and $y = \frac{100(3)}{\sqrt{3^2+1}} = \frac{300}{\sqrt{10}} = 30\sqrt{10}$. Using these two values of x and y, produces magnitude $= \sqrt{x^2 + y^2} = \sqrt{\left(10\sqrt{10}\right)^2 + \left(30\sqrt{10}\right)^2} = \sqrt{1000 + 9000} = \sqrt{10{,}000} = 100$ km

The direction is $\theta = \tan^{-1}\left(\frac{y}{x}\right) = \tan^{-1}\frac{30\sqrt{10}}{10\sqrt{10}} = \tan^{-1}3 = 71.6°$

(b) The horizontal component of this electron's velocity is $v_x = \frac{dx}{dt} = 100(t^2 + 1)^{-3/2}\left(-\frac{1}{2}\right)(2t) = \frac{-100t}{\left(t^2+1\right)^{3/2}}$. The vertical component of this electron's velocity is

$v_y = \frac{dy}{dt}$

$= \frac{100(t^2+1)^{1/2} - \frac{1}{2}(t^2+1)^{-1/2}(2t)(100t)}{t^2+1}$

$= \frac{100(t^2+1) - 100t^2}{\left(t^2+1\right)^{3/2}} = \frac{100}{\left(t^2+1\right)^{3/2}}$

(c) When $t = 3.0$, then $v_x = \frac{-100(3)}{\left(3^2+1\right)^{3/2}} = \frac{-300}{(10)^{3/2}} \approx$

-9.487 and $v_y = \frac{100}{(3^2+1)^{3/2}} = \frac{100}{(10)^{3/2}} \approx 3.162$. Us-

ing these two values of v_x and v_y, produces a magnitude of

$$\sqrt{(-9.487)^2 + (3.162)^2} = 10.0 \text{ km/s}$$

and the direction is $\theta = \tan^{-1}\left(\frac{3.162}{-9.487}\right) = -18.4°$.

(d) The horizontal component of this electron's acceleration is

$a_x = \dfrac{-100(t^2+1)^{3/2} + 100t\left(\frac{3}{2}\right)(t^2+1)^{1/2}(2t)}{\left(t^2+1\right)^3}$

$= \dfrac{-100(t^2+1) + 300t^2}{\left(t^2+1\right)^{5/2}} = \dfrac{200t^2 - 100}{\left(t^2+1\right)^{5/2}}$

The vertical component of this electron's acceleration is $a_y = 100\left(-\frac{3}{2}\right)(t^2+1)^{-5/2}(2t) = \frac{-300t}{\left(t^2+1\right)^{5/2}}$.

(e) When $t = 3.0$, then $a_x = \frac{200(3^2)-100}{\left(3^2+1\right)^{5/2}} \approx 5.376$

and $a_y = \frac{-300(3)}{\left(3^2+1\right)^{5/2}} \approx -2.846$. Using these two values of a_x and a_y, produces a magnitude of $\sqrt{5.376^2 + (-2.846)^2} \approx 6.083$ km/s^2 and a direction of $\theta = \tan^{-1}\left(\frac{-2.846}{5.376^2}\right) \approx -27.9°$.

25. (a) The horizontal component of this spacecraft's velocity is $v_x = \frac{dx}{dt} = 10\left(\frac{1}{2}\right)(t^4+1)^{-1/2} \cdot 4t^3 = \frac{20t^3}{\sqrt{t^4+1}}$. The vertical component of this spacecraft's velocity is $v_y = \frac{dy}{dt} = \left(\frac{3}{2}\right)40t^{1/2} = 60\sqrt{t}$.

(b) The magnitude of this spacecraft's velocity at time t is $\sqrt{v_x^2 + v_y^2} = \sqrt{\left(\frac{20t^3}{\sqrt{t^4+1}}\right)^2 + \left(60\sqrt{t}\right)^2} = 20\sqrt{\frac{t^6}{t^4+1} + 9t}$ m/s and the direction is

$\theta = \tan^{-1}\left(\frac{v_y}{v_x}\right) = \tan^{-1}\left(\dfrac{60\sqrt{t}}{\dfrac{20t^3}{\sqrt{t^4+1}}}\right)$

$= \tan^{-1}\left(\frac{60\sqrt{t}\sqrt{t^4+1}}{20t^3}\right) = \tan^{-1}\left(\frac{3\sqrt{t^4+1}}{t^{5/2}}\right)$.

(c) When $t = 5.0$ s, the magnitude is $20 \times \sqrt{\frac{5^6}{5^4+1} + 9(5)} \approx 167.28$ m/s and direction is $\theta = \tan^{-1}\left(\frac{3\sqrt{5^4+1}}{5^{5/2}}\right) \approx 53.3°$.

29.2 DIFFERENTIATION IN POLAR COORDINATES

1. Since $r = 3\sin\theta$, then $r' = 3\cos\theta$ which leads to $\frac{dy}{dx} = \frac{r'\sin\theta + r\cos\theta}{r'\cos\theta - r\sin\theta} = \frac{3\cos\theta\sin\theta + 3\sin\theta\cos\theta}{3\cos\theta\cos\theta - 3\sin\theta\sin\theta} = \frac{2\cos\theta\sin\theta}{\cos^2\theta - \sin^2\theta} = \frac{\sin 2\theta}{\cos 2\theta} = \tan 2\theta$

3. Since $r = 1 + \cos\theta$, then $r' = -\sin\theta$ and we have $\frac{dy}{dx} = \frac{r'\sin\theta + r\cos\theta}{r'\cos\theta - r\sin\theta} = \frac{-\sin\theta\sin\theta + (1+\cos\theta)\cos\theta}{-\sin\theta\cos\theta - (1+\cos\theta)\sin\theta} = \frac{-\sin^2\theta + \cos\theta + \cos^2\theta}{-2\sin\theta\cos\theta - \sin\theta} = \frac{\cos\theta + \cos 2\theta}{-\sin 2\theta - \sin\theta} = -\frac{\cos 2\theta + \cos\theta}{\sin 2\theta + \sin\theta}$.

5. We are given $r = 1 + \cos 3\theta$, and so $r' = -3\sin 3\theta$, which leads to $\frac{dy}{dx} = \frac{-3\sin 3\theta\sin\theta + (1+\cos 3\theta)\cos\theta}{-3\sin 3\theta\cos\theta - (1+\cos 3\theta)\sin\theta} = \frac{3\sin 3\theta\sin\theta - \cos 3\theta\cos\theta - \cos\theta}{3\sin 3\theta\cos\theta + \cos 3\theta\sin\theta + \sin\theta}$.

7. Here $r = \tan\theta$, which means that $r' = \sec^2\theta$, and so the desired derivative is $\frac{dy}{dx} = \frac{\sec^2\theta\sin\theta + \tan\theta\cos\theta}{\sec^2\theta\cos\theta - \tan\theta\sin\theta} = \frac{\tan\theta\sec\theta + \tan\theta\cos\theta}{\sec\theta - \tan\theta\sin\theta} \cdot \frac{\cos\theta}{\cos\theta} = \frac{\tan\theta + \tan\theta\cos^2\theta}{1 - \sin^2\theta} = \tan\left(\frac{1+\cos^2\theta}{\cos^2\theta}\right) = \tan\theta\left(\sec^2\theta + 1\right)$.

9. $r = \sin\theta$, $r' = \cos\theta$; $\frac{dy}{dx} = \frac{\cos\theta\sin\theta + \sin\theta\cos\theta}{\cos\theta\cos\theta - \sin\theta\sin\theta} = \frac{2\sin\theta\cos\theta}{\cos^2\theta - \sin^2\theta} = \frac{\sin 2\theta}{\cos 2\theta} = \tan 2\theta$. At the point $\left(1, \frac{\pi}{2}\right)$, $m = \tan 2\cdot\frac{\pi}{2} = \tan\pi = 0$. Thus, $x = r\cos\theta = 1\cos\frac{\pi}{2} = 0$ and $y = r\sin\theta = 1\sin\frac{\pi}{2} = 1$. Tangent: $y = 1$; Normal: $x = 0$

11. $r = 5\sin 3\theta$, $\left(\frac{5}{\sqrt{2}}, \frac{\pi}{12}\right)$, $r' = 15\cos 3\theta$; $\frac{dy}{dx} = \frac{15\cos 3\theta\sin\theta + 5\sin 3\theta\cos\theta}{15\cos 3\theta\cos\theta - 5\sin 3\theta\sin\theta}$. When $\theta = \frac{\pi}{12}$, then $m = \frac{dy}{dx} \approx 0.660254 \approx 0.6603$. Thus, $x = \frac{5}{\sqrt{2}}\cos\frac{\pi}{12} = 3.4151$ and $y = \frac{5}{\sqrt{2}}\sin\frac{\pi}{12} = 0.9151$. Tangent: $y - 0.9151 = 0.6603(x - 3.4151)$. Normal: $y - 0.9151 = -1.5146(x - 3.4151)$

13. We are given $r = 6\sin^2\theta$ and $\left(4.5, \frac{2\pi}{3}\right)$, and so $r' = 12\sin\theta\cos\theta$ which leads to the derivative $\frac{dy}{dx} = \frac{12\sin\theta\cos\theta\sin\theta + 6\sin^2\theta\cos\theta}{12\sin\theta\cos\theta\cos\theta - 6\sin^2\theta\sin\theta} = \frac{3\sin^2\theta\cos\theta}{\sin\theta\left(2\cos^2\theta - \sin^2\theta\right)} = \frac{3\sin\theta\cos\theta}{2\cos^2\theta - \sin^2\theta}$. At $\theta = \frac{2\pi}{3}$ we get $m = \frac{3\cdot\frac{\sqrt{3}}{2}\cdot\frac{-1}{2}}{2\cdot\frac{1}{4} - \frac{3}{4}} = \frac{\frac{-3\sqrt{3}}{4}}{-\frac{1}{4}} = 3\sqrt{3} \approx 5.1962$. At $\left(4.5, \frac{2\pi}{3}\right)$, we find that $x = 4.5\cos\frac{2\pi}{3} = \frac{-9}{4}$ and $y = 4.5\sin\frac{2\pi}{3} = \frac{9\sqrt{3}}{4}$. Tangent: $y - \frac{9\sqrt{3}}{4} = 3\sqrt{3}\left(x + \frac{9}{4}\right)$ or $4y - 9\sqrt{3} = 12\sqrt{3}x + 27\sqrt{3}$ or $4y - 12\sqrt{3}x = 36\sqrt{3}$ or $y - 3\sqrt{3}x = 9\sqrt{3}$. Normal: $y - \frac{9\sqrt{3}}{4} = \frac{-1}{3\sqrt{3}}\left(x + \frac{9}{4}\right)$,

or $4y - 9\sqrt{3} = \frac{-4}{3\sqrt{3}}x - \frac{9}{3\sqrt{3}}$, or $4y + \frac{4}{3\sqrt{3}}x = 8\sqrt{3}$, or $y + \frac{1}{3\sqrt{3}}x = 2\sqrt{3}$, or $9y + \sqrt{3}x = 18\sqrt{3}$.

15. $r = e^\theta$; $\left(2.8497, \frac{\pi}{3}\right)$; By exercise #8, we see that $\frac{dy}{dx} = \frac{\sin\theta + \cos\theta}{\cos\theta - \sin\theta}$, and at $\frac{\pi}{3}$, we obtain $m = \frac{\frac{\sqrt{3}}{2} + \frac{1}{2}}{\frac{1}{2} - \frac{\sqrt{3}}{2}} = \frac{\sqrt{3}+1}{1-\sqrt{3}} = \frac{(\sqrt{3}+1)(\sqrt{3}+1)}{(1-\sqrt{3})(1+\sqrt{3})} = \frac{4+2\sqrt{3}}{-2} = -2 - \sqrt{3}$. Thus, $x = 2.8497 \cdot \cos\frac{\pi}{3} = 1.4249$ and $y = 2.8497 \cdot \sin\frac{\pi}{3} = 2.4679$. Tangent: $y - 2.4679 = -(2+\sqrt{3})(x - 1.4249)$, or $y - 2.4697 = -3.7321x + 5.3178$, or $y + 3.7321x = 7.7875$. Normal: $y - 2.4679 = \frac{1}{2+\sqrt{3}}(x - 1.4249)$, or $y - 2.4679 = 0.2679x - 0.3818$, or $y - 0.2679x = 2.0861$.

17. Since $r = 3\cos 2\theta$ we have $r' = -6\sin 2\theta$, which gives the derivative $\frac{dy}{dx} = \frac{-6\sin 2\theta\sin\theta + 3\cos 2\theta\cos\theta}{-6\sin 2\theta\cos\theta - 3\cos 2\theta\sin\theta} = \frac{2\sin 2\theta\sin\theta - \cos 2\theta\cos\theta}{2\sin 2\theta\cos\theta + \cos 2\theta\sin\theta}$. If we set the numerator equal to 0 we get $2\sin 2\theta\sin\theta - \cos 2\theta\cos\theta = 0$ which expands to $2(2\sin\theta\cos\theta)\sin\theta - \left(1 - 2\sin^2\theta\right)\cos\theta = 0$; $\cos\theta\left(4\sin^2\theta - 1 + 2\sin^2\theta\right) = 0$. Which simplifies to $\cos\theta\left(6\sin^2\theta - 1\right) = 0$. Hence either $\cos\theta = 0$ or $\sin^2\theta = \frac{1}{6}$ or $\sin\theta = \pm\frac{\sqrt{6}}{6}$. This gives critical values of $\theta = \frac{\pi}{2}$, $\frac{3\pi}{2}$ and $\theta = 0.4205, 2.7211, 3.5621,$ and 5.8627.

Set the denominator to 0 and solve $2\sin 2\theta \times \cos\theta + \cos 2\theta\sin\theta = 0$. Using an expansion identity for $2\sin\alpha\cos\alpha$, we get $4\sin\theta\cos\theta\cos\theta + \left(2\cos^2\theta - 1\right)\sin\theta = 0$. Factoring, we get $\sin\theta \times \left[4\cos^2\theta + 2\cos^2\theta - 1\right] = 0$, so $\sin\theta = 0$ or $\cos^2\theta = \frac{1}{6}$ or $\cos\theta = \pm\frac{\sqrt{6}}{6}$. This gives critical values of $0, \pi, 1.1503, 1.9913, 4.2918,$ and 5.1329.

The graph of $r = 3\cos 2\theta$ is a four leaf rose. The first set of critical values gives the locations of horizontal tangents and the second set gives the locations of vertical tangents. Horizontal tangents at $\left(-3, \frac{\pi}{2}\right), \left(-3, \frac{3\pi}{2}\right), (2, 0.4205), (2, 2.7211),$ $(2, 3.5621),$ and $(2, 5.8627)$. Vertical tangents at $(3, 0), (3, \pi), (-2, 1.1503), (-2, 1.9913),$ $(-2, 4.2918),$ and $(-2, 5.1329)$.

19. $r = 1 + 2\cos\theta$, $r' = -2\sin\theta$. Hence, we obtain

$$\frac{dy}{dx} = \frac{-2\sin\theta\sin\theta + (1 + 2\cos\theta)\cos\theta}{-2\sin\theta\cos\theta - (1 + 2\cos\theta)\sin\theta}$$

$$= \frac{-2\sin^2\theta + \cos^2\theta + 2\cos\theta}{-2\sin\theta\cos\theta - \sin\theta + 2\cos\theta\sin\theta}$$

$$= \frac{-2(1 - \cos^2\theta) + \cos\theta + 2\cos^2\theta}{-4\sin\theta\cos\theta - \sin\theta}$$

$$= \frac{4\cos^2\theta + \cos\theta - 2}{-\sin\theta(4\cos\theta + 1)}$$

The numerator equals 0 when $4\cos^2\theta + \cos\theta - 2 = 0$ or $\cos\theta = (-1 \pm \sqrt{33})8$ or when $\theta \approx$ 0.9359, 5.3476, 2.5738, or 3.7094. The denominator equals 0 when $\sin\theta = 0$ or $\cos\theta = -\frac{1}{4}$ which yields $\theta = 0$, π, 1.8234, or 4.4597. This is a limaçon. Horizontal tangents are at (2.1861, 0.9359), (−0.6861, 2.5738), (−0.6861, 3.7094), and (2.1861, 5.3476). Vertical tangents are at (3, 0), (0.5, 1.8234), (−1, π), and (0.5, 4.4597).

21. **(a)** $\frac{dr}{d\theta} = \frac{10(3\sin\theta)}{(5 + 3\cos\theta)^2} = \frac{30\sin\theta}{(5 + 3\cos\theta)^2}$

(b) Critical values are 0, π, and 2π. Since r is increasing on the interval $(0, \pi)$ and decreasing on $(\pi, 2\pi)$, r has a maximum at $\theta = \pi$.

≡ 29.3 ARC LENGTH AND SURFACE AREA REVISITED

1. $x = 4t^3$; $y = 3t^2$ so $\frac{dx}{dt} = 12t^2$ and $\frac{dy}{dt} = 6t$.

The arc length is $L = \int_0^1 \sqrt{(12t^2)^2 + (6t)^2}\,dt =$ $\int_0^1 \sqrt{144t^4 + 36t^2}\,dt = \int_0^1 6t\sqrt{4t^2 + 1}\,dt$. Let $u = 4t^2 + 1$ and $du = 8t\,dt$. Substituting, we get $\frac{6}{8}\int\sqrt{u}\,du = \frac{3}{4}\frac{2}{3}u^{3/2} = \frac{1}{2}\left(4t^2 + 1\right)^{3/2}\Big|_0^1 = \frac{1}{2}\left[5^{3/2} - 1^{3/2}\right] \approx \frac{1}{2}(10.1803) = 5.0902$.

3. $x = 3\cos t$; $\frac{dx}{dt} = -3\sin t$; $y = 3\sin t$; $\frac{dy}{dt} = 3\cos t$. The arc length is

$L = \int_0^{2\pi}\sqrt{(-3\sin t)^2 + (3\cos t)^2}\,dt =$ $\int_0^{2\pi}\sqrt{9\sin^2 t + 9\cos^2 t}\,dt = \int_0^{2\pi} 3\,dt = 3t\big|_0^{2\pi} = 6\pi$. (Note: this curve is a circle of radius 3.)

5. $x = \cos t + t\sin t$; $\frac{dx}{dt} = -\sin t + t\cos t + \sin t = t\cos t$. $y = \sin t - t\cos t$; $\frac{dy}{dt} = \cos t + t\sin t - \cos t = t\sin t$. The arc length is

$L = \int_0^{\pi}\sqrt{(t\cos t)^2 + (t\sin t)^2}\,dt$

$= \int_0^{\pi}\sqrt{t^2\cos^2 t + t^2\sin^2 t}\,dt$

$= \int_0^{\pi} t\,dt = \frac{t^2}{2}\Big|_0^{\pi} = \frac{1}{2}\pi^2$.

7. Since $r = 1 + \cos\theta$ we have $r' = -\sin\theta$. Thus, the arc length is

$L = \int_0^{\pi}\sqrt{(1 + \cos\theta)^2 + (-\sin\theta)^2}\,d\theta =$

$\int_0^{\pi}\sqrt{1 + 2\cos\theta + \cos^2\theta + \sin^2\theta}\,d\theta =$

$\int_0^{\pi}\sqrt{2 + 2\cos\theta}\,d\theta = \sqrt{2}\int_0^{\pi}\sqrt{1 + \cos\theta}\,d\theta$. Since $\cos^2\frac{\theta}{2} = \frac{1 + \cos\theta}{2}$ we get $\sqrt{1 + \cos\theta} = \sqrt{2}\cos\frac{\theta}{2}$.

Now our integral is $\sqrt{2}\cos\int_2^{\pi}\sqrt{2}\cos\frac{\theta}{2}\,d\theta =$ $2\int_0^{\pi}\cos\frac{\theta}{2}\,d\theta$. Letting $u = \frac{\theta}{2}$ we have $du = \frac{1}{2}d\theta$ or $2\,du = d\theta$, and so, we get by substitution, $4\int\cos u\,du = 4\sin u = 4\sin\frac{\theta}{2}\Big|_0^{\pi} = 4\left(\sin\frac{\pi}{2} - \sin 0\right) = 4$.

9. Here $r = \sin^2\theta$, and so $r' = 2\sin\theta\cos\theta = \sin 2\theta$. The arc length is

$L = \int_0^{\pi/2}\sqrt{\sin^4\theta + (2\sin\theta\cos\theta)^2}\,d\theta$

$= \int_0^{\pi/2}\sqrt{\sin^4\theta + 4\sin^2\theta\cos^2\theta}\,d\theta$

$= \int_0^{\pi/2}\sin\theta\sqrt{\sin^2\theta + 4\cos^2\theta}\,d\theta$

$= \int_0^{\pi/2}\sin\theta\sqrt{1 + 3\cos^2\theta}\,d\theta$.

Letting $u = \sqrt{3}\cos\theta$, then $du = -\sqrt{3}\sin\theta$ and substituting we get $-\frac{1}{\sqrt{3}}\int\sqrt{1 + u^2}\,du$. By formula #30 in Appendix C we get

$-\frac{1}{\sqrt{3}}\left[\frac{u}{2}\sqrt{1 + u^2} + \frac{1}{2}\ln\left|u + \sqrt{1 + u^2}\right|\right] =$

$-\frac{1}{\sqrt{3}}\left[\frac{\sqrt{3}\cos\theta}{2}\sqrt{1 + 3\cos^2\theta} + \frac{1}{2}\ln\left|\sqrt{3}\cos\theta + \sqrt{1 + 3\cos^2\theta}\right|\right]_0^{\pi/2} = -\frac{1}{\sqrt{3}}\left(-\sqrt{3} - \frac{1}{2}\ln(\sqrt{3} + 2)\right) = 1 + \frac{\sqrt{3}}{6}\ln(2 + \sqrt{3}) \approx 1.3802$.

11. $x = t + 4$, so $\frac{dx}{dt} = 1$ and $y = t^3$, so $\frac{dy}{dt} = 3t^2$. Thus, the area of the surface of revolution is $S = 2\pi\int_0^2 t^3\sqrt{1 + (3t^2)^2}\,dt = 2\pi\int_0^2 t^3\sqrt{1 + 9t^4}\,dt$. Let $u = 1 + 9t^4$, and then $du = 36t^3\,dt$. Substituting we get $\frac{2\pi}{36}\int\sqrt{u}\,du = \frac{\pi}{18}\cdot\frac{2}{3}u^{3/2} = \frac{\pi}{27}\left(1 + 9t^4\right)^{3/2}\Big|_0^2$

$= \frac{\pi}{27}\left(145^{3/2} - 1\right) \approx 203.0436$.

13. Here $x = \cos t$, so $\frac{dx}{dt} = -\sin t$ and $y = \sin t$, so $\frac{dy}{dt} = \cos t$. Thus,

$S = 2\pi\int_0^{\pi/2}\sin t\sqrt{(-\sin t)^2 + (\cos t)^2}\,dt$

$$= 2\pi \int_0^{\pi/2} \sin t \sqrt{\sin^2 t + \cos^2 t}\, dt$$

$$= 2\pi \int_0^{\pi/2} \sin t\, dt$$

$$= 2\pi(-\cos t)\big|_0^{\pi/2} = 2\pi(0 + 1) = 2\pi.$$

15. $x = 1 + \sin t$, $\frac{dy}{dt} = \cos t$; $y = \cos t$; $\frac{dy}{dt} = -\sin t$. Thus, the area is
$S = 2\pi \int_0^{\pi/2} \cos t \sqrt{\cos^2 t + \sin^2 t}\, dt = 2\pi \int_0^{\pi/2} \cos t\, dt = 2\pi[\sin t]_0^{\pi/2} = 2\pi.$

17. $r = 1 + \cos\theta$; $\frac{dr}{d\theta} = -\sin\theta$. Thus, the area of the surface of revolution is

$$S = 2\pi \int_0^\pi r \sin\theta \sqrt{(1 + \cos t)^2 + (-\sin\theta)^2}\, d\theta$$

$$= 2\pi \int_0^\pi (1 + \cos\theta) \sin\theta$$
$$\times \sqrt{1 + 2\cos\theta + \cos^2\theta + \sin^2\theta}\, d\theta$$

$$= 2\pi \int_0^\pi (1 + \cos\theta) \sin\theta \sqrt{2 + 2\cos\theta}\, d\theta$$

$$= 2\sqrt{2}\pi \int_0^\pi (1 + \cos\theta)^{3/2} \sin\theta\, d\theta$$

Let $u = 1 + \cos\theta$, then $du = -\sin\theta\, d\theta$, and we get $-2\sqrt{2}\pi \int u^{3/2}\, du = -2\sqrt{2}\pi \cdot \frac{2}{5}u^{5/2} = \frac{-4\sqrt{2}\pi}{5}(1 + \cos\theta)^{5/2}\big|_0^\pi = \frac{-4\sqrt{2}\pi}{5}((1-1)^{5/2} - (1 - 1)^{5/2}) = \frac{-4\sqrt{2}\pi}{5}(2)^{5/2} = \frac{2^5}{5}\pi = \frac{32}{5}\pi \approx 20.1062.$

19. **(a)** $L = \int_0^{8\pi} \sqrt{r^2 + \left(\frac{dr}{d\theta}\right)^2}\, d\theta = \int_0^{8\pi} \sqrt{(b\theta)^2 + b^2}\, d\theta = b \int_0^{8\pi} \sqrt{\theta^2 + 1}\, d\theta$. Integrating by Formula 30, Appendix C, we obtain $L = b\left(\frac{1}{2}\right)\left[\theta\sqrt{\theta^2 + 1} + \ln(\theta + \sqrt{\theta^2 + 1})\right]_0^{8\pi}$ Here $d = 1$ cm, and so, $b = \frac{1}{2\pi}$. Substituting this value for b, we get $L \approx 50.56$ cm, **(b)** $50.56/5 = 10.11 = 10^+$ h

≡ 29.4 INTERSECTION OF GRAPHS OF POLAR COORDINATES

1. $r = 3\theta$, $r = \frac{\pi}{2}$: By direct substitution we get $\frac{\pi}{2} = 3\theta$ or $\theta = \frac{\pi}{6}$. This gives the one point of intersection $\left(\frac{\pi}{2}, \frac{\pi}{6}\right)$

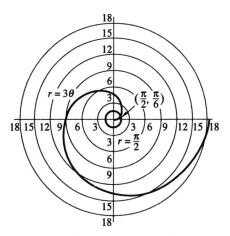

3. $r = \frac{1}{2}$; $r = \cos\theta$: By direct substitution we get $\cos\theta = \frac{1}{2}$ or $\theta = \frac{\pi}{3}$ and $\frac{5\pi}{3}$. This gives the two points of intersection $\left(\frac{1}{2}, \frac{\pi}{3}\right)$ and $\left(\frac{1}{2}, \frac{5\pi}{3}\right)$

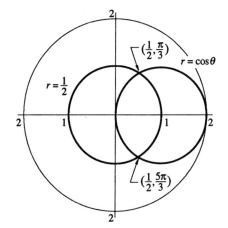

5. $r = 1 - \sin\theta$; $r = 1 + \cos\theta$: Direct substitution yields $1 - \sin\theta = 1 + \cos\theta$ or $-\sin\theta = \cos\theta$. This is equivalent to $\tan\theta = -1$. Hence we get $\theta = \frac{3\pi}{4}$ and $\frac{7\pi}{4}$ and the points $\left(1 - \frac{\sqrt{2}}{2}, \frac{3\pi}{4}\right)$ and $\left(1 + \frac{\sqrt{2}}{2}, \frac{7\pi}{4}\right)$. Looking at the graph we also see that the pole $(0, 0)$ is a common point.

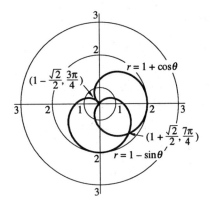

7. $r = \sqrt{3}$; $r = 2\cos\theta$. Direct substitution yields $2\cos\theta = \sqrt{3}$ or $\cos\theta = \frac{\sqrt{3}}{2}$. Hence we get $\theta = \frac{\pi}{6}$, $\frac{11\pi}{6}$ and the two points $\left(\sqrt{3}, \frac{\pi}{6}\right)$ and $\left(\sqrt{3}, \frac{11\pi}{6}\right)$.

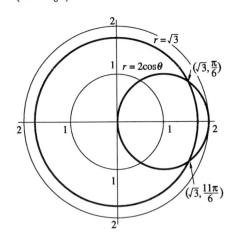

9. $r = \sin 2\theta$; $r = \sqrt{2}\sin\theta$: Direct substitution yields $\sin 2\theta = \sqrt{2}\sin\theta$ or using a trig identity $2\sin\theta\cos\theta = \sqrt{2}\sin\theta$. Hence $2\sin\theta\cos\theta - \sqrt{2}\sin\theta = 0$ or $\sin\theta(\cos\theta - \sqrt{2}) = 0$. Hence, $\sin\theta = 0 \Rightarrow \theta = 0$ or π and $\cos\theta = \frac{\sqrt{2}}{2} \Rightarrow \frac{\pi}{4}, \frac{7\pi}{4}$. Now, $\theta = 0$ and $\theta = \pi$ yield the same point $(0, 0)$, and so we have three solutions $(0, 0)$, $\left(1, \frac{\pi}{4}\right)$, and $\left(-1, \frac{7\pi}{4}\right)$.

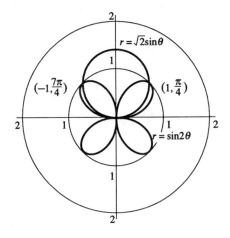

11. $r = 1 - \sin\theta$, $r = \frac{1}{1-\sin\theta}$. Direct substitution gives $1 - \sin\theta = \frac{1}{1-\sin\theta}$. Multiplying both sides by $1 - \sin\theta$ we get $(1 - \sin\theta)^2 = 1$ or $1 - \sin\theta = \pm 1$; $-\sin\theta = -1 \pm 1$ or $\sin\theta = 1 \pm 1$; $\sin\theta = 0$ or 2. Since 2 is not in the range of the sine function, $\theta = 0$ or π. This gives the two points $(1, 0)$ and $(1, \pi)$.

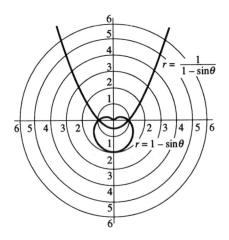

≡ 29.5 AREA IN POLAR COORDINATES

1. $r = 4\sin\theta$. This curve is complete when θ goes from 0 to π so the area is $\frac{1}{2}\int_0^\pi (4\sin\theta)d\theta = \frac{1}{2}\int_0^\pi 16\sin^2\theta\,d\theta = \frac{1}{2}\int_0^\pi 16\frac{1-\cos 2\theta}{2}d\theta = 4\int_0^\pi (1-\cos 2\theta)$ $\times\,d\theta = 4\left[\theta - \frac{1}{2}\sin\theta\right]_0^\pi = 4[\pi - 0 - 0 + 0] = 4\pi$

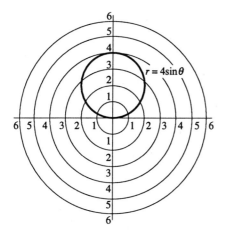

3. $r = 1 - \cos\theta$. For this curve we need to integrate from 0 to 2π with the result $\frac{1}{2}\int_0^{2\pi}(1-\cos\theta)^2\,d\theta = \frac{1}{2}\int_0^{2\pi}(1-2\cos\theta+\cos^2\theta)d\theta = \frac{1}{2}\int_0^{2\pi}\left(1-2\cos\theta + \frac{1+\cos 2\theta}{2}\right)d\theta = \frac{1}{2}\left[\frac{3}{2}\theta - 2\sin\theta + \frac{1}{4}\sin 2\theta\right]_0^{2\pi} = \frac{1}{2}[3\pi - 0 + 0 - 0 + 0 - 0] = \frac{3}{2}\pi$

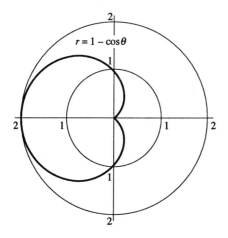

5. $r = \sin 2\theta$. This curve is a four leaf rose. By symmetry we can get the area of one leaf and multiply by 4. The result is $4 \cdot \frac{1}{2}\int_0^{\pi/2}(\sin 2\theta)^2 d\theta = 2\int_0^{\pi/2}\sin^2 2\theta\,d\theta = 2\int_0^{\pi/2}\frac{1-\cos 4\theta}{2}d\theta = \theta - \frac{1}{4}\sin 4\theta\Big|_0^{\pi/2} = \frac{\pi}{2} - 0 = \frac{\pi}{2}$

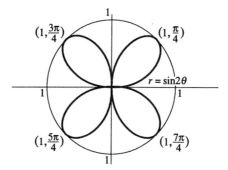

7. $r = 4 + \cos\theta$. Integrating from 0 to 2π, the area is
$$\frac{1}{2}\int_0^{2\pi}(4+\cos\theta)d\theta$$
$$= \frac{1}{2}\int_0^{2\pi}(16 + 8\cos\theta + \cos^2\theta)d\theta$$
$$= \frac{1}{2}\int_0^{2\pi}\left(16 + 8\cos\theta + \frac{1+\cos 2\theta}{2}\right)d\theta$$
$$= \frac{1}{2}\left[16\theta + 8\sin\theta + \frac{1}{2}\theta + \frac{1}{2}\sin 2\theta\right]_0^{2\pi}$$
$$= \frac{1}{2}[32\pi + 0 + \pi + 0] = \frac{33}{2}\pi$$

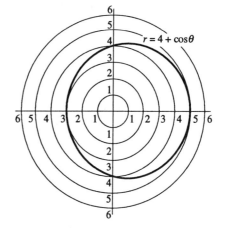

9. $r^2 = \sin\theta \Rightarrow r = \pm\sqrt{\sin\theta}$; $r = \sqrt{\sin\theta}$ gives the top half of the curve with $0 \le \theta \le \pi$. The bottom half is $r = -\sqrt{\sin\theta}$. By symmetry we can integrate $\sin\theta$ from 0 to π and double the answer.

$2 \cdot \frac{1}{2}\int_0^\pi (\sqrt{\sin\theta})^2 d\theta = \int_0^\pi \sin\theta\, d\theta = -\cos\theta\big|_0^\pi = -[(-1) - 1] = 2.$

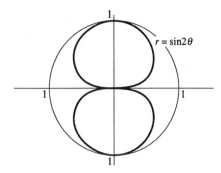

11. $\frac{1}{2}\int_0^{\pi/4}(3\cos\theta)^2 d\theta = \frac{1}{2}\int_0^{\pi/4} 9\cos^2\theta\, d\theta =$

$\frac{9}{2}\int_0^{\pi/4}\frac{1+\cos 2\theta}{2}d\theta = \frac{9}{2}\left[\frac{\theta}{2} + \frac{\sin 2\theta}{4}\right]_0^{\pi/4} = \frac{9}{3}\left[\frac{\pi}{8} + \frac{1}{4}\right] =$

$\frac{9}{16}[\pi + 2] \approx 2.8921.$

13. $\frac{1}{2}\int_0^{\pi/4}(\cos 2\theta)^2 d\theta = \frac{1}{2}\int_0^{\pi/4}\cos^2\theta\, d\theta =$

$\frac{1}{2}\int_0^{\pi/4}\frac{1+\cos 4\theta}{2}d\theta = \frac{1}{2}\left[\frac{1}{2}\theta + \frac{1}{4}\sin 4\theta\right]_0^{\pi/4} = \frac{1}{2}\left[\frac{\pi}{8}\right] =$

$\frac{\pi}{16} \approx 0.1963.$

15. $\frac{1}{2}\int_0^{\pi/2}\left(e^{2\theta}\right)^2 d\theta = \frac{1}{2}\int_0^{\pi/2}e^{4\theta}d\theta = \frac{1}{2}\left[\frac{1}{4}e^{4\theta}\right]_0^{\pi/2} =$

$\frac{1}{8}\left[e^{2\pi} - 1\right] \approx 66.8115$

17.   To get the area A of the region enclosed by one loop of $r = 4\cos 2\theta$ we can integrate from $-\frac{\pi}{4}$ to $\frac{\pi}{4}$ with the result $A = \frac{1}{2}\int_{-\pi/4}^{\pi/4}(4\cos 2\theta)^2 d\theta = 8\int_{-\pi/4}^{\pi/4}\frac{1+\cos 4\theta}{2}d\theta = 8\left[\frac{1}{2}\theta + \frac{\sin 4\theta}{8}\right]_{-\pi/4}^{\pi/4} = 8\left[\frac{\pi}{8} + \frac{\pi}{8}\right] = 2\pi.$

19. $r = 2\sin 3\theta$, and to get the area of the region enclosed by one loop if we integrate from 0 to $\frac{\pi}{3}$. Thus, if A is the area, we have $A = \frac{1}{2}\int_0^{\pi/3}(2\sin 3\theta)^2 d\theta = \int_0^{\pi/3} 2\sin 3\theta\, d\theta = \int_0^{\pi/3}(1 - \cos 6\theta)d\theta = \left[\theta - \frac{\sin 6\theta}{6}\right]_0^{\pi/3} = \frac{\pi}{3}.$

21.   Looking at the figure 29.23S we can determine that the graphs intersect at $\left(1, -\frac{\pi}{2}\right)$ and $\left(1, \frac{\pi}{2}\right)$ so we need to integrate as follows: $\frac{1}{2}\int_{-\pi/2}^{\pi/2}\left((1 + \cos\theta)^2 - 1^2\right)d\theta = \frac{1}{2}\int_{-\pi/2}^{\pi/2}\left(1 + 2\cos\theta + \cos^2\theta - 1\right)d\theta = \frac{1}{2}\int_{-\pi/2}^{\pi/2}\left(2\cos\theta + \frac{1+\cos 2\theta}{2}\right)d\theta = \left[\sin\theta + \right.$

$\frac{1}{4}\theta + \frac{1}{8}\sin 2\theta\Big]_{-\pi/2}^{\pi/2} = \left(1 + \frac{\pi}{8} + 0\right) - \left(-1 - \frac{\pi}{8} + 0\right) =$

$2 + \frac{\pi}{4} \approx 2.7854.$

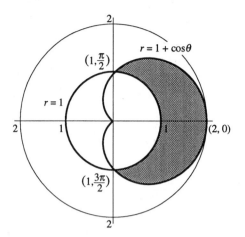

23.   First we need to solve for θ. Using direct substitution we get $1 - \sin\theta = 2\cos\theta$. Squaring both sides we have $1 - 2\sin\theta + \sin^2\theta = 4\cos^2\theta$. Since $\cos^2\theta = 1 - \sin^2\theta$ we have $1 - 2\sin\theta + \sin^2\theta = 4 - 4\sin^2\theta$ or $0 = 3 + 2\sin\theta - 5\sin^2\theta$. This factors into $(3 + 5\sin\theta)(1 - \sin\theta) = 0$. Hence $\sin\theta = 1$ or $\sin\theta = -\frac{3}{5}$, and so $\theta = \frac{\pi}{2}, \frac{3\pi}{2}$, 3.78509, or 5.63968. Checking we see that $\frac{\pi}{2}$ and 5.63968 are the answers we need. Let $5.63968 = T$. Then,

$\frac{1}{2}\int_{\pi/6}^{T}\left((1 - \sin\theta)^2 - (2\cos\theta)^2\right)d\theta$

$= \frac{1}{2}\int_{\pi/2}^{T}\left(1 - 2\sin\theta + \sin^2\theta - 4\cos^2\theta\right)d\theta$

$= \frac{1}{2}\int_{\pi/2}^{T}\left(1 - 2\sin\theta + \sin^2\theta - 4 + 4\sin^2\theta\right)d\theta$

$= \frac{1}{2}\int_{\pi/2}^{T}\left[-3 - 2\sin\theta + 5\left(\frac{1 - \cos 2\theta}{2}\right)\right]d\theta$

$= \frac{1}{2}\left[-3\theta + 2\cos\theta + \frac{5}{2}\theta - \frac{5\sin 2\theta}{4}\right]_{\pi/2}^{T}$

$= \left[-\frac{1}{4}\theta + \cos\theta - \frac{5\sin 2\theta}{8}\right]_{\pi/2}^{T}$

$\approx -0.009921 + 0.392699 = 0.382778$

So, the desired area is about 0.3828.

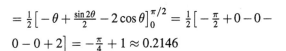

$$= \tfrac{1}{2}\left[-\theta + \tfrac{\sin 2\theta}{2} - 2\cos\theta\right]_0^{\pi/2} = \tfrac{1}{2}\left[-\tfrac{\pi}{2} + 0 - 0 -\right.$$
$$\left.0 - 0 + 2\right] = -\tfrac{\pi}{4} + 1 \approx 0.2146$$

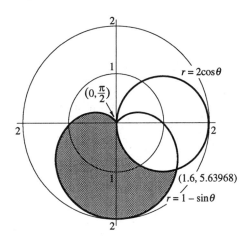

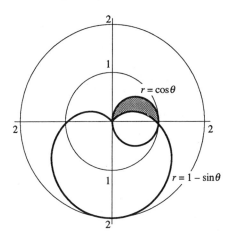

25. $\tfrac{1}{2}\int_0^{\pi/2}\left(\cos^2\theta - (1-\sin\theta)^2\right)d\theta = \tfrac{1}{2}\int_0^{\pi/2}\left(\cos^2\theta - 1 + 2\sin\theta - \sin^2\theta\right)d\theta = \int_0^{\pi/2}(\cos 2\theta - 1 + 2\sin\theta)d\theta$

CHAPTER 29 REVIEW

1. $x = t^2$, $y = t^3 + t$.

   (a) $\frac{dy}{dx} = \frac{dy/dt}{dx/dt} = \frac{3t^2+1}{2t}$; $\frac{d^2y}{dx^2} = \frac{2t(6t) - (3gt^2+1)2}{(2t)^3} =$
$\frac{12t^2 - 6t^2 - 2}{(2t)^3} = \frac{6t^2-2}{2(4t^3)} = \frac{3t^2-1}{4t^3}$

   (b) At $t = 1$, $\frac{dy}{dx} = \frac{3+1}{2} = 2$.

2.   **(a)** $\frac{dy}{dx} = \frac{dy/dt}{dx/dt} = \frac{-2t^{-2}}{-1t^{-2}} = 2$. Since $\frac{dy}{dx}$ is a constant, $\frac{d^2y}{dx^2} = 0$,

   (b) At $t = 1$, $\frac{dy}{dx} = 2$.

3.   **(a)** $\frac{dy}{dx} = \frac{\cos t}{2\cos t(-\sin t)} = -\tfrac{1}{2}\csc t$, and the second

   derivative is $\frac{d^2y}{dx^2} = \frac{\tfrac{1}{2}\csc t\cot t}{-2\cos t\sin t} = -\tfrac{1}{4}\csc^3 t$

   (b) At $t = 1$, $\frac{dy}{dx} = -\tfrac{1}{2}\csc 1 \approx -0.5942$

4.   **(a)** $\frac{dy}{dx} = \frac{-e^{-t}}{\tfrac{1}{2}t^{-1/2}} = -2\sqrt{t}e^{-t}$; $\frac{d^2y}{dx^2} =$
$\frac{2\sqrt{t}e^{-t} - t^{-1/2}e^{-t}}{\tfrac{1}{2}t^{-1/2}} = 4te^{-t} - 2e^{-t} = (4t-2)e^{-t}$

   (b) At $t = 1$, $\frac{dy}{dx} = -2e^{-1} \approx -0.7358$

5.   **(a)** $\frac{dy}{dx} = \frac{-2\sin\theta\sin\theta + (3 + 2\cos\theta)\cos\theta}{-2\sin\theta\cos\theta - (3 + 2\cos\theta)\sin\theta}$
$= \frac{-2\sin^2\theta + 3\cos\theta + 2\cos^2\theta}{-2\sin\theta\cos\theta - 3\sin\theta - 2\cos\theta\sin\theta}$
$= \frac{2(\cos^2\theta - \sin^2\theta) + 3\cos\theta}{-4\sin\theta\cos\theta - 3\sin\theta}$
$= \frac{2\cos 2\theta + 3\cos\theta}{-2\sin 2\theta - 3\sin\theta}$
$= -\frac{2\cos 2\theta + 3\cos\theta}{2\sin 2\theta + 3\sin\theta}$

   (b) If $\theta = \frac{\pi}{4}$, then
$$\frac{dy}{dx} = -\frac{2\cos\frac{\pi}{2} + 3\cos\frac{\pi}{4}}{2\sin\frac{\pi}{2} + 3\sin\frac{\pi}{4}}$$
$$= -\frac{2\cdot 0 + \frac{3\sqrt{2}}{2}}{2\cdot 1 + \frac{3\sqrt{2}}{2}} = -\frac{3\sqrt{2}}{4 + 3\sqrt{2}} \approx -0.5147.$$

6.   **(a)**
$$\frac{dy}{dx} = \frac{-12\sin 2\theta\sin\theta + 6\cos 2\theta\cos\theta}{-12\sin 2\theta\cos\theta - 6\cos 2\theta\sin\theta}$$
$$= \frac{2\sin 2\theta\sin\theta - \cos 2\theta\cos\theta}{2\sin 2\theta\cos\theta + \cos 2\theta\sin\theta}$$
$$= \frac{2(2\sin\theta\cos\theta)\sin\theta - (\cos^2\theta - \sin^2\theta)\cos\theta}{2(\sin\theta\cos\theta)\cos\theta + (\cos^2\theta - \sin^2\theta)\sin\theta}$$
$$= \frac{\cos\theta(4\sin^2\theta - \cos^2\theta + \sin^2\theta)}{\sin\theta(4\cos^2\theta + \cos^2\theta - \sin^2\theta)}$$
$$= \frac{\cos\theta(5\sin^2\theta - \cos^2\theta)}{\sin\theta(5\cos^2\theta - \sin^2\theta)}$$

(b) At $\theta = \frac{\pi}{4}$, $\frac{dy}{dx}$ is $\dfrac{\cos\frac{\pi}{4}\left(5\sin^2\frac{\pi}{4} - \cos^2\frac{\pi}{4}\right)}{\sin\frac{\pi}{4}\left(5\cos^2\frac{\pi}{4} - \sin^2\frac{\pi}{4}\right)} = 1$

7. $r^2 = \sin\theta$; $r = (\sin\theta)^{1/2}$; $r' = \frac{1}{2}(\sin\theta)^{-1/2}\cos\theta$

(a) $\dfrac{dy}{dx} = \dfrac{\frac{1}{2}\sin^{-1/2}\theta\cos\theta\sin\theta + \sin^{1/2}\theta\cos\theta}{\frac{1}{2}\sin^{-1/2}\theta\cos\theta\cos\theta - \sin^{1/2}\theta\sin\theta}$

$= \dfrac{\cos\theta\sin\theta + 2\sin\theta\cos\theta}{\cos^2\theta - 2\sin^2\theta}$

$= \dfrac{3\cos\theta\sin\theta}{\cos^2\theta - 2\sin^2\theta}$

(b) At $\theta = \frac{\pi}{4}$ we have $\dfrac{dy}{dx} = \dfrac{3\cdot\frac{\sqrt{2}}{2}\frac{\sqrt{2}}{2}}{\left(\frac{\sqrt{2}}{2}\right)^2 - 2\left(\frac{\sqrt{2}}{2}\right)^2} =$

$\dfrac{\frac{3}{2}}{-\frac{1}{2}} = -3$

8.   **(a)** If $r = \frac{8}{3+\cos\theta} = 8(3 + \cos\theta)^{-1}$, then $r' =$

$-8(3+\cos\theta)^{-2}(-\sin\theta) = \frac{8\sin\theta}{(3+\cos\theta)^2}$. So,

$\dfrac{dy}{dx} = \dfrac{\frac{8\sin\theta}{(3+\cos\theta)^2}\sin\theta + \frac{8}{3+\cos\theta}\cos\theta}{\frac{8\sin\theta}{(3+\cos\theta)^2}\cos\theta - \frac{8}{3+\cos\theta}\sin\theta}$

$= \dfrac{8\sin^2\theta + 8\cos\theta(3+\cos\theta)}{8\sin\theta\cos\theta - 8\sin\theta(3+\cos\theta)}$

$= \dfrac{\sin^2\theta + 3\cos\theta + \cos^2\theta}{\sin\theta\cos\theta - 3\sin\theta - \sin\theta\cos\theta}$

$= \dfrac{1 + 3\cos\theta}{-3\sin\theta}$.

(b) At $\theta = \frac{\pi}{4}$, then $\dfrac{dy}{dx} = \dfrac{1 + 3\cdot\frac{\sqrt{2}}{2}}{-3\cdot\frac{\sqrt{2}}{2}} \approx -1.4714$.

9. $\frac{dx}{dt} = 2t$; $\frac{dy}{dt} = 3t^2$ and $L = \int_0^2 \sqrt{(2t)^2 + \left(3t^2\right)^2}\,dt$ $= \int_0^2\sqrt{4t^2 + 9t^4}\,dt = \int_0^2 t\sqrt{4 + 9t^2}\,dt$. Let $u = 4+9t^2$, and then $du = 18t\,dt$, so we get $\frac{1}{18}\int u^{1/2}du$ $= \frac{1}{16}\cdot\frac{2}{5}u^{3/2} = \frac{1}{27}\left(4+9t^2\right)^{3/2}\Big|_0^2 = \frac{1}{27}\left(40^{3/2} - 4^{3/2}\right) \approx 9.0734$.

10. $\frac{dx}{dt} = 2t$; $\frac{dy}{dt} = 2$, Thus, $L = \int_0^3\sqrt{(2t)^2 + 2^2}\,dt =$ $\int_0^3\sqrt{4t^2 + 4}\,dt = 2\int_0^3\sqrt{t^2 + 1}\,dt =$ $2\left[\frac{t}{2}\sqrt{t^2+1} + \frac{1}{2}\ln\left|t + \sqrt{t^2+1}\right|\right]_0^3 =$ $\left(3\sqrt{10} + \ln\left|3 + \sqrt{10}\right|\right) \approx 11.3053$.

11. $\frac{dx}{dt} = 4\cos t$; $\frac{dy}{dt} = (-4\sin t)$, and as a result, $L =$ $\int_0^{\pi/2}\sqrt{(4\cos t)^2 + (-4\sin t)^2}\,dt = \int_0^{\pi/2} 4\ dt =$ $4t\big|_0^{\pi/2} = 2\pi$.

12. $r = e^{2\theta}$; $r' = 2e^{2\theta}$, then $L =$ $\int_0^\pi\sqrt{\left(e^{2\theta}\right)^2 + \left(2e^{2\theta}\right)^2}\,d\theta = \int_0^\pi\sqrt{e^{4\theta} + 4e^{4\theta}}\,d\theta =$ $\sqrt{5}\int_0^\pi e^{2\theta} = \frac{\sqrt{5}}{2}e^{2\theta}\Big|_0^\pi = \frac{\sqrt{5}}{2}\left(e^{2\pi} - 1\right) \approx 597.5798$.

13.   Since $r = \cos^2\left(\frac{\theta}{2}\right)$, we have $r' =$ $2\cos\frac{\theta}{2}\left(-\frac{1}{2}\sin\frac{\theta}{2}\right) = -\cos\frac{\theta}{2}\sin\frac{\theta}{2}$. Thus, $L = \int_0^\pi\sqrt{\cos^2\frac{\theta}{2} + \cos^2\frac{\theta}{2}\sin^2\frac{\theta}{2}}\,d\theta = \int_0^\pi\cos\frac{\theta}{2}\,d\theta$ $= 2\sin\frac{\theta}{2}\Big|_0^\pi = 2\cdot\sin\frac{\pi}{2} = 2$.

14. $r = 1 - \cos\theta$ so $r' = \sin\theta$

$L = \int_{-\pi}^0\sqrt{(1 - \cos\theta)^2 + \sin^2\theta}\,d\theta$

$= \int_{-\pi}^0\sqrt{1 - 2\cos\theta + \cos^2\theta + \sin^2\theta}\,d\theta$

$= \int_{-\pi}^0\sqrt{2 - 2\cos\theta} = \int_{-\pi}^0 2\sqrt{\frac{1-\cos\theta}{2}}\,d\theta$

$= 2\int_{-\pi}^0\sin\frac{\theta}{2}\,d\theta = -4\cos\frac{\theta}{2}\Big|_{-\pi}^0 = 4$

15. $\frac{dx}{dt} = 2t$; $\frac{dy}{dt} = 1$

$S = 2\pi\int_0^2 t\sqrt{(2t)^2 + 1^2}\,dt$

$= 2\pi\int_0^2 t\sqrt{4t^2 + 1}\,dt$

Let $u = 4t^2 + 1$, and then $du = 8t\,dt$ or $\frac{1}{8}du = t\,dt$. Substituting we get $\frac{1}{4}\pi\int\sqrt{u}\,du = \frac{\pi}{4}\cdot\frac{2}{3}u^{3/2} =$ $\frac{\pi}{6}\left(4t^2 + 1\right)^{3/2}\Big|_0^2 = \frac{\pi}{6}\left(17^{3/2} - 1\right) \approx 36.1769$.

16. $\frac{dx}{dt} = 2t$ and $\frac{dy}{dt} = \left(1 - t^2\right)$.

$S = 2\pi\int_0^1\left(t - \frac{t^3}{3}\right)\sqrt{(2t)^2 + \left(1-t^2\right)^2}\,dt$

$= 2\pi\int_0^1\left(t - \frac{t^3}{3}\right)\sqrt{4t^2 + 1 - 2t^2 + t^4}\,dt$

$= 2\pi\int_0^1\left(t - \frac{t^3}{3}\right)\sqrt{t^4 + 2t^2 + 1}\,dt$

$= 2\pi\int_0^1\left(t - \frac{t^3}{3}\right)\left(t^2 + 1\right)dt$

$= 2\pi\int_0^1\left(t^3 - \frac{t^3}{3} - \frac{t^5}{3} + t\right)dt$

$= 2\pi\left[\frac{2}{3}\cdot\frac{1}{4}t^4 - \frac{1}{3}\frac{1}{6}t^6 + \frac{t^2}{2}\right]_0^1$

$= 2\pi\left(\frac{1}{6} - \frac{1}{18} + \frac{1}{2}\right)$

$= 2\pi\left(\frac{3 - 1 + 9}{18}\right) = \frac{11\pi}{9} \approx 3.8397$

17. Here $\frac{dx}{dt} = 2t$ and $\frac{dy}{dt} = 3t^2$, so the surface is $S = 2\pi \int_0^2 t^2 \sqrt{(2t)^2 + (3+2)^2}\,dt = 2\pi \int_0^2 t^2 \times$

$\sqrt{4t^2 + 9t^4}\,dt = 2\pi \int_0^2 t^3 \sqrt{4 + 9t^2}\,dt$. Let $u = 4 + 9t^2$, then $t^2 = \frac{u-4}{9}$ and $du = 18t\,dt$ or $\frac{1}{18}du = t\,dt$.

Substituting, we get $2\pi \cdot \frac{1}{18} \int \left(\frac{u-4}{9}\right)u^{1/2}\,du = \frac{\pi}{81} \int (u^{3/2} - 4u^{1/2})\,du = \frac{\pi}{81}\left(\frac{2}{5}u^{5/2} - 4\frac{2}{3}u^{3/2}\right) = \frac{\pi}{81}\left[\frac{2}{5}\left(4+9t^2\right)^{5/2} -\right.$

$\left.\frac{8}{3}\left(u + 9t^2\right)^{3/2}\right]_0^2 = \frac{\pi}{81}\left[\frac{2}{5}40^{5/2} - \frac{8}{3}40^{3/2} - \frac{64}{5} + \frac{64}{3}\right] \approx 41.7485\pi \approx 131.1568$.

18. Since $x = e^t \sin t$, then $dx = \left(e^t + \cos t + e^t \sin t\right)dt$ and since $y = e^t \cos$, then $dy = \left(e^t \cos t - e^t \sin t\right)dt$. Thus, the desired surface area is given by

$$S = 2\pi \int_0^{\pi/2} e^t \sin t \sqrt{\left(e^t \cos t + e^t \sin t\right)^2 + \left(e^t \cos t - e^t \sin t\right)^2}\,dt$$

$$= 2\pi \int_0^{\pi/2} e^t \sin t \sqrt{2e^{2t} \cos^2 t + 2e^{2t} \sin^2 t}\,dt$$

$$= 2\pi \int_0^{\pi/2} e^t \sin t \sqrt{2}e^t\,dt = 2\sqrt{2}\pi \int_0^{\pi/2} e^{2t} \sin t$$

By formula #87, Appendix C, this integral is $2\sqrt{2}\pi \left(\frac{e^{2t}}{2^2+1^2}(2\sin t - \cos t)\right)\Big|_0^{\pi/2} = 2\sqrt{2}\pi\left(\frac{e^\pi}{5}\cdot 2 + \frac{1}{5}\right) = \frac{2\sqrt{2}\pi}{5}\left(2e^\pi + 1\right) \approx 84.0263$.

19. $r = 6\sin\theta$ and $r' = 6\cos\theta$; $y = 6\sin^2\theta$. Hence, we find that the desired surface area is
$S = 2\pi \int_0^\pi \sqrt{36\sin^2\theta + 36\cos^2\theta}\left(6\sin^2 6\right)dt = 2\pi \int_0^\pi 36\sin^2\theta\,d\theta = 36\pi \int_0^\pi (1 - \cos 2\theta)d\theta = 36\pi\left[\theta - \frac{\sin 2\theta}{2}\right]_0^\pi = 36\pi^2 \approx 355.3058$.

20. Here $r = 4 + 4\cos\theta$, $r' = -4\sin\theta$, and $y = 4\sin\theta + 4\sin\theta\cos\theta$. Hence, the desired surface area is

$$S = 2\pi \int_0^\pi (4\sin\theta + 4\sin\theta\cos\theta)$$
$$\times \sqrt{(4 + 4\cos\theta)^2 + (4\sin\theta)^2}\,d\theta$$
$$= 2\pi \int_0^\pi (4\sin\theta + 4\sin\theta\cos\theta)$$
$$\times 4\sqrt{1 + 2\cos\theta + \cos^2\theta + \sin^2\theta}\,d\theta$$
$$= 32\pi \int_0^\pi (\sin\theta + \sin\theta\cos\theta)\sqrt{2 + 2\cos\theta}\,d\theta$$
$$= 32\sqrt{2}\pi \int_0^\pi (1 + \cos\theta)\sqrt{1 + \cos\theta}\sin\theta\,d\theta$$
$$= 32\sqrt{2}\pi \int_0^\pi (1 + \cos\theta)^{3/2}\sin\theta\,d\theta$$

Let $u = 1 + \cos\theta$ and $du = -\sin\theta$. Substituting, we get $-32\sqrt{2}\pi\left[\int u^{3/2}du\right] = -32\sqrt{2}\pi\left[\frac{2}{5}u^{5/2}\right] = -32\sqrt{2}\pi\left[\frac{2}{5}(1+\cos\theta)^{5/2}\right]_0^\pi = \frac{-64\sqrt{2}}{5}\left(0 - 2^{5/2}\right) = \frac{64\sqrt{2}\pi}{5}(4\sqrt{2}) = \frac{2^9\pi}{5} \approx 102.4\pi \approx 321.6991$.

21. By direct substitution, $4 = 4 + 4\sin\theta$; $0 = \sin\theta$ so $\theta = 0$ or π; $(4, 0)$, $(4, \pi)$.

22. By direct substitution, $3 = 6\sin\theta$; $\sin\theta = \frac{1}{2}$; $\theta = \frac{\pi}{6}, \frac{5\pi}{6}$; $\left(3, \frac{\pi}{6}\right)$, $\left(3, \frac{5\pi}{6}\right)$.

23. By direct substitution, $4 + 4\cos\theta = 2$; $4\cos\theta = -2$; $\cos\theta = -\frac{1}{2}$; $\theta = \frac{2\pi}{3}, \frac{4\pi}{3}$. $\left(2, \frac{2\pi}{3}\right)$, $\left(2, \frac{4\pi}{3}\right)$.

24. By direct substitution, $2\sin 2\theta = 1$; $\sin 2\theta = \frac{1}{2}$, and so $2\theta = \frac{\pi}{6}, \frac{5\pi}{6}, \frac{13\pi}{6}$, and $\frac{17\pi}{6}$. Thus, $\theta = \frac{\pi}{12}, \frac{5\pi}{12}, \frac{13\pi}{12}$, and $\frac{17\pi}{12}$. Graphing yields symmetric points at $\frac{-\pi}{12}, \frac{-5\pi}{12}, \frac{-13\pi}{12}$, and $\frac{-17\pi}{12}$. The resulting points are $\left(1, \frac{\pi}{12}\right)$, $\left(1, \frac{5\pi}{12}\right)$, etc.

25. The area is $A = \frac{1}{2}\int_0^{2\pi}(1 + \cos\theta)^2\,d\theta = \frac{1}{2}\int_0^{2\pi}\left(1 + 2\cos\theta + \cos^2\theta\right)d\theta = \frac{1}{2}\int_0^{2\pi}\left(1 + 2\cos\theta + \frac{1+\cos\theta}{2}\right)d\theta = \frac{1}{2}\left[\frac{3}{2}\theta + 2\sin\theta + \frac{\sin 2\theta}{4}\right]_0^{3\pi} = \frac{3\pi}{2}$.

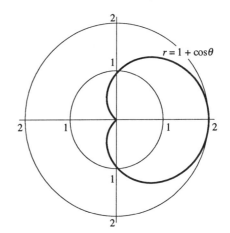
$r = 1 + \cos\theta$

26. The area is $A = \frac{1}{2}\int_0^{2\pi}(4+4\sin\theta)^2 d\theta = 8\int_0^{2\pi}(1+2\sin\theta+\sin^2\theta)d\theta = 8\int_0^{2\pi}\left(1+2\sin\theta+\frac{1-\cos\theta}{2}\right)d\theta = 8\left[\frac{3}{2}\theta - 2\cos\theta - \frac{\sin 2\theta}{4}\right]_0^{2\pi} = 24\pi - 16 + 16 = 24\pi.$

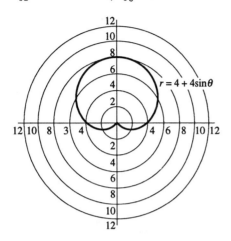

27. $A = \frac{1}{2}\int_0^{\pi}\left((4+4\sin\theta)^2 - 4^2\right)d\theta = \frac{1}{2}\int_0^{\pi}(16 + 32\sin\theta+16\sin^2\theta-16)d\theta = \frac{1}{2}\int_0^{\pi}\left[32\sin\theta+8(1-\cos 2\theta)\right]d\theta = \frac{1}{2}[8\theta - 32\cos\theta - 4\sin 2\theta]_0^{\pi} = 4\pi + 32 \approx 44.5664.$

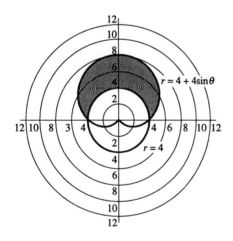

28. The area is $A = \frac{1}{2}\int_{\pi/6}^{5\pi/6}\left[(6\sin\theta)^2 - 3^2\right]d\theta = \frac{1}{2}\int_{\pi/6}^{5\pi/6}(36\sin^2\theta-9)d\theta = \frac{1}{2}\int_{\pi/6}^{5\pi/6}\left[18(1-\cos 2\theta)-9\right]d\theta = \frac{1}{2}[9\theta - 9\sin 2\theta]_{\pi/6}^{5\pi/6} = \frac{9}{2}\left[\frac{5\pi}{6} + \frac{\sqrt{3}}{2} - \frac{\pi}{6} + \frac{\sqrt{3}}{2}\right] = \frac{9}{2}\left[\frac{2\pi}{3} + \sqrt{3}\right] = 3\pi + \frac{9\sqrt{3}}{2} \approx 17.2190.$

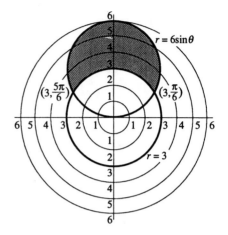

29. $\frac{1}{2}\int_{\pi/2}^{\pi}\left[(\sin\theta)^2-(1+\cos\theta)^2\right]d\theta = \frac{1}{2}\int_{\pi/2}^{\pi}\left[\sin^2\theta-(1+2\cos\theta+\cos^2\theta)\right]d\theta = \frac{1}{2}\int_{\pi/2}^{\pi}\left[\sin^2\theta-\cos^2\theta-1-2\cos\theta\right]d\theta = \frac{1}{2}\int_{\pi/2}^{\pi}[-\cos 2\theta - 1 - 2\cos\theta]d\theta.$ Evaluating this integral, we obtain $\frac{1}{2}\left[-\frac{\sin 2\theta}{2} - 0 - 2\sin\theta\right]_{\pi/2}^{\pi} = \frac{1}{2}\left[-\pi + \frac{\pi}{2} + 2\right] = 1 - \frac{\pi}{4} \approx 0.2146.$

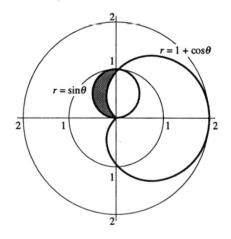

30. In problem #24 we solved for θ. We will use the first pair and multiply the answer by 4. $A = 4 \cdot \frac{1}{2} \int_{\pi/12}^{5\pi/12} \left[(2\sin 2\theta)^2 - 1^2 \right] d\theta = 2 \int_{\pi/12}^{5\pi/12} \left[4\sin^2 2\theta - 1 \right] d\theta = 2 \int_{\pi/12}^{5\pi/12} [2(1 - \cos 4\theta) - 1] d\theta = 2 \left[\theta - \frac{\sin 4\theta}{2} \right]_{\pi/12}^{5\pi/12} = 2 \left[\frac{5\pi}{12} + \frac{\sqrt{3}}{4} - \frac{\pi}{4} + \frac{\sqrt{3}}{4} \right] = \frac{2\pi}{3} + \sqrt{3} \approx 3.8264.$

≡ CHAPTER 29 TEST

1. $\frac{dy}{dx} = \frac{dy/dt}{dx/dt} = \frac{2t+5}{3t^2}$; $\frac{d^2y}{dx^2} = \frac{\frac{d}{dt}\left(\frac{dy}{dx}\right)}{dx/dt} = \frac{\left[(3t^2)(2) - (2t+5)(6t)\right]/(3t^2)^2}{3t^2} = \frac{-6t^2 - 30t}{(3t^2)^3} = \frac{-2t - 10}{9t^5}.$

2. The first derivative is $\frac{dy}{dx} = \frac{dy/dt}{dx/dt} = \frac{-2\cos t \sin t}{\cos t} = -2\sin t.$ When $t = \frac{\pi}{4}$, then $\frac{dy}{dx} = -\sqrt{2}$, so the slope is $-\sqrt{2}$. The tangent line is $y - \frac{1}{2} = \sqrt{2}\left(x - \frac{\sqrt{2}}{2}\right)$ or $y - \sqrt{2}x + \frac{3}{2}.$

3. We will use $\frac{dy}{dx} = \frac{r'\sin\theta + r\cos\theta}{r'\cos\theta - r\sin\theta}$. Differentiating $r = 4 - 2\sin\theta$ we obtain $r' = \frac{dr}{d\theta} = -2\cos\theta$ and so

$$\frac{dy}{dx} = \frac{-2\cos\theta\sin\theta + (4 - 2\sin\theta)\cos\theta}{-2\cos\theta\cos\theta - (4 - 2\sin\theta)\sin\theta}$$
$$= \frac{-4\sin\theta\cos\theta + 4\cos\theta}{-2\cos^2\theta - 4\sin\theta + 2\sin^2\theta}$$
$$= \frac{4\cos\theta - 4\sin\theta\cos\theta}{-2(1 - \sin^2\theta) - 4\sin\theta + 2\sin^2\theta}$$
$$= \frac{4\cos\theta(1 - \sin\theta)}{-2 + 2\sin^2\theta - 4\sin\theta + 2\sin^2\theta}$$
$$= \frac{4\cos\theta(1 - \sin\theta)}{4\sin^2\theta - 4\sin\theta - 2}$$
$$= \frac{2\cos\theta(1 - \sin\theta)}{2\sin^2\theta - 2\sin\theta - 1}$$

4. We find that $r' = 24\cos 3\theta$, so $\frac{dy}{dx} = \frac{r'\sin\theta + r\cos\theta}{r'\cos\theta - r\sin\theta} = \frac{24\cos 3\theta\sin\theta + 8\sin 3\theta\cos\theta}{24\cos 3\theta\cos\theta - 8\sin 3\theta\sin\theta}$. When $\theta = \frac{\pi}{4}$, we see that $\frac{dy}{dx} = \frac{1}{2}$, so the slope of the normal line at this point is -2. Using the relationships that $x = r\cos\theta$ and $y = r\sin\theta$, we see that the polar coordinate $\left(4\sqrt{2}, \frac{\pi}{4}\right)$ has the rectangular coordinate $(4, 4)$ and so the equation of the normal line is $y - 4 = -2(x - 4)$ or $y = -2x + 12.$

5. $L = \int_0^2 \sqrt{(e^t)^2 + (2e^t)^2}\,dt = \int_0^2 \sqrt{e^{2t} + 4e^{2t}}\,dt = \int_0^2 \sqrt{5}e^t\,dt = \sqrt{5}e^t\big|_0^2 = \sqrt{5}(e^2 - 1)$

6. Here $f(\theta) = 4\sin\theta$, $f'(\theta) = 4\cos\theta$, and $y = 4\sin^2\theta$. So the surface area is

$$S = 2\pi \int_0^{\pi/2} \left(4\sin^2\theta\right)\sqrt{(4\sin\theta)^2 + (4\cos\theta)^2}\,d\theta$$
$$= 8\pi \int_0^{\pi/2} \left(\sin^2\theta\right)\sqrt{16\sin^\theta + 16\cos^2\theta}\,d\theta$$
$$= 32\pi \int_0^{\pi/2} \left(\sin^2\theta\right) d\theta$$
$$= 32\pi \left(\frac{1}{2} - \frac{1}{4}\sin 2\theta\right)\Big|_0^{\pi/2}$$

(by Formula 56 in Appendix C). Evaluating this we obtain $32\pi\left(\frac{\pi}{4}\right) = 8\pi^2 \approx 78.9568$

7. Setting these equations equal to each other, we obtain $4\cos\theta = 1 - \cos\theta$ or $5\cos\theta = 1$ and so the points of intersection are when $\cos\theta = \frac{1}{5} = 0.2$ or when $\theta \approx 1.3694$ and -1.3694 and the points of intersection are $(0.2, 1.3694)$ and $(0.2, -1.3694) = (0.2, 4.9137)$

8. The points of intersection are $-\frac{\pi}{2}$ and $\frac{\pi}{2}$ so

$$A = \int_{-\pi/2}^{\pi/2} \frac{1}{2}\left(r_2^2 - r_1^2\right) d\theta$$
$$= \int_{-\pi/2}^{\pi/2} \frac{1}{2}\left[\left(1^2 - (1 - \cos\theta)^2\right)\right] d\theta$$
$$= \int_{-\pi/2}^{\pi/2} \frac{1}{2}\left[\left(1 - (1 - 2\cos\theta + \cos^2\theta)\right)\right] d\theta$$
$$= \frac{1}{2}\left[2\sin\theta - \frac{\sin 2\theta}{4} - \frac{\theta}{2}\right]_{-\pi/2}^{\pi/2} = 2 - \frac{\pi}{4}$$

CHAPTER
30
Partial Derivatives and Multiple Integrals

☰ 30.1 FUNCTIONS IN TWO VARIABLES

1. $f(x, y) = 3x + 4y - xy$:
(a) $f(1, 0) = 3 \cdot 1 = 3$;
(b) $f(0, 1) = 4 \cdot 1 = 4$;
(c) $f(2, 1) = 3 \cdot 2 + 4 \cdot 1 - 2 \cdot 1 = 6 + 4 - 2 = 8$;
(d) $f(x + h, y) = 3(x + h) + 4y - (x + h)y = 3x + 3h + 4y - xy - hy$;
(e) $f(x, y + h) = 3x + 4(y + h) - x(y + h) = 3x + 4y + 4h - xy - xh$

3. $j(x, y) = \sqrt{xy} - x + \frac{4}{y}$:
(a) $j(-1, -2) = \sqrt{2} + 1 + \frac{4}{-2} = \sqrt{2} - 1 \approx 0.4142$;
(b) $j(-1, -4) = \sqrt{4} + 1 - 1 = 2$;
(c) $j(4, 1) = \sqrt{4} - 4 + 4 = 2$;
(d) $j(x + h, y) = \sqrt{(x + h)y} - (x + h) + \frac{4}{y}$;
(e) $j(x, y + h) = \sqrt{x(y + h)} - x + \frac{4}{y + h}$

5. $f(x, y) = \frac{2xy - x^2}{y - x}$:
(a) $f(1, 0) = \frac{0 - 1}{-1} = 1$;
(b) $f(0, 1) = \frac{0}{1} = 0$;
(c) $f(2, 1) = \frac{2 \cdot 2 - 2^2}{1 - 2} = 0$;
(d) $f(1, 2) = \frac{4 - 1}{1} = 3$;
(e) the domain is all real ordered pairs (x, y) such that $x \neq y$.

7. $V = \frac{1}{3}\pi r^2 h$

9. $A = 2\pi rh + 2\pi r^2$

11. The area of the storage tank is $A = 2\pi r^2 + 2\pi rh$. For the cost we have $C_{\text{bottom}} + C_{\text{top}} = 200\left(\pi r^2\right) + 200\left(\pi r^2\right) = 400\pi r^2$ and $C_{\text{side}} = 1000(2\pi rh) = 2000\pi rh$. Thus, the total cost is $C_{\text{bottom}} + C_{\text{top}} + C_{\text{side}} = 400\pi r^2 + 2000\pi rh = 400\pi r(r + 5h)$.

13. $v(R, r) = c\left(R^2 - r^2\right)$; $v(0.0075, 0.0045) = \left(0.0075^2 - 0.0045^2\right) = 3.6 \times 10^{-5} = 0.000\,036$ cm/min

15. (a) $T = 20°F$ and $v = 200$ mph so we use the following part of the formula: $WCI = 91.4 - \frac{(10.45 + 6.69\sqrt{v} - 0.447v)(91.4 - F)}{22} = -10.6$ or $-11°F$,
(b) $T = 10°$, $v = 20$ mph, using the same formula as in part (a) we get the $WCI = -25°F$,
(c) Since $v = 4$ mph, $WCI = F = 10°F$,
(d) As in parts (a) and (b) $WCI = -15°F$

17. (a) The volume is the product of the box's length, width, and height, or $V(x, y, z) = xyz$.
(b) The amount of the material needed to construct the box is the same as the surface area of the box including the partition, $S(x, y, z) = xy + 2xz + 3yz$.

19. $S(2400, 65) = (0.000\,02)(2,400)(65)^2 = 202.8$ ft

21. (a) $A(84, 1.80) = 0.18215\left(84^{0.425}\right)\left(1.80^{0.725}\right) \approx 1.834$ m^2
(b) For this, we need to compute a new constant. Since 1 kg $\approx$ 2.2 lb, then $1^{0.425} \approx 2.2^{0.425} \approx 1.398$ and since 1 m = 39.37 in., we have $1^{0.725} = 39.37^{0.725} \approx 14.338$. We also need to convert m^2 to in.^{2}, and 1 m^2 = 39.37^2 in.^{2} ≈ 1550 in.^{2}. The new constant is $\frac{0.18215 \times 1550}{1.398 \times 14.338} \approx 14.085$. Hence, the formula $A(w, h) = 14.085w^{0.425}h^{0.725}$ can determine the surface area in in.^{2} of a person based on the person's weight, w, in lb and height, h, in in.
(c) $A(185, 71) = 14.085\left(185^{0.425}\right)\left(71^{0.725}\right) \approx 2,847.5$ in.^{2}. (Note that this answer is almost the same as part (a); that is, 1.834 m$^2 \approx 2,847.5$ in.^{2}.)

23. (a) Here $x = 300$ units of labor and $y = 50$ units of capital, we have $P(300, 50) = 400\left(300^{0.35}\right) \times \left(50^{0.65}\right) \approx 37,444$.
(b) Here $x = 600$ and $y = 100$, and we have $P(600, 100) = 400\left(600^{0.35}\right)\left(100^{0.65}\right) \approx 74,888$.

≡ 30.2 SURFACES IN THREE DIMENSIONS

1. $x + 2y + 3z - 6 = 0$: A plane whose intercepts are $(6, 0, 0)$, $(0, 3, 0)$ and $(0, 0, 2)$.

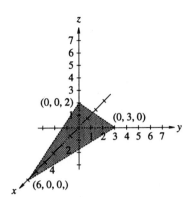

3. $x = 8y^2$: A parabolic cylinder perpendicular to the xy-plane.

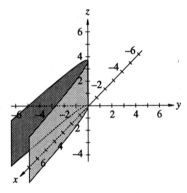

5. $x^2 + 2y^2 + z^2 = 4 \Rightarrow \frac{x^2}{4} + \frac{z^2}{4} = 1$: Ellipsoid with intercepts $(\pm 2, 0, 0)$, $(0, \pm\sqrt{2}, 0)$ and $(0, 0, \pm 2)$.

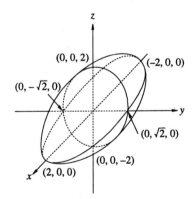

7. $2x^2 + y^2 + 2z^2 = 8 \Rightarrow \frac{x^2}{4} + \frac{y^2}{8} + \frac{z^2}{4} = 1$: Ellipsoid with intercepts $(\pm 2, 0, 0)$, $(0, \pm 2\sqrt{2}, 0)$, and $(0, 0, \pm 2)$.

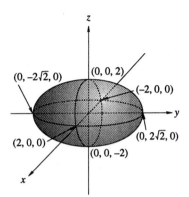

9. $4x^2 + y^2 - z^2 = -1$: Hyperboloid of two sheets with intercepts $(0, 0, \pm 1)$.

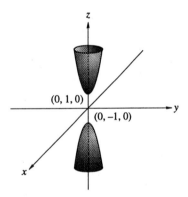

11. $3x - 4y - 6z = 12$: Plane with intercepts $(4, 0, 0)$, $(0, -3, 0)$, and $(0, 0, -2)$.

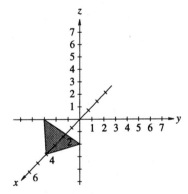

13. $z = x^2 + 4y^2 - 16$: This is an elliptic paraboloid with a z-translation of -16. The intercepts are $(\pm 4, 0, 0)$, $(0, \pm 2, 0)$, and $(0, 0, -16)$.

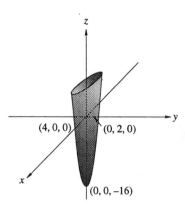

15. $36z = 4x^2 + 9y^2 \Rightarrow z = \frac{x^2}{9} + \frac{y^2}{4}$: Elliptic paraboloid with vertex $(0, 0, 0)$. Four other points are $(\pm 3, 0, 1)$ and $(0, \pm 2, 1)$.

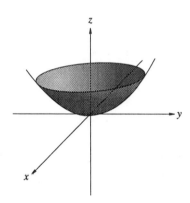

17. $36z = 4x^2 - 9y^2 \Rightarrow z = \frac{x^2}{9} - \frac{y^2}{4}$: Hyperbolic paraboloid with saddle point $(0, 0, 0)$. Four other points are $(\pm 3, 0, 1)$ and $(0, \pm 2, -1)$.

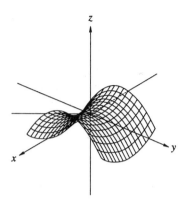

19. $2x = y^2 \Rightarrow x = \frac{1}{2}y^2$: Parabolic cylinder perpendicular to the xy-plane.

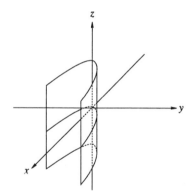

≡ 30.3 PARTIAL DERIVATIVES

1. $f(x, y) = x^2y + y^2x$; $f_x = 2xy + y^2$; $f_y = x^2 + 2xy$.

3. $f(x, y) = 3x^2 + 6xy^3$; $f_x = 6x + 6y^3$; $f_y = 18xy^2$

5. $f(x, y) = \frac{x^2 + y^2}{y} = \frac{x^2}{y} + y$; $f_x = \frac{2x}{y}$; $f_y = \frac{-x^2}{y^2} + 1 = \frac{y^2 - x^2}{y^2}$

7. $f(x, y) = e^y \cos x + e^x \sin y$; $f_x = -e^y \sin x + e^x \sin y$; $f_y = e^y \cos x + e^x \cos y$

9. $f(x, y) = e^{2x+3y}$; $f_x = 2e^{2x+3y}$; $f_y = 3e^{2x+3y}$

11. $f(x, y) = \sin\left(x^2 y^3\right)$; $f_x = 2xy^3 \cos\left(x^2 y^3\right)$; $f_y = 3x^2 y^2 \cos\left(x^2 y^3\right)$

13. $f(x, y) = \ln \sqrt{x^2 + y^2}$. So, $f_x = \dfrac{\frac{1}{2}\left(x^2 + y^3\right)^{-1/2} \cdot 2x}{\sqrt{x^2 + y^2}} = \frac{x}{x^2 + y^2}$ and $f_y = \dfrac{\frac{1}{2}\left(x^2 + y^2\right)^{-1/2}(2y)}{\sqrt{x^2 + y^2}} = \frac{y}{x^2 + y^2}$.

15. $f(x, y) = \sin^2(3xy)$; $f_x = 2\sin(3xy) \cdot \cos(3xy) \cdot 3y = 6y \sin(3xy)\cos(3xy)$; $f_y = 2 \sin 3xy \cos 3xy \cdot 3x = 6x \sin(3xy)\cos(3xy)$

17. $f(x, y) = x^3 y^2 - xy^5$; $f_x = 3x^2 y^2 - y^5$; $f_y = 2x^3 y - 5xy^4$; $f_{xx} = 6xy^2$; $f_{yx} = 6x^2 y - 5y^4$; $f_{yy} = 2x^3 - 20xy^3$; $f_{xy} = 6x^2 y - 5y^4$

19. $f(x, y) = 3e^{xy^3}$; $f_x = 3y^3 e^{xy^3}$; $f_y = 9xy^2 e^{xy^3}$; $f_{xx} = 3y^6 e^{xy^3}$; $f_{yx} = f_{xy} = 9e^{xy^3} y^2 \left(1 + xy^3\right)$; $f_{yy} = 27x^2 y^4 e^{xy^3} + 18xy e^{ey^3} = 9xy e^{xy^3} \left(3xy^3 + 2\right)$

21. $f(x, y) = \frac{2x}{y^5}$; $f_x = \frac{2}{y^5}$; $f_y = \frac{-10x}{y^6}$; $f_{xx} = 0$;

$f_{yx} = f_{xy} = \frac{-10}{y^6}$; $f_{yy} = \frac{60x}{y^7}$

23. $f(x, y) = \ln(x^3 y^5)$; $f_x = \frac{3}{x}$; $f_y = \frac{5}{y}$; $f_{xx} = \frac{-3}{x^2}$;

$f_{yx} = f_{xy} = 0$; $f_{yy} = \frac{-5}{y^2}$

25. $P = \frac{20 \cdot R \cdot 295}{V}$; $P_V = \frac{-5900R}{V^2}$ at $V = 3$; $P_V =$

$\frac{-5900R}{9}$ kg/m^2

27. $V = \frac{82.06T}{P}$; $V_P = \frac{-82.06T}{P^2}$ at $T = 300°$, $P =$

5 atm; $V = \frac{-82.06 \times 300}{5^2} = -984.72$ cm^3/atm

29. $z = 16 - x^2 - y^2$; $z_y = -2y$. At $(1, 3, 6)$, $z_y = -2 \cdot 3 = -6$.

31. **(a)** $\frac{dR}{dL} = \frac{k}{r^4}$,

(b) Rewriting the formula as $R = kLr^{-4}$, we have $\frac{dR}{dr} = -4kLr^{-5} = -\frac{4kL}{r^5}$.

33. **(a)** $A_w(w, h) = 5.9861w^{-0.575}h^{0.725}$ in.^{2}/lb; $A_h(w, h) = 10.2116w^{0.425}h^{-0.275}$ in.^{2}/in.

(b) Substituting in the formulas from (a) produces $A_w(185, 71) = 5.9861(185^{-0.575})(71^{0.725}) = 6.54$ in.^{2}/lb and $A_h(185, 71) = 10.2116(185^{0.425}) \times (71^{-0.275}) = 29.08$ in.^{2}/in.

(c) $A_w(185, 71) = 6.54$ in.^{2}/lb means that for a 185 lb person 5′11″ tall, the rate of change in surface area is 6.54 in.^{2} for each lb gained in weight if the height is fixed; $A_h(185, 71) = 29.07$ in.^{2}/in. means that for a 185 lb person 5′11″ tall, the rate of change in surface area is 29.07 in.^{2} for each in. gained in height if the weight remains the same.

35. **(a)** $S_w(w, r) = 0.000\,02r^2$ ft/lb,
(b) $S_w(2400, 65) = 0.00002(65^2) = 0.0845$ ft/lb,
(c) $S_r(w, r) = 0.000\,04wr$ ft/mph,
(d) $S_r(2400, 65) = 0.00004(2400)(65) = 6.24$ ft/mph,
(e) about 6.24 ft

≡ 30.4 SOME APPLICATIONS OF PARTIAL DERIVATIVES

1. $z = x^2 + y^2$; $z_x = 2x$; $z_y = 2y$; $dz = 2xdx + 2ydy$

3. $z = 3x^2 + 4xy - 2y^3$; $z_x = 6x + 4y$; $z_y = 4x - 6y^2$; $dz = (6x + 4y)dx + (4x - 6y^2)dy$

5. $z = \arctan\left(\frac{y}{x}\right)$; $z_x = \frac{-\frac{y}{x^2}}{1 + \frac{y^2}{x^2}} = \frac{-y}{x^2 + y^2}$; $z_y =$

$\frac{\frac{1}{x}}{1 + \frac{y^2}{x^2}} = \frac{x}{x^2 + y^2}$; $dz = \frac{-y}{x^2 + y^2}\,dx + \frac{x}{x^2 + y^2}\,dy$

7. $z = x\tan yx$; $z_x = \tan yx + xy\sec^2 yx$; $z_y = x^2\sec^2 yx$; $dz = (\tan yx + xy\sec^2 yx)\,dx + x^2\sec^2 yx\,dy$

9. $z = x^2 + y^2$; $x = te^t$; $y = t^2e^t$; $z_x = 2x$; $z_y = 2y$; $\frac{dx}{dt} = (t + 1)e^t$; $\frac{dy}{dt} = (t^2 + 2t)e^t$; $\frac{dz}{dt} = 2x(t + 1)e^t + 2y(t^2 + 2t)e^t = 2(te^t)(t + 1)e^t + 2(t^2e^t)(t^2 + 2t)e^t = (t + 1)2te^{2t} + (t^2 + 2t)2t^2e^{2t} = 2te^{2t}[1 + t + 2t^2 = t^3]$

11. $z = e^u\sin v$; $u = \sqrt{t}$; $v = \pi t$; $z_u = e^u\sin v$; $z_v = e^u\cos v$; $\frac{du}{dt} = \frac{1}{2\sqrt{t}}$; $\frac{dv}{dt} = \pi$; $\frac{dz}{dt} = e^u\sin v \cdot \frac{1}{2\sqrt{t}} + e^u\cos v \cdot \pi = \frac{e^{\sqrt{t}}\sin\pi t}{2\sqrt{t}} + \pi e^{\sqrt{t}}\cos\pi t$

13. $z = x^2 + 4y^2 + x + 8y - 1$; $z_x = 2x + 1$; $z_y = 8y + 8$; $z_x = 0 \Rightarrow x = -\frac{1}{2}$; $z_y = 0 \Rightarrow y = -1$; $z_{xx} = 2$;

$z_{yx} = 0$; $z_{xy} = 0$; $z_{yy} = 8$; $\Delta = \begin{vmatrix} z_{xx} & z_{xy} \\ z_{yx} & z_{yy} \end{vmatrix} = \begin{vmatrix} 2 & 0 \\ 0 & 8 \end{vmatrix} = 16 > 0$. Since z_{xx} is $2 > 0$ we have a local minimum at $\left(-\frac{1}{2}, -1\right)$.

15. $z = 20 + 12x - 12y - 3x^2 - 2y^2$; $z_x = 12 - 6x = 0 \Rightarrow x = 2$; $z_y = -12 - 4y = 0 \Rightarrow y = -3$; $z_{xx} = -6$; $z_{xy} = z_{yx} = 0$; $z_{yy} = -4$; $\Delta = \begin{vmatrix} -6 & 0 \\ 0 & -4 \end{vmatrix} = 24 > 0$. Since $z_{xx} = 2 > 0$ we have a local minimum at $(2, -3)$.

17. $z = x^2 + 2xy - y^2$; $z_x = 2x + 2y$; $z_y = 2x - 2y$; $2x + 2y = 0$ and $2x - 2y = 0 \Rightarrow x = 0$ and $y = 0$; $z_{xx} = 2$; $z_{xy} = z_{yx} = 2$; $z_{yy} = -2$. $\Delta\begin{vmatrix} 2 & 2 \\ 2 & -2 \end{vmatrix} = -8 < 0$, so saddle point at $(0, 0)$.

19. $z = x^3 + x^2y + y^2$; $z_x = 3x^2 + 2xy$; $z_y = x^2 + 2y$; $3x^2 + 2xy = 0$, and $x^2 + 2y = 0$. Solving for $2y$ in the second equation we get $2y = -x^2$. Substituting into the first equation we have $3x^2 - x^3 = 0$ or $x^2(3 - x) = 0$ so $x = 0$ or $x = 3$. Back substituting we get the ordered pairs $(0, 0)$ and $\left(3, -\frac{9}{2}\right)$. $z_{xx} = 6x + 2y$; $z_{xy} = z_{yx} = 2x$; $z_{yy} = 2$. $\Delta = \begin{vmatrix} 6x + 2y & 2x \\ 2x & 2 \end{vmatrix} = 12x + 4y - 4x^2$.

At $(0, 0)$ the test fails. At $\left(3, -\frac{9}{2}\right)$, $\Delta = 36 - 18 - 81 < 0$, so there is a saddle point at $\left(3, -\frac{9}{2}\right)$.

21. $A = bh$; $A_b = h$; $A_h = b$; $dA = hdb + bdh = 25(0.1) + 10(0.1) = 35$ cm^2

23. $R_{R_1} = \frac{(R_1+R_2)R_2+R_1R_2}{(R_1+R_2)^2}$; $R_{R_2} = \frac{(R_1+R_2)R_1+R_1R_2}{(R_1+R_2)^2} dR$

$= R_{R_1}dR_1 + R_{R_2}dR_2 = \frac{(500)400+40000}{(500)^2}1 +$

$\frac{500\cdot100+40000}{500^2}4 = 0.96+1.44 = 2.4\Omega$. (Note $dR_1 = 0.01 \times 100 = 1\Omega$; $dR_2 = 0.01 \times 400 = 4\Omega$.)

25. $PV = kT$ or $T = \frac{PV}{k}$. $\frac{dT}{dt} = \frac{\partial T}{\partial P}\frac{dP}{dt} + \frac{\partial T}{\partial V}\frac{dV}{dt} = \frac{V}{k}\cdot\frac{dP}{dt} + \frac{P}{k}\frac{dV}{dt} = \frac{1}{k}\left(V\frac{dP}{dt} + P\cdot\frac{dV}{dt}\right)$.

27. $W = \frac{Li^2}{2}$; $dw = \frac{\partial W}{\partial L}dL + \frac{\partial w}{\partial i}di = \frac{i^2}{2}(0.02) + Li(-0.1) = \frac{1^2}{2}(0.02)+30(1)(-0.1) = 0.01-0.3 = -0.29$.

29. $V = \ell\cdot w\cdot h$; $h = \frac{500}{\ell w}$. $A = \ell w+2\ell\cdot h+2\ell h = \ell w+ 2\cdot\frac{500}{w}+2\cdot\frac{500}{\ell}$. $A_\ell - w - \frac{1000}{w^2}$; $A_w = \ell - \frac{1000}{\ell^2}$; $A_\ell = A_w = 0$ so $w - \frac{1000}{\ell^2} = 0$ and $\ell - \frac{1000}{w^2} = 0$. Hence $\ell = \frac{1000}{w^2}$ and substituting we get $w - \frac{1000}{\left(\frac{1000}{w^2}\right)^2} = 0$

or $w - \frac{w^4}{1000} = 0$. Now $w\left(1 - \frac{w}{1000}\right) = 0$ so $w = 0$ or $w^3 = 1000 \Rightarrow w = 10$. But, w cannot be 0, so $w = 10$ and $\ell = \frac{1000}{10^2} = 10$; $h = \frac{500}{10\times 10} = 5$. To conclude that this is a minimum we compute $A_{\ell\ell} = \frac{2000}{\ell^3}$; $A_{ww} = \frac{2000}{w^3}$. $A_{w\ell} = A_{\ell w} = 1$;

$\Delta \begin{vmatrix} \frac{2000}{10^3} & 1 \\ 1 & \frac{2000}{10^3} \end{vmatrix} = 4 - 1 = 3 > 0$. $A_{\ell\ell} = 2 > 0$ so these dimensions yield a minimum; $h = 5$ cm, $\ell = 10$ cm, $w = 10$ cm.

31. **(a)** $V = \ell wh = 4$ or $h = \frac{4}{\ell w}$. Cost $= 1.25\ell w + 1.50(2\ell h+2\ell w) = 1.25\ell w+3\left(\frac{4}{w} + \frac{4}{\ell}\right) = 1.25\ell w + 12\left(\frac{1}{w} + \frac{1}{\ell}\right)$. $C_\ell = 1.25w - \frac{12}{\ell^2}$, $C_w = 1.25\ell - \frac{12}{w^2}$. Setting these both equal to zero we get $w = \frac{9.6}{\ell^2}$. Substituting, we have $1.25\ell - \frac{12}{\left(\frac{9.6}{\ell^2}\right)^2} = 0$ or

$1.25\ell - 0.1302\ell^4 = 0$, or $\ell\left(1.25 - 0.1302\ell^3\right) = 0$, so $\ell = 0$ or $\ell = \sqrt[3]{\frac{1.25}{0.1302}} = 2.125$. Since ℓ cannot be 0, we have $\ell = 2.125$ ft, $w = 2.125$ ft and $h = \frac{4}{(2.125)^2} = 0.886$ ft. This will yield a minimum cost.

(b) $C = 1.25(2.125)^2 + 6(2.125 \times 0.886) = \$16.94$

33. We begin by differentiating $R(x, y)$ with respect to x and with respect to y, with the following results.

$R_x(x, y) = 15y(2 - x - 4y) + 15xy(-1)$
$= 30y - 15xy - 60y^2 - 15xy$
$= 30y - 30xy - 60y^2$
$= 30\left(y - xy - 2y^2\right)$
$R_y(x, y) = 15x(2 - x - 4y) + 15xy(-4)$
$= 30x - 15x^2 - 60xy - 60xy$
$= 30x - 15x^2 - 120xy$
$= 15\left(2x - x^2 - 8xy\right)$

Setting $R_x = 0$, we obtain $y - xy - 2y^2 = 0$ or $xy = y - 2y^2$ or $x = 1 - 2y$, if $y \neq 0$. Substituting this value into R_y and setting it equal to 0, we obtain $x = \frac{1}{2}(2y + 1) = \frac{2y+1}{2}$. Substituting this value of x into R_y and setting it equal to 0, we obtain

$2(1 - 2y) - (1 - 2y)^2 - 8(1 - 2y)y = 0$
$2 - 4y - 1 + 4y - 4y^2 - 8y + 16y^2 = 0$
$12y^2 - 8y + 1 = 0$
$(6y + 1)(2y - 1) = 0$

Hence, $y = \frac{1}{6}$ or $y = \frac{1}{2}$. Using back substitution produces $x = 1-2\left(\frac{1}{6}\right) = \frac{2}{3}$ or $x = 1-2\left(\frac{1}{2}\right) = 0$. Since $x > 0$, the maximum reaction occurs when $x = \frac{2}{3}$ and $y = \frac{1}{6}$. This produces a reaction of $15\left(\frac{2}{3}\cdot\frac{1}{6}\right)\left(2 - \frac{2}{3} - 4\cdot\frac{1}{6}\right) = \frac{10}{9}$.

35. Rewriting $V(x)$ produces

$V(x) = \frac{Q}{4\pi\epsilon_0}\left(x^2 + a^2\right)^{-1/2}$. Differentiating this version, we get

$E_x = -\frac{\partial V}{\partial x} = -\frac{Q}{4\pi\epsilon_0}\left(-\frac{1}{2}\right)\left(x^2 + a^2\right)^{-3/2}(2x)$
$= \frac{Qx}{4\pi\epsilon_0\left(x^2 + a^2\right)^{3/2}}$

$\equiv$ 30.5 MULTIPLE INTEGRALS

1. $\int_0^1 \int_0^{x^2} xy\, dy\, dx = \int_0^1 \left[\frac{1}{2}xy^2\right]_0^{x^2} dx = \int_0^1 \frac{1}{2}x^5 dx = \left[\frac{1}{12}x^6\right]_0^1 = \frac{1}{12}$

3. $\int_0^\pi \int_{-\pi/2}^{\pi/2} \sin x \cos y\, dy\, dx = \int_0^\pi \left[-\sin x \sin y\right]_{-\pi/2}^{\pi/2} \times dx = \int_0^\pi \left[-\sin x(1 + 1)\right] dx = -2\cos x|_0^\pi = 2 + 2 = 4$

5. $\int_0^{\ln 2} \int_0^{\ln 5} e^{x+y} dy\, dx = \int_0^{\ln 2} \left[e^{x+y} \right]_0^{\ln 5} dx =$
$\int_0^{\ln 2} e^x(5-1) = 4e^x\big|_0^{\ln 2} = 4(2-1) = 4$

7. $\int_0^1 \int_x^{\sqrt{x}} (x+y) dy\, dx = \int_0^1 \left[xy + \frac{y^2}{2} \right]_x^{\sqrt{x}} dx = \int_0^1 \times$
$\left(x^{3/2} - \frac{x}{2} - x^2 - \frac{x^2}{2} \right) dx = \left[\frac{2}{5}x^{5/2} + \frac{x^2}{4} - \frac{x^3}{2} \right]_0^1$
$= \frac{2}{5} + \frac{1}{4} - \frac{1}{2} = \frac{8+5-10}{20} = \frac{3}{20}.$

9. $\int_0^1 \int_{x^3}^x (y-x) dy\, dx = \int_0^1 \left[\frac{y^2}{2} - xy \right]_{x^3}^x dx = \int_0^1 \times$
$\left(\frac{x^2}{2} - x^2 - \frac{x^6}{2} + x^4 \right) dx = \left[-\frac{x^3}{6} - \frac{x^7}{14} + \frac{x^5}{5} \right]_0^1 =$
$-\frac{1}{6} - \frac{1}{14} + \frac{1}{5} = \frac{-35-15+42}{210} = \frac{-8}{210} = \frac{-4}{105}.$

11. $\int_0^2 \int_0^3 (xy+x-y) dy\, dx = \int_0^2 \left(\frac{xy^2}{2} + xy - \frac{y^2}{2} \right)_0^3 dx$
$= \int_0^2 \left(\frac{9}{2}x + 3x - \frac{9}{2} \right) dx = \int_0^2 \left(\frac{15}{2}x - \frac{9}{x} \right) dx =$
$\left[\frac{15x^2}{4} - \frac{9x}{2} \right]_0^2 = 15 - 9 = 6.$

13. $\int_0^1 \int_0^{1-x} (x^2 y + xy^2)\, dy\, dx$
$= \int_0^1 \left[\frac{x^2 y^2}{2} + \frac{xy^3}{3} \right]_0^{1-x} dy$
$= \int_0^1 \left(\frac{x^2(1-x)^2}{2} + \frac{x(1-x)^3}{3} \right) dx$
$= \int_0^1 \left(\frac{x^2 - 2x^3 + x^4}{2} \right.$
$\left. + \frac{x - 3x^2 + 3x^2 - x^4}{3} \right) dx$
$= \int_0^1 \left(\frac{x}{3} - \frac{x^2}{2} + \frac{x^4}{6} \right) dx$
$= \left[\frac{x^2}{6} - \frac{x^3}{6} + \frac{x^5}{30} \right]_0^1 = \frac{1}{30}$

15. $\int_0^1 \int_{y^2}^y (xy + 1) dx\, dy = \int_0^1 \left[\frac{x^2 y}{2} + x \right]_{y^2}^y dy = \int_0^1 \times$
$\left(\frac{y^3}{2} + y - \frac{y^5}{2} - y^2 \right) dy = \left[\frac{y^4}{8} + \frac{y^2}{2} - \frac{y^6}{12} - \frac{y^3}{3} \right]_0^1 =$
$\frac{1}{8} + \frac{1}{2} - \frac{1}{12} - \frac{1}{3} =$
$\frac{3+12-2-8}{24} = \frac{5}{24}.$

17. For this region we can integrate with respect to y first from x to 1 and then with respect to x from 0 to 1. $\int_0^1 \int_x^1 (x^2 - y^2)\, dy\, dx = \int_0^1 \left[x^2 y - \frac{y^3}{3} \right]_x^1 dx$
$= \int_0^1 \left(x^2 - \frac{1}{3} - x^3 + \frac{x^3}{3} \right) dx = \int_0^1 \left(x^2 + \frac{2x^3}{3} - \frac{1}{3} \right)$
$\times dx = \left[\frac{x^3}{3} + \frac{x^4}{6} - \frac{x}{3} \right]_0^1 = \frac{1}{6}.$

19. $\int_0^8 \int_0^{\sqrt[3]{y}} (x^3 - y^3)\, dx\, dy = \int_0^8 \left(\frac{x^4}{4} - y^3 x \right)\Big|_0^{\sqrt[3]{y}} dy$
$= \int_0^8 \left(\frac{y^{4/3}}{4} - y^{10/3} \right) dy = \left[\frac{3y^{7/3}}{28} - \frac{3y^{13/3}}{13} \right]_0^8 = \frac{3 \cdot 2^7}{28} -$
$\frac{3 \cdot 2^{13}}{13} = -1876.7473.$

21. $\int_1^8 \int_{8/x}^{9-x} (x+y) dy\, dx$
$= \int_1^8 \left[xy + \frac{y^2}{2} \right]_{8/x}^{9-x} dx$
$= \int_1^8 \left(x(9-x) + \frac{(9-x)^2}{2} - 8 - \frac{32}{x^2} \right) dx$
$= \int_1^8 \left(9x - x^2 + \frac{81}{2} - 9x \right.$
$\left. + \frac{x^2}{2} - 8 - \frac{32}{x^2} \right) dx$
$= \int_1^8 \left(\frac{-x^2}{2} - \frac{32}{x^2} + \frac{65}{2} \right) dx$
$= \frac{-x^3}{6} + \frac{32}{x} + \frac{65x}{2}\Big|_1^8$
$= \frac{-8^3}{6} + \frac{32}{8} + \frac{65 \cdot 8}{2} + \frac{1}{6} - 32 - \frac{65}{2}$
$= \frac{343}{3} = 114\frac{1}{3}$

23. $\int_0^1 \int_{x^3}^x (x+y)^2 dy\, dx$
$= \int_0^1 \int_{x^3}^x (x^2 + 2xy + y^2) dy\, dx$
$= \int_0^1 \left[x^2 y + xy^2 + \frac{y^3}{3} \right]_{x^3}^x dx$
$= \int_0^1 \left[\left(x^3 + x^3 + \frac{x^3}{3} \right) \right.$
$\left. - \left(x^5 + x^7 + \frac{x^9}{3} \right) \right]_{x^3}^x dx$
$= \int_0^1 \left[\frac{7}{3}x^3 - x^5 - x^7 - \frac{1}{3}x^9 \right] dx$
$= \left[\frac{7}{12}x^4 - \frac{1}{6}x^6 - \frac{1}{8}x^8 - \frac{1}{30}x^{10} \right]_0^1$
$= \frac{7}{12} - \frac{1}{6} - \frac{1}{8} - \frac{1}{30}$
$= \frac{60 + 10 - 20 - 15 - 4}{120} = \frac{31}{120} \approx 0.2583$

25. The desired area is symmetric about the x-axis, so we need only to integrate the y-values from 0 to 3, and double that answer. Thus, we have
$\int_{-3}^3 \int_0^{y^2-9} xy\, dx\, dy = 2\int_0^3 \int_0^{y^2-9} xy\, dx\, dy =$
$2\int_0^3 \frac{x^2 y}{2}\Big|_0^{y^2-9} dy = \int_0^3 (y^2 - 9)^2 y\, dy = \int_0^3 (y^5 -$
$18y^3 + 81y)\, dy = \left[\frac{1}{6}y^6 - \frac{9}{2}y^4 + \frac{81}{2}y^2 \right]_0^3 = 121.5.$

27. $\bar{f} = \dfrac{1}{6} \displaystyle\int_0^2 \int_0^3 (x^2 + y^2)\, dy\, dx$

$= \dfrac{1}{6} \displaystyle\int_0^2 \left[\left(x^2 y + \dfrac{1}{3} y^3\right)\right]_0^3 dx$

$= \dfrac{1}{6} \displaystyle\int_0^2 [3x^2 + 9]\, dx$

$= \dfrac{1}{6} \left[x^3 + 9x\right]_0^2$

$= \dfrac{1}{6}[8 + 18] = \dfrac{26}{6} = \dfrac{13}{3}$

29. $\bar{f} = \dfrac{1}{3 \cdot 5} \displaystyle\int_1^4 \int_2^7 \dfrac{x}{y}\, dy\, dx$

$= \dfrac{1}{15} \displaystyle\int_1^4 [x \ln y]_2^7\, dx$

$= \dfrac{1}{15}[\ln 7 - \ln 2] \displaystyle\int_1^4 x\, dx$

$= \dfrac{1}{15} \left(\ln \dfrac{7}{2}\right) \left[\dfrac{1}{2} x^2\right]_1^4$

$= \dfrac{1}{30} \left(\ln \dfrac{7}{2}\right) [16 - 1] = \dfrac{1}{2} \ln \dfrac{7}{2} \approx 0.6264$

31. Solving $x + 2y + 2 = 4$ for z we get $z = 4 - x - 2y$. The x-intercept is 4 and the trace in the xy plane is $y = 2 - \frac{1}{2}x$. Thus the volume is

$V = \displaystyle\int_0^4 \int_0^{2 - \frac{1}{2}x} (4 - x - 2y)\, dy\, dx$

$= \displaystyle\int_0^4 \left[4y - xy - y^2\right]_0^{2 - \frac{1}{2}x} dx$

$= \displaystyle\int_0^4 \left[4\left(2 - \dfrac{1}{2}x\right) - x\left(2 - \dfrac{1}{2}x\right)\right.$

$\left. - \left(2 - \dfrac{1}{2}x\right)^2\right] dx$

$= \displaystyle\int_0^4 \left(8 - 2x - 2x + \dfrac{1}{2}x^2 - 4 + 2x - \dfrac{1}{4}x^2\right) dx$

$= \displaystyle\int_0^4 \left(4 - 2x + \dfrac{1}{4}x^2\right) dx$

$= \left[4x - x^2 + \dfrac{1}{12}x^3\right]_0^4$

$= 16 - 16 + \dfrac{1}{12}64 = \dfrac{16}{3} = 5\dfrac{1}{3}$

33. As in problem #28, $z = \pm\sqrt{2 - x^2}$. We can integrate from 0 to $\sqrt{2}$ and double the answer. Thus, $V = 2\int_0^{\sqrt{2}} \int_0^x \sqrt{2 - x^2}\, dy\, dx = 2\int_0^{\sqrt{2}} \sqrt{2 - x^2} \cdot x\, dx$. Let $u = 2 - x^2$, and $du = -2x\, dx$. Substituting, we get $V = -\int_0^{\sqrt{2}} u^{1/2} du = -\frac{2}{3}u^{3/2} = -\frac{2}{3}\sqrt{2 - x^2}^3 \big|_0^{\sqrt{2}} = \frac{2}{3}\sqrt{2}^3 = \frac{4\sqrt{2}}{3} \approx 1.8856$.

35. C_1 and C_2 are independent operations. Therefore, we will figure the averages separately and then add the results.

$\overline{C_1} = \dfrac{1}{100} \displaystyle\int_{300}^{400} (0.2x^2 + 40x + 3600)\, dx$

$= \dfrac{1}{100} \left[\dfrac{0.2}{3}x^3 + \dfrac{40}{2}x^2 + 3600x\right]_{300}^{400}$

$= 42{,}266.7$

$\overline{C_2} = \dfrac{1}{200} \displaystyle\int_{350}^{550} (0.4y^2 + 24y + 6{,}000)\, dy$

$= \dfrac{1}{200} \left[\dfrac{0.4}{3}y^3 + 12y^2 + 6{,}000y\right]_{350}^{550}$

$= 99{,}133.3$

Adding produces $C_1 + C_2 = 42{,}266.7 + 99{,}133.3 = 141{,}400$. The average cost of $C_1(x) + C_2(y)$ is \$141,400.

37. **(a)** By the Pythagorean theorem, $d^2 = x^2 + y^2$, so $C(x, y) = 100 - 15(x^2 + y^2) = 100 - 15x^2 - 15y^2$. **(b)** Since the plant is in the center of town, we integrate for $-1 \leq x \leq 1$ and $-2 \leq y \leq 2$.

$\overline{C} = \dfrac{1}{4 \cdot 2} \displaystyle\int_{-2}^2 \int_{-1}^1 (100 - 15x^2 - 15y^2)\, dx\, dy$

$= \dfrac{1}{8} \displaystyle\int_{-2}^2 [100x - 5x^3 - 15xy^2]_{-1}^1\, dy$

$= \dfrac{1}{8} \displaystyle\int_{-2}^2 [200 - 10 - 30y^2]\, dy$

$= \dfrac{1}{8} \displaystyle\int_{-2}^2 [190 - 30y^2]\, dy$

$= \dfrac{1}{8} \left[190y - 10y^3\right]_{-2}^2$

$= \dfrac{1}{8}[760 - 160] = \dfrac{1}{8}[600] = 75$

Thus, the average concentration of particulate matter throughout the town is 75 ppm.

▄▄ 30.6 CYLINDRICAL AND SPHERICAL COORDINATES

1. Here $x = 2\cos\frac{\pi}{4} = \sqrt{2}$ and $y = 2\sin\frac{\pi}{4} = \sqrt{2}$. Thus, the cylindrical coordinate $(2, \pi/4, 2)$ has the equivalent rectangular coordinate $(\sqrt{2}, \sqrt{2}, 2)$.

3. $x = 2\cos 0 = 2$, $y = 2\sin 0 = 0$. Thus, the cylindrical coordinate $(2, 0, 4)$ has the equivalent rectangular coordinate $(2, 0, 4)$.

5. Here $x = 5\cos\frac{5\pi}{4} = \frac{-5\sqrt{2}}{2}$ and $y = 5\sin\frac{5\pi}{4} = -\frac{5\sqrt{2}}{2}$, so the cylindrical coordinate $(5, 5\pi/4, 0)$ has the equivalent rectangular coordinate $\left(\frac{-5\sqrt{2}}{2}, \frac{-5\sqrt{2}}{2}, 0\right)$ or $(-3.5355, -3.5355, 0)$.

7. Here $r = \sqrt{x^2 + y^2} = \sqrt{2^2 + 2^2} = \sqrt{8} = 2\sqrt{2}$ and $\theta = \tan^{-1}\frac{y}{x} = \tan^{-1}\frac{2}{2} = \tan^{-1}1 = \frac{\pi}{4}$. This means that the rectangular coordinates $(2, 2, 5)$ have cylindrical coordinates $\left(2\sqrt{2}, \frac{\pi}{4}, 5\right) = (2.8284, 0.7854, 5)$.

9. Here $r = \sqrt{(-4)^2 + 3^2} = 5$ and $\theta = \tan^{-1}\frac{-3}{4} = 2.4981$. This means that the rectangular coordinates $(-4, 3, 2)$ have cylindrical coordinates $(5, 2.4981, 2)$.

11. Here $r = \sqrt{12^2 + (-5)^2} = 13$ and $\theta = \tan^{-1}\frac{-5}{12} = 5.8884$. Thus, the rectangular coordinates $(12, -5, -3)$ have cylindrical coordinates $(13, 5.8884, -3)$

13. Here $x = 4\sin\frac{\pi}{6}\cos\frac{\pi}{4} = 4 \cdot \frac{1}{2} \cdot \frac{\sqrt{2}}{2} = \sqrt{2}$ and $y = 4\sin\frac{\pi}{6}\frac{\pi}{4} = 4 \cdot \frac{1}{2} \cdot \frac{\sqrt{2}}{2} = \sqrt{2}$ and $z = 4\cos\frac{\pi}{6} = 4 \cdot \frac{3}{2} = 2\sqrt{3}$. Thus, the spherical coordinate $\left(4, \frac{\pi}{4}, \frac{\pi}{6}\right)$ has the equivalent rectangular coordinate $(\sqrt{2}, \sqrt{2}, 2\sqrt{3}) \approx (1.4142, 1.4142, 3.4641)$

15. Here $x = 3\sin\frac{5\pi}{3}\cos\frac{\pi}{2} = 3 \cdot \frac{-\sqrt{3}}{2} \cdot 0 = 0$, $y = \sin\frac{5\pi}{3}\sin\frac{\pi}{2} = \frac{-3\sqrt{3}}{2}$, and $z = 3\cos\frac{5\pi}{3} = \frac{3}{2}$. Thus, the spherical coordinate $\left(3, \frac{\pi}{2}, \frac{5\pi}{3}\right)$ has the equivalent rectangular coordinate $\left(0, \frac{-3\sqrt{3}}{2}, \frac{3}{2}\right) = (0, -2.5981, 1.5)$.

17. Here $x = 5\sin\frac{2\pi}{3}\cos\frac{7\pi}{6} = 5 \cdot \frac{-\sqrt{3}}{2} \cdot \frac{-\sqrt{3}}{2} = \frac{-15}{4} = -3.75$, $y = 5\sin\frac{2\pi}{3}\sin\frac{7\pi}{6} = 5 \cdot \frac{\sqrt{3}}{2} \cdot \frac{-1}{2} = \frac{-5\sqrt{3}}{4} \approx -2.1651$, and $z = 5\cos\frac{2\pi}{3} = 5 \cdot -\frac{1}{2} = -2.5$. Thus, the spherical coordinate $\left(5, \frac{7\pi}{6}, \frac{2\pi}{3}\right)$ has the equivalent rectangular coordinate $(-3.75, -2.1651, -2.5)$.

19. Here $\rho = \sqrt{4^2 + 3^2} = 5$, $\theta = \tan^{-1}\frac{3}{4} = 0.6435$, and $\phi = \cos^{-1}\frac{0}{5} = \frac{\pi}{2} \approx 1.5708$. Thus, the rectangular coordinate $(4, 3, 0)$ has the spherical coordinate $\left(5, 0.6435, \frac{\pi}{2}\right)$.

21. Here $\rho = \sqrt{2^2 + 1^2 + 2^2} = 3$, $\theta = \tan^{-1}\frac{1}{2} = 0.4636$, and $\phi = \cos^{-1}\frac{-2}{3} = 2.3005$. Thus, the rectangular coordinate $(2, 1, -2)$ has the spherical coordinate $(3, 0.4636, 2.3005)$.

23. Here $\rho = \sqrt{1^2 + 1^2 + \sqrt{2}^2} = 2$, $\theta = \tan^{-1}\frac{1}{1} = \frac{\pi}{4} = 0.7854$, and $\phi = \cos^{-1}\frac{\sqrt{2}}{2} = \frac{\pi}{4}$. Thus, the rectangular coordinate $(1, 1, \sqrt{2})$ has the spherical coordinate $\left(2, \frac{\pi}{4}, \frac{\pi}{4}\right) = (2, 0.7854, 0.7854)$.

25. The graph is a line that makes an angle of $\frac{\pi}{4}$ with the z-axis and whose image on the xy-plane makes an angle of $\frac{\pi}{4}$ with the x-axis.

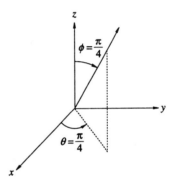

27. The region between the two spheres $\rho = 3$ and $\rho = 5$.

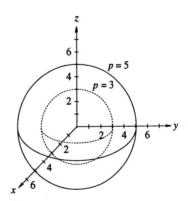

29. A cone shaped figure with spherical base.

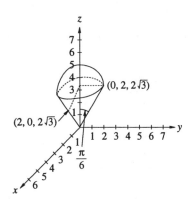

$(0, 2, 2\sqrt{3})$

$(2, 0, 2\sqrt{3})$

$\frac{\pi}{6}$

31. See *Computer Programs* in the main text.

30.7 MOMENTS AND CENTROIDS

1. $A = \int_0^3 \int_0^{2x+3} dy \, dx = \int_0^3 y|_0^{2x+3} dx = \int_0^3 (2x + 3)dx = x^2 + 3x|_0^3 = 18.$ $m = \rho A = 18\rho.$
$M_x = \int_0^3 \int_0^{2x+3} y\rho \, dy \, dx = \rho \int_0^3 \frac{y^2}{2}|_0^{2x+3} dx = \frac{1}{2}\rho \int_0^3 (2x + 3)^2 dx = \frac{1}{2}\rho \int_0^3 \left(4x^2 + 12x + 9\right) dx = \frac{1}{2}\rho \left[\frac{4}{3}x^3 + 6x^2 + 9x\right]_0^3 = 58.5\rho.$ $M_y = \int_0^3 \int_0^{2x+3} \times x\rho \, dy \, dx = \rho \int_0^3 x[y]_0^{2x+3} dx = \rho \int_0^3 \left(2x^2 + 3x\right) dx = \rho \left[\frac{2}{3}x^3 + \frac{3}{2}x^2\right]_0^3 = 31.5\rho.$ $\bar{x} = \frac{M_y}{m} = \frac{31.5\rho}{18\rho} = 1.75;$ $\bar{y} = \frac{M_x}{m} = \frac{58.5\rho}{18\rho} = 3.25$

3. $m = \rho \int_0^8 \int_0^{x^{1/3}} dy \, dx = \rho \int_0^8 x^{1/3} dx = \rho\frac{3}{4}x^{4/3}|_0^8 = 12\rho.$ $M_x = \rho \int_0^8 \int_0^{x^{1/3}} y \, dy \, dx = \rho \int_0^8 \frac{y^2}{2}|_0^{x^{1/3}} dx = \rho \int_0^8 \frac{1}{2}x^{2/3} dx = \rho \left[\frac{1}{2} \cdot \frac{3}{5}x^{5/3}\right]_0^8 = 9.6\rho.$ $M_y = \rho \int_0^8 \int_0^{x^{1/3}} x \, dy \, dx = \rho \int_0^8 x^{4/3} dx = \rho\frac{3}{7}x^{7/3}|_0^8 = 54.8571.$ $\bar{x} = \frac{M_y}{m} = \frac{32}{7} = 4.5714;$ $\bar{y} = \frac{M_x}{m} = 0.8$

5. $m = \rho \int_0^5 \int_0^{\sqrt{x+4}} dy \, dx = \rho \int_0^5 \sqrt{x + 4} \, dx = \rho \cdot \frac{2}{3}(x + 4)^{3/2}|_0^5 = \left(18 - \frac{16}{3}\right)\rho = \frac{38}{3}\rho = 12.6667\rho.$
$M_x = \rho \int_0^5 \int_0^{\sqrt{x+4}} y \, dy \, dx = \rho \int_0^5 \frac{y^2}{2}|_0^{\sqrt{x+4}} dx = \rho \int_0^5 \frac{x+4}{2} dx = \rho \left[\frac{x^2}{4} + 2x\right]_0^5 = \frac{65}{4}\rho = 16.25\rho.$
$M_y = \rho \int_0^5 \int_0^{\sqrt{x+4}} x \, dy \, dx = \rho \int_0^5 xy|_0^{\sqrt{x+4}} dx = \rho \int_0^5 x\sqrt{x + 4} \, dx.$ Let $u = x + 4$ and $du = dx.$ Then $x = u - 4,$ $x = 0 \Rightarrow u = 4,$ and $x = 5 \Rightarrow u = 9.$ Substituting, we obtain $\rho \int_4^9 (u - u)\sqrt{4} \, du = \rho \int_4^9 \left(u^{3/2} - 4u^{1/2}\right) = \rho \left[\frac{2}{5}u^{5/2} - \frac{8}{3}u^{3/2}\right]_4^9 = 33.7333\rho.$ $\bar{x} = \frac{M_y}{m} = 2.6632,$ $\bar{y} = \frac{M_x}{m} = 1.2829.$

7. $m = \rho \int_0^2 \int_{x^2}^{4x} dy \, dx = \rho \int_0^2 \left(4x - x^2\right) dx = \left(2x^2 - \frac{x^3}{3}\right)|_0^2 \rho = \left(8 - \frac{8}{3}\right)\rho = \frac{16}{3}\rho = 5.3333\rho.$
$M_x = \rho \int_0^2 \int_{x^2}^{4x} y \, dy \, dx = \rho \int_0^2 \frac{y^2}{2}|_{x^2}^{4x} dx = \rho \int_0^2 \times \left(8x^2 - \frac{x^4}{2}\right) dx = \rho \left[\frac{8}{3}x^3 - \frac{x^5}{10}\right]_0^2 = 18.1333\rho.$
$M_y = \rho \int_0^2 \int_{x^2}^{4x} x \, dy \, dx = \rho \int_0^2 x \left(4x - x^2\right) dx = \rho \int_0^2 \left(4x^2 - x^3\right) dx = \rho \left[\frac{4}{3}x^3 - \frac{x^4}{4}\right]_0^2 = 6.6667 = 6\frac{2}{3}\rho.$ $\bar{x} = \frac{M_y}{m} = 1.25.$ $\bar{y} = \frac{M_x}{m} = 3.4$

9. $m = \rho = \int_0^1 \int_{x^{3/2}}^{x} dy \, dx = \rho \int_0^1 \left(x - x^{3/2}\right) dx = \rho \left[\frac{x^2}{2} - \frac{2}{5}x^{5/2}\right]_0^1 = \rho \left(\frac{1}{2} - \frac{2}{5}\right) = \frac{1}{10}\rho.$ Thus, we have $M_x = \rho \int_0^1 \int_{x^{3/2}}^{x} y \, dy \, dx = \rho \int_0^1 \frac{y^2}{2}|_{x^{3/2}}^{x} dx = \rho \int_0^1 \left(\frac{x^2}{2} - \frac{x^3}{2}\right) dx = \rho \left[\frac{x^3}{6} - \frac{x^4}{8}\right]_0^1 = \frac{1}{24}.$ We also have $M_y = \rho = \int_0^1 \int_{x^{3/2}}^{x} x \, dy \, dx = \rho \int_0^1 \left(x^2 - x^{5/2}\right) dx = \rho \left[\frac{x^3}{3} - \frac{2}{7}x^{7/2}\right]_0^1 = \left(\frac{1}{3} - \frac{2}{7}\right)\rho = \frac{1}{21}\rho.$ Hence, $\bar{x} = \frac{M_y}{m} = \frac{1/21}{1/10} = \frac{10}{21} \approx 0.4762,$ and $\bar{y} = \frac{M_x}{m} = \frac{1/24}{1/10} = \frac{5}{12} \approx 0.4167.$

11. $y = x$ and $y = 12 - x^2$ intersect at $x = -4$ and $x = 3.$ First, we determine that $m = \rho \int_{-4}^3 \int_x^{12-x^2} \times dy \, dx = \rho \int_{-4}^3 \left(12 - x - x^2\right) dx = \rho \left(12x - \frac{x^2}{2} - \frac{x^3}{3}\right)_{-4}^3 = 57\frac{1}{6}\rho.$ Then, we have $M_x = \rho \int_{-4}^3 \int_x^{12-x^2} \times y \, dy \, dx = \rho \int_{-4}^3 \frac{y^2}{2}|_z^{12-x^2} dx = \frac{\rho}{2} \int_{-4}^3 (144 - 24x^2 + x^4 - x^2) dx = \frac{\rho}{2} \left[144x - \frac{25x^3}{3} + \frac{x^5}{5}\right]_{-4}^3 = \frac{(260.1+276.8)}{2}\rho = 251.5333\rho.$ We also have $M_y = \rho \int_{-4}^3 \int_x^{12-x^2} x \, dy \, dx = \rho \int_{-4}^3 \left[12x - x^3 - x^2\right] dx$

$= \rho \left[6x^2 - \frac{x^3}{3} - \frac{x^4}{4} \right]_{-4}^{3} = 28.5833\rho.$ Hence, $\bar{x} =$

$\frac{-28.5833}{57.1667} = -0.5$ and $\bar{y} = \frac{251.5333}{57.1667} = 4.4.$

13. Here $I_y = \rho \int_0^{\sqrt{2}} \int_{x^2}^{2} x^2 \, dy \, dx = \rho \int_0^{\sqrt{2}} x^2 y|_{x^2}^{2} dx =$

$\rho \int_0^{\sqrt{2}} (2x^2 - x^4) \, dx = \rho \left[\frac{2x^3}{3} - \frac{x^5}{5} \right]_0^{\sqrt{2}} =$

$\left(\frac{4}{3}\sqrt{2} - \frac{4}{5}\sqrt{2} \right) \rho = \frac{8\sqrt{2}}{15}\rho = 0.7542\rho.$ We also

find that $m = \rho \int_0^{\sqrt{2}} \int_{x^2}^{2} dy \, dx = \rho \int_0^{\sqrt{2}} (2 - x^2) \, dx$

$= \rho \left[2x - \frac{x^3}{3} \right]_0^{\sqrt{2}} = \rho \frac{4\sqrt{2}}{3}.$ Hence, $r_y = \sqrt{\frac{\frac{8}{15}}{\frac{4}{3}}} =$

$\sqrt{\frac{2}{5}} \approx 0.6325.$

15. $m = \rho \int_0^3 \int_0^5 dy \, dx = \rho \int_0^3 5 \, dx = 15;$ $I_x =$

$\rho \int_0^3 \int_0^5 y^2 dy \, dx = \rho \int_0^3 \frac{y^3}{3}|_0^5 dx = \rho \int_0^3 \frac{125}{3} \, dx =$

$125\rho;$ $r_x = \sqrt{\frac{125}{15}} = \sqrt{\frac{25}{3}} \approx 2.8868.$

17. $m = \rho \int_0^4 \int_0^{4x-x^2} dy \, dx = \rho \int_0^4 4x - x^2 \, dx =$

$\rho \left[2x^2 - \frac{x^3}{3} \right]_0^4 = \left(32 - \frac{64}{3} \right) \rho = \frac{32}{3}\rho;$ $I_x = \rho \int_0^4 \times$

$\int_0^{4x-x^2} y^2 dy \, dx = \rho \int_0^4 \frac{y^3}{3}|_0^{4x-x^2} dx = \frac{\rho}{3} \int_0^4 (64x -$

$48x^4 + 12x^5 - x^6) \, dx = \frac{\rho}{3} \left[16x^4 - \frac{48}{5}x^5 - 2x^6 - \right.$

$\left. \frac{x^6}{5} \right]_0^4 \approx 39.010$ or $\frac{4096}{105};$ $r_x = \sqrt{\frac{39.010}{10.667}} = 1.9124.$

19. $m = \rho \int_1^2 \int_{1/x}^{x^2} dy \, dx = \rho \int_1^2 [x^2 - x^1] \, dx = \rho \times$

$\left[\frac{x^3}{3} - \ln x \right]_1^2 = 1.6402\rho;$ $I_y = \rho \int_1^2 \int_{1/x}^{x^2} x^2 dy \, dx =$

$\rho \int_1^2 (x^4 - x) \, dx = \left[\frac{x^5}{5} - \frac{x^2}{2} \right]_1^2 \rho \left(\frac{32}{5} - 2 - \frac{1}{5} + \frac{1}{2} \right)$

$\times \rho = 4.7\rho;$ $r_y = \sqrt{\frac{4.7}{1.6402}} \approx 1.6928$

21. (a) $I_0 = \rho \int_0^{\sqrt{2}} \int_{x^2}^{2} (x^2 + y^2) \, dy \, dx$

$= \rho \int_0^{\sqrt{2}} \left[x^2 y + \frac{1}{3}y^3 \right]_{x^2}^{2} dx$

$= \rho \int_0^{\sqrt{2}} \left[2x^2 + \frac{8}{3} - x^4 - \frac{1}{3}x^6 \right] dx$

$= \rho \left[\frac{2}{3}x^3 + \frac{8}{3}x - \frac{1}{5}x^5 - \frac{1}{21}x^7 \right]_0^{\sqrt{2}}$

$= \rho \left[\frac{4\sqrt{2}}{3} + \frac{8\sqrt{2}}{3} - \frac{4\sqrt{2}}{5} - \frac{8\sqrt{2}}{21} \right]$

$= \frac{296\sqrt{2}}{105}\rho$

(b) As in Exercise 13, $m = \frac{4\sqrt{2}}{3}\rho,$ so $r_0\sqrt{\frac{I_0}{m}} =$

$\sqrt{\frac{296\sqrt{2}}{105} \cdot \frac{3}{4\sqrt{2}}} = \sqrt{\frac{74}{35}} \approx 1.45.$

23. (a) $I_0 = \rho \int_0^4 \int_0^{4x-x^2} (x^2 + y^2) \, dy \, dx$

$= \rho \int_0^4 \left[x^2 y + \frac{1}{3}y^3 \right]_0^{4x-x^2} dx$

$= \rho \int_0^4 \left[x^2 (4x - x^2) + \frac{1}{3} (4x - x^2)^3 \right] dx$

$= \rho \int_0^4 \left[4x^3 - x^4 + \frac{1}{3} \right.$

$\left. \times (64x^3 - 48x^4 + 12x^5 - x^6) \right] dx$

$= \rho \int_0^4 \left[\frac{76}{3}x^3 - 17x^4 + 4x^5 - \frac{1}{3}x^6 \right] dx$

$= \rho \left[\frac{19}{3}x^4 - \frac{17}{5}x^5 + \frac{2}{3}x^6 - \frac{1}{21}x^7 \right]_0^4$

$= \frac{9472}{105}\rho \approx 90.21\rho$

(b) As in Exercise 17, $m = \frac{32}{3}\rho,$ so $r_0 = \sqrt{\frac{I_0}{m}} =$

$\sqrt{\frac{9472}{105} \cdot \frac{3}{32}} = \sqrt{\frac{296}{35}} \approx 2.91$

CHAPTER 30 REVIEW

1. A plane whose intercepts are (6, 0, 0), (0, 2, 0), and (0, 0, 3)

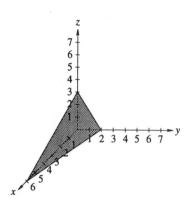

2. A plane whose intercepts are (3, 0, 0), (0, −4, 0) and (0, 0, −6)

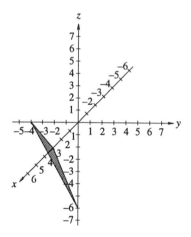

3. An elliptical cylinder whose axis is the x-axis, intercepts are (0, ±1, 0), (0, 0, ±2)

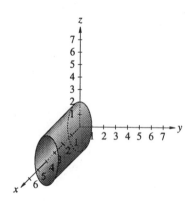

4. Sphere of radius 4.

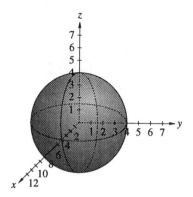

5. A parabolic cylinder whose axis is the y-axis and has the trace in the xy-plane of $y = 4x^2$

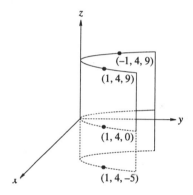

6. A hyperbolic cylinder whose axis is the x-axis and has the trace in the xy-plane of $9x^2 - 4y^2 = 1$.

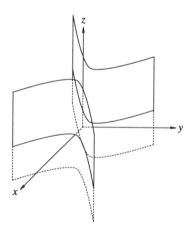

7. Hyperboloid of one-sheet. The trace in the yz-plane is the ellipse $\frac{x^2}{16} + \frac{y^2}{4} = 1$. The trace in the yz-plane is the hyperbola $\frac{y^2}{4} - \frac{z^2}{16} = 1$ and the trace in the xz-plane is the hyperbola $\frac{x^2}{16} - \frac{z^2}{16} = 1$.

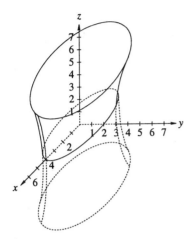

8. An elliptic cone

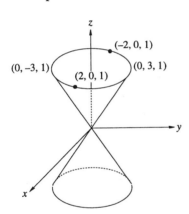

9. $z = 3x^2 + 6xy^3$; $\frac{\partial z}{\partial x} = 6x + 6y^3$; $\frac{\partial z}{\partial y} = 18xy^2$;

$\frac{\partial^2 z}{\partial x^2} = 6$; $\frac{\partial^2 z}{\partial y^2} = 36xy$; $\frac{\partial^2 z}{\partial x \partial y} = 18y^2$

10. $z = \frac{x^2 + y^2}{y} = x^2 y^{-1} + y$; $\frac{\partial z}{\partial x} = \frac{2x}{y}$; $\frac{\partial z}{\partial y} = \frac{-x^2}{y^2} + 1 = \frac{y^2 - x^2}{y^2}$; $\frac{\partial^2 z}{\partial x^2} = \frac{2}{y}$; $\frac{\partial^2 z}{\partial y^2} = \frac{2x^2}{y^3}$; $\frac{\partial^2 z}{\partial x \partial y} = \frac{-2x}{y^2}$

11. $z = e^x \cos y - e^y \sin x$; $\frac{\partial z}{\partial x} = e^x \cos y - e^y \cos x$;

$\frac{\partial z}{\partial y} = -e^x \sin y - e^y \sin x$; $\frac{\partial^2 z}{\partial x^2} = e^x \cos y + e^y \sin x$;

$\frac{\partial^2 z}{\partial y^2} = -e^x \cos y - e^y \sin x$; $\frac{\partial^2 z}{\partial x \partial y} = -e^x \sin y - e^y \cos x$

12. $z = \ln \sqrt{x^2 + y^3} + \sin^2(3xy)$;

$\frac{\partial z}{\partial x} = \frac{2x}{2(x^2 + y^3)} + 2\sin(3xy) \cdot \cos 3xy \cdot 3y$

$= \frac{x}{x^2 + y^3} + 6y \sin(3xy) \cos(3xy)$

$\frac{\partial z}{\partial y} = \frac{3y^2}{2(x^2 + y^3)} + 6x \sin 3xy \cos 3xy$

$\frac{\partial^2 z}{\partial x^2} = \frac{x^2 + y^2 - 2x^2}{(x^2 + y^3)^2} + 18y^2 \cos^2 3xy$
$\quad - 18y^2 \sin^2 3xy$

$= \frac{y^3 - x^2}{(x^2 + y^3)^2} + 18y^2(\cos^2 3xy - \sin^2 3xy)$

$= \frac{y^3 - x^2}{(x^2 + y^3)^2} + 18y^2 \cos 6xy$

$\frac{\partial^2 z}{\partial y^2} = \frac{2(x^2 + y^3)6y - 3y^2 \cdot 2 \cdot 3y^2}{4(x^2 + y^3)^2}$
$\quad + 18x^2 (\cos^2 3xy - \sin^2 3xy)$

$= \frac{6x^2 y - 3y^4}{2(x^2 + y^3)^2} + 18x \cos 6xy$

$\frac{\partial^2 z}{\partial x \partial y} = \frac{12xy^2}{4(x^2 + y^3)^2} + 6 \sin 3xy \cos 3xy$
$\quad + 18xy \cos^2 3xy - 18xy \sin^2 3xy$

$= \frac{3xy^2}{(x^2 + y^3)^2} + 6 \sin 3xy \cos 3xy$
$\quad + 18xy \cos 6xy$

13. $\int_0^2 \int_0^{x^3} xy \; dy \; dx = \int_0^2 \frac{xy^2}{2}\Big|_0^{x^3} dx = \int_0^2 \frac{x^7}{2} dx =$

$\frac{x^8}{18}\Big|_0^2 = 16$

14. $\int_0^{\pi/2} \int_0^{\pi} \sin x \cos y \; dy \; dx$

$= \int_0^{\pi/2} \sin x [\cos \pi - \cos 0] dx$

$= -2 \int_0^{\pi/2} \sin x \; dy = -2[\cos x]_0^{\pi/2}$

$= -2\left[\cos\left(\frac{\pi}{2}\right) - \cos 0 \right] = 2$

15. $\int_0^1 \int_0^4 (x+y)^{1/2} dy \; dx = \int_0^1 \frac{2}{3}(x+y)^{3/2}\Big|_0^4 dx =$

$\frac{2}{3} \int_0^1 \left[(x+4)^{3/2} - x^{3/2} \right] dx =$

$\frac{2}{3}\left[\frac{2}{5}(x+4)^{5/2} - \frac{2}{5}x^{5/2} \right]_0^1 = \frac{4}{15}\left(5^{5/2} - 1 - 4^{5/2} \right) =$

$\frac{4}{15}\left(5^{5/2} - 33 \right) \approx 6.1071$.

16. $\int_0^4 \int_0^9 xy\sqrt{x^2 + y^2} \; dy \; dx$: Let $u = x^2 + y^2$, and then $du = 2y \; dy$. When $y = 0 \Rightarrow u = x^2$, and when $y = 9 \Rightarrow u = x^2 + 9^2$. Substitution yields $\frac{1}{2} \int_0^4 \int_{x^2}^{x^2+9^2} x(u)^{1/2} du = \frac{1}{2} \int_0^4 \frac{2}{3}xu^{3/2}\Big|_{x^2}^{x^2+9^2} dx = \frac{1}{2} \int_0^4 (x(x^2 + 9^2)^{3/2} - x^4) dx = \frac{1}{3}\left[\frac{1}{2} \cdot \frac{2}{5}(x^2+9^2)^{5/2} - \frac{x^5}{5}\right]_0^4 = \frac{1}{15}\left[(4^2 + 9^2)^{5/2} - 4^5 - 9^5 \right] = 2172.9935$.

17. $\int_0^{\ln 4} \int_0^{\ln 10} e^{x+2y} dy\, dx = \int_0^{\ln 4} \frac{1}{2} e^{x+2y} \Big|_0^{\ln 10} dx =$

$\int_0^{\ln 4} \left(\frac{1}{2} e^{x+2\ln 10} - \frac{1}{2} e^x \right) dx = \int_0^{\ln 4} 49\frac{1}{2} e^x\, dx =$

$49.5 e^x \Big|_0^{\ln 4} = 49.5(4-1) = 148.5.$

18. $\int_0^1 \int_0^{x^2} x\, dy\, dx = \int_0^1 x^3\, dx = \frac{x^4}{4} \Big|_0^1 = \frac{1}{4}$

19. $\left(4, \frac{\pi}{6}, 2\right), x = 4\cos\frac{\pi}{6} = 2\sqrt{3} \approx 3.4641, y =$

$4\sin\frac{\pi}{6} = 2$, and $z = 2$; rectangular: $(2\sqrt{3}, 2, 2)$; ρ
$= \sqrt{(2\sqrt{3})^2 + 2^2 + 2^2} = \sqrt{12+4+4} = \sqrt{20}$
$= 2\sqrt{5} \approx 4.4721.$ $\theta = \tan^{-1} \frac{2}{2\sqrt{3}} = \frac{\pi}{6}, \phi =$

$\cos^{-1} \frac{2}{2\sqrt{5}} = 1.1071$; spherical $\left(2\sqrt{5}, \frac{\pi}{6}, 1.1071 \right)$

20. $x = 9\cos\frac{\pi}{6} = -\frac{9\sqrt{3}}{2} \approx -7.7942, y = 9\sin\frac{5\pi}{6} =$

$\frac{9}{2} = 4.5; z = -3$; rectangular: $(-7.7942, 4.5, -3)$.

$\rho = \sqrt{\left(\frac{9\sqrt{3}}{2}\right)^2 + \left(\frac{9}{2}\right)^2 + 3^2} = 9.4868; \theta = \tan^{-1}$

$\left(-\frac{1}{\sqrt{3}} \right) = \frac{5\pi}{6}, \phi = \cos^{-1} \frac{-3}{9.4868} = 1.8925$; spher-

ical: $\left(9.4868, \frac{5\pi}{6}, 1.8925 \right)$.

21. $x = 4\sin\frac{\pi}{6}\cos\frac{3\pi}{4} = 2 \cdot \cos\frac{3\pi}{4} = -\sqrt{2}, y =$

$4\sin\frac{\pi}{6}\sin\frac{3\pi}{4} = \sqrt{2}, z = 4\cos\frac{\pi}{6} = 2\sqrt{3}$; rectan-

gular: $(-\sqrt{2}, \sqrt{2}, 2\sqrt{3}).$ $r = \sqrt{(-\sqrt{2})^2 + \sqrt{2}^2} =$

$2, \theta = \tan^{-1}\frac{\sqrt{2}}{-\sqrt{2}} = \frac{3\pi}{4}$; cylindrical: $\left(2, \frac{3\pi}{4}, 2\sqrt{3} \right).$

22. $x = 5\sin\frac{2\pi}{3}\cos\frac{7\pi}{6} = 5 \cdot \frac{\sqrt{3}}{2} \cdot \frac{-\sqrt{3}}{2} = \frac{-15}{4} =$

$-3.75; y = 5\sin\frac{2\pi}{3}\sin\frac{7\pi}{6} = 5 \cdot \frac{\sqrt{3}}{2} \cdot \left(\frac{-1}{2}\right) =$

$\frac{-5\sqrt{3}}{4} \approx -2.1651; z = 5\cos\frac{2\pi}{3} = 5\left(\frac{-1}{2}\right) =$

-2.5; rectangular: $(-3.75, -2.1651, -2.5).$ $r =$

$\sqrt{(3.75)^2 + (-2.1651)^2} = 4.3301, \theta = \theta = \frac{7\pi}{6}, z$

$= z = -2.5$; cylindrical: $\left(4.3301, \frac{7\pi}{6}, -2.5 \right).$

23. $\overline{f} = \frac{1}{4 \cdot 4} \int_0^4 \int_{-2}^2 \left(x^2 + y^4 \right) dy\, dx$

$= \frac{1}{16} \int_0^4 \left[x^2 y + \frac{1}{5} y^5 \right]_{-2}^2 dx$

$= \frac{1}{16} \int_0^4 \left[4x^2 + \frac{64}{5} \right] dx$

$= \frac{1}{16} \left[\frac{4}{3} x^3 + \frac{64}{5} x \right]_0^4$

$= \frac{16}{3} + \frac{16}{5} = \frac{128}{15} \approx 8.53$

24. $\overline{f} = \frac{1}{4 \cdot 4} \int_1^5 \int_0^4 \left(x^3 + 2y \right) dy\, dx$

$= \frac{1}{16} \int_1^5 \left[x^3 y + y^2 \right]_0^4 dx$

$= \frac{1}{16} \int_1^5 \left[4x^3 + 16 \right] dx$

$= \frac{1}{16} \left[x^4 + 16x \right]_1^5$

$= \frac{1}{16} [(625 + 80) - (1 + 16)] = \frac{688}{16} = 43$

25. $V = \ell \cdot \omega \cdot h = 300$ so $h = \frac{300}{\ell \cdot \omega}.$ $A = \ell \cdot \omega +$

$2\ell h + 2\omega h = \ell \cdot \omega + \frac{600}{\omega} + \frac{600}{\ell}; A_\ell = \omega - \frac{600}{\ell^2} = 0;$

$A_w = \ell - \frac{600}{\omega^2} = 0.$ Solving the first equation for

ω we have $\omega = \frac{600}{\ell^2}.$ Substituting this into the

second equation, we get $\ell - \dfrac{600}{\left(\frac{600}{\ell^2}\right)^2} = \ell - \frac{\ell^4}{600} = 0$

or $\ell = \frac{\ell^4}{600} \Rightarrow \ell^3 = 600$ or $\ell = \sqrt[3]{600} \approx 8.434.$

in. Now $\omega = \frac{600}{\ell^2} = \frac{600}{\sqrt[3]{600^2}} = \sqrt[3]{600} \approx 8.434$ in.

$h = \frac{300}{\sqrt[3]{600}} = \frac{300\sqrt[3]{600}}{600} = \frac{\sqrt[3]{600}}{2} \approx 4.217$ in.

26. $R = \frac{R_1 R_2}{R_1 + R_1}; \frac{\partial R}{\partial R_1} = \frac{(R_1 + R_2) R_2 - R_1 R_2}{(R_1 + R_2)^2} = \frac{R_2^2}{(R_1 + R_2)^2}.$

$\frac{\partial R}{\partial_1 R_2} = \frac{R_1^2}{(R_1 + R_2)^2}, dR_1 = 300 \times 0.01 = 3, dR_2 =$

$600 \times 0.01 = 6.$ $dR = \frac{\partial R}{\partial R_1} dR_1 + \frac{\partial R}{\partial R_2} dR_2 =$

$\frac{R_2^2}{(R_1 + R_2)^2} dR_1 + \frac{R_1^2}{(R_1 + R_2)^2} dR_2 = \frac{600^2}{(900)^2} \times 3 + \frac{300^2}{900^2} \times 6 =$
$2\ \Omega.$

27. $\frac{\partial R}{\partial V} = \frac{1}{I}, \frac{\partial R}{\partial I} = -VI^{-2}, dR = \frac{\partial R}{\partial V} dV + \frac{\partial R}{\partial I} dI =$

$\frac{1}{I} dV - \frac{V}{I^2} dI = \frac{1}{2} \times 0.2 - \frac{116 \times 0.01}{2} = -0.19\ \Omega.$

28. $V = \ell \cdot \omega \cdot h = 25$, so $h = \frac{25}{\ell \omega}.$ $A = \ell\omega +$

$2\ell h + 2\omega h = \ell\omega + \frac{50}{\omega} + \frac{50}{\ell}.$ $A_\ell = \omega - \frac{50}{\ell^2}; A_\omega =$

$\ell - \frac{50}{\omega^2}; A_\ell = A_\omega = 0; \ell - \frac{50}{\omega^2} = 0 \Rightarrow \ell =$

$\frac{50}{\omega^2}; \omega - \dfrac{50}{\left(\frac{50}{\omega^2}\right)^2} = \omega - \frac{\omega^4}{50} = 0; \omega = \frac{\omega^4}{50}; 50 =$

ω^3 or $\omega = \sqrt[3]{50} \approx 3.684$ m. In like manner,

$\ell = \sqrt[3]{50} = 3.684$ m. Since $h = \frac{25}{\ell \cdot \omega},$ we get

$h = \frac{25}{\sqrt[3]{50^2}} = \frac{25\sqrt[3]{50}}{50} = \frac{\sqrt[3]{50}}{2} \approx 1.842$ m.

29. $PV = nRT \Rightarrow V = \frac{nRT}{P}; T = \frac{PV}{nR} = \frac{4 \times 1000}{8R} =$

$\frac{500}{R}; dV = \frac{\partial V}{\partial T} dT + \frac{\partial V}{\partial P} dP = \frac{nR}{P} dT - \frac{nRT}{P^2} dP =$

$\frac{8R}{4}(0.5) - \dfrac{8R \cdot \frac{500}{R}}{4^2}(0.4) = R - 100\ \text{cm}^2/\text{min}.$

30. $V = \frac{1}{3}\pi r^2 h$; $dV = \frac{\partial V}{\partial r}dr = \frac{1}{3}\pi r^2 \cdot dh + \frac{2}{3}\pi rh dr = \frac{1}{3}\pi(180)^2 15 + \frac{2}{3}\pi(180)(270)10 = 486{,}000\pi \text{ cm}^2/\text{s}$.

31. $m = \int_1^5 \int_0^{6x} dy\, dx = \int_1^5 6x\, dx = 3x^2\big|_1^5 = 75 - 3 = 72$, $M_x = \int_1^5 \int_0^{6x} y\, dy\, dx = \int_1^5 \frac{y^2}{2}\big|_0^{6x} dx = \int_1^5 18x^2 dx = 6x^3\big|_1^5 = 744$. $M_y = \int_1^5 \int_0^{6x} x\, dy\, dx = \int_1^5 6x^2\, dx = 2x^2\big|_1^5 = 248$, $\bar{x} = \frac{M_y}{m} = \frac{248}{72} = 3.4444$; $\bar{y} = \frac{M_x}{m} = \frac{744}{72} \approx 10.3333$.

32. $m = \int_1^2 \int_1^{x^4} dy\, dx = \int_1^2 x^4\, dx = \frac{x^5}{5}\big|_1^2 = \frac{31}{5}$, $M_x = \int_1^2 \int_0^{x^4} y\, dy\, dx = \int_1^2 \frac{y^2}{2}\big|_0^{x^4} dx = \int_1^2 \frac{x^8}{2}dx = \frac{x^9}{18}\big|_1^2 = \frac{511}{18}$, $M_y = \int_1^2 \int_0^{x^4} x\, dy\, dx = \int_1^2 x^5 dx = \frac{x^6}{6}\big|_1^2 = \frac{63}{6}$, $\bar{x} = \frac{63/6}{31/5} = \frac{105}{62} \approx 1.6935$, $\bar{y} = \frac{511/18}{31/5} = \frac{2555}{558} = 4.579$

33. $m = \int_0^1 \int_{x^3}^{\sqrt[3]{x}} dy\, dx = \int_0^1 (x^{1/3} - x^3)\, dx = \frac{3}{4}x^{4/3} - \frac{1}{4}x^4\big|_0^1 = \frac{1}{2}$; $M_x = \int_0^1 \int_{x^3}^{\sqrt[3]{x}} y\, dy\, dx = \int_0^1 \frac{y^2}{2}\big|_{x^3}^{\sqrt[3]{x}} dx = \frac{1}{2}\int_0^1 (x^{2/3} - x^6)\, dx = \frac{1}{2}\left[\frac{3}{5}x^{5/3} - \frac{1}{7}x^7\right]_0^1 = \frac{1}{2} \times \left(\frac{3}{5} - \frac{1}{7}\right) = \frac{8}{35}$; $M_y = \int_0^1 \int_{x^3}^{\sqrt[3]{x}} x\, dy\, dx = \int_0^1 (x^{4/3} - x^4)\, dx = \left[\frac{4}{7}x^{7/4} - \frac{1}{5}x^5\right]_0^2 = \frac{8}{35}$; $\bar{x} = \bar{y} = \frac{8/35}{1/2} = \frac{16}{35} \approx 0.4571$

34. These two graphs intersect when $x^2 = 9 - x^2$ or $2x^2 = 9 \Rightarrow x = \pm\frac{3/\sqrt{2}}{2} = \pm\frac{3}{\sqrt{2}} \approx \pm 2.1213$. $m = \int_{-3/\sqrt{2}}^{3/\sqrt{2}} \int_{x^2}^{9-x^2} dy\, dx = \int_{-3/\sqrt{2}}^{3/\sqrt{2}} (9 - 2x^2)\, dx = 9x - \frac{2}{3}x^3\big|_{-3\sqrt{2}}^{3/\sqrt{2}} = \frac{54}{\sqrt{2}} - \frac{9}{2\sqrt{2}} = 27\sqrt{2} - 9\sqrt{2} = 18\sqrt{2}$; $M_x = \int_{-3/\sqrt{2}}^{3/\sqrt{2}} \int_{x^2}^{9-x^2} y\, dy\, dx = \int_{-3/\sqrt{2}}^{3/\sqrt{2}} \frac{y^2}{2}\big|_{x^2}^{9-x^2} dx = \frac{1}{2}\int_{-3/\sqrt{2}}^{3/\sqrt{2}} (81 - 18x^2)\, dx = \frac{1}{2}\left[81x - 6x^3\right]_{-3/\sqrt{2}}^{3/\sqrt{2}} = \frac{243}{\sqrt{2}} - \frac{81}{\sqrt{2}} = 81\sqrt{2}$. $M_y = \int_{-3/\sqrt{2}}^{3/\sqrt{2}} \int_{x^2}^{9-x^2} x\, dy\, dx = \int_{-3/\sqrt{2}}^{3/\sqrt{2}} (9x - 2x^3)\, dx = 0$. This integral is 0 since $9x - 2x^3$ is an odd function and we are integrating from $-\frac{3}{\sqrt{2}}$ to $\frac{3}{\sqrt{2}}$. $\bar{x} = 0$, $\bar{y} = \frac{81\sqrt{2}}{18\sqrt{2}} = \frac{9}{2} = 4.5$. The answer is exactly what you would expect to get due to the symmetry of the region.

35. $I_x = \rho\int_0^5 \int_0^{6x} y^2 dy\, dx = \rho\int_0^5 \frac{y^3}{3}\big|_0^{6x} dx = \rho\int_0^5 72x^3\, dx = \rho 18x^4\big|_0^5 = 11{,}250\rho$. $m = \rho\int_0^5 dy\, dx = \rho\int_0^5 6x\, dx = \rho 3x^2\big|_0^5 = 75\rho$. $r_x = \sqrt{\frac{I_x}{m}} = \sqrt{\frac{11{,}250\rho}{75\rho}} = \sqrt{150} \approx 12.2474$

36. $I_y = \rho\int_0^4 \int_0^{4-x} x^2 dy\, dx = \rho\int_0^4 (4x^2 - x^3)\, dx = \rho\left[\frac{4}{3}x^3 - \frac{x^4}{4}\right]_0^4 = 21.3333\rho$ or $\frac{64}{3}\rho$. $m = \rho\int_0^4 \int_0^{4-x} dy\, dx = \rho\int_0^4 (4 - x)dx = \rho\left[4x - \frac{x^2}{2}\right]_0^4 = 8\rho$. $r_y = \sqrt{\frac{I_y}{m}} = \sqrt{\frac{\frac{64}{3}\rho}{8}} = \sqrt{\frac{8}{3}\rho} \approx 1.6330$.

37. $I_x = \rho\int_0^1 \int_0^{x^{2/3}} y^2 dy\, dx = \rho\int_0^1 \frac{y^3}{3}\big|_0^{x^{2/3}} dx = \rho\int_0^1 \frac{x^2}{3}dx = \rho\frac{x^3}{8}\big|_0^1 = \frac{1}{9}\rho$. $m = \rho\int_0^1 \int_0^{x^{2/3}} dy\, dx = \rho\int_0^1 x^{2/3}dx = \rho\frac{3}{5}x^{5/3}\big|_0^1 = \frac{3}{5}\rho$. $r_x = \sqrt{\frac{1/9\rho}{3/5\rho}} = \sqrt{\frac{5}{27}} \approx 0.4303$

38. $I_y = \rho\int_0^1 \int_{x^3}^{\sqrt[3]{x}} y^2 dy\, dx = \rho\int_0^1 \frac{y^3}{3}\big|_{x^3}^{\sqrt[3]{x}} dx = \rho\int_0^1 \left(\frac{x}{3} - \frac{x^9}{3}\right) dx = \rho\left[\frac{x^2}{6} - \frac{x^{10}}{30}\right]_0^1 = \left(\frac{1}{6} - \frac{1}{30}\right)\rho = \frac{4}{30}\rho = \frac{2}{15}\rho$. $m = \rho\int_0^1 \int_{x^3}^{\sqrt[3]{x}} dy\, dx = \rho\int_0^1 (x^{1/3} - x^3)\, dx = \rho\left(\frac{3}{4}x^{4/3} - \frac{x^4}{4}\right)\big|_0^1 = \frac{1}{2}\rho$. $r_y = \sqrt{\frac{\frac{2}{15}\rho}{\frac{1}{2}\rho}} = \sqrt{\frac{4}{15}} \approx 0.5164$

39. **(a)** $I_0 = \rho\int_0^8 \int_0^{10x} (x^2 + y^2)\, dy\, dx$
$= \rho\int_0^8 \left[x^2 y + \frac{1}{3}y^3\right]_0^{10x} dx$
$= \rho\int_0^8 \left[10x^3 + \frac{1000}{3}x^3\right] dx$
$= \rho\int_0^8 \left[\frac{1030}{3}x^3\right] dx$
$= \rho\left[\frac{1030}{12}x^4\right]_0^8 = \frac{1{,}054{,}720}{3}\rho$

(b) $m = \rho\int_0^8 \int_0^{10x} dy\, dx = \rho\int_0^8 10x\, dx = \rho\left[5x^2\right]_0^8 = 320\rho$, so $r_0 = \sqrt{\frac{I_0}{m}} = \sqrt{\frac{1{,}054{,}720}{3 \cdot 320}} = \sqrt{\frac{3{,}296}{3}} \approx 33.15$.

40. **(a)** $I_0 = \rho\int_0^6 \int_{x^2}^{6x} (x^2 + y^2)\, dy\, dx$
$= \rho\int_0^6 \left[x^2 y + \frac{1}{3}y^3\right]_{x^2}^{6x} dx$
$= \rho\int_0^6 \left[6x^3 + 72x^3 - x^4 - \frac{1}{3}x^6\right] dx$
$= \rho\left[\frac{3}{2}x^4 + 18x^4 - \frac{1}{5}x^5 - \frac{1}{21}x^7\right]_0^6$
$= \frac{363528}{35}\rho$

(b) $m = \rho \int_0^6 \int_{x^2}^{6x} dy\ dx = \rho \int_0^6 \left(6x - x^2\right) dx =$
$\rho \left[3x^2 - \frac{1}{3}x^3\right]_0^6 = 36\rho$, so $r_0 = \sqrt{\frac{I_0}{m}} = \sqrt{\frac{363528}{35\cdot 36}} =$
$\sqrt{\frac{10{,}098}{35}} \approx 16.99$.

41. $\int_0^2 \int_0^x \left(x^2 - y\right) dy\ dx + \int_2^4 \int_0^{4-x} \left(x^2 - y\right) dy\ dx =$
$\int_0^2 \left(x^2 y - \frac{y^2}{2}\right) \big|_0^x dx + \int_2^4 \left(x^2 y - y\frac{2}{2}\right) \big|_0^{4-x} dx =$
$\int_0^2 \left(x^3 - \frac{x^2}{2}\right) dx + \int_2^4 \left(4x^2 - x^3 - 8 + 4x - \frac{x^2}{2}\right)$
$\times dx = \left[\frac{x^4}{4} - \frac{x^3}{6}\right]_0^2 + \left[\frac{4}{3}x^3 - \frac{x^4}{4} - 8x + 2x^2 - \frac{x^3}{6}\right]_2^4$
$= 4 - \frac{8}{6} + \frac{256}{3} - 64 - 32 + 32 - \frac{64}{4} - \frac{32}{3} + 4 + 16 -$
$8 + \frac{8}{6} = 16$.

42. $\int_{-\sqrt{5}}^{\sqrt{5}} \int_{x^2+x-5}^{x} (x - 4y) dy\ dx = \int_{-\sqrt{5}}^{\sqrt{5}} (xy-$
$2y^2) \big|_{x^2+x-5}^{x} dx = \int_{-\sqrt{5}}^{\sqrt{5}} \left[x^2 - 2x^2 - (x^3 + x^2 -\right.$
$\left. 5x) + 2\left(x^2 + x - 5\right)^2\right] dx = \int_{-\sqrt{5}}^{\sqrt{5}} \left[-x^3 - 2x^2 +\right.$
$\left. 5x + 2\left(x^4 + 2x^3 - 9x^2 - 10x + 25\right)\right] dx = \int_{-\sqrt{5}}^{\sqrt{5}}$
$\times \left(2x^4 + 3x^3 - 20x^2 - 15x + 50\right) dx = \left[\frac{2}{5}x^5 +\right.$
$\left. \frac{3}{4}x^4 - \frac{20}{3}x^3 - \frac{15}{2}x^2 + 50x\right]_{-\sqrt{5}}^{\sqrt{5}} = 2\left(\frac{2}{5} \cdot 25\sqrt{5} -\right.$
$\left. \frac{20}{3} \cdot 5\sqrt{5} + 50\sqrt{5}\right) = 2\left(10\sqrt{5} - \frac{100}{3}\sqrt{5} + 50\sqrt{5}\right)$
$= \frac{160}{3}\sqrt{5} \approx 119.2570$ (Note: to evaluate from
$-\sqrt{5}$ to $\sqrt{5}$, we can double the odd terms and
cancel the even.)

CHAPTER 30 TEST

1. A plane whose intercepts are $(4, 0, 0)$, $(0, -8, 0)$, and $(0, 0, 2)$

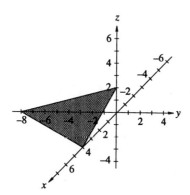

2. Sphere of radius 3

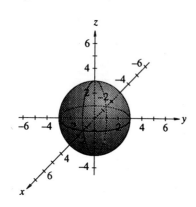

3. Hyperboloid of one-sheet. The trace in the xy-plane is the ellipse $\frac{x^2}{4} + \frac{y^2}{9} = 1$. The trace in the yz-plane is the hyperbola $\frac{y^2}{9} - \frac{z^2}{4} = 1$ and the trace in the xz-plane is the hyperbola $\frac{x^2}{4} - \frac{z^2}{4} = 1$.

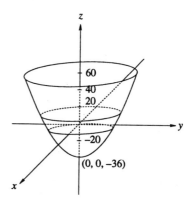

4. Elliptic paraboloid with intercepts $(\pm 2, 0, 0)$, $(0, \pm 3, 0)$, and $(0, 0, -36)$

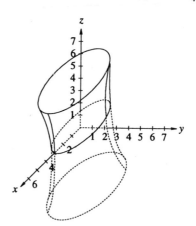

5. Treating y as a constant and differentiating z with respect to x, produces $\frac{\partial z}{\partial x} = 12x^2 - 10xy^4$. Treating x as a constant and differentiating z with respect to y, produces $\frac{\partial z}{\partial y} = -20x^2y^3$

6. Rewriting z as $z = \left(x^2 + 4y^3\right)^{1/2} + \ln\left(x^2 y\right)$, treating y as a constant and differentiating z with respect to x, produces $\frac{\partial z}{\partial x} = \frac{1}{2}\left(x^2 + 4y^3\right)^{-1/2}(2x) + \frac{2xy}{x^2 y}$ which simplifies to $\frac{\partial z}{\partial x} = x\left(x^2 + 4y^3\right)^{-1/2} + \frac{2}{x}$. Treating y as a constant and differentiating $\frac{\partial z}{\partial x}$ with respect to x produces $\frac{\partial^2 z}{\partial x^2} = x\left[-\frac{1}{2}\left(x^2 + 4y^3\right)^{-3/2}(2x)\right] + \left(x^2 + 4y^3\right)^{-1/2} - \frac{2}{x^2} = -x^2\left(x^2 + 4y^3\right)^{-3/2} + \left(x^2 + 4y^3\right)^{-1/2} - \frac{2}{x^2}$

7.
$$\int_0^4 \int_0^{\pi/4} x \sin y \, dy \, dx = \int_0^4 \left(-x \cos y\right)\Big|_{y=0}^{y=\pi/4} dx$$
$$= \int_0^4 -x\left(\frac{\sqrt{2}}{2} - 1\right) dx$$
$$= \left(1 - \frac{\sqrt{2}}{2}\right) \int_0^4 x \, dx$$
$$= \left(1 - \frac{\sqrt{2}}{2}\right) \frac{1}{2}x^2\Big|_0^4$$
$$= 8 - 4\sqrt{2} \approx 2.3431$$

8. $\int_0^2 \int_0^{x^2} 3xy^2 \, dy \, dx = \int_0^2 xy^3\big|_0^{x^2} dx = \int_0^2 x^7 dx = \frac{1}{8}x^8\big|_0^2 = \frac{256}{8} = 32$

9. First we will convert these coordinates from cylindrical coordinates to rectangular coordinates. A point with the cylindrical coordinates (r, θ, z), has the rectangular coordinates (x, y, z), where $x = r\cos\theta$, $y = r\sin\theta$, $z = z$. We begin with $x = r\cos\theta = 3\cos\frac{5\pi}{6} = 3\left(-\frac{\sqrt{3}}{2}\right) = -\frac{3\sqrt{3}}{2}$. Next, $y = r\sin\theta = 3\sin\frac{5\pi}{6} = 3\left(\frac{1}{2}\right) = \frac{3}{2}$. Finally, $z = z = 8$. So, the cylindrical coordinates $\left(3, \frac{5\pi}{6}, 8\right)$ are equivalent to the rectangular coordinates $\left(-\frac{3\sqrt{3}}{2}, \frac{3}{2}, 8\right)$. Next, we convert these rectangular coordinates to spherical coordinates. A point with the rectangular coordinates (x, y, z) has the spherical coordinates (ρ, θ, ϕ), where $\rho = \sqrt{x^2 + y^2 + z^2}$, $\tan\theta = \frac{y}{x}$, $\cos\phi = \frac{z}{\rho}$. First, we obtain $\rho = \sqrt{x^2 + y^2 + z^2} = \sqrt{\left(-\frac{3\sqrt{3}}{2}\right)^2 + \left(\frac{3}{2}\right)^2 + 8^2} = \sqrt{\frac{27}{4} + \frac{9}{4} + 64} = \sqrt{73}$. Next we determine $\tan\theta = \frac{y}{x} = \frac{3/2}{-\frac{3\sqrt{3}}{2}} =$

$-\frac{1}{\sqrt{3}}$ and so $\theta = \frac{5\pi}{6}$. Finally, we determine $\cos\phi = \frac{z}{\rho} = \frac{8}{\sqrt{73}}$ and so $\phi = \cos^{-1}\frac{8}{\sqrt{73}} \approx 0.3588$. [The value of θ is the same in both the cylindrical and spherical coordinate systems, so we could have used the given value $\theta = \frac{5\pi}{6}$.]

10. First we will convert these coordinates from spherical coordinates to rectangular coordinates. We have $\rho = 4$, $\theta = \frac{7\pi}{6}$ and $\phi = \frac{3\pi}{4}$. Using the formulas in (3), we have $x = \rho\sin\phi\cos\theta = 4\sin\frac{3\pi}{4} \times \cos\frac{7\pi}{6} = 4\left(\frac{\sqrt{2}}{2}\right)\left(-\frac{\sqrt{3}}{2}\right) = -\sqrt{6}$; $y = \rho\sin\phi\sin\theta = 4\left(\frac{\sqrt{2}}{2}\right)\left(-\frac{1}{2}\right) = -\sqrt{2}$; and $z = \rho\cos\phi = 4\left(-\frac{\sqrt{2}}{2}\right) = -2\sqrt{2}$. So, the spherical coordinates of $\left(4, \frac{7\pi}{6}, \frac{3\pi}{4}\right)$ are, in rectangular coordinates, $(-\sqrt{6}, -\sqrt{2}, -2\sqrt{2})$.

Next, we convert these rectangular coordinates to cylindrical coordinates. A point with the rectangular coordinates (x, y, z), has the cylindrical coordinates (r, θ, z), where $r = \sqrt{x^2 + y^2}$, $\tan\theta = \frac{y}{x}$, $z = z$. First, $r = \sqrt{x^2 + y^2} = \sqrt{(-\sqrt{6})^2 + (-\sqrt{2})^2} = \sqrt{6+2} = \sqrt{8} = 2\sqrt{2}$. Next, $\tan\theta = \frac{y}{x} = \frac{-\sqrt{2}}{-\sqrt{6}} = \frac{1}{\sqrt{3}}$. Since the coordinates of x and y are in the third quadrant of the xy-plane, we have $\theta = \pi + \tan^{-1}\frac{1}{\sqrt{3}} = \frac{7\pi}{6}$. [Notice that since θ is the same in both the spherical and cylindrical coordinate systems, we could have used the given value of *theta*]. Finally, $z = z = -2\sqrt{2}$. So, the spherical coordinates of $\left(4, \frac{7\pi}{6}, \frac{3\pi}{4}\right)$ are, in cylindrical coordinates, $\left(2\sqrt{2}, \frac{7\pi}{6}, -2\sqrt{2}\right)$.

11. Let $z = 0$. Then the base of the region in the xy-plane is the ellipse $x^2 + 4y^2 = 4$ or $\frac{x^2}{4} + y^2 = 1$. Solving this for y, we see that $y = \pm\sqrt{\frac{4-x^2}{4}}$ and so the bounds for y are $-\sqrt{(4-x^2)/4} \leq y \leq \sqrt{(4-x^2)/4}$ and $-2 \leq x \leq 2$. Thus, the volume is $V = \int_{-2}^2 \int_{-\sqrt{(4-x^2)/4}}^{\sqrt{(4-x^2)/4}} \left(4 - x^2 - 4y^2\right) dy \, dx = \int_{-2}^2 \left[4y - x^2 y - \frac{4}{3}y^3\right]_{-\sqrt{(4-x^2)/4}}^{\sqrt{(4-x^2)/4}} dx = \frac{2}{3}\int_{-2}^2 \left(4 - x^2\right)^{3/2} dx$. To integrate this, we use Formula #28, Appendix C, and obtain $\frac{2}{3}\left[\frac{x}{4}\left(x^2 - 10\right)\sqrt{4 - x^2} + \frac{3}{2}\sin^{-1}\frac{x}{2}\right]_{-2}^2 \approx 12.5664 = 4\pi$

12. The two functions intersect at the points $(-1, 4)$ and $(3, 4)$ as shown in Figure 30.0. So, we have $f(x) = 4$, $g(x) = x^2 - 2x + 1$, $a = -1$ and $b = 3$. The area of this region is $A = \int_{-1}^{3} \int_{x^2-2x+1}^{4} dy \, dx$ $= \int_{-1}^{3} y| \int_{x^2-2x+1}^{4} dx = \int_{-1}^{3} (3 + 2x - x^2) \, dx =$ $3x + x^2 - \frac{1}{3}x^3 \Big]_{-1}^{3} = 9 - \left(-\frac{5}{3}\right) = \frac{32}{3}$. So, the

mass is $\frac{32}{3}\rho$. Next, we find the first moments of this region. $M_y = \rho \int_{-1}^{3} \int_{x^2-2x+1}^{4} y \, dy \, dx =$ $\rho \int_{-1}^{3} (3x + 2x^2 - x^3) \, dx = \rho \left[\frac{3}{2}x^2 + \frac{2}{3}x^3 - \frac{1}{4}x^4\right]_{-1}^{3} = \frac{32}{3}\rho$ and $M_x = \rho \int_{-1}^{3} \int_{x^2-2x+1}^{4} y \, dy \, dx$ $= \rho \int_{-1}^{3} \int_{(x-1)^2}^{4} y \, dy \, dx = \rho \int_{-1}^{3} \frac{y^2}{2}\Big|_{(x-1)^2}^{4} dx =$ $\rho \int_{-1}^{3} \frac{1}{2} \left[16 - (x-1)^4\right] dx = \frac{\rho}{2} \left[16x - \frac{1}{5}\times\right.$ $\left.(x-1)^5\right]_{-1}^{3} = 25.6$. So, $\bar{x} = \frac{M_y}{m} = \frac{32}{3} / \frac{32}{3} = 1$ and $\bar{y} = \frac{M_x}{m} = 25.6/\frac{32}{3} = 2.4$. The centroid is $(\bar{x}. \bar{y}) = (1, 2.4)$.

13. $\bar{f} = \frac{1}{3 \cdot 4} \int_{-1}^{2} \int_{0}^{4} x^2 y^3 dy \, dx$

$= \frac{1}{12} \int_{-1}^{2} \left[\frac{1}{4}x^2 y^4\right]_0^4 dx = \frac{1}{12} \int_{-1}^{2} 64x^2 dx$

$= \frac{1}{12} \left[\frac{64}{3}x^3\right]_{-1}^{2} = 16$

14. **(a)** $I_0 = \rho \int_{0}^{2} \int_{x}^{6-x^2} (x^2 + y^2) \, dy \, dx$

$= \rho \int_{0}^{2} \left[x^2 y + \frac{1}{3}y^3\right]_x^{6-x^2} dx$

$= \rho \int_{0}^{2} \left[6x^2 - x^4 + \frac{1}{3}(216 - 108x^2\right.$

$\left. +18x^4 - x^6) - x^3 - \frac{1}{3}x^3\right] dx$

$= \rho \int_{0}^{2} \left[72 - 30x^2 - \frac{4}{3}x^3 + 5x^4 - \frac{1}{3}x^6\right] dx$

$= \rho \left[72x - 10x^3 - \frac{1}{3}x^4 + x^5 - \frac{1}{21}x^7\right]_0^2$

$= \frac{592}{7}\rho$

(b) $m = \rho \int_0^2 \int_x^{6-x^2} dy \, dx = \rho \int_0^2 (6 - x^2 - x) dx =$ $\rho \left[6x - \frac{1}{3}x^3 - \frac{1}{2}x^2\right]_0^2 = \frac{22}{3}\rho$, so $r_0 = \sqrt{\frac{I_0}{m}} =$ $\sqrt{\frac{592}{7} \cdot \frac{3}{22}} = \sqrt{\frac{888}{77}} \approx 3.40$.

CHAPTER

31

Infinite Series

≡ 31.1 MACLAURIN SERIES

1. We first find the derivatives as $f(x) = \sin x$, $f'(x) = \cos x$, $f''(x) = -\sin x$, $f'''(x) = -\cos x$, $f^{(4)}(x) = \sin x$, $f^{(5)}(x) = \cos x$, $f^{(6)}(x) = -\sin$ and $f^{(7)}(x) = -\cos x$. Then, we evaluate f and its derivatives at $x = 0$, with the following results: $f(0) = 0$, $f'(0) = 1$, $f''(0) = 0$, $f'''(0) = -1$, $f^{(4)}(0) = 0$, $f^{(5)}(0) = 1$, $f^{(6)}(0) = 0$, and $f^{(7)} = -1$. Hence, $a_0 = 0$, $a_1 = 1$, $a_2 = 0$, $a_3 = -1$, $a_4 = 0$, $a_4 = 0$, $a_5 = 1$, $a_6 = 0$, and $a_7 = -1$. Using these results we get $\sin x = x - \frac{x^3}{3!} + \frac{x^5}{5!} - \frac{x^7}{7!}$.

3. Finding the derivatives of $\cosh(x) = f(x)$, we obtain $f'(x) = \sinh x$, $f''(x) = \cosh x$, etc. Evaluating f and its derivatives at $x = 0$ produces $f(0) = 1$, $f'(0) = 0$, $f''(0) = 1$, $f'''(0) = 0$, $f^{(4)}(0) = 1$, etc. Thus we obtain the desired result $\cosh(x) = 1 + \frac{x^2}{2!} + \frac{x^4}{4!} + \frac{x^6}{6!}$.

5. Here $f(x) = e^{3x}$, $f'(x) = 3e^{3x}$, $f''(x) = 9e^{3x}$, and $f'''(x) = 27e^{3x}$. Thus, $f(0) = 1$, $f'(0) = 3$, $f''(0) = 9$, and $f'''(0) = 27$. Using these results, we obtain the Maclaurin series $e^{3x} = 1 + 3x + \frac{9x^2}{2!} + \frac{27x^3}{3!} = 1 + 3x + \frac{9}{2}x^2 + \frac{9}{2}x^3$.

7. Here we have

$$f(x) = \ln\left(1 + x^2\right), \qquad\qquad f(0) = 0$$

$$f'(x) = \frac{2x}{(x^2 + 1)}, \qquad\qquad f'(0) = 0$$

$$f''(x) = \frac{-2\left(x^2 - 1\right)}{\left(x^2 + 1\right)^2}, \qquad\qquad f''(0) = 2$$

$$f'''(x) = 4x\left(x^2 - 3\right)/\left(x^2 + 1\right)^3, \qquad\qquad f'''(0) = 0$$

$$f^{(4)}(x) = \frac{-12\left(x^4 - 6x^2 + 1\right)}{\left(x^2 + 1\right)^4}, \qquad\qquad f^{(4)}(0) = -12$$

$$f^{(5)}(x) = \frac{48x\left(x^4 - 10x^2 + 5\right)}{\left(x^2 + 1\right)^3}, \qquad\qquad f^{(5)}(0) = 0$$

$$f^{(6)}(x) = \frac{-240\left(x^6 - 15x^4 + 15x^2 - 1\right)}{\left(x^2 + 1\right)^6}, \qquad\qquad f^{(6)}(0) = 240$$

$$f^{(7)}(x) = \frac{1440\left(x^7 - 21x^5 + 35x^3 - 7x\right)}{\left(x^2 + 1\right)^7}, \qquad\qquad f^{(7)}(0) = 0$$

$$f^{(8)}(x) = \frac{-10{,}080\left(x^8 - 28x^6 + 70x^4 - 28x^2 + 1\right)}{\left(x^2 + 1\right)^8}, \qquad\qquad f^{(8)}(0) = -10{,}080$$

Hence, we have $\ln\left(1 + x^2\right) = \frac{2x^2}{2!} + \frac{12x^4}{4!} + \frac{240x^6}{6!} + \frac{-10{,}080x^8}{8!} = x^2 - \frac{x^4}{2} + \frac{x^6}{3} - \frac{x^8}{4} = x^2 - \frac{1}{2}x^4 + \frac{1}{3}x^6 - \frac{1}{4}x^8$.

9.   This time we have

$$f(x) = \cos x^2, \qquad\qquad\qquad\qquad\qquad\qquad f(0) = 1$$
$$f'(x) = -2x \sin x^2, \qquad\qquad\qquad\qquad\qquad\quad f'(0) = 0$$
$$f''(x) = -4x^2 \cos x^2 - 2 \sin x^2, \qquad\qquad\qquad f''(0) = 0$$
$$f'''(x) = 8x^3 \sin x^2 - 12x \cos x^2, \qquad\qquad\qquad f'''(0) = 0$$
$$f^{(4)}(x) = 16x^4 \cos x^2 - 12 \cos x^2 + 48x^2 \sin x^2, \qquad f^{(4)}(0) = -12$$
$$f^{(5)}(x) = 160x^3 \cos x^2 - 32x^5 \sin x^2 + 120x \sin x^2, \qquad f^{(5)}(0) = 0$$
$$f^{(6)}(x) = -64x^6 \cos x^2 + 720x^2 \cos x^2 - 480x^4 \sin x^2,$$
$$\qquad\qquad + 120 \sin x^2, \qquad\qquad\qquad\qquad\qquad\quad f^{(6)}(0) = 0$$
$$f^{(7)}(x) = -1{,}344x^5 \cos x^2 + 1{,}680x \cos x^2 + 128x^7 \sin x^2$$
$$\qquad\qquad - 3{,}360x^3 \sin x^2, \qquad\qquad\qquad\qquad\quad f^{(7)}(0) = 0$$
$$f^{(8)}(x) = 256x^8 \cos x^2 - 13{,}440x^4 \cos x^2 + 1{,}680 \cos x^2$$
$$\qquad\qquad + 3{,}584x^2 \sin x^2 - 13{,}440x^2 \sin x^2, \qquad\quad f^8(0) = 1680$$
$$f^{(9)}(x) = -512x^9 \sin x^2 + 9{,}216x^7 \cos x^2 + 48{,}384x^5 \sin x^2$$
$$\qquad\qquad - 80{,}640x^3 \cos x^2 - 30240x \sin x^2 \qquad\qquad f^{(9)}(0) = 0$$
$$f^{(10)}(x) = -1{,}024x^{10} \cos x^2 - 23{,}040x^8 \sin x^2 + 161{,}280x^6 \cos x^2$$
$$\qquad\qquad + 403{,}200x^4 \sin x^2 - 302{,}400x^2 \cos x^2 - 30{,}240 \sin x^2, \qquad f^{(10)}(0) = 0$$
$$f^{(11)}(x) = 2{,}048x^{11} \sin x^2 - 56{,}320x^9 \cos x^2 - 506{,}880x^7 \sin x^2$$
$$\qquad\qquad + 1774080x^5 \cos x^2 + 2{,}217{,}600x^3 \sin x^2 - 665{,}280x \cos x^2, \qquad f^{(11)}(0) = 0$$
$$f^{(12)}(x) = 4{,}096x^{12} \cos x^2 + 135{,}168x^{10} \sin x^2 - 1{,}520{,}640x^8 \cos x^2$$
$$\qquad\qquad - 7{,}096{,}320x^6 \sin x^2 + 13{,}305{,}600x^4 \cos x^2$$
$$\qquad\qquad + 7{,}983{,}360x^2 \sin x^2 - 665{,}280 \cos x^2, \qquad\qquad f^{(12)}(0) = -665{,}280$$

Hence, $\cos x^2 = 1 - \frac{12x^4}{4!} + \frac{1680x^8}{8!} - \frac{665{,}280x^{12}}{12!} = 1 - \frac{1}{2}x^4 + \frac{1}{24}x^8 - \frac{1}{720}x^{12} = 1 - \frac{1}{2!}x^4 + \frac{1}{4!}x^8 - \frac{1}{6!}x^{12}$.

11.   Here we have

$$f(x) = e^x \sin x, \qquad\qquad\qquad\qquad f(0) = 0$$
$$f'(x) = e^x \sin x + e^x \cos x, \qquad\qquad f'(0) = 1$$
$$f''(x) = e^x \sin x + e^x \cos x + e^x \cos x - e^x \sin x$$
$$\qquad = 2e^2 \cos x, \qquad\qquad\qquad\qquad f''(0) = 2$$
$$f'''(x) = 2e^x \cos x - 2e^x \sin x, \qquad\quad f'''(0) = 2$$
$$f^{(4)}(x) = 2e^x \cos x - 2e^x \sin x - 2e^x \sin x - 2e^x \cos x$$
$$\qquad = -4e^x \sin x, \qquad\qquad\qquad\quad f^{(4)}(0) = 0$$
$$f^{(5)}(x) = -4e^x \sin x - 4e^x \cos x, \qquad f^{(5)}(0) = -4$$

Thus, $e^x \sin x = x + \frac{2x^2}{2!} + \frac{2x^3}{3!} - \frac{4x^5}{5!} = x + x^2 + \frac{1}{3}x^3 - \frac{x^5}{30}$.

13.   This time we have

$$f(x) = x^2 e^{-x^2}, \qquad\qquad\qquad\qquad\qquad\qquad\qquad\qquad f(0) = 0$$
$$f'(x) = \left(2x - 2x^3\right)e^{-x^2}, \qquad\qquad\qquad\qquad\qquad\quad f'(0) = 0$$
$$f''(x) = \left(4x^4 - 10x^2 + 2\right)e^{-x^2}, \qquad\qquad\qquad\quad f''(0) = 2$$
$$f'''(x) = \left(-8x^5 + 36x^3 - 24x\right)e^{-x^2}, \qquad\qquad\quad f'''(0) = 0$$
$$f^{(4)}(x) = \left(16x^6 - 112x^4 + 156x^2 - 24\right)e^{-x^2}, \qquad f^{(4)}(0) = -24$$
$$f^{(5)}(x) = \left(-32x^7 + 320x^5 - 760x^3 + 360x\right)e^{-x^2}, \qquad f^{(5)}(x) = 0$$
$$f^{(6)}(x) = \left(64x^8 - 864x^6 + 3120x^4 - 3000x^2 + 360\right)e^{-x^2}, \qquad f^{(6)}(0) = 360$$
$$f^{(7)}(x) = \left(-128x^9 + 2240x^7 - 11424x^5 + 18{,}480x^3 - 6720x\right)e^{-x^2}, \qquad f^{(7)}(0) = 0$$
$$f^{(8)}(x) = \left(256x^{10} - 5632x^8 + 38{,}528x^6 - 94{,}080x^4 + 68{,}880x^2 - 6720\right)e^{-x^2}, \qquad f^{(8)}(0) = -6720$$

Hence, $x^2 e^{-x^2} = \frac{2x^2}{2!} - \frac{24x^4}{4!} + \frac{360x^6}{6!} - \frac{6720x^8}{8!} = x^2 - x^4 + \frac{1}{2}x^6 - \frac{1}{6}x^8 = x^2\left(1 - x^2 + \frac{1}{2}x^4 - \frac{1}{6}x^6\right)$.

15. Here we have

$$f(x) = x\sin 3x, \qquad\qquad f(0) = 0$$
$$f'(x) = 3x\cos 3x + \sin 3x, \qquad f'(0) = 0$$
$$f''(x) = 6\cos 3x - 9x\sin 3x, \qquad f''(0) = 6$$
$$f'''(x) = -27x\cos 3x - 27\sin 3x, \qquad f'''(0) = 0$$
$$f^{(4)}(x) = 81x\sin 3x- = 108\cos 3x, \qquad f^{(4)}(0) = -108$$
$$f^{(5)}(x) = 243x\cos 3x + 405\sin 3x, \qquad f^{(5)}(0) = 0$$
$$f^{(6)}(x) = 1458\cos 3x - 729x\sin 3x, \qquad f^{(6)}(0) = 1458$$
$$f^{(7)}(x) - 5103\sin 3x - 2187x\sin 3x \qquad f^{(7)}(0) = 0$$
$$f^{(8)}(x) = -17{,}496\cos 3x - 6561x\cos 3x \qquad f^{(8)}(0) = -17{,}496$$

Hence, we have $x\sin 3x = \frac{6x^2}{2!} - \frac{108x^4}{4!} + \frac{1458x^6}{6!} - \frac{17{,}496x^8}{8!} = 3x^2 - \frac{9}{2}x^4 + \frac{81}{40}x^6 - \frac{243}{560}x^8$.

17. As in example 31.5, let $f(t) = A'(t)$. Hence $f(t) = \frac{-60}{t^2+30}$, then $f(0) = -2$; $f'(t) = \frac{120t}{(t^2+30)^2}$, then $f'(0) = 0$, and $f''(t) = \frac{120(t^2+30)^2 - 120t(2)(t^2+30)2t}{(t^2+30)^4} = \frac{120t^2 + 120\cdot 30 - 480t^2}{(t^2+30)^3}$, then $f''(0) = \frac{4}{30} = \frac{2}{15}$. This gives the first two terms of the Maclaurin series for $A'(t)$ as $A'(t) = -2 + \frac{2}{15}\frac{t^2}{2} = -2 + \frac{t^2}{15}$. Thus, $A(t) = \int A'(t)dt = \int\left(-2 + \frac{t^2}{15}\right)dt = -2t + \frac{t^3}{45} + C$. Since the initial wound was 12cm^2, $C = 12$. Hence, $A(t) = 12 - 2t + \frac{t^3}{45}$, and so $A(2) = 12 - 2\cdot 2 + \frac{2^3}{45} = 12 - 4 + \frac{8}{45} = 8\frac{8}{45} \approx 8.178$ cm^2.

19.
$$E(x) = \sec x; \qquad\qquad E(0) = 1$$
$$E'(x) = \sec x\tan x; \qquad\qquad E'(0) = 0$$
$$E''(x) = \sec^3 x + \sec x\tan^2 x; \qquad E''(0) = 1$$
$$E'''(x) = 3\sec^3 x\tan x + \sec x\tan^3 x$$
$$\qquad + 2\sec^3 x\tan x$$
$$\qquad = 5\sec^3 x\tan x + \sec x\tan^3 x \qquad E'''(0) = 0$$
$$E^{(4)}(x) = 15\sec^3 x\tan^2 x + 5\sec^5 x$$
$$\qquad + \sec x\tan^4 x + 3\sec^3 x\tan^2 x; \qquad E^{(4)}(0) = 5$$
$$E(x) = 1 + \frac{1}{2!}x^2 + \frac{5}{4!}x^4 = 1 + \frac{1}{2}x^2 + \frac{5}{24}x^4.$$

☰ 31.2 OPERATIONS WITH SERIES

1. Since $e^x = 1 + x + \frac{x^2}{2!} + \frac{x^3}{3!}$, we see that $e^{3x} = 1 + 3x + \frac{(3x)^2}{2} + \frac{(3x)^3}{3!} = 1 + 3x + \frac{9x^2}{2} + \frac{9x^3}{2}$.

3. Since $\cos x = 1 - \frac{x^2}{2!} + \frac{x^4}{4!} - \frac{x^6}{6!}$, we can find $\cos\frac{x}{2} = 1 - \frac{x^2}{8} + \frac{x^4}{16\cdot 4!} - \frac{x^6}{2^6 6!} = 1 - \frac{x^2}{8} + \frac{x^4}{384} - \frac{x^6}{46{,}080}$

5. Since $\cos x = 1 - \frac{x^2}{2!} + \frac{x^4}{4!} - \frac{x^6}{6!}$, and so $\cos x^3 = 1 - \frac{x^6}{2!} + \frac{x^{12}}{4!} - \frac{x^{18}}{6!}$.

7. $\sin 2x^2 = 2x^2 - \frac{(2x^2)^3}{3!} + \frac{(2x^2)^5}{5!} - \frac{(2x^2)^7}{7!} = 2x^2 - \frac{4}{3}x^6 + \frac{2^5 x^{10}}{5!} - \frac{2^7 x^{14}}{7!} = 2x^2 - \frac{8x^6}{3!} + \frac{32x^{10}}{5!} - \frac{128x^{14}}{7!}$

9. $\sin x = x - \frac{x^3}{3!} + \frac{x^5}{5!} - \frac{x^7}{7!} + \frac{x^9}{9!} + \cdots$, and hence $\frac{d}{dx}\sin x = 1 - \frac{x^2}{2!} + \frac{x^4}{4!} - \frac{x^6}{6!} + \frac{x^8}{8!} + \cdots$, which is $\cos x$.

11. $e^{2x} = 1 + 2x + \frac{(2x)^2}{2!} + \frac{(2x)^3}{3!} + \frac{(2y)^4}{4!} = 1 + 2x + \frac{4x^2}{2!} + \frac{8x^3}{3!} + \frac{16x^4}{4!} + \cdots$. Hence, $\frac{d}{dx}e^{2x} = 2 + 4x + \frac{8x^2}{2!} + \frac{16x^3}{3!} + \cdots = 2\left(1 + 2x + \frac{4x^2}{2!} + \frac{8x^3}{3!} + \cdots\right) = 2\left(1 + 2x + \frac{(2x)^2}{2!} + \frac{(2x)^3}{3!} + \cdots\right) = 2e^{2x}$.

13. $\int_0^1 \sin x^2 dx \approx \int_0^1\left(x^2 - \frac{x^6}{3!} + \frac{x^{10}}{5!}\right)dx = \left[\frac{x^3}{3} - \frac{x^3}{7\cdot 3!} + \frac{x^{11}}{11\cdot 5!}\right]_0^1 = \frac{1}{3} - \frac{1}{42} + \frac{1}{1320} \approx 0.3103$.

15. $\int_0^{0.2}\sin\sqrt{x}\ dx = \int_0^{0.2}\left(x^{1/2} - \frac{x^{3/2}}{3!} + \frac{x^{5/2}}{5!}\right)dx = \left[\frac{2}{3}x^{3/2} - \frac{2x^{5/2}}{5\cdot 31} + \frac{2x^{7/2}}{7\cdot 5!}\right]_0^{0.2} = \frac{2}{3}(0.2)^{3/2} - \frac{1}{15}(0.2)^{5/2} + \frac{2}{7\cdot 5!}(0.2)^{7/2} \approx 0.05844$.

17. We know that $e^{-x} = 1 - x + \frac{x^2}{2!} - \frac{x^3}{3!} + \frac{x^4}{4!}$ and that $\cos x = 1 - \frac{x^2}{2!} + \frac{x^4}{4!} - \frac{x^6}{6!}$. Multiplying produces $e^{-x}\cos x = \left(1 - x + \frac{x^2}{2!} - \frac{x^3}{6} + \frac{x^4}{4!}\right)\left(1 - \frac{x^2}{2} + \frac{x^4}{4!} - \right.$

$\cdots) - 1 - x - \frac{x^2}{2} + \frac{x^2}{2} + \frac{x^3}{2} - \frac{x^3}{6} + \frac{x^4}{24} - \frac{x^4}{4} + \frac{x^4}{4!} = $

$1 - x + \frac{2x^3}{3!} - \frac{4x^4}{4!} + \cdots$.

19. $e^{-x^2} = 1 - x^2 + \frac{x^4}{2!} + \frac{x^6}{3!} + \frac{x^8}{4!} + \cdots$. Hence $x^2 e^{-x^2}$

$= x^2 - x^4 + \frac{x^6}{2!} - \frac{x^8}{3!} + \frac{x^{10}}{4!} - \cdots$

21. $5e^{0.8j} = 5(\cos 0.8 + j \sin 0.8) = 3.4835 + 3.5868j$

23. $6 \operatorname{cis} \frac{4\pi}{3} = 6e^{\frac{4\pi}{3}j} = 6e^{4\pi j/3}$

25. $y = e^{x^2} = 1 + x^2 + \frac{x^4}{2} + \cdots$. Hence, we obtain

$A = \int_0^1 \left(1 + x^2 + \frac{x^4}{2}\right) dx = \left[x + \frac{x^3}{3} + \frac{x^5}{10}\right]_0^1 = $

$1 + \frac{1}{3} + \frac{1}{10} = \frac{43}{30} = 1.4333$.

27. $x^2 e^2 = x^2 + x^3 + \frac{x^4}{2}$. Integrating, we obtain

$\int_0^{0.2} \left(x^2 + x^3 + \frac{x^4}{2}\right) dx = \left[\frac{x^3}{3} + \frac{x^4}{4} + \frac{x^5}{5}\right]_0^{0.2} = $

$\frac{(0.2)^2}{3} + \frac{(0.2)^4}{4} + \frac{(0.2)^5}{10} = 0.0031$.

29. $\cos 0.1 = \frac{(0.1)^2}{2!} + \frac{(0.1)^4}{4!} - \frac{(0.1)^6}{6!} + \frac{(0.1)^8}{8!} \approx 0.995004$.
This produces $15\cos(0.1) = 15 \times 0.9950 = 14.9251$.

31. **(a)** $i = \sin t^2 = t^2 - \frac{t^6}{3!} + \frac{t^{10}}{5!} - \frac{t^{14}}{7!} + \cdots$

(b) Charge $= \int_0^{0.02} \left(t^2 - \frac{t^6}{3!} + \frac{t^{10}}{5!} - \frac{t^{14}}{7!}\right) dx = $

$\left[\frac{t^3}{3} - \frac{t^7}{7 \cdot 3!} + \frac{t^{11}}{11 \cdot 5!} - \frac{t^{15}}{15 \cdot 7!}\right]_0^{0.02} = \frac{(0.02)^3}{3} - \frac{(0.02)^7}{7 \cdot 3!} + $

$\frac{(0.02)^{11}}{11 \cdot 5!} = 2.7 \times 10^{-6} = 0.0000027$.

33. **(a)** Since we know that $e^x = 1 + x + \frac{x^2}{2!} + \frac{x^3}{3!}$, then

$$e^{-2t} = 1 + (-2t) + \frac{(-2t)^2}{2!} + \frac{(-2t)^3}{3!}$$

$$= 1 - 2t + 2t^2 - \frac{4}{3}t^3$$

and $e^{-t} = 1 - t + \frac{t^2}{2} - \frac{t^3}{6}$

Hence,

$$P = P_0\left(e^{-2t} + e^{-t}\right)$$

$$= P_0\left(1 - 2t + 2t^2 - \frac{4}{3}t^3 + 1 - t + \frac{t^2}{2} - \frac{t^3}{6}\right)$$

$$= P_0\left(2 - 3t + \frac{5}{2}t^2 - \frac{3}{2}t^3\right)$$

(b) $\frac{dP}{dt} = P_0\left(-3 + 5t - \frac{9}{2}t^2\right)$.

≡ 31.3 NUMERICAL TECHNIQUES USING SERIES

1. $e^{-0.3} = 1 - 0.3 + \frac{0.3^2}{2!} - \frac{(0.3)^3}{3!} + \frac{(0.3)^4}{4!} = 1 - 0.3 + $
$0.045 - 0.0045 + 0.0003375$. Using the first four terms and rounding we get $e^{-0.3} = 0.7408$.

3. $\cos(0.1) = .1 - \frac{(0.1)^2}{2!} + \frac{(0.1)^4}{4!} - \frac{(0.1)^6}{6!} = 1 - 0.005 + $
0.00000417. Using the first two terms we get
0.9950

5. $5° = \frac{\pi}{36}$. Hence, $\sin 5° - \sin \frac{\pi}{36} = \frac{\pi}{36} - \frac{\left(\frac{\pi}{36}\right)^3}{3!} - $

$\frac{\left(\frac{\pi}{36}\right)^5}{5!} - 0.087266 - 0.00011 = 0.0872$

7. $\ln(0.97) = \ln(1 - 0.03) = -0.03 - \frac{(0.03)^2}{2} - \frac{(0.03)^3}{3}$
$- \frac{(0.03)^4}{4} = -0.03 - 0.00045 = -0.0305$

9. $\ln(0.5) = \ln(1 - 0.5) = -0.5 - \frac{(0.5)^2}{2} - \frac{(0.5)^3}{3} - $
$\frac{(0.5)^4}{4} - \cdots = -0.5 - 0.125 - 0.041667 - 0.015625 - $
$0.00625 - 0.002604 - 0.001116 - 0.000488 - $
$0.000217 - 0.000097 = -0.6931$

11. First we must find a series for $\frac{1 - \cos x}{x}$. Since
$\cos x = 1 - \frac{x^2}{2!} + \frac{x^4}{4!} \cdots$, we have then $1 - \cos x = $

$\frac{x^2}{2!} - \frac{x^4}{4!} + \frac{x^6}{6!}$, and $\frac{1 - \cos x}{x} = \frac{x}{2!} - \frac{x^3}{4!} + \frac{x^5}{6!}$. As a result, we get

$\int_0^{0.5} \frac{1 - \cos x}{x} dx$

$= \int_0^{0.5} \left(\frac{x}{2!} - \frac{x^3}{4!} + \frac{x^5}{6!} + \cdots\right) dx$

$= \frac{x^2}{4} - \frac{x^4}{4 \cdot 4!} + \frac{x^6}{6 \cdot 6!} - \frac{x^8}{8 \cdot 8!}\Big|_0^{0.5}$

$= \frac{(0.5)^2}{4} - \frac{(0.5)^4}{4 \cdot 4!} + \frac{(0.5)^6}{6 \cdot 6!} - \frac{(0.5)^8}{8 \cdot 8!}$

$= 0.0526 - 0.000651 + 0.0000036$

$= 0.0619$.

13. $\int_0^1 \frac{\sin x - x}{x^2} dx = \int_0^1 \frac{-x}{3!} + \frac{x^3}{5!} - \frac{x^5}{7!} dx = \left[\frac{-x^2}{12} + \frac{x^4}{4 \cdot 5!}\right.$
$\left. + \frac{x^6}{6 \cdot 7!}\right]_0^1 = -\frac{1}{12} + \frac{1}{480} - 0.000033 = -0.0813$.

15. $\int_0^{0.2} \frac{e^x - 1}{x}$

$= \int_0^{0.2} \left(1 + \frac{x}{2!} + \frac{x^2}{3!} + \frac{x^3}{4!} + \cdots\right) dx$

$= \left[x + \frac{x^2}{4} + \frac{x^3}{3 \cdot 3!} + \frac{x^4}{4 \cdot 4!}\right]_0^{0.2}$

$= 0.2 + 0.01 + 0.000444 + 0.0000167$

$= 0.2105$

17. **(a)** $\sin\theta = \theta - \frac{\theta^3}{3!} + \frac{\theta^4}{5!} - \cdots$, so $\sin\theta = \theta$ when $\frac{\theta^3}{3!}$ is negligible.

(b) If we want accuracy to 0.0001, then set $\frac{\theta^3}{3!} < 0.0001$ or $\theta^3 < 0.0006$ or $\theta < \sqrt[3]{0.0006} \approx 0.0843$.

19. The Maclaurin series for e^x is $1 + x + \frac{x^2}{2!} + \frac{x^3}{3!} + \cdots$, so $e^{-0.05t^2} = 1 - 0.05t^2 + \frac{0.0025t^4}{2!} - \frac{0.000125t^6}{3!} + \cdots$. Thus,

$$q = \int_0^{0.5} \left(1 - e^{-0.05t^2}\right)dt$$

$$= \int_0^{0.5} \left(0.05t^2 - \frac{0.0025t^4}{2!} + \frac{0.000125t^6}{3!} + \cdots\right)dt$$

$$= \left[\frac{0.05}{3}t^3 - \frac{0.0025t^5}{5\cdot 2!} + \frac{0.000125t^7}{7\cdot 3!} + \cdots\right]_0^{0.5}$$

$$\approx 0.002083 - .000008 + \cdots \approx 0.0021$$

This charge is about 0.0021 μC.

21. **(a)** The Maclaurin series for e^x is $1 + x + \frac{x^2}{2!} + \frac{x^3}{3!} + \cdots$, so the desired Maclaurin series is $V = 0.75e^{0.75t} = 0.75\left(1 + 0.75t + \frac{(0.75)^2}{2!}t^2 + \frac{(0.75)^3}{3!}t^3 + \cdots\right)$;

(b) When $t = 0.45$, this becomes $0.75(1 + 0.3375 + 0.05695 + 0.00641 + 0.00054\cdots) \approx 0.75(1.4014) \approx 1.05105$.

23. **(a)** $e^x = 1 + x + \frac{x^2}{2!} + \frac{x^3}{3!}$

$$e^{-t^2} = 1 + \left(-t^2\right) + \frac{\left(-t^2\right)^2}{2!} + \frac{\left(-t^2\right)^3}{3!}$$

$$= 1 - t^2 + \frac{1}{2}t^4 - \frac{1}{6}t^6$$

(b) $V(x) = V_0 \int_0^x e^{-t^2}dt$

$$= V_0 \int_0^x \left(1 - t^2 + \frac{1}{2}t^4 - \frac{1}{6}t^6\right)dt$$

$$= V_0 \left[t - \frac{1}{3}t^3 + \frac{1}{10}t^5 - \frac{1}{42}t^7\right]_0^x$$

$$= V_0 \left[x - \frac{1}{3}x^3 + \frac{1}{10}x^5 - \frac{1}{42}x^7\right]$$

(c) $V(1.4) = V_0\left[(1.4) - \frac{1}{3}(1.4)^3 + \frac{1}{10}(1.4)^5 - \frac{1}{42}(1.4)^7\right] \approx 0.7722V_0$.

(d) The next nonzero term in the series of e^x is $\frac{x^4}{4!}$ and so the next nonzero term in the series of e^{-t^2} is $\frac{(-t^2)^4}{4!} = \frac{t^8}{24}$ and $\int\frac{t^8}{24}dt = \frac{t^9}{9\cdot24}$. When $t = 1.4$, then $\frac{t^9}{9\cdot24}V_0 \approx 0.0957V_0$. The error is less than 0.0957 since this is an oscillating series.

≡ 31.4 TAYLOR SERIES

1. $f(x) = e^x$, $f(2) = e^2$, $f'(x) = e^x$, $f'(2) = e^2$, etc. $e^x = e^2 + e^2(x - 2) + e^2\frac{(x-2)^2}{2!} + e^2\frac{(x-2)^3}{3!} + \cdots = e^2\left[x - 1 + \frac{(x-2)^2}{2!} + \frac{(x-2)^3}{3!} + \cdots\right]$

3. $h(x) = \sin x$, $h\left(\frac{\pi}{4}\right) = \frac{\sqrt{2}}{2}$, $h'(x) = \cos x$, $h'\left(\frac{\pi}{4}\right) = \frac{\sqrt{2}}{2}$, $h''(x) = \sin x$, $h''\left(\frac{\pi}{4}\right) = -\frac{\sqrt{2}}{2}$, $h'''(x) = -\cos x$, $h'''\left(\frac{\pi}{2}\right) = -\frac{\sqrt{2}}{2}$. As a result we get

$$\sin x = \frac{\sqrt{2}}{2} = \frac{\sqrt{2}}{2}\left(x - \frac{\pi x}{4}\right) - \frac{\sqrt{2}}{2}\frac{\left(x - \frac{\pi}{2}\right)^2}{2!}$$

$$- \frac{\sqrt{2}}{2}\frac{\left(x - \frac{\pi}{4}\right)3}{3!} + \cdots$$

$$= \frac{\sqrt{2}}{2}\left[1 + \left(x - \frac{\pi}{4}\right) - \frac{\left(x - \frac{\pi}{4}\right)^2}{2!}\right.$$

$$\left. - \frac{\left(x - \frac{\pi}{4}\right)^3}{3!} + \frac{\left(x - \frac{\pi}{4}\right)^4}{4!} + \cdots\right]$$

5. $k(x) = \frac{1}{x}$, so $k(2) = \frac{1}{2}$, $k'(x) = -x^{-2}$, so $k'(2) = -\frac{1}{4} = \frac{-1}{2^2}$, $k''(x) = 2x^{-3}$, so $k''(2) = \frac{2}{2^3}$, and $k'''(x) = -6x^{-4}$, so $k'''(2) = \frac{-6}{2^4}$. Hence, we see that $\frac{1}{x} = \frac{1}{2} - \frac{(x-2)}{2^2} + \frac{2(x-2)^2}{2^32!} - \frac{6(x-2)^3}{2^43!} + \frac{4!(x-2)^4}{2^54!} - \cdots = \frac{1}{2}\left[1 - \frac{x-2}{2} + \frac{(x-2)^2}{2^2} - \frac{(x-2)^3}{2^3} + \frac{(x-2)^4}{2^4} - \cdots\right]$.

7. $f(x) = \frac{1}{(x+1)^2} = (x+1)^{-2}$, $f(0) = 1$, $f'(x) = -2(x+1)^{-3}$, $f'(0) = -2$ $f''(x) = 6(x+1)^{-4}$, $f''(0) = 6$, $f'''(x) = -24(x+1)^{-5}$, $f'''(0) = -24$. As a result, we obtain $\frac{1}{(x+1)^2} = 1 - 2(x) + \frac{6(x)^2}{2!} - \frac{24(x)^3}{3!} + \cdots = 1 - 2(x) + 3(x)^2 - 4(x)^3 + 5(x)^4 - \cdots$.

9. $h(x) = e^{-x}$, so $h(1) = e^{-1}$, and $h'(x) = -e^{-x}$, so $h'(1) = -e^{-1}$, and $h''(x) = e^{-x}$, so $h''(1) = e^{-1}\cdots$. Combining these results, we see that $e^{-x} =$

$e^{-1}e^{-1}(x-1) + e^{-1}\frac{(x-1)^2}{2!} - e^{-1}\frac{1}{3!}(x-1)^3 \cdots =$
$e^{-1}\left[1 - (x-1) + \frac{1}{2!}(x-1)^2 - \frac{1}{3!}(x-1)^3 + \cdots\right].$

11. $k(x) = \sqrt[3]{x},\ k(8) = 2,\ k'(x) = \frac{1}{3}x^{-2/3},\ k'(8) = \frac{1}{3}\cdot\frac{1}{2^2},\ k''(x) = \frac{-2}{9}x^{-5/3},\ k''(8) = \frac{-2}{9}\cdot\frac{1}{2^5} = \frac{-1}{9\cdot2^4},\ k'''(x) = \frac{10}{27}x^{-8/3}.$ Hence, we obtain $\sqrt[3]{x} = 2 + \frac{1}{3}\cdot\frac{1}{2^2}(x-8) - \frac{1}{9\cdot2^4}\cdot\frac{(x-8)^2}{2!} + \frac{10}{27}\cdot\frac{1}{2^8}\cdot\frac{(x-8)^3}{3!} + \cdots.$

13. Using Exercise #1, with $x = 2.1$, and $c = 2$, we see that $x - c = 0.2$. As a result, we obtain $e^{2.1} = e^2\left[1 + (0.1) + \frac{(0.1)^2}{2!} + \frac{(0.1)^3}{3!} + \cdots\right] = e^2[1.1052] = 8.1662.$

15. In this exercise we use the result from Exercise #3. We convert 43° to $\frac{43\pi}{180}$ radians. We have $x = \frac{\pi}{4}$ and $c = \frac{43\pi}{180}$, with the result that $x - c = \frac{43\pi}{180} - \frac{\pi}{4} = -\frac{\pi}{90}$. Thus, $\sin 43° = \sin \frac{43\pi}{180}\ \sin\left(-\frac{\pi}{90}\right) = \frac{\sqrt{2}}{2}\left[1 + \left(-\frac{\pi}{90}\right) - \frac{\left(-\frac{\pi}{90}\right)^2}{2!} - \frac{\left(-\frac{\pi}{90}\right)^3}{3!} + \frac{\left(-\frac{\pi}{90}\right)^4}{4!}\right] = \frac{\sqrt{2}}{2}[0.9645] = 0.6820.$

17. Using Exercise #5, with $x = 1.8$, $c = 2$, and $x - c = -0.2$, we get $\frac{1}{1.8} \approx \frac{1}{2}\left[1 - \frac{(-0.2)}{2} + \frac{(-0.2)^2}{2^2} - \frac{(-0.2)^3}{2^3} + \frac{(0.2)^4}{2^4}\right] = \frac{1}{2}[1 + 0.1 + 0.01 + 0.001 + 0.0001] = \frac{1}{2}[1.1111] = 0.5555$

19. $\frac{1}{1.12^2} = \frac{1}{(0.12+1)^2} = 1 - 2(0.12) + 3(0.12)^2 - 4(0.12)^3 = 1 - 0.24 + 0.0432 - 0.006912 + 0.0010368 - 0.000149 = 0.7972$

21. $e^{-0.8} \approx e^{-1}\left[1 - (-0.2) + \frac{1}{2!}(-0.2)^2 - \frac{1}{3!}(-0.2)^3\right] = e^{-1}[1.2214] = 0.4493.$

23. $\sqrt[3]{7.8} = \sqrt[3]{8 - 0.2} = 2 + \frac{1}{3}\cdot\frac{1}{2^2}(-0.2) - \frac{1}{9}\cdot\frac{1}{2^4}\frac{(-0.2)^2}{2} = 2 - 0.01667 - 0.000139 = 1.9832$

25. From Example 31.0, we know that $\frac{\sin \pi t}{t} = \pi\left(\frac{t-2}{t}\right) - \frac{\pi^3}{3!}\frac{(t-2)^3}{t} + \cdots$. When we let $t = 2.1$, we get $\frac{\sin \pi t}{t} = \pi\left(\frac{0.1}{2.1}\right) - \frac{\pi^3}{3!}\frac{(0.1)^3}{2.1} + \cdots \approx 0.14960 - (0.00246) + \cdots \approx 0.14714$. A calculator gives $\frac{\sin(2.1\pi)}{2.1} \approx 0.14715$.

≡ 31.5 FOURIER SERIES

1. $f(x) = \begin{cases} 2, & -\pi < x < 0 \\ -2, & 0 < x < \pi \end{cases}$, period 2π. $2L = 2\pi$, so $L = \pi$, First, $a_0 = \frac{1}{2\pi}\int_{-\pi}^{\pi} f(x)dx = \frac{1}{2\pi}\int_{-\pi}^0 (-2)dx + \frac{1}{2\pi}\int_0^\pi 2\,dx = \frac{1}{2\pi}[-2x]_{-\pi}^0 + \frac{1}{2\pi}[2x]_0^{-\pi} = \frac{-2\pi}{2\pi} + \frac{2\pi}{2\pi} = 0$. We have $a_k = \frac{1}{\pi}\int_{-\pi}^{\pi} f(x)\cos\frac{k\pi x}{\pi}dx = \frac{1}{\pi}\int_{-\pi}^0 2\cos kx\,dx + \int_0^\pi -2\cos kx\,dx = \frac{2}{\pi}\left[\frac{\sin kx}{d}\right]_{-\pi}^0 - \frac{2}{\pi}\left[\frac{\sin kx}{k}\right]_0^\pi = 0 - \frac{2}{k\pi}\sin(-k\pi) - \frac{2}{k\pi}\sin k\pi = 0$. We also have $b_k = \frac{1}{\pi}\int_{-\pi}^{\pi} f(x)\sin\frac{k\pi x}{\pi}dx = \frac{1}{\pi}\int_{-\pi}^0 2\sin kx\,dx + \frac{1}{\pi}\int_0^\pi = 2\sin kx\,dx = \frac{-2}{\pi}\left[\frac{\cos kx}{k}\right]_{-\pi}^0 + \frac{2}{\pi}\left[\frac{\cos kx}{k}\right]_0^\pi = \frac{-2}{k\pi}[1 - \cos k\pi] + \frac{2}{k\pi}[\cos k\pi - 1] = \frac{-4}{k\pi}(1 - \cos k\pi) = \begin{cases} 0 & \text{if } k \text{ is even} \\ \frac{-8}{k\pi} & \text{if } k \text{ is odd} \end{cases}$. Hence $f(x) = \frac{-8}{\pi}\sin x + \frac{-8}{3\pi}\sin 3x + \frac{-8}{5\pi}\sin 5x + \cdots = \frac{-8}{\pi}\left(\sin x + \frac{1}{3}\sin 3x + \frac{1}{5}\sin 5x + \cdots\right) = \frac{-8}{\pi}\sum_{k=1}^{\infty}\frac{1}{2k-1}\sin(2k-1)x.$

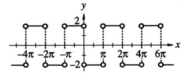

3. $h(x) = \begin{cases} 0, & -3 < x < 0 \\ 2, & 0 < x < 3 \end{cases}$, period 6 so $L = 3$. First, $a_0 = \frac{1}{6}\int_{-3}^3 h(x)dx = \frac{1}{6}\int_0^3 2\,dx = 1$. We have $a_k = \frac{1}{3}\int_{-3}^3 h(x)\cos\frac{k\pi x}{3}dx = \frac{1}{3}\int_0^3 2\cos\frac{k\pi x}{3} \times dx = \frac{2}{3}\frac{3}{k\pi}\left[\sin\frac{k\pi x}{3}\right]_0^3 = \frac{2}{k\pi}\sin k\pi = 0$ for all k. We also have $b_k = \frac{1}{3}\int_{-3}^3 h(x)\sin\frac{k\pi x}{3}dx = \frac{1}{3}\int_0^3 2\sin\frac{k\pi x}{3}dx = \frac{2}{3}\cdot\frac{3}{k\pi}\left[-\cos\frac{k\pi x}{3}\right]_0^3 = \frac{2}{k\pi}[-\cos k\pi + 1] = \begin{cases} 0 & \text{if } k \text{ is even} \\ \frac{4}{k\pi} & \text{if } k \text{ is odd} \end{cases}$. Hence, $h(x) = 1 + \frac{4}{\pi}\sin\frac{\pi x}{3} + \frac{4}{3\pi}\sin\frac{3\pi x}{3} + \frac{4}{5\pi}\sin\frac{5\pi x}{3} + \cdots.$

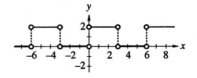

5. $f(x) = \begin{cases} -x & -\pi < x < 0 \\ x, & 0 < x < \pi \end{cases}$, period $= 2\pi$ so $L = \pi$. First, $a_0 = \frac{1}{2\pi} \int_{-\pi}^{\pi} f(x)dx = \frac{1}{2\pi} \int_{-\pi}^{0} -x\, dx + \frac{1}{2\pi} \int_0^{\pi} x\, dx = \frac{1}{2\pi} \frac{-x^2}{2} \Big|_{-\pi}^0 + \frac{1}{2\pi} \frac{x^2}{2} \Big|_0^{\pi} = \frac{\pi^2}{4\pi} + \frac{\pi^2}{4\pi} = \frac{\pi}{2}$. Next, $a_k = \frac{1}{\pi} \int_{-\pi}^{\pi} f(x) \cos kx\, dx = \frac{1}{\pi} \int_{-\pi}^{0} -x \cos kx\, dx + \frac{1}{\pi} \int_0^{\pi} x \cos kx\, dx = \frac{2}{\pi} \int_0^{\pi} x \cos kx\, dx = \frac{2}{\pi} \left[\frac{\cos kx}{k^2} + \frac{x \sin kx}{k} \right]_0^{\pi} = \frac{2}{\pi k^2} [\cos k\pi - 1] = \begin{cases} 0 & \text{if } k \text{ is even} \\ \frac{-4}{k^2 \pi} & \text{if } k \text{ is odd} \end{cases}$. Also, we see that $b_k = \frac{1}{2\pi} \times \int_{-\pi}^{\pi} f(x) \sin kx = 0$ since $f(x) \sin k\pi$ is an odd function. Hence, $f(x) = \frac{\pi}{2} - \frac{4}{\pi} \cos x - \frac{4}{9\pi} \cos 3x - \frac{4}{25\pi} \cos 5\pi - \cdots = \frac{\pi}{2} - \frac{4}{\pi} \sum_{k=1}^{\infty} \frac{1}{(2k-1)^2} \cos(2k-1)x$.

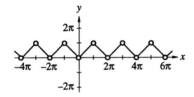

7. $h(x) = x$, $-\pi < x < \pi$, period 2π so $L = \pi$. First, $a_0 = \frac{1}{2\pi} \int_{-\pi}^{\pi} x\, dx = 0$. Next, $a_k = \frac{1}{\pi} \int_{-\pi}^{\pi} x \cos kx\, dx = 0$, since $[x \cos kx]$ is an odd function. Also, $b_k = \frac{1}{\pi} \int_{-\pi}^{\pi} x \sin kx\, dx = \frac{1}{\pi} \left[\frac{\sin kx}{k^2} - \frac{x \cos kx}{k} \right]_{-\pi}^{\pi} = \frac{1}{\pi} \left[\left(-0 - \frac{\pi \cos k\pi}{k} \right) - \left(0 + \frac{\pi \cos k\pi}{k} \right) \right] = \begin{cases} \frac{-2}{k} & \text{if } k \text{ even} \\ \frac{2}{k} & \text{if } k \text{ is odd} \end{cases}$. Hence, $h(x) = 2 \sin \pi - \frac{2}{2} \sin 2\pi + \frac{2}{3} \sin 3\pi - \cdots = 2 \sum_{k=1}^{\infty} \frac{(-1)^{k+1}}{k} \sin kx$.

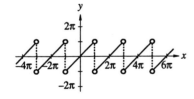

9. $f(x) = \begin{cases} 1, & 0 < x < \frac{2\pi}{3} \\ 0, & \frac{2\pi}{3} < x < \frac{4\pi}{3} \\ -1 & \frac{4\pi}{3} < x < 2\pi \end{cases}$, period $= 2\pi$, so $L = \pi$. First, $a_0 = \frac{1}{2\pi} \int_0^{2\pi} f(x)dx = \frac{1}{2\pi} \times \left[\int_0^{2\pi/3} 1\, dx + \int_{4\pi/3}^{2\pi} -1\, dx \right] = 0$. Then, we have $a_k = \frac{1}{\pi} \int_0^{2\pi} f(x) \cos kx\, dx = \frac{1}{2} \left[\int_0^{2\pi/3} \cos kx\, dx - \right.$

$\left. \int_{4\pi/3}^{2\pi} \cos kx\, dx \right] = \frac{1}{k\pi} \left[\sin kx \Big|_0^{2\pi/3} - \sin kx \Big|_{4\pi/3}^{2\pi} \right] = \frac{1}{\pi k} \left[\sin \frac{2k\pi}{3} - 0 - 0 + \sin \frac{4\pi k}{3} \right]$. This is always 0. Now, we have $b_k = \frac{1}{\pi} \int_0^{2\pi} f(x) \sin kx\, dx = \frac{1}{\pi} \int_0^{2\pi/3} \sin kx - \frac{1}{\pi} \int_{4\pi/3}^{2\pi} \sin kx = \frac{-1}{k\pi} \cos kx \Big|_0^{2\pi/3} + \frac{1}{k\pi} \cos kx \Big|_{4\pi/3}^{2\pi} = \frac{-1}{k\pi} \left[1 - \cos \frac{2k\pi}{3} + 1 - \cos \frac{4k\pi}{3} \right]$. Thus, $b_1 = \frac{3}{\pi}$, $b_2 = \frac{3}{2\pi}$, $b_3 = 0$, $b_4 = \frac{3}{4\pi}$, $b_5 = \frac{3}{5\pi}$, $b_6 = 0$, and we obtain $f(x) = \frac{3}{\pi} \sin x + \frac{3}{2\pi} \sin 2x + 0 \cdot \sin 3x + \frac{3}{4\pi} \sin 4x + \cdots \frac{3}{\pi} \sum_{k=1}^{\infty} \times \left(\frac{\sin(3k-2)x}{3k-2} + \frac{\sin(3k-1)x}{3k-1} \right)$.

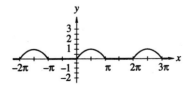

11. $h(x) = \begin{cases} 0, & \text{if } -\pi < x < 0 \\ \sin x, & \text{if } 0 < x < \pi \end{cases}$, period $= 2\pi$, so $L = \pi$. First, $a_0 = \frac{1}{2\pi} \int_{-\pi}^{\pi} h(x)dx = \frac{1}{2\pi} \int_0^{\pi} \sin x\, dx = \frac{1}{2\pi} [-\cos x]_0^{\pi} = \frac{1}{2\pi} [1+1] = \frac{1}{\pi}$. For a_1 we have $a_1 = \frac{1}{\pi} \int_0^{\pi} \sin x \cos x\, dx = \frac{1}{2\pi} \int_0^{\pi} 2 \sin x \cos x\, dx = \frac{1}{2\pi} \int_0^{\pi} \sin 2x\, dx = \left[-\frac{1}{4\pi} \cos 2x \right]_0^{\pi} = 0$. For a_k, $k > 1$, we have $a_k = \frac{1}{\pi} \int_{-\pi}^{\pi} h(x) \cos kx\, dx = \frac{1}{\pi} \int_0^{\pi} \sin x \cos kx\, dx = \frac{1}{\pi} \left[-\frac{\cos(1-k)x}{2(1-k)} - \frac{\cos(1+k)x}{2(1+k)} \right]_0^{\pi} = -\frac{1}{\pi} \left[\frac{\cos(1-k)\pi}{2(k-1)} + \frac{\cos(1+k)x}{2(1+k)} - \frac{1}{2(1-k)} - \frac{1}{2(1+k)} \right]$. If k is odd, $\cos(1-k)\pi = \cos(1+k)\pi = 1$, so $a_k = 0$. If k is even, $\cos(1-k)\pi = \cos(1+k)\pi = -1$, so $a_k = \frac{1}{\pi} \left[\frac{1}{1-k} + \frac{1}{1+k} \right] = \frac{1}{\pi} \frac{2}{1-k^2} = -\frac{2}{(k^2-1)\pi}$. In particular, $a_2 = -\frac{2}{(2^2-1)\pi} = -\frac{2}{3\pi}$, $a_4 = -\frac{2}{(4^2-1)\pi} = -\frac{2}{15\pi}$, $a_6 = -\frac{2}{35\pi}$. For b_1 we have $b_1 = \frac{1}{\pi} \int_0^{\pi} \sin^2 x\, dx = \frac{1}{\pi} \left[\frac{x}{2} - \frac{1}{4} \sin 2x \right]_0^{\pi} = \frac{1}{2}$. For $k > 1$, we have $b_k = \frac{1}{\pi} \int_0^{\pi} \sin x \sin kx\, dx = \frac{1}{\pi} \left[\frac{\sin(1-k)x}{2(1-k)} - \frac{\sin(1+k)x}{2(1+k)} \right]_0^{\pi} = 0$. Hence, $h(x) = \frac{1}{\pi} + \frac{1}{2} \sin x - \frac{2}{\pi} \left[\frac{1}{3} \cos 2x + \frac{1}{15} \cos 4x + \cdots \right] = \frac{1}{\pi} + \frac{1}{2} \sin x - \frac{2}{\pi} \sum_{k=1}^{\infty} \frac{\cos 2kx}{4k^2-1}$.

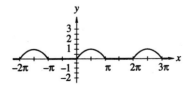

▤ CHAPTER 31 REVIEW

1. Since $\cos x = 1 - \frac{x^2}{2!} + \frac{x^4}{4!} - \frac{x^6}{6!} + \cdots$ we have $1 - \cos x = \frac{x^6}{2!} - \frac{x^4}{4!} + \frac{x^6}{6!} - \cdots$ and $\frac{1-\cos x}{x}$ equals $\frac{x}{2!} - \frac{x^3}{4!} + \frac{x^5}{6!} - \frac{x^7}{8!} + \frac{x^9}{10!} - \frac{x^{11}}{12!} + \cdots$.

2. Since $e^x = 1 + x + \frac{x^2}{2!} + \frac{x^3}{3!} + \cdots$ we get $e^{2x} = 1 - 2x + \frac{4x^2}{2!} - \frac{8x^3}{3!} + \cdots$ and so $xe^{-2x} = x - 2x^2 + \frac{4x^3}{2!} - \frac{8x^4}{3!} + \frac{16x^5}{4!} - \frac{32x^6}{5} + \cdots$

3. $\sin x \cos x = \left(x - \frac{x^3}{3!} + \frac{x^5}{5!} - \frac{x^7}{7!} + \cdots\right)\left(1 - \frac{x^2}{2!} + \frac{x^4}{4!} - \frac{x^6}{6!} + \cdots\right) = x - \frac{x^3}{2} - \frac{x^3}{3} + \frac{x^5}{4!} + \frac{x^5}{6} + \frac{x^5}{6!} - \frac{x^7}{3!4!} - \frac{x^7}{2\cdot 5!} - \frac{x^7}{6!} - \frac{x^7}{7!} + \cdots = x - \frac{5x^3}{3!} + \frac{16x^5}{5!} - \frac{64x^7}{7!} + \cdots$

4. $\ln(2 + x) = \ln(x + 1 + 1) = x + 1 - \frac{(x+1)^2}{2} + \frac{(x+1)^3}{3} - \frac{(x+1)^4}{4} + \cdots$

5. $f(x) = (1 + x)^{3/4}$, $f(0) = 1$; $f'(x) = \frac{3}{4}(1 + x)^{-1/4}$; $f'(0) = \frac{3}{4}$; $f''(x) = -\frac{3}{16}(1+x)^{-5/4}$; $f''(0) = -\frac{3}{16}$; $f'''(x) = \frac{15}{16}(1 + x)^{-9/4}$; $f'''(0) = \frac{15}{64}$. Hence $(1 + x)^{3/4} = 1 + \frac{3}{4}(1 + x) - \frac{3}{16}\frac{(1+x)^2}{2!} + \frac{15}{64}\frac{(1+x)^3}{3!} - \frac{135}{256}\frac{(1-x)^4}{4!} + \cdots$.

6. Since $e^x = 1 + x + \frac{x^2}{2!} + \frac{x^3}{3!} + \cdots$ we get $e^{-x^4} = 1 - x^4 + \frac{x^8}{2!} - \frac{x^{12}}{3!} + \frac{x^{16}}{4!} - \frac{x^{20}}{5!} + \cdots$

7. Since $\sin x = x - \frac{x^8}{3!} + \frac{x^5}{5!} - \frac{x^7}{7!} + \cdots$ we obtain $\sin x^2 = x^2 - \frac{x^6}{3!} + \frac{x^{10}}{5!} - \frac{x^{14}}{7!} + \frac{x^{18}}{9!} - \cdots$

8. $1 - e^x = -x - \frac{x^2}{2!} - \frac{x^3}{3!} - \frac{x^4}{4!} - \cdots$ and we get $\frac{1-e^x}{x} = -1 - \frac{x}{2!} - \frac{x^2}{3!} - \frac{x^3}{4!} - \frac{x^4}{5!} - \cdots$

9. First we obtain the Maclaurin series for $\frac{\sin x}{\sqrt{x}}$ which is $x^{1/2} - \frac{x^{5/2}}{3!} + \frac{x^{9/2}}{5!} - \cdots$. Now we integrate $\int_0^1 \left(x^{1/2} - \frac{x^{5/2}}{3!} + \frac{x^{9/2}}{5!} - \frac{x^{13/2}}{7!}\right) = \left[\frac{2}{3}x^{3/2} - \frac{2x^{3/2}}{7\cdot 3!} + \frac{2}{11}\frac{x^{11/2}}{5!} - \frac{2x^{15/2}}{157!}\right]_0^1 = \frac{2}{3} - \frac{2}{42} + \frac{2}{1320} - \frac{2}{75600} = 0.66667 - 0.04762 + 0.001515 - 0.000026 = 0.6205$

10. $x^3 e^{-x^3} = x^3 - x^6 + \frac{x^9}{2!} - \frac{x^{12}}{3!} + \frac{x^{15}}{4!}$. Hence, we obtain $\int_0^1 x^3 e^{-x^3} dx = \int_0^1 \left(x^3 - x^6 + \frac{x^9}{2!} - \frac{x^{12}}{3!} + \cdots\right) dx = \left[\frac{x^4}{4} - \frac{x^7}{7} + \frac{x^{10}}{20} - \frac{x^{13}}{13\cdot 13!} + \frac{x^{16}}{16\cdot 4!} - \frac{x^{19}}{19\cdot 5!} + \cdots\right]_0^1 = \frac{1}{4} - \frac{1}{7} + \frac{1}{20} - \frac{1}{78} + \frac{1}{384} - \frac{1}{2280} + \frac{1}{15840} = 0.1466$.

11. $e^x = 1 + x + \frac{x^2}{2!} + \frac{x^3}{3!} + \frac{x^4}{4!} + \cdots$. Hence, $e^{0.25} = 1 - 0.25 + \frac{0.25^2}{2!} - \frac{0.25^3}{3!} + \cdots = 1 - 0.25 + 0.03125 - 0.002604 + 0.000163 - 0.000008 = 0.7788$.

12. $\cos x^4 = 1 - \frac{x^8}{2!} + \frac{x^{16}}{4!} - \cdots$. Hence, $\int_0^1 \cos x^4 dx = \int_0^1 \left(1 - \frac{x^8}{2!} + \frac{x^{16}}{4!} - \frac{x^{24}}{6!} + \cdots\right) dx = \left(x - \frac{x^9}{18} + \frac{x^{17}}{17\cdot 4!} - \frac{x^{25}}{25\cdot 6!} + \cdots\right)\big|_0^1 = 1 - \frac{1}{18} + \frac{1}{408} - \frac{1}{18000} = 0.9468$.

13. $f(x) = \cos x$, $\cos \frac{\pi}{6} = \frac{\sqrt{3}}{2}$, $f'(x) = -\sin x$, $f'\left(\frac{\pi}{6}\right) = -\frac{1}{2}$, $f''(x) = -\cos x$, $f''\left(\frac{\pi}{6}\right) = -\frac{\sqrt{3}}{2}$, $f'''(x) = \sin x$, $f'''\left(\frac{\pi}{6}\right) = \frac{1}{2}$. Hence, we obtain $\cos x = \frac{\sqrt{3}}{2} - \frac{1}{2}\left(x - \frac{\pi}{6}\right) - \frac{\sqrt{3}}{2}\cdot\frac{\left(x - \frac{\pi}{6}\right)^2}{2!} + \frac{1}{2}\frac{\left(x - \frac{\pi}{6}\right)}{3!} + \cdots$

14. $f(x) = \sinh x$, $f(\ln 2) = \frac{3}{4}$, $f'(x) = \cosh x$, $f'(\ln(2)) = \frac{5}{4}$; $f''(x) = f(x)$, $f'''(x) = f'(x)$, etc. Hence, $\sinh x = \frac{3}{4} + \frac{5}{4}(x - \ln 2) + \frac{3}{4}\cdot\frac{(x-\ln 2)^2}{2!} + \frac{5}{4}\frac{(x-\ln 2)^3}{3!} + \cdots$

15. $f(x) = x^{1/2}$, $f(9) = 3$; $f'(x) = \frac{1}{2}x^{-1/2}$; $f'(9) = \frac{1}{6}$, $f''(x) = -\frac{1}{4}x^{-3/2}$; $f''(9) = \frac{-1}{4\cdot 27} = \frac{-1}{108}$; $f'''(x) = \frac{3}{8}x^{-5/2}$, $f'''(9) = \frac{3}{8\cdot 3^5}$. Thus, we obtain $\sqrt{x} = 3 + \frac{1}{6}(x - 9) - \frac{1}{4\cdot 3^3}\frac{(x-9)^2}{2!} + \frac{3}{8\cdot 3^5}\frac{(x-9)^3}{3!} - \cdots$.

16. $f(x) = \ln x$, $f(1) = 0$, $f'(x) = \frac{1}{x}$, $f'(1) = 1$, $f''(x) = -x^{-2}$, $f''(1) = -1$, $f'''(x) = 2x^{-3}$; $f'''(1) = 2$, $f^{(4)}(x) = -6x^{-4}$; $f^{(4)} = -6$. Thus, we get $\ln x = (x-1) - \frac{(x-1)^2}{2} + \frac{(x-1)^3}{3} - \frac{(x-1)^4}{4} + \cdots$.

17. $\cos 29° = \cos(30 - 1)^2 = \cos\left(\frac{\pi}{6} - \frac{\pi}{180}\right) = \frac{\sqrt{3}}{2} - \frac{1}{2}\left(-\frac{\pi}{180}\right) - \frac{\sqrt{3}}{4}\left(\frac{-\pi}{180}\right)^2 + \frac{1}{2}\left(\frac{-\pi}{180}\right)^3 + \cdots = 0.866025 + 0.008727 - 0.00013 = 0.8746$

18. $\sinh(\ln 1.9) = \sinh(\ln(2 - 0.1)) = \sinh\left(\frac{\ln 2}{\ln 0.1}\right)$; $\sinh(\ln 1.9) = \frac{3}{4} + \frac{5}{4}(\ln 1.9 - \ln 1) + \frac{3}{4}\frac{(\ln 1.9 - \ln 2)^2}{2} + \cdots = 0.75 - 0.06412 + 0.00099 - 0.00003 = 0.6868$.

19. $\sqrt{9.1} = 3 + \frac{1}{6}(.1) - \frac{1(0.1)^2}{4\cdot 3^2 2} = 3 + 0.016667 - 0.000046 = 3.0166$

20. $\ln(1.1) = 0.1 - \frac{0.1^2}{2} + \frac{0.1^3}{3} - \frac{0.1^4}{4} = 0.1 - 0.005 + 0.0003333 - 0.000025 = 0.09531$

21. $f(x) = \begin{cases} 1 & -\pi < x < 0 \\ 0 & 0 < x < \pi \end{cases}$, period $= 2\pi$ so $L = \pi$. First, $a_0 = \frac{1}{2\pi}\int_{-\pi}^{\pi} f(x)dx = \frac{1}{2\pi}\int_{-\pi}^{0} 1\, dx = \frac{1}{2\pi}x\big|_{-\pi}^{0} = \frac{1}{2}$. Next, we have $a_k = \frac{1}{\pi}\int_{-\pi}^{\pi} f(x)\cos kx\, dx = \frac{1}{\pi}\int_{-\pi}^{0}\cos kx\, dx = \frac{1}{k\pi}\sin kx\big|_{-\pi}^{0} = 0$. Also, $b_k = \frac{1}{\pi}\int_{-\pi}^{0}\sin kx\, dx = \frac{1}{k\pi}(-\cos kx)\big|_{-\pi}^{0} = \frac{1}{k\pi}(-1 + \cos(-k\pi)) = \begin{cases} 0, & \text{if } k \text{ is even} \\ \frac{-2}{k\pi} & \text{if } k \text{ is odd} \end{cases}$. Thus, $f(x) = \frac{1}{2} - \frac{2}{\pi}\sin x + \frac{2}{3\pi}\sin 3x - \frac{2}{5\pi}\sin 5x - \cdots = \frac{1}{2} - \frac{2}{\pi}\sum_{k=1}^{\infty}\frac{\sin(2k-1)x}{2k-1}$.

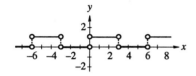

22. $g(x) = \begin{cases} 1 & -\pi < x < 0 \\ 2 & 0 < x < \pi \end{cases}$, period 2π so $L = \pi$. First, $a_0 = \frac{1}{2\pi}\int_{-\pi}^{\pi} g(x)dx = \frac{1}{2\pi}\int_{-\pi}^{0} 1\, dx + \frac{1}{2\pi}\int_{0}^{\pi} 2\, dx = \frac{1}{2\pi}(0 + \pi) + \frac{1}{2\pi}(2\pi) = \frac{1}{2} + 1 = \frac{3}{2}$. Next, $a_k = \frac{1}{\pi}\int_{-\pi}^{\pi} g(x)\cos kx\, dx = \frac{1}{\pi}\int_{-\pi}^{0}\cos hx\, dx + \frac{1}{\pi}\int_{0}^{\pi} 2\cos kx\, dx = \frac{1}{k\pi}\sin kx\big|_{-\pi}^{0} + \frac{2}{k\pi}\sin kx\big|_{0}^{\pi} = 0$. We also have $b_k = \frac{1}{\pi}\int_{-\pi}^{0}\sin kx\, dx + \frac{1}{\pi}\int_{0}^{\pi} 2\sin kx\, dx = \frac{-1}{k\pi}(\cos kx)\big|_{-\pi}^{0} + \frac{-2}{k\pi}\cos kx\big|_{0}^{\pi} = \frac{-1}{k\pi}(1 - \cos(-k\pi)) - \frac{2}{k\pi}(\cos k\pi - 1) = \begin{cases} 0 & \text{if } k \text{ is even} \\ \frac{2}{k\pi} & \text{if } k \text{ is odd} \end{cases}$. This gives $g(x) = \frac{3}{2} + \frac{2}{\pi}\sin x + \frac{2}{3\pi}\sin 3x + \cdots = \frac{3}{2} + \frac{2}{\pi}\sum_{k=1}^{\infty}\frac{\sin(2k-1)x}{2k-1}$.

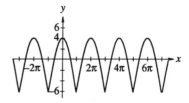

23. $h(x) = x^2$, $-\pi < x < \pi$, period 2π. First, $a_0 = \frac{1}{2\pi}\int_{-\pi}^{\pi} x^2 dx = \frac{x^3}{6\pi}\big|_{-\pi}^{\pi} = \frac{\pi^2}{3}$. Next, $a_k = \frac{1}{\pi}\int_{-\pi}^{\pi} x^2\cos kx\, dx = \frac{x^2}{\pi k}\sin kx - 2\int x\sin kx =$

$\frac{x^2}{k\pi}\sin kx + \frac{2x}{k^2\pi}\cos kx - \frac{2}{\pi}\int\cos kx = \Big[\frac{x^2}{k\pi}\sin kx + \frac{2x}{k^2\pi}\cos kx - \frac{1}{k^3\pi}\sin kx\Big]_{-\pi}^{\pi} = \begin{cases} -\frac{4}{k^2}, & k \text{ odd} \\ \frac{4}{k^2}, & k \text{ even} \end{cases}$

Also, $b_k = 0$ since $x^2\sin kx$ is an odd function. Hence, we obtain $h(x) = \frac{\pi^2}{3} - 4\cos x + \frac{4}{2^2}\cos\cos 2x - \frac{4}{3^2}\cos 3x + \frac{4}{4^2}\cos 4x - \cdots = \frac{\pi^2}{3} + 4\sum_{k=1}^{\infty}\frac{(-1)^k}{k^2}\cos kx$.

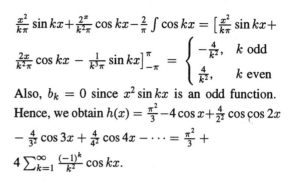

24. $j(x) = 4\cdot x^2$, $-\pi < x < \pi$, period $= 2\pi$. First, $a_0 = \frac{1}{2\pi}\int_{-\pi}^{\pi}(4 - x^2)dx = \frac{1}{2\pi}\Big(4x - \frac{x^3}{3}\Big)\big|_{-\pi}^{\pi} = \frac{1}{\pi}\Big(4\pi - \frac{\pi^3}{3}\Big) = 4 - \frac{\pi^2}{3}$. (Note: the function is even).

$a_k = \frac{1}{\pi}\int_{-\pi}^{\pi}(4 - x^2)kx\, dx$
$= \frac{1}{\pi}\int_{-\pi}^{\pi} 4\cos kx - \frac{1}{\pi}\int_{-\pi}^{\pi} x^2\cos kx\, dx$
$= \Big[\frac{4}{\pi k}\sin kx - \frac{x^2}{k\pi}\sin kx + \frac{x}{k^2\pi}\cos kx - \frac{1}{k^3\pi}\cos kx\Big]_{-\pi}^{\pi}$
$= \begin{cases} \frac{-4}{k^2} & k \text{ even} \\ \frac{4}{k^2} & k \text{ odd} \end{cases}$

$b_k = 0$ since $(4 - x^2)\sin kx$ is an odd function. Hence $j(x) = 4 - \frac{\pi^2}{3} + 4\sum_{k=1}^{\infty}\frac{(-1)^{k+1}}{k^2}\cos kx$.

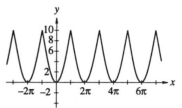

CHAPTER 31 TEST

1. Since $\sin x = x - \frac{x^3}{3!} + \frac{x^5}{5!} - \frac{x^7}{7!} + \cdots$, we get $\sin x^2 = x^2 - \frac{x^6}{3!} + \frac{x^{10}}{5!} - \frac{x^{14}}{7!} + \cdots$

2. Since $\cos x = 1 - \frac{x^2}{2!} + \frac{x^4}{4!} - \frac{x^6}{6!} + \cdots$, we obtain $\cos\sqrt{x} = 1 - \frac{x}{2!} + \frac{x^2}{4!} - \frac{x^3}{6!} + \cdots$

3. $h(x) = \frac{1}{x}$; $h(2) = \frac{1}{2}$; $h'(x) = -x^2$; $h'(2) = -\frac{1}{4}$; $h''(x) = 2x^{-3}$; $h''(2) = \frac{1}{4}$; $h'''(x) = -6x^{-4}$; $h'''(2) = \frac{-6}{2^4}$. Thus, we see that $h(x) = \frac{1}{x} = \frac{1}{2} - \frac{1}{4}(x-2) + \frac{1}{4} \cdot \frac{(x-2)^2}{2!} - \frac{6}{2^4} \cdot \frac{(x-2)^3}{3!} + \cdots = \frac{1}{2} - \frac{(x-2)}{4} + \frac{(x-2)^2}{8} - \frac{(x-2)^3}{16} + \cdots = \sum_{k=0}^{\infty} = \frac{(-1)^k (x-2)^k}{2^{k+1}}$.

4. $\frac{1}{2.1} = \frac{1}{2} - \frac{0.1}{4} + \frac{(0.1)^2}{8} - \frac{(0.1)^3}{16} = 0.5 - 0.025 + 0.00125 - 0.0000625 = 0.47619$

5. $j(x) = x^{1/3}$; $j(-1) = -1$; $j'(x) = \frac{1}{3}x^{-2/3}$; $y'(-1) = \frac{1}{3}$; $y''(x) = -\frac{2}{9}(x)^{-5/3}$; $y''(-1) = \frac{2}{9}$; $y'''(x) = \frac{10}{27}x^{-8/3}$; $y'''(-1) = \frac{10}{27}$. Hence, $j(x) = -1 + \frac{1}{3}(x+1) + \frac{2}{9} \cdot \frac{(x+1)^2}{2!} + \frac{10}{27}\frac{(x+1)^3}{3!} = -1 + \frac{(x+1)}{3} + \frac{(x+1)^2}{9} + \frac{5(x+1)^3}{81}$.

6. $\sqrt[3]{0.9} = -1 + \frac{(0.1)}{3} + \frac{(0.1)^2}{9} + \frac{5(0.1)^3}{81} = -1 + 0.03333 + 0.001111 + 0.000062 = -0.96549$

7. $f(x) = x$, $0 < x < 2\pi$, period is 2π so $L = \pi$. First, $a_0 = \frac{1}{2\pi}\int_0^{2\pi} x\, dx = \frac{1}{2\pi}\frac{x^2}{2}\Big|_0^{2\pi} = \pi$.
Next, we see that $a_k = \frac{1}{\pi}\int_0^{2\pi} x\cos kx\, dx = \frac{1}{k^2\pi}\int_0^{2\pi} kx\cos kx\, k\, dx = \frac{1}{k^2\pi}[\cos kx + kx\sin kx]_0^{2\pi} = 0$. We also determine that $b_k = \frac{1}{\pi}\int_0^{2\pi} x\sin kx\, dx = \frac{1}{k^2\pi}\int_0^{2\pi} kx\sin kx\, k\, dx = \frac{1}{k^2\pi}[\sin kx - kx\cos kx]_0^{2\pi} = \frac{1}{k^2\pi}[-2k\pi] = \frac{-2}{k}$. Hence, we obtain $f(x) = \pi - 2\left(\sin x + \frac{\sin 2x}{2} + \frac{\sin 3x}{3} + \cdots\right) = \pi - 2\sum_{k=1}^{\infty} \frac{\sin kx}{k}$.

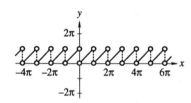

8. $g(x) = \begin{cases} -\cos x & -\pi < x < 0 \\ \cos x & 0 < x < \pi \end{cases}$, period is 2π so $L = \pi$. First, $a_0 = \frac{1}{2\pi}\int_{-\pi}^{\pi} g(x)dx = \frac{1}{2\pi}\int_{-\pi}^{0} -\cos x\, dx + \frac{1}{2\pi}\int_0^{\pi}\cos x\, dx = \frac{1}{2\pi}(-\sin x)\Big|_{-\pi}^{0} + \frac{1}{2\pi}(\sin x)\Big|_0^{\pi} = 0$. Then, we have $a_k = \frac{1}{\pi} \times \int_{-\pi}^{\pi} g(x)\cos kx\, dx = \frac{1}{\pi}\int_{-\pi}^{0} -\cos x\cos kx\, dx + \frac{1}{\pi}\int_0^{\pi}\cos x\cos kx\, dx$. These integrals add up to 0 since the cosine function is even and the product of the cosine functions is even. Hence, $\int_{-\pi}^{0} = \int_0^{\pi}$ and, since one of the integrals is negative, their sum is 0. We also have, for $k > 1$, $b_k =$

$\frac{1}{\pi}\int_{-\pi}^{0} -\cos x\sin kx\, dx + \frac{1}{\pi}\int_0^{\pi}\cos x\sin kx\, dx$. By form 72 this is $\frac{1}{\pi}\left[\frac{\cos(k-1)x}{2(k-1)} + \frac{\cos(k+1)x}{2(k+1)}\right]_{-\pi}^{0} + \frac{1}{\pi} \times \left[-\frac{\cos(k-1)x}{2(k-1)} - \frac{\cos(k+1)x}{k+1}\right]_0^{\pi} = \frac{1}{\pi}\left[\frac{-\cos(k-1)(-\pi)}{2(k-1)} - \frac{\cos(k+1)(-\pi)}{2(k+1)} + \frac{1}{2(k-1)} + \frac{1}{2(k+1)} - \frac{\cos(k-1)\pi}{2(k-1)} - \frac{\cos(k+1)\pi}{2(k+1)} + \frac{1}{2(k-1)} + \frac{1}{2(k+1)}\right]$. When k is odd, $\cos(k-1)(-\pi) = \cos(k+1)(-\pi) = \cos(k-1)\pi = \cos(k+1)\pi = 1$, and so $b_k = 0$. When k is even, $\cos(k-1)(-\pi) = \cos(k+1)(-\pi) = \cos(k-1)\pi = \cos(k+1)\pi = -1$, and so we get $b_k = \frac{1}{\pi}\left[\frac{4}{2(k-1)} + \frac{4}{2(k+1)}\right] = \frac{4k}{\pi(k^2-1)}$. In particular, $b_2 = \frac{4\cdot 2}{\pi(2^2-1)} = \frac{8}{3\pi}$, $b_4 = \frac{4\cdot 4}{(4^2-1)\pi} = \frac{16}{15\pi}$, $b_6 = \frac{24}{35\pi}$. Now, $k = 1$ yields $b_1 = \frac{1}{\pi}\int_{-\pi}^{0}(-\cos x\sin x\, dx) + \frac{1}{\pi}\int_0^{\pi}\cos x\sin x\, dx = 0$. Putting it all together, we get $g(x) = \frac{8}{\pi}\left(\frac{\sin 2x}{3} + \frac{2\sin 2x}{15} + \frac{3\sin 2x}{35} + \cdots\right) = \frac{8}{\pi}\sum_{k=1}^{\infty}\frac{k\sin 2x}{4k^2-1}$.

9. The Maclaurin series for $\cos x$ is $1 - \frac{x^2}{2!} + \frac{x^4}{4!} + \cdots$, so $\cos x^3 = 1 - \frac{(x^3)^2}{2!} + \frac{(x^3)^4}{4!} + \cdots = 1 - \frac{x^6}{2!} + \frac{x^{12}}{4!} + \cdots$. Thus, $\int_1^2 \cos x^3 dx = \int_1^2 \left(1 - \frac{x^6}{2!} + \frac{x^{12}}{4!} + \cdots\right)dx = \left[x - \frac{x^7}{7\cdot 2!} + \frac{x^{13}}{13\cdot 4!} + \cdots\right]_1^2 \approx 19.1136 - 0.9318 = 18.1818$.

10. The Maclaurin series for e^x is $1 + x + \frac{x^2}{2!} + \frac{x^3}{3!} + \cdots$, so $e^{-0.30t^2} = 1 - 0.30t^2 + \frac{0.09t^4}{2!} - \frac{0.027t^6}{3!} + \cdots$. Thus, $q = \int_0^{0.15} e^{-0.30t^2} dt = \int_0^{0.15}\left(1 - 0.30t^2 + \frac{0.09t^4}{2!} - \frac{0.027t^6}{3!} + \cdots\right)dt = \left[t - 0.10t^3 + 0.009t^5 - \frac{0.027t^7}{7\cdot 3!} + \cdots\right]_0^{0.15} \approx 0.149663\ \mu C$.

CHAPTER

32

First-Order Differential Equations

≡ 32.1 SOLUTIONS OF DIFFERENTIAL EQUATIONS

1. $y = e^x$; $y' = e^x$, $y'' = e^x$. Hence $y'' - y = e^x - e^x = 0$

3. $y = 2x^3$, $y' = 6x^2$. Hence $xy' = x6x^2 = 6x^3 = 3 \cdot 2x^3 = 3y$

5. $y = 5\cos x$; $y' = -5\sin x$. Hence $y' + y\tan x = -5\sin x + 5\cos x \tan x = -5\sin x + 5\sin x = 0$

7. $y = x^2 + 3x$; $y' = 2x + 3$; $y'' = 2$. Hence $xy' - 2y + 3xy'' - 3x = x(2x+3) - 2(x^3 + 3x) + 3x(2) - 3x = 2x^2 + 3x - 2x^2 - 6x + 6x - 3x = 0$

9. $y = 2e^x + 3e^{-x} - 4x$, $y' = 2e^x - 3e^{-x} - 4$, $y'' = 2e^x + 3e^{-x}$. Hence $y'' - y = (2e^x + 3e^{-x}) - (2e^x + 3e^{-x} - 4x) = 4x$

11. $y = 9x^2 + 6x - 54$; $y' = 18x + 6$; $y'' = 18$; $y''' = 0$

13. $y = x^2 + 4$; $y' = 2x$

15. $y = 5e^{-3x} + 2e^{2x}$, $y' = -15e^{-3x} + 4e^{2x}$, $y'' = 45e^{-3x} + 8e^{2x}$. Hence $y'' + y' - 6y = (45e^{-3x} + 8e^{2x}) + (-15e^{-2x} + 4e^{2x}) - 6(5e^{-3x} + 2e^{2x}) = (45 - 15 - 30)e^{-3x} + (8 + 4 - 12)e^{2x} = 0$

17. $y' = \int y'' dx = \int 6x^2 dx = 2x^3 + C_1$. Since $y'(1) = -2, 2(1)^3 + C_1 = 2 + C_1 = 2 + C_1 = -2$ or $C_1 = -4$ so $y' = 2x^3 - 4$. Next, we see that $y = \int y' dx = \int (2x^3 - 4) dx = \frac{x^4}{2} - 4x + C_2$. Since $y(1) = 3$, $\frac{1}{2} - 4 + C_2 = 3$ or $C_2 = 6\frac{1}{2}$. Hence $y = \frac{1}{2}x^4 - 4x + 6\frac{1}{2}$

19. $y = \int \frac{dy}{dx} dx = \int \sec^2 x \, dx = \tan x + C$. Since $y\left(\frac{\pi}{4}\right) = -6$, $\tan \frac{\pi}{4} + C = -6 \Rightarrow C = -7$ so $y = \tan x - 7$.

21. As in Example 32.0, we have $a(t) = -30$, so $v(t) = -30t + C_1$. We were given $v(0) = 65$ mph $= 95\frac{1}{3}$ ft/s and so $v(t) = -30t + 95\frac{1}{3}$ ft/s. By integrating, we obtain $s(t) = -15t^2 + 95\frac{1}{3}t + C_2$.

Since $s(0) = 0$, we see that $C_2 = 0$ and thus $s(t) = -15t^2 + 95\frac{1}{3}t$.

(a) When the car stops $v(t) = -30t + 95\frac{1}{3} = 0$. Solving for t, we see that the car will stop in about $t = 3.178$ s.

(b) Substituting $t = 3.178$ s in the equation for $s(t)$, we get $s(3.178) \approx 151.474$. The car takes about 151.474 ft to stop.

23. **(a)** The cars will strike when $s(t) = 151$. Solving $s(t) = -15t^2 + 95\frac{1}{3}t = 151$ or $15t^2 - 95\frac{1}{3}t + 151 = 0$ for t we get $t = 3$ s.

(b) When they strike $v(3) = -30(3) + 95\frac{1}{3} = 5\frac{1}{3}$.

So, the first car is going $5\frac{1}{3}$ mph.

25. **(a)** Acceleration is gravity, $g = -32$ ft/s². So, $v(t) = -32t + C_1$. Since $v(0) = 64$, we have $C_1 = 64$ and $v(t) = -32t + 64$. Thus, the ball's position is $s(t) = -16t^2 + 64t + C_2$. At $t = 0$, we are given $s(0) = 192$, so $C_2 = 192$ and $s(t) = -16t^2 + 64t + 192 = -16(t^2 - 4t - 12) = -16(t - 6)(t + 2)$. The ball will strike the ground at $t = 6$ s.

(b) At $t = 6$, $v(6) = -32(6) + 64 = -128$, so it is going 128 ft/s downward when it strikes the ground.

27. **(a)** Let $p(t)$ be the number of mg of pain reliever still in her system t hours after taking the tablet. We are given $\frac{dp(t)}{dt} = -75$ as the rate of metabolism. Integrating, we get $p(t) = -75t + C$. At $t = 0$, there are 500 mg in the system, so $C = 500$ and $p(t) = -75t + 500$.

(b) We want to find t so that $p(t) = -115t + 500 = 0$. Solving for t, we get $t = 6\frac{2}{3}$ h.

29. Let $c(t)$ be the number of mg of cortisone still in the patient's system t hours after taking the 5 mg dose. We are given $\frac{dc(t)}{dt} = -\frac{1}{2\sqrt{t}} = -\frac{1}{2}t^{-1/2}$

377

as the rate of metabolism. Integrating, produces $c(t) = -t^{1/2} + C$. At $t = 0$, there are 5 mg in the system, so $C = 5$ and $c(t) = -t^{1/2} + 5$. We want to find t so that $c(t) = -t^{1/2} + 5 = 0$. Solving for t, we get $\sqrt{t} = 5$ and so $t = 25$ h.

31. If $i = \frac{V}{R} + ke^{-Rt/L}$, then $\frac{di}{dt} = -\frac{kR}{L}e^{-Rt/L}$. Substituting this into the original differential equation, we obtain $L\left(-\frac{kR}{L}e^{-Rt/L}\right) + \left(\frac{V}{R} + ke^{-Rt/L}\right) = -kRe^{-Rt/L} + V + kRe^{-Rt/L} = V$, which was to be shown.

≡ 32.2 SEPARATION OF VARIABLES

1. $y' = \frac{dy}{dx} = \frac{2x}{y} \Rightarrow 2x\,dx - y\,dy = 0$. Integrating we get $x^2 - \frac{1}{2}y^2 = C_1$ or $\frac{1}{2}y^2 - x^2 = C_2$ or $y^2 - 2x^2 = C_3$.

3. $5x\,dx + 3y^2\,dy = 0$. Integrating we get $\frac{5}{2}x^2 + y^3 = C$

5. $5y\,dx + 3\,dy = 0$. Multiply by $I(x,y) = \frac{1}{y}$ to get $5\,dx + \frac{3}{y} = 0$. Now by integrating we get $5x + 3\ln|y| = C$

7. $4y^5\,dx - 3x^2y\,dy = 0$. Multiplying by $I(x,y) = \frac{1}{y^5x^2}$ we get $\frac{4}{x^2}\,dx - \frac{3}{y^4}\,dy = 0$. Now integration yields $-4x^{-1} + y^{-3} = C$.

9. $x^3(y^2 + 4) + yy' = 0$ is the same as $x^3(y^2 + 4)\,dx + y\,dy = 0$. Multiplying by $\frac{1}{y^2+4}$ yields $x^3\,dx + \frac{y}{y^2+4} = 0$. Now integration gives us $\frac{1}{4}x^4 + \frac{1}{2}\ln(y^2 + 4) = C_1$ or $\frac{1}{4}x^4 + \ln\sqrt{y^2 + 4} = C_1$ or $\frac{1}{2}x^4 + \ln(y^2 + 4) = C_2$.

11. $ye^{x^2}\,dy = 2x(y^2 + 4)\,dx$. The integrating factor is $I(x,y) = \frac{1}{(y^2+4)\,3^{x^2}}$. Multiplying by $I(x,y)$ yields $\frac{y}{y^2+4}\,dy = \frac{2x}{e^{x^2}}\,dx$. Now integration yields $\frac{1}{2}\ln(y^2 + 4) = -e^{-x^2} + C$ or $\ln\sqrt{y^2 + 4} = -e^{-x^2} + C$ or $\ln(y^2 + 4) = -2e^{-x^2} + C_2$.

13. $2y + e^{-3x}y' = 0 \Rightarrow 2y\,dx + e^{-3x}\,dy = 0 \Rightarrow 2e^{3x}\,dx + \frac{1}{y}\,dy = 0$. Integration yields $\frac{2}{3}e^{3x} + \ln|y| = C$

15. $(y^2 - 4)\cos x\,dx + 2y\sin x\,dy = 0$. Multiplying by the integrating factor yields $\frac{\cos x}{\sin x}\,dx + \frac{2y}{y^2-4}\,dy = 0$. Integration yields $\ln|\sin x| + \ln|y^2 - 4| = \ln C$. This simplifies to $|y^2 - 4||\sin x| = C$ or $|(y^2 - 4)\sin x| = C$.

17. $y' = \frac{x^2y - y}{y^2+3} = \frac{y(x^2-1)}{y^2+3} = \frac{dy}{dx}$. Hence $(x^2 - 1)\,dx = \frac{y^2+3}{y}\,dy$. Integrating we get $\frac{1}{3}x^3 - x = \frac{y^2}{2} + 3\ln|y| + C$ or $\frac{1}{2}y^2 + 3\ln|y| = \frac{1}{3}x^3 - x + C$.

19. $\sec 3x\,dy + y^2e^{\sin 3x}\,dx = 0$ is equivalent to $\frac{1}{y^2}\,dy + \cos 3x\,e^{\sin 3x}\,dx = 0$. Integration yields $-\frac{1}{y} + \frac{1}{3}e^{\sin 3x} = C$

21. $y' + y^2x^3 = 0$ becomes $\frac{1}{y^2}\,dy + x^3\,dx = 0$. Integration yields the general solution $-\frac{1}{y} + \frac{x^4}{4} + C$. Substituting -1 for y and 2 for x we get $1 + 4 = C$ so $C = 5$. The particular solution is $-\frac{1}{y} + \frac{1}{4}x^4 = 5$.

23. $\frac{dy}{dx} = \frac{3x+xy^2}{y+x^2y} = \frac{x(3+y^2)}{y(1+x^2)}$. This is equivalent to $\frac{y}{3+y^2}\,dy = \frac{x}{1+x^2}\,dx$. Integration yields the general solution $\frac{1}{2}\ln(3 + y^2) = \frac{1}{2}\ln(1 + x^2) + \frac{1}{2}\ln C$ or $(3+y^2)/(1 + x^2) = C$. Substituting 3 for y and 1 for x we get $\frac{12}{2} = 6 = C$. The particular solution is $\frac{3+y^2}{1+x^2} = 6$.

25. $y' = \frac{dy}{dx} = 3x^2y - 2y = y(3x^2 - 2)$. Hence we get $\frac{1}{y}\,dy = (3x^2 - 2)\,dx$. Integration yields the general solution $\ln|y| = x^3 - 2x + C$. Substituting $y = x = 1$ we get $\ln 1 = 1 - 2 + C$ or $C = 1$. The particular solution is $\ln|y| = x^3 - 2x + 1$

27. **(a)** As in Example 32.13, the general solution is $T = T_m + (T_0 - T_m)e^{-kt}$, where $T_m = 325°$F and $T_0 = 10°$F. Hence, $T = 325 + (-315)e^{-kt}$. After $t = 1$ hour, $T = 75°$F. Substituting, we solve for k

$$75 = 325 - 315e^{-k}$$
$$e^{-k} = \frac{75 - 325}{-315} \approx 0.793651$$
$$-k = \ln 0.79365$$
$$k \approx 0.231112$$

The particular solution is $T = 325 - 315e^{-0.231112t}$.

(b) Setting $T = 145$, we solve for t.

$$145 = 325 - 315e^{-0.231112t}$$
$$e^{-0.231112t} = \frac{145 - 325}{-315} \approx 0.5714286$$
$$-0.231112t = \ln 0.5714286 = -0.559616$$
$$t \approx 2.42$$

It will take about 2.42 hr $\approx$ 2 hr 25 min.

29.
$$\frac{dx}{dt} = \frac{5}{x} \text{ or } dx = 5\,dt$$
$$\int x\,dx = \int 5\,dt$$
$$\frac{1}{2}x^2 = 5t + C_1$$
$$x^2 = 10t + C$$
$$x = \sqrt{10t + C}$$

When $t = 0$ (6:00 A.M.), $x = 2$ and so, $C = 4$. We want to solve for t when $x = 6$.
$$6 = \sqrt{10t + 4}$$
$$36 = 10t + 4$$
$$10t = 32$$
$$t = 3.2$$

The ice will be 6 in. thick 3.2 hours or about 3 hr 24 min after 6:00 A.M., or at about 9:24 A.M.

31. **(a)** If i represents the number of people who are infected, then $\frac{di}{dt} = ki(500{,}000 - i)$

(b)
$$\frac{di}{dt} = ki(500{,}000 - i)$$
$$\frac{1}{i(500{,}000 - i)}di = k\,dt$$

By form #48 in Appendix C, $\int \frac{1}{i(500{,}000-i)}di = \frac{1}{500{,}000}\ln\left|\frac{i}{500{,}000-i}\right|$, so we get

$$\frac{1}{500{,}000}\ln\left|\frac{i}{500{,}000 - i}\right| = kt + C_1$$
$$\ln\left|\frac{i}{500{,}0000 - i}\right| = 500{,}000kt + C_2$$
$$\frac{i}{500{,}000 - i} = e^{(500{,}000kt + C_2)} = C_3 e^{500{,}000kt}$$
$$i = (500{,}000 - i)C_3 e^{500{,}000kt}$$
$$= 500{,}000Ce^{500{,}000kt} - iC_3 e^{500{,}000kt}$$
$$i\left(1 + C_3 e^{500{,}000kt}\right) = 500{,}000C_3 e^{500{,}000kt}$$
$$i = \frac{500{,}000C_3 e^{500{,}000kt}}{1 + C_3 e^{500{,}000kt}}$$
$$= \frac{500{,}000C_3}{e^{-500{,}000kt} + C_3}$$
$$= \frac{500{,}000}{Ce^{-500{,}000kt} + 1}$$
$$= \frac{500{,}000}{1 + Ce^{-500{,}000kt}} = \frac{500{,}000}{1 + Ce^{-Bt}}$$

(c) When $t = 0$, we know that $i = 100$, and so $100 = \frac{500{,}000}{1+Ce^{-B(0)}} = \frac{500{,}000}{1+C}$. Solving this equation for C, produces $C = 4{,}999$. Thus, $i = \frac{500{,}000}{1+4{,}999e^{-Bt}}$. When $t = 10$, we are given $i = 750$, and so $750 = \frac{500{,}000}{1+4{,}999e^{-B(10)}}$ or $750\left(1 + 4{,}999e^{-B(10)}\right) = 500{,}000$ or $(750)4{,}999e^{-10B} = 3{,}749{,}250e^{-10B} = 499{,}250$ and $e^{-10B} = \frac{499{,}250}{3{,}749{,}250}$ and $-10B = \ln\left(\frac{499{,}250}{3{,}749{,}250}\right)$ 2 which results in $B \approx 0.201620$. Thus, a particular solution is $i = \frac{500{,}000}{1+4{,}999e^{-0.201620t}}$.

(d) Evaluating $i = \frac{500{,}000}{1+4{,}999e^{-0.201620t}}$ when $t = 20$ produces $i \approx 5{,}578$. Thus, around 5,578 people will be infected after 20 days.

☰ 32.3 INTEGRATING FACTORS

1. No, $\frac{\partial(x^2-y)}{\partial y} = -1$, $\frac{\partial x}{\partial x} = 1$

3. Yes, $\frac{\partial(x^2+y^2)}{\partial y} = 2y = \frac{\partial 2xy}{\partial x}$

5. Yes, $\frac{\partial(x+y\cos x)}{\partial y} = \cos x = \frac{\partial \sin x}{\partial x}$

7. Yes, $\frac{\partial(2xy+x)}{\partial y} = 2x = \frac{\partial(y+x^2)}{\partial x}$

9. No, $\frac{\partial(x-y)}{\partial y} = -1$, $\frac{\partial(y+x)}{\partial x} = 1$

11. $y\,dx - x\,dy = 0$. By Table 32.1 we can use one of four Integrating factors. Choosing $-\frac{1}{xy}$ we get $-\frac{y\,dx - x\,dy}{xy} = 0$. Integration yields $\ln\frac{y}{x} = C_1$ or letting $C_2 = e^{C_1}$ we get $\frac{y}{x} = C_2$.

13. $y\,dx - x\,dy - 5y^4\,dy = 0$. Choose $I(x,y) = \frac{1}{y^2}$. Multiplying by $\frac{1}{y^2}$ we get $\frac{y\,dx - x\,dy}{y^2} - 5y^2\,dy = 0$. Integration yields $\frac{x}{y} - \frac{5}{3}y^3 = C$.

15. $y\,dx + y^2\,dx = x\,dy - x^2\,dx$ is equivalent to $y\,dx - x\,dy + x^2\,dx + y^2\,dx = 0$. This time we choose $I(x,y) = \frac{-1}{x^2+y^2}$ and multiplying we get $-\frac{y\,dx - x\,dy}{x^2+y^2} - \frac{x^2+y^2}{x^2+y^2}\,dx = 0$. Now integration yields $\arctan\frac{y}{x} - x = C$.

17. $y\,dx + x\,dy - 3x^3y\,dx = 0$. The only integrating factor is $\frac{1}{xy}$. Multiplying by this we get $\frac{y\,dx + x\,dy}{xy} - 3x^2\,dx = 0$. Integration yields $\ln(xy) - x^3 = C$.

19. $xy^2\,dx + x^2y\,dy = 0$. This does not fit a set form but using $I(x,y) = \frac{1}{x^2y^2}$ we get $\frac{1}{x}\,dx + \frac{1}{y}\,dy = 0$.

Integrating we get $\ln x + \ln y = C_1$ or $\ln(xy) = C_1$ or $xy = C_2$ where $C_2 = e^{C_1}$.

21. $2y\,dx + x\,dy - 3x\,dx = 0$. Using $I(x,y) = x$ we get $2xy\,dx + x^2\,dy - 3x^2\,dx = 0$. Integrating we have $x^2y - x^3 = C$.

23. $3x^2(x^2+y^2)\,dx + y\,dx - x\,dy = 0$. Using $I(x,y) = \frac{-1}{x^2+y^2}$ we get $-3x^2\,dx - \frac{y\,dx - x\,dy}{x^2+y^2} = 0$. Integration yields $-x^3 + \arctan\frac{y}{x} = C$.

25. $\tan(x^2+y^2)\,dy = x\,dx + y\,dy$. First we can multiply both sides by $\cot(x^2+y^2)$ giving us $dy = \cot(x^2+y^2)(x\,dx + y\,dy)$. Upon examination you will discover that $(x\,dx + y\,dy)$ is the derivative of x^2+y^2. Hence we can take the integral of both sides giving us $y = \ln\left|\sin(x^2+y^2)\right| + C$.

27. $2x\,dy + 2y\,dx = 3x^3y\,dx$. Multiplying by $\frac{1}{xy}$ we get $2\frac{x\,dy + y\,dx}{xy} = 3x^2\,dx$ so the general solution is $2\ln(xy) = x^3 + C$ or $\ln x^2y^2 = x^3 + C$ or $x^2y^2 = e^{x^3} + C$. Now letting $x = 1$ and $y = 3$ and solving for C we get $9 = e^1 + C$ or $C = 9 - e$. Hence the particular solution is $x^2y^2 = e^{x^3} + 9 - e$.

29. $(x^2 + y + y^2)\,dx = x\,dy$ is equivalent to $(x^2 + y^2)\,dx = x\,dy - y\,dx$. Now let $I(x,y) = \frac{-1}{x^2+y^2}$ and multiplying yield $-dx = \frac{dy\,d - x\,dy}{x^2+y^2}$. Integration yields $-x = -\arctan\frac{y}{x} + C$ or $\arctan\left(\frac{y}{x}\right) = x + C$. Now setting $y = \frac{\pi}{3}$ and $x = \frac{\pi}{3}$ we solve for C $\arctan 1 = \frac{\pi}{3} + C$ or $C = -\frac{\pi}{12}$. Hence the particular solution is $\arctan\frac{y}{x} = x - \frac{\pi}{12}$.

☰ 32.4 LINEAR FIRST-ORDER DIFFERENTIAL EQUATIONS

1. $dy + 2xy\,dx = 6x\,dx$ is equivalent to $\frac{dy}{dx} + 2xy = 6x$. The integrating factor is $I(x,y) = e^{\int 2x\,dx} = e^{x^2}$. Multiplying we get $\frac{dy}{dx}e^{x^2} + 2xye^{x^2} = 6xe^{x^2}$. Now taking the integral of both sides we get $ye^{x^2} = 3e^{x^2} + C$.

3. $y' - 2y = 4$ or $\frac{dy}{dx} - 2y = 4$. $I(x,y) = e^{-2x}$ and multiplying we get $\frac{dy}{dx}e^{-2x} - 2ye^{-2x} = 4e^{-2x}$. Integrating we get $ye^{-2x} = -2e^{-2x} + C$.

5. $\frac{dy}{dx} + \frac{y}{x} = x^2$. $I(x,y) = e^{\int x^{-1}\,dx} = e^{\ln x} = x$. Multiplying we get $\frac{dy}{dx}x + y = x^3$. Integration yields $xy = \frac{1}{4}x^4 + C$.

7. $y' - 6y = 12x$. $I(x,y) = e^{-6x}$ and multiplying we get $\frac{dy}{dx}e^{-6x} - 6ye^{-6x} = 12xe^{-6x}$. The left-hand side integrates to ye^{-6x}. The right-hand side we can integrate of parts with $u = 12x$, $du = 12\,dx$, $dv = e^{-6x}\,dx$, $v = \frac{-1}{6}e^{-6x}$. Hence

$\int 12xe^{-6x}\,dx = 12x\left(\frac{-1}{6}e^{-6x}\right) - \int \frac{-1}{6}e^{-6x} \cdot 12\,dx$
$= -2xe^{-6x} - \frac{1}{3}e^{-6x}$. Putting it all together we
get $ye^{-6x} = -\frac{1}{3}e^{-6x}(-6x - 1) + C$.

9. $\frac{dy}{dx} = e^x - \frac{y}{x}$ is equivalent to $\frac{dy}{dx} + \frac{y}{x} = e^x$.
$I(x,y) = e^{\int \frac{1}{x}dx} = e^{\ln x} = x$. Multiplying we
get $\frac{dy}{dx}x + y = xe^x$. Integration yields $xy = e^x(x - 1) + C$. (Use form 84 from Appendix C).

11. $y' - \frac{1}{x^2}y = \frac{1}{x^2}$. $I(x,y) = e^{\int \frac{-1}{x^2}dx} = e^{1/x}$. Now
multiplying by $e^{1/x}$ we get $\frac{dy}{dx}e^{1/x} - \frac{1}{x^2}ye^{1/x} = \frac{1}{x^2}e^{1/x}$. Integration yields $ye^{1/x} = -e^{1/x} + C$.

13. $y\,dy - 2y^2\,dx = y\sin 2x\,dx$. Dividing by $y\,dx$ we
get $\frac{dy}{dx} - 2y = \sin 2x$. $I(x,y) = e^{-2x}$ and multiplying we get $\frac{dy}{dx}e^{-2x} - 2ye^{-2x} = e^{-2x}\sin 2x$. Integrating using form 87 we get $ye^{-2x} = \frac{e^{-2x}}{8}(-2\sin 2x + 2\cos 2x) + C$ or $ye^{-2x} = -\frac{1}{4}e^{-2x}(\sin 2x + \cos 2x) + C$.

15. $4x\,dy - y\,dx = 8x\,dx$. Dividing by $4x\,dx$ we
get $\frac{dy}{dx} - \frac{y}{4x} = 2$. $I(x,y) = e^{-\int \frac{1}{4x}dx} = e^{-\frac{1}{4}\ln x} = e^{\ln x^{-\frac{1}{4}}} = x^{-\frac{1}{4}}$. Multiplying by this we get $\frac{dy}{dx}x^{-1/4} - \frac{y}{4x}x^{-1/4} = 2x^{-1/4}$. Now integrating we get $yx^{-1/4} = \frac{8}{3}x^{3/4} + C$.

17. $\frac{dy}{dx} - y\sec^2 x = \sec^2 x$. $I(x,y) = e^{\int -\sec^2 x\,dx} = e^{-\tan x}$. Multiplying we get $\frac{dy}{dx}e^{-\tan x} - y\sec^2 xe^{-\tan x} = \sec^2 xe^{-\tan x}$. Now integration yields $ye^{-\tan x} = -e^{-\tan x} + C$.

19. $x\,dy + (1 - 4x)y\,dx = 4x^2e^{4x}\,dx$. Dividing by
$x\,dx$ we obtain $\frac{dy}{dx} + \frac{1-4x}{x}y = 4xe^{4x}$. To get the
integrating factor we must find $\int \frac{1-4x}{x}dx = \left(\frac{1}{x} - 4\right)dx = \ln x - 4x$. Hence $I(x,y) = e^{\ln x - 4x}$. Multiplying by this we obtain the equation $\frac{dy}{dx}e^{\ln x - 4x} +$

$\frac{1-4x}{x}ye^{\ln x - 4x} = 4xe^{4x}e^{\ln x - 4x} = 4xe^{\ln x} = 4x^2$.
Now integration yields $ye^{\ln x - 4x} = \frac{4}{3}x^3 + C$. The
left side of this equation is $ye^{\ln x} \cdot e^{-4x} = yx \cdot e^{-4x} = \frac{yx}{e^{4x}}$. Hence the answer is $\frac{yx}{e^{4x}} = \frac{4}{3}x^3 + C$.

21. $\frac{dy}{dx} + \frac{2y}{x} = x$. $I(x,y) = e^{\int \frac{2}{x}} = e^{2\ln x} = e^{\ln x^2} = x^2$. Multiplying we obtain $\frac{dy}{dx}x^2 + 2yx = x^3$. Integration yields the general solution $yx^2 = \frac{1}{4}x^4 + C$. Substituting $y = 3$ and $x = 2$ we solve for C. $3 \cdot 4 = \frac{1}{4} \cdot 2^4 + C$ or $12 = 4 + C$ or $C = 8$. The
particular solution is $yx^2 = \frac{1}{4}x^4 + 8$.

23. $y' - \frac{2y}{x} = x^2\sin 3x$. The integrating factor is
$e^{\int \frac{-2}{x}} = e^{-2\ln x} = \frac{1}{x^2}$. Multiplying by $\frac{1}{x^2}$ yields
$\frac{dy}{dx} \cdot \frac{1}{x^2} - \frac{2y}{x^3} = \sin 3x$. Integrating we get the general solution $\frac{y}{x^2} = \frac{-1}{3}\cos 3x + C$. Substituting $y = \pi^2$ and $x = \pi$ we solve for C. $1 = \frac{-1}{3}\cos 3\pi + C$. $1 = \frac{1}{3} + C$ or $C = \frac{2}{3}$. The particular solution is
$yx^{-2} = -\frac{1}{3}\cos 3x + \frac{2}{3}$.

25.   **(a)** The general solution is $w(t) = \frac{2}{35}C + ke^{-0.005t}$. Substituting 160 for $w(0)$ and 2,100 for C, we
can solve for k. $160 = \frac{2}{35}(2100) + ke^{-0.005(0)} = 120 + k$, hence $k = 40$. The particular solution is
$w(t) = 120 + 40e^{-0.005t}$.
(b) $w(30) = 120 + 40e^{-0.005(30)} \approx 154.4$ lb
(c) $w(t) = 150 = 120 + 40e^{-0.005t}$
$\quad\quad 30 = 40e^{-0.005t}$
$\quad\quad 0.75 = e^{-0.005t}$
$\quad\quad \ln 0.75 = -0.005t$
$\quad\quad\quad t = \frac{\ln 0.75}{-0.005} \approx 57.5$

It will take about 58 days for this person to lose
10 lb.
(d) $\lim_{t\to\infty} w(t) = \lim_{t\to\infty}\left(120 + 40e^{-0.005t}\right) = 120 + 40(0) = 120$.
(e) This person's weight will approach 120 lb if
this diet is maintained for a long time.

≡ 32.5 APPLICATIONS

1. As in example 32.22, $\frac{dN}{dt} - kN = 0$ has general solution $N = Ce^{kt}$. At $t = 0$, $N = 100$ which gives us that $C = 100$. Now when $t = 15$ days, $N = 75$ g and we can solve for k. $75 = 100e^{15k}$, so $\frac{75}{100} = e^{15k}$ and taking natural logarithms we get $15k = \ln\left(\frac{3}{4}\right)$ or $k = -0.0192$. The equation is $N = 100e^{-0.0192t}$ where N is in grams and t in days.

3. If the half-life is 1000 years, we get $\frac{1}{2} = e^{k1000}$ or $k = \dfrac{\ln\frac{1}{2}}{1000} = -6.93147 \times 10^{-4}$. When 10% has decayed we have $0.9 = e^{-6.93147\times10^{-4}t}$ or $-6.93147 \times 10^{-4}t = \ln(0.9) \Rightarrow t = 152.0031$ yr.

5. When $t = 0$, $N = 1387$.

7. First we must find k in the equation $N = Ce^{kt}$. Since the half-life is 5600 years, $\frac{1}{2} = e^{5600k}$ or $k = -1.23776 \times 10^{-4}$. Now we solve for t in the equation $\frac{1}{500} = e^{-1.23776\times10^{-4}t}$, with the result $t = \dfrac{\ln\left(\frac{1}{500}\right)}{-1.23776\times10^{-4}} = 50208.4$ years ≈ 50210 years.

9. (a) As in example 32.23 we have $\frac{dT}{dt} + kT = 30k$. Solving this differential equation we get $Te^{kt} = 30e^{kt} + C$ or $T = 30 + Ce^{kt}$. When $t = 0$,

$T = 10°$, hence $C = -20$ or $T = 30 - 20e^{kt}$. Now when $t = 10$, $T = 15°$ or $15 = 30 - 20e^{-k10}$ or $\frac{-15}{-20} = e^{10k}$. Hence $k = \frac{\ln(0.75)}{-10} = 0.02877$. The equation is $T = 30 - 20e^{-0.02877t}$

(b) $22 = 30 - 20e^{-0.02877t}$, or $\frac{-8}{-20} = e^{-0.02877t}$, and so $t = \frac{\ln(0.4)}{-0.02877} = 31.85$ min.

(c) $t = 30 - 20e^{-0.02877\times60} = 26.44°C$

11. As in example 32.23, $T = 40 + Ce^{-kt}$. At $t = 0$, $T = 375°F$ so $C = 375 - 40 = 335°F$. Now, when $t = 15$ min, $T = 280°F$ so $280 = 40 + 335e^{-15k}$ or $\frac{240}{335} = e^{-15k}$, and so $k = 0.02223$. Thus, we get $T = 40 + 335e^{-0.02223t}$. Hence $75 = 40 + 335e^{-0.02223t}$, or $\frac{35}{335} = e^{-0.02223t}$, and as a result, $t = \dfrac{\ln\left(\frac{35}{335}\right)}{-0.02223} = 101.6$ min or 1 hr 41.6 min.

13. This is an RL circuit and so $\frac{dI}{dt} + \frac{R}{L}I = \frac{E}{L}$ or $\frac{dI}{dt} + \frac{10}{2}I = \frac{100}{2}$ or $\frac{dI}{dt} + 5I = 50$. The general solution is $I = Ce^{-5t} + 10$. At $t = 0$, $I = 0$ so $C = -10$. The particular solution is $I = -10e^{-5t} + 10$.

15. $\frac{dI}{dt} + 50I = 5$. The integrating factor is e^{50t} and we get $Ie^{50t} = \frac{1}{10}e^{50t} + C$. Hence $I = 0.1 + Ce^{50t}$. When $t = 0$, $I = 0$, so $C = -0.1$ and the particular solution is $I = 0.1 - 0.1e^{-50t}$

≡ 32.6 MORE APPLICATIONS

1. $x^2 + y^2 = C$ is a family of circles with center at the origin. Taking the derivative implicitly we get $2x + 2yy' = 0$ or $\frac{dy}{dx} = -\frac{x}{y} = f(x, y)$. The orthogonal trajectories satisfy the equation $\frac{dy}{dx} = \frac{-1}{f(x,y)}$ or $\frac{dy}{dx} = \frac{y}{x}$. Separating variables we get $\frac{dy}{y} = \frac{dx}{x}$. Integration gives $\ln y = \ln x + C_1$ or $y = e^{\ln x + C_1} = e^{C_1}x$. Letting $e^{C_1} = k$ we get the family of curves $y = kx$ or the set of lines through the origin.

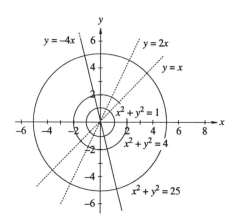

3. $y = Ce^x$ had derivative $\frac{dy}{dx} = Ce^x$. Substituting ye^{-x} for C we obtain $\frac{dy}{dx} = y$. The family of curves we seek satisfies the equation $\frac{dy}{dx} = -\frac{1}{y}$ or

$y\,dy = -dx$. Integrating we get $\frac{1}{2}y^2 = -x + C$ or $y^2 = -2x + k$ or $y^2 + 2x = k$.

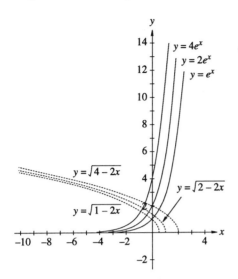

5. $x^2 + \frac{1}{2}y^2 = C^2$ has implicit derivative $2x + yy' = 0$ or $y' = \frac{-2x}{y}$. The orthogonal trajectories satisfy the equation $\frac{dy}{dx} = \frac{y}{2x}$. Separating variables we get $\frac{dy}{y} = \frac{dx}{2x}$. Integration yields $\ln y = \frac{1}{2}\ln x + C_1$ or $\ln y^2 = \ln x + C_2$ or $y^2 = kx$.

7. As in example 32.26 we start with the equation $\frac{dv}{dt} + \frac{kv}{m} = g$ or $\frac{dv}{dt} + \frac{0.2v}{10} = 9.8$. This is of linear form and the integrating factor is $e^{0.02t}$. Hence we obtain the equation $\frac{dv}{dt}\left(ve^{0.02t}\right) = 9.8e^{0.02t}$. Integration yields $ve^{0.02t} = 490e^{0.02t} + C_1$ or $v = 490 + C_1e^{-0.02t}$. Since the initial velocity is 0, $C_1 = -490$ and the equation becomes $v(t) = 490\left(1 - e^{-0.02t}\right)$ m/s or $v(t) = 490\left(1 - e^{-\frac{t}{50}}\right)$ m/s.

9. (a) Using the general solution from problem 7 and substituting 1 for v and 0 for t we find $C_1 = 489$ so the particular solution is $v(t) = 490 - 489e^{-\frac{t}{50}}$,

(b) $v(10) = 490 - 489e^{-\frac{1}{5}} = 89.64$ m/s,

(c) $v_L = \frac{10 \times 9.8}{0.2} = 490$ m/s. (Note: $\lim_{t \to \infty} v(t) = 490$).

11. As in Example 32.27, $\frac{dQ}{dt} + \frac{2}{50 + (2-2)t}Q = 10$ or $\frac{dQ}{Q} = -\frac{dt}{25}$. Separating variables we get $\frac{dQ}{Q} + \frac{dt}{25} = 0$. Integration yields $\ln Q = -\frac{1}{25}t + C$ or $Q = ke^{-t/25}$. When $t = 0$, $Q = 25$ so $k = 25$ and $Q = 25e^{-t/25}$. After 15 minutes $Q = 25e^{-15/25} = 13.72$ lb.

13. (a) $\frac{dQ}{dt} + \frac{4}{8}Q = 12$. This equation is linear and the integrating factor is $e^{t/2}$. Multiplying yields $\frac{dQ}{dt}\left(Qe^{t/2}\right) = 12e^{t/2}$. Integrating we get $Qe^{t/2} = 24e^{t/2} + C$ or $Q = 24 + Ce^{-t/2}$. Since $Q = 2$ when $t = 0$, $C = -22$ so the particular solution is $Q(t) = 24 - 22e^{-t/2}$
(b) $Q(8) = 24 - 22e^{-4} = 23.597$ lb.

15. (a) This is a mixture problem with $r_1 = r_2 = 10\text{m}^3/\text{hr}$, $V_0 = 200\text{m}^3$, $a = 10$ kg, and $b = 1$ kg/5 m$^3 = 0.2$ kg/m^3. Thus,

$$\frac{dQ}{dt} + \frac{10}{200 + (10 - 10)t}Q = (0.2)10$$
$$\frac{dQ}{dt} + \frac{1}{20}Q = 2$$

The integrating factor is $e^{t/200}$, so $\frac{d}{dt}\left(e^{t/200}Q\right) = 2e^{t/200}$, which leads to

$$e^{t/200}Q = \int 2e^{t/200}dt$$
$$e^{t/200}Q = 400e^{t/200} + C$$
$$Q = 400 + Ce^{-t/200}$$

Since $Q(0) = 10$, $C = -390$. Thus, the quantity of waste in the tank at any time t, in hr, is given by $Q(t) = 400 - 390e^{-t/200}$.
(b) When $t = 3$ hr, we have $Q(3) = 400 - 390e^{-3/200} \approx 15.806$ kg of waste in the tank.

CHAPTER 32 REVIEW

1. We have $y' = x + y$ or $\frac{dy}{dx} - y = x$. This is linear with integrating factor e^{-x}. Hence we get $\frac{dy}{dx}\left(ye^{-x}\right) = xe^{-x}$. Integrating with form 84 yields $ye^{-x} = e^{-x}(-x - 1) + C$ or $y = Ce^x - (x + 1)$.

2. $x\,dx - y^2\,dy = 0$ or $\frac{1}{2}x^2 - \frac{1}{3}y^3 = C$

3. $y' = y^2x^3$ or $\frac{dy}{y^2} = x^3\,dx$. Integrating produces $-y^{-1} = \frac{1}{4}x^2 + C_1$ or $x^4 + 4y^{-1} = C$.

4. $y' = 8y$ or $\frac{dy}{y} = 8\,dx$ and integrating gives us $\ln y = 8x + C$, or $y = ke^{8x}$.

5. $e^x \, dx - 2y \, dy = 0$ and integrating produces $e^x - y^2 = C$.

6. $(y^2 - y)dx + x \, dy = 0$ or $\frac{dx}{x} + \frac{dy}{y^2-y} = 0$. Integrating by Form 48 we get $\ln x - \ln\left|\frac{y}{y-1}\right| = C$ or $\ln \frac{x(y-1)}{y} = C_1$ or $-x + \frac{x}{y} = C$.

7. $(y - xy^3)dx + (x - 2x^2y^2)\, dy = 0$ is equivalent to $y \, dx + x \, dy - xy^3 \, dx - 2x^2y^2 \, dy = 0$. Now we multiply by $\frac{1}{xy}$ to get $\frac{y\,dx+x\,dy}{xy} - (y^2 \, dx + 2xy \, dy) = 0$. The first part has integral $\ln|xy|$. The second part has integral $-xy^2$. Putting it together we get $\ln|xy| - xy^2 = C$.

8. $(y+1)dx - x \, dy = 0$ is equivalent to $\frac{dx}{x} - \frac{dy}{y+1} = 0$. This has solution $\ln x - \ln(y+1) = \ln C$ or $\frac{x}{y+1} = C$ or $\frac{y+1}{x} = C$.

9. $y \, dx + (2-x)dy = 0$ is equivalent to $\frac{dx}{2-x} + \frac{dy}{y} = 0$. This has solution $-\ln|2 - x| + \ln y = \ln C$ or $\ln \frac{y}{x-2} = \ln C$ or $\frac{y}{x-2} = C$ or $\frac{x-2}{y} = C$.

10. $(y + x^3y)dx + x \, dy = 0$ or $y \, dx + x \, dy = -x^3 y \, dx$. Multiplying by the integrating factor $\frac{1}{xy}$, we get $\frac{y\,dx+x\,dy}{xy} = \frac{-x^3y}{xy} \, dx = -x^2 \, dx$. Integrating, we get $\ln xy = -\frac{1}{3}x^3 + C$

11. $y^2 \, dx + xy \, dy = 0$. Multiplying by $\frac{1}{xy^2}$ to separate variables we get $\frac{dx}{x} + \frac{dy}{y} = 0$. Integration yields $\ln|x| + \ln|y| = \ln C$ or $\ln|xy| = C$.

12. $dy - 3y \, dx = 6 \, dx$ is linear first-order and has integrating factor e^{3x}. Multiplying we get $e^{-3x}dy = 3ye^{-3x}dx = 6e^{-3x}dx$ and integrating we have $ye^{-3x} = -2e^{-3x} + C$ or $y = -2 + Ce^{3x}$

13. $y' - 4xy = x$ in linear first-order and has integrating factor e^{-2x^2}. Its solution is $ye^{-2x^2} = \int xe^{-2x^2}dx$ or $ye^{-2x^2} = \frac{-1}{4}e^{-2x^2} + C$ or $y = \frac{-1}{4} + Ce^{2x^2}$.

14. $y' + \frac{12y}{x} = x^{12}$ is linear first-order with integrating factor $e^{12 \ln x} = x^{12}$. Multiplying and integrating we get $yx^{12} = \int x^{24}dx$ or $yx^{12} = \frac{x^{25}}{25} + C$.

15. $y' + y = \sin x$ is linear first-order with integrating factor e^x. Multiplying and integrating we get $ye^x = \int e^x \sin x \, dx$. Using form 87 this becomes

$ye^x = \frac{e^x}{2}(\sin x - \cos x) + C$ or $y = \frac{1}{2}(\sin x - \cos x) + Ce^{-x}$.

16. $y' - 7y = e^x$ is linear first-order with integrating factor e^{-7x}. Multiplying and integrating we have $ye^{-7x} = \int e^{-6x}dx$ or $ye^{-7x} = \frac{1}{6}e^{-6x} + C$ which is equivalent to $y = -\frac{1}{6}e + Ce^{7x}$.

17. $\cos x \, dx + y \, dy = 0$ has general solution $\sin x + \frac{1}{2}y^2 = C$. Substituting $y = 2$ and $x = 0$ we find $C = \sin 0 + \frac{1}{2} \cdot 2^2 = 2$. The particular solution is $\sin x + \frac{1}{2}y^2 = 2$.

18. $(x^2 + y + y^2)dx - x \, dy = 0$ is equivalent to $y \, dx - x \, dy + (x^2 + y^2)dx = 0$. Using the integrating factor $\frac{-1}{x^2+y^2}$ we get the general solution $x + \arctan \frac{y}{x} = C$. Substituting $y = \frac{\pi}{3}$ and $x = \frac{\pi}{3}$ we solve for C. $C = -\frac{\pi}{3} + \arctan 1 = -\frac{\pi}{3} + \frac{\pi}{4} = -\frac{\pi}{12}$. The particular solution is $x + \arctan \frac{y}{x} = -\frac{\pi}{12}$.

19. $y' + \frac{2}{x}y = x$ is of linear first-order form with integrating factor x^2. Multiplying and integrating we get the general solution $yx^2 = \int x^3 \, dx$ or $yx^2 = \frac{1}{4}x^4 + C$. Substituting $x = 1$ and $y = 0$ we find that $C = -\frac{1}{4}$. The particular solution is $yx^2 = \frac{1}{4}x^4 - \frac{1}{4}$ or $4yx^2 = x^4 - 1$.

20. $dy + 6xy \, dx = 0$ This is linear first-order with integrating factor e^{3x^2}. Multiplying and integrating we get $ye^{3x^2} = C$. Substituting $y = 5$ and $x = \pi$ we get $C = 5e^{3\pi^2}$. Hence the particular solution is $ye^{3x^2} = 5e^{3\pi^2} \approx 3.6134 \times 10^{13}$.

21. (a) As in Example 32.22 we get $N = Ce^{kt}$. After 2 years $\frac{N}{C} = 0.9$ so $e^{2k} = 0.9$, and so $k = -0.05268$. $N(t) = Ce^{-0.0527t}$ where $C = N(0)$;
(b) $e^{-0.0527t} = \frac{1}{2} \Rightarrow -0.0527t = \ln \frac{1}{2}$ or $t = 13.15$ years or 13 years 57 days 12 hours.

22. $y = cx^2$ had derivative $\frac{dy}{dx} = 3cx^2$. Substituting $\frac{y}{x^3}$ for c we get $\frac{dy}{dx} = 3 \cdot \frac{y}{x^3} \cdot x^2 = \frac{3y}{x}$. The family we seek satisfies $\frac{dy}{dx} = -\frac{x}{3y}$ or $3y \, dy = -x \, dx$ or $\frac{3}{2}y^2 = -\frac{1}{2}x^2 + k$ or $\frac{3}{2}y^2 + \frac{1}{2}x^2 = k$ or $3y^2 + x^2 = C$.

23. As in example 32.33 we get $T = 72 + Ce^{-kt}$. When $t = 0$, $T = 0$ so $C = -72$ and hence

$T = 72 - 72e^{-kt}$. When $t = 15$ min, $T = 20°F$, so we get $20 = 72 - 72e^{-15t}$ or $e^{-15t} = \frac{52}{72}$ or

$$t = \frac{\ln \frac{52}{72}}{-15} = 0.021695.$$ This gives $T = 72\left(1 - e^{-0.021695t}\right)$.

(a) Substituting 50 for T and solving we get $1 - \frac{50}{72}e^{-0.021695t}$ or $t = \dfrac{\ln\left(\frac{20}{72}\right)}{0.021695} = 59.04$ min.

(b) $T(60) = 72\left(1 - e^{-0.021695 \times 60}\right) = 52.41°F$.

24. Using the formula $\frac{dI}{dt} + \frac{R}{L}I = \frac{E}{L}$ we get $\frac{dI}{dt} + 30I = \frac{25}{2}$. This equation is linear first-order with integrating factor e^{30t}. Multiplying and integrating we get $Ie^{30t} = \frac{25}{60}e^{30t} + C$ or $I = \frac{5}{12} + Ce^{-30t}$. The initial current is 0, so $C = -\frac{5}{12}$ and the particular solution is $I = \frac{5}{12}\left(1 - e^{-30t}\right)$.

25. Using the equation $\frac{dq}{dt} + \frac{1}{RC}q = \frac{E}{R}$ we get $\frac{dq}{dt} + 1.25q = 5\cos 2t$. This is linear first-order with integrating factor $e^{1.25t}$. Multiplying and integrating we get $qe^{1.25t} = 5\int e^{1.25t}\cos 2t$. Using form 88 we get $qe^{1.25t} = \frac{5e^{1.25t}}{17.5625}(1.25\cos 2t + 2\sin 2t) + C$ or $q = \frac{1}{3.5125}\left(1.25\cos 2t + 2\sin 2t + ke^{-1.25t}\right)$. When $t = 0$, $q = 0$ so $k = -1.25$. Finally, multiplying by $\frac{80}{80}$ to clear up fractions and decimals we get $q = \frac{1}{281}\left(100\cos 2t + 160\sin 2t - 100e^{-1.25t}\right)$.

26. First we convert 192 lb to slugs: $\frac{192}{32} = 6$ slugs. Since the limiting velocity is 16 ft/s, we get $k = \frac{192}{16} = 12$. Now using the equation $\frac{dv}{dt} + \frac{kv}{m} = g$, we have $\frac{dv}{dt} + \frac{12}{6}v = 32$. This is linear first-order with integrating factor e^{2t}. Multiplying and integrating we get $ve^{2t} = 16e^{2t} + C$ or $v = 16 + Ce^{-2t}$. Since $v_0 = 0$, $C = -16$ and $v(t) = 16\left(1 - e^{-2t}\right)$

(a) $v(1) = 13.83$ ft/s.

(b) $15 = 16\left(1 - e^{-2t}\right) \Rightarrow e^{-2t} = \frac{1}{16}$ or $t = \dfrac{\ln\left(\frac{1}{16}\right)}{-2} = 1.386$ s.

27. **(a)** As in example 32.28, $\frac{dQ}{dt} + \frac{100}{2000}Q = 0.035$. This is linear first-order with integrating factor $e^{0.05t}$. Multiplying and integrating we get $Qe^{0.05t} = 0.7e^{0.05t} + C$ or $Q = 0.7 + Ce^{-0.05t}$. The initial quantity of CO_2 is 2000×0.001 so $C = 1.3$. The particular solution is $Q(t) = 0.7 + 1.3e^{-t/20}$ ft^3,

(b) $Q(60) = 0.7547$ ft^3,

(c) $Q(180) = 0.7002$ ft^3.

28. **(a)** As in example 32.28, $\frac{dQ}{dt} + \frac{1}{250}Q = 0.0005$. The general solution is $Qe^{t/250} = 0.125e^{t/250} + C$ or $Q = 0.125 + Ce^{-t/250}$. Since $Q_0 = 2.5$, $C = 2.375$ and the solution is $Q(t) = 0.125 + 2.375e^{-t/250}$ ft^3,

(b) $Q(60) = 1.8807$ ft^3.

29. $\frac{dQ}{dt} + \frac{25}{100000}Q = 0.0125$ has solution $Qe^{t/4000} = 50e^{-t/4000} + C$ or $Q = 50 + ce^{-t/4000}$. Since $Q_0 = 100{,}000 \times 0.0001 = 10$, $C = -40$ and $Q(t) = 50 - 40e^{-t/4000}m^3$, **(b)** Solving $20 = 50 - 40e^{-t/4000}$ fot t we get $\frac{3}{4} = e^{-t/4000}$ or $t = -4000\ln\left(\frac{3}{4}\right) = 1151$ days or 3 years 56 days.

30. **(a)** $\frac{dQ}{dt} + \frac{500}{10^6}Q = 0.00074 \times 500$ or $\frac{dQ}{dt} + \frac{Q}{20000} = 0.37$ has solution $Qe^{t/20\,000} = \int 0.37e^{t/20\,000} = 7400e^{t/20\,000} + C$ or $Q = 7400 + Ce^{-\frac{t}{20\,000}}$. Since $Q_0 = 10^3$, $C = -6400$. Hence $Q(t) = 7400 - 6400e^{-t/20\,000}$,

(b) Solving $2000 = 7400 - 6400e^{-t/20\,000}$ for t we get $\frac{54}{64} = e^{-t20\,000}$ or $t = -20000\ln\frac{54}{64} = 3398$ days or 9 years 3.6 months,

(c) As t approaches infinity, Q approaches 7400 m^3.

☰ CHAPTER 32 TEST

1. Differentiating $y = 4e^{2x} + 2e^{-3x}$ we get $\frac{dy}{dx} = 8e^{2x} - 6e^{-3x}$ and $\frac{d^2y}{dx^2} = 16e^{2x} + 18e^{-3x}$. Substituting these in the given differential equation, we get $\left(16e^{2x} + 18e^{-3x}\right) + \left(8e^{2x} - 6e^{-3x}\right) - 6\left(4e^{2x} + 2e^{-3x}\right) = (16 + 8 - 24)e^{2x} + (18 - 6 - 12)e^{-3x} = 0.$

2. Here $M = y^2$ and $N = 2xy$. Both M and N are continuous. Since $\frac{\partial M}{\partial y} = 2y = \frac{\partial N}{\partial x}$, the equation is exact.

3. Here $y' = \frac{dy}{dx} = \frac{5x^4}{y^3}$ can be rewritten as $y^3\,dy = 5x^4\,dx$. Integrating, we obtain $\frac{1}{4}y^4 = x^5 + C$.

4. Dividing this equation by x^2, we obtain $\frac{x^2+1}{x^2}dx + y^2 dy = 0$ or $\left(1 + \frac{1}{x^2}\right)dx + y^2 dy = 0$. Integrating produces the solution $x - \frac{1}{x} + \frac{1}{3}y^3 = C$.

5. Rewriting the given equation as $\left(1 + y^2\right)dx + \left(x^2 + 1\right)dy = 0$ we see that the equation is separable. If the variables are separated by dividing by $\left(1+y^2\right)\left(x^2+1\right)$, we obtain $\frac{dx}{x^2+1} + \frac{y}{1+y^2}dy = 0$. Integrating we obtain the solution $\arctan x + \frac{1}{2}\ln\left(1+y^2\right) = C$.

6. Dividing by $\sin x$ we obtain $\frac{dy}{dx} = \frac{1}{\sin x} - 2y\frac{\cos x}{\sin x} = \csc x - 2y\cot x$, which can be written in the standard form for a linear first-order differential equation as $\frac{dy}{dx} + (2\cot x)y = \csc x$. This has an integrating factor of $e^{\int 2\cot x\, dx} = e^{\ln|\sin x|} = \sin x$ and the solution is $y\sin x = \int(\sin x)(\csc x)dx = \int dx = x + C$, and so the solution is $y\sin x = x + C$.

7. The equation contains the combination $x\,dy - y\,dx = -(y\,dx - x\,dy)$. Using integrating factor (4) from Table 32.1, we multiply the given equation by $-\frac{1}{x^2}$ to obtain $\frac{y\,dx - x\,dy}{x^2} = \frac{x^6}{x^2}dx = 0$ or $\frac{y\,dx - x\,dy}{x^2} + x^4\,dx = 0$. The left-hand term is the derivative of $-\frac{y}{x}$ so integrating produces the solution $-\frac{y}{x} + \frac{1}{5}x^5 = C$.

8. The given equation contains the combination $x\,dy + y\,dx$. Using integrating factor (1) from Table 32.1, we multiply the given equation by $I(x,y) = \frac{1}{xy}$ obtaining $4\left(\frac{x\,dy+y\,dx}{xy}\right) + \frac{6x^2 y}{xy}dx = 4\left(\frac{x\,dy+y\,dx}{xy}\right) + 6x\,dx = 0$. Integrating, we get $4\ln xy + 3x^2 = C$. We are given $x = 1$ when $y = e$. Substituting, we obtain $4\ln e + 3 = 7 = C$, so the particular solution is $4\ln xy + 3x^2 = 7$.

9. We have $F(x,\,y,\,c) = y - ce^{-2x}$. Differentiating the given equation, we get $y' = -2ce^{-2x}$. Solving the given equation for c, we obtain $c = ye^{2x}$ and so $y' = -2\left(ye^{2x}\right)e^{-2x} = -2y$. The orthogonal trajectories are the solutions of $\frac{dy}{dx} = \frac{-1}{-2y} = \frac{1}{2y}$ or $2y\,dy - dx = 0$. The solution of this differential equation is $y^2 - x = k$.

10. Let N represent the amount of the radioactive substance present at time t. Because the rate of decay is proportional to the amount of substance remaining we have $\frac{dN}{dt} = kN$ or $\frac{dN}{dt} - kN = 0$. Using the integrating factor e^{-kt} and integrating, we obtain $Ne^{kt} = C$ or $N = Ce^{-kt}$. If N_0 represents the amount when $t = 0$, the equation becomes $N = N_0 e^{-kt}$. We are told that $N = 0.70$ when $t = 40$ years, and so we get $0.7N_0 = N_0 e^{-40k}$ or $0.7 = e^{-40k}$ and so $k = \frac{\ln 0.7}{-40} \approx 0.008\,917$ and the desired equation is $N = N_0 e^{-0.008\,917t}$. To find the half-life, we solve $e^{-0.008\,917t} = \frac{1}{2}$ and obtain $t \approx 77.73$ years.

11. The body satisfies the equation $\frac{dv}{dt} + \frac{0.2v}{5} = 32$ or $\frac{dv}{dt} = 32 - 0.04v$. Separating variables, we obtain $\frac{dv}{32-0.04v} = dt$ and integrating produces $-25\ln(32 - 0.04v) = t + C_1$ which can be written as $\ln(32 - 0.04v) = -\frac{t}{25} + C$ where $-25\ln C = C_1$. This is equivalent to $32 - 0.04v = Ce^{-t/25}$. At $t = 0$, we are given $v = 0$ so $C = 32$ which makes the equation $32 - 0.04v = 32e^{-t/25}$ or $v = \frac{3s}{0.04}\left(1 - e^{-t/25}\right) = 800\left(1 - e^{-t/25}\right)$. When $t = 6$ s, $v = 800\left(1 - e^{-6/25}\right) \approx 170.70$ ft/s.

12. If T represents the temperature of the object at some time t minutes and T_m the temperature of the surrounding medium, then according to Newton's law of cooling, we have $\frac{dT}{dt} = -k\left(T - T_m\right)$. Here $T_m = 30°C$ and so $\frac{dT}{dt} + kT = 30k$. The solution of this differential equation is $Te^{kt} = 30e^{kt} + C$ or $T = 30 + Ce^{-kt}$. Since $T = 50°C$ when $t = 0$, we get $C = 20$ and the equation becomes $T = 30 + 20e^{-kt}$. When $t = 20$, we are given $T = 45$ which gives $45 = 30 + 20e^{-20k}$ or $0.75 = e^{-20k}$. Taking the natural logarithm we get $\ln = 0.75 = -20k$ or $k = \frac{0.75}{-20} \approx 0.0144$. The desired equation is $T = 30 + 20e^{-0.0144t}$.

CHAPTER
33
Higher-Order Differential Equations

≡ 33.1 HIGHER-ORDER HOMOGENEOUS EQUATIONS WITH CONSTANT COEFFICIENTS

1. $\left(D^2 - 7D + 10\right) y = 0$ has auxiliary equation $m^2 - 7m + 10 = 0$ which factors into $(m-2)(m-5) = 0$ and so has roots $m_1 = 2$ and $m_2 = 5$. The general solution is thus $y = c_1 e^{2x} + c_2 e^{5x}$.

3. $D^2 y - 8Dy + 12y = 0$ has auxiliary equation $m^2 - 8m + 12 = 0$ which has roots 2 and 6. The solution is $y = c_1 e^{2x} + c_2 e^{6x}$.

5. $2\frac{d^2y}{dx^2} + \frac{dy}{dx} - 5y = 0$ has auxiliary equation $2m^2 + 9m - 5 = 0$ which factors into $(2m-1)(m+5) = 0$. The roots are $\frac{1}{2}$ and -5 so the solution is $y = c_1 e^{\frac{1}{2}x} + c_2 e^{-5x}$.

7. $4D^2 y + 7Dy - 2y = 0$ has auxiliary equation $4m^2 - 7m - 2 = 0$. This factors into $(4m+1)(m-2) = 0$ and has roots $-\frac{1}{4}$ and 2. The solution is $y = c_1 e^{-\frac{1}{4}x} + c_2 e^{2x}$.

9. $y'' - 5y' + 4y = 0$ has auxiliary equation $m^2 - 5m + 4 = 0$. This has roots 1 and 4 so the solution is $y = c_1 e^x + c_2 e^{4x}$.

11. $y'' - y = 0$ has auxiliary equation $m^2 - 1 = 0$. The roots are 1 and -1 so the solution is $y = c_1 e^x + c_2 e^{-x}$.

13. $y'' = 6y$ has auxiliary equation $m^2 - 7 = 0$. This has roots $\sqrt{7}$ and $-\sqrt{7}$ so the solution is $y = c_1 e^{\sqrt{7}x} + c_2 e^{-\sqrt{7}x}$.

15. $2\frac{d^2y}{dx^2} + \frac{dy}{dx} - y = 0$ has auxiliary equation $2m^2 + m - 1 = 0$. This factors into $(2m-1)(m+1) = 0$ and has roots $\frac{1}{2}$ and -1. Hence the solution is $y = c_1 e^{x/2} + c_2 e^{-x}$.

17. $D^3 y - 6D^2 y + 11Dy - 6y = 0$ has auxiliary equation $m^3 - 6m^2 + 11m - 6 = 0$. This factors into $(m-1)(m-2)(m-3) = 0$ and has roots 1, 2 and 3. The solution is $y = c_1 e^x + c_2 e^{2x} + c_3 e^{3x}$.

19. $y''' + 6y'' + 11y' + 6y = 0$ has auxiliary equation $m^3 + 6m^2 + 11m + 6 = 0$. This has only negative roots. With synthetic division we can verify that the roots are -1, -2, and -3. Hence the solution is $y = c_1 e^{-x} + c_2 e^{-2x} + c_3 e^{-3x}$.

21. $D^2 y + 2Dy - 15y = 0$ has auxiliary equation $m^2 + 2m - 15 = 0$. This has roots 3 and -5 and so the general solution is $y = c_1 e^{3x} + c_2 e^{-5x}$. When $x = 0$ we have $y = 2$ so $2 = c_1 + c_2$ or $c_1 = 2 - c_2$ finding $y' = 3c_1, e^{3x} - 5c_2 e^{-5x}$ we also have that $y' = 6$ when $x = 0$. Hence $6 = 3c_1 - 5c_2$. Substituting $2 - c_2$ for c_1 we get $6 = 3(2-c_2) - 5c_2$ or $6 = 6 - 3c_2 - 5c_2$. Hence $c_2 = 0$ and $c_1 = 2$. The particular solution is $y = 2e^{3x}$.

23. $D^2 y + 2Dy - 8y = 0$ has auxiliary equation $m^2 + 2m - 8 = 0$. Its roots are -4 and 2. Hence the general solution is $y = c_1 e^{2x} + c_2 e^{-4x}$. When $x = 0$ we are given that $y = 0$ which produces

$$0 = c_1 + c_2 \qquad (1)$$

Differentiating that general solution gives $y' = 2c_1 e^{2x} - 4c_2 e^{-4x}$. When $x = 0$, we know that $y' = 6$, and so

$$6 = 2c_1 - 4c_2$$
$$\text{or} \quad 3 = c_1 - 2c_2 \qquad (2)$$

Subtracting equation (2) from equation (1) yields $3c_2 = -3$, and so $c_2 = -1$. Back-substitution into equation (1) gives $c_1 = 1$. The particular solution is $y = e^{2x} - e^{-4x}$.

25. The differential equation $y'' + 2y' - 15y = 0$ has auxiliary equation $m^2 + 2m - 15 = 0$. Its roots are 3 and -5. Hence, the general solution is $y = c_1 e^{3x} + c_2 e^{-5x}$. When $x = 0$ we are given that $y = 5$ which produces

$$c_1 + c_2 = 5 \qquad (1)$$

Differentiating the general solution gives $y' = 3c_1e^{3x} - 5c_2e^{-5x}$. When $x = 0$, we know that $y' = -1$, and so

$$3c_1 - 5_2 = -1 \qquad (2)$$

Multiplying equation (1) by 5 and adding it to equation (2) produces $8c_1 = 24$, $c_1 = 3$. Back-substitution gives $c_2 = 2$. The particular solution is $y = 3e^{3x} + 2e^{-5x}$.

27. The differential equation $y''' + y'' - 4y' - 4y = 0$ has auxiliary equation $m^3 + m^2 - 4m - 4 = 0$. Removing common factors gives $m^2(m+1) - 4(m+1) = (m^2 - 4)(m+1)$. So, the roots are -1, 2,

and -2. The general solution is $y = c_1e^{-x} + c_2e^{2x} + c_3e^{-2}$, its derivative is $y' = -c_1e^{-x} + 2c_2e^{2x} - 2c_3e^{-2}$, and its second derivative is $y'' = c_1e^{-x} + 4c_2e^{2x} + 4c_3e^{-2}$. When $x = 0$, we are given $y = 3$, $y' = 7$, and $y'' = -3$. Substituting these in the appropriate equations generate the system of equations

$$c_1 + c_2 + c_3 = 3$$
$$-c_1 + 2c_2 - 2c_3 = 7$$
$$c_1 + 4c_2 + 4c_3 = -3$$

The solutions of this system are $c_1 = 5$, $c_2 = 2$, and $c_3 = -4$. The particular solution is $y = 5e^{-x} + 2e^{2x} - 4e^{-2x}$.

≡ 33.2 AUXILIARY EQUATIONS WITH REPEATED OR COMPLEX ROOTS

1. $D^2y + 2Dy + y = 0$ has auxiliary equation $m^2 + 2m + 1 = 0$ This has -1 as a double root. Hence the general solution is $y = c_1e^{-x} + c_2xe^{-x}$ or $y = (c_1 + c_2x)e^{-x}$.

3. $D^2y - 4Dy + 4y = 0$ has auxiliary equation $m^2 - 4m + 4 = 0$. This has 2 as a double root and hence the solution is $y = (c_1 + c_2x)e^{2x}$.

5. $9y'' - 6y' + y = 0$ has auxiliary equation $9m^2 - 6m + 1 = 0$. $\frac{1}{3}$ is a double roots and the solution is $y = (c_1 + c_2x)e^{x/e}$.

7. $4y'' + 12y' + 9y = 0$ has auxiliary equation $4m^3 + 12m + 9 = 0$. This has $-\frac{3}{2}$ as a double root and hence the solution is $y = (c_1 + c_2x)e^{-3x/2}$.

9. $y'' + 4y' + 5y = 0$ has $m^2 + 4m + 5 = 0$ as its auxiliary equation. By the quadratic formula, this has $\frac{-4 \pm \sqrt{16-20}}{2} = -2 \pm j$ as its complex roots. Hence the solution is $y = e^{-2x}(c_1 \cos x + c_2 \sin x)$.

11. $D^2y + 9Dy = 0$ has auxiliary equation $m^2 + 9m = 0$. This has 0 and -9 for its roots and hence the solution is $y = c_1 + c_2e^{-9x}$.

13. $D^3y = y$ has auxiliary equation $m^3 - 1 = 0$. 1 is real solution and synthetic division yields $m^2 + m + 1$ as the depressed quotient. The other two roots are $-\frac{1}{2} \pm \frac{\sqrt{3}}{2}j$. Hence the solution is $y = c_1e^x + e^{-x/2}(c_2 \cos \frac{\sqrt{3}}{2}x + c_3 \sin \frac{\sqrt{3}}{2}x)$.

15. $\frac{d^3y}{dx^3} + 8y = 0$ has auxiliary equation $m^3 + 8 = 0$. We can see that -2 is one root. Using synthetic division we get the depressed quotient $m^2 - 2m + 4$

which has roots $1 \pm \sqrt{3}j$. Hence the solution is $y = c_1e^{-2x} + e^x(c_2 \cos \sqrt{3} + c_3 \sin \sqrt{3}x)$.

17. $y''' - 6y'' + 11y' - 6y = 0$ has auxiliary equation $m^3 - 6m^2 + 11m - 6 = 0$. Using the rational root theorem and synthetic division we find the roots are 1, 2 and 3. Hence the solution is $y = c_1e^x + c_2e^{2x} + c_3e^{3x}$.

19. $D^4y + 8D^3y + 24D^2y + 32Dy + 16y = 0$ has auxiliary equation $m^4 + 8m^3 + 24m^2 + 32m + 16 = 0$. This has -2 as a root four times. Hence the solution is $y = (c_1 + c_2x + c_3x^2 + c_4x^3)e^{-2x}$.

21. $(D - 1)^2(D + 2)^3y = 0$ has auxiliary equations with 1 as a double root and -2 as a triple root. Hence the solution is $y = (c_1 + c_2x)e^x + (c_3 + c_4x + c_5x^2)e^{-2x}$.

23. $(D-3)^2(D^2 - 6D - y)y = 0$ has auxiliary equation with 3 as a double root and irrational roots $\frac{6 \pm \sqrt{36+36}}{2} = 3 \pm 3\sqrt{2}$. Hence the solution is $y = (c_1 + c_2x)e^{3x} + c_3e^{(3+3\sqrt{3})x} + c_4e^{(3-3\sqrt{3})x}$.

25. $(3D^2 - 2D + 1)y = 0$ has $3m^2 - 2m + 1 = 0$ as its auxiliary equation. This equation has complex roots $\frac{2 \pm \sqrt{4-12}}{6} = \frac{1}{3} \pm \frac{\sqrt{2}}{3}j$. Hence the solution is $y = e^{x/3}(c_1 \cos \frac{\sqrt{2}}{3}x + c_2 \sin \frac{\sqrt{2}}{3}x)$.

27. $y'' - 2y' + 5y = 0$ has auxiliary equation $m^2 - 2m + 5 = 0$. This has complex roots $1 \pm 2j$. Hence the general solution is $y = e^x(c_1 \cos 2x + c_2 \sin 2x)$. When $x = 0$, y is 4 and so $c_1 = 4$. When $x = 0$, $y' = 7$ so $y' = e^x(c_1 \cos 2x + c_2 \sin 2x) + e^x(-2c_1 \sin 2x + 2c_2 \cos 2x) = e^x[(c_1 + c_2) \times \cos 2x + (c_2 - 2c_1) \sin 2x]$. Hence $c_1 + 2c_2 = 7$

or $4 + 2c_2 = 7$ or $c_2 = \frac{3}{2}$. The particular solution is $y = e^x\left(4\cos 2x + \frac{3}{2}\sin 2x\right)$.

29. $\left(4D^2 - 12D + 9\right)y = 0$ has auxiliary equation with double root $\frac{3}{2}$. Hence the general solution is $y = (c_1 + c_2 x)\,e^{3x/2}$. When $x = 0$, $y = 1$ so $c_1 = 1$ or $y = (1 + c_2 x)\,e^{3x/2}$, which means that $c_2 = -1$. Hence the particular solution is

$$y = (1 - x)e^{\frac{3}{2}x}.$$

31. The differential equation $y'' + 2y' + y = 0$ has auxiliary equation $m^2 + 2m + 1 = 0$. This equation has -1 as a double root. Hence, the general

solution is $y = c_1 e^{-x} + c_2 x e^{-x}$ and its derivative is $y' = -c_1 e^{-x} - c_2 e^{-x} - c_2 xe$. When $x = 0$, we are given $y = 1$ and $y' = -3$ which leads equations $c_1 = 1$ and $-c_1 + c_2 = -3$. Hence, $c_2 = -2$ and the particular solution is $y = e^{-x} - 2xe^{-x}$.

33. $\left(D^2 - 6D + 34\right)y = 0$ has the auxiliary equation $m^2 - 6m + 34 = 0$. This has the complex roots $3 \pm 5j$, so the general solution is $y = e^{3x}(c_1\cos 5x + c_2\cos 5x)$ and $y' = 3e^{3x}(c_1\cos 5x + c_2\sin 5x) + e^{3x}(-5c_1\sin 5x + 5c_2\cos 5x)$. When $x = 0$, we are given $y = 0.5$ and $y' = 1.5$ and so, $c_1 = 4$ and $3c_1 + 5c_2 = -3$. Hence, $c_2 = -3$ and the particular solution is $y = e^{3x}(4\cos 5x - 3\sin 5x)$.

33.3 SOLUTIONS OF NONHOMOGENEOUS EQUATIONS

1. $\left(D^2 - 10D + 25\right)y = 4$. To find y_c we solve the auxiliary equation $m^2 - 10m + 25 = 0$. This has double root 5. Hence $y_c = (c_1 + xc_2)e^{5x}$. For y_p, we use Case 1 and $y_p = A_0$, $y_p' = 0$. Substituting into the original equation we get $25A_0 = 4$ or $A_0 = \frac{4}{25}$. Hence the solution is $y = (c_1 + c_2 x)^{5x} + \frac{4}{25}$.

3. $\left(D^2 - 10D + 25\right)y = e^{3x}$ has the same complentary solution as problems 1 and 2, namely $y_c = (c_1 + c_2 x)e^{5x}$. To find the particular solution y_p, we use Case 2 with $y_p = A_0 e^{3x}$, $y' = 3A_0 e^{3x}$, and $y'' = 9A_0 e^{3x}$. Substituting into the original equation we get $9A_0 e^{3x} - 10 \cdot 3A_0 e^{3x} + 25A_0 e^{3x} = 3^{3x}$ or $4A_0 e^{3x} = e^{3x}$. Hence $A_0 = \frac{1}{4}$. The solution is

$$y = (c_1 + c_2 x)e^{5x} + \frac{1}{4}e^{3x}.$$

5. $\left(D^2 - 10D + 25\right)y = 10 + e^x$. Again $y_c = (c_1 + c_2 x)e^{5x}$. The y_p has two parts. Case 1 $y_{p1} = A_0$ and we get $25A_0 = 10$ or $A_0 = \frac{2}{5}$. Case 2, $y_{p2} = B_0 e^x$, $y_{p2}' = B_0 e^x$, $y_{p2}'' = B_0 e^x$. Substituting we get $16B_0 e^x = e^x$ or $B_0 = \frac{1}{16}$. The solution is

$$y = (c_1 + c_2 x)e^{5x} + \frac{2}{5} + \frac{1}{16}e^x.$$

7. $\left(D^2 - 10D + 25\right)y = 29\sin 2x$. Again $y_c = (C_1 + C_2 x)e^{5x}$. Using Case 3 we set $y_p = A\sin 2x + B\cos 2x$, and so $y_p' = 2A\cos 2x - 2B\sin 2x$, and $y_p'' = -4A\sin 2x - 4B\cos 2x$. Substituting into the original equation we get $(-4A + 20B + 25A)\sin 2x + (-4B - 20A + 25B)\cos 2x = 29\sin 2x$. Hence $21A + 20B = 29$ and $-20A + 21B = 0$. Eliminating B we get $(441 + 440)A =$

$29 \cdot 21$ or $A = \frac{609}{841} = \frac{21}{29}$. Also, $B = \frac{20A}{21} = \frac{20}{21} \cdot \frac{21}{29} = \frac{20}{29}$. The solution is $y = (C_1 + C_2 x)e^{5x} + \frac{21}{29}\sin 2x + \frac{20}{29}\cos 2x$.

9. $\left(D^2 - 4\right) = x^2 e^x - 3x$ has complementary auxiliary equation with roots ± 2. Hence $y_c = c_1 e^{3x} + c_2 e^{-2x}$. y_p will have 2 parts. $x^2 e^x$ is Case 2 so we set $y_{p1} = e^x(A_2 x^2 + A_1 x + A_0)$, with $y_{p1}' = e^x(A_2 x^2 + A_1 x + A_0) + e^x(2A_2 x + A_1) = A_2 e^x x^2 + (A_1 + 2A_2)e^x x + (A_0 + A_1)e^x$ and $y_{p1}'' = A_2 e^x x^2 + 2xA_2 e^x + (A_1 + 2A_2)e^x x + (A_1 + 2A_2)e^x + (A_0 + A_1)e^x = A_2 e^x x^2 + (2A_2 + A_1 + 2A_2)e^x x + (2A_2 + 2A_1 + A_0)e^x$. Substititing into the original equation yields $(A_2 - 4A_2)e^x x^2 + (4A_2 + A_1 - 4A_1)e^x x + (2A_2 + 2A_1 + A_0 - 4A_0)e^x = x^2 e^x$. Hence $-3A_2 = 1$ or $A_2 = -\frac{1}{3}$. Next, $-\frac{4}{3} - 3A_1 = 0$ or $A_1 = -\frac{4}{9}$ and finally $3A_0 = 2A_2 + 2A_1$ and so $A_1 = \dfrac{-\frac{2}{3} - \frac{8}{9}}{3} = \frac{-14}{27}$. For the second part of y_p, we have $y_{p2} = B_1 x + B_0$, $y_{p2}' = B$ and $y_{p2}'' = 0$. Hence $-4B_1 x = -3x$ and $B_1 = \frac{3}{4}$ $B_0 = 0$. The solution is $y = c_1 e^{2x} + c_2 e^{-2x} - \left(\frac{1}{3}x^2 + \frac{4}{9} + \frac{14}{29}\right)e^x + \frac{3}{4}x$.

11. $\left(D^2 + 9\right)y = 4\cos x + 2\sin x$. y_c has auxiliary equation with roots $\pm 3j$. Hence $y_c = c_1 \cos 3x - c_2 \sin 3x$. y_p fits Cases 3 and 4 so we set $y_p = A\cos x + B\sin x$, $y_p' = -A\sin x + B\cos x$ and $y_p'' = -A\cos x - B\sin x$. Substituting into the original equation we get $(-A\cos x - B\sin x) + 9(A\cos x + B\sin x) = 4\cos x + 2\sin x$ so $8A = 4$

and $8B = 2$ giving $A = \frac{1}{2}$ and $B = \frac{1}{4}$. The solution is $y = c_1 \cos 3x + c_2 \sin 3x + \frac{1}{2} \cos x + \frac{1}{4} \sin x$.

13. $y'' + y = 2e^{3x}$. y_c has auxiliary equation with roots $\pm j$. Hence $y_c = c_1 \cos x + c_2 \sin x$, y_p fits Case 2 so we set $y_p = Ae^{3x}$, $y_p' = 3Ae^{3x}$, and $y_p'' = 9Ae^{3x}$. Substituting we get $9Ae^{3x} + Ae^{3x} = 2e^{3x}$ or $10A = 2 \Rightarrow A = \frac{1}{5}$. The solution is $y = c_1 \cos x + c_2 \sin x + \frac{1}{5}e^{3x}$.

15. $y'' - 4y = 8x^2$. First, $y_c = c_1 e^{2x} + c_2 e^{-2x}$. Now set $y_p = A_2 x^2 + A_1 x + A_0$, $y_p' = 2A_2 x + A_1$, and $y_p'' = 2A_2$. Substituting we get $2A_2 - 4(A_2 x^2 + A_1 + A_0) = 8x^2$. Hence $-4A_2 = 8$ and $A_2 = -2$. Also, $A_1 = 0$. Finally, $2A_2 - 4A_0 = 0$, so $A_0 = \frac{1}{2}A_2 = -1$. The solution is $y = c_1 e^{2x} + c_2 e^{-2x} - 2x^2 - 1$.

17. $D^2 y + y = 4 + \cos 2x$. $y_c = c_1 \cos x + c_2 \sin x$. Set $y_{p1} = A$ and we get $A = 4$ so $y_{p1} = 4$. Now set $y_{p2} = A \sin 2x + B \cos 2x$, $y_{p2}' = 2A \cos 2x - 2B \sin 2x$, and $y_{p2}'' = -4A \sin x - 4B \cos 2x$. Hence, $-4A + A = 0$ and $-4B + B = 1$. Thus, we see that $A = 0$ and $B = -\frac{1}{3}$. The solution is $y = c_1 \cos x + c_2 \sin x - \frac{1}{3} \cos 2x + 4$.

19. $3D^2 y + 2Dy + y = 4 + 2x + 6x^2$. Here the auxiliary equation is $3m^2 + 2m + 1 = 0$ has complex roots $\frac{-1}{3} \pm \frac{\sqrt{2}}{3} j$ so $y_c = e^{-x/3}\left(c_1 \cos \frac{\sqrt{2}}{3}x + c_2 \sin \frac{\sqrt{2}}{3}x \right)$. As in problem 18 set $y_p = Ax^2 + Bx + C$ with $y_p' = 2Ax + B$ and $y_p'' = 2A$. Now substituting, we get $3 \cdot 2A = 2(2A + B) + (Ax^2 + Bx + C) = 4 + 2x + 6x^2$. Collecting like terms we obtain $Ax^2 + (4A+B)x + (6A+2B+C) = 4+2x+6x^2$. So $A = 6$, $4A + B = 2$, and so $B = 2 - 4A = 2 - 24 = -22$. We also obtain $6A + 2B + C = 4$ or $C = 4 - 6A - 2B = 4 - 36 + 44 = 12$. The solution is $y = e^{-x/3}\left(c_1 \cos \frac{\sqrt{2}}{3}x + c_2 \sin \frac{\sqrt{2}}{3}x \right) + 6x^2 - 22x + 12$.

21. $y'' - 2y' + x = x^2 - 1$. The auxiliary equation $m^2 - 2m + 1 = 0$ has 1 as a double root. Hence $y_c = (c_1 + c_2 x)e^x$. Now set $y_p = Ax^2 + Bx + C$, $y_p' = 2Ax + B$, and $y_p'' = 2A$. Substituting we get $2A - 2(2Ax + B) + Ax^2 + Bx + C = x^2 - 1$. Collecting like terms we have $Ax^2 + (-4A + B)x + (2A - 2B + C) = x^2 - 1$. Hence, $A = 1$, $-4A + B = 0$, and $2A - 2B + C = -1$. $B = 4A = 4$, and $C = -1 + 2B - 2A = -1 + 8 - 2 = 5$.

Thus, $y_p = x^2 + 4x + 5$. Now to find c_1 and c_2 of y_c when $x = 0$, $y = (c_1 + 0)1 + 5 = 7$ and so $c_1 = 2$. $y' = (2 + c_2 x)e^x + c_2 e^x + 2x + 4$ and when $x = 0$, $y' = 2 + c_2 + 4 = 15$, or $c_2 = 9$. Hence the particular solution is $y = (2 + 9x)e^x + x^2 + 4x + 5$.

23. $y'' - y' - 2y = \sin 2x$. $y_c = c_1 e^{2x} + c_2 e^{-x}$. Now set $y_p = A \sin 2x + B \cos 2x$, with $y_p' = 2A \cos 2x - 2B \sin 2x$ and $y_p'' = -4A \sin 2x - 4B \cos 2x$. Substituting we obtain the equation $(-4A \sin 2x - 4B \cos 2x) - (2A \cos 2x - 2B \sin 2x) - 2(A \sin 2x + B \cos 2x) = \sin 2x$. This yields $-4A + 2B - 2A = 1$ or $-6A + 2B = 1$ and $-4B - 2A - 2B = 0$ or $-2A - 6B = 0$. Solving we get $A = -\frac{3}{20}$, $B = \frac{1}{20}$. Hence $y = c_1 e^{2x} + c_2 e^{-x} - \frac{3}{20} \sin 2x + \frac{1}{20} \cos 2x$. When $x = 0$, $y = c_1 + c_2 + \frac{1}{20} = 1$, or $c_1 + c_2 = \frac{19}{20}$. Differentiating, we obtain $y' = 2c_1 e^{2x} - c_2 e^{-x} - \frac{6}{20} \cos 2x - \frac{1}{10} \sin 2x$, and when $x = 0$, we know that $y' = 2c_1 - c_2 - \frac{6}{20} = \frac{7}{4}$ or $2c_1 - c_2 = \frac{41}{20}$. So, $3c_1 = \frac{60}{20}$ and $c_1 = 1$; also $c_2 = \frac{19}{20} - 1 = -\frac{1}{20}$. The particular solution is $y = e^{2x} + \frac{1}{20}e^{-x} - \frac{3}{20} \sin 2x + \frac{1}{20} \cos 2x$.

25. The differential equation $y'' + 2y' + y = 3x^2 + 2x - 1$ has the auxiliary equation $m^2 + 2m + 1 = 0$. This has the double root -1, so $y_c = c_1 e^{-x} + c_2 x e^{-x} = (c_1 + c_2 x)e^{-x}$. This is Case 1, so $y_p = A_2 x^2 + A_1 x + A_0$, with $y_p' = 2A_2 x + A_1$, and $y_p'' = 2A_2$. Substituting, we get
$$2A_2 + 2(2A_2 x + A_1) + A_2 x^2 + A_1 x + A_0 = 3x^2 + 2x - 1$$
Eliminating parentheses and collecting like terms produces
$$A_0 x^2 + (4A_1 + A_1)x + 2A_2 + 2A_1 + A_0 = 3x^2 + 2x - 1$$
Hence, $A_2 = 3$, $4A_1 + A_1 = 2$ or $4(3) + A_1 = 2$ and $A_1 = -10$, and $2A_2 + 2A_1 + A_0 = -1$ or $2(3) + 2(-10) + A_0 = -1$ and $A_0 = 13$. This gives the general solution
$$y = (c_1 + c_2 x)e^{-x} + 3x^2 - 10x + 13$$
and its derivative is
$$y' = -(c_1 + c_2 x)e^{-x} + c_2 e^{-x} + 6x - 10$$
We are told that when $x = 0$, then $y = 3$ and $y' = 5$. Substituting these in the general solution and its derivative produce $c_1 + 13 = 3$ or $c_1 = -10$. Next, $10 + c_2 - 10 = 5$ or $c_2 = 5$. The particular solution is $y = (-10 + 5x)e^{-x} + 3x^2 - 10x + 13$ or $y = -10e^{-x} + 5xe^{-x} + 3x^2 - 10x + 13$.

27. The differential equation $y'' + 9y = 8\cos x$ has the auxiliary equation $m^2 + 9 = 0$ with roots $\pm 3j$. Hence, $y_c = c_1\cos 3x + c_2\sin 3x$. This is a Case 4 equation, so $y_p = A_0\sin x + B_0\cos x$, with $y_p' = A_0\cos x - B_0\sin x$ and $y_p'' = -A_0\sin x - B_0\cos x$. Substituting, we obtain the equation

$$-A_0\sin x - B_0\cos x + (A_0\sin x + B_0\cos x) = 8\cos x$$

Hence, $8A_0 = 0$ and $A_0 = 0$ as well as $8B_0 = 8$ and $B_0 = 1$. This leads to

$$y = y_c + y_p = c_1\cos 3x + c_2\sin 3x + \cos x$$
$$y' = -3c_1\sin 3x + 3c_2\cos 3x - \sin x$$

When $x = \frac{\pi}{2}$, we are given $y = -1$, so $-c_2 = -1$ and $c_2 = 1$. We are also told that when $x = \frac{\pi}{2}$, then $y' = 1$ and so $-3c_1 - 1 = 1$ and $c_1 = \frac{2}{3}$. Putting it all together, we see that the particular solution is $y = \cos x + \frac{2}{3}\cos 3x + \sin 3x$.

≡ 33.4 APPLICATIONS

1. Since $F = kx$, we have $k = \frac{F}{x} = \frac{20\text{lb}}{5\text{in.}} = 4\text{lb/in.} = 48$ lb/ft.

3. Here $F = mg$ and $F = kx$ or $k = \frac{F}{x} = \frac{mg}{x} = \frac{5\times 98}{0.2}\frac{\text{N}}{\text{m}} = 245$ N/m.

5. We know that $m = \frac{20}{32} = 0.625$ Slugs, $\ell = 2$ and $k = 48$, $2b = \frac{\ell}{m} = 3.2$, and $\omega^2 = \frac{k}{m} = \frac{48}{0.625} = 76.8$. So the auxiliary equation is $m^2 + 3.2m + 76.8 = 0$. This has solutions $-1.6 \pm \sqrt{1.6^2 - 76.8} = -1.6 \pm \sqrt{75.2}j$. Hence the general solution is $x = e^{-1.6t}\left(c_1\cos\sqrt{75.2}t + c_2 \times \sin\sqrt{75.2}t\right)$. Since $x(0) = 0$, $c_1 = 0$, and $x'(t) = -1.6c_2e^{-1.6t}\sin\sqrt{75.2}t + c_2\sqrt{75.2}e^{-1.6t}\cos\sqrt{75.2}t$ and since $x'(0) = c_2\sqrt{75.2} = 4$, then $c_2 = \frac{4}{\sqrt{75.2}}$. So, $x = \frac{4}{\sqrt{75.2}}e^{-1.6t}\sin\sqrt{75.2}t \approx 0.4613e^{-1.6t}\sin 8.6178t$.

7. $m = 5$, $\ell = 40$, and $k = 245$, so $2b = 8$ and $\omega^2 = 49$. The auxiliary equation is $m^2 + 8m + 49 = 0$ has solutions $-4 \pm \sqrt{16 - 49} = -4 \pm \sqrt{33}j$ and the general solution is $x = e^{-4t}\left(c_1\cos\sqrt{33}t + c_2\sin\sqrt{35}t\right)$. Since $x(0) = 0$, $c_1 = 0$, and we have $x'(t) = -4e^{-4t}c_2\sin\sqrt{33}t + c_2\sqrt{33}e^{-4t} \times \cos\sqrt{33}t$. Also, $x'(0) = c_2\sqrt{33} = 2$, so $c_2 = \frac{2}{\sqrt{33}}$. The particular solution is $x = \frac{2}{\sqrt{33}}e^{-4t} \times \sin\sqrt{33}t \approx 0.3482e^{-4t}\sin 5.7446t$.

9. $E(t) = L\frac{d^2q}{dt^2} + R\frac{dq}{dt} + \frac{q}{c}$ so $50\sin 100t = 0.1\frac{d^2q}{dt^2} + 6\frac{dq}{dt} + 100q$. q_c has auxiliary equation $0.1m^2 + 6m + 100 = 0$ with roots $\frac{-6\pm\sqrt{36-40}}{0.2} = -30 \pm 10j$. The general complementary solution is $q_c =$ $e^{-30t}\left(c_1\cos 10t + c_2\sin 10t\right)$. For q_p we get $q_p = A\sin 100t + B\cos 100t$, $q_p' = 100A\cos 100t - 100B\sin 100t$, and $q_p'' = -10^4A\sin 100t - 10^4\cos 100t$. Substituting we get $\left(-10^3A\sin 100t -10^3\cos 100t\right) + 6(100A\cos 100t - 100B\sin 100t) + 100(A\sin 100t + B\cos 100t) = 50\sin 100t$ so $-1000A - 600B + 100A = 50$ and $-1000B + 600A + 100B = 0$; $-900A - 600B = 50$, so $-18A - 12B = 1$. Also, $600A - 900B = 0$ or $2A - 3B = 0$. Hence, $A = -\frac{1}{26}$, $B = -\frac{1}{39}$. So $q_p = -\frac{1}{26}\sin 100t - \frac{1}{39}\cos 100t$. The general solution is $q = e^{-30t}\left(c_1\cos 10t + c_2\sin 10t\right) - \frac{1}{26}\sin 100t - \frac{1}{39}\cos 100t$. Since $q(0) = c_1 - \frac{1}{39} = 0$, then $c_1 = \frac{1}{39}$. Also, $i(t) = q'(t) = -30e^{-30t} \times \left(c_1\cos 10t + c_2\sin 10t\right) + e^{-30t}\left(-10c_1\sin 10t + 10c_2\cos 10t\right) - \frac{100}{26}\cos 100t + \frac{100}{39}\sin 100t$. Since $q'(0) = -30c_1 + 10c_2 - \frac{100}{26} = 0$, we have $c_2 = \frac{\frac{30}{39} - \frac{100}{26}}{10} = \frac{3}{39} - \frac{10}{26} = -\frac{24}{78} = -\frac{4}{13}$. The particular solution is $q(t) = e^{-30t}\left(\frac{1}{39}\cos 10t - \frac{4}{13}\sin 10t\right) - \frac{1}{26}\sin 100t - \frac{1}{39}\cos 100t$.

11. (a) $E(t) = L\frac{d^2q}{dt^2} + R\frac{dq}{dt} + \frac{q}{c}$ so we get $\frac{1}{10}\frac{d^2q}{dt^2} + \frac{100}{4}q = 180\cos 60t$. The auxiliary equation is $0.1m^2 + \frac{1000}{4} = 0$, so $m = \pm 50j$ and as a result, $q_c = c_1\cos 50t + c_2\sin 50t$. Now we get $q_p = A\sin 60t + B\cos 60t$, $q_p' = 60A\cos 60t - 60B\sin 60t$, and $q_p'' = -3600A\sin 60t - 3600 \times \cos 60t$. Substituting we get $-360A\sin 60t - 360B\cos 60t + 250A\sin 60t + 250B\cos 60t = 180\cos 60t$. Hence we have the equations $-360A + 250A = 0$

or $A = 0$ and $-360B + 259B = 180$ or $B = \frac{180}{-110} = -\frac{18}{11}$. The general equation is $q(t) = c_1 \cos 50t + c_2 \sin 50t - \frac{18}{11} \cos 60t$. Since $q(0) = 0$, $c_1 = \frac{18}{11}$. Also, since $q'(t) = -50c_1 \sin 50t + 50c_2 \cos 50t + \frac{18 \cdot 60}{11} \sin 6t$, and $q'(0) = 50c_1 = 0$, we find that $c_2 = 0$. The particular solution is $q(t) = \frac{18}{11} \cos 50t - \frac{18}{11} \cos 60t$

(b) The steady-state current is $i = \frac{dq_p}{dt} = \frac{d}{dt} \times \left(\frac{-18}{11} \cos 60t\right) = \frac{1080}{11} \sin 60t$.

13. The differential equation $\frac{d^2g}{dt^2} + 2\alpha \frac{dg}{dt} + \omega_0^2 g = 0$ has auxiliary equation $m^2 + 2\alpha m + \omega_0^2 = 0$. Completing the square, we get $m^2 + 2\alpha m + \alpha^2 = \alpha^2 - \omega_0^2$ and so $(m + \alpha)^2 = \alpha^2 - \omega_0^2$ and $m + \alpha = \pm\sqrt{\alpha^2 - \omega_0^2}$, which means that $m = -\alpha \pm$

$\sqrt{\alpha^2 - \omega_0^2}$. Since $\alpha^2 - \omega_0^2 < 0$, these are complex roots with $m = -\alpha \pm \sqrt{\omega_0^2 - \alpha^2}j$. Thus, the general solution is

$$g(t) = e^{-\alpha t}\left(c_1 \cos \sqrt{\omega_0^2 - \alpha^2}t + c_2 \sin \sqrt{\omega_0^2 - \alpha^2}t\right)$$

15. Substituting L, R, and C, produces the differential equation $1.2\frac{d^2q}{dt^2} + 5\frac{dq}{dt} + \frac{1}{0.025}q = 15$. This has auxiliary equation $1.2m^2 + 5m + 40 = 0$ with roots $-2.083 \pm 5.3845j$ or about $-2.1 \pm 5.4j$. Hence, $q_c = e^{-2.1t}\left(c_1 \cos 5.4t + c_2 \sin 5.4t\right)$. Here, $q_p = A_0$ and $40A_0 = 15$ so $A_0 = \frac{15}{40} = 0.375$. Hence, the general solution is $q(t) = e^{-2.1t}\left(c_1 \cos 5.4t + c_2 \sin 5.4t\right) + 0.375$.

CHAPTER 33 REVIEW

1. $D^2y = 0$ has auxiliary equation $m^2 = 0$ with 0 as a double root. Hence $y = \left(c_1 + c_2x\right)e^{0x}$ or more simply $y = c_1 + c_2x$.

2. $\left(D^2 - 9\right)y = 0$. The auxiliary equation $m^2 - 9 = 0$ has roots 3 and -3. Hence $y = c_1e^{3x} + c_2e^{-3x}$.

3. $\left(D^2 - 5D + 6\right)y = 0$. The auxiliary equation $m^2 - 5m + 6 = 0$ has roots 2 and 3. Hence $y = c_1e^{2x} + c_2e^{3x}$.

4. $\left(D^2 - 5D - 14\right)y = 0$. The auxiliary equation $m^2 - 5m - 14 = 0$ has roots 7 and -2. Hence the solution is $y = c_1e^{7x} + c_2e^{-2x}$.

5. $\left(D^2 - 6D + 25\right)y = 0$. The auxiliary equation $m^2 - 6m + 25 = 0$ has complex roots $3 + 4j$. Hence the solution is $y = e^{3x}\left(c_1 \cos 4x + c_2 \sin 4x\right)$

6. $\left(D^2 + 2D - 2\right)y = 0$. The auxiliary equation $m^2 + 2m - 2 = 0$ has roots $-1 \pm \sqrt{3}$. Hence $y = c_1e^{(-1+\sqrt{3})x} + c_2^{(-1-\sqrt{3})x}$

7. $\left(D^3 - 8\right)y = 0$; $m^3 - 8$ factors as $(m - 2)(m^2 + 2m + 4)$ and has roots 2, $-1 \pm \sqrt{3}j$. Hence $y = c_1e^{2x} + e^{-x}\left(c_2 \cos \sqrt{3}x + c_3 \sin \sqrt{3}x\right)$

8. $y''' + y'' - y' - y = 0$ has auxiliary equation $m^3 + m^2 - m - 1 = 0$. This factors by grouping into $\left(m^2 - 1\right)(m + 1) = 0$ and has roots -1 twice and 1. Hence, the solution is $y = c_1e^x + c_2e^{-x} + c_3xe^{-x}$

9. $\left(D^2 + 3D - 4\right)y = 9x^2$. The auxiliary equation is $m^2 + 3m - 4 = 0$ and it has roots -4 and 1. Hence

$y_c = c_1e^x + c_2e^{-4x}$. Now we set $y_p = Ax^2 + Bx + C$, $y_p' = 2Ax + B$, and $y_p'' = 2A$. Substituting we get $2A + 3(2A + B) - 4\left(Ax^2 + Bx + C\right) = 9x^2$ so $-4A = 9$ or $A = -\frac{9}{4}$, and also $6A - 4B = 0$ so $B = \frac{6A}{4} = \frac{-27}{8}$. Now, since $2A + 3B - 4C = 0$, we see that $C = \frac{2A+3B}{4} = \frac{-\frac{9}{2} - \frac{81}{8}}{4} = \frac{-117}{32}$. So $y_p = -\frac{9}{4}x^2 - \frac{27}{3}x - \frac{117}{32}$ and $y = y_c + y_p = c_1e^x + c_2e^{-4x} - \frac{9}{4}x^2 - \frac{27}{8}x - \frac{117}{32}$.

10. $\left(D^2 - 4D - 5\right)y = 3e^{2x}$. The auxiliary equation has roots 5 and -1 so $y_c = c_1e^{5x} + c_2e^{-x}$. Now we set $y_p = Ae^{2x}$, $y_p' = 2Ae^{2x}$ and $y_p'' = 4Ae^{2x}$. Substituting we get $4Ae^{2x} - 4\left(2Ae^{2x}\right) - 5\left(Ae^{2x}\right) = 3e^{2x}$. Hence $4A - 8A - 5A = 3$ or $-9A = 3$ or $A = -\frac{1}{3}$. The solution is $y = c_1e^{5x} + c - 2e^x - \frac{1}{3}e^{2x}$

11. $\left(D^2 + 7D + 12\right)y = \cos 5x$. The auxiliary equation $m^2 + 7m + 12 = 0$, has roots -3 and -4. So $y_c = c_1e^{-3x} + c_2e^{-4x}$. Now we set $y_p = A \sin 5x + B \cos 5x$, $y_p' = 5A \cos 5x - 5B \sin 5x$, and $y_p'' = -25A \sin 5x - 25B \cos 5x$. Substituting we get $(-25A \sin 5x - 25B \cos 5x) + 7(5A \cos 5x - 5B \sin 5x) + 12(A \sin 5x + B \cos 5x) = \cos 5x$. This yields the linear equations $-25A - 35B +$

$12A = 0$ and $-25B+35A+12B = 1$. These simplify to $-13A-35B = 0$ and $35A-13B = 1$. Using Cramer's Rule we get $A = \frac{35}{1394}$ and $B = \frac{-13}{1394}$. So $y_p = \frac{35}{1394}\sin 5x - \frac{13}{1394}\cos 5x$. The solution is $y = y_c + y_p = c_1 e^{-3x} + c_2 e^{-4x} + \frac{35}{1394}\sin 5x - \frac{13}{1394}\cos 5x$.

12. $\left(D^2 - 6D + 8\right)y = 6x^2 - 2$. The auxiliary equation has roots 2 and 4 so $y_c = c_1 e^{2x} + c_2 e^{4x}$. Now we set $y_p = Ax^2 + Bx + C$, $y_p' = 2Ax + B$, $y_p'' = 2A$. Substituting we get $2A - 6(2Ax + B) + 8\left(Ax^2 + Bx + C\right) = 6x^2 - 2$. Hence $8A = 6$ or $A = \frac{3}{4}$ and since $-12A + 8B = 0$, we obtain $B = \frac{12A}{8} = \frac{3}{2} \cdot \frac{3}{4} = \frac{9}{8}$, and since $2A - 6B + 8C = -2$, we find $C = \frac{-2-2A+6B}{8} = \frac{-2 - \frac{3}{2} + \frac{27}{4}}{8} = \frac{13}{32}$. So we get $y_p = \frac{3}{4}x^2 + \frac{9}{8}x + \frac{13}{32}$ and $y = y_c + y_p = c_1 e^{2x} + c_2 e^{4x} + \frac{3}{4}x^2 + \frac{9}{8}x + \frac{13}{32}$.

13. $\left(D^2 + 4D + 4\right)y = 4x + e^{2x}$. The auxiliary equation has double root at -2, so $y_c = \left(c_1 + c_2 x\right)e^{-2x}$. Now y_p has two parts. Set $y_{p1} = Ax + B$, with $y_{p1}' = A$, and $y_{p1}'' = 0$. Substituting we get $4A + 4Ax + 4B = 4x$. so $A = 1$ and $B = -1$. Next we set $y_{p2} = Ae^{2x}$ so $y_{p2}' = 2Ae^{2x}$ and $y_{p2}'' = 4Ae^{2x}$. Substituting we have $(4A + 8A + 4A)e^{2x} = e^{2x}$ or $A = \frac{1}{16}$. $y = y_c + y_{p1} + y_{p2} = \left(c_1 + c_2 x\right)e^{-2x} + x - 1 + \frac{1}{16}e^{2x}$.

14. $\left(D^2 + 4D - 9\right)y = x\sin 2x$. The auxiliary equation $m^2 + 4m - 9 = 0$ has solutions $-2 \pm \sqrt{13}$ and so $y_c = c_1 e^{(-2+\sqrt{13})x} + c_2 e^{(-2-\sqrt{13})x}$. Now we set $y_p = A_1\sin 2x + A_0\sin 2x + B_1 x\cos 2x + B_0\cos 2x$, $y_p' = 2A_1 x\cos 2x + A_1\sin 2x + 2A_0 \times \cos 2x - 2B_1 x\sin 2x + B_1\cos 2x - 2B_0\sin 2x$, and $y_p'' = 2A_1 x\cos 2x - 4A_1 x\sin 2x + 2A_1\cos 2x - 4A_0 x\sin 2x - 2B_1\sin 2x - 2B_1 x\cos 2x - 2B_1\sin 2x - 4B_0\cos 2x$. Substituting we get $y_p'' + 4y_p' - 9y = x\sin 2x$. Collecting like terms we have the following four linear equations.

$$-4A_1 + 4\left(-2B_1\right) - 9A_1 = 1 \quad x\sin 2x$$
$$-4B_1 - 4A_0 + 4 \times \left(-2B_0 + A_1\right) - 9A_0 = 0 \quad \sin 2x$$
$$-4B_1 + 4 \cdot 2A_1 - 9B_1 = 0 \quad x\cos 2x$$
$$-4B_0 + 4A_1 + 4\left(B_1 + 2A_0\right) - 9B_0 = 1 \quad \cos 2x$$

These simplify to

$$-13A_1 - 8B_1 = 1$$
$$4A_1 - 13A_0 - 4B_1 - 8B_0 = 0$$
$$8A_1 - 13B_1 = 0$$
$$4A_1 + 8A_0 + 4B_1 - 13B_0 = 0$$

Solving the first and third equations with Cramer's Rule yields $A_1 = \frac{-13}{233}$ and $B_1 = \frac{-8}{223}$. Substituting these results into the 2nd and 4th equations we get

$$13A_0 + 8B_0 = \frac{20}{233}$$
$$8A_0 - 13B_0 = \frac{84}{233}$$

Again using Cramer's Rule we have $A_0 = \frac{932}{54289}$ and $B_0 = \frac{932}{54289}$. Putting this all together we have $y = c_1 e^{(-2+\sqrt{13})x} + c_2 e^{(-2-\sqrt{13})x} - \frac{x}{233}(13\sin 2x + 8\cos 2x) + \frac{932}{54289}(\sin 2x + \cos 2x)$.

15. First we find the spring constant $k = \frac{mg}{x} = \frac{4 \cdot 9.8}{0.2} = 196$ N/m. So $m = 4$ kg, $k = 196$ N/m and $\ell = 10$. Hence we get auxiliary equation $4m^2 + 10m + 196 = 0$. This auxiliary equation has roots $\frac{-5}{4} \pm \frac{\sqrt{759}j}{4}$. Hence $x = e^{-5t/4}\left(c_1\cos\frac{\sqrt{759}}{4}t + c_2 \times \sin\frac{\sqrt{759}}{r}t\right)$. Assuming $x(0) = 0$, we obtain $c_1 = 0$. Now, differentiating, we see that $x'(t) = -\frac{5}{4}e^{-5t/4}\left(c_1\sin\frac{\sqrt{759}}{4}t\right) + c_2\frac{\sqrt{759}}{4}e^{-5t/4}\cos\frac{\sqrt{759}}{4}t$. Since $v(0) = x'(0) = c_2\frac{\sqrt{759}}{4} = 1$, we have $c_2 = \frac{4}{\sqrt{759}}$. Hence the solution is $x = \frac{4}{\sqrt{759}}e^{-5t/4}\sin\frac{\sqrt{759}}{4}t$.

16. **(a)** The given information yields the equation $\frac{d^2q}{dt^2} + 6\frac{dq}{dt} + 2500q = 16\cos 10t$. The auxiliary equation $m^2 + 6m + 2500$ has solutions $-3 \pm \sqrt{2491}j$. So, we see that $q_c = e^{-3t}\left(c_1\cos\sqrt{2491}t + c_2 \times \sin\sqrt{2491}t\right)$. To find q_p we set $q_p = A\sin 10t + B\cos 10t$, $q_p' = 10A\cos 10t - 10B\sin 10t$, and $q_p'' = 100A\cos 10t - 100B\sin 10t$. Substituting we get $(-100A\sin 10t - 100B\cos 10t) + 6(10A \times \cos 10t - 10B\sin 10t) + 2500(A\sin 10t + B\cos 10t) = 16\cos 10t$. As a result, $-100A - 60B + 2500A = 0$ and $-100B + 60A + 2500B = 16$. These simplify to $2400A - 60B = 0$ and $60A + 2400B = 16$. Using Cramer's Rule we get $A = \frac{960}{5763600}$ and $B = \frac{38400}{5763600}$ which reduce to $A = \frac{4}{24015}$ and $B = \frac{160}{24015}$. Thus, $q_p = \frac{1}{24015}(4\sin 10t + 160\cos 10t)$. Hence

the general solution is $q(t) = e^{-3t}\left(c_1 \cos\sqrt{2491}t\right.$
$\left. + c_2 \sin\sqrt{2491}t\right) + \frac{1}{24015}(4\sin 10t + 160\cos 10t)$.

Since $q(0) = c_1 + \frac{160}{24015} = 0$, $c_1 = \frac{-160}{24015}$. Also
$i(t) = q'(t) = -3e^{-3t}\left(c_1 \cos\sqrt{2491}t +\right.$
$c_2 \sin\sqrt{2491}t + e^{-3t}\left(-\sqrt{2491}c_1 \sin\sqrt{2491}t +\right.$
$\sqrt{2491}c_2 \cos\sqrt{(2491t)} + \frac{40}{24105}\cos 10t - \frac{1600}{24015}\sin 10t$.

So $q'(0) = -3c_1 + \sqrt{2491}c_2 + \frac{40}{24015} = 0$ and so
$c_2 = \frac{-520}{24015\sqrt{2491}}$. Putting it all together we get
$q(t) = \frac{1}{24015}\left[e^{-3t}\left(-160\cos\sqrt{2491}t - \frac{520}{\sqrt{2491}}\times\right.\right.$
$\sin\sqrt{2491}t\left.\right) + 4\sin 10t + 160\cos 10t\left.\right]$.

(b) The steady state current is $\frac{dq_p}{dt} = \frac{1}{24015}\times$
$[40\cos 10t - 1600\sin 10t]$.

CHAPTER 33 TEST

The auxiliary equation is formed by replacing the operators, D, with place-holders for the roots, usually m and replacing the variable, y, with 1. In this problem the resulting auxiliary equation is $5m^2 + 2m - 3 = 0$.

1. The auxiliary equation is $m^2 - 25 = 0$. This factors as $m^2 - 25 = (m - 5)(m + 5) = 0$ and has roots $m_1 = 5$ and $m_2 = -5$. So, the general solution of this differential equation is $y = c_1 e^{5x} + c_2 e^{-5x}$.

2. The auxiliary equation is $m^2 + 8m - 20 = 0$ which factors as $(m + 10)(m - 2) = 0$. The roots of the auxiliary equation are $m_1 = -10$ and $m = 2$. The solution of the differential equation is $y = c_1 e^{-10x} + c_2 e^{2x}$.

3. The auxiliary equation is $(m - 5)^3 \left(m^2 + 4\right) = 0$. This has the real root 5 with multiplicity 3 and the complex roots $\pm 2j$. So, the general solution is $y = \left(c_1 + c_2 x + c_3 x^2\right) e^{5x} + c_4 \cos 2x + c_5 \sin 2x$.

4. The auxiliary equation is $m^2 + 6m + 13 = 0$. Using the quadratic formula we get the roots $m = \frac{-6 \pm \sqrt{6^2 - 4(13)}}{2} = \frac{-6 \pm \sqrt{-16}}{2} = -3 \pm 2j$. So the general solution to this differential equation is $y = e^{-3x}\left(c_1 \cos 2x + c_2 \sin 2x\right)$.

5. The auxiliary equation is $m^2 + 6m - 7 = 0$ which factors as $(m + 7)(m - 1) = 0$. The roots of this auxiliary equation are $m = -7$ and $m = 1$. The general solution is $y = c_1 e^{-7x} + c_2 e^x$.

6. From Exercise 0 we have the complementary solution $y_c = c_1 e^{-7x} + c_2 e^x$. The given equation is a nonhomogeneous equation with $f(x) = 6e^{2x}$ and so this is a Case 2 problem using the Method of Undetermined Coefficients with $a = 2$ and $p_n = 6$ so $y_p = A_0 e^{2x}$. Differentiating, we obtain $y_p' = 2A_0 e^{2x}$ and $y_p'' = 4A_0 e^{2x}$. Substituting this in the given differential equation produces $4A_0 e^{2x} + 12A_0 e^{2x} - 7A_0 e^{2x} = 6e^{2x}$. Solving for A_0, we

determine that $A_0 = \frac{2}{3}$ and so the general solution is $y = c_1 e^{-7x} + c_2 e^x + \frac{2}{3}e^{2x}$.

7. Again the complementary solution is $y_c = c_1 e^{-7x} + c_2 e^x$. This is a nonhomogeneous equation with $f(x) = 3\cos x$, a Case 4 problem using the Method of Undetermined Coefficients with $a = 0$, $p_n = 3$, and $b = 4$, so $y_p = A_0 \sin 4x + B_0 \cos 4x$. Differentiating, produces $y_p' = 4A_0 \cos 4x - 4B_0 \sin 4x$ and $y_p'' = -16A_0 \sin 4x - 16B_0 \cos 4x$. Substituting in the given differential equation we get $(-16A_0 \sin 4x - 16B_0 \cos 4x) + 6(4A_0 \cos 4x - 4B_0 \sin 4x) - 7(A_0 \sin 4x + B_0 \cos 4x) = 3\cos 4x$. Multiplying and collecting terms produces $-23 \times (A_0 + B_0)\sin 4x + 24(A_0 - B_0)\cos 4x = 3\cos 4x$. Thus, $A_0 + B_0 = 0$ and $A_0 - B_0 = 3$. Solving these two equations we determine that $A_0 = \frac{3}{2}$ and $B_0 = -\frac{3}{2}$ and so the general solution is $y = c_1 e^{-7x} + c_2 e^x + \frac{3}{2}\sin 4x - \frac{3}{2}\cos 4x$.

8. Again the complementary solution is $y_c = c_1 e^{-7x} + c_2 e^x$. This is a nonhomogeneous equation with $f(x) = 8x$, a Case 1 problem using the Method of Undetermined Coefficients, so $y_p = A_1 x + A_0$. Differentiating, we obtain $y_p' = A_1$ and $y_p'' = 0$. Substituting these in the given differential equation produces $0 + 6(A_1) - 7(A_1 x + A_0) = 8x$. Multiplying and collecting terms produces $-A_1 x + A_0 = 8x$. Thus, $A_1 = -8$ and $A_0 = 0$ and so the general solution is $y = c_1 e^{-7x} + c_2 e^x - 8x$.

9. Once again the complementary solution is $y_c = c_1 e^{-7x} + c_2 e^x$. This is a nonhomogeneous equation with $f(x) = e^{4x}$, a Case 2 problem using the Method of Undetermined Coefficients with $a = 4$, and $p_n = 1$, so $y_p = A_0 e^{4x}$. Differentiating, we get $y_p' = 4A_0 e^{4x}$ and $y_p'' = 16A_0 e^{4x}$. Substituting these in the given differential equation produces $16A_0 e^{4x} + 6\left(4A_0 e^{4x}\right) - 7\left(A_0 e^{4x}\right) = e^{4x}$.

Collecting terms produces $33A_0e^{4x} = e^{4x}$. Thus, $A_0 = \frac{1}{33}$ and the general solution is $y = c_1e^{-7x} + c_2e^x + \frac{1}{33}e^{4x}$. To find the particular solution, we find $y' = -7c_1e^{-7x} + c_2e^x + \frac{4}{33}e^{4x}$. Substituting $x = 0$ and $y = \frac{4}{11}$ in the general solution produces $\frac{4}{11} = c_1 + c_2 + \frac{1}{33}$ and substituting $x = 0$ and $y' = \frac{5}{11}$ in the derivative gives $\frac{5}{11} - 7c_1 + c_2 + \frac{4}{33}$. Solving these two equations we find $c_1 = 0$ and $c_2 + \frac{1}{3}$. Thus, the particular solution of the given differential equation is $y = \frac{1}{3}e^x + \frac{1}{33}e^{4x}$.

10. Here $E(t) = 5\sin 10t$ and the desired differential equation is $5\sin 10t = 0.5\frac{d^2q}{dt^2} + 30\frac{dq}{dt} + \frac{q}{2\times 10^{-3}} = 0.5\frac{d^2q}{dt^2} + 30\frac{dq}{dt} + 500q$. The auxiliary equation is $0.5m^2 + 30m + 500 = 0$ or $m^2 + 60m + 1000 = 0$. Using the quadratic equation we get $m = -30 \pm 20j$. These are complex roots and so the complementary solution is $q_c = e^{-30t}(c_1\cos 20t + c_2 \times \sin 20t)$. For the particular solution we have a case 3 situation with $a = 0$, $b = 10$, and $p_n = 5$, and so $q_p = A_0\sin 10t + B_0\cos 10t$. Differentiating, we get $q'_p = 10A_0\cos 10t - 10B_0\sin 10t$ and $q''_p = -100A_0\sin 10t - 100B_0\cos 10t$. Substituting these in the differential equation, we have $0.5(100A_0\sin 10t - 100B_0\cos 10t) + 30(10A_0\times \cos 10t - 10B_0\sin 10t) + 500(A_0\sin 10t + B_0\times \cos 10t) = 5\sin 10t$. Multiplying and collecting terms, we get $(450A_0 - 300B_0)\sin 10t + (450B_0 + 300A_0)\cos 10t = 5\sin 10t$. Hence, $450A_0 - 300B_0 = 5$ and $450B_0 + 300A_0 = 0$. Solving these two equations simultaneously gives $A_0 = 0.02$ and $B_0 = \frac{1}{75}$. Thus, the general solution is $q(t) = e^{-30t}(c_1\cos 20t + c_2\sin 20t) + 0.02\sin 10t + \frac{1}{75}\times \cos 10t$. The steady-state solution is $q_p = 0.02\times \sin 10t + \frac{1}{75}\cos 10t$. Consequently, the steady-state current is $i = \frac{dq_p}{dt} = 0.2\cos 10t - \frac{10}{75}\sin 10t$.

CHAPTER
34
Numerical Methods and LaPlace Transforms

≡ 34.1 EULER'S OR THE INCREMENT METHOD

1. See *Computer Programs* in the main text.

3.

x	$h = 0, 1$	$h = 0.05$	Correct solution $y = \frac{1}{2}e^x + \frac{1}{2}e^{3x}$
0	0	0	0
0.5	–	0.5	0.05528
0.10	0.1	0.1101	0.12234
0.15	–	0.1818	0.2032
0.20	0.2405	0.2672	0.3004
0.25	–	0.3683	0.4165
0.30	0.4348	0.4878	0.5549
0.35	–	0.6285	0.7193
0.40	0.7002	0.7937	0.9141
0.45	–	0.9873	1.1446
0.50	1.0595	1.2138	1.4165
0.55	–	1.4784	1.7369
0.60	1.5422	1.7868	2.1138
0.65	–	2.1459	2.5566
0.70	2.1871	2.5636	3.0762
0.75	–	3.0488	3.6854
0.80	3.0046	3.6120	4.3988
0.85	–	4.2650	5.2337
0.90	4.1805	5.0218	6.2101
0.95	–	5.8980	7.3510
1.00	5.6807	6.9120	8.6836

5.

x	$h = 0.05$	Correct solution $y = \frac{1}{53}(55e^{7x} - 2\cos 2x - 7\sin 2x)$
0	1	1
0.05	1.35	1.4219
0.10	1.8275	2.0265
0.15	2.4770	2.8904
0.20	3.3588	4.1220
0.25	4.5538	5.8753
0.30	6.1717	8.3686
0.35	8.3600	11.9117
0.40	11.3182	16.9442
0.45	15.3154	24.0898
0.50	20.7149	34.2336

7.

x	y
0	1
0.1	1.1
0.2	1.222
0.3	1.3753
0.4	1.5735
0.5	1.8371
0.6	2.1995
0.7	2.7193
0.8	3.5078
0.9	4.8023
1.0	7.1895

9.

x	y
0	0
0.1	0.1
0.2	0.2010
0.3	0.3051
0.4	0.4147
0.5	0.5327
0.6	0.6633
0.7	0.8121
0.8	0.9887
0.9	1.2092
1.0	1.5062

11.

x	y
$\pi/2 \approx 1.57$	1
1.62	1.0270
1.67	1.0507
1.72	1.0711
1.77	1.0882
1.82	1.1022
1.87	1.1130
1.92	1.1209
1.97	1.1259
2.02	1.1281
2.07	1.1277
2.12	1.1249
2.17	1.1197
2.22	1.1123
2.27	1.1028
2.32	1.0913
2.37	1.0779

13. See *Computer Programs* in the main text.

15. Substituting $T_m(t) = 70°$ in the given formula, produces $\frac{dT}{dt} = k\left(T_m{}^4 - T^4\right) = 40^{-4}\left(70^4 - T^4\right)$. By Euler's method, we have $T_{n+1} = 40^{-4}\left(70^4 - T_n{}^4\right)h + T_n$

n	t	$T(t)$	
0	0	100	
1	0.1	97.0316	
2	0.2	94.5068	
$\vdots$			
10	1	82.6936	$T(1) \approx 82.694°$
11	1.1	81.8049	
$\vdots$			
20	2	76.4459	$T(2) \approx 76.446°$

34.2 SUCCESSIVE APPROXIMATIONS

1. $y' = x + 2$, $(1, 4)$, $x = 2$. Integrating we get $y = \int (x+2)dx = \frac{1}{2}x^2 + 2x + C$. Now substituting 1 for x and 4 for y we obtain $C = 4 - 2\frac{1}{2} = \frac{3}{2}$. Hence $y = \frac{1}{2}x^2 + 2x + \frac{3}{2}$, $y(2) = 7.5$ the exact value.

3. $y' = 2xy + 2x$. Substituting 1 for y we have $y' = 4x$ so $y = 2x^2 + C_1$. Substituting $(0, 1)$ we obtain $C_1 = 1$. The first approximation is $y = 2x^2 + 1$. Substituting this for y in the original equation we have $y' = 2x\left(2x^2 + 1\right) + 2x = 4x^3 + 4x$. Integration yields $y = x^4 + 2x^2 + C_2$. Substituting $(0, 1)$ again we get $C_2 = 1$ so the second approximation is $y = x^4 + 2x^2 + 1$. Substituting for y once more in the original equation we have $y' = 2x\left(x^4 + 2x^2 + 1\right) + 2x = 2x^5 + 4x^3 + 4x$. Integration yields $y = \frac{1}{3}x^6 + x^4 + 2x^2 + C_3$. Again $C_3 = 1$ so the third approximation is $y = \frac{1}{3}x^6 + x^4 + 2x^2 + 1$. $y(1) = 4\frac{1}{3}$ compares favorably with 4.4366, the correct solution.

5. $y' = x^2 + y^2$. Substituting 1 for y we have $y' = x^2 + 1$. By integration $y = \frac{1}{3}x^3 + x + C_1$. Substituting $(0, 1)$ we get $C_1 = 1$. The first approximation is $y = \frac{1}{3}x^3 + x + 1$. Substituting this for y in the original equation yields $y' = x^2 + \frac{1}{9}x^6 + \frac{2}{3}x^4 + \frac{2}{3}x^3 + x^2 + 2x + 1$. Integration yields $y = \frac{1}{63}x^7 + \frac{2}{15}x^5 + \frac{1}{6}x^4 + \frac{2}{3}x^3 + x^2 + x + C_2$. Again, $C_2 = 1$ so the second approximation is $y = \frac{1}{63}x^7 + \frac{2}{15}x^5 + \frac{1}{6}x^4 + \frac{2}{3}x^3 + x^2 + x + 1$.

7. $y' = 4 + xy$. Substituting 1 for y and integrating we have $y = 4x + \frac{1}{2}x^2 + C_1$. With the point $(0, 1)$ we get $C_1 = 1$ and $y = 4x + \frac{1}{2}x^2 + 1$. Substituting this for y in the original equation yields $y' = 4 + 4x^2 + \frac{1}{2}x^3 + x$. Integrating we get $y = \frac{1}{8}x^4 + \frac{4}{3}x^3 + \frac{1}{2}x^2 + 4x + C_2$. Again $C_2 = 1$ so the second approximation is $y = \frac{1}{8}x^4 + \frac{4}{3}x^3 + \frac{1}{2}x^2 + 4x + 1$.

9. $y' = \sin 2x + y$; $\left(\frac{\pi}{2}, 1\right)$. Substituting 1 for y and integrating we get $y = -\frac{1}{2}\cos 2x + x + C_1$. Using $\left(\frac{\pi}{2}, 1\right)$ we get $C_1 = 1 - \frac{1}{2} - \frac{\pi}{2} = \frac{1-\pi}{2}$. The first approximation is $y = -\frac{1}{2}\cos 2x + x + \frac{1-\pi}{2}$. Substituting this for y in the original equation we

get $y' = \sin 2x - \frac{1}{2}\cos 2x + x + \frac{1-\pi}{2}$. Integration yields $y = -\frac{1}{2}\cos 2x - \frac{1}{4}\sin 2x + \frac{1}{2}x^2 + \frac{1-\pi}{2}x + C_2$. Using $\left(\frac{\pi}{2}, 1\right)$ again we get $C_2 = 1 - \frac{1}{2} - \frac{\pi^2}{8} - \frac{1-\pi}{2} \cdot \frac{\pi}{2} = \frac{4-2\pi-\pi^2}{8}$. The second approximation is $y = -\frac{1}{2}\cos 2x - \frac{1}{4}\sin 2x + \frac{1}{2}x^2 + \frac{1-\pi}{2}x + \frac{4-2\pi-\pi^2}{8}$.

11. The Maclaurin series for e^{-x} is $1 - x + \frac{x^2}{2!} - \frac{x^3}{3!} + \frac{x^4}{4!} - \frac{x^5}{5!} + \cdots$. Substituting into $3e^{-x} + x - 1$ we obtain $3\left(1 - x + \frac{x^2}{2} - \frac{x^3}{6} + \frac{x^4}{24} - \frac{x^5}{120}\right) + x - 1 = 2 - 2x + \frac{3x^2}{2} - \frac{x^3}{2} + \frac{x^4}{8} - \frac{x^5}{40}$. The first four terms are the same.

13. $I'(t) = -2I(t) + \sin t$ is like $I' = -2I + \sin t$. Substitute 0 for I and integrate we get $I = -\cos t + C_1$. Using $(0, 0)$ we get $C_1 = 1$. The first approximation is $I = -\cos t + 1$. Substituting into the original equation to get $I' = -2(-\cos t + 1) + \sin t = 2\cos t + \sin t - 2$. Integrating again yields $I = 2\sin t - \cos t - 2t + C_2$. Using $(0, 0)$ again we obtain $C_2 = 1$. The second approximation is $I = 2\sin t - \cos t - 2t + 1$. Substituting once more we have $I' = -3\sin t + 2\cos t + 4t - 2$. Integrating we get $I = 3\cos t + 2\sin t + 2t^2 - 2t + C_3$. Using $(0, 0)$ we obtain $C_3 = -3$. The third approximation is $I(t) = 3\cos t + 2\sin t + 2t^2 - 2t - 3$. $I(0.5) \approx 0.0916$.

34.3 LAPLACE TRANSFORMS

1. $f(t) = n$, $L(f) = L(n) = \int_0^\infty e^{-st} n \, dt$

$= \lim_{b\to\infty} \int_0^b e^{-st} n \, dt$

$= \lim_{b\to\infty} \left[\frac{-n}{s} e^{-st}\right]_0^n$

$= \lim_{n\to\infty} \frac{-n}{b} e^{-sb} + \frac{n}{s} = \frac{n}{s}$

3. $f(t) = \cos at$, $L(f) = L(\cos at) = \int_0^\infty e^{-st} \times \cos at \, dt = \lim_{b\to\infty} \int_0^b e^{-st}\cos at \, dt$. By form 88, we see that this is $\lim_{b\to\infty}\left[\frac{e^{-st}}{s^2+a^2}(-s\cos at + a\sin at)\right]_0^b$
$= \lim_{b\to\infty} \frac{e^{-sb}}{s^2+a^2}(-s\cos ab + a\sin ab) - \frac{1}{s^2+a^2}(-s)$
$= \frac{s}{s^2+a^2}$.

5. $f(t) = t^3 = 3! \cdot \frac{t^3}{3!}$. $l(f) = 3!L\left(\frac{t^3}{3!}\right) = 3!\frac{1}{s^4} = \frac{3!}{s^4}$. (form 3)

7. $f(t) = \sin 6t$, $L(f) = \frac{6}{s^2+36}$. (form 8)

9. $f(t) = t^3 e^{5t}$, $L(f) = \frac{3!}{(s-5)^4}$. (form 7)

11. $f(t) = \sin 4t + 4t\cos 4t$, $L(f) = \frac{2\cdot4s^2}{\left(s^2+4^2\right)^2} = \frac{8s^2}{\left(s^2+16\right)^2}$. (form 14)

13. $L(\sin 5t + \cos 3t) = L(\sin 5t) + L(\cos 3t) = \frac{5}{s^2+25} + \frac{s}{s^2+9}$. (forms 8 and 9)

15. $L(3t\sin 5t - \cos 4t + 1) = 3L(t\sin 5t) - L(1 - \cos 4t) = \frac{30s}{\left(s^2+25\right)^2} - \frac{16}{s\left(s^2+16\right)}$. (forms 10 and 13)

17. $L\left(e^{-2t}\cos 3t - 2e^{5t}\sin t\right) = L\left(e^{-2t}\cos 3t\right) - 2L\times \left(e^{5t}\sin t\right) = \frac{s+2}{(s+2)^2+9} - \frac{2}{(s-5)^2+1}$. (forms 19 and 20)

19. $L(4 + 3t + 2e^t) = 4L(1) + 3L(t) + 2L(e^t) = \frac{4}{s} + \frac{3}{s^2} + \frac{2}{s-1}$. (forms 1, 2 and 5)

21. We need to use the trigonometric identity $\cos^2 \theta = \frac{1}{2}(1 + \cos 2\theta)$ with $\theta = 3t$. This leads to $\cos^2 3t = \frac{1}{2} + \frac{1}{2}\cos 6t$, and so we have $L(\cos^2 3t) = L(\frac{1}{2} + \frac{1}{2}\cos 6t$. By Table 34.1, $L\left(\frac{1}{2}\right) = \frac{1}{2}L(1) = \frac{1}{2}\left(\frac{1}{s}\right)$ and $L(\frac{1}{2}\cos 6t) = \frac{1}{2}L(\cos 6t) = \frac{1}{2}\left(\frac{s}{s^2+6^2}\right) = \frac{1}{2}\left(\frac{s}{s^2+36}\right)$. Thus, $L(\cos^2 3t) = L(\frac{1}{2} + \frac{1}{2}\cos 6t) = \frac{1}{2}\left(\frac{1}{s}\right) + \frac{1}{2}\left(\frac{s}{s^2+36}\right) = \frac{s^2+18}{s\left(s^2+36\right)}$.

≡ 34.4 INVERSE LAPLACE TRANSFORMS AND TRANSFORMS OF DERIVATIVES

1. $\frac{4}{s-5}$. $s - a$ appears in the denominator of form 5. $L^{-1}(F) = 4e^{5t}$.

3. $\frac{16}{s^3+16s} = \frac{16}{s\left(s^2+16\right)} = \frac{4^2}{s\left(s^2+4^2\right)}$ fits form 10. $L^{-1}(F) = 1 - \cos 4t$

5. $\frac{3s}{s^2+9} = 3 \cdot \frac{s}{s^2+3^2}$. $L^{-1}(F) = 3\cos 3t$. (form 9)

7. $\frac{5s}{s^2+9s+14} = \frac{5s}{(s+7)(s+2)}$. Using form 18 with $a = -2$ and $b = -7$ we get $-2e^{-2t} + 7e^{-7t}$

9. $\frac{2s+6s^2}{\left(s^2+25\right)^2} = \frac{1}{5} \cdot \frac{2 \cdot 5s}{\left(s^2+5^2\right)^2} = \frac{6}{10}\frac{2 \cdot 5s^2}{\left(s^2+5^2\right)^2}$. By forms 13 and 14, $L^{-1}(F) = \frac{1}{5}t\sin 5t + \frac{3}{5}(\sin 5t + 5t\cos 5t)$.

11. $L(f'') + 4L(f') = \left[s^2L(f) - sf(0) - f'(0)\right] + 4[sL(f) - f(0)] = \left(s^2 + 4s\right)L(f) - s - 4 = \left(s^2 + 4s\right)L(f) - (s+4)$

13. $2L(f'') - 3L(f') + L(f) = 2\left[s^2L(f) - sf(0) - f'(0)\right] - 3[sL(f) - f(0)] + L(f) = \left(2s^2 - 3s + 1\right)L(f) - (2s - 3)$

15. $L(f'') + 6L(f') - L(f) = \left[s^2L(f) - sf(0) - f'(0)\right] + 6[sL(f) - f(0)] - L(f) = \left(s^2 + 6s - 1\right)L(f) - s + 1 - 6 = \left(s^2 + 6s - 1\right)L(f) - (s+5)$

≡ 34.5 PARTIAL FRACTIONS

1. $\frac{1}{x(x+1)} = \frac{A}{x} + \frac{B}{x+1}$. Multiplying by $x(x+1)$ we get $1 = A(x+1) + Bx$. If $x = 0$ we get $A = 1$. If $x = -1$, we get $B = -1$. The partial fraction decomposition is $\frac{1}{x} + \frac{-1}{x+1}$.

3. $\frac{4}{x(x-1)(x+1)} = \frac{A}{x} + \frac{B}{x-1} + \frac{C}{x+1}$. Multiplying by the LCD we get $4 = A\left(x^2-1\right) + B\left(x^2+x\right) + C\left(x^2-x\right) = (A+B+C)x^2 + (B-C)x - A$. Hence $A = -4$. Substituting this we have $B + C - 4 = 0$ and $B - C = 0$. Solving we get $B = 2$ and $C = 2$. The partial fraction decomposition is $\frac{-4}{x} + \frac{2}{x-1} + \frac{2}{x+1}$.

5. $\frac{3x}{x^2-8+15} = \frac{3x}{(x-3)(x-5)} = \frac{A}{x-3} + \frac{B}{x-5}$. Multiplying by the LCD we get $3x = A(x-5) + B(x-3)$. Setting $x = 3$ yields $A = \frac{9}{-2} = -\frac{9}{2}$. Setting $x = 5$ yields $B = 15/2$. Hence the answer is $\frac{-9/2}{x-3} + \frac{15/2}{x-5}$.

7. $\frac{x-1}{x^2+2x+1} = \frac{x-1}{(x+1)^2} + \frac{A}{x+1} + \frac{B}{(x+1)^2}$. This is equivalent to $x - 2 = A(x+1) + B$. Setting $x = 1$ yields $B = -2$. Substituting this we have $x - 1 = Ax + A - 2$. Using the x-term we see $A = 1$. The answer is $\frac{1}{x+1} + \frac{-2}{(x+1)^2}$.

9. $\frac{x^3+x^2+x+2}{\left(x^2+1\right)\left(x^2+2\right)} = \frac{Ax+B}{x^2+1} + \frac{Cx+D}{x^2+2}$ is equivalent to $x^3 + x^2 + x + 2 = (Ax+B)\left(x^2+2\right) + (Cx+D)\left(x^2+1\right) = (A+C)x^3 + (B+D)x^2 + (2A+C)x + (2B+D)$. So $A + C = 1$ and $2A + C = 1$ yield $A = 0$ and $C = 1$. Also, $B + D = 1$ and $2B + D = 2$ yield $B = 1$ and $D = 0$. The partial fraction decomposition is $\frac{1}{x^2+1} + \frac{x}{x^2+2}$.

11. $L^{-1}\left[\frac{1}{s(s+1)}\right] = L^{-1}\left[\frac{1}{s}\right] + L^{-1}\left[\frac{-1}{s+1}\right] = 1 - e^{-t}$.

13. $L^{-1}\left[\frac{4}{s(s-1)(s+1)}\right] = L^{-1}\left[\frac{-4}{s}\right] + L^{-1}\left[\frac{2}{s-1}\right] + L^{-1}\left[\frac{2}{s+1}\right] = -4 + 2e^t + 2e^{-t}$.

15. $L^{-1}\left[\frac{3s}{s^2-8s+15}\right] = L^{-1}\left[\frac{-9/2}{s-3}\right] + L^{-1}\left[\frac{15/2}{s-5}\right] = -\frac{9}{2}e^{3t} + \frac{15}{2}e^{5t}$.

17. $L^{-1}\left[\frac{s-1}{s^2+s+1}\right] = L^{-1}\left[\frac{1}{s+1}\right] + L^{-1}\left[\frac{-2}{(s+1)^2}\right] = e^{-t} - 2te^{-t}$

19. $L^{-1}\left[\frac{s^3+s^2+s+2}{(s^2+1)(s^2+2)}\right] = L^{-1}\left[\frac{1}{s^2+1}\right] + L^{-1}\left[\frac{s}{s^2+2}\right] = \sin t + \cos\sqrt{2}t$.

21. First we must find the partial fraction decomposition of $\frac{2}{s^2(s-1)}$. We set it equal to $\frac{A}{s} + \frac{B}{s^2} + \frac{C}{s-1}$. Multiplying by the LCD we obtain $2 = As(s-1) + B(s-1) + Cs^2$. Setting $s = 0$ yields $B = -2$. Setting $s = 1$ yields $c = 2$. Expanding the s^2-term is $(A+C)s^2$ so $A = -C = -2$. $L^{-1}\left[\frac{2}{s^2(s-1)}\right] = L^{-1}\left[\frac{-2}{s}\right] + L^{-1}\left[\frac{-2}{s^2}\right] + L^{-1}\left[\frac{2}{s-1}\right] = -2 - 2t + 2e^t$.

23. $\frac{1}{(s+4)(s^2+9)} = \frac{A}{s+4} + \frac{Bs+C}{s^2+9}$ is equivalent to $1 = A(s^2+9) + (Bs+C)(s+4) = (A+B)s^2 + (4B+C)s + 9A + 4C$. Hence $A+B = 0$, $4B+C = 0$ and $9A+4C = 1$. Solving we find $A = \frac{1}{25}$, $B = -\frac{1}{25}$, and $C = \frac{4}{25}$. $L^{-1}\left[\frac{1}{(s+4)(s^2+9)}\right] = L^{-1}\left[\frac{1/25}{s+4}\right] + L^{-1}\left[\frac{-s/25+4/25}{s^2+9}\right] = \frac{1}{25}e^{-4t} - \frac{1}{25}\cos 3t + \frac{4}{25}\sin 3t$.

25. $\frac{1}{(s^2+1)(s-1)^2} = \frac{A}{s-1} + \frac{B}{(s-1)^2} + \frac{Cs+D}{s^2+1}$. Multiplying by the LCD we obtain $1 = A(s-1)(s^2+1) + B(s^2+1) + (Cs+D)(s-1)^2$. Setting $s = 1$ we get $B + \frac{1}{2}$. Expanding the right-hand side of the equation we get $As^3 - As^2 + As - A + Bs^2 + B + Cs^3 - 2Cs^2 + Cs + Ds^2 - 2Ds + D$. Collecting like terms we have $(A+C)s^3 + (-A+B-2C+D)s^2 + (A+C-2D)s + (A+B+D)$. So $A + C = 0$, $(-A+B-2C+D) = 0$, $(A+C-2D) = 0$, and $(-A+B+D) = 1$. Solving we get $A = -\frac{1}{2}$, $B = \frac{1}{2}$, $C = \frac{1}{2}$, and $D = 0$. Hence $L^{-1}\left[\frac{1}{(s^2+1)(s-1)^2}\right] = L^{-1}\left[\frac{-1/2}{s-1}\right] + L^{-1}\left[\frac{1/2}{(s-1)^2}\right] + L^{-1}\left[\frac{s/2}{s^2+1}\right] = -\frac{1}{2}e^t + \frac{1}{2}te^t + \frac{1}{2}\cos t$.

≡ 34.6 USING LAPLACE TRANSFORMS TO SOLVE DIFFERENTIAL EQUATIONS

1. $y' - y = 1$, $y(0) = 0$, $L(y') = sL(y) - y(0) = sL(y)$. Hence the Laplace Transform is $sY - Y = \frac{1}{s}$ or $Y(s-1) = \frac{1}{s}$ or $Y = \frac{1}{s(s-1)}$. The partial fraction decomposition for $\frac{1}{s(s-1)} = \frac{A}{s} + \frac{B}{s-1}$ or $1 = A(s-1) + Bs$. Setting $s = 0$ yields $A = -1$. Setting $s = 1$ yields $B = 1$ so $Y = \frac{-1}{s} + \frac{1}{s-1}$ and $L^{-1}(Y) = L^{-1}\left(\frac{-1}{s}\right) + L^{-1}\left[\frac{1}{s-1}\right]$ or $y = -1 + e^t$.

3. $y' + 6y = e^{-6t}$, $y(0) = 2$. The Laplace Transform is $sY - y(0) + 6Y = \frac{1}{s+6}$ or $(s+6)Y - 2 = \frac{1}{s+6}$. So $(s+6)Y = \frac{1}{s+6} + 2 = \frac{1+2s+12}{s+6} = \frac{2s+13}{s+6}$. Hence $Y = \frac{2s+13}{(s+6)^2} = \frac{A}{s+6} + \frac{B}{(s+6)^2}$. Multiplying by the LCD we have $2s + 13 = A(s+6) + B = As + 6A + B$. Hence $A = 2$ and $6A + B = 13$ or $B = 1$, $y = L^{-1}\left[\frac{2}{s+6}\right] + L^{-1}\left[\frac{1}{(s+6)^2}\right] = 2e^{-6t} + te^{-6t}$.

5. $y' + 5y = te^{-5t}$, $y(0) = 3$. The Laplace Transform is $sY - y(0) + 5Y = \frac{1}{(s+5)^2}$. Hence $(s+5)Y = \frac{1}{(s+5)^2} + 3 = \frac{1+3s^2+30s+75}{(s+5)^2} = \frac{3s^2+30s+76}{(s+5)^2}$. This yields $Y = \frac{3s^2+30s+76}{(s+5)^3} = \frac{A}{s+5} + \frac{B}{(s+5)^2} + \frac{C}{(s+5)^3}$ or $3s^2 + 30s + 76 = A(s+5)^2 + B(s+5) + C$. Letting $s = -5$ we get $C = 1$. The s^2-term gives $A = 3$. The s-term yields $10A + B = 30$, so $B =$

$30 - 10A = 30 - 30 = 0$. Hence $Y = \frac{3}{s+5} + \frac{1}{(s+5)^3}$ and $y = L^{-1}\left[\frac{3}{s+5} + \frac{1}{(s+5)^3}\right] = 3e^{-5t} + \frac{1}{2}t^2e^{-5t}$.

7. $y'' + 4y = 0$, $y'(0) = 3$, $y(0) = 2$. The Laplace Transform is $s^2Y - sy(0) - y(0) + 4Y = 0$ or $(s^2+4)Y - 2s - 3 = 0$. Hence $Y = \frac{2s+3}{s^2+4} = 2\frac{s}{s^2+2^2} + \frac{3}{s^2+2^2}$ and $y = 2\cos 2t + \frac{3}{2}\sin 2t$.

9. $y'' + 4y = e^t$, $y(0) = y'(0) = 0$. The Laplace transform is $s^2Y + 4y + \frac{1}{s-1}$ or $Y = \frac{1}{(s-1)(s^2+4)} = \frac{A}{s-1} + \frac{Bs+C}{s^2+4}$. Multiplying by the LCD we get $A(s^2+4) + (Bs+C)(s-1) = 1$. Setting $s = 1$ we get $A = \frac{1}{5}$. Expanding we get $(A+B)s^2 + (-B+C)s + (4A-C) = 1$. Hence $B = -A = -\frac{1}{5}$, and $C = B = -\frac{1}{5}$. So the Laplace becomes $Y = \frac{1}{5}\left(\frac{1}{s-1} + \frac{-s-1}{s^2+4}\right) = \frac{1}{5}\left(\frac{1}{s-1} - \frac{s}{s^2+2^2} - \frac{1}{s^2+2^2}\right)$ and the inverse Laplace transform yields $y = \frac{1}{5}\left(e^t - \cos 2t - \frac{1}{2}\sin 2t\right)$.

11. $y'' - 6y' + 9y = t$, $y(0) = 0$, $y'(0) = 1$. The Laplace Transform is $s^2Y - sy(0) - y'(0) - 6(sY - y(0)) + 9 = \frac{1}{s^2}$ or $(s^2 - 6s + 9)Y - 1 = \frac{1}{s^2}$ or $(s^2 - 6s + 9)Y = \frac{1}{s^2} + 1 = \frac{1+s^2}{s^2}$. Hence $Y =$

$\frac{s^2+1}{s^2(s-3)^2} = \frac{A}{s} + \frac{B}{s^2} + \frac{C}{s-3} + \frac{|Q}{(s-3)^2}$. Multiplying by the LCD we have $s^2 + 1 = As(s-3)^2 + B(s-3)^2 + C(s-3)s^2 + Ds^2$. Setting $s = 0$, we find $B = \frac{1}{9}$, and setting $s = 3$, we get $D = \frac{10}{9}$. The s-term yields $9A - 6B = 0$ so $A = \frac{6B}{9} = \frac{6}{81} = \frac{2}{27}$. The s^3-term yields $A + C = 0$ so $C = -A = -\frac{2}{27}$. $Y = \frac{2}{27}\frac{1}{s} + \frac{1}{9}\frac{1}{s^2} - \frac{2}{27}\frac{1}{s-3} + \frac{10}{9} \cdot \frac{1}{(s-3)^2}$. The inverse Laplace Transform yields $y = \frac{2}{27} + \frac{1}{9}t - \frac{2}{27}e^{3t} + \frac{10}{9}te^{3t}$.

13. $y'' + 6y' + 13y = 0$, $y(0) = 1$, $y''(0) = -4$. The Laplace Transform is $s^2Y - s \cdot 1 + 4 + 6(sY - 1) + 13Y = 0$ which simplifies to $(s^2 + 6s + 13)Y = s + 2$, so $Y = \frac{s+2}{s^2+6s+13} = \frac{s+2}{s^2+6s+9+4} = \frac{s+3-1}{(s+3)^2+2^2} = \frac{s+3}{(s+3)^2+2^2} - \frac{1}{2}\frac{2}{(s+3)^2+2^2}$. The inverse Laplace Transform yields $y = e^{-3t}\left(\cos 2t - \frac{1}{2}\sin 2t\right)$.

15. $y'' - 4y = 3\cos t$, $y(0) = y'(0) = 0$. The Laplace Transform is $s^2Y - 4Y = \frac{3s}{s^2+1}$ so $Y = \frac{3s}{(s^2+1)(s^2-1)} = \frac{3s}{(s^2+1)(s+2)(s-2)} = \frac{A}{s+2} + \frac{B}{s-2} + \frac{Cs+D}{s^2+1}$. Multiplying by the LCD we get $3s = A(s-2)(s^2+1) + B((s+2)(s^2+1)) + (Cs+D)(s^2-4)$. Setting $s = -2$, we get $A = \frac{-6}{(-4)(5)} = \frac{3}{10}$. Setting $s = 2$, we get $B = \frac{6}{4 \cdot 5} = \frac{3}{10}$. The s^3-term of the expansion is $A + B + C = 0$ so $C = -\frac{6}{10} = -\frac{3}{5}$. The constant term of the expansion is $-2A + 2B - 4D = 0$ so $D = \frac{-2A+2B}{4} = 0$. Hence $Y = \frac{3}{10} \cdot \frac{1}{s+2} + \frac{3}{10} \cdot \frac{1}{s-2} - \frac{3}{5} \cdot \frac{s}{s^2+1}$ and the inverse Laplace Transform is $y = \frac{3}{10}e^{-2t} + \frac{3}{10}e^{2t} - \frac{3}{5}\cos t$.

17. $y'' + 2y' + 5y = 3e^{-2t}$, $y(0) = y'(0) = 1$. The Laplace transform is $(s^2Y - s - 1) + 2sY - 2 + 5Y = \frac{3}{s+2}$. This is equivalent to $(s^2 + 2s + 5)Y - (s + 3) = \frac{3}{s+2}$. So $(s^2 + 2s + 5)Y = \frac{3}{s+2} + (s + 3) = \frac{3+(s^2+5s+6)}{s+2} = \frac{s^2+5s+9}{s+2}$ Hence $Y = \frac{(s^2+5s+9)}{(s+2)(s^2+2s+5)} = \frac{A}{s+2} + \frac{Bs+C}{s^2+2s+5}$. This gives $(s^2 + 5s + 9) = A(s^2 + 2s + 5) + (Bs+C)(s+2) = (A + B)s^2 + (2A + 2B + C)s + 5A + 2C$. And we have linear equations $A + B = 1$, $2A + 2B + C = 5$ and $5A + 2C = 9$. Solving we obtain $A = \frac{3}{5}$, $B = \frac{2}{5}$, and $C = 3$. So $Y = $

$\frac{3}{5} \cdot \frac{1}{s+2} + \frac{\frac{2}{5}s + 3}{s^2 + 2s + 5}$. The second fraction is $\frac{2}{5}\frac{s+\frac{15}{2}}{s^2+2s+1+4} = \frac{2}{5}\frac{s+1+\frac{13}{2}}{(s+1)^2+2^2} = \frac{2}{5}\frac{s+1}{(s+1)^2+2^2} = \frac{13}{10}\frac{2}{(s+1)^2+2^2}$. The inverse Laplace transform is $y = \frac{3}{5}e^{-2t} + e^{-t}\left(\frac{2}{5}\cos 2t + \frac{13}{10}\sin 2t\right)$.

19. $y'' + 2y' - 3y = te^{2t}$, $y(0) = 2$, $y'(0) = 3$. The Laplace transform is $(s^2Y - 2s - 3) + 2sY - 4 - 3Y = \frac{1}{(s-2)^2}$. Regrouping we get $(s^2+2s-3)Y - (2s+7) = \frac{1}{(s-2)^2}$. Hence $(s^2+2s-3)Y = \frac{1}{(s-2)^2} + (2s+7) = \frac{2s^3-s^2-20s+29}{(s-2)^2}$ or $Y = \frac{2s^3-s^2-20s+29}{(s-2)^2(s+3)(s-1)} = \frac{A}{s+3} + \frac{B}{s-1} + \frac{C}{s-2} + \frac{|Q}{(s-2)^2}$. Multiplying by the LCD we get $2s^3 - s^2 - 20s + 29 = A(s-1)(s-2)^2 + B(s+3)(s-2)^2 + C(s+3)(s-1)(s-2) + D(s+3)(s-1)$. Setting $s = -3$ we get $A = \frac{26}{-100} = \frac{-13}{50}$. Setting $s = 1$ we get $B = \frac{10}{4} = \frac{5}{2}$. Setting $s = 2$, we get $D = \frac{1}{5}$. the s^3-term yields $A + B + C = 2$ or $C = 2 - A - B = \frac{6}{25}$. Hence we have $Y = -\frac{13}{50} \cdot \frac{1}{s+3} + \frac{5}{2} \cdot \frac{1}{s-1} - \frac{6}{25} \cdot \frac{1}{s-2} + \frac{1}{5} \cdot \frac{1}{(s-2)^2}$. The inverse Laplace Transform gives the answer $y = -\frac{15}{50}e^{-3t} + \frac{5}{2}e^t - \frac{6}{25}e^{2t} + \frac{1}{5}te^{2t}$.

21. $y'' + 2y + 5y = 10\cos t$, $y(0) = 2$, $y'(0) = 1$. The Laplace transform is $(s^2Y - 2s - 1) + (2sY - 4) + 5Y = \frac{10s}{s^2+1}$. This is equivalent to $(s^2 + 2s + 5)Y - (2s + 5) = \frac{10s}{s^2+1}$ or $(s^2 + 2s + 5)Y = \frac{10s}{s^2+1} + (2s + 5) = \frac{2s^3+5s^2+12s+5}{s^2+1}$ or $Y = \frac{2s^3+5s^2+12s+5}{(s^2+1)(s^2+2s+5)} = \frac{As+B}{s^2+1} + \frac{Cs+D}{s^2+2s+5}$. Multiplying by the LCD we get $2s^3 + 5s^2 + 12s + 5 = (As+B)(s^2+2s+5) + (Cs+D)(s^2+1) = As^3 + (2A+B)s^2 + (5A+2B)s + 5B + Cs^3 + ds^2 + Cs + D$. The yields the linear equations $A + C = 2$, $2A + B + D = 5$, $5A + 2B + C = 12$, and $5B + D = 5$. Solving this system we get $A = 2$, $B = 1$, $C = 0$, and $D = 0$. Hence the partial fraction decomposition is $Y = \frac{2s+1}{s^2+1} = 2\frac{s}{s^2+1} + \frac{1}{s^2+1}$. The inverse Laplace transform yields the answer $y = 2\cos t + \sin t$. (Note: $2s^3 + 5s^2 + 12s + 5$ factors into $(2s+1)(s^2+2s+5)$ so we could have reduced $\frac{2s^3+5s^2+12s+5}{(s^2+1)(s^2+2s+5)}$ to $\frac{2s+1}{s^2+1}$ and not used partial fractions.)

23. We start with the equation $L\frac{d^2q}{dt^2} + R\frac{dq}{dt} + \frac{1}{C}q = E(t)$ which becomes $q'' + 20q' + 200q = 150$. The Laplace transform is $s^2Q + 20sQ + 200Q = \frac{150}{s}$. Hence $(s^2 + 20s + 200)Q = \frac{150}{s}$ or $Q = \frac{150}{s(s^2+20s+200)} = \frac{A}{s} + \frac{Bs+C}{s^2+20s+200}$. Multiplying by the LCD we get $10 = A(s^2 + 20s + 200) + (Bs + C) = (A + B)s^2 + (20A + C)s + 200A$. The constant term yields $A = \frac{150}{200} = \frac{3}{4}$. The s^2-term yields $A + B = 0$ so $B = -\frac{3}{4}$. The s-term yields $20A + C = 0$ or $C = -15$. Hence $Q = \frac{3}{4} \cdot \frac{1}{s} + \frac{-\frac{3}{4}s - 15}{s^2 + 20s + 200} = \frac{3}{4} \cdot \frac{1}{s} + \frac{-\frac{3}{4}(s + 20)}{s^2 + 20s + 100 + 100} = \frac{3}{4} \cdot \frac{1}{s} - \frac{3}{4} \cdot \frac{s+10+10}{(s+10)^2+10^2} = \frac{3}{4}\frac{1}{s} - \frac{3}{4}\frac{s+10}{(s+10)^2+10^2} - \frac{3}{4}\frac{10}{(s+10)^2+10^2}$. The inverse Laplace transform yields the answer $q(t) = \frac{3}{4} - \frac{3}{4}e^{-10t}(\cos 10t + \sin 10t)$. $i(t) = q'(t) = \frac{15}{2}e^{-10t}(\cos 10t + \sin 10t) - \frac{15}{2}e^{-10t} \times (-\sin 10t + \cos 10t) = 15e^{-10t}\sin 10t$. The steady state current is 0.

25. We begin with the equation $0.9q'' + 6q' + 50q = 6$. The Laplace transform is $0.1s^2Q + 6sQ + 50 = \frac{6}{s}$ so $Q = \frac{6}{s(0.1s^2+6s+50)}$ or $Q = \frac{60}{s(s^2+60s+500)}$ or $Q = \frac{60}{s(s+10)(s+50)} = \frac{A}{s} + \frac{B}{s+10} + \frac{C}{s+50}$. Multiplying by the LCD we get $60 = A(s + 10)(s + 50) + Bs(s + 50) + Cs(s + 10)$. Setting $s = 0$, we get $A = \frac{60}{500} = \frac{3}{25}$. Setting $s = -10$, we get $B = \frac{60}{-400} = -\frac{3}{20}$. Setting $s = -50$, yields $C = \frac{60}{2000} = \frac{3}{100}$. Hence we get $Q = \frac{3}{25} \cdot \frac{1}{s} - \frac{3}{20} \cdot \frac{1}{s+10} + \frac{3}{100} \cdot \frac{1}{s+50}$. The inverse Laplace transform yields the answer $q(t) = \frac{3}{25} - \frac{3}{20}e^{-10t} + \frac{3}{100}e^{-50t}$. Thus, $i(t) = q'(t) = \frac{3}{2}e^{-10t} - \frac{3}{2}e^{-50t} = \frac{3}{2}(e^{-50t} - e^{-50t})$.

27. First, a 16 pound weight has a mass of $\frac{1}{4}$ slug. The differential equation is $\frac{1}{2}\frac{d^2x}{dt^2} = -4x - 64\frac{dx}{dt}$, we need not bother changing to slugs. The differential equation is $16\frac{d^2x}{dt^2} = -4x - 64\frac{dx}{dt}$ which is equivalent to $x'' + 128x' + 8x = 0$. This has auxiliary equation $m^2 + 128m + 18 = 0$, which has roots $\frac{-128\pm\sqrt{128^2-32}}{2} = -64\pm\sqrt{4088} = -64\pm2\sqrt{1022}$. Hence the general solution is $x = c_1e^{(-64+2\sqrt{1022})t} + c_2e^{(-64-2\sqrt{1022})t}$. When $t = 0$, then $x = 2$, so $c_1 + c_2 = 2$. When $t = 0$, $x' = 0$, so $(-64 + 2\sqrt{1022})c_1 + (-64 - 2\sqrt{1022})c_2 = 0$. Substituting $2 - c_2$ for c_1, we get $(-64+2\sqrt{1022})(2-c_2) + (-64-2\sqrt{1022})c_2 = 0$ or $-128+4\sqrt{1022}+64c_2 - 2\sqrt{1022}c_2 - 64c_2 - 2\sqrt{1022} = 0$ or $-4\sqrt{1022}c_2 = 128 - 4\sqrt{1022}$, and so $c_2 = 1 - \frac{32}{\sqrt{1022}}$. Thus, $c_1 = 2 - c_2 = 2 - \left(1 - \frac{32}{\sqrt{1022}}\right) = 1 + \frac{32}{\sqrt{1022}}$. Thus, the final solution is $x = \left(1 + \frac{32}{\sqrt{1022}}\right)e^{(-64+2\sqrt{1022})t} + \left(1 - \frac{32}{\sqrt{1022}}\right)e^{(-64-2\sqrt{1022})t}$.

29. The Laplace transform of the left-hand side of the differential equation $y'' + 2y' + 17y = e^{-t}\cos 4t$ is $s^2Y - sy(0) - y'(0) + 2[sY - y(0)] + 17Y = s^2Y + 2sY + 17Y = Y(s^2 + 2s + 17)$. The Laplace transform of the right-hand side is $\frac{s+1}{(s+1)^2+4^2} = \frac{s+1}{s^2+2s+17}$. Hence, $Y(s^2 + 2s + 17) = \frac{s+1}{s^2+2s+17} = \frac{s+1}{(s+1)^2+4^2}$ or $Y = \frac{s+1}{(s^2+2s+17)^2} = \frac{s+1}{[(s+1)^2+4^2]^2} = \frac{s}{[(s+1)^2+4^2]^2} + \frac{1}{[(s+1)^2+4^2]^2}$. Using formulas 21 and 22 in Table 34.1, we obtain

$$y = L^{-1}\left(\frac{s+1}{[(s+1)^2+4^2]^2}\right)$$

$$= L^{-1}\left(\frac{s+1}{[(s+1)^2+4^2]^2}\right)$$

$$+ L^{-1}\left(\frac{s+1}{[(s+1)^2+4^2]^2}\right)$$

$$= \frac{1}{128}e^{-t}[4t\cos(4t) + 16t\sin(4t) - \sin(4t)]$$

$$+ \frac{1}{128}e^{-t}[\sin(4t) - 4t\cos(4t)]$$

$$= \frac{1}{8}te^{-t}\sin 4t$$

1.

	y	
x	$h = 0.1$	$h = 0.05$
0	2	2
0.05	–	1.8
0.10	1.6	1.6299
0.15	–	1.4838
0.20	1.3184	1.3572
0.25	–	1.2467
0.30	1.1098	1.1495
0.35	–	1.0636
0.40	0.9497	0.9873
0.45	–	0.9190
0.50	0.8234	0.8578
0.55	–	0.8026
0.60	0.7217	0.7527
0.65	–	0.7074
0.70	0.6384	0.6661
0.75	–	0.6284
0.80	0.5691	0.5938
0.85	–	0.5621
0.90	0.5108	0.5329
0.95	–	0.5059
1.00	0.4612	0.4809

2.

	y	
x	$h = 0.1$	$h = 0.05$
0	1	1
0.05	–	1.1
0.10	1.2	1.2052
0.15	–	1.3157
0.20	1.4218	1.4313
0.25	–	1.5520
0.30	1.6649	1.6778
0.35	–	1.8088
0.40	1.9288	1.9448
0.45	–	2.0859
0.50	2.2133	2.2321
0.55	–	2.3833
0.60	2.5183	2.5395
0.65	–	2.7001
0.70	2.8436	2.8671
0.75	–	3.0385
0.80	3.1892	3.2149
0.85	–	3.3963
0.90	3.5549	3.5827
0.95	–	3.7742
1.00	3.9408	3.9706

3.

	y
x	$h = 0.05$
0	60
0.05	59.9
0.10	59.8
0.15	59.6990
0.20	59.5961
0.25	59.4905
0.30	59.3812
0.35	59.2678
0.40	59.1496
0.45	59.0263
0.50	58.8975
0.55	58.7630
0.60	58.6228
0.65	58.4767
0.70	58.3249
0.75	58.1675
0.80	58.0047
0.85	57.8366
0.90	57.6635
0.95	57.4857
1.00	57.3034

4.

	y
x	$h = 0.01$
0	0
0.01	0.06
0.02	0.1200
0.03	0.1800
0.04	0.2400
0.05	0.2999
0.06	0.3598
0.07	0.4197
0.08	0.4796
0.09	0.5394
0.10	0.5991

5. $y' = xy + 4$, $(0, 1)$. Substituting 1 for y we get $y' = x + 4$. Integrating we have $y = \frac{1}{2}x^2 + 4x + C_1$. Solving for C_1 we get $C_1 = 1$, so the first approximation is $y = \frac{1}{2}x^2 + 4x + 1$. Now we substitute this for y and get $y' = \frac{1}{2}x^3 + 4x^2 + x + 4$. Integrating this we get $y = \frac{1}{8}x^4 + \frac{4}{3}x^3 + \frac{1}{2}x^2 + 4x + C_2$. Again $C_2 = 1$ so the second approximation is $y = \frac{1}{8}x^4 + \frac{4}{3}x^3 + \frac{1}{2}x^2 + 4x + 1$.

6. $y' = 2x + y$. Substituting 0 for y we have $y' = 2x$. Integration yields $y = x^2 + C_1$. $C_1 = 0$ so the first approximation is $y = x^2$. Substituting into the original equation we get $y' = 2x + x^2$. Integrating, we have $y = x^2 + \frac{1}{3}x^3 + C_2$. Again $C_2 = 0$ so the second approximation is $y = x^2 + \frac{1}{3}x^3$.

7. $y' = y\sqrt{x-9}$, $(10, 1)$. Substituting 1 for y we get $y' = \sqrt{x-9} = (x-9)^{1/2}$. Integration yields $y = \frac{2}{3}(x-9)^{\frac{3}{2}} + C_1$. Substituting $(10, 1)$ we get $1 = \frac{2}{3}(1) + C_1$ or $C_1 = \frac{1}{3}$. The first approximation is $y = \frac{2}{3}(x-9)^{3/2} + \frac{1}{3}$. Now substituting this for y in the original equation we obtain $y' = \left(\frac{2}{3}(x-9)^{3/2} + \frac{1}{3}\right)(x-9)^{1/2} = \frac{2}{3}(x-9)^2 + \frac{1}{3}(x-9)^{1/2}$.

Integrating again we get $y = \frac{2}{9}(x-9)^3 + \frac{2}{9}(x-9)^{\frac{3}{2}}$. Solving for C_2 we have $1 - \frac{2}{9} - \frac{2}{9} = \frac{5}{9}$. The second approximation is $y = \frac{2}{9}(x-9)^3 + \frac{2}{9}(x-9)^{\frac{3}{2}} + \frac{5}{9}$.

8. $y' = y + \cos x$, $\left(\frac{\pi}{2}, 0\right)$. Substituting 0 for y we have $y' = \cos x$ so $y = \sin x + C_1$. Using $\left(\frac{\pi}{2}, 0\right)$ we obtain $C_1 = -1$ so $y = \sin x - 1$. Substituting this in the original equation we get $y' = \sin x + \cos x - 1$. Integration yields $y = -\cos x + \sin x - x + C_2$. $C_2 = \frac{\pi}{2} - 1$ and the second approximation is $y = -\cos x + \sin x - x + \frac{\pi}{2} - 1$.

9. $2y' - y = 4$, $y(0) = 1$ has Laplace transform $2sY - 2 - Y = \frac{4}{s}$ or $(2s-1)Y = \frac{4}{s} + 2 = \frac{4+2s}{s}$. So we get $Y = \frac{4+2s}{s(2s-1)} = \frac{A}{s} + \frac{B}{2s-1}$. Multiplying by the LCD we have $4 + 2s = A(2s-1) + Bs$. Setting $s = 0$, we obtain $A = \frac{4}{-1} = -4$. Setting $s = \frac{1}{2}$, we obtain $B = \frac{5}{1/2} = 10$. So we have $Y = -4\frac{1}{s} + 10 \cdot \frac{1}{2s-1} = -4\frac{1}{s} + 5\frac{1}{s-\frac{1}{2}}$. The inverse Laplace transformation yields the solution $y = -4 + 5e^{t/2}$.

10. $3y' + y = t$, $y(0) = 2$. The Laplace transform is $3sY - 6 + Y = \frac{1}{s^2}$ so $(3s+1)Y = \frac{1}{s^2} + 6 = \frac{1+6s^2}{s^2}$ and $Y = \frac{1+6s^2}{s^2(3s+1)} = \frac{A}{3s+1} + \frac{B}{s} + \frac{C}{s^2}$. Multiplying by the LCD we have $1 + 6s^2 = As^2 + Bs(3s+1) + C(3s+1)$. Setting $s = -\frac{1}{3}$, we obtain $A = \frac{5/3}{1/9} = 15$. Setting $s = 0$, we obtain $C = 1$. The s^2-term yields $6 = A + 3B$ so $B = \frac{6-A}{3} = \frac{-9}{3} = -3$. So the partial fraction decomposition

is $Y = \frac{15}{3s+1} - \frac{3}{s} + \frac{1}{s^2} = \frac{5}{s+\frac{1}{3}} - 3 \cdot \frac{1}{s} + \frac{1}{s^2}$. The inverse Laplace transform yields the answer $y = 5e^{-t/3} - 3 + t$.

11. $y' + 2y = e^t$, $y(0) = 1$. The Laplace transform is $sY - 1 + 2Y = \frac{1}{s-1}$ or $(s+2)Y = \frac{1}{s-1} + 1 = \frac{s}{s-1}$. So $Y = \frac{s}{(s-1)(s+2)} = \frac{A}{s-1} + \frac{B}{s+2}$. This gives $s = A(s+2) + B(s-1)$. Setting $s = 1$, we get $A = \frac{1}{3}$. Setting $s = -2$, we get $B = \frac{2}{3}$. Hence $Y = \frac{1}{3} \cdot \frac{1}{s-1} + \frac{2}{3} \cdot \frac{1}{s+2}$. The inverse Laplace transform yields $y = \frac{1}{3}e^t + \frac{2}{3}e^{-2t}$.

12. $y' + 5y = 0$, $y(0) = 1$. The Laplace transform is $sY - 1 + sY = 0$ or $(s+5)Y = 1$ or $Y = \frac{1}{s+5}$. The inverse Laplace transform yields $y = e^{-5t}$.

13. $y'' - y = 0$, $y(0) = y'(0) = 1$. The Laplace transform is $s^2Y - s - 1 - Y = 0$ or $(s^2 - 1)Y = s + 1$. Hence $Y = \frac{s+1}{s^2-1} = \frac{1}{s-1}$. The inverse Laplace transform yields $y = e^t$.

14. $y'' - y = e^t$, $y(0) = y'(0) = 0$. The Laplace transform is $s^2Y - Y = \frac{1}{s-1}$ or $Y = \frac{1}{(s-1)(s^2-1)} = \frac{1}{(s-1)^2(s+1)}$. Getting partial fractions we have $\frac{1}{(s-1)^2(s+1)} = \frac{A}{s-1} + \frac{B}{(s-1)^2} + \frac{C}{s+1}$. Multiplying we get $1 = A(s^2-1) + B(s+10) + C(s-1)^2$. Setting $s = 1$, we obtain $B = \frac{1}{2}$. The s^2-term yields $A + C = 0$. The constant term yields $-A + B + C = 1$ or $-A + C = \frac{1}{2}$. Solving, we get $C = \frac{1}{4}$ nd $A = -\frac{1}{4}$. Hence $Y = -\frac{1}{4}\frac{1}{s-1} + \frac{1}{2} \cdot \frac{1}{(s-1)^2} + \frac{1}{4} \cdot \frac{1}{s+1}$. The inverse Laplace transform yields $y = -\frac{1}{4}e^t + \frac{1}{2}te^t + \frac{1}{4}e^{-t}$.

15. $y'' + 2y' + 5y = 0$, $y(0) = 1$, $y'(0) = 0$. The Laplace transform is $s^2Y - s + 2sY - 2 + 5y = 0$ or $(s^2 + 2s + 5)Y = s + 2$. So $Y = \frac{s+2}{s^2+2s+5} = \frac{s+2}{(s+1)^2+2^2} = \frac{s+1}{(s+1)^2+2^2} + \frac{1}{2}\frac{2}{(s+1)^2+2^2}$. The inverse Laplace transform yields the answer $y = e^{-t}\cos 2t + \frac{1}{2}e^{-t}\sin 2t$.

16. $y'' + y' + y = 0$, $y(0) = 4$, $y'(0) = -2$. The Laplace transform is $(s^2Y - 4s + 2) + (sY - 4) + Y = 0$ or $(s^2 + s + 1)Y = 4s + 2$. Hence

$Y = \frac{4s+2}{s^2+s+1} = \frac{4s+2}{s^2+s+\frac{1}{4}+\frac{3}{4}} = 4\frac{s+\frac{1}{2}}{\left(s+\frac{1}{2}\right)^2+\frac{3}{4}}$.

The inverse Laplace transform gives

$y = 4e^{-t/2}\cos\frac{\sqrt{3}}{2}t$.

17. $y'' + 2y' + 5y = 3e^{-2t}$, $y(0) = y'(0) = 1$. The Laplace transform is $(s^2Y - s - 1) + 2sY - 2 + 5 = \frac{3}{s+2}$ or $(s^2 + 2s + 5)Y - (s + 3) = \frac{3}{s+2}$. Hence $(s^2 + 2s + 5)Y = \frac{3}{s+2} + s + 3 = \frac{s^2+5s+9}{s+2}$ and $Y = \frac{s^2+5s+9}{(s+9)(s^2+2s+5)} = \frac{A}{s+2} + \frac{Bs+C}{s^2+2s+5}$. Multiplying we obtain $s^2 + 5s + 9 = A(s^2 + 2s + 5) + (Bs + C)(s + 2) = (A + B)s^2 + (2A + 2B + C)s + (5A + 2C)$. Hence we get the linear equation $A + B = 1$, $2A + 2B + C = 5$, and $5A + 2C = 9$. Solving we obtain $A = \frac{3}{5}$, $B = \frac{2}{5}$, and $C = 3$. Thus we have $Y = \frac{3}{5} \cdot \frac{1}{s+2} + \frac{2}{5} \cdot \frac{s}{s^2+2s+5} + \frac{3}{s^2+2s+5} = \frac{3}{5} \cdot \frac{1}{s+2} + \frac{2}{5}\frac{s+1}{(s+1)^2+2^2} + \frac{13}{10}\frac{2}{(s+1)^2+2^2}$. The inverse Laplace transform yields the answer $y = \frac{3}{5}e^{-2t} + \frac{2}{5}e^{-t}\cos 2t + \frac{13}{10}e^{-t}\sin 2t$.

18. $y'' + 2y' - 3y = 5e^{2t}$, $u(0) = 2$, $y'(0) = 3$. The Laplace transform is $(s^2Y - 2s - 3) + (2sY - 4) - 3Y = \frac{5}{s-2}$ or $(s^2 + 2s - 3)Y - (2s + 7) = \frac{5}{s-2}$ so $(s^2 + 2s - 3)Y = \frac{5}{s-2} + (2s + 7) = \frac{2s^2+3s-9}{s-2}$. This yields $Y = \frac{2s^2+3s-9}{(s-2)(s^2+2s-3)} = \frac{(2s-3)(s+3)}{(s-2)(s+3)(s-1)} = \frac{2s-3}{(s-2)(s+1)} = \frac{A}{s-2} + \frac{B}{s-1}$. Multiplying we have $2s - 3 = A(s - 1) + B(s - 2)$. Setting $s = 2$ we get $A = 1$. Setting $s = 1$, we get $B = 1$. The partial fraction decomposition of Y is $\frac{s}{s-2} + \frac{1}{s-1}$. The inverse Laplace transform yields the answer $y = e^{2t} + e^t$.

19. $y'' + 4 = \sin t$, $y(0) = y'(0) = 0$. The Laplace transform is $s^2Y + \frac{4}{s} = \frac{1}{s^2+1}$. So $s^2Y = \frac{1}{s^2+1} - \frac{4}{s}$ and $Y = \frac{1}{(s^2+1)s^2} - \frac{4}{s^3}$. The inverse Laplace transform (by forms 11 and 3) is $y = t - \sin t - 2t^2$.

20. $y'' + 4 = t$, $y(0) = -1$, $y'(0) = 0$. The Laplace transform is $s^2Y + s + \frac{4}{s} = \frac{1}{s^2}$. So $s^2Y = \frac{1}{s^2} - \frac{4}{s} - s$ and $Y = \frac{1}{s^4} - \frac{4}{s^3} - \frac{1}{s}$. The inverse Laplace transform is $y = \frac{t^3}{6} - 2t^2 - 1$.

21. We begin with the equation $\frac{1}{2}q'' + 10q' + 1001 = 10\sin t$. The Laplace transform is $\frac{1}{2}(s^2Q - 10) + 10(sQ) + 100Q = \frac{10}{s^2+1}$ or $\left(\frac{1}{2}s^2 + 10s + 100\right)Q = \frac{10}{s^2+1} + 5 = \frac{5s^2+15}{s^2+1}$. Multiplying by 2 to clear fractions we have $(s^2 + 20s + 200)Q = \frac{10s^2+30}{s^2+1}$. Hence $Q = \frac{10s^2+30}{(s^2+1)(s^2+20s+100)} = \frac{As+B}{s^2+1} + \frac{Cs+D}{s^2+20s+200}$. Multiplying by the LCD we obtain the equation $10s^2 + 30 = (As + b)(s^2 + 20s + 200) + (Cs + D)(s^2 + 1) = (A + C)s^3 + (20A + B + D)s^2 + (200A + 20B + C)s + 200B + D$. Hence we have linear equations $A + C = 0$, $20A + B + D = 10$, $200A + 20B + C = 0$, and $200B + D = 30$. Solving this system, we obtain $A = \frac{-400}{40,001}$, $B = \frac{3980}{40,001}$, $C = \frac{400}{40,001}$ and $D = \frac{404,030}{40,001}$. Substituting, we obtain $Q = \frac{10}{40,001}\left[\frac{-40s+398}{s^2+1} + \frac{40s+40,403}{s^2+20s+200}\right]$. Working each part separately, we have: $\frac{-40s+398}{s^2+1} = -40\frac{s}{s^2+1} + 398\frac{1}{s^2+1}$. The inverse Laplace transform yields $q = -40\cos t + 398\sin t$. For $\frac{40s+40,403}{s^2+20s+200}$, we complete the square in the denominator, getting $\frac{40s+40,403}{s^2+20s+200} = \frac{40s+40,403}{s^2+20s+100+100} = \frac{40s+40,403}{(s+10)^2+10^2} = 40\frac{s+10}{(s+10)^2+10^2} + \frac{40,003}{(s+10)^2+10^2} = 40\frac{s+10}{(s+10)^2+10} + \frac{40,403}{10} \cdot \frac{10}{(s+10)^2+10^2}$. The inverse Laplace transform yields $40e^{-10t}\cos 10t + \frac{40,003}{10}e^{-10t}\sin 10t$. Hence, we get

$q(t) = \frac{10}{40,001}\Big(-40\cos 5 + 398\sin 5$

$\qquad + 40e^{-10t}\cos 10t + \frac{40,003}{10}e^{-10t}\sin 10t\Big)$

Thus, we have

$i(t) = q'(t)$

$\quad = \frac{10}{40,001}\Big(40\sin t + 398\cos t$

$\qquad - 400e^{-10t}\cos 10t$

$\qquad - 400e^{-10t}\sin 10t - 40,003e^{-10t}\sin 10t$

$\qquad + 40,003e^{-10t}\cos 10t\Big)$

$\quad = \frac{10}{40,001}\Big(40\sin t + 398\cos t$

$\qquad + 39,603e^{-10t}\cos 10t$

$\qquad - 40,403e^{-10t}\sin 10t\Big)$

The steady-state current is $\frac{10}{40,001}(40\sin t + 398\cos t)$.

22.   The initial differential equation is $20x'' + 90x' + 700x = 5\sin t$. This is equivalent to $2x'' + 9x' + 70x = \frac{1}{2}\sin t$ with auxiliary equation $2m^2 + 9m + 70 = 0$. This has solutions $\frac{-9 \pm \sqrt{81-560}}{4} = -\frac{9}{4} \pm \frac{\sqrt{479}}{4}j$. Hence, the general complementary solution is $x_c = e^{-9t/4}\left(C_1\cos\frac{\sqrt{479}}{4}t + C_2\sin\frac{\sqrt{479}}{4}t\right)$. For x_p we set $x_p = A\sin t + B\cos t$. Thus, $x_p' = A\cos t - B\sin t$. Substituting these expressions into the equivalent equation above, we have $2(-A\sin t - B\cos t) + 9(A\cos t - B\sin t) + 70(A\sin t + B\cos t) = \frac{1}{2}\sin t$. Collecting like terms produces $-2A - 9B + 70A = \frac{1}{2}$ and $-2B + 9A + 70B = 0$. These equations are equivalent to $68A - 9B = \frac{1}{2}$ and $9A + 68B = 0$. Solving, we get $A = \frac{68}{9410} = \frac{34}{4705}$ and $B = -\frac{9}{9410}$. This produces $x_p = \frac{1}{9410}(68\sin t - 9\cos t)$. Hence, the general solution is $x = x_c + x_p = e^{-9t/4}\left(c_1\cos\frac{\sqrt{479}}{4}t + c_2\sin\frac{\sqrt{479}}{4}t\right) + \frac{1}{9410}(68\sin t - 9\cos t)$. Now, $x(0) = 0$ yields $c_1 = \frac{9}{9410}$. We find that $x'(t) = -\frac{9}{4}e^{9t/4} \times$
$\left(\frac{9}{9410}\cos\frac{\sqrt{479}}{4}t + c_2\sin\frac{\sqrt{479}}{4}t\right) + e^{-9t/4}\left(-\frac{9\sqrt{479}}{9410\cdot 4} \times \right.$
$\left.\sin\frac{\sqrt{479}}{4}t + \frac{\sqrt{479}}{4}c_2\cos\frac{\sqrt{479}}{4}t\right) + \frac{1}{9410}(68\cos t +$
$9\sin t)$. Now, using the fact that $v(0) = -1$ or $x'(0) = -1$, (recall, down is positive),we obtain $-1 = -\frac{9}{4}\cdot\frac{9}{9410} + \frac{\sqrt{479}}{4}c_2 + \frac{68}{9410}$ and we obtain $c_2 \approx -0.183692$. The particular solution is

$$x(t) = e^{-9t/4}\left(\frac{9}{9410}\cos\frac{\sqrt{479}}{4}t\right.$$
$$\left. - 0.1837\sin\frac{\sqrt{479}}{4}t\right)$$
$$+ \frac{1}{9410}(68\sin t - 9\cos t)$$
$$\approx e^{-2.25t}(0.0009564\cos 5.4715t$$
$$- 0.1837\sin 5.4715t) + 0.007226\sin t$$
$$- 0.0009564\cos t$$

CHAPTER 34 TEST

1.   From Entry 5 in Table 34.1, we see that the Laplace transform of e^{at} is $\frac{1}{s-a}$. Here $a = -3$, so the Laplace transform is $5\left(\frac{1}{s+3}\right) = \frac{5}{s+3}$.

2.   From Entry 3, we have the Laplace transform of $2t^3 = 6\left(\frac{t^3}{3}\right)$ as $6\left(\frac{1}{s^4}\right)$ and from Entry 8, the Laplace transform of $\sin 8t$ is $\frac{8}{s^2+8^2}$. Adding these together, we get the Laplace transform of $2t^3 + \sin 8t$ as $\frac{6}{s^4} + \frac{8}{s^2+64}$.

3.   From Entry 20, we have $a = -5$ and $b = 4$, so the Laplace transform is $\frac{s-5}{(s-5)^2+4^2}$.

4.   From Entry 7, we have $n - 1 = 9$ and $a = 5$, so the Laplace transform is $8\left(\frac{9}{(s-5)^{10}}\right) = \frac{72}{(s-5)^{10}}$.

5.   We see that this is of the form in Entry 19 of Table 34.1 with $a = 6$ and $B = \sqrt{7}$. So, $L^{-1}(F) = e^{-6t}\sin\sqrt{7}t$.

6.   Completing the square of the denominator we get $F(s) = \frac{s}{(s-6s+9)+4} = \frac{s}{(s-3)^2+2^2}$. This does not satisfy any of the entries in Table 34.1, but it is close to Entry 20. If we rewrite $F(s)$ by adding $0 = -3 + 3$ to the numerator and regrouping, we obtain $F(s) = \frac{s-3+3}{(s-3)^2+2^2} = \frac{s-3}{(s-3)^2+2^2} + \frac{3}{(s-3)^2+2^2} = \frac{s-3}{(s-3)^2+2^2} + \frac{3}{2}\left(\frac{2}{(s-3)^2+2^2}\right)$. Applying Entries 20 and 19 of Table 34.1, respectively, we ggt $L^{-1}(F) = e^{-3t}\cos 2t + \frac{3}{2}e^{-3t}\sin 2t$.

7.   Using partial fraction decomposition we have $\frac{5}{(s+3)(s^2+1)} = \frac{A}{s+3} + \frac{Bs+C}{s^2+1}$. Multiplying by $(s+3)(s^2+1)$ produces$5 = A(s^2+1) + (Bs+C)(s+3) = As^2 + A + Bs^2 + Cs + 3Bs + 3C = (A+B)s^2 + (3B+C)s + (A+3C)$. Solving the simultaneous equations $A + B = 0$, $3B + C = 0$, and $A + 3C = 5$ we obtain $A = 0.5$, $B = -0.5$, and $C = 1.5$. So we can rewrite $F(s)$ as $\frac{0.5}{s+3} + \frac{-0.5s+1.5}{s^2+1} = \frac{0.5}{s+3} - \frac{0.5s}{s^2+1} + \frac{1.5}{s^2+1}$. From Entries 5, 9, and 8 of Table 34.1, respectively, we obtain $L^{-1}(F) = 0.5e^{-3t} - 0.5\cos t + 1.5\sin t$.

8. Taking Laplace transforms of both sides produces $L(y') + 5L(y) = L(0)$. Since $L(y') = sY - y(0) = sY - 2$, this equation becomes $sY - 2 + 5Y = 0$ or $Y(s + 5) = 2$ and so $Y = \frac{2}{s+5}$. Taking the inverse Laplace transform of both sides we obtain $y = L^{-1}\left(\frac{2}{s+5}\right) = e^{-5t}$.

9. Taking the Laplace transform of the given equation produces $L(y'') + 2L(y) = L\left(e^{-2t}\right)$. Since $L(y'') = s^2Y - sy(0) - y'(0)$, $y(0) = 0$ and $y'(0) = 0$ and $y'(0) = 0$, we can rewrite this as $s^2Y - s + 2Y = \frac{1}{s+2}$ or $\left(s^2 + 2\right)Y = s + \frac{1}{s+2}$ and so $Y = \frac{s}{s^2+2} + \frac{1}{(s^2+2)(s+2)}$. Using partial fractions, we determine that $\frac{1}{(s^2+2)(s+2)} = \frac{1/6}{s+2} + \frac{-s/6+1/3}{s^2+2} = \frac{1}{6}\left(\frac{1}{s+2}\right) - \frac{1}{6}\left(\frac{s}{s^2+2}\right) + \frac{1}{3}\left(\frac{1}{s^2+2}\right)$. Combining this partial fraction with the first $\frac{s}{s^2+2}$, we obtain $Y = \frac{1}{2}\left(\frac{1}{s+2}\right) + \frac{5}{6}\left(\frac{s}{s^2+2}\right) + \frac{1}{3}\left(\frac{1}{s^2+2}\right)$. So, $y = \frac{1}{6}L^{-1}\left(\frac{1}{s+2}\right) + \frac{5}{6}L^{-1}\left(\frac{s}{s^2+2}\right) + \frac{1}{3}L^{-1}\left(\frac{1}{s^2+2}\right) = \frac{1}{6}L^{-1}\left(\frac{1}{s+2}\right) + \frac{5}{6}L^{-1}\left(\frac{s}{s^2+2}\right) + \frac{1}{3\sqrt{2}}L^{-1}\left(\frac{\sqrt{2}}{s^2+2}\right)$. Using Entries 5, 9, and 8 in Table 34.1 we determine that $y = \frac{1}{6}e^{-2t} + \frac{5}{6}\cos\sqrt{2}t + \frac{1}{3\sqrt{2}}\sin\sqrt{2}t$.

10. Taking the Laplace transform of the given equation produces $L(y'') + 2L(y') + L(y) = 0$. Since $L(y') = sY - y(0) = sY^{-1}$ and $L(y'') = s^2Y - sy(0) - y'(0) = s^2Y - s + 1$, we can rewrite this as $\left(s^2Y - s + 1\right) + 2sY + Y = 0$ or $\left(s^2 + 2s + 1\right)Y = s + 1$ and so $Y = \frac{s+1}{s^2+2s+1} = \frac{1}{(s+1)}$ and $y = L^{-1}(Y) = L^{-1}\left(\frac{1}{(s+1)}\right)$. By form 5, we get $y = e^{-t}$.

11. Using the formula $L\frac{d^2q}{dt^2} + R\frac{dq}{dt} + \frac{q}{C} = E(t)$ we have $0.2q'' + 8q' + \frac{q}{0.0025} = 0.2q'' + 8q' + 400q = 12\sin 20t$. Taking the Laplace transform of this equation, we get $0.2L(q'') + 8L(q') + 400L(q) = L(12\sin 20t)$. Since $L(q') = sQ - q(0) = sQ$, $L(q'') = s^2Q - sq(0) + q'(0) = s^2Q$, and $L \times (12\sin 20t) = 12\left(\frac{20}{s^2+20^2}\right) = \frac{240}{s^2+400}$, we can rewrite our equation as $0.2s^2Q + 8sQ + 400 = \frac{240}{s^2+400}$ or $\left(0.2s^2 + 8s\right)Q = -400 + \frac{240}{s^2+400}$ and so $Q = \frac{-400}{0.2s^2+8s} + \frac{240}{\left(0.2s^2+8s\right)\left(s^2+20^2\right)} = \frac{-2000}{s^2+40s} + \frac{1200}{\left(s^2+40s\right)\left(s^2+20^2\right)}$. Using partial fraction decomposition, we can rewrite this as $Q = \left(\frac{-50}{s} + \frac{50}{s+40}\right) + \left(\frac{0.075}{s} + \frac{-0.015}{s+40} - \frac{0.06s+0.6}{s^2+400}\right) = \frac{-49.925}{s} + \frac{49.985}{s+40} - \frac{0.06s}{s^2+400} - \frac{0.6}{s^2+400}$. Taking the inverse Laplace transform, we obtain $q = L^{-1}(Q) = -49.925 + 49.985e^{-40t} - 0.06\cos 20t - 0.03\sin 20t$.

12. Using Euler's method where $y_{n+1} = y_n + f(x_n, y_n)h$ we have $f(x, y) = 2x + y$. Thus we get the formula $y_{n+1} = y_n + 0.2(2x_n + y_n)$. We are given $x_0 = 0$, $y_0 = 2$, and $h = 0.2$. Using this data and applying the results successively, we obtain the following:

x	y $h = 0.2$
0	2
0.2	2.4
0.4	2.96
0.6	3.712
0.8	4.6944
1.0	5.95328

Solutions for Additional Computer Exercises

Technical Mathematics contains 32 exercises which ask students to write a computer program and *Technical Mathematics with Calculus* contains an additional 6 such exercises. This appendix consists of programs written in GWBASIC that serve as answers to each of those exercises. Just as the accompanying textbook increases in sophistication and complexity as you progress from beginning to end, the computer programs in this Program Book also increase in sophistication and complexity.

The name of each program corresponds to its location in the text. Each name consists of the three letters PRG followed immediately by five digits. The first two digits represent the chapter in the text. Thus, Chapter 9 is 09 and chapter 25 is 25. The next (third) digit denotes the section in the chapter; and the last two digits, the exercise in that section where the program was assigned.

Using the above description, the program for Chapter 9, Section 5, Exercise 43, has the name PRG09543. Similarly, the computer program for Chapter 17, Section 3, Exercise 22, is named PRG17322.

* * * * *

You may wish to write your own versions of these programs. If you do, then use the following programs as guidelines. In some cases, we have added additional steps to the programs to make them easier to use and more "user friendly." You may have some better ideas on how this might be done.

However, it is important that anyone programming a computer remember that a computer must be told what to do and how it is done. If the directions are wrong, the computer will give a wrong answer.

There are several special commands, statements, and functions that are used in BASIC. In fact, most of these, with some variation such as SQRT rather than SQR, are used in other computer program-

ming languages. This is not intended as a programming course, and so we assume that you are familiar with BASIC.

Commands, such as LOAD, RUN, LIST, and SAVE are not part of a program, but allow you to tell the computer to do something with a program.

Program *statements* used in these programs include PRINT, PRINT USING, INPUT, REM, GOTO (or GO TO), DEF FN, DIM, FOR...NEXT, IF...THEN, GOSUB...RETURN, CLS, CLEAR, and END.

Functions used in these programs include ABS, ATN, COS, EXP, INT, LEFT$, LOG, SGN, SIN, GQR, TAB, and TAN. You should remember that a computer evaluates trigonometric functions expressed in radians and not degrees. You will need to include conversion factors in any programs where you want to enter, or have the computer print, angles given in degrees.

You may also have noticed that BASIC includes only the sine, cosine, tangent, and arctangent functions. If your programs include any other trigonometric formulas, you will have to define them in that particular program. Naturally, since these are programs involving mathematics, you will need to use the symbols for the mathematical *operations*. The computer uses $+$, $-$, $*$, and $/$ for addition, subtraction, multiplication, and division, respectively. For exponentiation, these BASIC programs use $\wedge$. (Some computers and some languages use either $**$ or $\uparrow$.)

* * * * *

Not everyone wishes to take the time to write a computer program. Thus, each of these programs is available on an IBM formatted disk in either $5\frac{1}{4}''$ or $3\frac{1}{2}''$ sizes.

At the end of these program listings is an exercise set of 30 exercises. Some of the exercises ask you to modify a program so that it will do a better job of performing the desired task or so that it can be used in some additional situations. The remaining exercises ask you to use one of the programs with a given set of numbers or equations.

A solution to each of these additional exercises is given in the *Instructor's Solutions Guide*; solutions to the odd numbered exercises are in the *Student's Solutions Guide*. Whenever possible, each solution contains a copy of the exact wording and format used when the computer printed each solution.

1. If you select $x = -1.9$, you get $x = -1.90625$, $y = 1.160156$. If you select $y = 1.16$, you get $x = -1.90625$, $y = 1.1625$.

3. **(a)** $= -12.000000$,
    (b) If you let $\pi = 3.1416$, $\frac{4}{3} = 1.3333$, $-\frac{17}{6} = -2.8333$, $1 + \sqrt{2} = 2.4142$, and $\sqrt{11} = 3.3166$, you get the result is 13.879020.

5. **(a)** THIS SYSTEM IS DEPENDENT.
    (b) THE SOLUTIONS ARE: X = 0, Y = 0, Z = 0.
    (c) THIS SYSTEM IS INCONSISTENT.

7. **(a)** THE ROOTS ARE: 1.75 AND .5
    (b) THE ROOTS ARE: 3.651635 AND - 6.480089
    (c) THE ROOTS ARE BOTH THE SAME: 2.3 AND 2.3
    (d) THE ROOTS ARE NOT REAL NUMBERS.
    THE ROOTS ARE: 1.109375 + .6523321J
    AND 1.109375 - .6523321J

9. **(a)** THE HORIZONTAL VECTOR HAS
    MAGNITUDE: 5.17638
    AND DIRECTION: 0 RADIANS
    OR 0 DEGREES
    THE VERTICAL VECTOR HAS
    MAGNITUDE: 19.31852
    AND DIRECTION 1.570796 RADIANS
    OR 90 DEGREES.
    (b) THE HORIZONTAL VECTOR HAS
    MAGNITUDE: 13.56877
    AND DIRECTION: 3.141593 RADIANS
    OR 180 DEGREES
    THE VERTICAL VECTOR HAS
    MAGNITUDE: 8.478708
    AND DIRECTION 4.712389 RADIANS
    OR 270 DEGREES.
    (c) THE HORIZONTAL VECTOR HAS
    MAGNITUDE: 4.687785
    AND DIRECTION: 0 RADIANS
    OR 0 DEGREES
    THE VERTICAL VECTOR HAS
    MAGNITUDE: 17.79283

AND DIRECTION 4.712389 RADIANS
OR 270 DEGREES.
(d) THE HORIZONTAL VECTOR HAS
MAGNITUDE: 14.32789
AND DIRECTION: 3.141593 RADIANS
OR 180 DEGREES
THE VERTICAL VECTOR HAS
MAGNITUDE: 18.8786
AND DIRECTION 1.570796 RADIANS
OR 90 DEGREES.

11. **(a)** -4.96 + 2.25 J,
     (b) -5.709999 + .8699999 J

13. **(a)** THE QUOTIENT IS:
     -.0450621 + -.7665063 J
     (b) THE QUOTIENT IS:
     .1509434 + .5283019 J

15. **(a)** THE TOTAL IS -198.1568 + 165.6024
     OR
     258.2446 CIS 2.445451 RADIANS
     258.2446 CIS 140.114 DEGREES
     (b) THE TOTAL IS
     18.89179 + -10.47188 J
     OR
     21.6 CIS
     5.77704 RADIANS
     21.6 CIS 331 DEGREES
     (c) THE TOTAL IS
     -17.64838 + -20.43367 J
     OR
     27 CIS 4 RADIANS
     27 CIS 229.1831 DEGREES

17. **(a)** THE TOTAL IS
     -28.31983 + -153.408 J
     OR
     156.0001 CIS 4.52984 RADIANS
     156.0001 CIS 259.5407 DEGREES
     (b) THE FIRST ROOT IS
     1.271921 $+ -$.2530007 J
     OR
     1.29684 CIS 6.086836 RADIANS
     1.29684 CIS 348.75 DEGREES
     THE 2ND ROOT IS
     .2530007 + 1.271921 J
     OR
     1.29684 CIS 1.374447 RADIANS
     1.29684 CIS 78.75 DEGREES
     THE 3RD ROOT IS
     -1.271921 + .2530008 J
     OR
     1.29684 CIS 2.945243 RADIANS
     1.29684 CIS 168.75 DEGREES
     THE 4TH ROOT IS

```
-.253001 + -1.271921 J
OR
1.29684 CIS 4.51604 RADIANS
1.29684 CIS 258.75 DEGREES
```

19. (a) THE ANSWER IS
```
      7.7000   -2.0000
     -2.9000   10.0000
     11.1000    6.0000
```
(b) THE ANSWER IS
```
      2.4000   -1.3000    3.5000
      5.7000    8.2000   -2.0000
```

21. THE ANSWER IS
```
     42.8600    94.1100   -206.0900
    108.8600   -23.3600    -64.0500
```

23. (a) THE MATRIX IS SINGULAR. THIS
SYSTEM HAS NO SOLUTION.
(b) THE INVERTED MATRIX IS
```
     -.067   -.333    .267
     -.067    .667   -.233
      .333   -.333    .167
```
THE ANSWER IS
```
     x( 1 ) = 2
     x( 2 ) = -3
     x( 3 ) = 3
```

25. THE TOTAL NUMBER IN THE SAMPLE IS:
150
THE MEAN IS: 2.833333
THE VARIANCE IS:
1.938889
AND THE STANDARD DEVIATION IS:
1.39244
THE NUMBER OF SCORES WITHIN ONE
STANDARD DEVIATION OF THE MEAN IS:
109 OR 72.66666 PERCENT OF THE
TOTAL VALUE.

AND THE NUMBER WITHIN TWO S.D. OF
THE MEAN IS: 144 OR 96 PERCENT.

27. ENTER THE LEFT END POINT, A=?   0

NEXT ENTER THE RIGHT POINT,
B =?   3.1416

AND FINALLY, THE NUMBER OF SEGMENTS,
N= ?   50
```
A   B        N
0   3.1416   50
```
THE AREA IS 3.915187

29. THE AREA UNDER F(X) OVER THE INTER-
VAL FROM 0 TO 3.1416 IS
3.91427 BY SIMPSON'S RULE AND
3.912377 BY THE TRAPEZOIDAL RULE